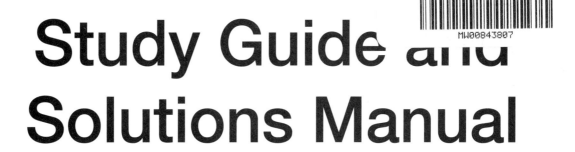

Study Guide and Solutions Manual

to Accompany

ORGANIC CHEMISTRY
Seventh Edition

Jim Parise
Department of Chemistry and Biochemistry
University of Notre Dame

Marc Loudon
Department of Medicinal Chemistry
and Molecular Pharmacology
Purdue University (Emeritus)

Lisa Bonner
Department of Chemistry and Biochemistry
Eckerd College

macmillan
learning

Austin • Boston • New York • Plymouth

Macmillan Learning
120 Broadway
New York, NY 10271
www.macmillanlearning.com

In 1946, William Freeman founded W. H. Freeman and Company and published Linus Pauling's *General Chemistry*, which revolutionized the chemistry curriculum and established the prototype for a Freeman text. W. H. Freeman quickly became a publishing house where leading researchers can make significant contributions to mathematics and science. In 1996, W. H. Freeman joined Macmillan and we have since proudly continued the legacy of providing revolutionary, quality educational tools for teaching and learning in STEM.

Contents

Preface

The purpose of this preface is to give students and faculty using this manual an overview of the organization and features of this *Study Guide and Solutions Manual* so that you can use it more effectively.

Organization

Each chapter corresponds to the chapter of the text with the same number and title. The following sections are found within each chapter, each headed by its own icon. (Some chapters do not contain all sections. For example, Chapters 13 and 14, which cover spectroscopy, do not have Reaction Reviews.)

CHAPTER SUMMARY

The *Chapter Summary* is a conceptual summary of the major points of the chapter. These summaries appeared in the 6th edition of this text as "Key Ideas."

STUDY GUIDE LINKS

Study Guide Links are short supplements to text material that are called out in the margins of the text by the icon depicted above. Study Guide Links provide hints for study, shortcuts, approaches to problem-solving, or additional explanations of topics that typically cause difficulty. For example, Study Guide Link 5.1, "How to Study Organic Reactions," on p. 108 of this manual, is called out with a margin icon at a point in the text (p. 257) at which some students begin to have difficulty learning reactions. It provides detailed hints on how to study and learn organic reactions efficiently.

FURTHER EXPLORATIONS

Further Explorations are short discussions that provide additional depth on certain topics; these are called out in the margins of the text by the icon above. For example, Further Exploration 14.2, called out with a margin icon on text p. 784, gives a more in-depth discussion of Fourier-Transform NMR, a modern method used for obtaining NMR spectra.

REACTION REVIEW

Reaction Reviews present the reactions in the chapter in detailed outline form. This summary consolidates the reactions of the chapter in one place and presents not only the reactions themselves but also the essential features of their mechanisms and stereochemistry for cases in which these issues are discussed in the text. Studying and learning reactions is a skill that should be cultivated, and Study Guide Link 5.1 on page 108 of this manual will help you to use these reaction summaries to best advantage.

 # SOLUTIONS TO PROBLEMS

Undoubtedly, your instructor will stress the importance of problem solving for the mastery of organic chemistry. This text provides over 2000 problems of varying difficulty levels, many with multiple parts. The solutions to many of the problems are worked-out examples in which not only is the answer given, but also the detailed approach to the solution is discussed.

First are the problems presented within the text itself, called *Focused Problems*. These problems for the most part focus on the material just presented in the foregoing section of the text. These are designed to test your understanding of the material in that section.

At the conclusion of each chapter you will find *Skills Objectives*. Each objective directs your attention to a specific skill that you should have developed when you master the current chapter, and the associated *Skills Objectives Problems* ask questions that are focused on a particular skill.

The section entitled *Integrated Problems* provides problems that test your ability to integrate what you have learned in the current chapter with what you have learned in other chapters, as well as what you have learned in other courses, such as general chemistry and mathematics.

It is important not to rely too heavily on the solutions presented in this manual. You learn organic chemistry by *solving problems on your own*. Many students develop a false sense of understanding by trying a problem, giving up too soon, and then reading the detailed solution. The solution seems to make sense to them, they feel like they understand the concept, and they move on. What they've actually done is short-circuited the learning process, and they're simply agreeing what the solution says; they haven't developed the capacity to solve problems independently.

Here is a better approach if you get stuck on a problem. First, attempt to solve it by reviewing the text or your class notes, or by talking to a classmate, TA, or your instructor. Be sure to ask for *guidance* only, not the answer. If you're still stuck, glance quickly at the solution in this manual without reading for detail. Focus on the references within the solution—pages, equations, sections, etc.—where the answer might be found. Jot those down next to your work, and close the manual. Refer back to the text and see if you can solve the problem after referring to relevant sources. It may take longer to solve problems this way, but your time will be much better spent if you take this approach.

Solutions Icon Comments

Within the solutions, you will occasionally find two types of comments written in a special font and marked with icons.

 Comments marked with this icon provide additional information or depth about the accompanying solution.

 Comments marked with this icon provide cautionary warnings—things to be particularly careful about or common traps.

Acknowledgments

We acknowledge Dr. Joseph Stowell of Purdue University for his contributions to previous editions of this manual. We are also indebted to our proofreaders, Mr. Luke Witmer, an undergraduate at Notre Dame, and Dr. Animesh Aditya, Lecturer of Medicinal and Organic Chemistry at Kennesaw State University.

We have tried very hard to eliminate as many errors in both the textbook and this manual as possible. Nevertheless, if errors are found, we would like to know about them so that we can correct them on reprint if possible. You can send these or any other comments to us by email at *loudonm@purdue.edu*. A list of both textbook errors and study guide and solutions manual errors can be found at the link below, which is also repeated within the footer of most pages of this manual.

https://web.ics.purdue.edu/~loudonm/teaching/

Errors are corrected with each subsequent printing of both the text and this manual.

We also enjoy hearing from students with comments and suggestions about the text or the solutions manual. We hope that you find this manual very useful in your study of organic chemistry.

Jim Parise, Marc Loudon, and Lisa Bonner
May 2021

About the Authors

Jim Parise

Jim Parise received his BS in chemistry from the State University of New York at Oswego in 2000 and his PhD in organic chemistry in 2007 from Duke University, where he worked with Professor Eric Toone. After a postdoctoral fellowship with Professor David Lawrence at the University of North Carolina at Chapel Hill, Dr. Parise joined the Department of Chemistry at Duke University, where he taught organic chemistry lecture and coordinated the laboratory courses. In 2011, he joined the faculty in the Department of Chemistry and Biochemistry at the University of Notre Dame. He teaches organic chemistry to pre-professional students, engineers, and science majors. Additionally, he works within a recently developed program designed to support underprepared and first-generation students. He has received a number of teaching awards including, most recently, the 2021 Shilts/Leonard Award for Sustained Excellence in Teaching. In addition to teaching classes, Dr. Parise has worked on developing more effective teaching methods for large lecture classes and on improving organic chemistry pedagogy. He was a pioneer in introducing technology and digital learning into the chemistry curriculum, including the use of iPads, online assessments, and original Lightboard videos that supplement and enhance lecture courses. Dr. Parise has authored laboratory manuals and a peer-reviewed book chapter on laboratory writing, and he is co-author of the accompanying textbook, *Organic Chemistry, Seventh Edition.*

Marc Loudon

Marc Loudon received his BS (magna cum laude) in chemistry from Louisiana State University, Baton Rouge, in 1964, and his PhD in organic chemistry in 1968 from the University of California, Berkeley, where he worked with Professor Don Noyce. After a two-year postdoctoral fellowship with Professor Daniel E. Koshland, Jr., at Berkeley, Dr. Loudon joined the Department of Chemistry at Cornell University, where he taught organic chemistry to both pre-professional students and science majors. In 1977, he joined the Department of Medicinal Chemistry and Pharmacognosy (now the Department of Medicinal Chemistry and Molecular Pharmacology) at Purdue University in the School of Pharmacy and Pharmaceutical Sciences, where he taught organic chemistry to pharmacy and pre-pharmacy students. Dr. Loudon developed and implemented cooperative-learning methods for teaching organic chemistry to large classes. He has received numerous teaching awards. Dr. Loudon was named the Gustav E. Cwalina Distinguished Professor in 1996, and was named Distinguished Professor Emeritus following his retirement from Purdue in 2015. Dr. Loudon is the author of numerous research articles, a co-author of an in-house laboratory manual, and he is the co-author of the accompanying textbook, *Organic Chemistry, Seventh Edition.* The first edition was published in 1984.

Lisa Bonner

Lisa Bonner grew up in Baltimore, received her BS (summa cum laude) in biochemistry and chemistry from Loyola College in Maryland, and has wanted to be an organic chemistry professor since taking the course herself. She earned her PhD from the Department of Medicinal Chemistry and Molecular Pharmacology at Purdue University, synthesizing novel dopaminergic agonists in the laboratory of Dr. David Nichols. While at Purdue, she was fortunate enough to work as a Teaching Assistant for Dr. Marc Loudon's organic chemistry courses for five years, earning valuable pedagogical experience and several teaching awards. She has taught undergraduate chemistry for over 10 years, and is currently an Assistant Professor at Eckerd College in St. Petersburg, Florida. Through her work as the Director of the Natural Science Peer Tutoring Program, Dr. Bonner established active learning communities where chemistry students could work problems in small groups in order to improve their outcomes in their courses. Recognizing the importance of engaging students inside and outside of the classroom, she is the Faculty Advisor for the Student Affiliates Chapter of the American Chemical Society and the President of Eckerd College's Phi Beta Kappa Chapter.

Chemical Bonding and Chemical Structure

CHAPTER 1 SUMMARY

- **Organic chemistry** is the chemistry of carbon compounds.

- The atomic number of an atom in the periodic table is equal to both the number of electrons and the number of protons in the neutral atom.

- When atoms form ions they tend to gain or lose electrons so as to give the ion with the same number of valence electrons as the nearest noble gas. (Octet rule for atoms)

- Electrons in atoms are arranged in orbitals, which describe the wave motions of electrons. The orbitals (and therefore the electrons within them) are characterized by their energy and their shape.

  - Each orbital is designated by a principal quantum number, which is an integer 1, 2, 3, … and by a sublevel quantum number, s, p, d …

  - Orbitals with principal quantum number n have n subtypes. An $n = 1$ orbital is always a 1s orbital. The $n = 2$ level has one 2s and three 2p sublevels. The $n = 3$ level has one 3s, three 3p, and five 3d orbitals.. Orbital energies increase with n and, within a quantum level n, in the order $s < p < d$ (Fig. 1.2, Sec. 1.2C).

  - The sublevel quantum numbers s, p, d … govern the orbital shape. All s orbitals are spheres; all p orbitals have two lobes (Figs. 1.3, 1.4, 1.7, 1.9, Sec. 1.2E).

  - The different orbitals with the same n and the same sublevel type have the same shape and energy but are oriented in different directions. The three 2p orbitals are oriented along the three Cartesian axes x, y, and z (Fig. 1.8, Sec. 1.2E).

- Because orbitals describe the wave motion of an electron, orbitals contain nodes, peaks and troughs. Orbitals have $n-1$ nodes, which can be planar or spherical. A 1s orbital has no nodes; a 2p orbital has a planar node separating its two lobes, a 3s orbital has two spherical nodes; and a 3p orbital has both a planar node and a spherical node. A node separates a wave peak and a wave trough.

- Electrons in atoms and ions occupy orbitals from lowest to highest energy (aufbau principle) subject to the Pauli principle and Hund's rules.

  - Pauli principle: no more than two electrons can occupy each orbital, and these must have opposite spin.

  - Hund's rules: orbitals of the same energy are occupied with single electrons before they are doubly occupied; and these single electrons have the same spin.

- The number of electrons in the highest principal energy level of an atom are the valence electrons—the electrons involved in the chemistry of the atom. The number of valence electrons in an atom is equal to the group number in the "A" group of the periodic table.

- Chemical compounds can contain two types of bonds: ionic and covalent. In ionic compounds, ions are held together by electrostatic attraction (the attraction of opposite charges). In covalent compounds, atoms are held together by the sharing of electrons.

- Both the formation of ions and bonding in covalent compounds tend to follow the octet rule: each atom is surrounded by eight valence electrons (two electrons for hydrogen). All bonding and nonbonding electrons are counted. The most common bonding arrangements consistent with the octet rule are given in Sec. 1.3C.

■ Ions can consist of covalently bonded atoms. The formal-charge rules (Sec. 1.4B, Study Problem 1.4) are used to assign charge to specific atoms within an ion.

■ The electron-counting procedures for formal charge and octet assessment are different.

- Formal-charge electron counting involves half of the bonding electrons plus all nonbonding electrons.

- Determining whether an atom in a covalent compound follows the octet rule involves counting all electrons surrounding an atom—both bonding and nonbonding.

■ Molecules that are not adequately described by a single Lewis structure are represented as resonance hybrids, which are weighted averages of two or more fictitious Lewis structures. Resonance hybrids are more stable than any of their contributing resonance structures.

■ The structure of a molecule is determined by its connectivity and its geometry. The molecular geometry of a molecule is determined by its bond lengths, bond angles, and dihedral angles.

- Bond lengths are governed, in decreasing order of importance, by the period of the periodic table from which the bonded atoms are derived; by the bond order (whether the bond is single, double, or triple); and by the column (group) of the periodic table from which the atoms in the bond are derived.

- Approximate bond angles can be predicted by assuming that the groups surrounding a given atom are as far apart as possible; nonbonding electron pairs (as in the nitrogen of ammonia) are treated as groups for purposes of predicting geometry. The common geometries are tetrahedral, trigonal pyramidal, trigonal planar, and linear. (Table 1.1, Sec. 1.6D)

- Dihedral angles are required to specify the spatial relationship of bonds on adjacent atoms.

■ Covalent bonds between atoms can be described in two ways: bond orbitals and molecular orbitals.

- Bond orbitals are formed by the overlap of atomic orbitals or hybrid orbitals on two atoms so that electrons in the bond are shared between atoms.

- Molecular orbitals (MOs) are orbitals of an entire molecule formed by combining atomic orbitals of the constituent atoms arranged in the proper geometry. Electrons are then added to the resulting MOs following the Pauli principle and Hund's rules. Mixing j atomic orbitals gives j MOs. Typically both bonding and antibonding MOs are formed; when bonding MOs are occupied with electrons, there is a bonding advantage corresponding to the energies of the electrons in the occupied MOs. When antibonding MOs are occupied, there is a bonding disadvantage corresponding to the energies of the electrons in the antibonding MOs.

■ Atomic orbitals used in bond orbitals can be produced by hybridization of atomic orbitals. Hybridization is used to account for molecular geometry (Table 1.2, Sec. 1.7E). Atomic orbitals are mixed (or rehybridized) so that fractional s character (f_s) and fractional p character (f_p) are distributed across several bonds. Hybrid orbitals involving s and p orbitals are notated as sp^n, where $n = f_p / f_s$. Hybrid orbitals overlap with hybrid orbitals or atomic orbitals on other atoms to form covalent bonds.

■ Tetrahedral bonding geometry about a central atom (bond angles = 109.5°) involves four sp^3-hybridized orbitals (Fig. 1.11, Sec. 1.7B). Each hybrid orbital can overlap with a hybrid orbital or atomic orbital from another atom. Methane has an sp^3-hybridized carbon atom (Display 1.61, Sec. 1.7B).

- Trigonal-planar bonding geometry about a central atom (bond angles = 120°) involves three sp^2-hybridized orbitals, with one 2p orbital unhybridized (Fig. 1.12, Sec. 1.7C). Double bonds are formed by the overlap of sp^2 orbitals from different atoms to form a sigma (σ) bond and overlap of two 2p orbitals to form a pi (π) bond. The carbons of ethylene are sp^2-hybridized (Display 1.65, Sec. 1.7C).

- Linear bonding geometry (bond angle = 180°) about a central atom involves formation of two sp-hybridized orbitals, with two 2p orbitals remaining unhybridized (Fig, 1.13, Sec. 1,7D).

Triple bonds are formed by the overlap of sp orbitals from different atoms to form a σ bond and the overlap of two 2p orbitals to form two π bonds. The carbons of acetylene are sp-hybridized (Display 1.70, Sec. 1.7D).

- Departures from exact tetrahedral, trigonal-planar, or linear geometry can be accounted for by fractional hybridization of the central atom—that is, non-integer n values in sp^n. Situations involving fractional hybridization result from the tendency of atomic orbitals containing nonbonding electron pairs to assume a greater fraction of s character.

- In polar covalent bonds, electrons are shared unequally between bonded atoms with different electronegativities. This unequal sharing results in a bond dipole moment. The dipole moment of a molecule is the vector sum of its individual bond dipole moments. The local charge distribution in a molecule can be described graphically with an electrostatic potential map (EPM).

- In ions consisting of covalently bonded atoms, the actual charge on each atom of the ion is not necessarily the same as the computed formal charge. The actual charge can be qualitatively assessed by using the relative electronegativities of the constituent atoms.

STUDY GUIDE LINKS

1.1   A Refresher on Vector Addition

Vectors can be added together to produce a new vector, called a *resultant.* There are two ways to add two vectors. The first way is to line up the two vectors with their tails touching and construct a parallelogram. Form the resultant by connecting the vertex of the two vectors to be added to the opposite side of the parallelogram, as shown on the right.

resultant = R

Alternatively, let the head of one vector touch the tail of the second, and form the resultant by connecting the tail of the first to the head of the second to form a triangle, as shown on the right.

In either case, trigonometry can be used to calculate the magnitude of the resultant as well as its angle. (The law of cosines can also be used.)

To line up the vectors, you are allowed to move a vector in *any* direction parallel to, or along, its path. In other words, you can move a vector anywhere as long as its length (magnitude) doesn't change, and its angle (direction) relative to any arbitrary axis doesn't change, as shown for the three equivalent vectors on the right.

All three vectors are equivalent because they have the same magnitude and direction.

FURTHER EXPLORATIONS

1.1   Electron Density Distribution in Orbitals

The orbitals in the text (Figs. 1.3, 1.4, 1.7, and 1.9) are drawn as closed geometrical figures designed to encompass about 90% of the electron density within the orbital. These figures do not show the distribution of the density within the orbital. The purpose of this Further Exploration is to provide a better idea of this distribution.

In the text we showed that the wave motion of an electron in an orbital is described by a *wavefunction*, symbolized by the Greek letter psi (ψ). The electron density at any point in space of an electron in an orbital ψ is proportional to the *square* of the wavefunction, ψ^2. Consequently, by plotting the square of the wavefunction as a function of spatial coordinates we can obtain a description of the electron density at any point in space. Because ψ is a function of three spatial dimensions, such a plot would require four dimensions, or at least a contour plot in three dimensions. A simpler plot that has all the information we need is obtained by plotting the electron density ψ^2 as a function of the radial distance r along any straight line originating at the nucleus. Consider, for example, the 1s orbital. Because the orbital is a sphere, we would get the same plot in any direction. The result for a 1s orbital is the plot in **Fig. SG1.1a** (dark line). From this plot we find that the electron density is greatest at the nucleus and then falls off exponentially. There are no nodes—no regions of zero electron density—in this plot.

The plot just described gives the electron density at a *point*. A different view of electron density is obtained if we plot the probability of finding the electron in a *spherical shell* at some distance r from the nucleus. The appropriate function for this probability is $r^2\psi^2$. The reason this makes sense is that the further away we are from the nucleus, the more surface area there is in a spherical shell. Another way of thinking about this is to imagine the dimples on two basketballs, one large and one small. Assume that the dimples on both are the same size. The large basketball, then, has more dimples. Now think of each dimple as one probability unit for finding an electron. The larger the sphere (that is, the greater is r), the larger is the probability of finding an electron. Since the surface area of a sphere increases as the square of the radius, that is, as r^2, then multiplying the probability ψ^2 at one point by r^2 gives the probability of finding the electron in a spherical shell of radius r. The gray curve in Fig. SG1.1a is a plot of this probability. This product has a maximum at 52.9 pm (0.529 Å). This is called the *Bohr radius*. It is exactly the same as the radial dimension of the electron orbit calculated in an earlier, planetary model of the atom. This model said that the electron was in a fixed orbit with a radius of 52.9 pm; the quantum model says that the *maximum likelihood* of finding an electron is on a spherical shell 52.9 pm from the nucleus; but there is a finite probability at *any* distance from the nucleus.

A similar plot of electron density for the 2s orbital is given in **Fig. SG1.1b**. In the plot of ψ^2 the node—the region of zero electron density—is at 110 pm. There is a small maximum in the spherical-shell electron probability at the Bohr radius, but the largest maximum is in the "outer ball" of electron density is at about 280 pm. If we contrast the value of r at which this maximum occurs with that for the 1s orbital we get a very graphic sense of the different sizes of the two orbitals.

Figure SG1.1   Plots of wavefunctions (ψ^2) and their probability distribution functions ($r^2\psi^2$) for (a) a 1s orbital; (b) a 2s orbital. The square of the wavefunction (ψ^2) is scaled so that this quantity can be shown on the same plot. The "sizes" of the orbitals can be approximated by the distances at which $r^2\psi^2$ is a maximum. For example, the 1s orbital has a maximum at 52.9 pm; the 2s orbital has a maximum at about 300 pm.

Figure SG1.2   Plots of wavefunctions (ψ^2) and their probability distribution functions ($r^2\psi^2$) for (a) a 2p orbital; (b) a 3p orbital.

Plots of electron density for the 2p and 3p orbitals are given in **Fig. SG1.2a** and **Fig. SG1.2b**, respectively. These plots can be imagined to be taken on a line running through the nucleus and along the axis of each orbital. Both positive and negative values of r are shown to stress the lobe character of the orbitals. Both orbitals have planar nodes at the nucleus that separate the orbital into two lobes. The $3p$ orbital has, in addition, a spherical node that separates the inner and outer regions of each lobe. This node is shown in Fig. SG1.2b as two dashed lines, which represent the $+r$ and $-r$ values—that is, the radius—of the spherical node. The maximum in the spherical-shell electron probability $r^2\psi^2$ for the 2p orbital (grey line, Fig. SG1.2a) is at about 4 Bohr radii (210 pm), whereas the largest maximum in the 3p orbital occurs at about 12 Bohr radii (630 pm).

Again, notice the larger size of the orbital with the larger quantum number.

These orbital pictures are derived from the *hydrogen atom*. Orbitals of atoms with higher atomic number have basically the same shape and nodal characteristics, but their sizes differ.

 ## 1.2   Fractional Hybridization

If the unshared electron-pair hybrid orbital has more s character than an sp³ hybrid orbital, how do we express its hybridization in sp^n notation? The point to remember is that the sp^n notation doesn't refer directly to the number of s and p orbitals in the hybrid, but rather to a *ratio* of s and p character in a hybrid orbital (Eq. 1.58, Sec. 1.7B.)

$$n = f_p/f_s \qquad\qquad \text{(SG1.1)}$$

In an sp³ hybrid orbital, the fraction s character is 0.25, or 25%. Let's say that in the nonbonding-pair hybrid orbital of ammonia there is slightly more s character, because nonbonding electrons prefer orbitals of greater s character. For example, suppose that there is 27% s character in this orbital. Then there must be 73% p character in the orbital. From Eq. SG1.1, $n = 73/27 = 2.70$. In other words, the hybrid orbital containing the nonbonding electron pair has sp²·⁷ hybridization. Because we've used up 27% of an s orbital to form this hybrid orbital, 73% of an s orbital is available to the other three hybrid orbitals—the ones used to bond with the hydrogens. When we distribute this s character equally between the remaining three hybrid orbitals, each one must contain 24.3% s character. This means that these orbitals each contain $1 - 0.243 = 0.757$ fraction p character, or 75.7%. The number n in the sp^n notation is then $0.757/0.243 = 3.12$. In other words, the hybrid orbitals involved in the N—H bonds of ammonia would be sp³·¹² hybrid orbitals. What we see from this calculation is that n in the sp^n notation *does not have to be a whole number*; orbitals can have "fractional hybridization."

1.3   Dipole Moments and Charge Separation

As noted in the text, the mathematical definition of the dipole moment $\boldsymbol{\mu}$ is as follows:

$$\boldsymbol{\mu} = q\mathbf{r} \tag{SG1.2}$$

where q is the magnitude of the separated charge and $\mathbf{r}$ is a vector from the site of positive charge to the site of negative charge. When $\boldsymbol{\mu}$ is a bond dipole, as in H—Cl, the magnitude of $\mathbf{r}$ is simply the bond length; in the case of H—Cl, it is the length of the H—Cl bond. In the HCl molecule the dipole moment vector $\boldsymbol{\mu}$ lies along the H—Cl bond, and it points from the H (the site of partial positive charge) to the Cl (the site of partial negative charge), as shown in the text. That is, $\boldsymbol{\mu}$ and $\mathbf{r}$ *have the same direction.*

The dimensions of the dipole moment, as suggested by Eq. SG1.2 above, correspond to charge × length. As the text indicates, the units of the dipole moment are called *debyes.* A dipole moment of one debye results when opposite charges, each with a magnitude of 1×10^{-10} electro–static unit (esu), are separated by one Ångstrom (or 100 pm). That is,

$$1 \text{ debye} = 1 \times 10^{-10} \text{ esu Å} = 1 \times 10^{-8} \text{ esu pm}$$

These units were established prior to the current trend toward the use of standard international units. To put these units on more familiar ground, the charge on an electron is 4.8×10^{-10} esu, and one Ångstrom = 10^{-10} meter = 100 picometers. (In standard international (SI) units, in which the unit of length is the meter and the unit of charge is the coulomb (C), 1 D = 3.34×10^{-30} C m, or coulomb meter.)

An actual calculation of the charge separation in H—Cl can provide a more quantitative sense of the meaning of the dipole moment. A table of experimentally measured dipole moments gives $\boldsymbol{\mu}$ =1.08 D for HCl. A table of bond lengths gives the length of the H—Cl bond as 127.4 pm; this is $\mathbf{r}$ in Eq. SG1.2. The only unknown remaining in Eq. SG1.2 is q. Because $\boldsymbol{\mu}$ and $\mathbf{r}$ have the same direction, q can be calculated by dividing the magnitude of $\boldsymbol{\mu}$ by that of $\mathbf{r}$.

$$q = |\boldsymbol{\mu}| \div |\mathbf{r}| = 1.08 \times 10^{-8} \text{ esu pm} \div 127.4 \text{ pm} = 0.848 \times 10^{-10} \text{ esu}$$

As noted above, the charge on an electron is 4.8×10^{-10} esu. Consequently, the charge separation in HCl in electronic charge units is

$$q \text{ (in electrons)}\quad = \quad (0.848 \times 10^{-10} \text{ esu}) \div (4.8 \times 10^{-10} \text{ esu per electron})$$

$$= \quad 0.18 \text{ electron}$$

In other words, the dipole moment of HCl, 1.08 D, means that a partial negative charge of 0.18 units ("18% of an electron") is on the chlorine of HCl, and a partial positive charge of 0.18 units is on the hydrogen.

The definition of dipole moment in Eq. SG1.2 contains two elements: the *amount* of charge separated (q) and the *distance* between the separated charges ($\mathbf{r}$). A smaller amount of charge separated by a long bond can result in as significant a dipole moment as a larger amount of charge separated by a shorter bond. For example, the dipole moment of fluoromethane (H_3C—F, 1.82 D) is somewhat less that of chloromethane (H_3C—Cl, 1.94D) even though fluorine is much more electronegative than fluorine. The reason is that the carbon–chlorine bond length (178 pm) is greater than the carbon–fluorine bond length (139 pm). If you perform a calculation on the methyl halides like the one above on HCl (ignoring the C—H bond dipoles in the halomethanes), you will find that there is more charge separation in the C—halogen bond of fluoromethane despite its smaller dipole moment.

SOLUTIONS TO PROBLEMS

Solutions to Focused Problems

1.1   (a)   Because boron is in Group 3A, it has three valence electrons
        (b)   Sulfur is in Group 6A and therefore has six valence electrons.
        (c)   Phosphorus is in Group 5A and has five valence electrons.
        (d)   Calcium is in Group 2A and has 2 valence electrons.
        (e)   Iodine, being in Group 7A, has 7 valence electrons.

1.2   Dot structures of the neutral elements:

   (a)   (b)   (c)   (d)   (e)

   $\cdot\overset{\cdot\cdot}{B}$ $\overset{\cdot\cdot}{\underset{\cdot}{:S}}\cdot$ $\cdot\overset{\cdot\cdot}{\underset{\cdot}{P}}\cdot$ $Ca:$ $\overset{\cdot\cdot}{\underset{\cdot\cdot}{:I}}\cdot$

   Electrons occupying a single orbital are paired. For example, in (b), sulfur has fully a occupied 3s and 3p orbital and therefore two sets of paired electrons. The other two electrons, by Hund's rules, are unpaired, with one each in the two remaining 2p orbitals.

1.3   (a)   The ion formed by less of a valence electron from a bromine atom has six valence electrons and a positive charge:

   $:\overset{\cdot\cdot}{\underset{\cdot}{Br}}^{+}$

   (b)   The ion formed when sulfur gains two valence electrons has eight valence electrons and a charge of 2–.

   $:\overset{\cdot\cdot}{\underset{\cdot\cdot}{S}}:^{2-}$

1.4   In displaying electron configurations with the orbital/superscript notation, you don't need to separately list the x, y, and z 2p or 3p orbitals unless they are not full [as in the nitrogen example in part (a)]. When all of the orbitals are fully occupied, you can use sp6 or 3p6, as in parts (c) through (f) below. However, it's not wrong to separate them out any time.

   (a)

   nitrogen, N: $(1s)^2(2s)^2(2p_x)^1(2p_y)^1(2p_z)^1$
         or
   $[He](2s)^2(2p)^3$

   | ENERGY | | |

   2p $\underline{\uparrow}$ $\underline{\uparrow}$ $\underline{\uparrow}$ } valence orbitals and valence electrons
   2s $\underline{\uparrow\downarrow}$
   1s $\underline{\uparrow\downarrow}$ } core electrons

   (b)

   oxygen, O: $(1s)^2(2s)^2(2p_x)^2(2p_y)^1(2p_z)^1$
         or
   $[He](2s)^2(2p)^4$

   2p $\underline{\uparrow\downarrow}$ $\underline{\uparrow}$ $\underline{\uparrow}$ } valence orbitals and valence electrons
   2s $\underline{\uparrow\downarrow}$
   1s $\underline{\uparrow\downarrow}$ } core electrons

(c)

chloride ion, Cl⁻: $(1s)^2(2s)^2(2p)^6(3s)^2(3p)^6$
or
$[Ne](3p)^6$
or
$[Ar]$

(d)

phorphorus, P: $(1s)^2(2s)^2(2p)^6(3s)^2(3p_x)^1(3p_y)^1(3p_z)^1$
or
$[Ne](3s)^2(3p_x)^1(3p_y)^1(3p_z)^1$

(e)

magnesium, Mg: $(1s)^2(2s)^2(2p)^6(3s)^2$
or
$[Ne](3s)^2$

(f)

magnesium ion, Mg²⁺: $(1s)^2(2s)^2(2p)^6$
or
$[Ne]$

In this case, the magnesium atom, when it loses all of its valence electrons to become the dipositive Mg²⁺ ion, has no remaining valence electrons. In principle, the level-2 electrons could then be viewed as valence electrons of the ion; however, these electrons are so tightly held that Mg²⁺ never loses an electron to become Mg³⁺. Therefore, it is more chemically reasonable to view the 2s and 2p electrons as core electrons.

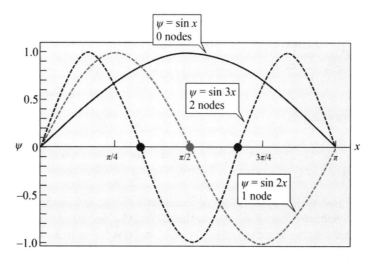

Figure SG1.3   A plot of the function $\psi = \sin nx$ for $n = 1, 2,$ and 3 for the solution to Focused Problem 1.6. The nodes are given by a gray filled circle (for $\psi = \sin 2x$) and black filled circles (for $\psi = \sin 3x$).

1.5   Both N and P have the same number of valence electrons. This is expected because they are in the same group of the periodic table (group 5A). However, their valence electrons are in different principal quantum levels.

1.6   **Fig. SG1.3** shows a plot of $\psi = \sin nx$ for the different values of n. Evidently, the number of nodes in this domain is one less than the "quantum number" n; that is, nodes $= n - 1$.

 Orbitals are described by mathematical functions of three variables, whereas $\psi = \sin nx$ is a function of one variable. However, the quantum numbers of orbitals have the same effect as n in this simple equation: they control the number of nodes.

1.7   A 4s orbital should consist of four concentric spheres and a total of three spherical nodes, as shown in the following diagram. In contrast, a 3s orbital has two spherical nodes and is smaller.

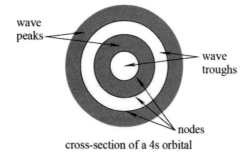

cross-section of a 4s orbital

1.8   (a)   (b)   (c)   (d)

1.9   (a)   Phosphorous acid   (b)   Dimethyl sulfoxide (DMSO)

The two octet structures for phosphorous acid are in equilibrium; the structure with the P—H bond is favored. However, either octet structure is a correct answer for this problem.

1.10   The electrostatic law (Eq. 1.25 in the text) shows that the interaction energy is proportional to the product of the charges on the two ions. If we assume that K^+ and Ca^{2+} are about the same size, then the attraction of Ca^{2+} for Cl^- is stronger than the attraction of K^+ for a Cl^-, because the charge on Ca^{2+} is twice that of K^+. Actually, Ca^{2+} has a considerably smaller ionic radius than K^+. This means that the distance between Ca^{2+} and Cl^- (r in the denominator of Eq. 1.25) is smaller than the distance between K^+ and Cl^- in their respective crystals. The smaller distance between the Ca^{2+} ion and Cl^- also contributes to a greater attractive energy. In summary, the ionic bond in $CaCl_2$ is stronger than the ionic bond in KCl. (The reported *lattice energy* of crystalline $CaCl_2$—the energy required to convert crystalline $CaCl_2$ into its constituent ions in the gas phase—is 2195 kJ mol^{-1}, and the lattice energy of KCl is 715 kJ mol^{-1}.)

1.11   (a)   Carbon with three bonds and one nonbonding electron pair has a negative charge and a complete octet. It conforms to the octet rule.

(b)   Carbon with three bonds and no nonbonding electrons has a positive charge. Its electronic structure is allowed, but it is short of an octet by two electrons.

(c)   Nitrogen with four single bonds and no nonbonding electrons has a +1 charge and a complete octet.

(d)   Nitrogen with two bonds and two pair of nonbonding electrons has a −1 charge and a complete octet.

1.12   (a)   (b)   (c)   (d)

total charge –2   total charge 0   total charge –1   total charge 0

1.13   (a)   (b)   (c)   (d)   (e)

[Only structure (d) does not require adding additional electrons.]

1.14   (a)   (b)   (c)

chloroform   **ammonium ion**   **ethoxide ion**   **methoxymethyl anion**
(trichloromethane)

(d)

ethylammonium ion   **dimethylammonium ion**

The first structure under (c), ethoxide ion, is a common species that we'll see often. The second structure, although a rather unusual ion, is nevertheless an allowed structure.

1.15   Two structures for diazomethane:

H₂C=N̈=N̈:   and   H₂C̈—N≡N̈

diazomethane

1.16   Two resonance structures for benzene:

benzene

Each bond is a single bond in one structure and a double bond in the other. On average, each bond has a bond order of 1.5.

1.17   (a)   Structures of the allyl anion:

$$[H_2C{=}CH{-}\overset{..}{C}H_2 \longleftrightarrow H_2\overset{..}{C}{-}CH{=}CH_2]$$

resonance structures of the allyl anion

(b)   Each terminal (CH$_2$) carbon bears 0.5 of a negative charge; the central carbon has no charge. To calculate the fraction of the charge on a given carbon, we add up the charge on that carbon in both structures (–1) and divide by the number of resonance structures (2). Similarly, to calculate the bond order of a given carbon–carbon bond, we add up the bond orders of that bond in all resonance structures (2 + 1) and divide by the number of resonance structures (2) to give a bond order of 1.5 for each carbon–carbon bond. This process assumes that all resonance structures contribute equally to the structure of the hybrid.

(c)   A hybrid structure:

$$H_2\overset{\delta-}{C}\text{----}CH\text{----}\overset{\delta-}{C}H_2$$

hybrid structure of the allyl anion

1.18   (a)   The tetrafluoroborate anion is tetrahedral—that is, it has F—B—F bond angles of 109.5°.
(b)   The formaldehyde molecule has trigonal planar geometry. Thus, both the H—C—H bond angle and the H—C=O bond angle are about 120°.
(c)   In acetonitrile, each of the H—C—H bond angles, as well as the H—C—C bond angle, are tetrahedral (about 109.5°). The C—C≡N bond angle is linear (180°).
(d)   Chloramine, like ammonia, is trigonal pyramidal.
(e)   The H—C—H geometry of methylene is bent, and the bond angle is about 120° if we consider the nonbonding electrons to be a "group." (For reasons given in Sec. 1.7E, the H—C—H angle should be considerably smaller.)

1.19   The line-and-wedge structures corresponding to the models of dichloromethane:

1.20   (a)   (b)

molecule (a)   molecule (d)   molecule (b)

1.21   Bond angles: *aa, ab, bc, bd, cd, de, df,* and *ef* are all about 120°, because all are centered on atoms with trigonal planar geometry; *fg* is predicted to have the tetrahedral value of 109.5. The bond lengths increase in the order

$$a < g < e < b < d < f < c$$

(In Chapter 4, we show that C—H bonds attached to carbons of double bonds are shorter than C—H bonds attached to carbons of single bonds; this is why $a < g$. A single bond that connects carbons of double bonds is shorter than a single bond that connects the carbon of a double bond to the carbon of another single bond. This is why $d < f$.)

1.22   If two atoms X and Y are bonded, then one or more dihedral angles are required when there are bonds on *both* X and Y to other atoms. In molecule *A*, the dihedral angles between C—H bonds and N—H bonds would be required to specify geometry fully. In molecule *D*, dihedral angles between the C—H bonds on each carbon and the O—C bond to the opposite carbon would be required. The geometry of molecules *B* and *C* can be specified without dihedral angles.

1.23   First we hybridize the orbitals of a *neutral* nitrogen atom and populate them with electrons.

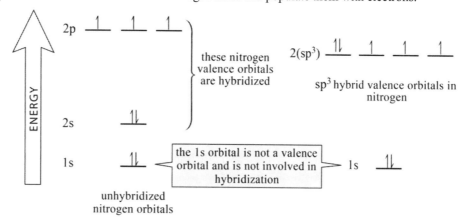

Then we remove a valence electron to give the hybrid orbitals of N⁺. The result is four sp^3 hybrid orbitals, each containing one electron. Each orbital can combine with a hydrogen atom to give the four electron pair bonds in $\overset{+}{N}H_4$.

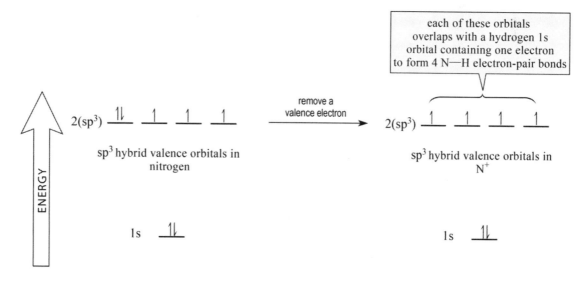

1.24   The hybridization diagram for boron (B) in BF_3 and the structure of BF_3:

The 2p orbital is empty in an sp^2-hybridized boron atom. Each of the singly-occupied sp^2 hybrid orbitals overlaps with the orbital of a fluorine atom containing one electron to give three B—F σ bonds.

1.25   Beryllium (atomic number 4) has two 1s electrons and two 2s electrons. Because BeH_2 is a linear molecule, the Be is sp-hybridized. The hybrid-orbital diagram for Be is much like the one for carbon in acetylene (Display 1.67, Sec. 1.7D of the text).

A perspective sketch of the orbitals involved and a description of the bonds:

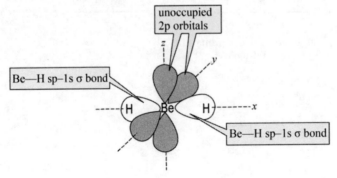

1.26   (a)   Because the central carbon of allene is bonded to two groups, its geometry is linear, and its hybridization is sp.

(b)   The 2p orbitals of allene are shown with the structure in perspective. The $2p_z$ orbitals are shaded gray and the $2p_y$ orbitals are unshaded.

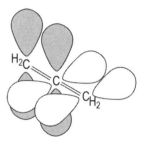

The two 2p orbitals on the central sp-hybridized carbon are perpendicular; these are the two unhybridized 2p orbitals. The two 2p orbitals on the end carbons are necessarily perpendicular, because each forms a π bond with a different 2p orbital on the central carbon.

(c)   The two sets of perpendicular 2p orbitals form two π orbitals. [The shading differentiates the π orbitals and corresponds to the shading in part (b); it is not meant to convey peaks and troughs.]

(d)   Because the two π orbitals are perpendicular, the two H—C—H planes must also be perpendicular.

1.27   (a)   Because oxygen is surrounded by three "groups" (the carbon, the hydrogen, and a nonbonding electron pair) it should be sp^2-hybridized.

valence orbitals in an sp^2-hybridized O^+

(Because an O^+ has five valence electrons, the hybrid-orbital diagram is identical to the one for nitrogen in text Display 1.78.)

(b)   The carbon and oxygen are both sp^2-hybridized. However, the orbital containing the nonbonding electron pair on oxygen will have slightly more 2s character because nonbonded electrons prefer s orbitals. The other hybrid orbitals will have slightly less 2s character and more 2p character to compensate.

(c)   A sketch of the orbitals involved:

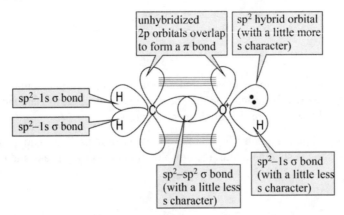

(d)   The geometry is basically trigonal planar at both the oxygen and the carbon atoms. However, because the orbital containing the nonbonded electron pair has more 2s character, and the other hybrid orbitals on oxygen have slightly more 2p character, the C—O—H bond angle is somewhat smaller than 120°.

1.28   (a)   The nitrogen and the carbon are both sp-hybridized in acetonitrile.

(b)   The nonbonding electron pair on nitrogen is in an sp hybrid orbital. [But see part (c).]

(c)   A sketch of the orbitals involved:

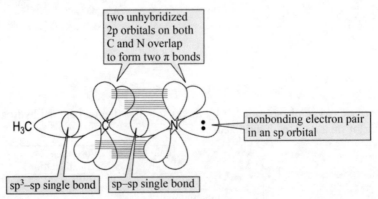

(The sp^3 hybrid orbitals on carbon used to form bonds to the hydrogens are not shown.) The orbital housing the nonbonding electron pair on nitrogen will assume somewhat greater 2s character, and the hybrid orbital used to form the bond to carbon to compensate will have somewhat less 2s character and more 2p character.

1.29   The pictures and energy levels of the molecular orbitals for parts (a), (b), (c), and (d) are essentially the same as they are for the dihydrogen molecular orbitals in Display 1.83 of the text; they differ in electron occupancy.

(a)   The He_2^+ ion contains three electrons. This can be conceived as the combination of a He^+ ion with a He atom. By the aufbau principle, two electrons occupy the bonding molecular orbital, and one occupies the antibonding molecular orbital. This is shown in the electron-occupancy diagram (a) of **Fig. SG1.4**.

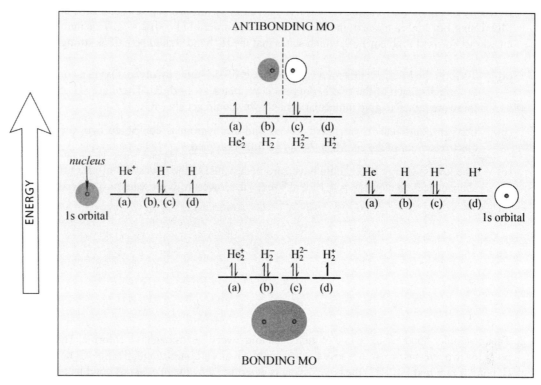

Figure SG1.4   Electron-occupancy diagrams for the solutions to Problems 1.29a-d. The electrons are represented by single-barbed arrows.

Because the bonding molecular orbital contains a greater number of electrons than the antibonding molecular orbital, this species is stable.

(b)   The H_2^- ion contains three electrons. This can be conceived as the combination of a hydrogen atom with a hydride ion (H^-). By the aufbau principle, two occupy the bonding molecular orbital, and one occupies the antibonding molecular orbital. This is shown in the electron-occupancy diagram (b) of Fig. SG1.4. Because the bonding molecular orbital contains a greater number of electrons than the antibonding molecular orbital, this species is stable. Notice that in terms of electron occupancy, H_2^- and He_2^+ are identical.

(c)   The H_2^{2-} ion can be conceived to result from the combination of two hydride ions (H^-). This species contains four electrons; two occupy the bonding molecular orbital, and two occupy the antibonding molecular orbital. This is shown in the electron-occupancy diagram (c) of Fig. SG1.4. In this species, the energetic advantage of the electrons in the bonding molecular orbital is cancelled by the energetic disadvantage of the same number of electrons in the antibonding molecular orbital. This species consequently has no energetic advantage over two dissociated hydride (H^-) ions, and therefore it readily dissociates.

(d)   The H_2^+ ion is conceptually derived from the combination of a hydrogen atom with a proton. It contains one electron in a bonding molecular orbital. This is shown in the electron-occupancy diagram (d) of Figure SG1.4. Because the bonding molecular orbital contains a greater number of electrons than the antibonding molecular orbital, this species is stable.

1.30   Because the H_2^+ ion has one electron in a bonding molecular orbital, it should have about half the stability of H_2 itself relative to dissociated fragments. Therefore, 218 kJ mol^{-1} (52 kcal mol^{-1}) is an estimate of the bond dissociation energy of this species.

1.31   Referring to Figure SG1.4d, the H_2^+ ion has 1 electron in the bonding MO and no electrons in the antibonding MO. Using Eq. 1.87 in the text, the H_2^+ ion has a bond order of $(1 - 0)/2 = \frac{1}{2}$. This number corresponds with the answer to Focused Problem 1.30, which states that the H_2^+ bond should be half as strong as the H_2 bond.

1.32   (a)   To form the MOs, we mix the valence orbitals of the atoms involved. Oxygen contributes one 2s and three 2p orbitals; each of the two hydrogens contributes a 1s orbital. There are six atomic orbitals, and when these are mixed to give molecular orbitals we should get six MOs.

   (b)   Oxygen contributes 6 valence electrons and the hydrogens contribute two, for a total of eight valence electrons. Four of the six MOs should be fully occupied.

   (c)   The lowest-energy MO should have zero nodes. Just as the lowest-energy MO for methane is a methane-shaped electron cloud (Fig. 1.14 in the text), the lowest MO for water is a water-shaped electron cloud.

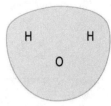

1.33   The most polar bond has to be one of the bonds to the very electronegative chlorine. The Si—C bond is the most polar for two reasons. First, Si is less electronegative than carbon. Second, the Si—Cl bond is longer than the C—Cl bond. From text Eq. 1.92, the bond dipole is determined by the *product* of bond length and charge separation.

1.34   The carbon with the most partial positive charge is the one bonded to the two electronegative atoms O and Cl.

the carbon with the most partial positive charge

1.35   Because we can move any vector along any parallel path (Study Guide Link 1.1), bond dipole vectors for identical bond types pointing in opposite directions must have a net zero resultant. Therefore, molecules *B* and *C* have zero dipole moment. Molecules *A* and *D* have a nonzero dipole moment.

1.36   (a)   Because nitrogen is more electronegative than H, most of the positive charge actually resides on the hydrogens.
   (b)   Because nitrogen is more electronegative than hydrogen, most of the negative charge actually resides on the nitrogen. Therefore, the formal charge gives a qualitatively accurate picture of actual charge.
   (c)   Because oxygen is much more electronegative than either hydrogen or carbon, the positive charge in this molecule is shared by the carbon and the directly attached hydrogens. Therefore, formal charge does *not* show the concentration of actual charge.

1.37   Oxygen is more electronegative than carbon, so the formal-charge orientation (which would result in a bond dipole oriented toward carbon) opposes the orientation of the relative electronegativities (which would result in a bond dipole oriented toward oxygen). The two effects almost cancel, so the actual bond dipole of CO is small.

Solutions to Skills Objectives Problems

1.38   Because the two atoms in question are in "A" groups of the periodic table, the number of valence electrons is given by the group number: 3 for Al and 7 for I.

1.39   Atoms on the far left of the periodic table tend to lose electrons to give a positive ion with the octet configuration of the noble gas from the previous period. Therefore, Mg (period 3, group 2) tends to lose its two valence electrons to form Mg^{2+}, which has the same electronic configuration as neon, the period-2 noble gas. This configuration is $1s^2 2s^2 2p^6$.

1.40   (a)   The first principal quantum level at which d orbitals first appear is $n = 3$. Therefore, the 2d orbital doesn't exist.

(b)   The orbital of highest energy is the 6p orbital, because it has the highest principal quantum number. The orbital of lowest energy is the 2p orbital.

1.41   The electronic configuration of Br:

bromine atom, Br: $(1s)^2(2s)^2(2p)^6(3s)^2(3p)^6(4s)^2(3d)^{10}(4p_x)^2(4p_y)^2(4p_z)^1$
or
$[Ar](4s)^2(3d)^{10}(4p_x)^2(4p_y)^2(4p_z)^1$

In this case, the energy of the 4s orbital lies below that of the 3d orbital. However, the 4s and 4p electrons are the valence electrons because they are the ones involved in the usual chemistry of the bromine atom.

1.42   (a)   A 2p orbital has one planar node. (Cross section in **Fig. SG1.5a**.)
(b)   A 3s orbital has two spherical nodes. (Cross section in **Fig. SG1.5b**.)
(c)   A 1s orbital has no nodes.
(d)   A 3p orbital has a spherical node and a planar node. (Cross section in text Fig. 1.9b.)

1.43   All atoms except boron and sulfur have octets. (The H has a "duet," which we'll classify as an "honorary octet" because it has the nearest-noble-gas number of electrons.) The sulfur has an expanded octet because it is surrounded by 6 pairs of bonding electrons. The boron is surrounded by 6 bonding electrons and is therefore short of an octet.

1.44   Structure *B* is the one that does not have all common bonding patterns. The oxygen in this structure, though it has an octet, is missing a formal charge of 2+; and a dipositive oxygen, because of its electronegativity, is never encountered in ordinary chemistry.

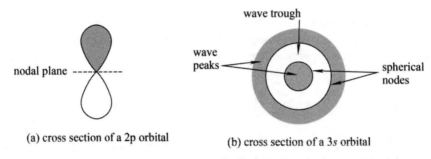

(a) cross section of a 2p orbital   (b) cross section of a 3s orbital

Figure SG 1.5   Orbital cross sections for Problem 1.42.

1.45   (a)   (b)

:N̈=N⁺=N̈:⁻   H—N̈=C=Ö   or   H—N⁺≡C—Ö:̈⁻
azide ion   **isocyanic acid**
                               (two possible structures)

(c)   (d)

butane   **"isobutane"**   **ketene**

1.46   (a)

perchlorate   **methylene**   **ethyl radical**   **trimethylamine oxide**
overall charge: –1   *overall charge = 0*   *overall charge = 0*   *overall charge = 0*

(b)

:C̈≡C—CH₃   :N̈=N⁺=N̈—CH₃   **formate   trichloromethyl
propyn-1-ide   **methyl azide   anion**
anion**

1.47   (a)   Because sulfur and carbon have comparable electronegativities, we expect the positive charge to be shared by sulfur and the carbons, as well as by the hydrogens, which have even lower electronegativity.

(b)   Because oxygen is considerably more electronegative than iodine, we expect the bulk of the negative charge to be on the oxygen as the formal-charge assignment suggests.

1.48   Structures *A* are not equivalent because, in the structure on the left, the positive charge is on a carbon bonded to only one other carbon, whereas in the other structure the positive charge is on a carbon bonded to three other carbons. Structures *B* are equivalent.

1.49   The three structures for the carbonate ion are completely equivalent; each accounts for one-third of the properties of the actual hybrid ion.

(a)   Because each oxygen has a negative charge in two of the three structures, there is $^2/_3$ of a negative charge on each oxygen.

(b)   Each C—O bond is a double bond in one of the three structures and a single bond in two of the three structures. Therefore, the bond order of each C—O bond is $^4/_3$.

1.50   Each bond in ozone has a bond order of $^3/_2$; the "half bond" in the resonance hybrid is represented with dashed lines. The central oxygen has a positive charge in both structures, so it has a full positive charge in the hybrid; and the terminal oxygens each have a negative charge in one structure and no charge in the other, so their charge is represented as $\delta-$ in the hybrid.

To depict nonbonding electrons in a hybrid structure, we typically show only the nonbonded electrons that are invariant across the different structures. The others are inferred from the partial charges. They of course would be shown explicitly in the actual resonance structures (as in the structures of ozone in the problem statement in the text).

1.51   The bond lengths increase in the following order:

$$c \approx a < d < b \approx f < e < h < g$$

The rules in Sec. 1.6B are listed in order of importance. Therefore, the row of the periodic table is paramount in determining bond length. At this point in our study of organic chemistry, we have no reason to differentiate the lengths of bonds a and c. However, we will show in Chapter 4 that a C—H bond involving an sp^2-hybridized carbon is shorter than a C—H bond involving an sp^3-hybridized carbon. Bond b is between two period-2 elements (C and C), whereas bond f is between a period-1 and a period-3 element. The "average period" for both is 2. In fact, the two bonds have about the same length.

1.52   The molecular shapes at each labeled atom:

a: The geometry at nitrogen is trigonal pyramidal.
b: The geometry at carbon is tetrahedral.
c: The geometry at the carbon of the triple bond is linear.
d: The geometry at the carbon of the double-bond is trigonal-planar.
e: The geometry at the carbon is tetrahedral.

For determining bond lengths, invoke the rules for determining relative bond lengths in Sec. 1.6B. The shortest bond is the N—H bond. The longest bond is the C—Br bond.

1.53   With four bonds to Al and no nonbonding electrons on Al, the geometry at the aluminum must be tetrahedral.

tetrachloroaluminate ion

1.54   (a)   The asterisked carbon is sp^2-hybridized. For the oxygen, we take our cue from the similar situation at nitrogen (text Display 1.79) and the hybridization of oxygen in water (text Display 1.80 and discussion) to determine that the oxygen is also sp^2-hybridized This hybridization results in trigonal-planar geometry, which allows the best overlap of an unshared electron pair on oxygen with the carbonyl π bond.

(b)   The geometry at the carbon is trigonal-planar with bond angles about 120°, and the geometry at the oxygen is roughly trigonal-planar with bond angles somewhat smaller than 120°.

(c)   The structures with dihedral angles of 0° and 180°:

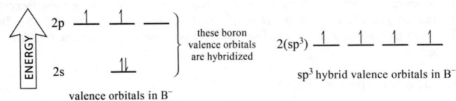

1.55   The boron in tetrafluoroborate is tetrahedral and must be sp^3-hybridized. A hybrid-orbital diagram starts with a negatively charged boron (which has *four* valence electrons) and hybridizes the valence orbitals to form four sp^3 hybrid orbitals. (In the following diagram below we don't show the core 1s electrons, which are not involved in hybridization.) Each of these can combine with a fluorine sp orbital containing a single electron to make four equivalent B—F bonds.

For H_2S, we assume that sulfur is hybridized approximately like the oxygen in H_2O. (Follow the diagram below.) We form the sulfur sp^2 hybrid. One of the sp^2 hybrid orbitals contains the nonbonding electron pair, and the other two sp^2 hybrid orbitals contain single electrons that will pair with the 1s orbitals of hydrogen atoms to make the S—H bonds. We then add s character to the hybrid orbitals containing the nonbonding pair because orbitals containing nonbonded pairs adopt as much s character as possible. A corresponding amount of p character must be added to the singly-occupied hybrid orbitals. The more p character is added, the closer the H—S—H bond angle will approach 90°. (The actual H—S—H bond angle is 92.1°.)

1.56   (a)   The idealized hybridization of oxygen (*a*) of ozone should be trigonal planar with an O—O—O angle of 120°. The orbital containing the nonbonding electron pair on this oxygen will assume somewhat greater s character, and the other hybrid orbitals will assume increasing p character, with the result that the O—O—O bond angle should be smaller than 120°. (The actual bond angle in ozone is 116.8°.)

(b)   The hybridization of sulfur (*b*) in methanethiol, if we follow the water model, should be sp^2 with a C—S—H bond angle of 120°. In this model, one electron pair is in an sp^2 hybrid orbital. This orbital assumes increased s character, which imparts greater p character to the orbitals used to form the bonds to C and H. This should compress the C—S—H angle; the amount of compression depends on the amount of p character; the limit would be a bond angle of 90°. (The actual C—S—H angle in methanethiol is 96°.)

(c)   The hybridization of carbon *c* in methanethiol should be—and is—sp³, with H—C—H and H—C—S bond angles close to tetrahedral.

(d)   The hybridization of the central carbon *d* of 2,4-dimethyl-2,3-pentadiene is sp, with a C—C—C bond angle of 180°.

(e)   The hybridization of carbon *e* of 2,4-dimethyl-2,3-pentadiene is sp² with bond angles close to 120°.

(f)   The nitrogen of the *N*-methylamide anion has an electronic structure similar to that of water—two bonds, two nonbonding electron pairs, and eight valence electrons. Its hybridization should then be sp² with a C—H—N bond angle of 120°. As in H_2O, the electron pair in the hybrid orbital will assume as much s character as possible, thereby giving more p character to the orbitals used to form the bonds to carbon and hydrogen. This will compress the C—N—H angle to be smaller than 120°, possibly close to the C—O—H angle in methanol (H_3C—O—H), which is 109°.

1.57   Dilithium, Li_2, is derived from the combination of two lithium atoms. It contains two bonding electrons derived from two 2s valence electrons, one from each lithium atom. Each Li atom also has two 1s electrons, but these are not valence electrons and therefore are not involved in bond formation and are not shown in the interaction diagram (**Fig. SG1.6**, p. 24). The interaction of the two 2s orbitals gives a bonding and an antibonding MO. The nodes present in the individual 2s orbitals are retained in the MOs. This is a stable species because the bonding molecular orbital contains a greater number of electrons than the antibonding molecular orbital.

1.58   (a)   Assuming that we mix the valence orbitals of the atoms involved to form the molecular orbitals, there are three 1s orbitals from H, a 2s orbital from C, and the three 2p orbitals from C, for a total of seven MOs.

(b)   There are three valence electrons from the hydrogens and three valence electrons from the C^+. The six electrons will fill three of the seven MOs—the three MOs of lowest energy.

1.59   (a)   Each nitrogen contributes 5 valence electrons, so there are 10 electrons (5 pairs) to distribute among the MOs. The distribution is shown in **Fig. SG1.7** on p. 24.

(b)   Using Eq. 1.87 in Sec. 1.8B, we have

$$\text{bond order} = \frac{\text{electrons in bonding MOs} - \text{electrons in antibonding MOs}}{2} = \frac{8-2}{2} = 3$$

(c)   Because the bond order is 3, a corresponding Lewis structure has a triple bond. The two nonbonding electron pairs are the ones in the 2sσ and 2sσ*.

:N≡N:

dinitrogen Lewis structure

1.60   (a)   Because C is considerably more electronegative than Mg, the bond dipoles point away from Mg and toward C.

$$H_3\overset{\delta-}{C} - \overset{\delta+}{Mg} - \overset{\delta-}{C}H_3$$

(b)   The C—Mg—C bond angle is 180°, because there are no nonbonding electrons. The H—C—Mg bond angle and the H—C—H bond angles are close to tetrahedral (109.5°).

(c)   Because the carbon–magnesium bond dipoles are oriented in opposite directions, the dipole moment is zero. (We ignore the very small C—H bond dipoles.)

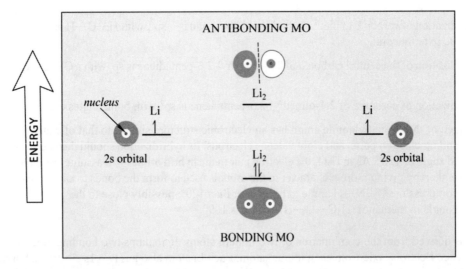

Figure SG1.6   Electron-occupancy diagram for the solution to Problem 1.57.

Figure SG1.7   Molecular orbital occupancy diagram for N_2 to accompany Problem 1.59.

Solutions to Integrated Problems

1.61   The bond angles are different because one of the unshared pairs in sulfur can assume more s character than the analogous pair in oxygen. This difference is possible because the bonds to sulfur are longer than the bonds to oxygen, so the bond angle at sulfur can get closer to the 90° limit required for higher p character of the bond orbitals. (See the answer to Problem 1.55 and 1.56 for a more detailed analysis of this hybridization effect,)

1.62   (a)   (b)   (c)

methyl isocyanate   **methyl cyanate**   **methyl fulminate**

1.63   (a)   The allyl cation is a hybrid of the two structures. Because the structures are equally important, each structure contributes equally to the structure of the hybrid. Because each carbon–carbon bond is a single bond (bond order = 1) in one resonance structure and a double bond (bond order = 2) in the other, the *average* bond order is (1 + 2)/2 = 1.5. In other words, each carbon–carbon bond is equivalent to a single bond plus "half" of a double bond.

(b)   By analogous reasoning, each of the terminal carbons bears +1 charge in one structure and 0 charge in the other. Therefore, the charge on each carbon of the ion is the average of these two, or (+1 + 0)/2 = +0.5. The charge on the central carbon is zero in all resonance structures and is therefore zero overall.

(c)   The hybridization of each carbon is sp^2, and the bond angles at each carbon are 120°.

(d)   A hybrid structure for the allyl cation:

(e)   The resonance structure on the right for the cation shown in (e) violates the octet rule at nitrogen. (There are five bonds to nitrogen.)

This example illustrates that a positive charge does not necessarily mean that an atom lacks an octet. This point will be considered further in Chapter 3

1.64   A linear water molecule would have zero dipole moment, because the O—H bond dipoles would be oriented in opposite directions and would cancel. Hence, the nonzero dipole moment for water shows that it is a bent molecule. Out with Professor Szents!

1.65   To solve this problem, we first have to know the geometry of CCl$_4$. It's tetrahedral, because the carbon is bonded to four identical groups. Now, as suggested in the hint, carry out a vector addition for the bond dipoles of each Cl—C—Cl pair. (Use a dipole arrow of arbitrary length as long as it is the same for both. We'll use an arrow with the length of the C—Cl bonds.) Here is the result of the vector addition, which we've done graphically, for the first pair. Notice that, because the C—Cl bonds are identical, the resultant must *bisect* the angle between the two bonds.

Now derive the resultant for the second pair. Because the C—Cl bonds are identical to those of the first pair, it follows that the resultant is the same length, and it again bisects the Cl—C—Cl bond angle. (We've rotated the molecule about an axis along the dipole so we are viewing the other two chlorines from the perspective we used to view the first two—in the plane of the page.)

The two resultants we have just obtained are oriented at an angle of 180°; that is, they are pointed in opposite directions. Because they are identical in magnitude, they cancel. Therefore, the dipole moment of CCl$_4$ is zero.

resultant dipole = 0

1.66   (a)   When the O—H bonds lie at a dihedral angle of 180° their bond dipoles cancel; consequently, a hydrogen peroxide molecule in this conformation would have a dipole moment of zero. This conformation is *ruled out* by the observation of a significant dipole moment.

the O—H bond dipoles cancel if the dihedral angle is 180°

(b)   The following construction for the O—O—H angle is useful for solving this problem:

The *x* components of the two O—H bond dipoles point in opposite directions regardless of the bond angle or the dihedral angle, so the *x* components cancel. Therefore, we can deal with the *y* component of the O—H bond dipole exclusively. First, we have to calculate the *y* component. This is just 1.52 cos (6.5°) = 1.52 (0.9935) = 1.51 D.

Note that 96.5° is not the dihedral angle, but the bond angle. The idea is to calculate the dihedral angle. From the analysis in the previous paragraph, we know that the resultant dipole moment of the molecule (given as 2.13 D) is derived from vector addition of the *y* components of the two O—H bond dipoles, which we now know are 1.51 D. All we have to do is determine the angle θ that will give 2.13 D as the resultant from vector addition of two 1.51 D dipoles.

1.51 D θ 1.51 D θ 2.13 D   angle = 180° − θ

The proper construction is the arrangement on the left; because we can move a vector at will so long as its magnitude and direction remain unchanged, we derive the construction on the right. The resultant, 2.13D, is the vector that completes a head-to-tail connection of the two bond dipoles. From the law of cosines, this vector is

$$(2.13 \text{ D})^2 = (1.51 \text{ D})^2 + (1.51 \text{ D})^2 - 2(1.51 \text{ D})(1.51 \text{ D}) \cos (180° - \theta)$$

or

$$4.54 \text{ D}^2 = 2.28 \text{ D}^2 + 2.28 \text{ D}^2 - (4.56 \text{ D}^2) \cos (180° - \theta)$$

from which we have

$$\frac{-0.02 \text{ D}^2}{-4.56 \text{ D}^2} = \cos (180° - \theta) = 0.00439$$

Using the "arccos" function on a calculator, we find that 180° − θ = 89.8° or θ, the desired dihedral angle is 180° − 89.8° = 90.2°. Therefore, it appears that the dihedral angle in hydrogen peroxide is about 90°. (The actual structure of hydrogen peroxide determined by electron diffraction shows a dihedral angle of 93.8°, so the use of dipole moments comes very close to predicting the correct dihedral angle.) The conformation of hydrogen peroxide is best represented as follows:

dihedral angle ~ 90°

1.67 The solution to this problem requires a construction similar to the one used in the solution to Problem 1.66 in which the resultant is known and the individual O—H bond dipoles are unknown.

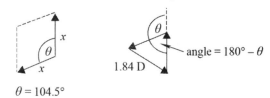

$\theta = 104.5°$

Applying the law of cosines, with x as the O—H bond dipoles,

$$1.84^2 = x^2 + x^2 - 2x^2[\cos(180° - \theta)]$$

$$3.39 = 2x^2(1 - \cos 75.5°) = 2x^2(1 - 0.25) = 1.50\,x^2$$

$$x = \sqrt{\frac{3.39}{1.50}} = 1.50\ \text{D}$$

This is the calculated value of the O—H bond dipole, which is very similar to the value assumed in Problem 1.66.

1.68 (a) The bonding molecular orbital is derived by the constructive overlap of wave peaks. The antibonding molecular orbital is derived by changing the peak to a trough and the trough to a peak in one of the two $2p$ orbitals. The destructive overlap of a peak with a trough gives a node. The resulting molecular orbitals are shown in **Fig. SG1.8**.

(b) The nodes are shown in Fig. SG1.8. Both of the MOs retain the nodes of the original 2p orbitals. The antibonding MO has an additional node that results from the destructive overlap of a peak and a trough.

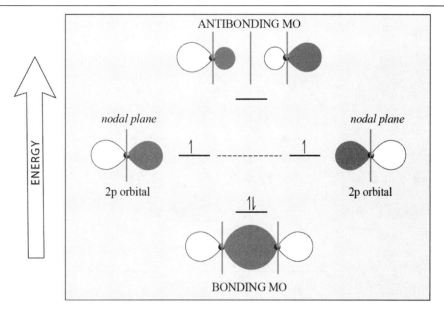

Figure SG1.8   Molecular orbital interaction diagram for Problem 1.68a

(c)   Figure SG1.8 is the required interaction diagram.

    (d)   The resulting bond is a σ bond because it looks the same no matter how it is rotated about the orbital axis—that is, it is *cylindrically symmetrical* about the bond axis.

1.69   (a)   The bonding molecular orbital is derived by the constructive "side-to-side" overlap of peaks with peaks and troughs with troughs. The antibonding molecular orbital is derived by changing the peak to a trough and the trough to a peak in one of the two 2p orbitals. Destructive overlap of troughs with peaks and peaks with troughs gives an additional node in the antibonding orbital. These molecular orbitals are shown in **Fig. SG1.9**.

    (b)   The nodes are also shown in Fig. SG1.9. Both of the MOs have the nodes of the original 2p orbitals, which merge into a single node because the nodal plane is common to both orbitals. The antibonding MO has in addition a node between the original orbitals that results from the destructive overlap of a peak and a trough.

    (c)   Fig. SG1.9 is the required interaction diagram.

    (d)   The resulting bond is *not* a σ bond because it does *not* look the same when it is rotated about the internuclear axis—that is, it I *not* cylindrically symmetrical about the bond axis.

 This type of bond, called a pi (π) bond, is important in the carbon–carbon double bond. Pi bonds are discussed in Chapter 4.

1.70   (a)   See the solution to Problem 1.57 and **Fig. SG1.10a**.

    (b)   See the solution to Problem 1.68 and **Fig. SG1.10b**.

    (c)   See the solution to Problem 1.69 and **Fig. SG1.10c**.

    (d)   Bringing the two $2p_z$ orbitals together results in two new MOs, one bonding ($2p\pi_z$) and one antibonding ($2p\pi_z{}^*$), as shown in **Fig. SG1.10d**. These orbitals are identical to those obtained from overlapping the $2p_y$ orbitals as shown in part (c), except that they are oriented on the z-axis, rotated 90° from the y-axis.

    (e)   The atomic orbitals of oxygen are combined according to parts (a–d) of this problem and are ordered in increasing energy as shown in **Fig. SG1.11**. Each oxygen atom has six valence electrons, two in the 2s orbital, and four in the three 2p orbitals. These electrons are placed in molecular orbitals according to Hund's rules.

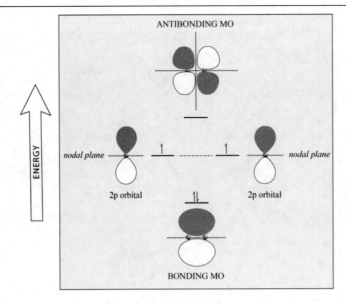

Figure SG1.9   Orbital interaction diagram for Problem 1.69. MOs resulting from side-to-side bonding are called pi (π) orbitals.

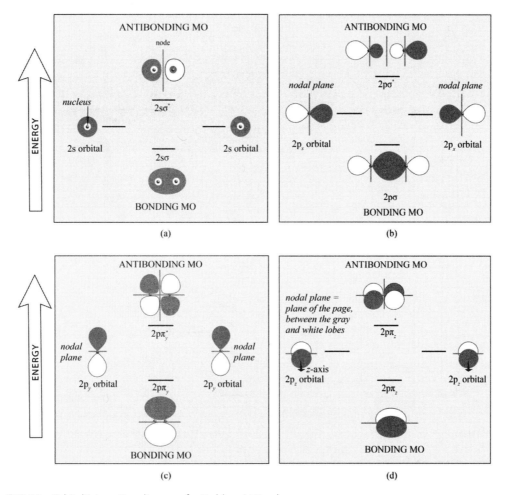

Figure SG1.10   Orbital interaction diagrams for Problem 1.70a–d.

(f)   The oxygen molecule has a net (or, nonzero) spin, and is magnetic. An unpaired electron generates a magnetic field due to its spin. The two unpaired electrons in the $2p\pi_y^*$ and $2p\pi_z^*$ orbitals do not have electrons of opposing spin to cancel them out, so the oxygen molecule is "paramagnetic" and liquid oxygen can be trapped between poles of a magnet.

(g)   We eliminate structure *A* because it does not have unpaired electrons. We also eliminate structure *D* because it has less than 12 total electrons. According to text Eq. 1.87 (Sec. 1.8B), oxygen has a bond order of $(8 - 4)/2 = 2$. However, none of the remaining structures contains a double bond. Yet we can use an "average structure", or resonance hybrid, of the molecule containing the single bond (*B*) and the one containing a triple bond (*C*). This average would have a bond order = 2.

$$\left[\cdot \ddot{\underset{\cdot\cdot}{O}} \!-\! \ddot{\underset{\cdot\cdot}{O}} \cdot \quad \longleftrightarrow \quad \cdot \ddot{O} \!\equiv\! \ddot{O} \cdot \right]$$

You may notice that the second structure above violates the octet rule. But, as you perhaps learned when you worked Focused Problems 1.29a and b, MO theory takes us beyond conventional Lewis structures and allows this violation provided that there is net bonding. However, there is an energetic cost: the bond energy of O_2 is about the same as that of a conventional *single* bond, but the bond length is about that of a conventional triple bond. This dichotomy requires the use of resonance structures if we insist on describing O_2 with Lewis structures!

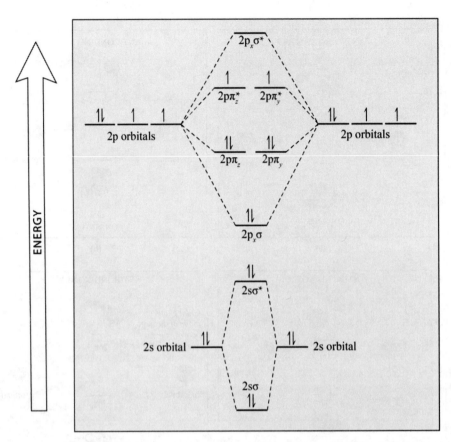

Figure SG1.11   An orbital interaction diagram for dioxygen to accompany Problem 1.70e and f showing electron occupancies for the MOs.

Alkanes and Organic Nomenclature

CHAPTER 2 SUMMARY

- Hydrocarbons are compounds that contain only carbon and hydrogen. There are two broad classes of hydrocarbons.

  - *Aliphatic hydrocarbons*, of which there are three sub-classes: alkanes, compounds with only carbon–carbon single bonds; alkenes, compounds with one or more carbon–carbon double bonds; and alkynes, compounds with one or more carbon–carbon triple bonds.

  - *Aromatic hydrocarbons*, which contain benzene or other aromatic rings.

- The complete description of molecular structure consists of the "three C's:" *composition* (molecular formula; *constitution*, or *connectivity* (the way carbons are bonded together); and *configuration*, the arrangement of atoms in three-dimensional space.

- Alkanes can contain branched chains, unbranched chains, or rings. A noncyclic alkane with n carbons has the composition C_nH_{2n+2}. A cyclic alkane with n carbons and m rings has the composition $C_nH_{2n+2-2m}$.

- Alkanes have sp³-hybridized carbon atoms with tetrahedral geometry.

- The structures of hydrocarbons are typically represented as condensed structures or skeletal structures. Skeletal structures of noncyclic alkanes are drawn as "zig-zag" lines with a carbon understood to be at each end and each vertex. Hydrogens need not be written explicitly, but are understood to be present on each carbon to fill out the carbon valence of 4. Skeletal structures of cyclic alkanes are drawn as regular polygons in which a carbon and its complement of hydrogens are assumed to be at each vertex.

- Carbons are classified as primary (three attached hydrogens), secondary (two attached hydrogens, *tertiary* (one attached hydrogen) or *quaternary* (no attached hydrogens). The corresponding hydrogens are also classified as primary, secondary, or tertiary.

- Isomers are different compounds with the same molecular formula. Compounds that have the same molecular formula but differ in their constitution (or connectivities) are called *constitutional isomers*.

- Alkanes are named systematically according to the substitutive nomenclature rules of the IUPAC. The name of a compound is based on its principal chain, which, for an alkane, is the longest continuous carbon chain in the molecule. The names of the parent unbranched alkanes with 1–12 carbon atoms are listed in Table 2.1 (Sec. 2.2B).

- An alkane name is constructed from the principal chain and the location of substituents using rules 1–10 in Sec. 2.4C. Common substituents are listed in Tables 2.3 and 2.4 in Sec. 2.4C.

- The conformation of an alkane is known when all of its dihedral angles are specified. Newman projections are often used to show dihedral angles. (Sec. 2.5A and B) A Newman projection depicts the dihedral angles about one carbon–carbon bond. Many Newman projections might be required to show the conformations of a molecule.

- Along with Newman projections, sawhorse projections and line-and-wedge structures can be used to draw molecular conformations about specific bonds. A single line-and-wedge structure can be used to show the conformations about several sequential carbon–carbon bonds provided that these bonds can all be drawn in the plane of the page.

■ Alkanes exist in various conformations that rapidly interconvert at room temperature.

   ● The stable conformations of alkanes have staggered bonds on adjacent carbons (dihedral angles are multiples of 60°).

   ● The staggered conformation that minimizes van der Waals repulsions has the lowest energy and is the predominant one. In butane, the major conformation is the anti conformation; the gauche conformations exist to a lesser extent.

■ The boiling points of alkanes and many other types of organic compounds increase within a homologous series by 20–30 °C per carbon atom.

■ The melting points of alkanes and many other types of organic compounds increase with the number of carbons within a homologous series. Within this trend, however, alkanes with odd and even numbers of carbons lie on separate curves in which the melting points of even-numbered compounds have the higher values.

■ Combustion is the most widely used reaction of alkanes. It finds practical application in the generation of much of the world's energy. The metabolism of carbohydrates is a form of combustion that supplies energy for living organisms.

■ Organic compounds are classified by their functional groups. Different compounds containing the same functional groups undergo the same types of reactions.

■ The "R" notation is used as a general abbreviation for substituent groups; Ph is the abbreviation for a phenyl group, and Ar is the abbreviation for an aryl (substituted phenyl) group.

■ Alkanes are derived from petroleum and are used mostly as fuels; however, they are also important as raw materials for the industrial preparation of other organic compounds.

STUDY GUIDE LINKS

2.1   Nomenclature of Simple Branched Compounds

The IUPAC recognizes as valid the following older names for branched alkane isomers of four, five, and six carbon atoms because of their common historical usage.

H_3C
$CH-CH_3$ or $(CH_3)_3CH$
H_3C
isobutane

H_3C
$CH-CH_2CH_3$ or $(CH_3)_2CHCH_2CH_3$
H_3C
isopentane

H_3C
$CH-CH_2CH_2CH_3$ or $(CH_3)_2CHCH_2CH_2CH_3$
H_3C
isohexane

CH_3
$H_3C-C-CH_3$ or $(CH_3)_4C$
CH_3
neopentane

CH_3
$H_3C-C-CH_2CH_3$ or $(CH_3)_3CCH_2CH_3$
CH_3
neohexane

The distinguishing structural feature of an alkane named with the "iso" prefix is a pattern of two methyl branches at the end of a carbon chain. The distinguishing structural feature of an alkane with a "neo" prefix is a pattern of three methyl branches at the end of a carbon chain.

Table 2.4 in Sec. 2.4C of the text shows that the names of the common branched alkyl groups are derived from the names shown above for the corresponding alkanes.

FURTHER EXPLORATIONS

2.1   Atomic Radii and van der Waals Repulsion

The *atomic radius* is the effective size of an atom when it is *covalently bonded to another atom*. In this text, this term means the same thing as *covalent radius*. The van der Waals radius, in contrast, measures an atom's effective size when it is *not covalently bonded* to another atom.

The covalent radius of an atom is much smaller than its van der Waals radius. The reason is that because electrons are shared in a chemical bond, nuclei are drawn closer together than they would be in the absence of such sharing. In other words, electrons in a bond are attracted to both nuclei, and this attraction pulls the nuclei together. *Nonbonded* atoms do not share their electrons, by definition.

The covalent bond in the hydrogen molecule provides a simple example. The bond length of H_2 (H—H, or dihydrogen) is 74.6 pm (0.746 Å). If we define the covalent radius as half of the separation of the two hydrogen nuclei, then the covalent radius of hydrogen is (74.6 pm)/2 = 37.8 pm. The *van der Waals radius* of hydrogen is 120 pm. This comparison shows that a covalently bonded hydrogen is separated much less from its bonded partner than it is from a nonbonded atom.

If nonbonded atoms are pushed together so that their electron clouds overlap, they resist. This resistance is called *van der Waals repulsion*. Pushing atoms together can result from applying pressure. We know that if we apply pressure to a balloon, the instant we release the pressure, the balloon springs back to its original shape. This is because the molecules of gas inside the balloon resist being pushed together. In some structures, atoms that are not bonded to each other are forced to violate each other's van der Waals radii; *gauche*-butane is such a structure. The resulting van der Waals repulsions cause the molecule to have a greater internal energy as compared with *anti*-butane.

REACTION REVIEW

I.   REACTION OF HYDROCARBONS

A.   COMBUSTION

1.   Combustion is the most widely utilized reaction of alkanes.

2.   Complete combustion affords carbon dioxide and water as the only combustion products.

$$C_nH_{2n+2} \ + \ \frac{3n+1}{2}O_2 \ \longrightarrow \ n\,CO_2 \ + \ (n+1)\,H_2O$$

3.   Combustion can be used in elemental analysis, which is the quantitative determination of elemental compositions.

   a.   The mass percents of the elements obtained from a combustion analysis of a compound can be used, together with the molecular mass, to determine the molecular formula. (See Integrated Problem 2.55 in the text.)

SOLUTIONS TO PROBLEMS

Solutions to Focused Problems

2.1 (a) The number of hydrogens in an alkane (branched or unbranched) is $2n + 2$, where n is the number of carbons. Consequently, there are $2(18) + 2 = 38$ hydrogens in the alkane with eighteen carbons.

(b) No; all alkanes—indeed, all hydrocarbons—must have an even number of hydrogens. If $2n + 2 = 23$, then n would have to be 10.5—an impossible number of carbons.

2.2 Skeletal formulas of tridecane:

$$CH_3CH_2CH_2CH_2CH_2CH_2CH_2CH_2CH_2CH_2CH_2CH_2CH_3$$

structural formula of tridecane   skeletal structure of tridecane

Don't forget that there is a carbon at each end of the skeletal structure as well as at each vertex.

2.3 (a) There are 8 carbons and 18 hydrogens: (b) A structure showing all of the atoms:

octane

2.4 (a) Every ring decreases the number of hydrogens by 2 from the fully saturated number of $2n + 2$ (n = the number of carbons). Therefore, the number of hydrogens is $2n + 2 - 4 = 2n - 2$.

(b) By the reasoning in part (a), the formula is $2n - 4$.

(c) By the reasoning in part (a), the formula is number of hydrogens $(N_H) = 2n + 2 - 2m = 2(n + 1 - m)$.

2.5 (a) Use the formula in part (c) of the solution to the previous problem with $n = 8$. With the number of hydrogens $= 10$, then $10 = 2(8 + 1 - m)$. Solving for m gives $m = 4$. The compound has 4 rings. By the same process, a cycloalkane C_7H_{12} has 2 rings.

(b) The compound has 14 carbons and has number of hydrogens $(N_H) = 2(14 + 1 - 4) = 22$. The compound has 22 hydrogens.

2.6 (a) 2,3,5-Trimethylhexane. Numbering of the chain is determined by pairwise comparison of the possible numbering schemes: (2,3,5) versus (2,4,5). The first scheme has the lower number at first point of difference (Rule 8).

(b) 2,4-Dimethylhexane

(c) 4-Ethyl-5-propyloctane. In this case, alphabetical citation governs the numbering of the principal chain, because numbering is ambiguous otherwise (Rule 10).

(d) 3,3,6-Trimethyloctane. Rule 8 rules out 3,6,6-trimethyloctane. (This is a substituted octane because the "Et" group is part of the principal chain.)

2.7   (a)

ethylcyclopentane

(b)

2,6-diethyl-1,1-dimethylcyclohexane

2.8   (a)   The name is 1-ethyl-2,4-dimethylcyclopentane. Rule 10 (alphabetical citation) and Rule 8 dictate the numbers of the substituents.

(b)   The carbon with two branches receives the number 1 by Rule 8. The name is 1,1,2,4-tetraethyl-3-methylcyclobutane. Tetraethyl begins with the letter *e* for citation purposes; that is, the prefix "tetra" is ignored in the alphabetical citation of prefixes.

(c)   The name is 1-isopropyl-4-methylcyclohexane because *iso* receives citation preference over *methyl*, and, in this case, the first-cited substituent determines numbering (Rule 10).

2.9   The compound is 3-ethyl-4-propylheptane, and the expanded structure is as follows:

$H_3C-CH_2-CH_2-CH-CH-CH_2-CH_3$
$H_3C-CH_2-CH_2$ CH_2-CH_3

skeletal structure

Your structures might look different from these. Any structures with the same connectivity as these are also correct.

2.10   (a)   The Lewis and skeletal structures.

$$\begin{array}{ccc} p & p & p \\ CH_3 & CH_3 & CH_3 \end{array}$$
$$\begin{array}{c} p \\ CH_3CHCH_2C-CHCH_2CH_3 \\ t \quad s^2 \qquad t \quad s^2 \end{array} \quad p$$
$$\begin{array}{c} CH(CH_3)_2 \\ t \quad p \end{array}$$

4-isopropyl-2,4,5-trimethylheptane

(b)   The primary carbons are identified with a *p*, the secondary carbons with an *s*, the tertiary carbons with a *t*, and the quaternary carbon with a *q* on the skeletal structure.

(c)   The primary hydrogens are identified with a *p*, the secondary hydrogens with an *s*, and the tertiary hydrogens with a *t* on the Lewis structure.

2.11   (a)   The structure of 3-isopropyl-1,1-dimethylcyclohexane (at right).

(b)   The primary, secondary, tertiary, and quaternary carbons are labeled with *p*, *s*, *t*, and *q*, respectively, on the structure.

(c)   The primary hydrogens are on the carbons labeled *p*; the secondary hydrogens are on the carbons labeled *s*; and the tertiary hydrogens are on the carbons labeled *t*.

3-isopropyl-1,1-dimethylcyclohexane

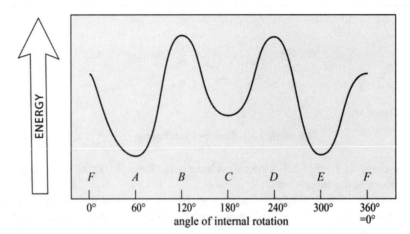

Figure SG2.1   A diagram of potential energy versus angle of internal rotation in isopentane to accompany the solution to Problem 2.12. The conformations identified by letter are shown in the solution to part (a).

2.12   (a)   The staggered conformations of isopentane are *A, C,* and *E;* the eclipsed conformations are *B, D,* and *F.*

A	*B*	*C*	*D*	*E*	*F*

(b)   The curve of potential energy versus angle of rotation is shown in **Fig. SG2.1**. The eclipsed conformations *B* and *D* have the highest energy because they have eclipsed methyl groups.

(c)   Conformations *A* and *E* have the lowest energy because they have one less interaction in which methyl groups are "close together"—that is, the C—CH$_3$ bonds have a 60° dihedral angle—than conformation *C*; consequently, either of these conformations is present in greater concentration than conformation *C*. Conformations *A* and *E* are present in equal amounts because they have the same energy.

2.13   (a)   The projected bond is the carbon–carbon bond at the end of the butane molecule:

projected bond
↓

H$_3$C—CH$_2$CH$_2$CH$_3$

There are three identical staggered conformations and three identical eclipsed conformations.

A	*B*	*C*	*D*	*E*	*F*

(b)   The curve of potential energy versus angle of internal rotation is essentially identical to the curve for ethane (text Fig. 2.3, Sec. 2.5A), except that the energy difference between staggered and eclipsed forms is slightly greater. That is, the energies of all staggered forms lie at equal minima, and the energies of all eclipsed forms lie at equal maxima.

(c)   All staggered conformations are present in equal amount because they are identical.

There is a subtlety here. Problem 2.13 has considered the conformation about the C1–C2 bond in isolation. Any staggered conformation about the C1–C2 bond contains a mixture of gauche and anti conformations about the C2–C3 bond.

2.14 The line-and-wedge and sawhorse structures for the two gauche conformations of butane.

sawhorse projections { ... } line-and-wedge structures

2.15 Structures *B* and *C* are drawn correctly: two bonds in the plane, one solid wedge coming towards you and one dashed wedge going away. In structure *A*, the H—C—H bond angle appears to be 180°, or linear, which is inconsistent with the 109.5° bond angles in an sp^3-hybridized atom. In structure *D*, the wedge and dashed wedge are on opposite sides of the H—C—Br bonds, which is not possible for a tetrahedral molecule. You should build a model and prove to yourself that arrangements *A* and *D* cannot exist.

2.16 (a) Observing the sawhorse structure from the perspective shown, the viewer would observe the compound as the Newman projection shown to the right. The propyl group is abbreviated as C_3H_7.

(b) Rotating C2, the front carbon, by 120° in the clockwise direction gives the structures shown below. View a model in each of the orientations shown so that you can relate the three-dimensional structures to your drawings.

Newman projection from part (a)   C2 rotated clockwise 120°   sawhorse structure   line-and-wedge structure

(c)   Rotating C2, the front carbon, by 180° gives the structures shown below. These conformations are eclipsed.

| Newman projection from part (a) | C2 rotated clockwise 180° | sawhorse structure | line-and-wedge structure |

2.17 (a) Pentane < 2-methylhexane < 2-methylheptane. These are arranged essentially by atomic mass. Branching will lower the boiling points of the last two compounds somewhat, but not enough to compensate for the 50°–60° expected for the different atomic mass. (The actual boiling points are 36, 90, and 116 °C.)

(b) Benzene < toluene < ethylbenzene. These are arranged in order of increasing molecular mass. (The actual boiling points are 80.1. 110.6, and 136 °C.)

2.18 (a) 2,2,3,3-Tetramethylbutane, because of its highly symmetrical structure, has the highest melting point of +100.8 °C. Octane is the next most symmetrical compounds and has the melting point of –56 °C. 2-Methylheptane has the lowest melting point of –109 °C.

(b) Benzene, being more symmetrical, has the higher melting point of +5.5 °C despite its lower molecular mass.

2.19 Water is not very effective in distinguishing a gasoline fire because the gasoline is not "smothered" by the water. Because it is less dense than water, gasoline always rises to the surface of water, where the air supports further combustion (see text Sec. 8.6D for further discussion).

2.20 (a) For the complete combustion of an alkane with n carbons, n mol of CO_2 is formed. For an alkane with $2n$ + 2 hydrogens there must be $[(2n + 2)$ hydrogens ÷ 2 hydrogens per water$] = n + 1$ water molecules. Add up the oxygens from the CO_2 and H_2O formed to get the O_2 required.

$$C_nH_{2n+2} + \frac{3n+1}{2}\,O_2 \longrightarrow n\,CO_2 + (n+1)\,H_2O$$

(b) Follow the same process used in part (a).

$$C_nH_{2n} + \frac{3n}{2}\,O_2 \longrightarrow n\,CO_2 + n\,H_2O$$

2.21 Use the balanced equation in the solution to Focused Problem 2.21a with $n = 8$. First, calculate the moles of octane in 15 gallons of octane. (The molecular mass of octane is 114.)

$$\text{moles } C_8H_{18} = (15\text{ gal})(3.785\text{ L gal}^{-1})(692\text{ g L}^{-1})([1/114]\text{ mol g}^{-1}) = 345\text{ moles}$$

As shown in the solution to Focused Problem 2.21a, eight moles of CO_2 are formed per mole of octane burned. Therefore, $(345)(8) = 2{,}760$ moles of CO_2 are formed.

Next, convert 2,760 moles of CO_2 into pounds of CO_2 to obtain the answer:

$$\text{lb } CO_2 = (2{,}760\text{ mol})(44\text{ g mol}^{-1})(0.001\text{ kg g}^{-1})(2.20\text{ lb kg}^{-1}) = 267\text{ lbs}$$

2.22 (a) The only carboxylic acid with this formula is acetic acid (at right).

acetic acid

(b)   Any structure with five carbons, an —OH group, and *either* a ring or a double bond is correct. Two of many examples are the following:

cyclopentanol H_2C=$CHCH_2CH_2CH_2$—OH

4-penten-1-ol

2.23   With 12 hydrogens and 5 carbons, this compound cannot have any double bonds or rings. Because amides, carboxylic acids, phenols, and esters each contain one or more double bonds, these groups are ruled out. An amide is also ruled out because it contains nitrogen, which is not present in the formula. The remaining possibilities are a compound with two ether groups, a compound with two alcohol groups, or a compound with both an ether and an alcohol. Examples of each are the following:

HO—$CH_2CH_2CH_2CH_2CH_2$—OH   HO—$CH_2CH_2CHCH_3$ CH_3O—CH_2CHCH_3

a compound with   a compound with   a compound with
two alcohol groups   an alcohol and an ether group   two ether groups

(Other correct structures are possible.)

2.24   (a)   Structure *B*   (b)   Structure *A*   (c)   Structure *B*   (d)   Structure *C*

For part (c), the two carbons of the ring are considered to be alkyl substituents even though they are "tied together" in a ring.

Solutions to Skills Objectives Problems

2.25   Compound *A* is an aromatic hydrocarbon; compound *B* is an alkyne; compound *C* is an alkene; and compound *D* is an alkyne, an alkene, and an aromatic hydrocarbon. None of the structures are "pure" alkane, although each structure has alkyl groups.

2.26   (a)   The composition of butane is given by its formula, C_4H_{10}.
(b)   The constitution of butane is given by its structural formula: four carbons in an unbranched chain with three hydrogens on the terminal carbons and two hydrogens on the internal carbons: $CH_3CH_2CH_2CH_3$.
(c)   Each carbon has tetrahedral geometry; that is, the bonds to each carbon point to the corners of a tetrahedron.

(d)   The possible staggered conformations are the gauche and anti conformations. In addition to these staggered conformations, there are two types of eclipsed conformation—one in which the carbon–methyl bonds eclipse each other, and two others in which the carbon–methyl bonds eclipse carbon–hydrogen bonds.

gauche   eclipsed 1   anti   eclipsed 1   gauche   eclipsed 2

2.27   (a)   The unbranched alkane with ten carbons is decane. $CH_3(CH_2)_8CH_3$

decane

      (b)   The unbranched alkane with five carbons is pentane.

$CH_3(CH_2)_3CH_3$

pentane

2.28   (a)   There are $2n + 2$ hydrogens with $n = 15$—that is, 32 hydrogens.

      (b)   In this case, $2n + 2 = 22$, so $n = 10$. There are 10 carbons in this hydrocarbon.

      (c)   With no rings there would be $2(30) + 2$ hydrogens, or 62 hydrogens. Each ring results in two fewer hydrogens, so, with 4 rings, there are 8 fewer hydrogens. There are $62 - 8 = 54$ hydrogens in this hydrocarbon.

2.29   Structure *C* has nine eight carbons, whereas the other structures have nine carbons. Therefore, structure *C* is not an isomer of structures *A*, *B*, or *D*.

To determine the relationship of the remaining structures, consider a distinguishing feature in one molecule and see how it relates to the connectivity of the same feature in the others. For example, in both *A* and *D*, the isobutyl group is connected to the penultimate carbon of a five-carbon chain, as shown at right. Therefore, structures *A* and *D* are identical. Structure *B* has an isopropyl group attached to a six-carbon chain rather than an isobutyl group attached to a five-carbon chain. Therefore, structure *B* has a different constitution and is therefore a constitutional isomer of structure *A* and *D*.

2.30   (a)   (b)   (c)

methylbutane   **2,2,3-trimethylbutane**   **3,4-dimethylhexane**
"isopropylethane"   "2-*tert*-butylpropane"   "2-*sec*-butylbutane"

Although methylbutane could be called 2-methylbutane, the numerical designation is superfluous because there is only one possible position of a methyl substituent on a four-carbon unbranched carbon chain.

2.31   Compound *A* is 5-ethyl-4-methylnonane. (Notice that the principal chain is not drawn straight across the page.) The substitution pattern in compound *B* is a 4,5,5,6-substituted nonane. Ethyl receives numerical precedence because it is first cited in the name (Rule 10) and there is no other basis for deciding precedence. Therefore, the name is 4-ethyl-5,5,6-trimethylnonane.

2.32   (a)   (b)

4-isobutyl-2,5-dimethylheptane   **5-neopentyldecane**

(c)

(d)

Me   Me   Et

1-cyclopropyl-2,4-dimethylcyclohexane

5-ethyl-3,3-dimethyloctane

2.33   A Lewis structure and a skeletal structure for 3,5-diethyl-4-(2-ethylpropyl)heptane:

CH₃CH₂

CH₃CH₂—CH   H

C

CH₃CH₂—CH   CH—CH₂CH₃

CH₃CH₂   CH₂CH₃

2.34   (a)

HO   OH

Me ·····C — C ◄ Et

Me   Et

(b)

Cl   Et

H—C — C ·····Cl

Et   H

(c)

Cl   Me

Cl ·····C — C ·····F

Me   Br

2.35   (a)   Newman projections of the three staggered conformations of 2,3,3,4-tetramethylpentane: (*i*Pr = isopropyl)

Me

Me   Me

Me   *i*Pr

H

A

Me

H   Me

Me   *i*Pr

Me

C

Me

Me   H

Me   *i*Pr

Me

E

(b)   To rank the energies of these conformations, list the destabilizing gauche interactions. We assume that any interaction with a hydrogen is negligible.

For *A*:   3 Me–Me, 1 Me–*i*Pr

For *C*:   2 Me–Me, 2 Me–*i*Pr

For *E*:   3 Me–Me, 1 Me–*i*Pr

The energies of *A* and *E* are equal. The energy of *C* is higher, because the methyl–*i*Pr interaction is more severe than a Me–Me interaction.

(c)   The three eclipsed conformations:

Me Me

Me H   Me

*i*Pr

B

Me H

Me Me   Me

*i*Pr

D

Me Me

H Me   H

*i*Pr

F

(d)   The energies of *B* and *D* are the same and are higher than the energy of *F*. All three conformations have methyl–methyl and methyl–hydrogen eclipsing interactions, but *B* and *D* have an isopropyl–methyl eclipsing interaction instead of an isopropyl–hydrogen interaction. The van der Waals repulsion between the two large groups raises the energy of conformations *B* and *D*.

(e)   The sketch of the conformational energy versus angle of rotation is in **Fig. SG2.2**. (The relative shape of the curve is the same the one in Fig. SG2.1.)

(f)   The conformations *A* and *E* are present in the same amount, and in greater amount than *C*, in the conformational equilibrium, because lower-energy conformations are always present in greater amounts than higher-energy ones.

2.36   (a)   Classification of carbons in *tert*-butylcyclopentane: (*p*= primary; *s* = secondary; *t* = tertiary; *q* = quaternary)

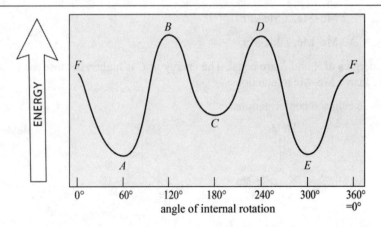

tert-butylcyclopentane

(b)   The hydrogens on carbons labeled *p* are primary hydrogens; the hydrogens on carbons labeled *s* are secondary hydrogens; and the hydrogen on the carbon labeled *t* is a tertiary hydrogen.

2.37   Add 25 °C per additional carbon or subtract 25 °C for each fewer carbon.

(a)   For 2-methylheptane the estimate is 149 °C – 25 °C = 124 °C. (The reported boiling point is 117–118 °C.)

(b)   For 2-methyldecane the estimate is 149 °C + 50°C = 199 °C. (The reported boiling point is 189–190 °C.)

2.38   The highly symmetrical 2,2,3,3,4,4-hexamethylpentane has the highest melting point (+65 °C). The less symmetrical undecane has the next highest melting point (–26 °C). The compound 2-methyldecane has the lowest melting point (–45 °C).

2.39   (a)   For octane, descriptions 2, 6 and 8 are applicable.

(b)   For propane, descriptions 4 and 8 are applicable.

2.40   (a)   Because the only source of the hydrogens in water is the hydrogens in the hydrocarbon, and because the cyclic hydrocarbon has fewer hydrogens per mol, one mol of the cycloalkane produces fewer moles of water.

Figure SG2.2   The conformational energy of 2,3,3,4-tetramethylpentane to accompany the solution to Problem 2.35.

(b)   Because the same number of carbons are in both the cyclic and noncyclic alkane, they produce the same amount of carbon dioxide per mole.

The conclusions of both parts can be verified with a comparison of the balanced equations in the solution to Focused Problem 2.20.

2.41   (a)   Ketone
        (b)   Nitrile
        (c)   Alcohol
        (d)   Ether

2.42   The partial structures show substitution around the carbon with the OH group; the tetrasubstituted carbon is irrelevant to this substitution. Patterns *A* and *D* are the applicable R notations.

Solutions to Integrated Problems

2.43   (a)   Add 50 °C for the two additional carbons to obtain 178 °C as the estimate. (The actual boiling point is 172 °C.)

        (b)   Because these two compounds have the same functional group (ketone) and the same number of carbons, they should have about the same boiling point. (The actual boiling point is reported as 147–150 °C.)

2.44   (a)   Take a systematic approach by starting with all of the trimethylpentanes:

2,3,4-trimethylpentane   **2,2,4-trimethylpentane**   **2,2,3-trimethylpentane**   **2,3,3-trimethylpentane**

Then we consider the possible pentanes with one ethyl and one methyl substituent:

3-ethyl-2-methylpentane   **3-ethyl-3-methylpentane**

These structures are the only possibilities. Notice that if we put an ethyl substituent on the second carbon of pentane, then we have actually formed a hexane, which doesn't meet the criterion that the principal chain contain 5 carbons.

        (b)   We have to consider all possible dimethylhexanes and ethylhexanes.

2,3-dimethylhexane

(continued)

$$CH_3$$
$$CH_3CCH_2CH_2CH_2CH_3$$
$$CH_3$$

2,2-dimethylhexane

$$CH_3$$
$$CH_3CH_2CCH_2CH_2CH_3$$
$$CH_3$$

3.3-dimethylhexane

$$CH_2CH_3$$
$$CH_3CHCHCH_2CH_2CH_3$$

3-ethylhexane

2.45 (a) Neopentane (dimethylpropane) has five carbons, and all hydrogens are primary:

dimethylpropane

(b) Cyclopentane has five carbons and only secondary hydrogens:

cyclopentane

(c) The structures that fit this description have several rings fused together. Among these are the following, along with their trivial names.

prismane   **cubane**   **tetrahedrane**

2.46 (a) The principal chain has eight carbons and is numbered in the structure at right. There are two methyl substituents at carbon 4 and an ethyl substituent at carbon 5. The name is 5-ethyl-4,4-dimethyloctane.

(b) The principal chain has 8 carbons with substituents at carbons 4 and 5. The ethyl group receives precedence in numbering because it is cited first. The name is 4-ethyl-5-methyloctane.

(c) The principal chain has nine carbons with methyl substituents at carbons 2 and 3. The name is 2,3-dimethylnonane. (Remember that there is a carbon at the end of each line as well as at the vertices.)

(d) The name is 4-isopropyloctane.

(e) The proper numbering scheme (by Rule 8) is 1,1,2,5. The name is 2,5-diethyl-1,1-dimethylcyclopentane.

2.47 (a)

$$CH(CH_3)_2$$
$$CH_3-CH_2-CH-CH-CH-CH_2-CH_3$$
$$CH_2-CH_3 \quad CH_3$$

3-ethyl-4-isopropyl-5-methylheptane

(b)

2,3,5-trimethyl-4-propylheptane

(c)

5-*sec*-butyl-6-*tert*-butyl-2,2-dimethylnonane

2.48 The procedure is to draw the structure with the incorrect (but unambiguous) name, and then name the structure correctly.

(a)   "2-Ethyl-2,4,6-trimethylheptane" actually has an 8-carbon principal chain and is properly named 2,4,6,6-tetramethyloctane.

2,4,6,6-tetramethyloctane

(b)   5-Neopentyldecane is named correctly.

$$CH_2—C(CH_3)_3$$
$$CH_3(CH_2)_3CH(CH_2)_4CH_3$$

5-neopentyldecane

(c)   4-Ethyl-1,2-diisopropylcyclohexane is named correctly. Note application of the "first point of difference" rule.

*i*Pr

Et —   —*i*Pr

4-ethyl-1,2-diisopropylcyclohexane

(d)   "3-Butyl-2,2-dimethylhexane" should be named 4-*tert*-butyloctane.

4-*tert*-butyloctane

2.49   (a)   The systematic IUPAC name of "marathane" is 3,3,4-triethylhexane.

(b)   There are several possible candidates for the principal chain that have 7 carbons. Rule 3 covers this situation: choose the chain with the greatest number of branches. The correct choice has 6 branches. (The other possibilities have 3 and 4 branches, respectively.) This is not named as a substituted cyclohexane because the principal chain has more than 6 carbons.

3,5-dicyclohexyl-5-ethyl-2,2-dimethyl-4-propylheptane

(c)   The systematic name for "pullane" is 1,1-di-*tert*-butyl-2,4-diethylcyclopentane.

(d)   The 11-carbon cycloalkane is cycloundecane, and "puh-lane" has substituents that will be numbered 1,1,3,10 to give the name 1-ethyl-1-methyl-3,10-dipropylcycloundecane. The "first point of difference" rule (Rule 8) dictates which of the several possible numbering schemes is chosen.

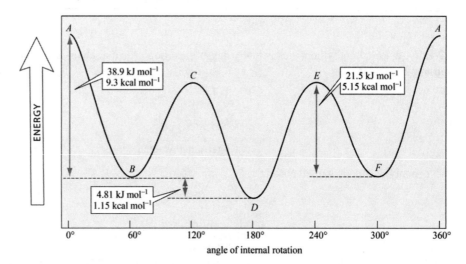

Figure SG2.3   The conformations of 1,2-dichloroethane to accompany the solution to Problem 2.50.

2.50   (a)   The sketch of potential energy versus dihedral angle is shown in **Fig. SG2.3**. The conformations are coded in the figure as follows:

(b)   Conformation D is present in greatest amount because it has the lowest energy.

2.51   Compound A should have the greatest energy difference between staggered and eclipsed forms. If we take the energy of the staggered form in each case as an arbitrary zero of energy, the methyl groups in the eclipsed conformation are much closer together in compound A than in compound B because of the smaller bond length of the intervening bond (C—C vs. Si—Si). The van der Waals repulsions between the methyl groups in compound A are therefore greater than the repulsions in compound B. These repulsions raise the energy of the eclipsed form of compound A more than they raise the energy of the eclipsed form of compound B.

2.52   Because the C—O bond length in compound A is smaller than the C—C bond length in B, the methyl groups in compound A have more severe van der Waals repulsions in the gauche conformation than those in compound B. If we assume that the effect on the anti conformation is miniscule, the gauche conformation of A then has a higher energy relative to its anti conformation than the gauche conformation of B.

2.53   (a)   The expected dipole moment for the anti conformation is zero, because the polar C—Br bonds are oriented in opposite directions and their bond dipoles cancel. Likewise, corresponding to every C—H bond is another C—H bond pointing in the opposite direction.

(b)   It follows from part (a) that the observed dipole moment is due solely to the relatively small amount of gauche conformation present. Hence,

$$\mu = \mu_{anti} N_{anti} + \mu_{gauche} N_{gauche}$$

or, with $\mu_{anti} = 0$,

$$1.0 = (0)(0.82) + \mu_{gauche}(0.18) = \mu_{gauche}(0.18)$$

Solving for μ_{gauche} gives μ_{gauche} = 5.56 D. This is a very large dipole moment. But this is what would be expected for a compound in which two C—Br bond dipoles (as well as the smaller C—H bond dipoles) are oriented in nearly the same direction.

2.54 The Oxhide family burns (12,000 miles/25 miles gal^{-1}) = 480 gallons of gasoline per year. The result from Focused Problem 2.21 is based on 15 gallons; scale the result from that solution to 480 gallons by multiplying it by the factor (480/15) = 32. This gives a carbon footprint of 8,544 pounds per year of CO_2 for the Oxhide family car.

For about 285,000,000 vehicles in the United States traveling 12,000 miles yr^{-1} with an average mileage of 18 miles gal^{-1}, the annual gasoline consumption is 190 billion gallons with a carbon footprint of about 3.4×10^{12} (3.4 *trillion*) pounds of CO_2!

2.55 (a) The mass of carbon in the sample equals the mass of carbon in the CO_2 product.

mass of carbon in CO_2 = (mass of the CO_2 sample)(fraction of the CO_2 mass that is carbon)
= (24.60 mg)(12.01/44.01) = 6.71 mg

The mass of hydrogen in the sample equals the mass of hydrogen in the H_2O product:

mass of hydrogen in H_2O = (mass of the H_2O sample)(fraction of the H_2O mass that is hydrogen)

= (11.51 mg)(2.016/18.016) = 1.29 mg

The total mass of C and H is 8.00 mg, as it must be if C and H are the only elements present in the sample.

(b) The millimoles (mmol) of C in the sample is 6.71/12.01 = 0.559, and the mmol of H is 1.29/1.008 = 1.28. Therefore, there are (1.28/0.559) = 2.29 moles of hydrogen per mole of carbon in the sample, or $C_1H_{2.29}$.

(c) Multiplying 2.29 by 7 gives 16.0 hydrogens. Therefore, the formula of X is C_7H_{16}.

(d) The molecular mass for the formula C_7H_{16} is 100.20. Therefore, this is the actual formula for the compound.

2.56 The functional groups in acebutolol are identified and the compound classes are labeled in the following structure.

ketone

amide

amine

$CH_3CH_2CH_2$—C—ṄH—   ÖCH₂—CH—CH₂—ṄH—CH(CH₃)₂

ether

aromatic
hydrocarbon   :ÖH

alcohol

acebutolol

2.57 (a) The side chain of asparagine contains an amide functional group.

(b) The side chain of tyrosine contains a phenol functional group.

(c) The side chain of threonine contains an alcohol functional group.

2.58 (a) Two constitutionally isomeric amides with the formula C_4H_9NO containing an isopropyl group:

:O: :O:

$(CH_3)_2CH$—C—ṄH₂   H—C—ṄH—CH(CH₃)₂

(b) Three constitutionally isomeric amides with the formula C_4H_9NO *not* containing an isopropyl group:

$$H_3C - \overset{\overset{\displaystyle :O:}{\|}}{C} - \overset{..}{N}H - CH_2CH_3 \qquad H - \overset{\overset{\displaystyle :O:}{\|}}{C} - \overset{..}{N}H - CH_2CH_2CH_3 \qquad CH_3CH_2 - \overset{\overset{\displaystyle :O:}{\|}}{C} - \overset{..}{N}H - CH_3$$

(Other structures are also possible.)

(c)   An isomer X of the amide in parts (a) and (b) with the formula C_4H_9NO that is *not* an amide and that contains an amino group and an alcohol group: (These are four of several possible structures.)

(d)   This compound could *not* be a nitrile. A nitrile has a triple bond, which would require two fewer hydrogens (that is, C_4H_7NO).

The Curved-Arrow Notation, Resonance, Acids and Bases, and Chemical Equilibrium

CHAPTER 3 SUMMARY

■ Two fundamental types of reactions involve electron pairs. All electron-pair reactions can be classified as one of these two reaction types or as combinations of them.

- *Lewis acid–base association reactions*, in which a nucleophile donates a pair of electrons to an electron-deficient electrophile. The reverse of a Lewis acid–base association is a Lewis acid–base dissociation, in which a leaving group departs from the electrophile with a pair of electrons.

- *Electron-pair displacement reactions*, in which a nucleophile donates a pair of electrons to an electrophile that is not electron-deficient. In this case, a bond between the electrophilic center and a leaving group becomes an unshared pair on the leaving group, and the electrophilic atom is transferred from an atom of the leaving group to the nucleophilic center. The reverse of an electron-pair displacement reaction is another electron-pair displacement in which the roles of the nucleophile and leaving group in the forward reaction are reversed.

■ The curved-arrow notation is an important symbolism for depicting the flow of electron pairs in chemical reactions. In this notation, the tail of a curved arrow indicates a source of electrons, and the head indicates the destination.

■ Several terms relating to electron movement in chemical reactions are very important:

- A *nucleophile* donates an electron pair in a chemical reaction. The atom of a nucleophile that actually donates the electrons is a *nucleophilic center*.

- An *electrophile* receives the donated electron pair from the nucleophile, and the atom of the electrophile that actually receives the electrons is the *electrophilic center*.

- In electron-pair displacement reactions, the electrophile also loses an electron-pair bond, which becomes a nonbonding electron pair on an atom of the *leaving group*. The electrophilic center is transferred from the leaving group to the nucleophile.

- In a Lewis acid–base association reaction, there is no leaving group. In a Lewis acid–base dissociation reaction, there is a leaving group but no nucleophile.

■ The curved-arrow notation can be used to derive resonance structures.

■ Resonance structures are typically drawn in the following situations:

- Two atoms with significantly different electronegativities are connected by a π bond, which can be delocalized onto the more electronegative atom.

- An atom with a nonbonding electron pair is connected to an electron-deficient atom.

- An electron-deficient atom is connected to one atom involved in a π bond.

- An atom with a nonbonding electron pair is connected to one atom involved in a π bond.

- Three alternating single and double bonds occur within a six-membered ring.

- The relative importance of resonance structures can be assessed with the following guidelines:

  - Structures in which every atom has a complete octet are more important than structures containing electron-deficient atoms.

  - If structures contain charges, the more important structures have negative charges on atoms of greater electronegativity and positive charges on atoms of lesser electronegativity.

  - Neutral structures and structures that delocalize charge are more important than structures that separate charge. An important exception is the greater importance of octet structures with separated charges than hypervalent neutral structures.

- A Brønsted acid–base reaction is a special case of an electron-pair displacement reaction in which the electrophilic center is a proton; a proton is transferred in the reaction, just as an electrophilic center is transferred in a general electron-pair displacement reaction. The electron donor is called a Brønsted base, and the species that donates the electrophilic proton is called the Brønsted acid.

- The strength of a Brønsted acid is indicated by the magnitude of its dissociation constant K_a. Because dissociation constants for various acids can differ by many orders of magnitude, a logarithmic pK_a scale is used, in which $pK_a = -\log K_a$. The strength of a Brønsted base is inferred from the K_a (or pK_a) of its conjugate acid.

- The equilibrium constant K_{eq} for a reaction is related to the standard free-energy difference $\Delta G°$ between products and starting materials by the relationship $\Delta G° = -2.3RT(\log K_{eq})$. Reactions with positive $\Delta G°$ values have $K_{eq} < 1$ and favor starting materials at equilibrium. Reactions with negative $\Delta G°$ values have $K_{eq} > 1$ and favor products at equilibrium.

- The logarithm of the equilibrium constant for any Brønsted acid–base reaction can be calculated by subtracting the pK_a of the acid on the reactant side from the pK_a of the acid on the product side. A Brønsted acid–base equilibrium always favors formation of the weaker acid and weaker base.

- The dissociation state of an acid with dissociation constant K_a is determined by the difference $pH - pK_a$. When $pH << pK_a$, an acid is undissociated. When $pH >> pK_a$, an acid is fully dissociated. When $pH = pK_a$, half of the acid molecules are dissociated; the concentrations of an acid and its conjugate base are equal. The dissociation state is quantitatively described by the fraction dissociated, f_A, or by the fraction undissociated, f_{AH}, given in 3.67a–b in Sec. 3.6E.

- Acidity, basicity, and other chemical properties vary with structure. Five structural effects on Brønsted acidity are the *charge effect*, the *element effect*, the *hybridization effect*, the *resonance effect*, and the *polar effect*.

  - *Charge effect:* An acid is stronger when the element bearing the acidic proton is more positively charged.

  - *Element effect:* Brønsted acidity increases as the atom is further to the right in the "A" groups of the periodic table, and, within a group, toward higher-numbered periods.

  - *Hybridization effect:* Proton acidity increases with increasing s character in the bond to hydrogen.

  - *Resonance effect:* Resonance stabilization of a base increases acidity of its conjugate acid; resonance stabilization of an acid decreases its acidity.

  - *Polar effect:* Electronegative groups near the ionizable hydrogen increase acidity; electropositive groups decrease acidity. The effect is distance-dependent—the closer the groups the bigger the effect.

- The pK_a of an acid is proportional to its standard free energy of ionization $\Delta G°_a$. The process for analyzing the effect of structure on acidity is to assess the effect of structure on the energy of the acid and the base and then to consider how the resulting energies affect the pK_a.

STUDY GUIDE LINKS

3.1   Rules for Use of the Curved-Arrow Notation

Now that you've seen several examples of the curved-arrow notation, let's summarize its use with six basic rules. If you follow these you will always use the curved-arrow notation correctly. The first two rules reiterate points that have been made repeatedly in the text. Examples of the last four rules are provided below.

Rules for use of the curved-arrow notation:

The first two rules are very important and general:

1.   Curved arrows indicate the flow or "movement" of electron pairs, not atoms or nuclei.
2.   A curved arrow is always drawn with its tail at the source of electrons (a bond or unshared pair) and its head at the destination (an atom).

The second two rules deal with the manipulation of nonbonding electron pairs, illustrated by the equation at right below:

3.   When an atom donates a nonbonding electron pair, a curved arrow must indicate that the electron pair is used to form an additional bond to the *same atom.*
4.   When an atom accepts an unshared electron pair, a curved arrow must show that the electron pair originates from a bond to the *same atom.*

> the oxygen that donates the electron pair is involved in the newly-formed bond
>
> $CH_3\ddot{O}H$ + H—$\ddot{\underset{..}{C}l}$: ⟶ $CH_3\overset{H}{\underset{|}{\overset{+}{O}}}H$ + :$\ddot{\underset{..}{C}l}$:
>
> the additional electron pair on the chlorine came from a bond to the same chlorine

The last two rules deal with the manipulation of bonds:

5.   When a bond is broken, a curved arrow must show its conversion either into a nonbonding electron pair on one of the two bonded atoms, or into a new bond involving one of the two original bonded atoms.
6.   When a bond is formed, a curved arrow must show that the bond originates from either a nonbonding electron pair on one of the two bonded atoms, or from a bond involving one of the two bonded atoms.

The following example illustrates these rules:

> the bond broken to Br becomes a nonbonding electron pair on the same Br
>
> $H\ddot{O}$:⁻ $\rightarrow$ H—$\ddot{O}$—CH_2—$\ddot{B}r$: ⟶ $H\ddot{O}$—H + $\ddot{O}$=CH_2 + :$\ddot{B}r$:⁻
>
> the bond between O and H becomes a new bond between one of the atoms (O)

Don't try to memorize these rules. Instead, go back in the text and look at about three or four equations that show the curved-arrow notation and demonstrate to yourself that these rules are applied correctly. Then, the next few times you use the notation, use your intuition to answer the problem, but then check your result against these rules before you look at the answer. If you do this three or four times, you're not likely to misuse the notation.

3.2   Identification of Acids and Bases

This section stresses an important subtlety in dealing conceptually with acids and bases. An acid or a base—whether of the Lewis type or the Brønsted type—is defined by a *reaction,* and not by a *structure.* For example, some compounds react by donating a proton in certain reactions, in which case they are Brønsted acids; in other reactions, they react in other ways, in which case, they are *not* Brønsted acids.

You can sometimes tell from a structure whether a compound *might* react as an acid or a base. When someone looks at a structure and says, "That compound is an acid," what that person really means is "That compound is *capable of acting as an acid.*" It *is* an acid only in reactions in which it is acting like one. Thus, a compound with an unshared electron pair *might* react as a Lewis base; but if the same compound has an acidic hydrogen, it might also react as a Brønsted acid. The water molecule, with its amphoteric character, is an example of exactly this situation. Whether water is an acid or a base depends on the reaction. An electron-deficient compound certainly has the capability to react as a Lewis acid, and we might say, "That compound is a Lewis acid," but it is not a Lewis acid until it reacts like one. It might undergo other types of reactions in which it does not act as a Lewis acid. You have to see what happens to a compound in a reaction before you can classify the compound as an acid or base, and only then is the acid or base character defined for that particular reaction.

Consider the following analogy. Suppose you have a friend who has just purchased a sleek, red convertible. They ask you, "Is that a fast car, or what?" Logically, you can't answer the question until you see the car perform, even though you suspect that it may have the capability of running at great speed. In the Indianapolis 500, your friend's car may *not* be fast at all. On the back roads of rural Indiana, it may be the fastest car anyone has ever seen. When someone shows you a structure and asks, "Is this a strong Brønsted acid?" your logical answer should be, "Show me the reaction and I'll tell you."

SOLUTIONS TO PROBLEMS

Solutions to Focused Problems

3.1   (a)   The product results from donation of an unshared electron pair from the oxygen of H_2O to the electron-deficient carbon of the cation. (The curved arrow in the product is for the reverse reaction.)

$$H_3C-\overset{CH_3}{\underset{CH_3}{\overset{|}{\underset{|}{C^+}}}} \quad :\ddot{O}H_2 \quad \longrightarrow \quad H_3C-\overset{CH_3}{\underset{CH_3}{\overset{|}{\underset{|}{C}}}}-\overset{+}{\ddot{O}}H_2$$

Lewis base

Lewis acid

(b)   The nitrogen of ammonia donates an electron pair to the electron-deficient boron of BF_3. (The curved arrow in the product is for the reverse reaction.)

$$:\ddot{F}-\overset{:\ddot{F}:}{\underset{:\ddot{F}:}{\overset{|}{\underset{|}{B}}}} \quad :NH_3 \quad \longrightarrow \quad :\ddot{F}-\overset{:\ddot{F}:}{\underset{:\ddot{F}:}{\overset{|}{\underset{|}{B^-}}}}-\overset{+}{N}H_3$$

Lewis base

Lewis acid

3.2   (a)

HÖ—CH₂   :Cl:⁻
        |
        CH₃

(b)

        H
        |
H₃C   C⁺—C—CH₃   :Br:⁻
H₃C |
            CH₃

(c)

H₂C   O:
    \ / :O:
H₂C   O:

3.3

electrophilic center   nucleophilic center

leaving group

        CH₃
        |
H₃C—S⁺—CH₃   +   :ÖH   ⟶   H₃C—S: +   CH₃—ÖH
                                      |
                                      CH₃

electrophile   nucleophile

3.4   (a)   To complete the reaction, we recognize that the positively charged carbon is electron-deficient and therefore can act as an electrophile; and the oxygen, with its nonbonding electrons, can act as a nucleophile in a Lewis acid–base association reaction.

(b)   The curved-arrow notation for the reverse reaction is shown on the product.

(c)   The nucleophilic center and electrophilic center for the forward reaction and the leaving group for the reverse reaction are also identified.

electrophilic center   nucleophilic center   leaving group

        CH₃   CH₃
        | |
H₃C—C⁺   HÖCH₃   ⇌   H₃C—C—ÖCH₃
        | |
        CH₃   CH₃ H

3.5   (a)   The forward direction of this example illustrates Situation #2, and the reverse direction illustrates Situation #1.

        H₃C   CH₃   H₃C   CH₃
            \N/   \N⁺/

            ⁺   ⟷

(b)   Both forward and reverse directions illustrate Situation #3.

(c)   This example represents two successive applications of Situation #4 in both directions. The resonance structures could be interconverted in two successive steps. Another resonance structure not included in the problem is shown on the right.

(d)   This example illustrates Situation #1 in the forward direction and Situation #2 in the reverse direction. This also illustrates the conversion of an uncharged hypervalent (octet-expanded) structure into an octet structure with formal charges.

sulfate ion

3.6 (a) This is an example of Situation #4. The nonbonding electrons on a negatively charged oxygen are delocalized to the other oxygen.

acetate ion

(b) This is another example of Situation #4, in this case involving a triple bond.

acetonitrile-*N*-oxide

3.7 The resonance structure indicated by the curved-arrow notation:

3.8 (a) The curved-arrow notation for interconversion of the resonance structures:

acetic anhydride
A

carbonate ion
B

C

trimethyl phosphate
D

(b) The more important structure in each case is marked with an asterisk. (The structures are equivalent and equally important for *B*.) In *A*, the second structure is less important because it separates charge. In *C*, the second structure is more important because every atom has an octet. In *D*, the second structure is more important (despite charge separation) because every atoms has an octet.

3.9 (a) The Brønsted acids and bases and (b) the curved-arrow notation, are both shown in the following equation.

H₃C—O: + H—S—CH₃ ⇌ H₃C—O—H + :S—CH₃
Brønsted base   Brønsted acid   Brønsted acid   Brønsted base

3.10 (a)

(b)

3.11   Let H_2O be the acid on the left. Then CH_3OH is the acid on the right. (If you chose CH_3OH as the acid on the left, then switch reactants and products in the reaction that follows.)

$$\begin{array}{c} \boxed{\text{conjugate base–acid pair}} \\ \\ H\ddot{O}\text{—}H \ + \ \ddot{:}\ddot{O}CH_3 \ \rightleftharpoons \ H\ddot{O}\ddot{:}^- \ + \ H\text{—}\ddot{O}CH_3 \\ \\ \boxed{\text{conjugate acid–base pair}} \end{array}$$

3.12   (a)   The reverse of reaction 3.37a is a Lewis acid–base association reaction; there is no leaving group.

an electrophile;
the positively charged
carbon is the
electrophilic center

a nucleophile
and the nucleophilic center

(b)   The reverse of reaction 3.37b is a Lewis acid–base dissociation reaction. There is no nucleophile or electrophile.

leaving group

(c)   Like the forward reaction, the reverse of reaction 3.37c is a Brønsted acid–base reaction.

Brønsted base;
oxygen is a
nucleophilic center

hydrogen is the
electrophilic center

leaving group

$$\begin{array}{ccccccc} \underset{\substack{CH_3 \\ | \\ CH_3}}{H_3C\text{—}C\text{—}\overset{+}{\underset{\cdot\cdot}{O}}\text{—}H} & + & \ddot{O}H_2 & \rightleftharpoons & \underset{\substack{CH_3 \\ | \\ CH_3}}{H_3C\text{—}C\text{—}\ddot{O}\text{:}} & \cdots\,H\text{—}\overset{+}{\underset{\cdot\cdot}{O}}H_2 \\ & & F & & G & H \end{array}$$

E
conjugate acid

conjugate
base

conjugate base

Brønsted acid;
conjugate
acid

3.13   (a)   Two curved arrows with the proton of HBr as an electrophile makes this a Brønsted acid–base reaction.

nucleophilic
center

electrophilic center

leaving group

$$H_2C\text{=}CH_2 \quad H\text{—}\ddot{B}r\text{:} \quad \longrightarrow \quad H_2\overset{+}{C}\text{—}CH_2 \overset{\displaystyle H}{\underset{}{}} \quad \text{:}\ddot{B}r\text{:}^-$$

Brønsted base   Brønsted acid

(b)   This is a Lewis acid–base association reaction. (Nonbonding electron pairs are omitted for the oxygens that don't participate in the reaction.)

3.14   Setting [B] = 1000 = 10^3, [A] = 1/5.7 = 0.175. Of course, we can't have less than one molecule of A, but this is simply a ratio. We could equally well say that for every molecule of A there are 5700 molecules of B.

3.15   Setting [D] = 1000, there are $(1/7) \times 10^6$ molecules of C, or 1.43×10^5 molecules of C.

3.16   When A and B are present in equal amounts at equilibrium, their standard free energies are equal, $K_{eq} = 1$, and therefore $\Delta G° = 0$. The free-energy diagram is as follows:

3.17   Reactions (a) and (b) are unfavorable and reactions (c) and (d) are favorable.

3.18   In reactions (a) and (b) the products have a greater standard free energy than the reactants. In reactions (c) and (d) the reactants have a greater standard free energy than the products.

3.19   (a)   Because every 5.71 kJ mol^{-1} changes the equilibrium concentration ratio by a factor of 10, then twice 5.71 = 11.4 kJ mol^{-1} changes the equilibrium ratio by a factor of 100. Because the $\Delta G°$ is negative, the equilibrium constant [products]/[reactants] = 100. (This can be verified from Table 3.1.)

(b)   For $\Delta G° = -5.71 \log K_{eq} = -11.4$ kJ mol^{-1}, then $\log K_{eq} = 2$, or $K_{eq} = 100$.

(c)   Every factor of 3 corresponds to a log difference of 0.5, and a factor of 300 corresponds to a log difference of 2.5. Therefore, the free energy difference will be (2.5)(−5.71) = about −6 × 2.5 = −15 kJ mol^{-1} (about 3.6 kcal mol^{-1}). [The exact amount is −14.3 kJ mol^{-1} (−3.41 kcal mol^{-1}).]

3.20 (a) To complete this calculation, first calculate the equilibrium constant, and then use it to do a mass-action calculation.

$$K_{eq} = 10^{-\Delta G°/2.30RT} = 10^{-(-2.93)/5.71} = 10^{0.513} = 3.26$$

	A	B	C
Initially	0.1	0.2	0
Change:	$-x$	$-x$	$+x$
At equilibrium:	$0.1 - x$	$0.2 - x$	x

$$K_{eq} = \frac{[C]}{[A][B]} = \frac{x}{(0.1-x)(0.2-x)} = 3.26$$

$$x^2 - 0.607x + 0.02 = 0$$

$$x = \frac{0.607}{2} \pm \frac{\sqrt{0.368 - 0.080}}{2} = 0.303 \pm 0.268$$

Only $x = 0.035$ M is physically meaningful, because the alternate value, 0.571 M, is larger than the starting values of the reagents. Choose the physically meaningful value of x to obtain

$$[C] = 0.035 \ M, [B] = 0.2 - x = 0.165 \ M, \text{ and } [A] = 0.1 - x = 0.065 \ M.$$

  (b) If the $\Delta G°$ is positive, the equilibrium constant is less than unity and the reaction is less favorable; consequently, there should be much less C and more A and B at equilibrium.

An actual calculation shows that $[C] = 0.0061$. Then $[A] = 0.1 - x = 0.0939 \ M$, and $[B] = 0.2 - x = 0.1939$. The concentration of C is 10% of that in part (a), a result consistent with the statement in part (b).

3.21 (a) A factor of 10 is 1 log unit; so, applying Eq. 3.48a, we find that the difference in standard free energy is 5.71 kJ mol^{-1}, or 1.36 kcal mol^{-1}.

  (b) Applying Eq. 3.48b, we find that a 1 kJ mol^{-1} difference in standard free energy corresponds to a factor of $10^{0.175} = 1.50$ in equilibrium constant. Every 1 kcal mol^{-1} difference in standard free energy corresponds to a factor of $10^{0.735} = 5.43$ in K_{eq}.

3.22 (a) A stronger acid has a lower pK_a, so chloroacetic acid, the strongest acid, has p$K_a = 2.85$, formic acid has p$K_a = 3.75$, and propionic acid, the weakest acid, has p$K_a = 4.87$.

  (b) The dissociation constant for each acid is given $K_a = 10^{-pK_a}$. For chloroacetic acid, $K_a = 1.41 \times 10^{-3}$; for formic acid, $K_a = 1.78 \times 10^{-4}$; and for propionic acid, $K_a = 1.35 \times 10^{-5}$.

3.23 (a) The reaction: (b) The equilibrium constant expression:

$$H_2O + H_3O^+ \rightleftharpoons H_3O^+ + H_2O$$

$$K_{eq} = \frac{[H_3O^+][H_2O]}{[H_2O][H_3O^+]} = 1$$

With the unit activity reference state for water as both a reactant and a product, then $K_{eq} = K_a$. Because $K_a = 1$, then p$K_a = -\log K_a = 0$.

The pK_a value for H_3O^+, previously cited incorrectly as -1.7 in other sources (as well as previous editions of this text), comes from treating reactant water as 55.6 M and ignoring the concentration of the product water (which is equivalent to assigning the unit-activity standard state to it). We can't have two different standard states for water in the same reaction. The pK_a of 0 for H_3O^+ used in this text uses a single standard state for water.

3.24   (a)   Because acetic acid is the weakest acid, sodium acetate is the strongest base of the sodium salts. Its equilibrium will lie farthest to the right.

(b)   Because sodium chloroacetate is the weakest base of the three salts, less NaOH will be formed from a solution of this salt in water than would be formed with the other two salts. Therefore, sodium chloroacetate will give the solution with the *least* amount of NaOH and therefore the lowest pH.

3.25   (a)   The acid–base reaction is as follows:

$$H—CN + \ddot{N}H_3 \rightleftharpoons :\bar{C}N + \overset{+}{N}H_4$$

$\log K_{eq} = 9.25\text{–}9.4 = -0.15$, from which we obtain $K_{eq} = 10^{-0.15} = 0.71$.

(b)   The acid–base reaction is as follows:

$$H—CN + :\ddot{\underset{..}{F}}:^- \rightleftharpoons :\bar{C}N + H—\ddot{\underset{..}{F}}:$$

$\log K_{eq} = 3.2\text{–}9.4 = -6.2$, from which we obtain $K_{eq} = 10^{-6.2} = 6.31 \times 10^{-7}$.

3.26   (a)   Ammonia acting as a base towards the acid water:

$$\ddot{N}H_3 + H_2\ddot{O}: \rightleftharpoons \overset{+}{N}H_4 + {}^-:\ddot{O}H$$
$$\text{base} \qquad \text{acid}$$

Water is the acidic species on the left side of this equation; its pK_a is 14.0. The ammonium ion is the acid on the right; its pK_a is 9.25. The K_{eq} for this reaction is $10^{-(9.25-14)} = 10^{-4.75}$.

(b)   Ammonia acting as an acid towards the base water:

$$\ddot{N}H_3 + H_2\ddot{O}: \rightleftharpoons :\bar{N}H_2 + H_3\overset{+}{O}$$
$$\text{acid} \qquad \text{base}$$

Ammonia is the acidic species on the left side of this equation; its pK_a is about 35. Hydronium ion is the acidic species on the right ; its pK_a is 0. The K_{eq} for this reaction is $10^{(0-35)} = 10^{-35}$.

Reaction (a) has, by far, the greater equilibrium constant, and it is therefore the most important acid–base reaction in aqueous ammonia.

3.27   Using Eq. 3.61c we have $\Delta G_a^{\circ} = 5.71(pK_a) = 27.2 \text{ kJ mol}^{-1}$ ($6.50 \text{ kcal mol}^{-1}$)

3.28   Because less free energy is required to dissociate formic acid, it is a stronger acid.

3.29   The derivation of Eq. 3.67a:

$$f_A = \frac{10^{-pK_a}}{10^{-pK_a} + 10^{-pH}} = 10^{-pK_a} \frac{1}{10^{-pK_a} + 10^{-pH}} = \left(\frac{1}{10^{pK_a}}\right)\left(\frac{1}{10^{-pK_a} + 10^{-pH}}\right)$$

$$= \frac{1}{(10^{pK_a} \times 10^{-pK_a}) + (10^{pK_a} \times 10^{-pH})} = \frac{1}{10^{pK_a - pK_a} + 10^{pK_a - pH}} = \frac{1}{1 + 10^{-(pH - pK_a)}}$$

$$= \frac{1}{1 + 10^{-\Delta}}$$

The derivation of Eq. 3.67b uses similar manipulations:

$$f_{HA} = \frac{10^{-pH}}{10^{-pK_a} + 10^{-pH}} = \left(\frac{1}{10^{pH}}\right)\left(\frac{1}{10^{-pK_a} + 10^{-pH}}\right)$$

$$= \frac{1}{(10^{pH} \times 10^{-pK_a}) + (10^{pH} \times 10^{-pH})} = \frac{1}{10^{pH-pK_a} + 10^{pH-pH}} = \frac{1}{10^{(pH-pK_a)} + 1}$$

$$= \frac{1}{1 + 10^{\Delta}}$$

3.30 (a) This part of the question is asking us to determine [HA] and [A]. The pK_a of ibuprofen is 4.43, which is less than the pH of the solution, pH = 5.0; therefore, we expect more ibuprofen dissociated than undissociated, or [A] > [HA]. Use Eq. 3.67a to solve for the fraction dissociated: $f_A = 0.78$. In a 10^{-4} M solution, 78% dissociated makes [A] = 7.8×10^{-5} M, while the remaining is undissociated, and [HA] = 2.2×10^{-5} M.

(b) Because the pK_a > pH, we predict that more ibuprofen is undissociated, or [A] < [HA]. Using Fig. 3.1 in the text, we might predict that $f_{HA} > 0.99$ and that $f_A < 0.01$. When we use Eq. 3.67a we can precisely calculate $f_A = 0.00370$, or 0.370 %, which is consistent with our prediction.

(c) Here, since pH > pK_a, there will be much more ibuprofen dissociated than undissociated—nearly all dissociated if we consult text Fig. 3.1. Solving for f_A gives 0.999, meaning that ibuprofen is over 99% dissociated in the bloodstream.

3.31 We want to again use Eq. 3.67a to solve for f_A. However, in order to solve for Δ, we need the pH, which is not given. So, first, we must figure out the pH of a 0.1 M solution of acetic acid in water. We have the pK_a, which is the negative log of the K_a. We can use $K_a = 1.8 \times 10^{-5}$ to solve for the $[H_3O^+]$ using Eq. 3.51 in Sec. 3.6A of the text. You may recall from your general or high school chemistry courses the "ICE" method for solving problems such as these, using the approximation that x is very small in determining your answer.

$$K_a = 1.8 \times 10^{-5} = \frac{[H_3O^+][A^-]}{[HA]} = \frac{(x)(x)}{(0.1-x)}$$

Assuming x is very small,

$$K_a = 1.8 \times 10^{-5} = \frac{x^2}{0.1} \quad \text{or} \quad 1.8 \times 10^{-6} = x^2$$

from which we find that $x = [H_3O^+] = 1.34 \times 10^{-3}$. The pH is $-\log[H_3O^+] = 2.87$. Finally, we will employ Eq. 3.67a to calculate $f_A = 0.0129$, meaning that a 0.1 M solution of acetic acid is only about 1.3% dissociated in water, a result consistent with our classification of acetic acid as a weak acid.

3.32 (a) Consider each dissociation individually, and construct a plot for each as shown in **Fig. SG3.1** on the following page. First, at low pH values, H_2B (dotted line) is present in high concentrations. When the pK_a = pH (3.13), H_2B is 50% dissociated. At a pH about 2 units greater than this (5.13), $[H_2B]$ is nearly 0. Considering next HB (solid line), it begins to appear as H_2B disappears, such that the sum of $f_{H_2B} + f_{HB}$ always equals 1. At pH = 8.02, HB is 50% dissociated, and at pH = 10.2, it its concentration is nearly 0. Finally, considering B (gray line), it is the product of dissociation of HB, and is present at 50% concentration at pH = 8.02. At pH = 10, it is present at nearly 100%.

(b) HB is a maximum at a pH exactly halfway between the pK_a values of HB and H_2B, as shown by the pointer labeled (b).

(c) In the bloodstream (pH = 7.4), the major species would be HB, as shown by pointer labeled (c).

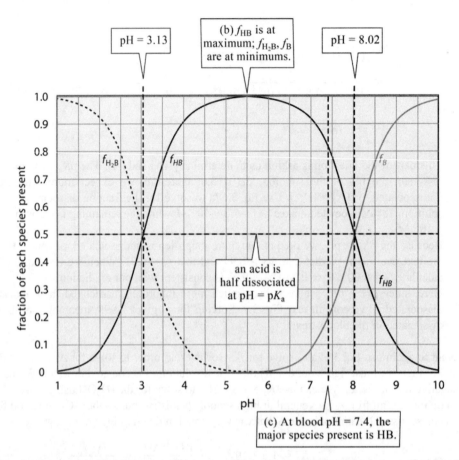

Figure SG3.1   Fraction of various species in the nicotine acid–base equilibrium to accompany the solution to Problem 3.32.

3.33   (a)   The acidity order is $C < D < A < B$. (The pK_a order is the reverse.) The order of A and B might be confusing because the comparison mixes different rows and periods. This order is reasoned as follows. From text Fig. 3.2 we see that HF is more acidic than H_2S. It then seems reasonable that H_2F^+ should be more acidic than H_3S^+ and more acidic than $CH_3\overset{+}{S}H_2$ (because methyl and hydrogen have comparable electronegativities).

   (b)   The pK_a of their conjugate acids (methanol, protonated methanol, ammonia) is in the order $B < A < C$. This is the order of basicity of the bases themselves.

3.34   (a)   According to the hybridization effect, the protonated nitrile ($CH_3C\equiv\overset{+}{N}$—H) is more acidic than the protonated amine ($CH_3CH_2\overset{+}{N}H_3$); there is much more s character in the sp–1s N—H bond of the protonated nitrile than in the sp^3–1s N—H bond of the protonated amine. The pK_a of the protonated nitrile is –10, and the pK_a of the protonated amine is 10.6.

   (b)   There is more s character in the sp^2–1s C—H bond of the second compound (protonated formaldehyde) than in the (roughly) sp^3–1s O—H bond of the first compound (protonated methanol). Therefore, protonated formaldehyde is more acidic; its pK_a is –6. The pK_a of the protonated alcohol is –2.5.

3.35   The conjugate acid A has an sp^2–1s σ bond, whereas the conjugate acid B has an sp^3–1s σ bond. Therefore, conjugate acid A, with more s character in the bond to the acidic proton, is more acidic. (The pK_a values are, for A, 3.13; and for B, 8.02.; see Focused Problem 3.32.)

conjugate acid *A*
(more acidic)

conjugate acid *B*
(less acidic)

3.36   (a)   The curved arrows for the resonance structures of phenoxide:

(b)   Statement (1) is true, because, by stabilizing the conjugate base, resonance lowers the pK_a of the conjugate acid (phenol). Statement (3) is also true because it is essentially another way of making Statememt (1). Statement (5) is also true. Lowering the pK_a of phenol decreases the basicity of its conjugate base phenoxide.

(c)   The conjugate base of phenol is resonance-stabilized; the conjugate base of cyclohexanol has no resonance stabilization. Therefore phenol is the more acidic compound. (The different in acidity is about 7 pK_a units.)

As Sec. 3.8 that follows in the text shows, resonance stabilization of phenoxide is only part of the explanation for the lower pK_a of phenol. The benzene ring, because of its sp^2-hybridized carbons, has a significant polar effect that also enhances the acidity of phenol.

3.37   (a)   Compound *A* is the stronger acid because the polar effect of the chlorine substituent stabilizes the conjugate base.

(b)   Compound *B* is the stronger acid, because the positively charged nitrogen has a greater polar effect than the uncharged nitrogen in *A*. This polar effect *destabilizes* the conjugate acid and therefore increases its acidity.

(c)   The order of acidity is *C* > *B* > *A* (which is the reverse of the pK_a order) because the polar effect of sp^2-hybridized carbons stabilizes the conjugate base. Compound *C* has two double bonds, more sp^2-hybridized carbons, and therefore a larger polar effect.

3.38   (a)   In terms of acidity, the order is *A* > *B* > *C*; the order of pK_a is therefore the reverse, that is, *C* > *B* > *A*. Compounds *A* and *B* both have chlorine substituents and so are more acidic than *C* by the polar effect. The thiol *A* is more acidic than the alcohol *B* by the element effect.

(b)   The order of decreasing pK_a is *B* > *A* > *C*. Both acids *A* and *C* have electronegative oxygen substituents and so are more acidic than *B* by the polar effect. Acid *C* has two of these substituents, so the polar substituent effect is greater and consequently *C* is most acidic.

(c)   By the charge effect, acids *A* and *B* are much stronger acids than *C*; between *A* and *B*, compound *B* is the stronger acid because of the polar effect of the chlorine. Therefore, the order of decreasing pK_a is *C* > *A* > *B*.

3.39   Range 13–16: *E*; range 9–11: *C*, *D*, and *F*; range 3–5: *A*; range –3 to 0: *B*.

Solutions to Skills Objectives Problems

3.40   A compound is electron-deficient when it has an atom with six or fewer electrons. Compounds (a), (d), and (e) each have one such atom: (a) has an electron-deficient carbon; (d) has an electron-deficient boron; and (e) has an electron-deficient carbon. The carbon in (e) is electron-deficient even though it has a pair of nonbonding electrons because its *total* electron count is 6.

3.41   The general approach is to allow a nucleophile to react with an electron-deficient atom; then compute the resulting formal charges.

(a)

(b)

(c)

Don't forget that carbon chains are flexible and can undergo internal rotations. Groups that appear far apart in structures that are drawn linearly can in many cases be brought into proximity by internal rotations. If this idea is not clear, construct a model for this case, and convince yourself that the OH group can easily be brought within bonding distance of the positively charged carbon.

(d)

(e)

3.42   (a)

(b)

3.43   (a)

(b)

Any time the leaving group is part of a ring, it remains part of the molecule when it "leaves," as in part (b). That is, the ring is opened to a noncyclic product instead of forming two separate products [as in part (a)].

3.44   (a)

(b)

(c)

3.45   (a)

(b)

(c)

(d)

(e)

We can also draw additional resonance structures in which each oxygen, in turn, has a double bond; this is implied by the hybrid structure. However, the octet structure is most important.

3.46   (Provide the curved arrows for extra practice.)

(a)

(b)

(c)

3.47   In all cases the hybrid structure has lower energy than any of the resonance contributors.

(a)   Structure *B* is more important because every atom has an octet. This importance is mitigated somewhat by the presence of the positive charge on the more electronegative atom.

(b)   Structure *A* is the most important structure because structures *B* and *C* have charge separation. Structures *B* and *C* are equivalent.

3.48   Reactions (b) and (c) are Brønsted acid–base reactions. Although a hydrogen is transferred in the first reaction, it is transferred as a *hydride*—a hydrogen with its two electrons. You can tell this at a glance because, if a proton had been transferred, the aluminum would have taken on a second negative charge.

(a)

(b)

conjugate base–acid pair

conjugate acid–base pair

(c)   (The designations *A1*, *B1*, *A2*, and *B2* are for the solution to Problem 3.49.)

conjugate base–acid pair

conjugate acid–base pair

3.49   We use the designations shown in Problem 3.49c. The equilibrium constant expression for the reaction in Problem 3.48c is $K_{eq} = ([A2][B2])/([A1][B1])$, where the quantities in brackets are the equilibrium concentrations of the various species.

3.50   (a)   Using the equation $K_{eq} = 10^{-\Delta G°/2.3RT}$, with $2.3RT$ at 298 K = 5.71 kJ mol^{-1}, we have $K_{eq} = 10^{-(-28.5)/5.71} = 10^5$.

(b)   A standard free energy scale is shown at right.

(c)   Because the products have the same (zero) initial concentration and the reactants have equal initial concentrations, we know that whatever the final concentrations of each, the two products will have the same concentration and the two reactants will have the same concentration. Therefore, we can write $[P]^2/[R]^2 = 10^5$, or $[P]/[R] = \sqrt{10^5} = 316$. The total concentration of products and reactants must be 2 M. Therefore, $2[P] + 2[R] = (2)(316[R])$ $+2[R] = 634[R] = 2$ M. From this we obtain $[R] = 2/634 = 0.00315$ M. Both reactants together are then present at $2(0.00315) = 0.00630$ M. $2[P]$ is then $2 - 2[R]$, or $[P] = 1 - [R] = 0.9937$ M. These are the concentrations of *each* product and *each* reactant. As a check, if we take the ratio $[P]^2/[R]^2$ we find that this in fact equals 10^5. So, each product is about 316 times the concentration of each reactant. In other words—as suggested by the large equilibrium constant—the reaction essentially goes to completion.

3.51   (a)   The equation for dissociation of trimethylammonium ion:

$$Me_3\overset{+}{N}\!-\!H \;+\; H_2O \;\rightleftharpoons\; Me_3\ddot{N} \;+\; H_3\overset{+}{O}$$

(b)   The expression for the dissociation constant uses the *unit activity convention* for the solvent water (Sec. 3.6A).

$$K_a = \frac{[Me_3\ddot{N}][H_3\overset{+}{O}]}{[Me_3\overset{+}{N}\!-\!H]}$$

3.52   (a)   The dissociation constant $K_a = 10^{-pK_a} = 10^{-9.76} = 1.74 \times 10^{-10}$.

(b)   The standard free energy of dissociation $\Delta G_a° = 2.3RT(pK_a)$. At 298 K, $2.3RT = 5.71$ kJ mol^{-1} (1.36 kcal mol^{-1}), so $\Delta G_a° = 55.7$ kJ mol^{-1} (13.3 kcal mol^{-1}).

3.53   (a)   Table 3.2 gives the pK_a of HCN as 9.4. Use text Eq. 3.59b (Sec. 3.6C) to obtain

$$K_{eq} = 10^{(9.76-9.4)} = 10^{0.36} = 2.3.$$

(b)   With the pK_a of H_2O = 14, application of the equation used in part (a) gives

$$K_{eq} = 10^{(14.0-10.5)} = 10^{3.5} = 3.16 \times 10^3$$

3.54   We can convert any two pieces of data into the same basis as the third. We arbitrarily choose to convert *HB* and *HC* into pK_a data. For the pK_a of *HB*, $pK_a = \Delta G_a°/2.3RT = 22.8/5.71 = 4$. For the p$K_a$ of *HC*, $pK_a = -\log(1.6 \times 10^{-7}) = 6.8$. In order of increasing acid strength, $HC < HB < HA$. The order of increasing base strength for the conjugate bases is the order of increasing pK_a values of the conjugate acids, which is $A < B < C$.

3.55   We use Eq. 3.67a or b with $\Delta = \text{pH} - pK_a = 0.19$. Because the pH is higher than the pK_a, we expect that more than 50% of the acid is in its conjugate-base form. The fraction of conjugate base is $1/(1 + 10^{-\Delta}) = (1/(1 + 10^{-0.19}) = 0.61$. Therefore 61% of the acid is in its conjugate-base form and 39% is in its conjugate-acid form. The conjugate-base concentration is $0.61 \times 10^{-4} = 6.1 \times 10^{-5}$ M, and the concentration of the conjugate acid is 3.9×10^{-5} M.

3.56 (a) In order of increasing acid strength, $C < B < A$, which follows the number and position of the chlorine substituents, which are electron-withdrawing and therefore acid-strengthening.

(b) Because the acid-strengthening hybridization of carbon substituents is in the order sp > sp^2 > sp^3, we have the order of increasing acid strength as $A < C < D < B$, with the order of C and D following from the relative distance of the sp^2-hybridized carbon from the acidic proton.

(c) The phenyl substituent is acid-strengthening because of the polar effect of its sp^2-hybridized carbon atoms, and the $^+NH_3$ substituent is very acid strengthening. The acidity order is $A < C < B$.

3.57 (a) Ranking the basicity of the conjugate bases follows by ranking the acidity of the conjugate acids. In terms of conjugate-acid pK_a, $CH < AH < BH$. The conjugate-base basicity is therefore in the same order: $C < A < B$.

(b) All of the compounds protonate on nitrogen. By the hybridization effect, B and A are both less basic than C, and the electron-withdrawing chlorine substituent makes A less basic than B. Therefore, the basicity order is $A < B < C$.

3.58 The order of increasing acidity is $D < E \approx C < A < F < B$. The order $A < F$ is a consequence of the element effect. The charge effect and the element effect work in opposite directions in determining the relative pK_a values of C and E with the result that the relative pK_a values are not very different.

Solutions to Integrated Problems

3.59 As suggested by the hint, the strategy is to examine the two possible conjugate acids and determine which one is resonance-stabilized. Protonation on the double-bonded oxygen gives a resonance-stabilized conjugate acid, whereas protonation on the other oxygen does not. Therefore, protonation on the double-bonded oxygen is preferred. Notice that the resonance effect outweighs the hybridization effect, which, by itself, would make the double-bonded oxygen *less* basic.

resonance-stabilized conjugate acid   not resonance-stabilized

3.60 (a) The curved arrow is backwards, and there is also no curved arrow to complete the proton transfer. This is the type of curved-arrow error discussed in the caution below text Eq, 3.34b, Sec. 3.4A.

(b) Although the curved arrow shown is correct, a second curved arrow to the leaving group is missing.

(c) The three curved arrows are all backwards; their directions should be reversed.

3.61 (a) The situation is that there is a nonbonding electron pair on an atom connected to a double bond.

(b)   The situation here is the cyclic shift of three double bonds. All of the structures are equivalent.

(c)   The three situations here are a cyclic shift of three double bonds, nonbonding electrons on an atom attached to a double bond, and a double bond between two atoms of differing electronegativity.

 In Sec. 15.7 we show that structures with three double bonds alternating with three single bonds in a cyclic arrangement are especially stable. This special stability (called *aromaticity*) makes structures *A* and *B* the most important structures. We also show in Chapter 15 that structures with two double bonds with a single bond in between them (called *conjugated double bonds*) are also particularly stable. This consideration makes structures *D* and *E* somewhat more important than structure *C*. However, not having studied Chapter 15 at this point, you would not be expected to invoke these additional considerations.

3.62   (a)   Electrons are flowing from the nitrogen to the oxygen shown by three curved arrows:

$$(CH_3)_2\ddot{N}H \;+\; H_2C\!=\!CH\!-\!\overset{\overset{\displaystyle :O:}{\|}}{C}\!-\!OEt \;\longrightarrow\; (CH_3)_2\overset{+}{N}H\!-\!CH_2\!-\!CH\!=\!\overset{\overset{\displaystyle :\ddot{O}:^-}{|}}{C}\!-\!OEt$$

(b)   Electrons are flowing from the negatively charged oxygen to the bromine with the breaking of two σ bonds:

$$:\!\ddot{B}r\!-\!CH_2\!-\!CH_2\!-\!\overset{\overset{\displaystyle :O:}{\|}}{C}\!-\!\ddot{O}:^- \;\longrightarrow\; :\!\ddot{B}r:^- \;+\; H_2C\!=\!CH_2 \;+\; \ddot{O}\!=\!C\!=\!\ddot{O}$$

3.63   For simplicity, let dimethylpropane be represented by D and pentane by P. First convert the standard free-energy difference into an equilibrium constant with the equation $K_{eq} = [D]/[P] = 10^{-\Delta G^\circ/2.3RT} = 10^{6.86/5.71} = 10^{1.20} = 15.9$. Therefore, [D] at equilibrium = 15.9[P]. Taking the total hydrocarbon as 1.00 mol, and noting from above that $[D] = 15.9[P]$,

$$1.00 \quad = \quad [D] + [P] = 15.9[P] + [P] = 16.9[P]$$

or [P] = 1.00/16.9 = 0.0592 and [D] = 1.000 − 0.059 = 0.941. Therefore, the equilibrium mixture contains 5.92% P and 94.1% D.

3.64   (a)   Any of the structures below are valid structures for the N_5^+ cation because they are resonance structures of the same molecule. They are also the answers to part (b).

3.65   (a)   For the first ionization ($pK_a = 2.86$), either of the two equivalent carboxylic acid groups can ionize:

$$HO-\overset{O}{\underset{\|}{C}}-CH_2-\overset{O}{\underset{\|}{C}}-OH \ + \ H_2O \ \rightleftharpoons \ HO-\overset{O}{\underset{\|}{C}}-CH_2-\overset{O}{\underset{\|}{C}}-O^- \ + \ H_3O^+$$

For the second ionization ($pK_a = 5.70$), the remaining carboxylic acid group ionizes:

$$HO-\overset{O}{\underset{\|}{C}}-CH_2-\overset{O}{\underset{\|}{C}}-O^- \ + \ H_2O \ \rightleftharpoons \ {}^-O-\overset{O}{\underset{\|}{C}}-CH_2-\overset{O}{\underset{\|}{C}}-O^- \ + \ H_3O^+$$

   (b)   Use a strategy similar to that used to solve Focused Problem 3.32 and sketch a plot similar to the one in that solution. At low pH, when protons are abundant, both carboxylic acid groups are undissociated, and AH_2 is present at nearly 100%. As the pH increases, AH_2 decreases and AH^- increases until the curves meet at pH = 2.86 (50/50 AH_2/AH^-). At pH = 4.3, at 1.44 pH units above the first pK_a, AH_2 has dropped to about 3.5% and AH^- has grown to 96.5%. Because pH 4.3 is 1.4 pH units below the second pK_a, it is nearly insignificant, and we ignore it. (An exact calculation shows that the second ionization affects the amount of AH^- by less than 0.1%.) Therefore, the ionization state at pH = 4.3, below the pK_a of AH_2 and above the pK_a of AH^-, is 96.5% AH^- with only small amounts of AH_2 and A^{2-}.

   (c)   As reasoned in part (b), at pH = 4.3, malonic acid is essentially in the AH^- form, or in other words, one of its carboxylic acid groups has been almost completely ionized. If we take the fraction of AH^- as 96.5%, then it takes 0.965 mole of sodium hydroxide per mole of malonic acid to get to this pH.

   (d)   The first pK_a is lower than that of acetic acid because, in the product of the first ionization, the electron-withdrawing polar effect of the un-ionized carboxylic acid group stabilizes the conjugate-base anion on the other. (The two electronegative oxygens and the C=O bond dipole are responsible for the polar effect of the carboxylic acid group.) The second pK_a is greater than that of acetic acid because one negative charge in the di-anionic product of the second ionization interacts unfavorably (repulsively) with the second negative charge, and the energy of the dianion is thereby raised.

   (e)   As n becomes larger, the polar groups become more remote. Hence, *both* polar effects discussed in part (d) become less significant, and *both* pK_a values approach that of acetic acid; that is, the pK_a values of the two ionizations become more similar. (In terms of the hint, as r in the denominator of the electrostatic law (text Eq. 3.84) becomes larger, the energy of interaction of one charge with another becomes smaller.)

Even if n is very large, the two pK_a values are not exactly equal. Theoretically, the first pK_a should be 0.6 smaller than the second pK_a (that is, the first K_a should be 4 times larger than the second K_a). The reason is that either of *two* equivalent acidic hydrogens can undergo the first ionization, but only one can undergo the second; this means that the first ionization is twice as probable statistically. Furthermore, either of *two* equivalent carboxylate ions can be protonated in the conjugate base of the monoacid, but only one is available in the conjugate base of the diacid. Thus, protonation of the dianion is twice as probable as protonation of the monoanion, which means that the second ionization is half as probable as the first—or again, that the first ionization is twice as probable as the second. Both of these factors multiply; thus, the first ionization is four times as likely as the second on purely statistical grounds; $\log (4) = 0.6$.

3.66   The primary effect of a polar substituent is on the stability of a charged species. The charged species in these equilibria are the cations. The question then becomes, how does a fluorine substituent affect the stability of a cation? As text Eq. 3.88 (Sec. 3.7E) shows, the polar effect of the fluorine de*stabilizes positive charge* because the positive ends of the C—F dipoles interact repulsively with the positive charge of the cation. Consequently, the cation product of the second equilibrium is destabilized relative to the neutral starting material, as shown in the following diagram. Because this cation has a higher energy, the energy required to form it is larger, and the equilibrium constant for its formation is smaller. In other words, Eq. (1) has the equilibrium constant most favorable to the right.

 It is worth mentioning that in this energy diagram, we are *not* comparing the free energies of the trifluoro alcohol on the right and the alcohol on the left. The point is to compare the two $\Delta G°$ values indicated by the arrows. The implicit assumption in diagrams like this is that the substituent effect operates primarily on one species in each equilibrium. In this particular case, the assumption is that the fluorine substituents have the major effect on the charged species and not the neutral species. The two neutral alcohols are arbitrarily placed at the same free energy to make the effect of the fluorine substituents easier to see.

3.67 (a) Protons with pK_a values more than about 2 units below the pH are completely dissociated, and those with pK_a values more than about 2 units above the pH are completely associated. So, at pH = 7.4, ascorbic acid has a net −1 charge. Proton *a* is the more acidic proton because the resulting mono-anion conjugate base has more important resonance structures than the mono-anion formed by ionization of proton *b*. The structure of the mono-anion is shown in part (b).

(b) The mono-anion conjugate base of ascorbic acid and its resonance structures:

3.68 (a) Remember that decreasing pK_a is increasing acidity. The descending pK_a order is $B > A > C > D$.

(b) The decreasing pK_a order is $C > D > A > B$. The element effect determines the first three, and the charge effect the last two.

(c) The descending pK_a order is $A > B > C$. The order is determined by the polar effect.

3.69 The pK_a of 10.42 is in the range for a protonated amine, and the other two pK_a values are typical for carboxylic acids, although the carboxylic acid with $pK_a = 2.16$ must have significant electron-withdrawing substituents nearby. Imagine starting with a solution of kainic acid in its fully undissociated form at very low pH. As we raise the pH, the first group to dissociate will be the carboxylic acid with $pK_a = 2.16$. (See the equation below.) This must be the carboxylic acid group closest to the nitrogen, because the positively charged nitrogen exerts a significant acid-strengthening polar effect. The other carboxylic acid group, being undissociated, is a polar electron-withdrawing group and also has a pK_a-lowering effect. One carboxylic acid is now ionized. Its negative charge has a pK_a-raising effect on the remaining carboxylic acid, but the protonated nitrogen continues to exert a pK_a-lowering polar effect. The two effects working in opposite directions tend to cancel so that the remaining carboxylic acid pK_a of 4.57 is fairly typical. Finally, as the pH is raised further, the proton on the nitrogen dissociates. Its $pK_a = 10.42$ is at the high end of amine pK_a values; the favorable interaction of the negative charges of the two ionized carboxylic acid groups with the positive charge of the protonated amine and stabilize it and raise its pK_a.

3.70   The equilibrium constant K_{eq} for the reaction of text Eq. 3.58 is given by

$$K_{eq} = \frac{[HB][A^-]}{[HA][B^-]}$$

Following the hint, first show that this equals K_{AH}/K_{BH}. We do that by applying the definition of K_{AH} and K_{BH}.

$$\frac{K_{HA}}{K_{HB}} = \frac{[H_3O^+][A^-]}{[HA]} \times \frac{[HB]}{[H_3O^+][B^-]} = \frac{[HB][A^-]}{[HA][B^-]} = K_{eq}$$

or $K_{eq} = K_{HA}/K_{HB}$. Then take the logarithms of both sides of this equation and apply the definition of pK_a:

$$\log K_{eq} = \log K_{HA} - \log K_{HB} = pK_{HB} - pK_{HA}$$

Taking antilogs of both sides proves Eq. 3.59b:

$$10^{(\log K_{eq})} = K_{eq} = 10^{(pK_{HB} - pK_{HA})}$$

3.71   The $\Delta G°$ for the reaction is found by taking the logarithm of the equilibrium constant and multiplying it by $-2.3RT$. From the solution to Problem 3.70,

$$\Delta G°_{eq} = -2.3RT \log K_{eq} = -2.3RT(pK_{HB} - pK_{HA}) = \Delta G°_{a,HA} - \Delta G°_{a,HB}$$

3.72 (a) The K_{eq} for this reaction is calculated in the solution to Focused Problem 3.53a as $K_{eq} = 2.3$ and $\log K_{eq} = 0.36$. Using the equation $\Delta G°_{eq} = -2.3RT \log K_{eq}$ we calculate, with $2.3RT = 5.71$ kJ mol^{-1} at 298 K (25 °C), $\Delta G°_{eq} = (-5.71)(0.36) = -2.06$ kJ mol^{-1} (0.491 kcal mol^{-1}).

  (b) If the ions are each present initially at 0.1 M and the compounds on the left are present initially at 0 M, then at equilibrium, when x M of each ion has reacted, the concentration of each ion is $(0.1 - x)$ M and the concentration of each compound on the left is x M. Therefore, the equilibrium constant expression is

$$K_{eq} = 2.3 = \frac{(0.1 - x)(0.1 - x)}{x^2} \quad \text{or} \quad 1.3x^2 + 0.2x - 10^{-2} = x^2 + 0.154x - 0.00769 = 0$$

Solving for x using the quadratic expression, we find that $x = 0.0396$ M and $0.1 - x = 0.0603$ M. These are the concentrations of each neutral compounds and each of the ions, respectively.

3.73   The relevant acid–base equation is:

Using the text Eq. 3.59b (the same equation that was derived in the solution to Problem 3.70), the K_{eq} for this reaction is $10^{(0.23-4.76)} = 10^{-4.53} = 2.95 \times 10^{-5}$, so the reaction lies far to the left. We start with equal amounts of all four compounds. Therefore, at equilibrium the concentrations of the two compounds on the left (L) are equal, and the concentrations of the two compounds on the right (R) are equal. Therefore, the solution contains mostly the two compounds on the left at almost 0.2 M each and only a miniscule amount of the compounds on the right.

More precisely, the equilibrium-constant expression is $[L]^2/[R]^2 = 2.95 \times 10^{-5}$, so that $[L]/[R] = \sqrt{2.95 \times 10^{-5}} = 0.0543$. In other words, each compound on the right is present at about 184 times the concentration of its conjugate-acid or conjugate-base component on the left. Because $2[L] + 2[R] = 0.4\ M$, and $[L] = 0.0543[R]$, then $[R] = 0.19\ M$ and $[L] = 0.01\ M$.

3.74   (a)   Because the nitrogen bonded to the acidic proton is sp^3-hybridized in A and sp^2-hybridized in B, compound B should be more acidic by the hybridization effect.

       (b)   Acid B is resonance-stabilized:

Resonance stabilization of the acid B weakens its acidity as shown in the following diagram:

       (c)   Because cation B is the weaker acid, it appears that the resonance effect on acidity outweighs the hybridization effect.

> The conjugate base of B also has resonance structures, but these structures *separate* charge rather than *disperse* charge, and are therefore less important and are ignored in discussing the resonance effect.

Introduction to Alkenes and Alkynes
Structure and Reactivity

CHAPTER 4 SUMMARY

- Alkenes are compounds containing carbon–carbon double bonds. Alkene carbon atoms, as well as other trigonal planar atoms, are sp^2-hybridized.

- Alkynes are compounds containing carbon–carbon triple bonds. Alkyne carbon atoms, as well as other atoms with linear bonding geometry, are sp-hybridized.

- The carbon–carbon double bond consists of a σ bond and a π bond. The carbon–carbon triple bond consists of a σ bond and two π bonds. π Electrons are more reactive than σ electrons and can be donated to Brønsted or Lewis acids.

- The overlap of 2p orbitals results in two molecular orbitals, a π (bonding) and π^* (antibonding). The bonding π MO *is* the π bond. The π^* MO has an additional planar node between the bonded atoms.

- Relative to sp^3-hybridized carbon atoms, carbon atoms with sp^2 and sp hybridization are electronegative.

- Because rotation about the alkene double bond does not occur under normal conditions, some alkenes can exist as double-bond stereoisomers. These are named using the *E,Z* priority system.

- If the positional interchange of two groups bonded to an atom gives rise to stereoisomers, the atom is a stereocenter. The carbons of a double bond in an alkene with *E,Z* stereoisomers are stereocenters.

- In the IUPAC substitutive nomenclature of alkenes and alkynes, the principal chain, which is the carbon chain containing the greatest number of double and triple bonds, is numbered so that the double and triple bonds receive the lowest numbers. If there is ambiguity, double bonds take numerical precedence.

- The unsaturation number of a compound, which is equal to the number of rings plus π bonds in the compound, can be calculated from the molecular formula by Eq. 4.18 in Sec. 4.3.

- Heats of formation (enthalpies of formation) can be used to determine the relative stabilities of various bonding arrangements within a set of isomers. Heats of formation reveal that alkenes and alkynes with more alkyl substituents at their double or triple bonds are more stable than isomers with fewer alkyl groups; this greater stability is due to hyperconjugation. In most cases, trans alkenes are more stable than their cis isomers because of van der Waals repulsions between substituents in the cis isomers.

- Dipolar molecules such as H—Br and H—OH add to alkenes and alkynes in a regioselective manner so that the hydrogen adds to the carbon of the double bond with the most hydrogens, and the electronegative group to the carbon of the double or triple bond with the greater number of alkyl groups. Addition of water requires acid catalysis because water itself is too weak an acid to protonate the π bond.

- Reactants are converted into products through unstable species called transition states. A transition state of a reaction step can be approximated as a structure that is intermediate between reactants and products, and it can be drawn by using dashed bonds and partial charges.

- A reaction rate is determined by the standard free energy of activation $\Delta G^{\circ\ddagger}$, which is the standard free energy difference between the transition state and the reactants. Reactions with smaller standard free energies of activation are faster.

- The rates of reactions increase with increasing temperature.

- The rates of multistep reactions are determined by the rate of the slowest step, called the rate-limiting step. This is the step with the transition state of highest standard free energy.

- According to Hammond's postulate, the structures and energies of transition states for reactions involving unstable intermediates (such as carbocations) resemble the structures and energies of the unstable intermediates themselves.

- The regioselectivity observed in the addition reactions of hydrogen halides or water to alkenes is a consequence of two facts: (1) the rate-limiting transition states of the two competing reactions resemble carbocations; and (2) the relative stability of carbocations is in the order tertiary > secondary > primary. Application of Hammond's postulate leads to the conclusion that the reaction involving the more stable carbocation is faster.

- Reactions of alkenes involving carbocation intermediates, such as hydrogen halide addition and hydration, show rearrangements in some cases. Unstable alkyl carbocations rearrange to more stable ones by a shift of an alkyl group, aryl group, or hydrogen to the electron-deficient carbon from an adjacent carbon. The group that moves brings along its bonding electron pair. As a result, the adjacent carbon becomes electron-deficient.

- Addition of HBr to alkynes involves vinylic carbocations. A second addition of HBr can occur to give a geminal dibromide.

- A catalyst increases the rate of a reaction without being consumed in the reaction. A catalyst does not affect the equilibrium constant for a chemical equilibrium. Catalysts accelerate the forward and reverse reactions of an equilibrium equally.

- Catalysts are of two types: heterogeneous and homogeneous. Catalytic hydrogenation of alkenes and alkynes are important examples of heterogeneous catalysis; acid-catalyzed hydration of alkenes involves homogeneous catalysis.

- Catalytic hydrogenation of alkynes gives cis-alkenes. Hydrogenation of alkynes can be stopped at the cis-alkene by employing catalysts poisoned with such things as lead salts or amines.

- Enzymes are biological catalysts that accelerate biological reactions by many orders of magnitude.

STUDY GUIDE LINKS

4.1   Different Ways to Draw the Same Structures

The discussion in the text requires you to realize that there are several different ways to draw the same alkene stereoisomer. For example, two equivalent ways to represent *cis*-2-butene in the plane of the page are the following:

two ways of drawing *cis*-2- butene

If this is not clear, think of the structure as an object. Rotate the structure 180° about a horizontal axis through the C═C bond, as follows:

When we perform such a rotation, the atom labels, for example, CH_3, are maintained "right side up"; that is, they are not inverted.

Similarly, two equivalent ways of drawing *trans*-2-butene are as follows:

different ways of drawing *trans*-2-butene

Their equivalence can be demonstrated by a 180° rotation about a vertical axis.

The two structures in text Eq. 4.10 (Study Problem 4.1) are examples of identical structures drawn differently. Some students think that structures such as these are stereoisomers, but in fact they represent the same molecule. As these structures demonstrate, the ultimate test of identity of two structures is their congruence—the superimposability of each atom of one structure and an identical atom of the other.

Don't forget that we can also rotate structures 90° out of the plane about a horizontal axis by using line–wedge structures. This was explained in Display 1.54, Sec.1.6D, text p. 36.

4.2   Drawing Structures from Names

When asked to draw a complicated structure from a name, as in Problem 4.6(b), you should always strive to be *systematic*. The natural tendency is to try to write the finished structure immediately. Instead, you should take it one step at a time:

1. Write the carbon skeleton of the principal chain; do not be concerned with stereochemistry.
2. Add the substituents to the principal chain.
3. Add the hydrogens.
4. Decide on relative group priorities at the double bonds to which an *E* or *Z* configuration must be assigned. Write them on your structure.
5. Redraw your structure with proper stereochemistry. It helps to maintain proper 120° bond angles at the double bonds.

Let's illustrate with an example. Suppose you are asked to draw the structure of (2*E*,4*Z*)-3-isobutyl-2,4-hexadiene. Follow the above steps in order.

1. The principal chain contains six carbons with double bonds at carbons 2 and 4:

$$C—C=C—C=C—C$$

2. There is an isobutyl group at carbon 3. Add the hydrogens within this group, since you won't have to manipulate this group further.

$$C-C=C-C=C-C$$
$$\underset{\text{isobutyl group}}{|}$$
$$CH_2CH(CH_3)_2$$

3. Rewrite, adding the missing hydrogens.

$$H_3C-CH=C-CH=CH-CH_3$$
$$|$$
$$CH_2CH(CH_3)_2$$

4. Indicate the relative priorities of the groups at each double bond. For example, focusing on the C2–C3 double bond (carbons in boldface below), the relative priorities of the attached groups are as follows:

(You should complete the priorities for the C4–C5 double bond.)

5. Redraw the structure with proper stereochemistry.

(2E,4Z)-3-isobutyl-2,4-hexadiene

You should work Problem 4.6(b) using this step-by-step procedure.

In many cases, it is easier to work a complicated problem if it can be broken into smaller "chunks." When you learn to bypass the anxiety created by a complex problem and adopt this approach, you have made significant progress toward becoming a good problem solver.

4.3   Solving Structure Problems

A number of problems ask you to deduce structures that are initially unknown by piecing together chemical data. Problem 4.76 is of this type. This study guide link illustrates a systematic approach to this type of problem by beginning the solution to Problem 4.76.

1. If the formulas of any of the unknowns are given, deduce all the information you can from the formulas. Begin with the unsaturation numbers (Sec. 4.3).

   In Problem 4.76, compound X has an unsaturation number $U = 2$, and compound Y has $U = 1$. This means that X has one ring and one double bond, or two double bonds. (It has to have at least one double bond because it undergoes addition.) Compound Y results from addition of HBr because the formula of Y is equal to that of X plus the elements of HBr.

2. Write all the information in the problem in equation form. This process gives you the entire problem at a glance.

(continued)

3. If there is a structure given explicitly anywhere in the problem, *even at the end of the problem*, deduce what you can from this structure.

   1,1-Dimethylcyclopentane is given explicitly in the problem as the catalytic hydrogenation product of *X*. Barring rearrangements, the structure provides the real key to solving the problem: it provides *the carbon skeleton of compound X*. It follows that one of the degrees of unsaturation of *X* must be a ring, and the other, as previously deduced, is a double bond. Thus, compound *X* must be a five-membered ring containing a double bond with two methyl groups on one carbon. The carbon bearing the methyl groups cannot be part of the double bond (why?).

4. Write out *all possibilities* and choose between (or among) them from the evidence presented, if possible.

   This step is one that meets great resistance from many students for some reason. But it is essential. Suppose a detective arrests the first person he or she meets that fits the description of a burglar; chances are, the detective will make a lot of false arrests. Instead, he or she ideally should list *all* the suspects and decide which among them fit all the data. The only "suspects" for *X* are the following:

   But what about the following structure?

   It is the same as *X1*! Don't forget that most structures can be drawn in different ways.

   To finish solving the problem, notice that compound *Y* is a *single* compound. Only one of the "suspects" above would give a *single compound* on addition of HBr; the other would give a mixture. What is the solution? Which compound is *X*?

FURTHER EXPLORATIONS

4.1 Relationship between Free Energy and Enthalpy

If you want to calculate an equilibrium constant, the *standard free energies* ($\Delta G°$ values) of the reactants and products are the relevant quantities. If you want to ask which of two molecules has a less energetic (or more stable) arrangement of bonds, then you need the *standard enthalpies of formation* ($\Delta H_f°$ values) of the two compounds. This section discusses why two different types of energy are needed for different purposes.

The discrepancy between the enthalpy change for any process (such as a reaction) and the free-energy change lies in something called the *entropy change* $\Delta S°$. The relationship between the free-energy change, the enthalpy change, and the entropy change is deceptively simple:

$$\Delta G° = \Delta H° - T\Delta S° \qquad \text{(FE4.1)}$$

where T is the absolute temperature in kelvins.

This equation is essentially a statement of the division of energy into heat and work. $\Delta H°$ is the total energy change in a reaction (at constant temperature and pressure); $\Delta G°$ is the amount of work we could obtain from the reaction. A simple way to think about "chemical work" is charging a battery. If we could charge a battery and then harness it do some work—for example, by running an automobile (all with 100% efficiency)—then the $\Delta G°$ is the amount of work we could get from the reaction. The quantity $T\Delta S°$ is the amount of heat that would be produced when the reaction is harnessed with 100% efficiency to do work.

We can interpret the entropy change in more useful and descriptive terms than simply a discrepancy between total energy and work. Entropy measures molecular randomness, or *freedom of motion*. For example, consider the following reaction:

cyclobutane $\longrightarrow$ $H_2C={CHCH_2CH_3}$
1-butene

The experimentally determined $\Delta H°$ for the reaction at 25 °C (or 298 K) is –26.53 kJ mol^{-1} (–6.34 kcal mol^{-1}), and the $\Delta G°$ for the reaction at 25 °C is –40.13 kJ mol^{-1} (–9.59 kcal mol^{-1}). Solving Eq. FE4.1, $T\Delta S°$ at $T = 298$ K is 13.60 kJ mol^{-1} (3.25 kcal mol^{-1}); that is, $\Delta S°$ is positive. A positive $\Delta S°$ means that there is more *randomness* or *freedom of motion* in the products than the reactants.

What is it about 1-butene that is more "random" than cyclobutane? 1-Butene has two carbon–carbon single bonds. Internal rotation can occur readily about these bonds, just as in butane. However, in cyclobutane, *internal rotations cannot occur* because the various carbon atoms are constrained into a small ring. (If this is not clear to you, make a model of cyclobutane and convince yourself that internal rotation about its carbon–carbon bonds cannot occur.) So, there is *greater freedom of motion* in 1-butene than in cyclobutane. Because of the internal rotations in 1-butene, the hydrogens on the carbons connected by single bonds can move through a greater volume of space than the same hydrogens in cyclobutane—that is, *their positions are more random.* Consequently, the randomness—or *entropy*—in 1-butene is greater than that in cyclobutane.

Because of entropy—the "randomness factor"—the energy that controls the position of a chemical equilibrium ($\Delta G°$) is different from the energy stored in chemical bonds ($\Delta H°$). Eq. FE4.1 shows that an increase of randomness in a reaction—a positive $\Delta S°$—gives the product an additional advantage in a chemical equilibrium *over and above* that which results from the formation of more stable bonds. In terms of the heat–work interpretation, we can get more work from this reaction than is available from bond changes, because heat is absorbed from the surroundings that goes toward the increased randomness of the product.

4.2   Sources of Heats of Formation

Heats of formation are not obtained by direct measurement. Rather, they are calculated from more readily available data by applying Hess's law. Two types of data that are commonly used are heats of combustion and heats of hydrogenation. To illustrate, suppose that we want to calculate the heat of formation of *trans*-2-butene from the following known heats of combustion: hydrogen, –241.8 kJ mol^{-1}; carbon, –393.5 kJ mol^{-1}; and *trans*-2-butene, –2530.0 kJ mol^{-1}. The combustion reaction (Sec. 2.7) is the reaction of each species with oxygen to give liquid water (in the combustion of hydrogen), carbon dioxide (in the combustion of carbon), and both liquid water and carbon dioxide (in the combustion of hydrocarbons). Although water produced by combustion is a gas, the "standard state" of water at 298 K (25 °C) is the liquid state. All heats of combustion have been corrected to this standard state.

Hess's law allows us to express the formation of *trans*-2-butene from its elements as the sum of three combustion reactions. Identical species on opposite sides of the equations cancel, and the resulting equations and their enthalpies are added algebraically:

Equations: $\Delta H°$ (kJ mol^{-1})

$$4\,H_2 + 2\!\!\!/O_2 \longrightarrow 4\!\!\!/H_2O \qquad\qquad 4(-285.8) = -1143.20$$

$$4\,C + 4\,O_2 \longrightarrow 4\!\!\!/CO_2 \qquad\qquad 4(-393.5) = -1574.0$$

$$4\!\!\!/CO_2 + 4\!\!\!/H_2O \longrightarrow 6\!\!\!/O_2 + \textit{trans-2-butene} \qquad\qquad +2706$$

Sum: $4\,C + 4\,H_2 \longrightarrow$ *trans*-2-butene $\Delta H_f° = -11.2$ kJ mol^{-1}

Because four moles of both H_2 and C are required, their respective enthalpies of combustion must be multiplied by 4. Note also that the combustion of *trans*-2-butene must be written in reverse so that the formation equation comes out with *trans*-2-butene on the right. Consequently, the sign of the enthalpy of this combustion is also reversed. See Focused Problem 4.16 for the calculation of a heat of formation from combustion data.

As this example illustrates, heats of formation derived from combustion are in many cases *small differences between large numbers* and are therefore subject to considerable uncertainty. Another reaction that has been used for obtaining heats of formation is *catalytic hydrogenation*. In this reaction, hydrogen is added to an alkene double bond in the presence of a catalyst (Sec. 4.10A). For *trans*-2-butene, this reaction is as follows:

trans-2-butene

The overall $\Delta H°$ of this reaction can be measured with excellent precision, and is typically much smaller than a heat of combustion. (For example, the $\Delta H°$ of hydrogenation of *trans*-2-butene is -115.5 kJ mol^{-1}.) The heats of formation of simple alkanes are known with good precision. For example, the heat of formation of butane is -127.1 kJ mol^{-1}. The heat of formation of H_2 is zero by definition. Hence, it is possible to calculate the heat of formation of the alkene (try it!). Focused Problem 4.18 illustrates the calculation of heats of formation from hydrogenation data.

Fortunately, calculations like these are not necessary every time a heat of formation is needed, because such calculations have already been done to provide the data available in standard tables of heats of formation. A good, readily available source of heats of formation is the NIST Chemistry WebBook (https://webbook.nist.gov/chemistry/). (NIST is the National Institute of Standards and Technology operated under the auspices of the Department of Commerce of the United Stated Government.)

4.3   Molecular Orbital Description of Hyperconjugation

A molecular orbital treatment shows why hyperconjugation is a stabilizing effect. If you understand the molecular orbital treatment of π bonds (Sec. 4.1C), you should be able to follow the treatment that follows.

As indicated by the margin note near Eq. 4.27c, the source of hyperconjugation is the overlap between the electrons of the C—H (or C—C) σ bond with the *antibonding* (π*) MO of the alkene. (The bonding π MO is fully occupied and cannot accommodate any more electrons.) We follow the addition–subtraction protocol used for forming the MOs of the hydrogen molecule (Secs. 1.8A, 4.1C). The interacting orbitals are the sp^3–1s bond orbital of the C—H bond and the π* MO of the alkene. The bonding (in-phase) combination provides an MO of lower energy, and the antibonding (out-of-phase) combination provides the unoccupied MO of higher energy:

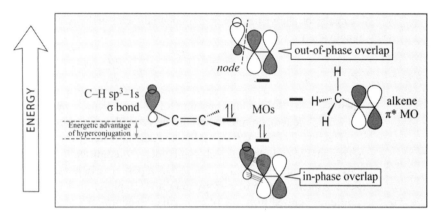

The lowering of the system energy provided by the occupied MO is hyperconjugation.

It is not strictly correct to use a bond orbital as a starting orbital for this treatment, because a bond orbital is not a proper molecular orbital of the alkene. In fact, a full-fledged molecular-orbital treatment of the alkene would start with all the carbons and hydrogens in the proper geometry and allow all their valence orbitals to interact. Adopting such an approach gives a more rigorous and complete molecular orbital picture, but, as noted in Sec 1.8C discrete bonds disappear and the descriptive value of the treatment is diminished. However, in such a full-fledged molecular-orbital approach, the overlap described here can be discerned as a component of the complete molecular orbitals.

 ## 4.4   Activation Energy

You may have learned in general chemistry about the *Arrhenius equation,* which treats rates in terms of an energy barrier called the *activation energy, E_a.*

$$k_{rate} = Ae^{-E_a/RT} = A(10^{-E_a/2.3RT}) \qquad \text{(FE4.2)}$$

In this equation, k_{rate} is the rate constant, which is the rate measured under standard conditions of 1 *M* concentration of all reactants and catalysts. (We discuss the rate constant in Chapter 9; but, for now, think of it as the rate.) The factor *A* is called the *preexponential factor.* The activation energy and the preexponential factor are determined experimentally from a plot of log k_{rate} vs. $1/T$. The purpose of this Further Exploration is to relate the activation energy to the standard free energy of activation.

The Arrhenius equation originated as an empirical way to describe the variation of rate with temperature. The view was that the activation energy is an energy barrier, but its description in thermodynamic terms was not provided. Eyring theory, or transition-state theory, actually envisioned a barrier characterized by a structure—the transition state. The transition state can in principle be described in terms of its thermodynamic properties, such as free energy, enthalpy, and entropy.

As transition-state theory emerged as an important theoretical framework for discussing reaction rates, it became of interest to relate the variables in the Arrhenius equation to those in transition-state theory. It can be shown that the activation energy of the Arrhenius equation can be related to the *enthalpy* of activation:

$$E_a = \Delta H^{\circ\ddagger} + RT \qquad \text{(FE4.3)}$$

The enthalpy of activation is the enthalpy of the transition state minus that of the starting materials. Because many typical organic reactions have $\Delta H^{\circ\ddagger}$ values that are substantially greater than RT (which equals about 2.5 kJ mol^{-1}), then, for these reactions,

$$E_a \approx \Delta H^{\circ\ddagger} \qquad \text{(FE4.4)}$$

It can also be shown that the preexponential factor in the Arrhenius equation is related to the *entropy* of activation:

$$A = \frac{k_B T}{h} e^{\Delta S^{o\ddagger}/R}$$

(FE4.5)

where k_B is the Boltzmann constant (= 1.38×10^{-23} J K^{-1}=R/N, where N = Avogadro's constant), T is the absolute temperature in kelvins, and h = Planck's constant = 6.63×10^{-34} J sec.

Because $\Delta G^{o\ddagger} = \Delta H^{o\ddagger} - T\Delta S^{o\ddagger}$, the standard free energy of activation can be calculated from $\Delta H^{o\ddagger}$ and $T\Delta S^{o\ddagger}$, both of which can be derived empirically from the Arrhenius equation as shown above. The entropy of activation $\Delta S^{o\ddagger}$ is a particularly interesting aspect of transition-state theory. (See the interpretation of entropy in Further Exploration 4.1.) The $\Delta S^{o\ddagger}$, like ΔS^{o} in an equilibrium, can be interpreted in terms of molecular freedom, or randomness. This interpretation provides a "window" on the structure of the transition state.

REACTION REVIEW

I.   ADDITION REACTIONS OF ALKENES

(More alkene additions are presented in Chapter 5.)

A.   ADDITION OF HYDROGEN HALIDES TO ALKENES

1.   Hydrogen halides add to alkenes in a regioselective manner so that the hydrogen adds to the carbon of the double bond with fewer alkyl substituents and the halide group to the carbon of the double bond with more alkyl substituents (Markovnikov's rule).

2.   The products of these reactions are alkyl halides.

an alkyl halide

3.   The addition of a hydrogen halide to an alkene is a regioselective reaction that occurs in two successive steps.

   a.   In the first step, protonation of the alkene double bond occurs at the carbon with the fewer alkyl substituents so that the more stable carbocation is formed—the one with the greater number of alkyl substituents at the electron-deficient carbon.

   b.   In the second step, the halide ion reacts at the electron-deficient carbon.

4.   Hydrogen halide addition to an alkene is a regioselective reaction because addition involves the transition state that resembles the more stable of the two possible carbocation intermediates.

5. The formation of a carbocation from an alkene is a Brønsted acid–base reaction—that is, an electron-pair displacement reaction in which the π bond acts as a Brønsted base towards the Brønsted acid H—X.

6. The reaction of the halide ion with the carbocation is a Lewis acid–base association reaction.

7. In hydrogen halide addition to alkenes, rearranged products are formed if a rearrangement-prone carbocation intermediate is involved—for example, a secondary carbocation that could rearrange to a tertiary carbocation.

B. *CATALYTIC HYDROGENATION OF ALKENES*

1. Catalytic hydrogenation is an addition of hydrogen in the presence of a catalyst; it is one of the best ways to convert alkenes into alkanes.

    a. Many of the common hydrogenation catalysts are insoluble in the reaction solution; these are examples of heterogeneous catalysts.

    b. Benzene rings are inert to conditions under which normal double bonds react readily. (Under conditions of high temperature and pressure they can be hydrogenated.)

C. *HYDRATION OF ALKENES*

1. The addition of water to the alkene double bond, called alkene hydration, is an acid-catalyzed reaction.

    a. In many cases, the catalyzing acid is soluble in the reaction solution. Soluble catalysts are called homogeneous catalysts.

    b. Hydration is a regioselective reaction. A proton is added to the carbon of the double bond with fewer alkyl substituents, and the hydroxy (OH) group is added to the carbon of the double bond with more alkyl substituents.

2. Alkene hydration is a multistep reaction.

    a. In the first step of the reaction (the rate-limiting step), a Brønsted acid–base reaction, the double bond is protonated to give a carbocation.

    b. In the next step of the hydration reaction, the carbocation reacts with the Lewis base water in a Lewis acid–base association reaction.

    c. Finally, a proton is lost to solvent in another Brønsted acid–base reaction to give an alcohol product and regenerate the catalyzing acid.

3. In alkene hydration, rearranged hydration products are formed if a rearrangement-prone carbocation intermediate is involved—for example, a secondary carbocation that could rearrange to a tertiary carbocation.

4. Alkene hydration is a reversible reaction.

    a. Although hydration is used industrially, dehydration of alcohols to alkenes is the more common laboratory use of this reaction.

b. The mechanism of alcohol dehydration is the reverse of the mechanism of alkene hydration (principle of microscopic reversibility).

II.   ADDITION REACTIONS OF ALKYNES

(More alkyne additions are presented in Chapter 5.)

A.   *CATALYTIC HYDROGENATION OF ALKYNES*

1. Alkynes undergo catalytic hydrogenation.

    a. Addition of one molar equivalent of H_2 to an alkyne gives an alkene.

    b. Addition of a second molar equivalent of H_2 gives an alkane.

$$R-C\equiv C-R \xrightarrow{\text{H}_2,\text{ catalyst}} \underset{\text{a cis alkene}}{\overset{\text{H}\qquad\text{H}}{\underset{R\qquad R}{C=C}}} \xrightarrow{\text{H}_2,\text{ catalyst}} R-CH_2-CH_2-R$$

an alkyne   an alkane

2. Hydrogenation of an alkyne can be stopped at the alkene stage if the reaction mixture contains a catalyst poison.

    a. A catalyst poison is a compound that disrupts the action of a catalyst. Some poisons can selectively block the hydrogenation of alkenes without preventing the hydrogenation of alkynes to alkenes.

    b. Useful catalyst poisons are salts of Pb^{2+} and certain nitrogen compounds such as pyridine, quinoline, or other amines.

    c. The Lindlar catalyst [$Pd/CaCO_3$ that has been washed with $Pb(OAc)_2$] is a commonly used poisoned catalyst.

$$R-C\equiv C-R \xrightarrow[\substack{Pd/CaCO_3 \\ \text{[washed with Pd(OAc)}_2]}]{H_2} \underset{\text{a cis alkene}}{\overset{\text{H}\qquad\text{H}}{\underset{R\qquad R}{C=C}}}$$

an alkyne

3. When carried out with a poisoned catalyst, hydrogenation of alkynes is one of the best methods for preparing cis alkenes. (The stereochemistry of additions is discussed in Sec. 7.8A.)

 # SOLUTIONS TO PROBLEMS

Solutions to Focused Problems

4.1   The order of increasing bond length is $h < d < a < g < e < f < c < b$. The three carbon-hydrogen bonds are shorter than any of the carbon-carbon bonds, so h, d, and a are shorter than any of the others. These are ranked in order of increasing p character (or decreasing s character). Bonds g and e are ranked by decreasing bond order. The carbon–carbon single bonds are also ranked in order of increasing p character.

4.2   (a)   1-Pentene cannot exist as stereoisomers.

(b)   3-Hexene can exist as both cis and trans isomers; both carbons of the double bond (carbons 3 and 4) are stereocenters.

(c)   1,3-Pentadiene can exist as cis and trans isomers at the bond between carbons 3 and 4; both of these carbons are stereocenters. (Cis-trans isomerism is *not* possible at the double bond between carbon-1 and carbon-2.)

stereoisomers of 1,3-pentadiene

(d)   2-Methyl-2-pentene cannot exist as cis, trans isomers because two of the groups at one carbon of the double bond are the same (the methyl groups). This molecule contains no stereocenters.

(e)   In principle, cyclobutene, which is a cis alkene, might also exist as a trans isomer. However, attempting to build a model of *trans*-cyclobutene should convince you that such an isomer is far too strained to exist. The carbons of the double bond in cyclobutene are stereocenters if you allow for the theoretical possibility of double-bond isomerism.

4.3   (a)   (b)   (c)

1,3-dimethylcyclohexene

4-methyl-1,3-hexadiene

1-isopropenylcyclopentene

(d)

5-(3-pentenyl)-1,3,6,8-decatetraene

4.4   (a)   1,3-Dimethylcyclopentene   (b)   3-Heptene
          (c)   5-allyl-1,8-nonadiene (Remember, the 2-propenyl group is called the *allyl group*.)

4.5   (a)   First, name the compound without considering stereochemistry: 3-isopropyl-2-hexene. Then, determine the stereochemistry at the double bond between carbons 2 and 3. The isopropyl group is higher priority than the propyl group. Since the isopropyl group is on the same side of the double bond as the higher-priority methyl group, then the name of the compound is (Z)-3-isopropyl-2-hexene.

          (b)   First, draw out the compound completely, expanding abbreviations like *i*-Bu so you can see the bonds required for determining stereochemistry. Next, name the compound without considering stereochemistry: 4-isobutyl-1,4-hexadiene. Then, determine the stereochemistry at the double bond between carbons 4 and 5. (*E,Z* isomerism is not possible at carbon-1). Start by writing the structure with the double bond between carbons 1 and 2 replicated, and decide on the path of highest priority within each group (shown in boldface in the following structure):

As the diagram above shows, the first point of difference occurs at the third carbon out; thus, the allyl group has higher priority than the isobutyl group. Because the groups of higher priority are on the same side of the double bond, the stereochemistry at this double bond is *Z*, and the full name is therefore (*Z*)-4-isobutyl-1,4-hexadiene. No number before the *Z* is necessary when only one double bond has a stereochemical designator.

4.6 (a)

(*E*)-4-allyl-1,5-octadiene

(b)

(2*E*,7*Z*)-5-[(*E*)-1-propenyl]-2,7-nonadiene

4.7 (a)

(b)

(c)

(d)

4.8 (a)

(CH₃)₂CH—C≡C—H
isopropylacetylene

(b)

cyclononyne

(c)

(CH₃)₂CHCH₂—C≡C—H
4-methyl-1-pentyne

(d)

1-ethynylcyclopentene

(e)

H—C≡C—CH₂CH₂—C≡C—H
1,5-hexadiyne

4.9 (a) 1-Hexyne (*Common name:* butylacetylene)   (b) 5-decyne (*Common name*: dibutylacetylene)
       (c) (*E*)-3-propyl-5-hepten-1-yne   (d) 3-Ethynyl-1,4-pentadiene

4.10 (a) Use Eq. 4.18 to obtain $U = 1$.   (b) $U = 3$ (c) $U = 4$

4.11 (a) With 1 ring and no double bonds, methylcyclohexane has $U = 1$.

(b)   With three double bonds and no rings, 2,4,6-octatriene has $U = 3$.

4.12   (a)   Formula A cannot be correct, because a compound with an odd number of nitrogens must have an odd number of hydrogens. Formula C is not possible because the maximum number of hydrogens in a molecule containing 10 carbons and three nitrogens is 25. The other formulas are possible.

(b)   There are twelve isomeric alcohols (counting E,Z isomers) and one ketone with the formula C_4H_8O. The structures identified as enols are in equilibrium with aldehydes or ketones; the three marked with an asterisk (*) are both in equilibrium with the ketone shown. As we show in Chapter 5, the ketone form is the dominant form at equilibrium. The other two enols are in equilibrium with an aldehyde (not shown). The alcohol marked with a dagger (†) can exist as stereoisomers. You'll learn about these types of stereo-isomers in Chapters 6 and 7.

4.13   (a)   In *cis*-2-butene, the bond dipoles of the H_3C—C bonds add vectorially, with a slight reduction from the H—C bond dipoles. (See construction in text Display 4.20.). In contrast, the dipole moment of *trans*-2-butene is zero because corresponding bonds are oriented in opposite directions, and their bond dipoles consequently cancel.

(b)   Let d be the resultant of a H_3C—C bond dipole m and the H—C bond dipole c pointing in the opposite direction. In propene the two alkene C—H bond dipoles a and b cancel. This leaves a molecular bond dipole of d. In methylpropene, there are two resultants d at an angle of 120°. From the law of cosines, the resultant is $\sqrt{2d^2 - 2d^2 \cos 60°} = \sqrt{2d^2 - d^2} = d$, where d is the magnitude of d. This shows that the dipole moment of methylpropene should be the same as that of propene, although oriented in a different direction.

In fact, the dipole moment of methylpropene is 0.49 D, and that of propene is 0.35 D. The vector analysis above assumes idealized bond angles of 120°. Minor deviations from these angles and experimental errors could easily account for the difference.

(c)   The dipole moment of 2-butyne is zero. The dipole moment of 1-butyne is the resultant of the bond dipoles m and c, which are not equal. Therefore, the dipole moment of 1-butyne is greater.

4.14   (a)   The enthalpy of reaction is obtained by subtracting the ΔH_f° of the reactant, 1-butene, from that of the product, methylpropene. Therefore, $\Delta H^\circ = -17.30 - (-0.30) = -17.0$ kJ mol^{-1} (-4.06 kcal mol^{-1}).

(b)   The butene with the lower ΔH_f° is more stable. Since the conversion of 1-butene to methylpropene is "energetically downhill"—that is, negative—methylpropene is more stable

4.15   (a)   The overall ΔH° of the process, +15.3 kJ mol^{-1}, must equal the difference in the ΔH_f° values of product and reactant.

$$+15.3 = \Delta H_f^\circ \text{ (1-hexene)} - \Delta H_f^\circ \text{ (2-ethyl-1-butene)}$$
$$+15.3 = -40.5 - \Delta H_f^\circ \text{ (2-ethyl-1-butene)}$$

Solving, ΔH_f° (2-ethyl-1-butene) $= -15.3 - 40.5 = -55.8$ kJ mol^{-1} (-13.3 kcal mol^{-1}).

(b)   Because the ΔH_f° of 2-ethyl-1-butene is more negative than that of 1-hexene, 2-ethyl-1-butene is the more stable isomer.

4.16   The combustion reaction for 1-heptene is shown at right. The given ΔH° for this reaction is -4693.1 kJ mol^{-1}. This is equal to the heats of formation of products minus those of the reactants. The heats of formation of CO_2 and liquid H_2O are the same as the heats of combustion of C and H_2, respectively, because these combustion reactions form CO_2 and H_2O from their elements.

$$H_2C\!=\!CHCH_2CH_2CH_2CH_2CH_3 \; + \; 10.5\,O_2$$
$$\downarrow$$
$$7\,CO_2 \; + \; 7\,H_2O$$

$$-4693.1 \text{ kJ mol}^{-1} = 7\Delta H_f^\circ \text{ (CO}_2) + 7\Delta H_f^\circ \text{ (H}_2\text{O)} - \Delta H_f^\circ \text{ (1-heptene)} - 10.5\Delta H_f^\circ {}^\circ\text{(O}_2)$$

Letting ΔH_f° (1-heptene) $= x$,

$$-4693.1 \text{ kJ mol}^{-1} = 7(-393.51) + 7(-285.83) - x - 10.5(0)$$

Solving,

$$x = \Delta H_f^\circ \text{ (1-heptene)} = -62.28 \text{ kJ mol}^{-1} \text{ } (-14.89 \text{ kcal mol}^{-1})$$

 This calculation illustrates the point made in Further Exploration 4.2, that heats of formation derived from heats of combustion are small difference between large numbers and are therefore subject to some uncertainty.

4.17   (a)   The order of stability is $B < A$. Compound A is more stable than compound B because the double bond in A has an additional alkyl (methyl) substituent.

(b)   The order of stability is $A < B$. Compound B is more stable than compound A, because compound A, which is a cis alkene, is destabilized by significant van der Waals repulsions between the *tert*-butyl group and the methyl group. Because compound B is a trans alkene, these destabilizing repulsions are absent.

4.18   Hydrogenation of either (E)- or (Z)-3-hexene gives the same product, hexane. This allows us to use the heats of hydrogenation in a Hess's law calculation:

$$\begin{array}{c} CH_3CH_2 \quad\quad CH_2CH_3 \\ \\ C\!=\!C \quad\quad + \quad H_2 \quad \longrightarrow \quad CH_3CH_2CH_2CH_2CH_2CH_3 \quad\quad \Delta H^\circ = -121.6 = -167.2 - x \\ H \quad\quad\quad H \\ \textbf{hexane} \\ \textbf{(Z)-3-hexene} \quad\quad\quad \Delta H_f^\circ = -167.2 \\ \Delta H_f^\circ = x \end{array}$$

From this we obtain the ΔH_f° of (Z)-3-hexene as -45.6 kJ mol^{-1} (-10.9 kcal mol^{-1}). A similar calculation for (E)-3-hexene gives the ΔH_f° as -49.3 kJ mol^{-1} (-11.8 kcal mol^{-1}). The E isomer is more stable. (Why?)

4.19   The *least* stable compound releases more energy on combustion because it is higher on the energy scale from the common products CO_2 and H_2O. From the relative heats of formation given in this section of the text, 1-hexyne is the least stable of the three hydrocarbons, and therefore releases more energy on combustion.

4.20   (a)

$$H_2C{=}CH_2 + HI \longrightarrow CH_3CH_2I$$

(b)

$$HC{\equiv}CH + Br_2 \longrightarrow Br{-}HC{=}CH{-}Br$$

Without knowing more about the reaction, the 1,2-dibromoethene product of (b) could be the cis isomer, the trans isomer, or a mixture of both.

(c)

or $B(CH_2CH_3)_3$

(d)

4.21   In (a), the chlorine goes to the carbon of the alkene with the methyl branches. In (b), the bromine goes to the ring carbon with the methyl branch.

(a) $(CH_3)_3C{-}Cl$

(b)

4.22   (a)   Cation A is primary, B is tertiary, and C is secondary. Cation B is most stable because it has the greatest number of alkyl substituents at the electron-deficient carbon.

Don't be confused by the quaternary carbon in cation A. It's the number of alkyl substituents on the *cationic* (electron-deficient) carbon that determines stability, not the number of substituents on carbons elsewhere in the molecule.

(b)   Cation A is primary, cation B is secondary, and cation C is tertiary. Cation C is most stable because it has the greatest number of alkyl substituents at the electron-deficient carbon.

4.23   The alkene is protonated at carbon-1 so as to form the tertiary carbocation at carbon-2 along with bromide ion. The bromide ion reacts in a Lewis acid–base association reaction to give the product, 2-bromo-2-methylpentane. Notice in the first step that, even though a proton is transferred, the curved-arrow notation shows the *flow of electrons*.

2-methyl-1-pentene   *a tertiary carbocation*   **2-bromo-2-methylpentane**

4.24   The two alkenes must have a double bond that includes the carbon bonded to Br in the product and an adjacent carbon.

(a)

(b)

4.25 The carbocations in parts (a) and (c) are secondary and can rearrange to tertiary carbocations by the paths shown below. The carbocation in part (b) is tertiary and cannot rearrange to a more stable carbocation.

(a) The rearrangement of the carbocation in part (a):

(c) The rearrangement of the carbocation in part (c):

4.26 Protonation of the double bond gives a secondary carbocation:

This secondary carbocation can react by two pathways: (*a*) rearrangement to a more stable tertiary carbocation followed by reaction of the rearranged carbocation with Br⁻; (*b*) reaction with Br⁻.

4.27 If these alkyl halides could be prepared from alkenes, the alkene starting materials would have to be *A1, B1* or *B2,* and *C1* for *A, B,* and *C,* respectively.

Reaction of *A1* with HBr would not give *A* because the bromine would end up on the carbon of the double bond with the alkyl substituent. Reaction of HBr with *either B1* or *B2* would give *B*. Addition of HBr to alkene *C1* would not give solely compound *C*, but would also give one or more rearrangement products. (Be sure to draw out the rearrangement and derive these products.

4.28 (a) This is an electron-pair displacement reaction.

transition state

(b) This is a Lewis acid–base dissociation reaction.

transition state

4.29 (a) The curved-arrow notation and the transition state for the reverse of Eq. 4.51:

(b) The transition state for any reaction and its reverse are the same.

4.30 (a) Apply Eq. 4.54b in Sec. 4.8B with $\Delta G_B^{\circ\ddagger} - \Delta G_A^{\circ\ddagger} = -15$ kJ mol^{-1}.

$$\log\left(\frac{\text{rate}_A}{\text{rate}_B}\right) = \frac{\Delta G_B^{\circ\ddagger} - \Delta G_A^{\circ\ddagger}}{2.3RT} = \frac{-15}{5.71} = -2.63$$

$$\frac{\text{rate}_A}{\text{rate}_B} = 10^{-2.63} = 2.36 \times 10^{-3}$$

Therefore, the rate of reaction A is 0.00236 times that of reaction B, or reaction B is (1/0.00236) or 424 times as fast as reaction A.

(b) The text states that a reaction rate approximately doubles for every 10° rise in temperature. The question is, then how many "doublings" in rate correspond to the rate factor 424 above?

$$\begin{aligned} 424 &= 2^n \\ \log 424 &= n \log 2 \\ 2.63 &= 0.301n \\ n &= 8.74 \end{aligned}$$

Evidently, a rate factor of 424 corresponds to almost 9 doublings of the rate—8.74, to be exact. The temperature at which reaction A is run would have to be raised an estimated 87° to make it as fast as reaction B.

4.31 (a) Apply Eq. 4.54b:

$$\log\left(\frac{\text{rate}_A}{\text{rate}_B}\right) = \log(10^{-6}) = -6 = \frac{\Delta G_B^{\circ\ddagger} - \Delta G_A^{\circ\ddagger}}{5.71} = \frac{\Delta G_B^{\circ\ddagger} - 90}{5.71}$$

Solving for $\Delta G_B^{\circ\ddagger}$ gives $\Delta G_B^{\circ\ddagger} = 55.7$ kJ mol^{-1} (13.3 kcal mol^{-1}).

(b) The reaction free-energy diagram for the two reactions is shown in **Fig. SG4.1** on the next page..

(c) The standard free energy of activation of a reverse reaction is the height of the energy barrier measured from the standard free energy of the products. In each case, this is 10 kJ mol^{-1} greater than the standard free energy of activation of the forward reaction: 100 kJ mol^{-1} for reaction A, and 66 kJ mol^{-1} for reaction B, as shown in the Figure.

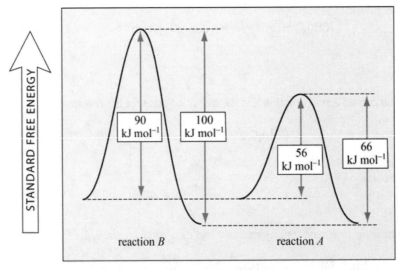

Figure SG4.1   Reaction free-energy diagrams to accompany the solution to Problem 4.31(b).

4.32, 4.33   The diagrams for these problems follow:.

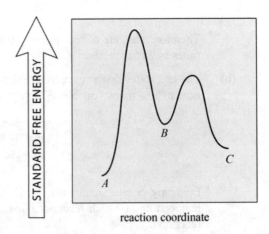

Diagram for the Solution to Problem 4.32

Diagram for the Solution to Problem 4.33

4.34   Because the transition states of the two reactions should resemble the respective carbocation intermediates (Hammond's postulate), examine the structures of the two carbocations (at right) and determine which should be more stable. The carbocation intermediate in the addition of HBr to methylpropene is tertiary, and it is more stable than the carbocation intermediate in the addition of HBr to *trans*-2-butene, which is secondary. If the transition states for the two addition reactions resemble the carbocation intermediates, the transition state for the reaction of methylpropene is also more stable. The relative rates of the two reactions are governed by the *differences* between the standard free energies of the transition states and their respective starting materials. Therefore, we have to consider the free energies of *both* the alkenes and the transition states. *If the two alkenes do not differ appreciably in energy, then the relative rates of the two reactions are governed mostly by the relative free energies of the two transition states.* (This is a fairly good assumption, because both alkenes have two alkyl branches.) Under this assumption, addition of HBr to methylpropene is faster, because the transition state for this reaction has lower standard free energy.

carbocation intermediate from HBr addition to 2-methylpropene

carbocation intermediate from HBr addition to *trans*-2-butene

4.35 (a)   The possible products from addition of HCl to 2-hexyne:

$$H_3C—C\equiv C—CH_2CH_2CH_3 + HCl \longrightarrow$$
2-hexyne

Compounds *B* and *D* are constitutional isomers of compounds *A* and *C*. Compounds *A* and *C* are stereoisomers; compounds *B* and *D* are stereoisomers.

(b)   Ion *AC* is the carbocation intermediate leading to compounds *A* and *C*; ion *BD* is the carbocation intermediate leading to compounds *B* and *D*.

4.36 (a)   The hydrogenation product of 1-pentene is pentane, $CH_3CH_2CH_2CH_2CH_3$.
(b)   The hydrogenation product of (*E*)-1,3-hexadiene is hexane, $CH_3CH_2CH_2CH_2CH_2CH_3$.

4.37 (a)   The formula restricts us to considering only alkenes with one double bond. Any alkene with a six-carbon unbranched carbon skeleton will give hexane as a product. Five alkenes meet this criterion.

(b)   Any alkene with the carbon skeleton of methylcyclopentane will undergo hydrogenation to give this cycloalkane. There are four such alkenes.

4.38 (a)   (b)   (c)   (d)
$CH_3(CH_2)_5CH=CH_2$ $CH_3(CH_2)_5CH_2CH_3$

1-octene   **octane**

The alkyne in part (c) contains a pyridine ring. Because pyridine is a catalyst poison, the reaction stops at the alkene stage. That is, the alkyne (or the alkene product) poisons its own hydrogenation.

4.39 In the first step of the reaction, the double bond is protonated by the acid catalyst in a Brønsted acid–base reaction to give a carbocation *A*, which rearranges to a tertiary carbocation *B* by a hydride shift. Carbocation *B* reacts with water in a Lewis acid–base association reaction, and the resulting adduct loses a proton to another water molecule in a Brønsted acid–base reaction to form the product alcohol and regenerate the catalyzing acid.

4.40 (a) The alkenes result from loss of the two chemically nonequivalent protons on the carbons adjacent to the positively charged carbon in carbocation *B*:

(b) In the solution to Focused Problem 4.39, the rearrangement of carbocation *A* to carbocation *B* is completely irreversible because of the much greater stability of *B*. Once this carbocation is formed, the only alkenes that can form are those that are derived from it—the ones shown in part (a).

4.41 The intermediate in the hydration of ethylene is a primary carbocation, so the transition state resembles such a cation. Because primary carbocations are *much* less stable than secondary and tertiary carbocations, transition states that resemble primary carbocations are also unstable. An unstable transition state means that the transition state has a very high energy. When a reaction has a transition state of very high energy, the reaction is slow.

4.42 The hydration of propene involves the usual mechanistic steps of alkene protonation to form a carbocation followed by the Lewis acid–base association reaction of water with the carbocation. The structure of isopropyl alcohol follows from the structure of the carbocation. A secondary carbocation rather than a primary carbocation is involved. (Why?).

Solutions to Skills Objectives Problems

4.43 *a:* sp^3–sp^3 *b:* sp^3–$1s$ *c:* sp^3–sp^2 *d:* sp^2–$1s$ *e:* sp^2–sp^2
f: sp^2–sp *g:* sp–sp *h:* sp–sp^3

4.44 The three principles involved are that bonds involving atoms from lower-numbered periods are shorter; bonds with higher bond order are shorter; and bonds with more s character are shorter. So, the C—H bonds are shorter than C—C bonds; for carbon–carbon bonds, triple bonds are shorter than double bonds, which are shorter than single bonds; and, for bonds between the same two atoms, bonds with more s-character are shorter. The s-character of the bonds follows from the answers to Problem 4.43.

Ranking: $d < b < g < e < f < h < c < a$

4.45 (a) The π and π^* molecular orbitals of methylpropene are very similar to those of ethylene. In other words, the presence of the methyl group has very little effect on these molecular orbitals.

(b) The π and π^* molecular orbitals of acetylene are much like those of ethylene, except that there are two degenerate (equal-energy) π MOs and two degenerate π^* MOs. These are shown separately aligned on the y and z axes.

Sketches for (a) and (b) are given in **Fig. SG4.2a–b** at the bottom of this page.

4.46 (a) 3-Ethylcyclopentene   (b) 6-Methyl-1-heptene
(c) 2-Isopropyl-1-methyl-1,3-cyclohexadiene   (d) 1-Buten-3-yne
(e) 2-Ethynyl-1,3-cyclopentadiene

4.47 (a)   (b)   (c)

$CH_3CH{=}CHCH_2CH_2CH_3$

2-hexene

4-methylcycloheptene

3,4-diethyl-1,3,5-hexatriene

(d)   (e)

$CH_3C{\equiv}C{-}C{\equiv}CCH_2CH_2CH_2CH_3$

2,4-nonadiyne

4-(3-methyl-1-butynyl)cyclohexene

4.48 Compounds *B* and *C* can exist as *E,Z* isomers. The two cyclobutenyl groups in *C* are *not* identical because they are attached at different ring carbons to the double bond. (Although the double bond in each four-membered ring could in principle exist as double-bond isomers, such isomers would be too strained to exist, so these possibilities are ignored.)

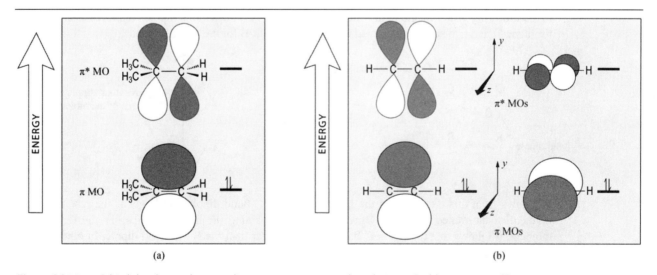

Figure SG4.2   Orbital sketches and energy diagrams to accompany the solution to Problems 4.45a and b.

4.49 (a) On the basis of atomic number, Br has higher priority than Cl, and I has higher priority than Br. The stereochemistry is *E*. We might have been tempted to assign *Z* stereochemistry to this alkene because the bromine substituents are on the same side of the double bond. As the above analysis shows, this is not correct.

(b) On the basis of atomic mass, D has a higher priority than H and CD_3 has a higher priority than CH_3. The stereochemistry is *Z*.

(c) The ring carbon with the methyl substituent is on the opposite side of the double bond from the methyl substituent on the other carbon. The stereochemistry is *E*.

(d) The four-membered rings are numbered *A*, *B*, and *C* in the diagram at right. Ring *A* is higher priority than ring *C* because the level-1 substitution pattern in ring *C* is C(C,C,H) at the point of attachment to the double bond, whereas, in ring *A*, the level-1 analysis is C(C,C,C). The analysis of the relative priorities of rings *A* and *B* also starts at the carbon attached to the double bond (bold), which is labeled as level 1 in each case. These carbons each have three carbons attached, so we must proceed to the next level to continue the analysis. In ring *B*, all paths outward are equivalent. In ring *A* we must choose the path of highest priority (labeled with the arrow), which leads to the level-2 carbon. Notice the replication of the double bond. Comparison of the level-2 carbons shows the substitution pattern C(C,C,H) for ring *A*, and C(C,H,H) for ring *B*. This establishes that ring *A* has the higher priority. Because ring *A* has the higher priority at both carbons, and because this substituent is on opposite sides of the double bond at the two alkene carbons, this alkene stereochemistry is *E*.

4.50 (a) With two nitrogens, seven carbons, and a chlorine, the maximum number of hydrogens is 17. The actual number is 13, so $U = (17-13)/2 = 2$. The compound could contain one triple bond.

(b) 17α-Ethynylestradiol has four rings, 3 double bonds, and one triple bond for $U = 9$. With twenty carbons, there must be $2(20) + 2 - 2(9) = 24$ hydrogens.

(c) With $U = (2N_C + 2 - 8)/2 = N_C - 3 = 6$, then $N_C = 9$. The structure of **aspirin (2-acetoxybenzoic acid)** aspirin is shown at right; check the answer for yourself.

4.51 (a) We assume that the C—Cl bond dipole is oriented from carbon to chlorine in both molecules. As shown in the following vector diagram, in which R = CH_3 or R = H, the parallel dipoles reinforce to give a larger resultant **A**; and two such resultants add to give a resultant **B** for the entire molecule.

It follows from this diagram that the larger is the R—C bond dipole, the larger is the resultant dipole moment of the molecule. Because the compound containing the methyl groups has the larger dipole moment, it follows that the H_3C—C bond dipole is larger than the H—C bond dipole. In other words, a methyl group is more electron donating than a hydrogen towards a double bond.

(b)   Assume that the major contributors to the molecular dipole moment are the H_3C—C and the C—Cl bond dipoles, as indicated in part (a). These bond dipoles reinforce in the E isomer, and are nearly opposed in the Z isomer. Consequently, the E isomer has the greater dipole moment.

4.52   Because an sp-hybridized carbon is more electronegative than an sp^3-hybridized carbon, the H_3C—$C\equiv$ bond dipole is oriented from H_3C to $C\equiv$. The $\equiv C$—Cl bond dipole is oriented toward the Cl. Therefore, the two bond dipoles in A reinforce. In B the two $\equiv C$—Cl bond dipoles oppose each other, and the dipole moment is zero. In C, the H_3C—$C\equiv$ and H—$C\equiv$ bond dipoles are oriented toward the —$C\equiv$, and oppose each other; however, because these bond dipoles are not equal, there is a net dipole in C oriented from H_3C to $C\equiv$. In other words, the order of increasing dipole moments is $\mu_B < \mu_C < \mu_A$.

4.53   (a)   The relative stability following from the number of alkyl substituents on the sp^2-hybridized carbons. Compound B has 3 substituents, compound A has two, and compound C has one. Therefore, the relative stability is $C < A < B$.

(b)   In molecule B the cis *tert*-butyl groups suffer significant van der Waals repulsions that destabilize the molecule. Molecule A can avoid severe van der Waals repulsions by rotating the branched carbon away from the *tert*-butyl group, but we assume that, despite this restricted rotation, this group acts as if it is larger than a methyl group. In molecule C, the two *tert*-butyl groups are trans and have no unfavorable van der Waals repulsions. The order of increasing stability is $C < A < B$.

(c)   The two dienes A and B have lower energy than the alkyne C as the heats of formation in Sec. 4.5C show. Of the two dienes, alkene B with the trans-substituted double bond is more stable than alkene A, which has a cis-substituted double bond.

(d)   3-Hexyne ($CH_3CH_2C\equiv CCH_2CH_3$) with two alkyl substituents on its triple bond, is more stable than 1-hexyne ($HC\equiv CCH_2CH_2CH_2CH_3$), which has one alkyl substituent on its triple bond.

4.54   (a)   The enthalpies of the three alkenes and their hydrogenation product 2-methylbutane are shown in the following energy diagram. (The shadowed pointers are heats of hydrogenation.)

(b)   Because all three alkenes are converted into the same alkane (2-methylbutane) by catalytic hydrogenation, we can use the heat liberated in the hydrogenation of each to determine their relative stabilities. Since 2-methyl-2-butene gives off the least amount of heat (-112.6 kJ mol^{-1}), it is the most stable. 3-Methyl-1-butene gives off the most heat, so it is the least stable.

(c)   Because 3-methyl-1-butene, 2-methyl-1-butene, and 2-methyl-2-butene are isomers, they are formed from the same proportions of carbon and hydrogen (C_5H_{10}). The difference in heats of hydrogenation must equal the difference in their heats of formation. We don't know what the individual heats of formation are, but we know that they differ by the amounts shown in the energy diagram.

(d) As discussed in Sec. 4.5B, the more alkyl groups that are attached to the carbons of the double bond, the more stable the alkene.

4.55 The structure of 4-octene (ignoring stereochemistry) is $CH_3CH_2CH_2CH{=}CHCH_2CH_2CH_3$. The reagent given add across the carbon–carbon double bond. Because the alkene is symmetrical, there are no issues of constitutional isomerism in the products.

(a)　　　　　　　　(b)　　　　　　　(c)　　　　　　(d)

The product of part (c) results from each B—H bond adding across the double bond of a different 4-octene molecule.

4.56 Additions to 3-hexyne ($CH_3CH_2C{\equiv}CCH_2CH_3$):

(a)　　　　　　　　　　　　(b)　　　　　　　　　　　(c)

4.57 (a)　　　　(b)　　　　(c)

4.58 (a)　　　　　　　　　(b)　　　　　　　(c)

(from A)　　　(from B)　　(from A and B assumes an excess of H_2)　　(from A)　　(from B)

4.59 (a)　　　　　　　　　(b)

$=CH_2$ and ——CH_3　　　　　and

(c) There are no alkynes that would undergo catalytic hydrogenation to give 2,4-dimethylpentane (why?), but there are two alkenes with one double bond (shown at right).

(d)　　　　　(e)　　　　　(f)　　　　　(g)

HC≡C——Me Me　　　　　　　and　　$CH_3CH_2C{\equiv}CCH_2C(CH_3)_3$

4.60   (a)   Reaction step *a* is both a Brønsted acid–base reaction as well as an electron-pair displacement reaction. Reaction step *b* is a Lewis acid–base association reaction.

$$(CH_3)_2C\!=\!CH\!-\!CH_3 \xrightarrow{(a)} (CH_3)_2\overset{+}{C}\!-\!CH_2\!-\!CH_3 \xrightarrow{(b)} (CH_3)_2C\!-\!CH_2\!-\!CH_3$$

(b)   Reaction steps *a* and *c* are both Brønsted acid–base reactions as well as electron-pair displacement reactions. Reaction step *b* is a Lewis acid–base association reaction.

$$(CH_3)_2C\!=\!CH\!-\!CH_3 \xrightarrow{(a)} (CH_3)_2\overset{+}{C}\!-\!CH_2\!-\!CH_3 \xrightarrow{(b)} (CH_3)_2C\!-\!CH_2\!-\!CH_3 \overset{(c)}{\rightleftharpoons} (CH_3)_2C\!-\!CH_2\!-\!CH_3 + H_3\overset{+}{O}:$$

4.61   (a)   Bonds involved in hyperconjugation are marked with an asterisk:

(b)   Overlap of a C—H bond orbital with the empty 2p orbital of the carbocation:

(c)   Overlap of a C—C bond orbital with the empty 2p orbital of the carbocation:

(d)   Resonance structures that show the overlap in parts (b) and (c):

C—H hyperconjugation for part (b)   C—C hyperconjugation for part (c)

4.62   (a)   Cation *B*, the tertiary carbocation, is most stable; carbocation *C* is least stable.
        (b)   Cation *B*, the allylic cation, is most stable because it is resonance stabilized. The vinylic cation *A* is least stable.

4.63   Carbocation *A* can rearrange in one way, and carbocation *C* in two ways, the second of which is a ring expansion by migration of a ring carbon. [Corresponding carbons are labeled with an asterisk (*) and a double-dagger (‡) to help in following the rearrangement.]

Carbocation *A* could also rearrange to a tertiary carbocation by a ring *contraction* to a four-membered ring, but this is not likely to happen because such a ring is highly strained (as we show in Sec. 7.5B).

4.64   (a)   *A* corresponds to starting materials, *D* to products, *B* and *C* to reactive intermediates, and ‡ to transition states.

   (b)   Because compound *D* has the lowest standard free energy, it is present in greatest amount when the reaction comes to equilibrium. In other words, the reaction proceeds to the right. Because compound *C* has the highest standard free energy, it is present in least amount when the reaction comes to equilibrium.

   (c)   The rate-limiting step of the reaction is *B* ⟶ *C*, because that step is the one with the transition state of highest free energy.

   (d)   The standard free energy of activation for the overall reaction is the energy of the transition state of highest standard free energy minus the standard free energy of the reactants. This is labeled with the vertical arrow.

   (e)   The reaction *C* ⟶ *D* is faster than the reaction *C* ⟶ *B*, because the faster reaction has the smaller standard free energy of activation.

This problem illustrates the sometimes confusing point that the energy barrier of the rate-limiting step can be smaller than other energy barriers in the mechanism. In these diagrams of *standard* free energy, the various compounds are present at 1*M* concentration. If everything were present at 1*M*, the rate of the *B* ⟶ *C* step would indeed be faster than *A* ⟶ *B*. However, in the actual reaction, *B* is present at *much* lower concentration than *A*. Because rates are proportional to concentration, the rate of the *B* ⟶ *C* step is actually the slowest (and therefore rate-limiting) step in the reaction. It is always true that the rate-limiting step is the step with the highest standard free energy of activation.

4.65   (a)   The reaction *A* ⟶ *B* is faster because it has a smaller standard free energy of activation. To calculate the rate factor, we use Eq. 4.54a in Sec. 4.8B and obtain

$$\frac{\text{rate}_A}{\text{rate}_B} = 10^{-(\Delta G_B^{\circ\ddagger} - \Delta G_A^{\circ\ddagger})/2.3RT} = 10^{-(81.5-87.2)/5.71} = 10^1 = 10$$

   (b)   To get a rate factor of $100 = 10^2$, the standard free energies of activation must differ by $2 \times 2.3RT$, or $2(5.71) = 11.4$ kJ mol^{-1} at 298 K (2.72 kcal mol^{-1}).

4.66   (a)   The product from hydration of methylenecyclobutane:

methylenecyclobutane   1-methylcyclobutanol
                              (compound *X*)

(b)   The reactive intermediate in the rate-limiting step is the carbocation, as we deduce from the curved-arrow notation:

carbocation
intermediate

(c)   The transition state for the rate-limiting step:

(d)   If the rate-limiting step is protonation of the double bond in the hydration (forward) direction, then the rate-limiting step in the reverse direction is the reverse of the same step: deprotonation of the carbocation intermediate to give the alkene. The transition state is identical to the one in part (c), because a reaction and its reverse have the same transition state under the same conditions.

4.67   The reaction of B with HBr involves a tertiary carbocation intermediate, whereas the same reaction of A involves a less stable secondary carbocation intermediate. From Hammond's postulate, we infer that the more stable reactive intermediate comes from the transition state of lower standard free energy. This means that the reaction of B is faster.

4.68   The catalyzed reaction involves a smaller standard free energy of activation without affecting the standard free energies of reactants and products. Therefore, Diagram 2 corresponds to the catalyzed reaction.

4.69   (a)   The stepwise curved-arrow mechanism for the hydration of (E)-3-methyl-2-pentene:

(b)   By the principle of microscopic reversibility, the mechanism for the reverse reaction under the same conditions has to be the exact reverse mechanism:

Solutions to Integrated Problems

4.70

(a) cyclobutene

(b) 3-methyl-1-octene

(c) 5,5-dimethyl-1,3-cycloheptadiene

(d) 1-vinylcyclohexene

(e) 4-allyl-2-methyl-1-octen-5-yne

4.71 (a) (*E*)-2-Butyl-1,3-pentadiene   (b) 1-(1-Cyclopentenyl)cyclopentene
      (c) (*E*)-5-Allyl-1,6-octadien-3-yne

4.72 (a) N (b) C (c) N (d) I (e) C (f) S

4.73   Both *cis-* and *trans*-3-hexene react with H—I to give 3-iodohexane and undergo catalytic hydrogenation to give hexane:

4.74   Any one of the following compounds, when allowed to react with hydrogen in the presence of a catalyst, will give propylcyclohexane.

any ONE of these

4.75   (a)   A five-carbon alkyne containing a ring: (b)   A six-carbon trans alkene containing a ring:

4.76   The process to be used in working this problem is thoroughly explained in Study Guide Link 4.3 in this chapter of this manual. The structures of X and Y are as follows:

4.77   Hydration, because of its mechanism, yields an alcohol in which the —OH group is positioned at the carbon of the double bond with *more alkyl substituents*. Al Keyhall's proposal would require a different regioselectivity, and therefore would not work. (As we show in Sec. 5.6B, there is a different way to prepare this alcohol from 3-methyl-1-butene.)

4.78   (a)   Before drawing the reaction-free energy diagram, convert K_{eq} into a free energy. (Assume 25 °C or 298 K.)

$$\Delta G° = -2.3RT \log K_{eq} = -5.71 \log(150) = -12.4 \text{ kJ mol}^{-1} (-2.97 \text{ kcal mol}^{-1})$$

This number is the energy difference between A and B. Compound B has the lower energy, because it is favored in the equilibrium. The reaction free-energy diagram is shown at right.

(b)   The standard free energy of activation for the reaction in the B to A direction is 12.4 kJ/mol + 96.0 kJ/mol = 108.4 kJ mol^{-1} (25.9 kcal mol^{-1}). This follows geometrically from the diagram in part (a); it is the difference between the standard free energy of B and the standard free energy of the transition state.

$\Delta G°^{\ddagger} = 96.0 \text{ kJ mol}^{-1}$

$\Delta G°^{\ddagger}$ (reverse reaction) = 96.0 + 12.4 = 108.4 kJ mol^{-1}

$\Delta G° = 12.4 \text{ kJ mol}^{-1}$

reaction coordinate

4.79   This hydration reaction, like others in this chapter, begins with protonation of the double bond of 3,3-dimethyl-1-butene to give the more stable carbocation. This cation rearranges by a methyl shift to give a more stable carbocation, which then undergoes a Lewis acid–base reaction with water to give 2,3-dimethyl-2-butanol, the rearranged product alcohol.

(reaction scheme)

3,3-dimethyl-1-butene

2,3-dimethyl-2-butanol

4.80   (a)   The bonding arrangement in (*E*)-1,3-pentadiene is more stable because this diene has the smaller heat of formation.

When two double bonds are separated by one single bond, as in (*E*)-1,3-pentadiene, the double bonds are said to be *conjugated.* The conjugated arrangement of double bonds is a particularly stable arrangement. The reason for this enhanced stability is discussed in Chapter 15.

(b)   The relevant combustion equation is

$$(E)\text{-1,3-pentadiene} + 7\,O_2 \longrightarrow 5\,CO_2 + 4\,H_2O$$
$$(C_5H_8)$$

| heat of formation (kJ mol^{-1}) | 75.8 | 0 | 5(−393.5)   4(−285.8) |

(Note that the heats of formation of CO_2 and H_2O are by definition identical to the heats of combustion of carbon and hydrogen, respectively.) The $\Delta H°$ for this reaction is

$$4(−285.8) + 5(−393.5) − 75.8 = −3186.5 \text{ kJ mol}^{-1}$$

The heat of combustion (always expressed as a positive number for heat liberated) is therefore 3186.5 kJ mol^{-1} (761.6 kcal mol^{-1}).

4.81   When we attempt build models of these two cyclic alkenes we find that (*E*)-cycloheptene is very difficult or impossible to build (depending on the flexibility of your models), but (*Z*)-cycloheptene can be built without difficulty. This exercise suggests that (*E*)-cycloheptene is very strained and very unstable, and certainly the least stable of the two alkenes. The source of the problem is that the distance between two trans carbons is too great to be bridged by only three other carbons without significant strain.

this distance is difficult to bridge with only 3 carbons

4.82   The mechanism is a sequence of two Brønsted acid–base reactions. In the first (step *a*), the double bond is protonated to give a tertiary carbocation; in the second (step *b*), the carbocation loses a different proton to give the alkene. The equilibrium is driven to the right by the greater stability of the alkene product. (Why is it more stable?)

(reaction scheme)

4.83 This is an acid-catalyzed addition to the double bond much like hydration, except that an alcohol rather than water is the nucleophile that reacts with the carbocation intermediate. Notice that the ionized form of H_2SO_4 in methanol is protonated methanol ($CH_3\overset{+}{O}H_2$).

4.84 Follow the suggestions in the problem. First, we label corresponding atoms (at right). Next, consider the fate of the various functional groups. The double bond has disappeared. Carbon-3 has gained a proton, and carbon-4 has been joined to the oxygen. In other words, an *addition* of the O—H group to the double bond has occurred. We have seen two other additions involving acids: addition of HBr, and acid-catalyzed hydration. Because a catalyzing acid is present in this reaction, and an oxygen is involved in the addition, it is reasonable to think of acid-catalyzed hydration as a possible model for this reaction. Carbon-3 can gain a proton if the double bond reacts as a Brønsted base with the Brønsted acid H_3O^+ (step *a* in the mechanism below). This forms a tertiary carbocation, which then reacts in a Lewis acid–base association reaction with the oxygen *within the same molecule* (step *b*). Loss of a proton to solvent H_2O in another Brønsted acid–base reaction (step *c*) regenerates the catalyzing acid and forms the product.

4.85 Protonation occurs at carbon-5 (see numbered structure below) to form a tertiary carbocation. (Protonation at carbon-1 would give a secondary carbocation.) Because the tertiary carbocation is more stable, it is formed more rapidly (Hammond's postulate). The tertiary carbocation resulting from protonation then adds to the other double bond (the only choice) to give a secondary carbocation (step *b*). This step can be viewed as a Lewis acid–base association reaction in which the π electrons of the double bond act as a nucleophile. The result is a secondary carbocation. A Lewis acid–base reaction of this carbocation with water (step *c*) followed by loss of a proton to another solvent water molecule (step *d*, another Brønsted acid–base reaction) gives the product.

4.86 Because protonation of alkene *A* gives a secondary carbocation and protonation of alkene *B* gives a (more stable) tertiary carbocation, we put the carbocation from *B* at the lower standard free energy. In this diagram, there is a negative sign on –2.3*RT*pK_a because the acid has higher free energy than the alkene base. (For weak acids, the acid is at the lower free energy.) Because carbocations have negative pK_a values (they are strong acids), then the negative pK_a

and the negative sign combine to give a positive standard free energy for conversion of the alkenes to their respective carbocations, as the diagram shows. From the diagram,

$$-2.3RTpK_{AH^+} > -2.3RT\,pK_{BH^+}$$

Dividing through by –2.3*RT*, $\qquad\qquad$ $pK_{AH^+} < pK_{BH^+}$

(Remember that dividing an equality by a negative number reverses the inequality.) This says that carbocation AH^+ is a stronger acid than carbocation BH^+. Therefore, alkene *A* is a weaker base than alkene *B*. In other words, alkene *B* is a stronger base because it protonates to give the more stable carbocation.

4.87 (a) We refer to the alkene on the left, 2,3-dimethyl-1-butene, as *A*, and the alkene on the right, 2,3-dimethyl-2-butene, as *B*. Compound *B* is favored at equilibrium because it has the lower ΔG_f°. The equilibrium constant is obtained from Eq. 3.48b (Sec. 3.5B). Define K_{eq} as [*B*]/[*A*], and use –3.10 kJ mol^{-1} as the ΔG° difference between the two alkenes:

$$K_{eq} = 10^{-\Delta G^\circ/2.3RT} = 10^{-(-3.10/5.71)} = 10^{0.543} = 3.49$$

(b) The equilibrium constant tells us *nothing* about how rapidly this interconversion takes place. For example, in the absence of a catalyst, the rate of this interconversion is essentially zero. However, this reaction does occur at a convenient rate in the presence of an acid catalyst, as suggested by Problem 4.82.

(c) The equilibrium favors the alkene with more alkyl substituents on the double bond, which is the source of the greater stability.

4.88 (a) First draw the structures of the two compounds (at right). Methylpropene has two alkyl substituents on its double bond, whereas 1-butene has only one. Because alkyl substituents on the double bond stabilize an alkene, methylpropene is the more stable of the two alkenes.

$$H_2C=CHCH_2CH_3$$

1-butene
A

$$H_2C=C\begin{smallmatrix}CH_3\\CH_3\end{smallmatrix}$$

methylpropene
B

(b) The reaction with the smaller standard free energy of activation ($\Delta G^{\circ\ddagger}$) is faster. Consequently, hydration of methylpropene is faster.

(c) The following diagram shows what we know about the two reactions.

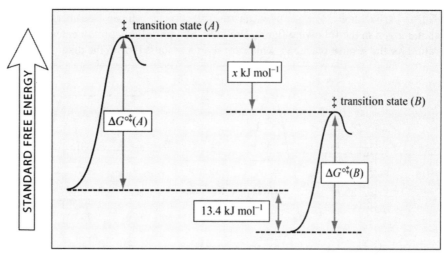

(d)   Let x represent the difference in transition-state free energies, as shown in the diagram. The diagrams in this figure require the following equality:

$$13.4 + \Delta G^{\circ\ddagger}(A) = x + \Delta G^{\circ\ddagger}(B)$$

Rearranging, and recognizing that the difference between the standard free energies of activation (given in the problem) is 22.8 kJ mol^{-1}, we have

$$\Delta G^{\circ\ddagger}(A) - \Delta G^{\circ\ddagger}(B) = 22.8 = x - 13.4$$

Solving for x gives

$$x = 22.8 + 13.4 = 36.2 \text{ kJ mol}^{-1}$$

In other words, the transition state for hydration of 1-butene is 36.2 kJ mol^{-1} *less stable* than the transition state for hydration of methylpropene. From Sec. 4.10C, the mechanism of hydration involves a carbocation intermediate, and the first step, protonation of the double bond to form this intermediate, is rate-limiting. In the case of methylpropene, the carbocation intermediate is tertiary; in the hydration of 1-butene, it is secondary. Tertiary carbocations are more stable than secondary carbocations. Because the transition states resemble the respective carbocations in structure and energy (Hammond's postulate), the transition state for formation of the more stable carbocation is the more stable transition state, as observed.

4.89   This problem can be framed in terms of two reaction coordinate diagrams. To do this for methylenecyclobutane requires a knowledge of the standard free energy of activation for its hydration. Letting the hydration of methylenecyclobutane be reaction A, and the hydration of methylpropene be reaction B, we use Eq. 4.54b in Sec. 4.8B to obtain

$$\log\left(\frac{\text{rate}_A}{\text{rate}_B}\right) = \log(0.6) = -0.222 = \frac{\Delta G_B^{\circ\ddagger} - \Delta G_A^{\circ\ddagger}}{2.3RT} = \frac{\Delta G_B^{\circ\ddagger} - \Delta G_A^{\circ\ddagger}}{5.71}$$

or $\Delta G_A^{\circ\ddagger} - \Delta G_B^{\circ\ddagger} = 1.27$ kJ mol^{-1}. With $\Delta G_B^{\circ\ddagger} = 91.3$ kJ mol^{-1}, then $\Delta G_A^{\circ\ddagger} = 92.6$ kJ mol^{-1}.

Next, we need to convert the relative hydration equilibrium constants for reactions A and B into relative standard free energies. A similar equation applies.

$$\log\left(\frac{K_A}{K_B}\right) = \log(250) = 2.40 = \frac{\Delta G_B^{\circ} - \Delta G_A^{\circ}}{2.3RT} = \frac{5.56 - \Delta G_A^{\circ}}{5.71}$$

or $\Delta G_A^\circ = -19.3$ kJ mol^{-1}. We can now construct reaction coordinate-free energy diagrams for the two reactions, which are given in the following figure. We have enough information to calculate the standard free energies of activation for the reverse reactions, which are shown in italic type in the diagram.

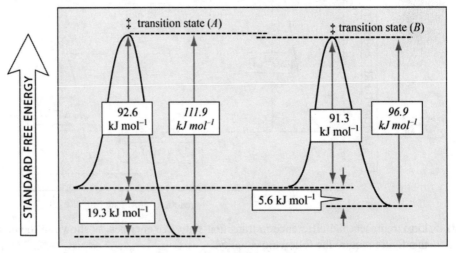

Because the standard free energy of activation of the reverse of reaction A is larger, it is the slower reaction. The relative rates are calculated from Eq. 4.54b:

$$\log\left(\frac{\text{rate}_B}{\text{rate}_A}\right) = \frac{\Delta G_A^{\circ\ddagger} - \Delta G_B^{\circ\ddagger}}{2.3RT} = \frac{15}{5.71} = 2.63$$

The reverse of reaction B is therefore $10^{2.63}$, or 427, times faster than the reverse of reaction A.

Addition Reactions of Alkenes and Alkynes

CHAPTER 5 SUMMARY

- A characteristic reaction of alkenes and alkynes is addition to the double or triple bond.

- Many addition reactions of alkenes and alkynes are *electrophilic additions,* in which the first step of the reaction mechanism is the reaction of an electrophile with a π bond of the alkene or alkyne. (Catalytic hydrogenation is an exception, and other addition mechanisms are considered in Chapter 10.)

- Electrophilic addition mechanisms can be further characterized as follows:

    - Mechanisms involving carbocation intermediates (hydrogen halide addition, hydration)

    - Mechanisms involving cyclic-ion intermediates (halogen addition, oxymercuration)

    - Concerted mechanisms (hydroboration, ozonolysis)

- Some useful transformations of alkenes involve additions followed by other transformations. These include oxymercuration–reduction and hydroboration–oxidation, which give alcohols; and ozonolysis followed by treatment with $(CH_3)_2S$ or H_2O_2, which gives aldehydes, ketones, or carboxylic acids by cleavage of the double bond.

- Electrophilic additions of alkynes include hydration, which is conceptually related to both alkene hydration and alkene oxymercuration, and hydroboration–oxidation. Hydration of alkynes (except for acetylene itself) gives ketones, and hydroboration–oxidation gives aldehydes from 1-alkynes and ketones from internal alkynes.

- Enols are intermediates in both the hydration of alkynes and hydroboration–oxidation of alkynes.

Enols are spontaneously transformed into the isomeric aldehydes or ketones.

- The constitutional isomer of the product formed in unsymmetrical additions to a π bond (additions in which the two groups that add are different) depends on the number of alkyl substituents at each carbon of the π bond. Reactions in which one constitutional isomer is strongly favored are said to be *regioselective.*

    - In electrophilic hydrogen-halide addition, the halogen adds to the carbon of the π bond with more alkyl substituents, and the hydrogen to the carbon with fewer alkyl substituents.

    - When halogens in a nucleophilic solvent (such as water) add to the double bond of a 1-alkene in which there are two alkyl substituents on one carbon, the cyclic-ion intermediate (a halonium ion) reacts with the solvent to give a halohydrin in which the halogen is on the terminal carbon and the solvent-derived species (such as OH from water) is on the carbon with two substituents.

    - In oxymercuration of alkenes, the solvent water reacts as a nucleophile at the carbon of the mercurinium-ion intermediate bonded to the greater number of alkyl substituents. Removal of the mercury by reduction with sodium borohydride ($NaBH_4$) gives an alcohol in which the OH group ends up on the carbon of the alkene with the greater number of substituents.

    - In the hydration of alkynes, an enol is formed in which the OH group of water ends up on the more substituted carbon of the alkyne. The enol is spontaneously converted into a ketone (except in the hydration of acetylene itself, which forms an aldehyde). Alkyne hydration is useful with a 1-alkyne or a symmetrical alkyne; otherwise, a mixture of products is formed.

• In hydroboration of an alkene or alkyne, the hydrogen goes to the carbon of the π bond with the greater number of alkyl substituents, and the boron goes to the carbon with fewer alkyl substituents. In the case of alkene addition, oxidation converts the resulting organoborane into an alcohol in which the OH group is positioned on the alkene carbon with fewer substituents. In the case of 1-alkyne

hydroboration, an aldehyde is formed at the terminal carbon of the triple bond. Like alkyne hydration, alkyne hydroboration is useful with 1-alkynes or symmetrical alkynes.

■ A significant part of the chemical economy is based on simple alkenes such as ethylene, propene, and methylpropene. Examples are polyethylene, a widely used polymer produced from ethylene, and methylpropene, which is used as a starting material for high-octane gasoline.

STUDY GUIDE LINKS

5.1   How to Study Organic Reactions

You have now studied several organic reactions, and perhaps it is apparent that you are going to study many more. How can you keep all these reactions straight? How can you recall just the right reaction to use in solving a particular problem?

There are four keys to learning reactions; each of these points is discussed further below.

1. Be active when you read the text.
2. Be organized.
3. Review frequently.
4. Study in small chunks.

Be active when you read the text. To see what this means, let's go through one section of the text right now. Open your text to Section 5.4A, "Oxymercuration–Reduction of Alkenes." This section is fairly typical of those in the text that describe important reactions. First comes a statement of the reaction. When you read this, be active! Try to make a hypothesis about where the different atoms come from. (The use of color in the equations is designed to help you with this.) For example, from where does the —OH group in the product of text Eq. 5.19 originate? Where do the two —OAc groups end up? If you think you have the right idea, write the same reaction using a different alkene starting material.

Next, the text provides a few practical facts about the reaction, for example, some information about the solvent or the reaction conditions. These details can be very important, even crucial (in which case the text will say so), or they might be added for completeness or for your later use as reference material. In most cases, you should read them but not bother to memorize them, because the object of our first pass through the chapter is to understand the reaction itself.

Next follows a discussion of the *mechanism* of the reaction. What should you do with the mechanism? What you *should not* do is memorize it. What you *should do* is to follow each step of the mechanism to see whether your hypothesis about where the various atoms come from in the product is correct, and how they get there.

Here's an important point about curved-arrow mechanisms—a point we have already introduced in Chapter 3. Each step in a curved-arrow mechanism *must* be one of three types.

1. A Lewis acid–base association (Sec. 3.1B)
2. A Lewis acid–base dissociation (Sec. 3.1B)
3. An electron-pair displacement reaction (Sec. 3.2A)

a. A Brønsted acid–base reaction (Sec. 3.4A)
b. Another electron-pair displacement (Sec. 3.2A)
c. A combination of (a) and (b)

Now is a good time to review Secs. 3.1 and 3.4A. You will rely on these sections *a lot* in studying reaction mechanisms.

Reactions of type (1) and (2) require *single* curved arrows. Reactions of type (3) require *two or more* curved arrows. Reactions of type (3a) involve, by definition, electron-pair displacements on a *proton*. As you examine a mechanism, *classify each mechanistic step* using the list above. Then *look for points in common with other mechanisms*. For example, *all* of the alkene reactions you have studied so far start with the π electrons of an alkene reacting as a Lewis base (a nucleophile or a Brønsted base) with an electrophile. Then consider the reactive intermediates, if any (for example, carbocations). Ask where you have seen such intermediates before.

One purpose of mechanisms is to help you see certain unifying ideas that seemingly different reactions have in common. (You may have noticed, for example, that carbocation theory keeps occurring in the discussions of many of the addition reactions you have studied.) As you go further along in the text, you might even try to write a mechanism on your own before looking at the one in the text. (Use the instructions in Problem 4.84 to help you.) Comparing your mechanism to the one in the text will help you to refine your mechanism-writing skills. Finally, after studying the mechanism, *be active!* Start with a different alkene starting material and write the mechanism on your own.

Once again: you should *not* have to memorize the mechanism. After analyzing each step as suggested, and after writing a few mechanisms on your own, you will begin to gain an intuition about mechanisms that will help you to avoid memorizing—and will help you eventually to predict reactions!

Sometimes there are further applications of the reactions or additional facts—sort of a "wrap-up." For example, Eq. 5.25 (Sec. 5.4B) illustrates the point that no rearrangements are observed in oxymercuration.

Finally come the problems. *Working problems is the most important thing you can do to master organic chemistry.* The answers to the problems in the text are provided in this manual. You might be tempted to read the answers before you work the problems. Instead, work "Focused Problems" (problems within the chapter) immediately when you come to them, and work them *with this manual closed!* These provide a valuable test over your understanding of the chemistry involved. Put a mark by the ones you find difficult; re-work these problems before an exam.

As you read the section, you may come across certain unfamiliar terms. For example, in the very first sentence of Section 5.4A is the term "addition." Do you know what this means? *The great temptation is to skip over unfamiliar terms, hoping that they will become clear in context.* This strategy often works when reading a novel, but you *can't* do this in scientific reading! You *must* understand each term before proceeding to the next. Ask yourself, "What is an addition reaction?" If you don't know, don't feel that you're intellectually deficient—just find out the answer. *Everyone* requires a number of repetitions in order to learn new ideas. (The author can't keep repeating the definition, because it would make the book even larger than it is!) How do you find out? *Use the index!* Or, in some cases, a cross-reference to an earlier section will be given. Look it up! If you get in the habit of taking this approach, you will find yourself retaining more, and you ultimately will have to do less work. If an unfamiliar term is one you have had to look up before, make yourself a glossary: Write the term and its definition *in your own words*—*not* necessarily the words of the text—and review it periodically. One prominent theory of learning says that students construct their own knowledge. What this means is that until you have processed something in *your own mind* you don't know it. Hence, when you see a new term, *read* the definition; *process it* (that is, frame it terms that *you* can understand, no matter how unsophisticated you might think your description is), and then *write it* (that is, be active). In summary, here's what you should be doing:

1. *Read*
2. *Process*

3. *Write*

If you do what has been suggested, you'll notice that you have been very active. You have a piece of scratch paper on your desk, and it should be filled with notes and structures.

Two other suggestions: First, *do not* underline (or highlight) large sections of the text! This represents an effort to be active, when in fact it represents in many cases *physical* activity coupled with *mental* passivity. If there is something you really want to emphasize, then by all means, highlight away! Otherwise, don't bother. It's a waste of time and effort. Second, if you outline the chapter, don't go into an "autopilot" mode; think about what you write, and don't passively copy sections of the text. Pretend you are a teacher and that you're writing the outline for someone who can't understand the text at all. This Study Guide has also provided outlines of the reactions for you. Use these for a quick review, but don't let them substitute for your own efforts to organize the material.

Be organized. After you have studied a reaction, write it down in a general form for later review. One good way is to make flash cards. Write the reactants on one side of the card and the products on the other side. Write two arrows, one on each side. Use "R" groups or "dangling bonds" to make the reaction as general as possible. Let's illustrate. Flash cards for oxymercuration–reduction might look as shown below:

Cards 1 and 2 summarize the individual steps of oxymercuration and reduction. Card 3 is a summary card for both steps. Card 4 is used to test whether you know the conditions for bringing about the reaction; or, in reverse, whether you know the transformation for which a given set of reagents is used. Card 5 is a "term" card, which you may want to separate out with other cards containing definitions of terms. Each time you learn a new reaction, make a new group of flash cards. Notice above that not only the individual reactions of oxymercuration and reduction have been entered, but also a summary reaction. The choice of one "R" group is arbitrary; you might instead use an alkene starting material $R_2C\!\!=\!\!CH\!\!-\!\!R$. The point here is to use enough "R" groups that the regioselectivity of the reaction is apparent. Notice also that the reactions are not balanced; the goal is to focus on what happens to the organic compound.

Each time you finish a study session, look at one side of each card, and complete the other side on a piece of scratch paper. Then, with the other side in view, complete the first side. (Notice that the arrow bearing the reaction conditions is repeated on both sides of the card for this purpose.) You should find that this process helps you learn reactions in both directions. This knowledge will become particularly valuable when you start combining reactions in multistep syntheses. Next, ask yourself why the reaction is reasonable. For the oxymercuration step, you should recognize that the reaction is an addition. Go through the mechanism. What is the electrophile? Why does it react with the alkene π electrons? Why is the reaction regioselective? (Note that the text does not cover the mechanism of the reduction step.) Then write a specific example of the reaction using R groups of your choice. Make examples that use cyclic compounds.

If you do this *every time you read,* and *if you understand what you read,* then what you are doing here is *not* memorizing the reactions, at least not in a rote fashion, because you will have already learned them by careful study. Rather, you are cataloging them for use as a future study tool. If you can't complete one side or the other of a reaction, look at the missing side. If it looks unfamiliar, return to the text and focus exclusively on the part that is unfamiliar.

Review frequently. Review your reaction sheet *every time you study!* Go back about three assignments for each review, dropping older material from your review. Then review the older material again once every week or two, and right before an exam or quiz. (This gives you something *active* to do in preparing for an examination.) Do the same thing with the list of terms. You will find that relatively few reactions and terms are covered in each assignment, and that learning each day's assignment takes relatively little time. If you get behind, however, the number of reactions and terms that must be learned will grow rapidly! A nice feature of flash cards is that they can be shuffled, so that you can review them out of context. (You might want to number your cards so you can return them to their original order.)

Many students and professors suffer from adherence to a theory that one of the author's colleagues calls the "immunization theory of learning." The student's version of this theory is: "I've studied this material once, and therefore I should never have to see it again ... *ever.*" The professor's version of the theory is, "Students have studied this material once, and therefore they should automatically know it. There is no need for me to review, and a student should have total recall of everything he/she has ever been taught." Those who adhere to the "immunization theory" deny a fundamental caveat about learning: *Continued reinforcement is one of the best ways to learn.* The author finds that students are sometimes frustrated because they didn't learn something on the first pass. You should *expect* to forget material, and you should expect to re-learn it, probably more than once. Each relearning, however, takes less time and brings with it deeper understanding. What would happen to the college basketball team if the coach taught them a play *once* and they never reviewed it before a game? What would happen to a musician who practiced a piece once before a concert? Why should studying chemistry (or anything else, for that matter) be any different?

Study in small chunks. Set realistic goals for each study session and pursue them regularly. If you can only afford six hours per week, it is far better to study six times per week for one hour than one time for six hours. If you can allow twelve hours per week, then break it into six two-hour sessions. Notice how this strategy is built into the discussion of the flash cards above: you should study your flash cards over and over again, but confine your concentration to relatively small numbers of them except for general reviews.

If you have ever played an organized sport, or studied a musical instrument seriously, you should understand the "small chunk" strategy well. Does the coach of a winning team hold one ten-hour practice per week, or five two-hour practices? Why? Because the mental and physical learning capacities of most individuals erode significantly as fatigue sets in. Does a skilled performer on a musical instrument practice once per week for twelve hours, or six times per week for two hours?

What is remarkable is that when you *stay organized* and *study in small chunks*, your capacity for further study increases! Just as an athlete's endurance increases with regular workouts, your ability to concentrate for longer periods of time will increase. Just as an athlete's speed and skill increase with each workout, you will also find that you will become more efficient: you will accomplish much more in less time!

The suggestions given here for learning reactions are not the only ones that will work. But if you are at a loss about how to study organic chemistry on your own, try these suggestions, and then refine them to suit your own needs. One of the authors (ML) has actually tested these suggestions on his class, and has determined that student grades are significantly improved when students follow these suggestions.

5.2   How to Solve Structure Problems

Before working Problem 5.43 in the text, be sure you have read Study Guide Link 4.3 (also called "Solving Structure Problems").

Problem 5.43 contains a lot of information. The temptation is to solve this problem from the top down. It bears repeating that this is not usually the best way to deal with this type of problem. What is known about *A*? Its formula indicates an unsaturation number of 3. Its reactions with Br_2, H_2/catalyst, and O_3 all suggest that *A* is an alkene. Immediately you should look for a *structure*. Two structures are given in the problem, one in words, the other explicitly. The hydrogenation product of *A* is 1-isopropyl-4-methylcyclohexane:

1-isopropyl-4-methylcyclohexane

This tells us *immediately* the carbon skeleton of *A*. Only the double bonds are missing. This structure further implies that *A* contains a ring. Remember that the ring accounts for one degree of unsaturation. How many double bonds does compound *A* have?

The second structure given in the problem is the ozonolysis product. Remember that ozonolysis produces *two* C=O groups for every double bond in the molecule. However, in this ozonolysis product, there are *three* C=O groups. But there are also *nine carbons*. Evidently, one carbon and one C=O group are missing. This means that the missing carbon is part of the missing C=O group, and that a one-carbon fragment was liberated from an alkene =CH$_2$ group by ozonolysis as a second product which was evidently not recovered. This conclusion is consistent with the unsaturation number, which indicates that there are two double bonds. Two double bonds, after ozonolysis, gives four C=O groups.

(1)   *(2)*

Because *A* contains two double bonds, one additional double bond must be provided to convert either *(1)* or *(2)* into possible structures for *A*. From the ozonolysis product, this double bond must be *within* the ring, because ozonolysis opens the ring. Now it is time to *write all the possibilities*.

(Recall that this point was stressed in Study Guide Link 4.3.) From *(1)* the only possibilities for *A* are the following:

(1-1)   *(1-2)*   *(1-3)*

From *(2)* the only possibilities are the following:

(2-1)   *(2-2)*

There are two ways to finish the problem. One is to apply ozonolysis to each of the five structures and see which one gives the observed product. Notice that the observed product contains two acetyl groups:

$$\overset{O}{\overset{\|}{-C}}-CH_3$$

an acetyl group

Which structures would give *two* of these groups after ozonolysis? You should finish the problem by deciding between these.

$$H_3C-\overset{O}{\overset{\|}{\underset{1}{C}}}-CH_2CH_2-\overset{\underset{\underset{CH_3}{\|}}{\underset{3}{C}=O}}{CH}-CH_2-\overset{O}{\overset{\|}{\underset{2}{C}}}-OH \qquad H-\overset{O}{\overset{\|}{\underset{4}{C}}}-OH$$

the one-carbon fragment

(The missing carbon, which evidently gives formic acid, is referred to as carbon-4.) Once again, *write out all possibilities.* All possible connections are:

1. C1 with C2, C3 with C4
2. C1 with C3, C2 with C4
3. C1 with C4, C2 with C3

All but one of these possibilities can be ruled out by recalling that compound *A* has to contain a six-membered ring. The only structure for *A* that has a six-membered ring is obtained by connecting C1 to C2. Hence, C3 and C4 are also connected. You should finish reconstructing compound *A*. (The final solution is in the "Solutions to Integrated Problems" part of this chapter.

FURTHER EXPLORATIONS

5.1 Mechanism of Organoborane Oxidation

Although the oxidation step of hydroboration–oxidation may seem strange, it is actually a series of Brønsted and Lewis acid–base reactions.

First, the hydroperoxide anion is produced by ionization of H_2O_2. As you can see from the pK_a values in the following equation, this ionization is quite favorable when $^-$OH is used as the base:

$$HO:^{\frown}H^{\frown}O{-}OH \longrightarrow HO{-}H + :O{-}OH$$

pK_a = 11.6 $\qquad$ pK_a = 14.0 $\qquad$ **hydroperoxide anion**

Next, this anion reacts with the boron of the organoborane in a Lewis acid–base association reaction. (The alkyl groups on boron are abbreviated as R groups.)

$$R_3B \longleftarrow :O{-}OH \longrightarrow R_3\bar{B}{-}O{-}OH$$

hydroperoxide anion

An alkyl group on boron moves from boron to oxygen with breaking of the (very weak) bond between the two oxygens and loss of the ⁻OH group:

$$R_2\bar{B}{-}O^{\frown}OH \longrightarrow R_2B{-}O{-}R + :OH$$
$$\;\;\;\;| \atop R$$

This is the step that replaces a carbon–boron bond with a carbon-oxygen bond. Why would such a reaction occur?

Let's use the device of imagining that this process occurs in a stepwise manner. The O—O bond is very weak, because two electronegative atoms are "pulling" at the bonding electrons. Imagine that ⁻OH is expelled, leaving an *electron-deficient oxygen:*

$$R_2\bar{B}{-}O^{\frown}OH \longrightarrow R_2\bar{B}{-}O^+ + :OH$$
$$\;\;\;\;| \atop R \qquad\qquad\qquad | \atop R$$

an electron-deficient oxygen

The electron-deficient oxygen is like a carbocation, except with *much* higher energy. (Notice: this is not the same situation that occurs in $H_3\overset{+}{O}$:, which is *not* electron deficient!) This very unstable cation can rearrange by taking electrons from the adjacent C—B bond:

$$R_2\bar{B}{-}O^+ \longrightarrow R_2B{-}O{-}R$$
$$\;\;\;\;| \atop R$$

The driving force for this rearrangement is the flow of electrons from the relatively electropositive boron to the very electronegative oxygen of ⁻OH, as well as the neutralization of charge.

In subsequent steps of the reaction, the bonds between the other R groups are converted in a similar manner into R—O bonds:

the same mechanism

$$R_2B{-}O{-}R \xrightarrow{H_2O_2,\ ^-OH} RO{-}\underset{|}{\overset{OR}{B}}{-}OR$$

a trialkyl borate

Finally, the resulting trialkyl borate undergoes hydrolysis to the alcohol product. The final hydrolysis breaks the boron–oxygen bonds but leaves the oxygen-carbon bonds unaffected.

$$4 :OH + RO{-}\underset{OR}{\overset{OR}{B}}{-}OR \xrightarrow{H_2O} 3 H{-}OR + \bar{B}(OH)_4$$

Try to write a stepwise mechanism for this hydrolysis. The first step is the Lewis acid–base association of ⁻OH with the boron of the trialkyl borate.

5.2   Mechanism of Ozonide Conversion into Carbonyl Compounds

The conversion of ozonides into carbonyl compounds can also be rationalized by a series of electron-pair displacement reactions. Dimethyl sulfide acts as a nucleophile toward an oxygen electrophile in the O—O bond, displacing the other oxygen. A series of electron-pair displacements eventually liberates dimethyl sulfoxide, an oxidation product of dimethyl sulfide.

A similar series of reactions can occur with water as the nucleophile.

Hydrogen peroxide is produced as a by-product of this reaction. (Water is literally oxidized by its reaction with the ozonide.) Aldehydes are easily oxidized by hydrogen peroxide (or other oxidants, such as the ozonide) to carboxylic acids. This involves the nucleophilic reaction of hydrogen peroxide at the aldehyde carbonyl carbon.

The key step in the oxidation is the migration of the aldehyde hydrogen to an oxygen. This is much like the migration step in organoborane oxidation, which, as shown in Further Exploration 5.1, can be viewed as a migration to an electron-deficient oxygen. The hydrogen rather than the R group migrates because the resulting "carbocation" (which is simply a resonance structure of the carboxylic acid) is stabilized by the alkyl substituent.

REACTION REVIEW

I.   CONVERSION OF ALKENES INTO HALOGEN-CONTAINING COMPOUNDS

(More alkene additions are presented in Chapters 4, 10, 11, and 12.)

A.   ADDITION OF CHLORINE AND BROMINE

1.   Bromine and chlorine undergo addition to alkenes to give compounds with halogens on adjacent carbons called vicinal dihalides. (X = Br or Cl)

$$X_2 \; + \; \diagdown_{\diagup}C=C_{\diagdown}^{\diagup} \longrightarrow X-\overset{|}{\underset{|}{C}}-\overset{|}{\underset{|}{C}}-X$$

2.   The rapid disappearance of the red bromine color during bromine addition is a useful qualitative test for alkenes.

3.   The equilibrium for addition of I_2 favors the alkene. Addition of F_2 occurs, but the reaction is violent and is accompanied by other reactions.

4.   Bromine addition to alkenes in many cases involves a reactive intermediate called a bromonium ion.

   a.   In a bromonium ion, a positively charged bromine with an octet of electrons is bonded to two other atoms in a three-membered ring.

$$C=C \longrightarrow C-C$$

a bromonium ion

   b.   Reaction of the bromide ion as a nucleophile at either of the carbons bound to bromine in the bromonium ion completes the addition of bromine.

B.   FORMATION OF HALOHYDRINS

1.   When an alkene reacts with Cl_2, Br_2, or I_2 in a solvent containing a large amount of water, water reacts as a nucleophile with the cyclic ion. (X = Cl, Br, or I).

2. Loss of a proton from the oxygen in a subsequent Brønsted acid–base reaction with the solvent gives a compound containing both an OH and a halogen (X) called a halohydrin.

a. Halohydrins are compounds that contain both a halogen (X) and an OH group. In vicinal halohydrins, the OH and halogen are on adjacent carbons.

b. The reaction is completely regioselective when one carbon of the alkene is bonded to two alkyl substituents and the other is bonded to two hydrogens; in such cases, the OH group goes to the carbon with two alkyl substituents, and the halogen goes to the carbon bearing the hydrogens.

II.   CONVERSION OF ALKENES INTO ALCOHOLS

A.   OXYMERCURATION–REDUCTION OF ALKENES

1. Alkenes react with mercuric acetate [Hg(OAc)$_2$] in aqueous solution to give addition products in which an —HgOAc (acetoxymercuri) group and a —OH (hydroxy) group derived from water have added to the double bond; this reaction is called oxymercuration.

a. The first step of the reaction mechanism involves the formation of a cyclic ion called a mercurinium ion. This is a two-electron three-center ion.

a mercurinium ion

b. The mercurinium ion has resonance structures in which structure *A* is more important than structure *B* because positive charge and electron deficiency are on the carbon with more alkyl substituents:

resonance structures of the mercurinium on

c. The solvent water reacts as a nucleophile with the mercurinium ion (shown as resonance structure *A*) exclusively at the carbon with the greater number of alkyl substituents.

d. The addition is completed by a Brønsted acid–base reaction in which a proton is transferred to the acetate ion that is formed in the previous step.

2. Oxymercuration adducts are converted into alcohols by treatment with the reducing agent sodium borohydride ($NaBH_4$) in base. In this reaction, the carbon–mercury bond is replaced by a carbon–hydrogen bond.

3. The overall reaction sequence, called oxymercuration–reduction, results in the net addition of the elements of water (H and OH) to an alkene double bond in a regioselective manner.

   a. The OH group is added to the carbon of the double bond bearing more alkyl substituents, and the H is added to the carbon of the double bond bearing fewer alkyl substituents.

   b. The product is free of rearrangements and other side reactions that are encountered in hydration. (See Eq. 5.25, Sec. 5.4B.)

B. HYDROBORATION–OXIDATION OF ALKENES

1. Borane (BH_3) adds regioselectively to alkenes so that the boron goes to the carbon of the double bond bearing fewer alkyl substituents, and the hydrogen goes to the carbon of the double bond bearing more alkyl substituents.

   a. The addition of BH_3 to alkenes, called hydroboration, occurs in a single step without intermediates—that is, by a concerted mechanism.

   b. In many cases, one borane molecule can add to three alkene molecules; the products in these situations are trialkylboranes.

a trialkylborane

2. The regioselectivity of the reaction is governed by the presence of partial positive charge in the transition state on the carbon of the alkene with more substituents (Eq. 5.34a, Sec. 5.6A), and by the smaller van der Waals repulsions when the boron forms a bond to the carbon with fewer alkyl substituents (Eq. 5.34b).

3. Trialkylboranes are oxidized to alcohols with basic hydrogen peroxide, H_2O_2. The net result of this transformation is replacement of the boron by an —OH in each alkyl group.

$$\left(H-\overset{\displaystyle |}{\underset{\displaystyle |}{C}}-\overset{\displaystyle |}{\underset{\displaystyle |}{C}}\right)_{\!3}\!\!B \xrightarrow{H_2O_2,\ ^-OH} 3\ H-\overset{\displaystyle |}{\underset{\displaystyle |}{C}}-\overset{\displaystyle |}{\underset{\displaystyle |}{C}}-OH\ +\ \text{borate salts}$$

4. The overall reaction sequence, called hydroboration–oxidation, results in the net addition of the elements of water (H and OH) to an alkene in a regioselective manner.
   a. The —OH group is added to the carbon atom of the double bond with fewer alkyl substituents; a hydrogen is added to the carbon of the double bond with the greater number of alkyl substituents.
   b. The product is free of rearrangements and other side reactions that are encountered in hydration.

$$H_3C-\overset{\displaystyle CH_3}{\underset{\displaystyle CH_3}{C}}-CH{=}CH_2 \xrightarrow[\text{2) } H_2O_2/OH^-]{\text{1) } BH_3} H_3C-\overset{\displaystyle CH_3}{\underset{\displaystyle CH_3}{C}}-CH_2-\overset{\displaystyle OH}{CH_2}$$

C. COMPARISON OF METHODS FOR THE SYNTHESIS OF ALCOHOLS FROM ALKENES

1. Hydration of alkenes is a useful industrial method for preparing a few alcohols, but it is not a good laboratory method.
2. Oxymercuration–reduction is a good laboratory method that gives an alcohol in which —OH has been added to the carbon of the double bond with more alkyl substituents and —H has been added to the carbon of the double bond with fewer alkyl substituents.
3. Hydroboration–oxidation is a good laboratory method that gives an alcohol in which —OH has been added to carbon of the double bond with fewer alkyl substituents and —H to the carbon of the double bond with more alkyl substituents.

$$R-CH{=}CH_2 \xrightarrow[\text{of H and OH}]{\text{net addition}} \begin{cases} R-\underset{\displaystyle OH}{CH}-\underset{\displaystyle H}{CH_2} & \text{oxymercuration–reduction} \\ \\ R-\underset{\displaystyle H}{CH}-\underset{\displaystyle OH}{CH_2} & \text{hydroboration–oxidation} \end{cases}$$

III. HYDROBORATION–OXIDATION OF ALKYNES

A. HYDROBORATION–OXIDATION OF INTERNAL ALKYNES

1. The hydroboration of internal alkynes is analogous to the same reaction of alkenes.
2. When alkynes react with diborane (B_2H_6), the elements of BH_3 are added to the triple bond.
   a. Oxidation of the organoborane with alkaline hydrogen peroxide yields the corresponding "alcohol," an enol.
   b. The enol spontaneously reacts further to give the corresponding aldehyde or ketone.

$$R-C{\equiv}C-R \xrightarrow{BH_3} \left(R-\overset{\displaystyle H}{C}{=}\underset{\displaystyle R}{C}\right)_{\!3}\!B \xrightarrow[^-OH]{H_2O_2} \underset{\underset{\displaystyle R\qquad R}{}}{\overset{\displaystyle H\quad OH}{C{=}C}} \longrightarrow R-\overset{\displaystyle H}{\underset{\displaystyle H}{C}}-\overset{\displaystyle O}{\underset{\displaystyle R}{C}}$$

an enol   a ketone

   c. A mixture of ketones is formed if the two substituents on the triple bond are different—that is, for alkynes of the form R—C≡C—R′,

B. *HYDROBORATION–OXIDATION OF 1-ALKYNES*

   1. The hydroboration of 1-alkynes can be stopped after a single addition provided that an organoborane containing highly branched groups is used instead of BH_3. The branching prevents more than one alkyne from reacting with the borane by a steric effect.

   2. One reagent developed for this purpose is called disiamylborane.

   3. Boron adds to the unbranched carbon atom of the triple bond, and hydrogen adds to the branched carbon.

an enol   an aldehyde

IV. CONVERSION OF ALKENES INTO ALDEHYDES, KETONES, OR CARBOXYLIC ACIDS

A. *OZONOLYSIS OF ALKENES*

   1. The reaction of an alkene with ozone to yield products of double-bond cleavage is called ozonolysis.

   2. Ozone, O_3, adds to alkenes at low temperature in a concerted cycloaddition reaction to yield an initial cyclization product (sometimes called a *molozonide*).

a molozonide

   3. This initial cycloaddition product is unstable and spontaneously forms the ozonide.

an ozonide

   4. If the ozonide is treated with dimethyl sulfide, $(CH_3)_2S$, the ozonide is split into products that contain a carbonyl (C=O) group.

     a. If a carbon of the double bond in the starting material bears a hydrogen, an aldehyde is formed.

     b. If a carbon of the double bond bears no hydrogens, a ketone is formed.

an aldehyde   a ketone

   5. If the ozonide is treated with water and hydrogen peroxide (H_2O_2), carboxylic acids are formed instead of aldehydes.

The reaction scheme at top:

$$\text{(cyclic peroxide structure)} \xrightarrow{H_2O_2,\ H_2O} R-C\!\!\begin{smallmatrix}OH\\ \\O\end{smallmatrix} + \begin{smallmatrix}R\\ \\C-R\\O\end{smallmatrix}$$

a carboxylic acid   a ketone

SOLUTIONS TO PROBLEMS

Solutions to Focused Problems

5.1   (a)   The iodine is the electrophile, because it adds to the carbon of the double bond with fewer alkyl substituents. This result does seem to fit the electronegativity pattern for electrophilic addition, as nitrogen is considerably more electronegative than iodine.

   (b)   The less electronegative group (I) adds to the carbon of the double bond with fewer substituents, and the more electronegative group (Br) adds to the carbon of the double bond with more substituents. (Structure at right.)

$$(CH_3)_2C\!-\!CH_2$$
$$\qquad\ \ |\qquad |$$
$$\qquad\ Br\quad\ I$$

5.2   (a)   The reaction of Br_2 with 2-methyl-1-hexene follows the pattern in Eqs. 5.10a and b of the text.

$$H_2C\!=\!C(CH_2)_3CH_3 \longrightarrow H_2C\!-\!C(CH_2)_3CH_3 \longrightarrow H_2C\!-\!C(CH_2)_3CH_3$$

2-methyl-1-hexene

   (b)   The mechanism involves a bromonium ion intermediate, formed as shown in part (a). Water reacts as a nucleophile at the carbon of the bromonium ion that has the greater number of substituents.

$$H_2C\!=\!C(CH_2)_3CH_3 \longrightarrow H_2C\!-\!C(CH_2)_3CH_3 \longrightarrow H_2C\!-\!C(CH_2)_3CH_3 \longrightarrow H_2C\!-\!C(CH_2)_3CH_3$$

2-methyl-1-hexene

$$+\ H_3O^+\ :\!\overset{..}{\underset{..}{Br}}\!:^-$$

   (c)   Because iodine is the electrophile in iodine azide [as discussed in the solution to Problem 5.1(a)], an iodonium ion is formed. An azide ion reacts with this cyclic ion to give the product.

2-methyl-1-hexene

$$(^{-}\ddot{N}_3 = \ddot{:}N = \overset{+}{N} = \ddot{N}\ddot{:})$$

structure of the azide ion

5.3 (a) The alkene that could be used to form chlorohydrin *B* is methylenecyclohexane:

methylenecyclohexane   *B*

(b) The curved-arrow mechanism for formation of *B*:

5.4 (a) In this part, the double bond is symmetrically placed in the molecule; so, regioselectivity is not an issue.

cyclohexene   **cyclohexanol**

(b) The OH group ends up on the carbon with the greater number of alkyl substituents:

$(CH_3)_2C = CHCH_2CH_3$ $\xrightarrow[\text{2) NaBH}_4\text{, NaOH}]{\text{1) Hg(OAc)}_2\text{, H}_2\text{O}}$ $(CH_3)_2\overset{\text{OH}}{\underset{|}{C}}CH_2CH_2CH_3$

2-methyl-2-pentene   **2-methyl-2-pentanol**

(c) In this case, two constitutional isomers are formed because each carbon of the double bond has the same number of substituents—one.

$\xrightarrow[\text{2) NaBH}_4\text{, NaOH}]{\text{1) Hg(OAc)}_2\text{, H}_2\text{O}}$ $CH_3\overset{\text{OH}}{\underset{|}{C}}HCH_2CH(CH_3)_2$ + $CH_3CH_2\overset{\text{OH}}{\underset{|}{C}}HCH(CH_3)_2$

trans-4-methyl-2-pentene

4-methyl-2-pentanol   **2-methyl-3-pentanol**

(d) As in part (a), the double bond is symmetrically substituted; so, regioselectivity is not an issue.

cis-3-hexene

$$\xrightarrow[\text{2) NaBH}_4\text{, NaOH}]{\text{1) Hg(OAc)}_2\text{, H}_2\text{O}}$$

$$\underset{\text{3-hexanol}}{\text{CH}_3\text{CH}_2\text{CHCH}_2\text{CH}_2\text{CH}_3}$$
OH

5.5 (a)

CH₃CH₂CCH₂CH₃   or   CH₃CH₂C=CHCH₃
   ‖CH₂ |CH₃

2-ethyl-1-butene   **3-methyl-2-pentene**
                         (either *E* or *Z* stereoisomer)

(b)

cyclopropylethylene

5.6 The only alkynes that give hydration products free of isomers are 1-alkynes, which give methyl ketones, and symmetrical alkynes.

(a)
$$HC\equiv CCH_2CH_2CH_3$$
1-pentyne

(b)
$$(CH_3)_3C-C\equiv CH$$
3,3-dimethyl-1-butyne

(c)

5-decyne

5.7 (a) Hydration of 1-propyne ($CH_3C\equiv CH$) would give acetone [$(CH_3)_2C=O$], not propanal, the compound shown in the problem.

(b) One carbon bonded to the carbonyl group must have at least two hydrogens if the carbonyl compound is to come from an alkyne hydration.

(c) This ketone (cyclohexanone) would come from hydration of cyclohexyne, which is too unstable to exist (why?).

5.8 Compounds *A* and *D* are enols that will spontaneously form ketones. Compound *B* is an alkene and an alcohol, so it might be referred to as an "enol." However, we usually reserve the word "enol" for a compound in which an OH group is bonded to a carbon of a double bond. The ketones formed spontaneously from *A* and *D* are shown at right.

from *A*

from *D*

5.9 (a) In this part and in part (d), the double bond is symmetrically placed in the molecule; so, regioselectivity is not an issue.

cyclohexene

$$\xrightarrow[\text{2) H}_2\text{O}_2\text{, }^-\text{OH}]{\text{1) BH}_3\text{, THF}}$$

(b) Because the alkene has a different number of alkene substituents on each carbon of the double bond, the reaction is regiospecific and only one constitutional isomer is formed in the reaction.

$$(CH_3)_2C=CHCH_2CH_3 \xrightarrow[\text{2) H}_2\text{O}_2\text{, }^-\text{OH}]{\text{1) BH}_3\text{, THF}} (CH_3)_2CHCHCH_2CH_3$$
OH

2-methyl-2-pentene

(c) In this case the double bond is not symmetrical, so two constitutional isomers are formed. In addition, the two carbons of the double bond have the same number of alkyl substituents.

trans-4-methyl-2-pentene

(d)   Because the alkene is symmetrical, only one product is formed.

cis-3-hexene

5.10   The two symmetrical alkenes in parts (a) and (d) give the same alcohols with either hydroboration–oxidation of oxymercuration–reduction. The alkene in part (c) gives a mixture of the same two alkenes with either method, although the percentages of each will likely differ with the two methods. The alkene in part (b) gives relatively pure samples of different constitutional isomers with the two methods.

5.11   (a)   (b)

methylenecyclohexane   **3-methyl-2-pentene (E or Z)**

5.12   (a)   Because the OH is placed symmetrically in the alcohol, either reaction would work; cyclopentene is the starting alkene.

cyclopentene

(b)   The only possible alkene starting material is 5-methyl-1-hexene; only hydroboration–oxidation will put the OH on the carbon of the double bond with fewer substituents.

5-methyl-1-hexene

(c)   This alcohol has to come from 3-ethyl-2-pentene; only oxymercuration–reduction will bring about the desired conversion, because the OH group must end up on the carbon of the double bond with the greater number of substituents.

3-ethyl-2-pentene

5.13   (a)   Because *cis*-2-butene is a symmetrical alkene, it reacts to give the same product in either oxymercuration-reduction or hydroboration-oxidation.

cis-2-butene

(b)   Because 1-methylcyclohexene has different numbers of alkyl substituents at the carbons of its double bond, it gives different products in the two reactions. (Remember that a ring carbon bonded to a carbon of the double bond is an alkyl substituent.)

1-methylcyclohexene

5.14   (a)   Cyclohexylacetylene gives different products from hydroboration–oxidation and Hg^{2+}-catalyzed hydration.

cyclohexylacetylene

(b)   2-Butyne gives the same product from the two reactions.

5.15   The initially formed cycloaddition product from 2 pentene can decompose to an aldehyde oxide in two different ways in which the curved arrows move in opposite directions.

initial cycloaddition product   *W*   *X*   initial cycloaddition product   *Y*   *Z*

By the mechanism of text Eq. 5.47b, *W* can react with *X* and *Y* can react with *Z* to give product *B*; *W* can react with *Z* to give product *A*; and *X* can react with *Y* to give product *C*. Product *B* is twice as probable because it can be formed in two ways.

5.16 (a)

$$CH_3CH=CCH_2CH_3 \xrightarrow{O_3} \xrightarrow{(CH_3)_2S} CH_3CH=O \ + \ O=CCH_2CH_3$$
with CH_3 substituent on the middle carbon; products: CH_3CH=O and O=CCH_2CH_3 with CH_3 substituent

3-methyl-2-pentene

(b)

methylenecyclohexane + O_3, (CH_3)_2S → cyclohexanone (ring =O) + O=CH_2

methylenecyclohexane

(c)

cyclooctene + O_3, (CH_3)_2S → dialdehyde (HC=O ... CH=O)

cyclooctene

(d)   2-Methylpentane is an alkane and, with no π bond, it does not react with ozone.

5.17 (a)

$$CH_3CH=CCH_2CH_3 \xrightarrow{O_3} \xrightarrow{H_2O_2} CH_3\overset{O}{\overset{\|}{C}}OH \ + \ O=CCH_2CH_3$$
with CH_3 substituents

3-methyl-2-pentene

(b)

methylenecyclohexane + O_3, H_2O_2 → cyclohexanone (ring =O) + HC(=O)—OH

methylenecyclohexane

(c)

cyclooctene + O_3, H_2O_2 → diacid (HOC=O ... COOH)

cyclooctene

(d)   2-Methylpentane does not react with ozone because it has no π bond.

5.18   (a)

$CH_3CH_2CH_2CH = CHCH_2CH_2CH_3$

***cis*- or *trans*-4-octene**

(b)

1-methylcycloheptene

(c)

5.19   As the solution to Problem 5.18(a) illustrates, *E* and *Z* isomers of an alkene give the same ozonolysis products. Therefore, ozonolysis cannot be used to determine the stereochemistry at the alkene double bond.

5.20   (a)   Aldehydes can be prepared from 1-alkynes by hydroboration–oxidation.

$(CH_3)_3CCH_2CH_2C\equiv CH$ $\xrightarrow[\text{2) } H_2O_2/^-OH]{\text{1) }(\text{BH})_2}$ $(CH_3)_3CCH_2CH_2CH_2CH$

5,5-dimethyl-1-hexyne   **5,5-dimethylhexanal**

(b)   One of the best ways to prepare a cis-alkene is by catalytic hydrogenation of an alkyne with a poisoned catalyst. (Remember that benzene rings do not undergo catalytic hydrogenation under ordinary conditions.)

$\xrightarrow[\text{Lindlar catalyst}]{H_2}$

1,2-diphenylacetylene (tolan)   ***cis*-stilbene**

Solutions to Skills Objectives Problems

5.21   (a)

$\underset{CH_2CHCH_2CH_3}{Br\ Br}$

(b)

CH_2CH_3

(c)

$H_2C=O$
+
$O=CHCH_2CH_3$

(d)

$\underset{HCOH}{O}$ + $\underset{HO}{\overset{O}{C}}CH_2CH_3$

(e)

CO_2
+ H_2O

(f)

$\underset{CH_2CHCH_2CH_3}{H\ Br}$

(g)

$\underset{CH_2CHCH_2CH_3}{|\ OH}$
+
$\underset{CH_2CHCH_2CH_3}{OH\ |}$

(h)

$CH_3CH_2CH_2CH_3$

(i)

$B(CH_2CH_2CH_2CH_3)_3$

(j)

$HOCH_2CH_2CH_2CH_3$

(k)

$\underset{CH_2CHCH_2CH_3}{AcOHg\ OH}$

(l)

$\underset{CH_3CHCH_2CH_3}{OH}$

(m)

$\underset{CH_3CHCH_2CH_3}{|}$

5.22 (a) (b) (c) (d) (e) CO_2 $+ H_2O$ (f)

(g) (h) (i) (j)

(k) (l) (m)

5.23 (a) (b) $CH_3CH_2CH_2CH_2CH_3$ (c) $CH_3CH_2CH_2CH=CH_2$

(d) (e) (f)

5.24 (a) (b) (c)

(d) (e) (f)

5.25 (a) or (b) or (c)

(d)

(e)

(f)

(g)

(h)

and

(i)

(j)

(k)

and many others

5.26 (a) The tertiary carbocation intermediate undergoes a Lewis acid–base association reaction with its counter-ion chloride.

(b) The tertiary carbocation intermediate, identical to the one in part (a), undergoes a Lewis acid–base association with a solvent water molecule, and the resulting ion loses a proton to solvent to give the product alcohol.

(c) The iodonium ion intermediate reacts with solvent water at the carbon with two alkyl substituents, and the resulting ion loses a proton to solvent to give the product iodohydrin.

(d)   The mercurinium ion intermediate reacts with solvent water at the carbon with more alkyl substituents, and the resulting ion loses a proton to solvent to give the product.

There is no simple way to show the curved-arrow notation for the reaction of the 3-center 2-electron bond in the mercurinium ion with a nucleophile. For simplicity we assume that there is a full positive charge at the carbon that reacts with the water molecule. This corresponds to the most important resonance contributor of the mercurinium ion structure. (See Eq. 5.20b, Sec. 5.4A in the text.)

5.27   The transition state for the first addition in the hydroboration of (*E*)-3-methyl-2-pentene:

transition state that gives the observed product

positive charge on the carbon with more alkyl substituents

steric repulsion

transition state that gives the product that is not formed

5.28   (a)   (b)   and   (c)   and others

5.29   (a)

1) BH$_3$, THF
2) H$_2$O$_2$, $^-$OH

OH

(b)

Br$_2$, H$_2$O

HO   Br

(c)

or   or   or   C≡CH

H$_2$
Pd/C catalyst

(d)

Me Me   C≡CH

1) ( )$_2$ BH
2) H$_2$O$_2$/$^-$OH

Me Me

O
‖
CH

(e)

Because the alkyne is symmetrical, hydroboration–oxidation (BH_3, then H_2O_2/NaOH) would also work.

(f)

Solutions to Integrated Problems

5.30   When 3-methyl-1-butene undergoes oxymercuration–reduction, addition to the double bond occurs *without* rearrangement, because the mercurinium ion intermediate does not rearrange. In contrast, rearrangement occurs in the hydration reaction because the initially formed secondary carbocation intermediate can rearrange by a hydride shift to a tertiary carbocation.

5.31   (a)   (b)

(c)   (d)

5.32   The products after the first, second, and third addition of diborane to 2-methyl-1-butene:

The transition state for the third addition:

5.33 (a) (or the Z isomer) (b) (c) (d)

(e) (f) (g) (h)

5.34 (a)

(b)

(c)

(d)

(e)

5.35 (a) This transformation can be carried out by hydroboration–oxidation in which the source of deuterium is BD_3.

4-methyl-1-pentene

(b) An oxymercuration–reduction on the alkene used in part (a) with $NaBD_4$ as the source of deuterium gives the desired compound.

$$H_2C = CHCH_2CH(CH_3)_2 \xrightarrow[\text{2) NaBD}_4,\text{ NaOH}]{\text{1) Hg(OAc)}_2,\text{ H}_2\text{O}} H_2C - CHCH_2CH_2(CH_3)_2$$

4-methyl-1-pentene

with D and OH substituents

5.36   The three isotopically substituted compounds can be prepared by a choice of catalysts and either H_2 or D_2. In the preparation of compound A, D_2 can be used in the first reaction if H_2 is used in the second with the same catalysts.

$CH_3CH_2C \equiv CH$
1-butyne

reacting via:
- H_2, Lindlar catalyst, then D_2, Pd/C catalyst → compound A
- D_2, Lindlar catalyst → compound B
- D_2, Pd/C catalyst → compound C

5.37   The product depends on the nucleophile (Nuc:) that reacts with the bromonium ion or iodonium ion as follows (X = halogen):

$$H_2C = CCH_2CH_3 + X_2 \longrightarrow$$
with CH_3 substituent
2-methyl-1-butene

halonium ion from 2-methyl-1-butene

(a)   The nucleophile is a molecule of the solvent water:

$$H_2C - CCH_2CH_3 + H - \overset{+}{O}H_2 \quad Br^-$$
with :Br:, CH_3, and HO: substituents
(ionized form of HBr in water)

(b)   The nucleophile is a molecule of the solvent methanol:

$$H_2C - CCH_2CH_3 + CH_3\overset{+}{O}H_2 \quad I^-$$
with :I:, CH_3, and CH_3O: substituents
(ionized form of HI in methanol)

5.38   The reasoning is ‡similar to that used in the solution to Problem 5.37. The product depends on the nucleophile that reacts with the mercurinium ion.

$$H_2C = CH(CH_2)_3CH_3 \xrightarrow{\text{Hg(OAc)}_2}$$
1-hexene

mercurinium ion with $\overset{\delta+}{Hg}$—OAc, and $Ac\overset{..}{O}\overset{-}{:}$

(a)   This is the usual oxymercuration–reduction product (shown at right), in which water reacts with the mercurinium ion. (The mechanism is shown in Eqs. 5.21 in Sec. 5.4A.)

$$H_3C - CH(CH_2)_3CH_3$$
with :OH substituent

(b) Isopropyl alcohol is the nucleophile that reacts with the mercurinium ion, and the ultimate product is an ether. The reasoning follows from the mechanism, which follows.

(c) In this case, the nucleophile is the either the acetate ion ($^-$OAc), the counter-ion of the mercurinium ion, which is the conjugate base of acetic acid, or a molecule of the solvent, acetic acid itself. The product is an ester. (We leave it to you to complete the mechanism.)

5.39 (a) The regioselectivities of both the first and second additions are discussed in Sec. 4.9. (See Eq.4.63b and associated discussion for the regioselectivity of the second addition.)

(b) Reason by analogy to the same reaction of alkenes, which gives bromohydrins. Two isomeric bromohydrins are formed initially; but, because they are enols, they spontaneously form the corresponding bromo ketones.

5.40 The two large alkyl substituents on the boron atom in disiamylborane amplify the steric effect. In the two products, the position of the OH group serves as a marker for the position of the boron in the organoborane. In product *B*, the OH group (and therefore the boron) is closer to the carbon branch than it is in product *A*. Therefore, the steric repulsion between the "disiamyl" groups of the borane and the alkyl branch derived from the alkene is larger for the transition state leading to compound *B*. Consequently, the percentage of compound *B* is reduced and the percentage of compound *A* is increased when disiamylborane is used. The transition states leading to the two organoborane intermediates are as follows:

severe steric repulsion
raises the energy of this
transition state

transition state that leads to
the product (*A*) formed in
greater amount

transition state that leads to
the product (*B*) that is formed
in smaller amount

5.41   Compound *A* has the connectivity of octane. Because it has an unsaturation number $U = 1$ and reacts with bromine, it is an alkene. The ozonolysis results show that compound *A* is 4-octene, $CH_3CH_2CH_2CH{=}CHCH_2CH_2CH_3$. Because the double bond is located symmetrically, only one ozonolysis product is formed. Ozonolysis cannot determine whether the alkene is cis or trans.

5.42   The unsaturation number ($U = 4$) requires two triple bonds, a triple bond and two double bonds, or four double bonds. The hydrogenation result in the absence of a catalyst poison to give hexane defines the carbon chain as an unbranched chain of six carbons. To accommodate six carbons, the ozonolysis result requires that *two* formic acid molecules are formed for each succinic acid molecule. This result, in turn, defines compound *B* as 1,5-hexadiene. Because 1,5-hexadiene is formed from *A* in the presence of a catalyst poison, and hexane is formed in the absence of a catalyst poison, compound *A* can only be 1,5-hexadiyne. The following scheme summarizes these results:

5.43   A detailed discussion of the approach to solving this problem is found in Study Guide Link 5.2. Compound *A* is limonene, a natural product obtained from the oils of lemons and oranges. (Structure at right.)

5.44   The addition of ozone to *trans*-3-hexene gives an initial cycloaddition product, as illustrated by Eq. 5.44, Sec. 5.8A. This product can fragment into propanal and propanal oxide, similar to the reaction in Eq. 5.47a and shown below.

These two products can then recombine to form an ozonide, similar to Eq. 5.47b. However, the problem states that there is an excess of ^{18}O-labeled acetaldehyde. (An *excess* of a compound means that there are many more equivalents of it present relative to anything else.) So, it is more likely that the propanal oxide will react with the ^{18}O-labeled acetaldehyde, because there are many more equivalents of it, rather than the propanal. Note the final position of the isotope label in the ozonide product.

5.45   (a)   Protonation of the double bond gives a tertiary carbocation, which undergoes a Lewis acid–base association reaction with one of the oxygens within the same molecule. As suggested in the problem, corresponding atoms in the product and starting material are numbered. (Why isn't the other double bond protonated first?)

  (b)   The nitrogen within the same molecule reacts as a nucleophile with the bromonium ion intermediate, as shown below. The bromonium ion is formed by the mechanism shown in Eq. 5.10a in Sec. 5.2A of the text.

  (c)   In the first part of the mechanism, a sulfonium ion is formed by a nucleophilic reaction of the π electrons of one double bond with the sulfur in sulfur dichloride. (This process is analogous to the formation of a bromonium ion.) The chloride ion liberated in this step then reacts at a ring carbon to open the sulfonium ion. The process is then repeated, except that the process is intramolecular.

An isomer of the product shown could also be formed. What is it? Consider the last step. Once you've tried to work it out for yourself, see part (d) that follows for a similar situation.

(d)   Adduct *A* below is first formed by the mechanism shown in Eqs. 5.20a and 5.21 (Sec. 5.4A) of the text. Then the mercurinium ion *B* forms at the second double bond by the mechanism of Eq. 5.20a. The oxygen of a hydroxy group introduced in the oxymercuration reaction of the first double bond serves as the nucleophile in the opening of the mercurinium ion formed at the second double bond within the same molecule. Two products are formed because the oxygen can react with the mercurinium ion at either of two carbons.

[The acetate ion is a by-product of the first step of oxymercuration; its reaction with the protonated ether is shown only for *(b)*.] Treatment with $NaBH_4$ replaces the mercury with a hydrogen:

Principles of Stereochemistry

CHAPTER 6 SUMMARY

■ Stereoisomers are molecules with the same atomic connectivity but a different arrangement of their atoms in space.

■ Two types of stereoisomers are:

- enantiomers—molecules that are related as object and noncongruent mirror image;

- diastereomers—stereoisomers that are not enantiomers.

■ A molecule that has an enantiomer is said to be chiral. Chiral molecules lack certain symmetry elements, such as a plane of symmetry or center of symmetry.

■ An asymmetric carbon is a carbon bonded to four different groups. All asymmetric carbons are stereocenters, but not all stereocenters are asymmetric carbons.

■ The *R,S* system is used for designating absolute configuration. The system is based on the clockwise or counterclockwise arrangement of group priorities when a molecule is viewed along a bond from the asymmetric atom to the group of lowest priority. The priorities are assigned as in the *E,Z* system (Sec. 4.2B).

■ Two enantiomers have the same physical properties except for their optical activities—that is, the way they rotate the plane of plane-polarized light. The optical rotations of a pair of enantiomers have equal magnitudes but opposite signs.

■ A mixture of enantiomers is characterized by the enantiomeric excess (EE), which is the difference in the percentages of the two enantiomers, and by the enantiomeric ratio (ER), which is the ratio of the two enantiomers in the mixture.

■ An equimolar mixture of enantiomers is called a racemate, or racemic mixture. A mixture of enantiomers containing different percentages of the two is called a scalemic mixture.

■ The absolute configurations of some compounds can be determined experimentally by correlating them chemically with other chiral compounds of previously known absolute configuration.

■ Diastereomers in general differ in their physical properties.

■ Meso compounds are achiral compounds that have chiral diastereomers. The most common cases of meso compounds are achiral compounds with two or more asymmetric centers. Meso compounds can be recognized by applying the criteria in Sec. 6.7B.

■ Some chiral molecules contain no asymmetric atoms. Examples are the gauche conformation of butane and certain substituted derivatives of allene. These are also examples of twist chirality.

■ Enantiomers can be separated or differentiated only by allowing them to interact with an enantiomerically pure chiral agent (the principle of enantiomeric differentiation).

■ The separation of enantiomers is called an enantiomeric resolution. All enantiomeric resolutions involve the temporary formation of diastereomers, which have different properties. Two methods of enantiomeric resolution are chiral chromatography and diastereomeric salt formation. (A third method, selective crystallization, is discussed in Chapter 8.)

■ Chiral compounds in which a central atom is bonded to three groups and a nonbonding electron pair (such as amines) can undergo inversion, a process that interconverts enantiomers. When the central atom is from the second period, inversion is so fast that the enantiomers cannot be isolated. When the central atom is from the third and greater periods, stereochemical inversion is so slow that it does not occur at room

temperature. The enantiomers of such compounds (such as sulfonium ions and phosphines) can be isolated.

STUDY GUIDE LINKS

6.1   Stereocenters and Asymmetric Atoms

Stereochemical terminology changes as chemists acquire a more rigorous understanding of the geometrical aspects of stereochemistry. The purpose of this section is to alert you to a problem that still exists with the term *stereocenter*. The definition used in this text—"an atom at which interchange of two groups gives a stereoisomer"—is the one accepted by all experts in the field of stereochemistry. A number of texts, however, have defined *stereocenter* to be an atom with four different groups attached. This is the same definition that is used in this text for *asymmetric carbon*. Thus, some texts do not distinguish between the terms *asymmetric carbon* and *stereocenter*. As shown in the text, however, all asymmetric carbons are stereocenters, but the converse is not true: not all stereocenters are asymmetric carbons. The cases discussed in Sec. 4.1D of the text demonstrate this point: the carbons of the double bond in *E* and *Z* isomers are stereocenters but are not asymmetric carbons. Because so many texts mis-define this term, you should be alert to the fact that you may see it misused.

A number of chemists have objected to the use of the term *asymmetric carbon* because a carbon atom is not "asymmetric." In fact, many chiral *molecules* themselves are not necessarily "asymmetric;" many have certain types of symmetry. Perhaps a better term for asymmetric carbon would be *asymmetrically substituted carbon*. Other terms you might hear are "chiral center" and "chiral carbon," but objections have been raised to these also. The term *asymmetric carbon* has been so widely used that everyone knows what it means, and this text has continued to use it. The important point is that, whatever we call it, an asymmetric carbon is *not* the same as a stereocenter; rather, asymmetric atoms constitute only one type of stereocenter.

One last point about the definition of asymmetric carbon: When we say that an asymmetric carbon has "four different groups" attached, we might have occasion to ask, "How different do the groups have to be for a carbon to be labeled *asymmetric*?" The differences among attached groups used to define "asymmetric carbon" generally refer to *constitutional* (that is, compositional) or *connectivity differences. Any* constitutional or connectivity difference between two attached groups makes them different. However, it is also possible for two groups to have the same connectivity but differ only in their *stereochemical configurations*. Problem 6.54, especially parts d and e, on text p. 324 explores how we deal with such a situation. However, this is a relatively rare situation for beginning students.

FURTHER EXPLORATIONS

6.1 Terminology of Racemates

Nowadays the terms *racemic mixture* and *racemate* are often used interchangeably to indicate *any* equimolar mixture of enantiomers. However, the term *racemic mixture* also has a more precise use. This term is sometimes used to indicate a *solid state* of a racemate that is a mixture of two crystal types, one for each enantiomer. In other words, a racemic mixture is a conglomerate of two enantiomeric crystal forms.

The term *racemic compound* is sometimes used, and it, too, has a more precise definition. This term indicates a *single* crystal form of a racemate that contains equal numbers of molecules of each enantiomer. It is not a conglomerate, but rather a single crystal that contains equal amounts of each enantiomer within its crystal structure.

Racemic mixtures and racemic compounds can be distinguished by an analysis of their melting behavior as a function of composition.

6.2 Isolation of Conformational Enantiomers

Although butane undergoes internal rotation too rapidly for its enantiomeric gauche forms to be isolated, chemists have succeeded in preparing other compounds in which internal rotation about a single bond is so slow that conformational enantiomers can be isolated and purified. The following compound is an example:

conformational enantiomers
can be isolated at room temperature

(How might you separate the enantiomers of this compound?) The two benzene rings are forced to occupy perpendicular planes; if they were to occupy the same plane, substituents on one ring would "bump into" those on the other, and very severe van der Waals repulsions would result. The conformation of the compound shown above is chiral. (You can verify its chirality by drawing or constructing its mirror image and showing that it is noncongruent.) Interconversion of the two enantiomeric conformations requires a rotation about the central bond, indicated by the arrows, which takes the rings into the same plane. Because of the van der Waals repulsions just noted, the energy barrier to this process is so high that rotation cannot occur at room temperature. In contrast, the compound in which the —CO_2H groups are replaced by hydrogens and the chlorine substituents are replaced by fluorines *cannot* be separated into its enantiomers at room temperature. In this case, racemization requires a planar conformation in which two small atoms (F and H) are close. Because van der Waals repulsions are much smaller here, this process requires less energy, and racemization occurs readily at room temperature.

SOLUTIONS TO PROBLEMS

Solutions to Focused Problems

6.1   The asymmetric carbons are indicated with asterisks. (In part (d), there are no asymmetric carbons.)

6.2   (a)   This compound is chiral.
(b)   This compound is achiral.
(c)   This compound is achiral.
(d)   This compound is chiral; the asymmetric atom is the nitrogen.

6.3   (a)   A shoe is chiral, because it has a noncongruent mirror image (the shoe of the opposite foot).
(b)   A book is achiral, if we neglect the writing.
(c)   A human body is achiral, if we neglect the internal organs and any superficial characteristics. However, if the hair is parted to one side or the other, or if there are other features that differentiate left from right, the body is chiral. If we "burrow down" to the molecular level, the amino acid and DNA building blocks of the body are chiral. Thus, the body itself is chiral.
(d)   A cone is achiral.
(e)   The pair of shoes is achiral. Again, however, if we "burrow down" to the molecular level, we would find that the biomolecules that make up leather are chiral, and this would make the pair itself chiral.
(f)   Chiral. (If you're right-handed, try using scissors to cut something with your left hand.)

6.4   (a)   Planes of symmetry in methane bisect one set of H—C—H bonds and contain the other H—C—H bonds. (There are four such planes.)
(b)   One plane of symmetry in water bisects the H—O—H atoms and bonds.
(c)   One plane of symmetry in ethylene is the plane of the page; the two others are the planes perpendicular to the page. The center of symmetry is the point in the center of the C=C bond.
(d)   The plane of symmetry is the plane of the page, and the center of symmetry is the point in the center of the C=C bond.
(e)   The plane of symmetry in *cis*-2-butene is the plane of the page and the plane perpendicular to the page that bisects the C=C bond.
(f)   The plane of symmetry contains the three C—C bonds, and the center of symmetry is at the center of the C2–C3 bond.

6.5   Remember that there are many different ways to draw a correct line-and-wedge structure. If your structures don't look like these, and if you're not sure whether yours is correct, make a model of both and check them for congruency. (When possible, we often adopt a "standard" representation in which the bond to the atom of lowest priority—hydrogen in (a) and (b)—is the dashed wedge bond (going away), and it is placed to the left of the asymmetric carbon. This makes it very easy to determine configuration. However, this "standard" representation is not necessary nor is it always possible.)

(a)   (b)   (c)

6.6　(a)　The asymmetric carbon in the given stereoisomer of alanine has the *S* configuration.

　　　(b)　The asymmetric carbon in the given stereoisomer of malic acid has the *R* configuration.

　　　(c)　The asymmetric atom is a nitrogen, and it has the *S* configuration.

　　　(d)　The asymmetric carbon in ibuprofen has the *R* configuration.

6.7

propranolol　　　**methadone**　　　**pregabalin (Lyrica)**　　　**captopril**

6.8　(a)　Use Biot's law, Eq. 6.13:

$$\alpha = [\alpha]cl = \frac{(66.1 \text{ degrees mL g}^{-1} \text{ dm}^{-1})(1 \text{ dm})(5 \text{ g})}{100 \text{ mL}} = 3.31 \text{ degrees}$$

　　　(b)　The specific rotations of enantiomers must have the same magnitude but the opposite sign. Therefore, the enantiomer of sucrose must have specific rotation = –66.1 degrees mL g^{-1} dm^{-1}.

6.9　We are given the observed rotation (α), concentration (c), and path length (l). Use Biot's law, Eq. 6.13, and rearrange to solve for the specific rotation, [α].

$$[\alpha] = \frac{\alpha}{cl} = \frac{-9.2 \text{ degrees}}{(0.3 \text{ g mL}^{-1})(1 \text{ dm})} = -31 \text{ degrees mL g}^{-1} \text{ dm}^{-1}$$

6.10　(a), (b)　The asymmetric carbon of thalidomide (marked with an asterisk, *) and the configuration of the *S* enantiomer:

(*S*)-thalidomide

6.11 **(a)** A concentration of 0.1 M is the same as a concentration of (0.1 mol L^{-1})(150 g mol^{-1})(0.001 L mL^{-1}) = 0.015 g mL^{-1}. Using this number for c in Eq. 6.13.

$$[\alpha] = \frac{\alpha}{cl} = \frac{+0.20 \text{ degrees}}{(0.015 \text{ g mL}^{-1})(1 \text{ dm})} = +13.3 \text{ degrees mL g}^{-1} \text{ dm}^{-1}$$

(b) Because the two enantiomers are present in equal amounts, their rotations cancel, and the observed rotation is zero.

(c) Because the concentration is halved, the observed rotation is also halved. Thus, it is +0.10 deg.

(d) The specific rotation is the same, because by definition this parameter is independent of concentration.

(e) The specific rotation of L is –13.3 deg mL g^{-1} dm^{-1}; enantiomers have specific rotations of equal magnitudes and opposite signs. Again, the specific rotation is the same regardless of concentration.

(f) In this solution, the 0.005 mol of L cancels the rotation of 0.005 mol of D. The resulting rotation is as if only 0.005 mol, or a concentration of 0.05 mol L^{-1}, of D were present. This amount of D corresponds to a concentration of (0.05 mol L^{-1})(150 g mol^{-1})(0.001 L mL^{-1}) = 0.0075 g mL^{-1}. Since this concentration is half of the original concentration, the observed rotation is also half of the original—that is, +0.10 deg.

(g) Using Eq. 6.19a, we can calculate the EE by figuring out the % of the major enantiomer. Because there is 0.01 mol of D and 0.005 mol of L, then D is the major enantiomer, and the solution is (0.01 D/0.015 $D + L$) = 66.7% D. Substituting into Eq. 6.19a we find that the EE of D is 33%.

(h) From part (g), the percent of the major enantiomer D is 67%. The remaining 33% is L. The ER is about 2; there is twice as much D to L.

6.12 **(a)**

(S)-ibuprofen

(b) Because the specific rotation of the mixture, $[\alpha]_{mixture}$, and the EE is given, use Eq. 6.17 and solve for $[\alpha]_{pure}$, which works out to +57.4 degrees mL g^{-1} dm^{-1}. The pure R enantiomer is then –57.4 degrees mL g^{-1} dm^{-1}.

(c) Using Eq. 6.19b, solving for the percentage of the major, S-enantiomer, gives 95%. Thus the minor, R-enantiomer, must be present in the remaining 5%.

6.13 The racemate has no effect on the observed rotation other than to dilute the sample. Therefore, after addition of the racemate, the concentration of the excess (R)-2-butanol is 0.75 M. This corresponds to (0.75 mol L^{-1})(74.12 g mol^{-1})(0.001 L mL^{-1}) = 0.055 g mL^{-1}. Use this as the value of c in Eq. 6.13 with $[\alpha]$ = –14.7 degrees mL g^{-1} dm^{-1}:

$$\alpha = (-14.7 \text{ degrees mL g}^{-1} \text{ dm}^{-1})(0.0556 \text{ g mL}^{-1})(1 \text{ dm}) = -0.817 \text{ degrees}$$

6.14 Begin with the perspective structure of the reactant. In the product, corresponding groups are in corresponding positions, because the reaction breaks none of the bonds to the asymmetric carbon.

6.15 **(a)** I   **(b)** E   **(c)** C   **(d)** D   **(e)** D   **(f)** N

In part (a), there are two methyl groups on the carbon bonded to the OH group. Don't be confused by the fact that one is part of the skeletal structure and the other is drawn out explicitly.

6.16 (a) For a molecule to have a meso stereoisomer, it must have more than one asymmetric atom, and it must be divisible into constitutionally identical halves (that is, halves that have the same connectivities relative to the dividing line). By these criteria, compound (a) can exist as a meso stereoisomer.

(b) This compound does not have a meso stereoisomer. This compound has two asymmetric carbons, but no symmetry between them. A meso compound must have asymmetric carbons but be achiral overall.

(c) Although achiral, *trans*-2-hexene cannot exist as a meso stereoisomer because it has no asymmetric atoms (and no chiral diastereomers).

6.17 As suggested by the hint, the internal mirror plane passes through the central carbon atom, the hydrogen, and the OH group. This carbon can have either of two configurations. The only requirement for a meso compound (in addition to the identical connectivities of the two half-structures relative to a dividing line) is that the asymmetric carbons at the two ends of the structure must have opposite configurations.

the internal mirror plane passes through these three atoms   the internal mirror plane passes through these three atoms

➡️ The central carbon is an example of a *pseudoasymmetric carbon.* Two of the groups differ only by their configuration and not by their constitution. (See a further explanation in the note following the solution to Problem 6.54.)

6.18 (a)

A is chiral

B is achiral (plane of symmetry through the atoms/groups indicated by ◊)

C is chiral

enantiomer of *A*

enatiomer of *C*

(b) If you interchange two groups attached to an atom and get a stereoisomer, that atom is a stereocenter.

A   *B*   *C*   * these atoms are stereocenters

(c) Tetrahedrally-substituted atoms with four different groups attached are asymmetric centers. None of these molecules have asymmetric centers.

6.19 In *anti*-butane, the two internal carbons are stereocenters, because interchange of a CH_3 group and an H at either carbon gives a stereoisomer (*gauche*-butane).

anti-butane   *gauche*-butane

Remember that we are just imagining these structures as if they were objects that cannot interconvert. Of course, butane has free rotation about the C–C bonds so these carbons are *not* stereocenters in butane under normal circumstances.

6.20 (a) The three conformations of *meso*-3,4-hexanediol:

A	*B*	*C*
achiral; has a	chiral	chiral
center of symmetry		

enantiomers

Conformation *A* is achiral; it has a center of symmetry and is congruent to its mirror image. (See Displays 6.3 and 6.4 in text Sec. 6.1C.) Conformations *B* and *C* are enantiomers. This relationship can be seen from the following manipulation of conformation *C*:

enantiomers

(b) Because its conformations interconvert rapidly, *meso*-3,4-hexanediol cannot be optically active. As the text indicates, molecules that consist of rapidly interconverting enantiomers are said to be achiral. However, at very low temperatures near absolute zero, conformations *B* and *C* could in principle be isolated; each would be optically active, and the two conformations would have rotations of equal magnitudes and opposite signs.

6.21 (a) All staggered conformations of propane are achiral (and identical).

(b) Like butane, 2,3-dimethylbutane, $(CH_3)_2CH$—$CH(CH_3)_2$, contains two enantiomeric staggered conformations about the C2–C3 bond and one achiral staggered conformation. The two conformations in which the hydrogens are gauche are the chiral conformations.

enantiomeric conformations of 2,3-dimethylbutane   the achiral conformation
                                                          of 2,3-dimethylbutane

(c)   Like ethane, 2,2,3,3-tetramethylbutane, $(CH_3)_3C$—$C(CH_3)_3$ consists entirely of achiral (and identical) staggered conformations.

6.22   (a)   The carbon stereocenter is fixed in the S configuration, but the nitrogen stereocenter undergoes inversion between R and S configurations. Consequently, the molecule is a mixture of C(S),N(S) and C(S),N(R) conformations; that is, it is a rapidly equilibrating mixture of diastereomers.

(b)   Each of the rapidly interconverting species in part (a) has an enantiomer: C(S),N(S) has an enantiomer C(R),N(R), and C(S),N(R) has an enantiomer C(R),N(S). Because inversion of the nitrogen stereocenter does not affect the configuration of the carbon stereocenter, it would be possible to resolve the racemate of this compound into enantiomeric sets of rapidly interconverting *diastereomers*.

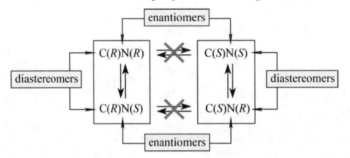

In other words, the set of compounds in one box could be resolved from the set in the other box.

6.23   Arsines should invert at an even slower rate than either amines or phosphines, for the same reason that phosphines invert at a slower rate than amines. The nonbonding electrons on arsenic are in an orbital with an even larger degree of s character, meaning that it takes a very large amount of energy to convert them into a p orbital, which is required for inversion.

6.24   (a)   If the enantiomer of the CSP were used, the elution order of R- and S-Nirvanol would be reversed—that is, the S-enantiomer would elute first.

(b)   If the racemate of the CSP were used, the two enantiomers would not separate by this method—they would elute at the same time.

6.25   The resolving agent must be *chiral,* and it must be *enantiomerically pure.* Compounds A and C meet both of these criteria.

6.26   Pasteur isolated the chiral stereoisomers, that is, the (2*R*,3*R*) and the (2*S*,3*S*) stereoisomers. (Because they are optically active, they must be chiral.) Yet to be isolated was the meso stereoisomer. Like all meso compounds, *meso*-tartaric acid is achiral and therefore is optically inactive. (The structures are drawn in eclipsed conformations for ease in seeing the symmetry relationships.)

(2*S*,3*S*)-tartaric acid   (2*R*,3*R*)-tartaric acid   *meso*-tartaric acid

6.27   The "resolving agent" was the first crystal that Pasteur separated. The handedness of each subsequent crystal was either "like" that of the first one or "opposite" to it.

Solutions to Skills Objectives Problems

6.28   This compound exists as four stereoisomers: (2*Z*,4*S*), (2*E*,4*S*), (2*Z*,4*R*), and (2*E*,4*R*).
   (a)   The carbon stereocenters are carbons 2, 3, and 4, indicated with diamonds (◊).
   (b)   Carbon 4 is an asymmetric carbon, indicated with an asterisk (*).

3,4-dimethyl-2-hexene

6.29   (a) S, also E and D   (b) R   (c) E   (d) SM   (e) D   (f) M
   (g) C   (h) no match   (i) R   (j) SM

6.30   (a) the same   (b) different (in theory, may be similar or very close)   (c) equal magnitude, opposite directions
   (d) unrelated   (e) meso compounds are optically inactive   (f) unrelated   (g) the same

6.31   (a)   (b)   (c)

6.32   (a)   (b)   (c)

6.33   (a)   The structures in *A* are both (*S*)-3-methylhexane, so they are identical; I. The structures in *B* are the *S* and
         R enantiomers of 3-methylhexane; E. The structures in *C* are 3-ethylpentane and 2-ethylpentane; C. The
         structures in *D* are stereoisomers, but the first is meso (note the plane of symmetry) and the second is
         chiral; D. The allenes in *E* are chiral, and mirror images of each other; E. If you rotate one of the compounds
         in *F* 180° along the long axis of the molecule it is congruent with the other, so they are identical; I.
   (b)

enantiomer of compound in 6.31 (a)   enantiomer of compound in   diastereomer of compound in   diastereomer of compound in
(only one stereocenter;   6.31 (b)   6.31 (b)   6.31 (b)
no diasteromer possible)

6.34   *Trans*-3-hexene, *cis*-3-hexene, and *trans*-2-pentene all have planes of symmetry in the plane of the page, passing
   through all of the carbon atoms in each structure. *Cis*-3-hexene has a second plane of symmetry, and meso-1,3-
   cyclohexanediol one plane of symmetry, perpendicular to the page.

planes of symmetry

cis-**3-hexene**   *meso*-**1,3-cyclohexanediol**

6.35 (a) A 0.5 M solution of this alkene contains $(0.5 \text{ mol L}^{-1})(146.2 \text{ g mol}^{-1})(0.001 \text{ L mL}^{-1}) = 0.0731 \text{ g mL}^{-1}$. Using Eq. 6.13 in Sec. 6.3B of the text, with $l = 0.5$ dm, the observed rotation is

$$(+79 \text{ degrees mL g}^{-1} \text{ dm}^{-1})(0.0731 \text{ g mL}^{-1})(0.5 \text{ dm}) = +2.89 \text{ degrees}$$

(b) The rotation of the (–)-enantiomer cancels half of the rotation of the (+)-enantiomer, and the concentration of the (+)-enantiomer is, in addition, halved. Therefore the rotation is 25% of that in part (a)—that is, +0.72 degrees.

6.36 You're given the observed rotation (α), density (c), and path length (l). Use Biot's law, Eq. 6.13, and rearrange to solve for the specific rotation, $[\alpha]$.

$$[\alpha] = \frac{\alpha}{cl} = \frac{-58.7 \text{ degrees}}{(0.962 \text{ g mL}^{-1})(1 \text{ dm})} = -61.0 \text{ degrees mL g}^{-1} \text{ dm}^{-1}$$

6.37 (a) Because the specific rotation of the mixture is negative, the major enantiomer is the (–)-enantiomer. The specific rotation of the (+)-enantiomer is given, so the specific rotation of the (–)-enantiomer is known as well. Using Eq. 6.17 in Sec. 6.4A of the text, we find that the EE of the (–)-enantiomer is 68%.

(b) Using Eq. 6.19b, the % major enantiomer is 84%, so the % minor enantiomer is 16%. The ER is 5.25.

(c) Because there is twice as much of the major enantiomer, the % major enantiomer is about 66.6% to 33.3% of the minor enantiomer. Using Eq. 6.19a, we get:

$$\text{EE} = 2(\% \text{ major enantiomer}) - 100\% = 33.4\%$$

You can check your work by solving for the specific rotation, $[\alpha]$, of the mixture in Problem 6.35, part (b), then using Eq. 6.17 to solve for the EE.

6.38 (a,c) These compounds have symmetry, but no asymmetric atoms, so by definition, they cannot exist as a meso compound. In compound (c), the branched carbon is bonded to two methyl groups—one is part of the skeletal structure and the other is indicated by "Me."

(b) If you're not convinced that this is a meso compound, try rotating it about the C3–C4 bond.

6.39 First, draw the (S)-enantiomer of the starting material using line and wedge bonds. Then, reduce the vinyl group to an ethyl group, but make sure the product substituents are drawn in the same relative positions as they were in the starting material. When the vinyl group is reduced to an ethyl group, it goes from being the second priority group to being the third priority group. This serves to reverse the absolute configuration of the product; it is the (R)-enantiomer. This is an example of a rare case where, even though there was no chemical reaction at the asymmetric carbon, the reaction caused the attached groups to swap priorities. (The relative priorities are the circled numbers.)

6.40 Hydrogenation of (S)-3-methyl-1-hexene does not break any of the bonds to the asymmetric carbon, and we are given the experimental result that the product, 3-methylhexane, has (–) rotation. Therefore, we can deduce the absolute configuration of (–)-3-methylhexane. Notice that *the relative priorities of the groups at the asymmetric carbon change as a result of the reaction.*

(S)-(+)-3-methyl-1-hexene   (R)-(–)-3-methylhexane   (S)-(+)-3-methylhexane

6.41   The Newman projections of 2-methylbutane about the C2–C3 bond (viewed from the C3 direction):

A   B   C

The sawhorse projections of 2-methylbutane:

A   B   C

(a)   Conformations *A* and *B* are chiral; they are conformational enantiomers.

(b)   Conformations *A* and *B* are conformational enantiomers. Conformations *A* and *C* are conformational diastereomers, as are conformations *B* and *C*.

(c)   These two conformations interconvert so rapidly that they cannot be isolated. As discussed in Sec. 6.9A, a molecule that consists of rapidly interconverting enantiomers is considered to be achiral.

(d)   The heats of formation of *A* and *B* are identical, and are lower than the heat of formation of *C*, because *C* contains two *gauche*-methyl interactions, whereas *A* and *B* each have only one. The heats of formation of *A* and *B* are equal because enantiomers have identical physical properties.

6.42   (a)

A   B   C   D

(b)   Compounds *A* and *D* invert into enantiomeric conformations. After inversion, compound *B* is identical. Compound *C* inverts into a diastereomer, as there are two asymmetric centers (the nitrogen and the indicated carbon). The nitrogen changes its configuration after inversion whereas the carbon configuration remains unchanged.

(c)   Even though compound *A* is chiral, if we assume that inversion at the nitrogen is rapid at room temperature, compound *A* would be optically inactive because it would be a mixture of rapidly interconverting enantiomeric conformations. There is no stereocenter in compound *B*; therefore it is not chiral, and it is not optically active. If we assume that the *carbon stereocenter* in compound *C* has a single specific configuration (as indicated in the problem), the nitrogen changes configuration on inversion but the carbon stereocenter does not. Therefore, compound C could be optically active. If we assume that inversion at phosphorus is slow, compound *D* would be stable in either the conformation given or in the inverted conformation, and either one of those conformations would be optically active.

6.43   (a)   Either pure (R)- or pure (S)-α-phenethylamine could be used to resolve racemic ibuprofen. If, for example, the pure (R)-amine was used, then the reaction with racemic ibuprofen would give a mixture of diastereomeric salts: the (R)-amine and (R)-ibuprofen and the (R)-amine and (S)-ibuprofen. Benzylamine is achiral and would just give a racemic mixture of the amine/ibuprofen salt. The racemic phenethylamine is also achiral overall so wouldn't be a good resolving agent.

(b)   Using the (R)-amine, two diastereomeric salts would be formed. In theory, these two diastereomeric salts would have different solubilities in ethanol, so one might crystallize preferentially (you probably couldn't predict *which* one, but you could predict *that* one (or the other) would.

You could also use the (S)-amine to form an "(S,S)-salt" and "(R,S)-salt", which would also be diastereomers and would also have different physical properties.

Solutions to Integrated Problems

6.44   The asymmetric carbons are indicated with asterisks (*).

(a)

CH₃CH₂CH—*⬡*
 |
CH₃   CH₃

(b)

CH₃CH₂C*H—⬡—CH₃
 |
CH₃

(c)

(d) none

(e)

6.45   In answering this question, notice that, in deciding priorities at carbon-1, carbon-2 has a higher priority than the phenyl (Ph) group, because carbon-2 has an attached nitrogen.

(a)   If you put the methyl group on the left and the phenyl on the right, the positions of the other groups will be reversed from the following ones.

ephedrine

pseudoephedrine

(b)   Sawhorse and Newman projections:

ephedrine

pseudoephedrine

(c)   These two compounds are not constitutional isomers because they have the same atomic connectivity. Because they are not identical, they are stereoisomers. If these two compounds were enantiomers, they would have opposite configurations at *both* asymmetric carbons. Because they are stereoisomers but not enantiomers, they are diastereomers.

(d)   In principle, their melting points should be different. In fact, they are. Ephedrine has a melting point of 38–39 °C; pseudoephedrine has a melting point of 116–119 °C.

(e)   One can say two things about the optical activities of these two compounds. First, they are both optically active, because they are chiral. Second, because they are diastereomers, their optical activities have no simple relationship—that is, their magnitudes are expected to be different, and one cannot say anything about their signs.

6.46   (a)   The sawhorse projections of ephedrine: (The projections labeled with double letters are eclipsed; the projections labeled with single letters are staggered.)

(b)   Each of these conformations is chiral. Had any one of them been achiral, ephedrine would not be chiral.

6.47   Either of the following heptane isomers is a correct answer:

3-methylhexane

2,3-dimethylpentane

6.48   The following stereoisomer of 1,2-dimethylcyclopropane is chiral. If you're not convinced, build a model of it and another model of its mirror image and test them for congruence.

H⸺ ⟨triangle⟩ ⸺CH₃
H₃C   H

6.49   (a)   True. 2-Chlorohexane and 3-chlorohexane are examples of chiral constitutional isomers. (Note that while they are chiral, they are not enantiomers.)

   (b)   True by definition.

   (c)   False. Every object—chiral or achiral—has a mirror image. Only an achiral molecule is congruent to its mirror image. (See Fig. 6.1, text Sec. 6.1A, for an example.)

   (d)   True by definition.

   (e)   False. For example, 2-chlorohexane has no diastereomer.

   (f)   False. Some E,Z isomers are not chiral (e.g., (E)- and (Z)-2-butene). Likewise, any meso compound is an achiral diastereomer of a compound containing asymmetric carbons.

   (g)   False. Meso compounds have asymmetric carbons and are not chiral. (A compound containing one *and only one* asymmetric carbon is chiral, however.)

   (h)   False, because some stereoisomerism is not associated with chirality—for example, E,Z stereoisomerism (double-bond stereoisomerism).

   (i)   True. By definition, interchanging two groups at a stereocenter generates a stereoisomer.

   (j)   False. Molecules with a mirror-image relationship must be either enantiomeric or identical.

   (k)   False. Every chiral compound is optically active, although it is possible in some cases for the optical activity to be too small to be detected experimentally.

   (l)   False. Optical rotation has no general relationship to R and S configuration.

   (m)   False. A chiral structure must have neither a plane of symmetry nor a center of symmetry. Therefore, a structure might not have a plane of symmetry and yet be achiral because it has a center of symmetry. (For example, see the solution to Problem 6.20(a), conformation A.)

   (n)   True, because the presence of a plane of symmetry is sufficient (although not necessary) to eliminate chirality.

   (o)   True. An asymmetric carbon is one type of stereocenter.

   (p)   False. Not all stereocenters are asymmetric atoms. For example, cis-2-butene contains two stereocenters, but these carbons are not asymmetric.

6.50   Compound B can undergo amine inversion, which rapidly interconverts its enantiomeric forms, as shown in Fig. 6.9, text Sec. 6.9B. Compound A cannot undergo inversion because it does not have nonbonding electron pair; therefore, its two enantiomeric forms cannot interconvert and can be isolated.

6.51   Any compound that is chiral can in principle be resolved into enantiomers. Compound A possesses a phosphorous atom containing four "groups" (the electron pair counts as a different group in this case), so it is chiral and could be resolved. Elements in the third period do not undergo rapid inversion (Sec. 6.9B). Compound B also contains a phosphorous atom with 4 different groups, so it can be resolved. The sulfur atom in compound C is attached to 4 groups and, like phosphorous, is too slow to interconvert, but two of its attached groups are identical. Thus, it is achiral and doesn't have an enantiomer. Compounds D and E can be resolved because they contain asymmetric carbons. Compound F can be resolved into enantiomers for the same reasons as compounds A and B.

6.52   (a)   Sulfur is the asymmetric center in omeprazole. It has four different "groups" attached, if we include the nonbonding electron pair. Because sulfur does not undergo rapid inversion (Sec. 6.9B) the two isomers can be isolated.

   (b)   A simplified structure of the S enantiomer of omeprazole is below. The cyclic groups are abbreviated "Ar". Priority 1 is O, 2 is Ar, 3 is CH₂Ar, and 4 is the electron pair.

Ar⸺S⁺⸺CH₂—Ar
     |
     ⁻O

6.53   Hydrogenation makes two of the branches to the asymmetric carbon of the starting alkene identical. Therefore, the product is achiral and is optically inactive.

3-methyl-1-pentene
(either enantiomer)

3-methylpentane
(achiral and therefore
optically inactive)

6.54   (a)   The two meso stereoisomers of 2,3,4-trichloropentane:

two meso stereoisomers of 2,3,4-trichloropentane

(b)   Each compound has a plane of symmetry that passes through carbon-3 as well as the C—Cl and C—H bonds to that carbon:

planes of symmetry

(c)   These two stereoisomers must be diastereomers, because they are not enantiomers.

(d)   Carbon-3 is a stereocenter, because interchanging the H and the Cl gives a stereoisomer—specifically, the other meso compound.

(e)   To specify the configuration of carbon-3, we would need a rule to specify the relative priorities of groups that differ only in their stereochemical configuration. If we arbitrarily let an *R* carbon take precedence over *S* (that is, precedence is alphabetical), then carbon-2 has precedence over carbon-4. This means that the stereoisomer on the left in parts (a) and (b) has the 3*S* configuration, and the one on the right has the 3*R* configuration.

 Asymmetric carbons are generally diagnosed using connectivity relationships. Therefore, this type of asymmetric carbon is unusual, because the connectivity of two of the attached groups is the same. In recognition of this situation, a carbon that is "asymmetric" only because of the differing configurations of the attached groups is called a *pseudoasymmetric carbon*. The sequence rule governing pseudoasymmetric carbons is that *R* takes precedence over *S*.

(f)   The two stereoisomers in parts (a) and (b) are the meso stereoisomers. There are also two enantiomers that have, respectively, the 2*R*,4*R* and the 2*S*,4*S* configurations. (Convince yourself that in these enantiomers, carbon-3 is *not* a stereocenter.) These are the only four stereoisomers.

6.55   (a)   Compounds *A* and *B* are diastereomers; both are meso. (See the solution to Problem 6.54 for a similar case.) Compound *C* is a diastereomer of both *A* and *B* because it differs in configuration at some, but not all, of the asymmetric carbons. The two asymmetric carbons in compounds *C* and *D* have the *R* configuration, so it appears that these are the same compound. To show this convincingly, first carry out an internal rotation in *D*, then rotate *D* 180° about an axis in the page:

H OH   H CH$_2$OH
HOCH$_2$ ⟍ ⟋ ⟍ ⟋ OH   120° internal
                                    rotation
H OH
D

H OH   HO H
HOCH$_2$ ⟍ ⟋ ⟋ ⟍ CH$_2$OH   180° rotation
                                       of entire molecule
H OH
D

H OH   HO H
HOCH$_2$ ⟍ ⟋ ⟍ ⟋ CH$_2$OH
HO H
C

Compound *E* is the enantiomer of *D* (and *C*), because both asymmetric carbons have the opposite (*S*) configuration. To summarize:

A and *B*:	diastereomers	*A* and *C*:	diastereomers
A and *D*:	diastereomers	*A* and *E*:	diastereomers
B and *C*:	diastereomers	*B* and *D*:	diastereomers
B and *E*:	diastereomers	*C* and *D*:	identical
C and *E*:	enantiomers	*D* and *E*:	enantiomers

(b)   Compounds *A* and *B* are both meso; they are diastereomers.

(c)   Compounds *C* (and its identity *D*) and *E* are chiral and therefore optically active.

6.56   Salts that are either identical or enantiomers should have identical solubilities. The solution to the problem then hinges on determining the relationship between each pair of salts. If they have the same configurations at corresponding asymmetric carbons, they are identical and have identical solubilities. If they have different configurations at *both* of the corresponding asymmetric carbons, they are enantiomers and have identical solubilities, because enantiomers have identical properties. If they have different configurations at one asymmetric carbon and the same configuration at the other, they are diastereomers and have different solubilities, because diastereomers have different properties.

Salt *A* has the *S* configuration at both asymmetric carbons, and salt *D* has the *R* configuration at both asymmetric carbons. Therefore, salts *A* and *D* are enantiomers and have identical solubilities. Salt *B* has the *S* configuration at the asymmetric carbon of the ammonium salt and the *R* configuration at the asymmetric carbon of the carboxylate. Salt *C* has the opposite configuration at both of these carbons, that is, (*R,S*). Therefore, salts *B* and *C* are enantiomers and have identical solubilities. However, the solubilities of *A* and *D* are different in principle from the solubilities of *B* and *C*.

Once you have deduced the configuration of one salt, the configuration of the corresponding carbon in the others can be determined by interchanging groups to reproduce the original. Each interchange changes the configuration; even numbers of interchanges leave the configuration unchanged. Alternatively, you can rotate about a single bond freely; this does not change the configuration. Thus, interchanging the H and the Ph gives the ammonium ion of *C* from that of *A*; hence, the configuration is different, and it must therefore be *R*. An internal rotation of the asymmetric carbon 120° counterclockwise about the methyl—C bond converts the ammonium ion of *A* into that of *B*; therefore, these two ions must be identical.

6.57   (a)   We should know by now that if the geometry is tetrahedral, there are two enantiomers:

F   F
| |
Cl·····C—Br   Br—C·····Cl
| |
I   I

enantiomers

(b)   For square-planar geometry, there are three diastereomers. These are found by making all possible pairwise switches between adjacent groups and ruling out identities. Note that, because of the square planar geometry, mirror images must be congruent, and therefore identical. There can be no enantiomers.

(c)   For pyramidal geometry, there are three diastereomeric sets of enantiomers (six stereoisomers total):

6.58   In a compound of the form X_2ZY_2 with square-planar geometry, there are two ways to arrange groups X and Y about atom Z: with like groups in adjacent corners, or with like groups in opposite corners, as shown. Because these are stereoisomers, and they are not enantiomers, they must be diastereomers. Tetrahedral compounds of the form X_2ZY_2 (for example, H_2CCl_2) *cannot* exist as stereoisomers. Therefore, the fact that $Cl_2Pt(NH_3)_2$ exists as stereoisomers with different properties shows that these stereoisomers are diastereomers, and their geometry about platinum is square-planar.

6.59   The ultimate test for chirality is to make the mirror image and test it for congruence:

Because the structure is not congruent to its mirror image, the compound is chiral.

6.60   (a)   Using the equation given and a ratio of R to $S = r = 0.064$, solving for t gives an age of 205 years.

(b)   Using Eq. 6.19a (text Sec. 6.9A), the EE is 88%.

(c)   First, rearrange Eq. 6.17 to find the specific rotation of the mixture, $[\alpha]_{mixture}$. Then use this value, +21.56 degrees, in Eq. 6.13 (text Sec. 6.3B) with a concentration $c = 0.01 \text{ g mL}^{-1}$. Solving for the observed rotation, α, gives $\alpha = +0.216$ degrees.

6.61   Each compound contributes to the optical activity in proportion to the fraction present. If the fraction of the α form is P_α, and that of the β form is P_β,

$$\text{specific rotation} = 52.7° = P_\alpha(+112°) + P_\beta(+18.7°)$$

Because $P_\alpha + P_\beta = 1.0$, then

$$52.7° = P_\alpha(+112°) + (1.00 - P_\alpha)(+18.7°)$$

Solving for P_α gives $P_\alpha = 0.36$, or 36%, and the fraction of $P_\beta = 64\%$.

6.62 (a) The optical rotations and melting points show that Fischer obtained enantiomers as a result of this transformation. Consider what each of the two possible geometries predicts for an interchange of the two groups. For tetrahedral carbon, the interchange of any two groups at an asymmetric carbon gives enantiomers. Consequently, the experimental result is consistent with tetrahedral geometry.

If the molecule were pyramidal, and if Fischer *happened* to be dealing with a stereoisomer in which the —CO₂H and —CONH₂ groups were at opposite (rather than adjacent) corners of the pyramid, he would also have obtained enantiomers from his experiment.

But if Fischer *happened* to be dealing with a stereoisomer in which the —CO₂H and —CONH₂ groups were at adjacent corners, the exchange should have given diastereomers. Thus, the conversion of one stereoisomer into its enantiomer could have been the result of two things occurring simultaneously: (1) pyramidal geometry, and (2) the fortuitous choice of a particular stereoisomer. Consequently, Fischer's result was consistent with either tetrahedral or pyramidal geometry.

(b) Neither result taken alone rules out pyramidal geometry. However, for an atom with two identical groups and *pyramidal* geometry to be achiral, the two carboxy groups in the product, and therefore, the carboxy and carboxamido group in the starting material, must be at the opposite corners of the pyramid.

If these two groups had been at opposite corners of the pyramid in the *first* experiment, then in the *second experiment with the same starting material,* the carboxy group and the methyl group must have been at *adjacent* corners, and this transformation would have yielded a chiral compound.

(Be sure to convince yourself that the molecule on the right is chiral. Do this by drawing (or making a model of) its mirror image and show that the two are noncongruent.) On the other hand, if the starting compound had tetrahedral geometry, then making *any* two groups identical would give an achiral compound. Since this was the experimental result, pyramidal geometry was ruled out, and tetrahedral geometry thus remained the only (and, as it turned out, correct) possibility.

7 Cyclic Compounds and Reaction Stereochemistry

CHAPTER 7 SUMMARY

- All of the cycloalkanes except for cyclopropane have puckered carbon skeletons.

- Of the cycloalkanes containing relatively small rings, cyclohexane is the most stable because it has no angle strain and it can adopt a conformation in which all bonds are staggered.

- The most stable conformation of cyclohexane is the chair conformation. In this conformation, hydrogens or substituent groups assume axial or equatorial positions. Cyclohexane and substituted cyclohexanes rapidly undergo the chair interconversion, in which equatorial groups become axial, and vice versa. The twist-boat conformation is a less stable conformation of cyclohexane derivatives. Twist-boat conformations are interconverted through boat transition states.

- A cyclohexane conformation with an axial substituent is typically less stable than a conformation with the same substituent in an equatorial position because of unfavorable van der Waals interactions (1,3-diaxial interactions) between the axial substituent and the two axial hydrogens on the same face of the ring. The 1,3-diaxial interaction of an axial methyl group and an axial ring hydrogen is very similar to the interaction of the two methyl groups in *gauche*-butane.

- Cyclopentane exists in an envelope conformation. Cyclopentane has a greater heat of formation per CH_2 than cyclohexane because of eclipsing between hydrogen atoms.

- Cyclobutane and cyclopropane contain significant angle strain because their bonds are forced to deviate significantly from the ideal tetrahedral angle. Cyclopropane has bent carbon–carbon bonds. Cyclobutane and cyclopropane are the least stable cycloalkanes.

- Cycloalkanes can be represented by planar polygons in which the stereochemistry of substituents is indicated by dashed or solid wedges. Planar line-and-wedge structures show the configurations of carbon stereocenters but contain no conformational information.

- Spirocyclic compounds contain two rings with a single common atom.

- Bicyclic compounds contain two rings joined at two common atoms, called bridgehead atoms. If the bridgehead atoms are adjacent, the compound is a fused bicyclic compound; if the bridgehead atoms are not adjacent, the compound is a bridged bicyclic compound. Either cis or trans ring fusion is possible.

  - Trans fusion, which avoids 1,3-diaxial interactions, is the most stable way to connect larger rings.

  - Cis fusion, which minimizes angle strain, is the most stable way to connect smaller rings.

  - Polycyclic compounds can contain many fused or bridged rings (or both).

- Steroids are important polycyclic compounds that are built on the carbon skeleton shown in Sec. 7.6D.

- Cycloalkenes with trans double bonds within rings containing fewer than eight atoms are too unstable to exist under normal circumstances. Bicyclic compounds with a double bond at a bridgehead atom within a ring of fewer than eight atoms are also unstable (Bredt's rule).

- The following fundamental principles govern reactions involving stereoisomers:

- A pair of enantiomers have identical reactivities unless the reaction conditions cause them to be involved in diastereomeric interactions (for example, a chiral catalyst, a chiral solvent, and so on).

- Chiral products are always formed as racemates in a chemical reaction involving achiral starting materials, unless the reaction conditions create diastereomeric interactions (for example, a chiral catalyst, a chiral solvent, and so forth).

- Diastereomers in general have different reactivities.

- Diastereomeric products of chemical reactions are formed at different rates and in unequal amounts.

◼ Addition reactions can occur with syn or anti stereochemistry. Substitution reactions can occur with retention or inversion of configuration. The stereochemistry of a reaction is determined by comparing the stereochemistry of the reactants and the products. Each carbon at which a chemical change occurs must be a stereocenter in the product in order for the stereochemistry of the reaction to be determined.

◼ In a stereoselective reaction, some stereoisomers of the product are formed in large excess over others.

A stereospecific reaction is a highly stereoselective reaction in which each stereoisomer of the reactant gives a different stereoisomer of the product. All stereospecific reactions are stereoselective, but not all stereoselective reactions are stereospecific.

◼ Bromine addition to simple alkenes is a stereospecific anti addition in which the syn addition of one bromine to give a bromonium ion is followed by the nucleophilic reaction of bromide ion at a carbon of the bromonium ion with inversion of configuration.

◼ The hydroboration of alkenes is a stereospecific syn addition, and the subsequent oxidation of organoboranes is a substitution that occurs stereospecifically with retention of configuration. Thus, hydroboration–oxidation of alkenes is an overall stereospecific syn addition of H and OH to alkenes.

◼ Catalytic hydrogenation is a stereospecific syn addition.

◼ Oxymercuration is a stereospecific anti addition. The subsequent reduction step in oxymercuration–reduction is not generally stereoselective (and therefore is not stereospecific), because the replacement of mercury with hydrogen can occur with mixed stereochemistry.

STUDY GUIDE LINKS

7.1 Relating Cyclohexane Conformations

Although you can often resort to models to visualize the stereochemical relationships between chair conformations of substituted cyclohexanes, it is useful to be able to carry out on paper manipulations like the ones used in the text. To gain greater facility with such manipulations, try the following exercises, using *cis*-1,2-dimethylcyclohexane as an example. Draw the result of each manipulation, and then check your drawing against the results shown. *Be sure to follow each manipulation with a model if there is any question about it!*

First, turn the structure 180° about a horizontal ("*x*") axis:

Now turn the original structure 180° about a "*y*" axis:

Practice turning the structure in 60° increments about the same axis:

When you rotate a structure 60°, remember that the ring perspective changes. (See text Display 7.4b, Sec. 7.2A.) Finally, turn the original structure 180° about an axis perpendicular to the page (a *"z"* axis):

Let's use these skills to solve the following problem. What is the stereochemical relationship between the two chair conformations of *trans*-1,3-dimethylcyclohexane?

(SG7.1)

There are three possible relationships between these conformations: they could be conformational diastereomers; they could be conformational enantiomers; or they could be identical. The strategy here is to orient one of the methyl groups in one structure in the same way as a methyl group in the other structure. Then compare the relative positions of the remaining methyl group. One way to perform this manipulation is to rotate the second structure about an axis perpendicular to the page, as follows:

Finally, rotate the resulting structure 60° about the *y* axis, as follows:

The resulting structure is identical to the first conformation in Eq. SG7.1. Consequently, the chair interconversion in this case results in *identical molecules.*

It is worth noting that some people, even after practice, have difficulty with manipulations like these because they have inherent difficulty with spatial perception. (Don't assume you are such a person until you have practiced! In fact, working through manipulations like this can help you develop improved spatial perception.) This has nothing to do with your intelligence! If you are such a person, don't hesitate to use models. The answer you get may take a little more time, but it is just as correct.

7.2   Reactions of Chiral Molecules

Sec. 7.7A of the text deals only with reactions of *achiral molecules* that give enantiomeric pairs of products. What happens when *chiral* molecules react to give stereoisomeric products? No general principles cover this situation. In some cases, stereoisomers of the product may be formed in equal amounts; in other cases, one stereoisomer may predominate. The result can be determined by the reaction mechanism—that is, the stereochemical nature of chemical changes that occur at stereo-centers. (A situation of this sort is discussed in Sec. 7.8B.) In other cases, reactants or intermediates are racemized in the reaction, in which case the principles of Sec. 7.7A apply to each enantiomer of the racemate.

It is even possible for chiral molecules to react in some cases to give achiral products even when a stereocenter does not undergo a chemical change. For example, the following hydrogenation reaction involving a chiral alkene gives an achiral alkane because the asymmetric carbon in the alkene is no longer an asymmetric carbon in the alkane product:

$$H_2C=CH\text{-----}\overset{\overset{\displaystyle CH_2CH_3}{|}}{\underset{\underset{\displaystyle H}{|}}{C}}\diagdown CH_3 \;+\; H_2 \;\xrightarrow{\text{catalyst}}\; CH_3CH_2\text{-----}\overset{\overset{\displaystyle CH_2CH_3}{|}}{\underset{\underset{\displaystyle H}{|}}{C}}\diagdown CH_3$$

chiral and optically active   achiral and optically inactive

Optical activity (and chirality) is lost in this reaction because the product is not a chiral molecule. In *this particular case,* the reaction causes a plane of symmetry to occur within the structure of the product that does not exist in the reactant. Optical activity would *not* be lost in the same reaction of a different alkene in which the product is chiral:

$$H_2C=CHCH_2\text{-----}\overset{\overset{\displaystyle CH_2CH_3}{|}}{\underset{\underset{\displaystyle H}{|}}{C}}\diagdown CH_3 \;+\; H_2 \;\xrightarrow{\text{catalyst}}\; CH_3CH_2CH_2\text{-----}\overset{\overset{\displaystyle CH_2CH_3}{|}}{\underset{\underset{\displaystyle H}{|}}{C}}\diagdown CH_3$$

chiral and optically active   chiral and optically active

7.3   Analysis of Reaction Sterechemistry

A given type of reaction might yield diastereomers with one starting material, enantiomers with another, and no stereoisomers at all with a third. For example, the bromine addition to cyclohexene in Study Problem 7.5 gives a pair of diastereomers, and one of these is formed as a pair of enantiomers. In bromine addition to propene, only a pair of enantiomers is formed. In bromine addition to ethylene, no stereoisomers at all are formed. For a given reaction, no single stereochemical result applies to all compounds. Each reaction must be analyzed individually.

7.4   Stereoselective and Stereospecific Reactions

Some texts define the term *stereospecific reaction* to mean "100% stereoselective within limits of detection." This idea can be conveyed by saying simply that a reaction is "highly stereoselective." This definition of *stereospecific* is not generally accepted; be careful not to confuse the terms *stereospecific* and *stereoselective* in this way.

7.5   When Stereoselectivity Matters

The stereoselectivity of a reaction (or lack of it) is of no concern if stereochemically different modes of a reaction give the same products. Let's see how this statement would apply in the case of bromine addition to alkenes.

The addition of bromine to ethylene does not give stereoisomers; hence, it makes no difference in the outcome of the reaction whether the reaction is a syn addition, an anti addition, or a mixture of the two.

$$H_2C=CH_2 \xrightarrow{Br_2} Br-CH_2-CH_2-Br$$

Furthermore, we cannot even tell whether the addition is syn or anti. It is often assumed (reasonably) that additions which occur in one particular way on certain compounds occur in the same way on similar compounds for which the stereoselectivity cannot be determined. So, because addition of bromine is an anti addition with other alkenes, we assume that it is an anti addition for ethylene as well. In fact, this assumption has been tested: the stereochemistry of bromine addition to isotopically substituted ethylenes was investigated in 1998:

$$\underset{D}{\overset{H}{\diagdown}}C=C\underset{D}{\overset{H}{\diagup}} \quad \text{and} \quad \underset{D}{\overset{H}{\diagdown}}C=C\underset{H}{\overset{D}{\diagup}}$$

This experiment established that anti-addition does in fact occur. Most chemists would accept the anti addition of bromine to the deuterium-substituted ethylenes as proof that ethylene itself undergoes anti addition of bromine.

In the addition of bromine to propene, stereoisomers are formed:

$$H_3C-CH=CH_2 \xrightarrow{Br_2} H_3C-\underset{\underset{Br}{|}}{CH}-CH_2-Br$$

Two *enantiomers* are possible, but they *must* be formed in the same amounts, because the reagents are achiral. Because carbon-1 of the product is not a stereocenter, the question of a syn or anti addition is irrelevant to the composition of the product mixture in the addition of bromine to propene.

Always remember a point made in Sec. 7.8A (following Eq. 7.50b) of the text: the stereochemistry of an addition is relevant only when *both* carbons of the double bond become stereocenters as a result of the addition. Thus, one can study the mode of bromine addition to *cis*- and *trans*-2-butene, as in Sec. 7.8C, because the two carbons of the double bond in the addition product are stereocenters. In a practical sense, bromine addition would not be as useful when applied to the 2-butenes or to cyclohexene if it were not stereoselective, because a mixture of diastereomers would be formed.

In contrast, the text shows that oxymercuration–reduction is not a stereoselective reaction sequence. The lack of stereoselectivity matters only when it is applied to an alkene that would give a mixture of diastereomeric products. Many reactions that are not stereoselective are useful as long as their use is confined to situations in which stereoselectivity is not an issue. (Focused Problem 7.30 illustrates this point.)

FURTHER EXPLORATIONS

 ### 7.1 Other Ways of Designating Relative Configuration

The use of *cis* and *trans* for designating relative stereochemistry in disubstituted rings has been adopted by *Chemical Abstracts,* the index to the world's chemical literature.

Chemical Abstracts adopted a somewhat different system for designating absolute configuration of chiral cyclic compounds than the IUPAC-based system used in the text. The *Chemical Abstracts* system involves the use of relative configurations. Consider the following compound:

The IUPAC name for this compound is (1*S*,2*R*)-1-chloro-2-methylcyclohexane. The *Chemical Abstracts* name is (1*S-cis*)-1-chloro-2-methylcyclohexane. In this system, the absolute configuration is given for the stereocenter of lower number (in this case the carbon bearing the chlorine), and then the configuration of the other stereocenter is given with the relative descriptor *cis* or *trans.*

As the text states, the cis–trans system is cumbersome when a ring contains more than two substituents. Although various extensions of this system have been proposed, *Chemical Abstracts* uses the terms α and β to describe relative stereochemistry. The term α means "on the same side of the ring in a planar projection," and β means "on the opposite side of the ring in a planar projection." The positions of all substituents are compared to the position of the substituent on a reference carbon, which, in the simplest cases, is the lowest-numbered carbon. The substituent at the reference carbon by definition is in the α position. Consider the following example:

[1*S*-(1α,2β,4α)]-1,4-dichloro-2-methylcyclopentane

The chlorine at carbon-1 is α by definition. Since it is "up", the other "up" substituent (the other chlorine) is also α, and the "down" substituent (the methyl group) is β. If we did not know the absolute configuration, but only the relative configuration, we would leave off the initial "1*S*" and name the compound as (1α,2β,4α)-1,4-dichloro-2-methylcyclopentane.

The following two cases are examples of achiral compounds with three stereocenters:

(1α,3α,5α)-1,3,5-trimethylcyclohexane   **(1α,3α,5β)-1,3,5-trimethylcyclohexane**

In the compound on the left, the three methyl groups are all designated as α because they are on the same "side" of the ring—that is, all "down."

One other way of designating relative configuration is to use an *R*,S** notation. If we don't know the actual absolute configurations, but we do know the *relative* configurations, we refer to the configuration of the lowest-numbered carbon as *R**, and assume that it has the *R* configuration. For example, if we had in hand one of the enantiomers of the following compound, but didn't know which one, we would refer to carbon 1 as *R**. Then carbon 2 would also be *R**, and carbon 4 would be *S**.

Because we don't know which enantiomer we have, the compound would be named (1*R**,2*R**,4*S**-1,4-dichloro-2-methylcyclopentane. The asterisks let us know that we have one of the two enantiomers, but we don't know which one. The *racemate* of this compound (that is, a 1:1 mixture of the two enantiomers) would be named (1*RS*,2*RS*,4*SR*)-1,4-dichloro-2-methylcyclopentane, with carbon-1 being arbitrarily assigned the "*RS*" configuration. Carbons with the same configuration as carbon-1 (carbon-2 in this example) are also assigned the *RS* configuration, and carbons with a different configuration (carbon-4 in this example) are assigned the *SR* configuration.

The *Chemical Abstracts* system has additional layers of complexity for more complicated cases, but this introduction covers most of the cases that would normally be encountered.

7.2 Alkenelike Behavior of Cyclopropanes

The large amount of strain in cyclopropanes causes them to be more reactive than ordinary cycloalkanes. For example, cyclopropanes can be hydrogenated, although not so readily as alkenes.

$$CH_2 \quad \overset{Pt/C}{\underset{50\ °C}{\overset{1\ atmosphere}{\longrightarrow}}} \quad H_2C—CH_2 + H_2 \longrightarrow CH_3CH_2CH_3$$

Although reactions like this suggest that cyclopropanes react like alkenes, an alternative view is that alkenes are cycloalkanes with "two-membered rings."

$$H_2C \quad CH_2$$

ethylene viewed as a "cyclic alkane"

In fact, it can be argued from hybrid-orbital considerations that a view of the alkene double bond as two "bent bonds" (similar to those in cyclopropane) is just as valid as the picture adopted by this text in which the two bonds are different, one a σ bond and one a π bond. (Some model sets represent double bonds as "bent bonds.")

If ethylene indeed is visualized as a "two-membered ring"—that is, *cycloethane*—it would be expected to be very strained, and that this strain would be reflected in a very high heat of formation. Indeed, the heat of formation of ethylene per CH_2 is even greater than that of cyclopropane.

	cyclobutane	cyclopropane	ethylene (cycloethane)
ΔH_f° per carbon atom, kJ mol^{-1}	7.1	17.8	26.2
kcal mol^{-1}	1.7	4.25	6.27

In the conventional model of the alkene double bond, the high energy of the bonds in ethylene resides completely in the relatively weak π bond. In the bent-bond model, the two bonds are equivalent. They are stronger than a π bond, but weaker than a σ bond.

7.3 Optical Activity

The interaction of polarized light with a chiral molecule is an important example of the *principle of enantiomeric differentiation.* Plane-polarized light is actually the *vector sum of right and left circularly polarized light,* two forms of light in which the electric field vector propagates as a right- and left-handed helix, respectively:

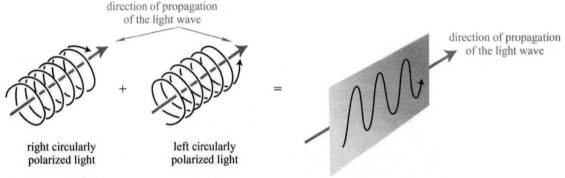

These two forms of light are noncongruent mirror images—that is, they are enantiomers. In other words, plane-polarized light is "enantiomeric light." Let's call the right-handed form of light *R,* and the left-handed form *S.* Suppose that plane polarized light—consisting of equal amounts of *R* and *S* forms—is passed through an enantiomerically pure sample that has the *S* configuration. When the light interacts with the sample, two types of interaction occur: *S* sample with *S* light, and *S* sample with *R* light. *These two interactions are diastereomeric,* because an *S,S* combination is the diastereomer of an *S,R* combination. Because diastereomers have different energies, these two interactions have different energies—one is stronger than the other. The form of light with the stronger interaction is retarded in its passage through the sample. This in turn causes the plane of the polarized light to rotate—that is, it causes optical activity. *The enantiomeric forms of light differ when they interact with a chiral object.*

7.4 Stereochemistry of Organoborane Oxidation

The mechanism of organoborane oxidation was discussed in Further Exploration 5.1 in this manual. Retention of stereochemistry in this reaction can be understood in terms of this mechanism. An alkyl group on boron moves from boron to oxygen in such a way that the configuration of the migrating carbon is unaffected:

As the second part of this equation shows, this process occurs at each of the carbon–boron bonds to yield ultimately a trialkyl borate. Thus, a carbon–boron bond is replaced by a carbon–oxygen bond without changing the stereochemical configuration of the carbon.

Finally, the resulting trialkyl borate undergoes hydrolysis to the alcohol product. The final hydrolysis breaks the boron–oxygen bonds but leaves the oxygen–carbon bonds unaffected. Because these bonds are not affected, the configuration at carbon does not change.

 SOLUTIONS TO PROBLEMS

Solutions to Focused Problems

7.1 First, the half-chair conformation is strained because the C—C—C bond angles in the flattened part of the molecule are widened to nearly 120°, whereas the optimal angle is 109.5°. (These angles are indicated with arcs in the following structure. Your model may pop back from a half-chair to a boat or chair if you try to place it on a table without holding it.) Second, there is significant eclipsing between some of the hydrogens. As we learned when we studied the conformations of ethane, eclipsing interactions are destabilizing.

half-chair cyclohexane
(eclipsed hydrogens are shown)

7.2 We apply text Eq. 3.47b in Sec. 3.5B, with $\Delta G° = 7.4$ kJ mol^{-1}.

$$K_{eq} = 10^{-\Delta G°/2.30RT} = 10^{-7.4/5.71} = 10^{-1.30} = 5.1 \times 10^{-2}$$

Letting the axial form be A and the equatorial form be E,

$$K_{eq} = \frac{[A]}{[E]} = 0.051 \text{ or } [A] = 0.051[E]$$

Recognizing that, at 1 mol L^{-1} total concentration, $[A] + [E] = 1.0$,

$$1.051[E] = 1.0$$

from which we obtain $[E] = 0.95$ and $[A] = 0.05$. A sample of methylcyclohexane contains 95% equatorial and 5% axial conformation.

7.3 Following the procedure in the solution to Focused Problem 7.2, we first calculate the K_{eq} for *tert*-butylcyclohexane:

$$K_{eq} = 10^{-\Delta G°/2.30RT} = 10^{-20/5.71} = 10^{-3.5} = 3.1 \times 10^{-4}$$

From this we calculate, in a total concentration of mol L^{-1}, $[A] = (3.1 \times 10^{-4})[E] \approx 0.00031$. There is about $0.051/0.00031 = 168$ times more of the axial conformation of methylcyclohexane than there is of the axial conformation of *tert*-butylcyclohexane per mole.

7.4 (a) Because the axial conformation of fluorocyclohexane has *two* fluorine–hydrogen 1,3-diaxial interactions (just as the axial conformation of methylcyclohexane has two methyl–hydrogen 1,3-diaxial interactions), the energetic cost of a single 1,3-diaxial interaction in fluorocyclohexane is 0.50 kJ mol^{-1} (0.13 kcal mol^{-1}).

(b)   In the case of methylcyclohexane, the energy cost of an axial methyl group is twice that of the methyl–methyl interaction in *gauche*-butane. If we take the same approach with fluorocyclohexane and 1-fluoropropane, the energy cost of an axial fluorine in fluorocyclohexane [1.0 kJ mol^{-1}, from part (a)] should be about twice the energy cost of a methyl–fluorine interaction in the gauche conformation of 1-fluoropropane. Therefore, such a gauche interaction should be about 0.50 kJ mol^{-1}.

7.5   (a)   The two chair conformations of *cis*-1,3-dimethylcyclohexane:

cis-1,3-dimethylcyclohexane

(b)   The conformations of *trans*-1-ethyl-4-isopropylcyclohexane:

trans-1-isopropyl-4-ethylcyclohexane

   Don't forget that a substituted cyclohexane can be drawn in many valid ways. If the structures you drew look different from those above, either verify that they are correct with models, or try to manipulate your structures mentally to see whether they can be made to look identical to the ones above. (Study Guide Link 7.1 provides some exercises that should be helpful.)

7.6   Boat conformations of the compounds in Focused Problem 7.6; remember that there are several valid ways to draw these structures:

(a)

cis-1,3-dimethylcyclohexane
(a boat conformation)

(b)

trans-1-isopropyl-4-ethylcyclohexane
(a boat conformation)

7.7   (a)

and

cis-1-bromo-3-methylcyclohexane
(enantiomers)

(b)

(1R,2S,3R)-2-chloro-1-ethyl-3-methylcyclohexane

Again, remember that there are other correct ways to draw these structures.

7.8   (a)

(b)

7.9   (a)   (b)

(a)   180°   120°

7.10   (a)

(a)   180°

(b)

180°

(b)

7.11   (a)   Compound *A* is achiral and undergoes the chair interconversion to give conformational enantiomers:

120°

conformational enantiomers

(b)   Compound *B* is chiral and undergoes the chair interconversion to give identical conformations. If the compound is chiral, then the individual chair conformations *must* be chiral.

(c)   Compound *C* is achiral and undergoes the chair interconversion to give conformational diastereomers:

7.12   (a)   *Trans*-1,4-dimethylcyclohexane does *not* have any asymmetric carbons.

(b)   *Trans*-1,4-dimethylcyclohexane has two stereocenters, identified with (*). (This is an example of stereocenters that are not asymmetric atoms.)

trans-1,4-dimethylcyclohexane

(c)   *Trans*-1,4-dimethylcyclohexane is achiral because its mirror image is congruent.

(d)   The two chair conformations of *trans*-1,4-dimethylcyclohexane are diastereomers. (They are stereoisomers, but not mirror images; therefore, they are conformational diastereomers.

7.13   (a)   First, draw the two chair conformations of *trans*-1,4-dimethylcyclohexane:

two 1,3-diaxial interactions

two 1,3-diaxial interactions

B

trans-1,4-dimethylcyclohexane

Conformation *A* has no 1,3-diaxial methyl-hydrogen interactions. Conformation *B* has four—two for each methyl group. At 3.7 kJ mol^{-1} (0.9 kcal mol^{-1}) each, these interactions make conformation *B* 14.8 kJ mol^{-1} (3.54 kcal mol^{-1}) less stable than conformation *A*.

(b)   The more stable conformations of the two 1,4-dimethylcyclohexanes:

trans-1,4-dimethylcyclohexane   **cis-1,4-dimethylcyclohexane**

The cis isomer has the same number of 1,4 methyl–hydrogen interactions—two—as the axial conformation of methylcyclohexane itself, and thus has a destabilizing contribution of 7.4 kJ mol^{-1} (1.8 kcal mol^{-1}). There are no destabilizing interactions in the trans isomer.

7.14   (a)   The larger *tert*-butyl group assumes the equatorial position. (Structure *A* below)

(b)   The most stable conformation is the one that has the greater number of groups in the equatorial position. (Structure *B* below)

(c)   The larger isopropyl group assumes the equatorial position. The presence of oxygen does not alter the fact that the ring exists in a chair conformation. (Structure *C* below)

7.15   (a)   If *trans*-1,3-dibromocyclobutane were planar, the carbon–bromine bonds would be directed in opposite directions; in this case, their bond dipoles would cancel, and the net dipole moment of the molecule would be zero. In puckered *trans*-1,3-dibromocyclobutane, the bond dipoles do not cancel.

planar *trans*-1,3-dibromocyclobutane: puckered *trans*-1,3-dibromocyclobutane:
C—Br bond dipoles cancel   C—Br bond dipoles do not cancel

(b)   Cyclobutane undergoes an interconversion of puckered forms analogous to the chair interconversion of cyclohexane. This interchanges axial and equatorial groups. Therefore, one conformation of *trans*-1,2-dimethylcyclobutane—the more stable conformation—has diequatorial substituents, and the other—the less stable conformation—has diaxial substituents. The diaxial conformation is destabilized by two 1,3-diaxial interactions.

more stable conformation

trans-1,2-dimethylcyclobutane

7.16 (a) *Cis*-1,2-dimethylcyclopropane is achiral; it is a meso compound.

(b) *Trans*-1,2-dimethylcyclopropane is chiral.

cis-1,2-dimethylcyclopropane

trans-1,2-dimethylcyclopropane
noncongruent mirror images

7.17 (a) Bicyclo[2.2.2]octane, a bridged bicyclic compound
(b) Bicyclo[3.2.0]heptane, a fused bicyclic compound

7.18 A fused bicyclic compound, by definition, has a "zero-carbon bridge" and so must have a zero in the number associated with the name. Consequently, compound *B* is a fused bicyclic compound; compound *A* is a bridged bicyclic compound.

7.19 The two indicated carbons in the diagram at right of *cis*-decalin are axial substituents in the other ring; consequently, the hydrogens on these carbons are involved in 1,3-diaxial interactions (double-headed arrows in the following structure.). There are, as in other axially-substituted alkylcyclohexanes, two 1,3-diaxial interactions per axial carbon, except that one is common to both carbons; hence, there are a total of three 1,3-diaxial interactions. Because there are no 1,3-diaxial interactions in *trans*-decalin, it is more stable than *cis*-decalin by $3 \times 3.7 = 11.1$ kJ mol^{-1} (2.65 kcal mol^{-1}).

7.20 (a) The model of *cis*-bicyclo[3.1.0]hexane is much more easily built. The smaller is a ring, the closer is the dihedral angle of the trans bonds to 180°. Forcing this dihedral angle to much smaller values introduces strain within the ring. Such a large dihedral angle causes the ends of trans bonds to be too far apart to be readily bridged by only one carbon.

cis-bicyclo[3.1.0]hexane

trans bonds

this distance is too large
to be bridged by only one carbon

 Be sure to build the models called for in the two parts of this problem. Models are much more revealing, and provide you with much greater insight, than simply reading verbal descriptions.

(b) The model of *trans*-bicyclo[5.3.0]decane is easier to build. The larger is a ring, the easier it is to compress the dihedral angle of the trans bonds at the ring junction without introducing significant strain in the ring. This angle must be very close to 0° in order to accommodate a fused cyclopropane, that is, to bridge the ends of trans bonds with only one carbon. (In this Newman projection, the bridgehead hydrogens are labeled and shown as H*.)

a dihedral angle close to 0°
is required for the cyclopropane ring

bridgehead
hydrogen

strain in the larger ring
decreases with
increasing ring size

bridgehead
hydrogen

7.21   (a)   Compound *B* is less stable because it has a bridgehead double bond within a small ring. Consequently, it would have the greater (more positive or less negative) heat of formation.

(b)   Although both molecules have bridgehead double bonds, the double bond in compound *B* is more twisted, and a model of this molecule is more difficult to build. Consequently, compound *B* is less stable and therefore would have the greater (more positive or less negative) heat of formation. (The double bond in *A* is trans within a 10-membered ring, which is large enough to accommodate a trans double bond.)

7.22   (a)   Because a nail and a hammer are achiral objects, all interactions of these objects with any chiral object (such as a hand) and its enantiomer are identical. Hence, assuming equal strength, a right and a left hand should be equally adept at driving a nail. (This assumes, of course, that the person driving the nail doesn't accidentally bend it into some chiral shape.) In contrast, when it drives a screw, a chiral hand interacts with the helical (and chiral) threads of the screw. For this reason, the right and left hands of the same person should differ in their ability to drive a screw.

(b)   Premeds could take out their stethoscopes and determine which side of each person the heart is on. Or, you could offer to shake hands. Mr. L would extend what he calls his right hand, but to us it would be his left. Or, you could ask them to smell a spearmint leaf and describe the odor. (*R*)-Carvone, the active principle of spearmint, has a spearmint odor, but its enantiomer (*S*)-carvone smells like caraway (the odor of rye bread). (*R*)-Carvone would smell like caraway to Mr. L.

(*R*)-(–)-carvone

In any case, the point is that the two mirror images are distinguished by comparing a chiral reference element (our bodies, our right hands, or the odors or tastes of enantiomers) with the corresponding elements of the two people.

7.23   (a)   Because enantiomers react with achiral compounds at the same rate, enantiomers are formed in equal amounts.

enantiomers; formed at the same rate

(b)   Enantiomers react at different rates with chiral reagents, so the products are diastereomers formed in unequal amounts.

diastereomers; formed at the different rates

7.24 Three stereoisomers of 2,3-dibromobutane—(2*S*,3*S*), (2*R*,3*R*), and meso— could form in bromine addition to *cis*-2-butene. The enantiomers, (2*S*,3*S*) and (2*R*,3*R*), must be formed in identical amounts (that is, as a racemic mixture) because the starting materials are achiral. The meso diastereomer is formed in a different amount from the racemate.

2,3-dibromobutane

7.25 Hydroboration-oxidation of *trans*-2-butene gives racemic 2-butanol—that is, two enantiomers formed in equal amounts.

1) BH_3, THF
2) H_2O_2, ^-OH

racemic 2-butanol

7.26 Write all possible diastereomers resulting from addition to one enantiomer of the starting material. These are in principle formed in different amounts. (Problem 7.28 considers this case further.)

A   *B*   *C*   *D*

Each of these structures has an enantiomer; drawing these gives four additional diastereomers, which are also formed in different amounts. (Diastereomers are *always* formed in different amounts in principle.)

A′   *B′*   *C′*   *D′*

Each point in the process that gives *A* (that is, each reactive intermediate, each transition state, etc.) is enantiomeric to the corresponding point in the process that gives *A′*; therefore the two processes must occur with identical rates. Because the starting alkene is racemic, the same amounts of the enantiomeric alkenes are involved in each process; therefore, the enantiomers of each pair are formed in identical amounts. Thus, *A* and *A′* are formed in identical amounts; *B* and *B′* are formed in identical amounts; and so on.

7.27 Because the bromonium-ion mechanism involves anti addition, bromine addition to cyclohexene must give *trans*-1,2-dibromocyclohexane as a product.

Br_2

(1*R*,2*R*)-1,2-dibromocyclohexane

Reaction of the bromide ion at the other carbon of the bromonium ion gives the enantiomer of *trans*-1,2-dibromocyclohexane, and it is formed in equal amounts.

(1S,2S)-1,2-dibromocyclohexane

7.28   (Refer to the structures in the solution to Problem 7.26.) Because the bromines of the product are trans, products *B* and *C*, and their respective enantiomers *B′* and *C′*, are the major products. Compounds *B* and *B′* are formed in identical amounts, because they are enantiomers, and compounds *C* and *C′* are formed in identical amounts for the same reason. Compounds *B* and *C*, as well as compounds *B′* and *C′*, are formed in different amounts, because they are diastereomers.

7.29   (a)   Because hydroboration-oxidation is a net syn addition, the H and OH are added to the same face of the alkene. This addition can occur in two different (and enantiomeric) ways to give enantiomeric products, which are formed in identical amounts. For ease in seeing the mode of addition, the alkene is shown in a plane perpendicular to the page, and each product, resulting from addition to the top face of the alkene, is shown in an eclipsed conformation. (If the addition occurs from the bottom face of the alkene, we get the same two products.)

(2S,3S)   (2R,3R)

(b)   It is always true that, for a stereospecific reaction, use of a diastereomeric starting material gives the diastereomeric set of enantiomeric products, which are formed in identical amounts. As in part (a), the alkene is shown in a plane perpendicular to the page, and each product is shown in an eclipsed conformation. One product results from addition to the top face of the alkene and the other from addition to the bottom face.

(2S,3R)   (2R,3S)

7.30   Alkenes *A* and *B* give a single compound; these products are achiral and have no stereocenters. Alkene *C* gives two diastereomers. (Each compound is a mixture of two enantiomers, which are formed in equal amounts because the overall reaction is not stereospecific.)

product from *A*   product from *B*   products from *C*
(each one is racemic)

Alkene *D* gives two constitutional isomers, each of which is a mixture of two diastereomers.

constitutional isomer #1
(a mixture of diastereomers)

constitutional isomer #2
(a mixture of diastereomers)

7.31   The cis isomer of the starting alkene was used. This is shown by starting with the cis isomer and imagining the syn addition of the two OH groups to the same face of the double bond.

addition to a symmetrical
cis alkene

a meso stereoisomer

$R— = CH_3(CH_2)_7—$

 By a similar analysis, you can show that syn addition to the trans isomer of the alkene gives the racemate.

Solutions to Skills Objectives Problems

7.32   (a)   (b)   (c)

7.33   (a)

120°

enantiomeric conformations

The conformations of both (b) and (c) are conformational diastereomers, because the two conformations in each case have different numbers of axial and equatorial substituents; therefore, they can't be enantiomers or identical.

(b)   (c)

7.34   (a)   (b)   (c)

7.35   (a)   (b)

7.36   A simple way to approach this, problem is to work from planar projections of each compound, because planar projections remove the complication of the chair interconversion. If the planar projections of the two compounds are mirror images, then the two are actual enantiomers. If the two planar projections are identical, then the two conformations must be conformations of the same molecule. In that case, we have to look at the chair interconversion to see the relationships of the two conformations.

(a)   The planar projections are identical. Although drawing them from the chair conformations shows them as mirror images, as shown at right, you can demonstrate that they are congruent (and therefore the same compound) by rotating either projection 180° about an axis perpendicular to the page. However, the *chair conformations are not congruent.* That they are enantiomers can be demonstrated by rotating either one 180° about a vertical axis to see the mirror-image relationship more clearly:

conformational enantiomers

(b)   The planar projections are actual enantiomers, as shown below.

enantiomers

As to the relationship of the two conformations, the first has 1,4-diequatorial and 2,3-diaxial methyl groups, whereas the other has 1,4-diaxial and 2,3-diequatorial methyl groups. Therefore they can be neither enantiomers or identical. Because they are stereoisomers, they must be diastereomers. In other words, the two conformations are diastereomeric conformations of enantiomers.

(c) The first chair conformation is the same as in part (a). The corresponding planar projections are identical once we rotate either one about a horizontal axis as shown:

The two chair conformations are identical. We can demonstrate this by rotating the second one 180° about an axis perpendicular to the page (a z axis) as shown:

7.37 The process is to draw the mirror image of the planar projection and determine whether it is congruent. Any structure for which an enantiomer can be drawn can be separated into enantiomers at room temperature.

Therefore, compounds A, C, and E can be separated into enantiomers at room temperature.

7.38 Solve the problem by drawing the planar projection of each compound and seeing whether it is congruent to its mirror image.

(a) *Trans*-1,2-dimethylcyclohexane is chiral and it can be isolated in optically active form. (It undergoes the chair interconversion to give a diastereomeric conformation.)

(b) 1,1-Dimethylcyclohexane is not chiral and therefore cannot be isolated in optically active form.

(c) *Cis*-1-Ethyl-4-methylcyclohexane is not chiral and therefore cannot be isolated in optically active form.

(d) *Cis*-1-Ethyl-3-methylcyclohexane is chiral and can be isolated in optically active form.

7.39 (a) Notice that compound X, *trans*-1,2-diethylcyclohexane, is chiral, and it has the S configuration at both asymmetric carbons. Conformations A, B, and D also have the S configuration at all asymmetric carbons and are possible boat conformations of X. Boat conformation C is derived from the *enantiomer* of X, and therefore is not a possible boat conformation. Boat conformation E has *cis* ethyl groups and therefore is achiral and is not a possible boat conformation of X. (Can you show how conformation D can be formed from X?)

(b)   Of the conformations *A*, *B*, and *D*, conformation *A* has the lowest energy because the bonds to the ethyl groups are staggered with bonds to hydrogens on adjacent carbons. In conformation *B*, the carbon–ethyl bonds are eclipsed with carbon–hydrogen bonds, and in conformation *D*, the two ethyl groups are in more energetic axial positions.

7.40   (a)   Of the two chair conformations, the one with the *tert*-butyl group in the equatorial position has the lower energy.

lower energy

(b)   The conformation with the fewest axial methyl groups has lower energy.

Conformation *A* has four 1,3-diaxial methyl–H interactions (two for each axial methyl group) for an energy contribution of 4×3.7 kJ mol^{-1}, and one *gauche*-butane methyl–methyl interaction (3.7 kJ mol^{-1}) for a total of 18.5 kJ mol^{-1} (4.43 kcal mol^{-1}). Conformation *B* has one *gauche*-butane methyl–methyl interaction (3.7 kJ mol^{-1}) and two 1,3-diaxial methyl–H interactions (2×3.7 kJ mol^{-1}) for a total energy contribution of 11.1 kJ mol^{-1} (2.65 kcal mol^{-1}). The energy difference between the two conformations is therefore 7.4 kJ mol^{-1} (1.8 kcal mol^{-1}), with conformation *B* having the lower energy, as we concluded qualitatively above on the basis of the relative number of axial methyl groups.

7.41   (a)   From the equilibrium constant (2.07), we use $\Delta G° = -2.303RT\log K_{eq} = -5.71 \times \log (2.01) = -1.73$ kJ mol^{-1} (–0.41 kcal mol^{-1}). The conformation with the equatorial chlorine is more stable.

(b)   From $\log K_{eq} = -\Delta G°/2.303RT = -(-9.2/5.71) = 1.61$, from which we have $K_{eq} = 40.8$. In other words, there is about 41 times as much of the equatorial conformation of isopropylcyclohexane as there is of the axial conformation.

7.42   (a)   Compound *A*: none of the above (it is spirocyclic)
Compound *B*: none of the above, because the two rings do not share one or more common atoms.
Compound *C*: bridged bicyclic
Compound *D*: fused bicyclic

(b)   The name of compound *C* is bicyclo[3.2.2]nonane. The name of compound *D* is bicyclo[5.3.0]decane.

7.43   Compound *C* violates Bredt's rule. Compound *B* does not because the double bond is "cis" within the fused ring and is not twisted.

7.44   (a)   Compounds *A* and *B* have cis ring junctions, whereas compound *C* has a trans ring junction.

A B C

(b)   To draw an envelope cyclopentane, draw a chair cyclohexane, clip off one of the carbon, and join the resulting ends:

Here are the chair/envelope conformations requested:

A B C

In B we have to use wedges if we want to convey the point of the envelope explicitly. (The wedges can be omitted when drawing such structures ordinarily.) In C, the bonds to the H's at the ring junction are parallel; they are distorted slightly so they can both be seen.

7.45   (a)   (1)   The products:

$$CH_3CH_2\overset{CH_3}{\underset{R}{CH}}-\overset{CH_3}{\underset{R}{CH}}-CH_2OH \quad + \quad CH_3CH_2\overset{CH_3}{\underset{R}{CH}}-\overset{CH_3}{\underset{S}{CH}}-CH_2OH$$

(2)   The two products are diastereomers.
(3)   The two products are formed in different amounts.
(4)   The two products have different boiling points and melting points, and could therefore be separated (in principle) by fractional distillation or fractional crystallization.

(b)   (1)   The products:

$$CH_3CH_2CH_2\overset{Br}{\underset{CH_2CH_3}{C}}CH_3 \quad \text{mixture of } R \text{ and } S$$

(2)   The two products are enantiomers.
(3)   The two products are formed in identical amounts.
(4)   The two products have identical boiling points and identical melting points. An enantiomeric resolution would be required to separate the two compounds.

(c)   (1)   The products:

$$CH_3CH_2\overset{Br}{CH}CH_2Br \quad \text{mixture of } R \text{ and } S$$

(2)   The two products are enantiomers.
(3)   The two products are formed in identical amounts.

(4)   The two products have identical boiling points and identical melting points. An enantiomeric resolution would be required to separate the two compounds.

(d)   (1)   The starting alkene is racemic, and is therefore an equimolar mixture of two enantiomers. To predict the products, make the prediction for each enantiomer separately and then combine the results.

$$
\underset{A}{\overset{Ph}{\underset{S}{|}}}\quad
$$

Ph
| *S*
CH₃CHCHCH₂Br + CH₃CHCHCH₂Br + CH₃CHCHCH₂Br + CH₃CHCHCH₂Br
S | *S* | *R* | *R* |
Br   Br   Br   Br
        A   *B*   *C*   *D*

        ‾‾‾‾‾‾‾‾‾‾‾‾‾‾‾‾‾‾‾‾‾‾‾‾   ‾‾‾‾‾‾‾‾‾‾‾‾‾‾‾‾‾‾‾‾‾‾‾‾
            from the 3*S* alkene   from the 3*R* alkene

(2)   Compounds *A* and *B*, which come from the *S* enantiomer of the starting material, are diastereomers, as are compounds *C* and *D*, which come from the *R* enantiomer of the starting material. Compounds *A* and *C* and compounds *B* and *D* are also diastereomers. Compounds *A* and *D*, as well as compounds *B* and *C*, are enantiomers.

(3)   The diastereomers are formed in different amounts; the enantiomers are formed in identical amounts.

(4)   Any pair of diastereomers have different melting points and boiling points; any pair of enantiomers have identical melting points and boiling points.

(e)   (1)   The product:

(2), (3), (4)   Only one product is formed.

7.46   (a)   The bromonium ion is symmetrical; we assume first that it forms on the lower face of the alkene. The oxygen of the water undergoes opposite-side substitution at a carbon of the bromonium ion at the upper face; because the two carbons are equivalent, reaction at either carbon is equally likely. We obtain the two structures on the left, which are identical. (Be sure to demonstrate this to yourself.)

alkene is drawn in
a plane perpendicular to the page
(see Display 1.54, Sec. 1.6)

        the same compound   the same compound

                        enantiomers

If the bromonium ion is formed at the upper face of the alkene, we obtain the two identical structures on the right, which are the enantiomers of the ones on the left. All structures are equally probable. Therefore, we obtain enantiomers, and they are formed in equal amounts.

(b)   In this case, the water reacts with the bromonium ion at the carbon bearing the ethyl group, because reaction of the nucleophile is preferred at the carbon with more alkyl substituents. Two enantiomers are formed in identical amounts.

7.47   (a)   Imagine both syn and anti additions to fumarate with the OD entering from the face of the alkene that leads to the product malate-3-*d* in which carbon-2 has the *S* configuration. A syn addition will lead to (2*S*,3*S*)-malate-3-*d*, whereas an anti addition will lead to the observed (2*S*,3*R*)-malate-3-*d*. Therefore, the reaction is an anti-addition.

> There are experimental techniques—for example, NMR spectroscopy, which is the subject of Chapter 14—that can make the subtle stereochemical distinctions required to solve stereochemical problems like this.

(b)   The use of D_2O allows us to differentiate the protons of the solvent from those of the starting material. The stereochemistry of the addition cannot be determined without this distinction.

7.48   Because it is given that the reaction is an anti addition, we carry out an internal rotation of the product to place the OH groups in an anti relationship. Fortunately, this relationship already exists in the product structure that is given; so, all we need to do is mentally reverse the addition to reveal the alkene and its stereochemistry:

7.49   (a)   Chlorine and cyanide both have the highest priority relative to methyl, hydrogen, and ethyl in the Cahn–Ingold–Prelog system. Therefore, we can use the relative configurations of the reactant and product to determine whether an inversion or retention of stereochemistry has occurred. The configuration of the reactant is *S* and the configuration of the product is *R*; therefore, the reaction has occurred with inversion of configuration.

(b)   The curved-arrow mechanism:

7.50   In the following, each of the chiral products is racemic.

(a)   (b)   (c)   (d)

(formed in different amounts)

(e)   (f)   (g)

Solutions to Integrated Problems

7.51   (a)

bicyclo[2.2.0]hexane

bicyclo[2.1.1]hexane
(two of several possibilities)

(b)

(*S*)-4-cyclobutylcyclohexene

7.52   Methylcyclohexane and (*E*)-4-methyl-2-hexene are *constitutional isomers*. Hence, they have the same molecular mass but different properties.

methylcyclohexane

(*E*)-4-methyl-2-hexene
(* = asymmetric carbon)

(a)   A molecular mass determination would not distinguish between the two compounds because their molecular masses are identical.

(b)   Uptake of H_2 over a catalyst would distinguish between the two compounds because the alkene would take up hydrogen in the presence of a catalyst, whereas the cycloalkane would not.

(c)   The reaction with Br_2 would distinguish between the two compounds because the alkene would decolorize Br_2 solution (because alkenes react with Br_2), whereas the alkane would not.

(d)   Constitutional isomers by definition have the same molecular formula, so determination of the molecular formula would not differentiate between the two compounds.

(e)   Because constitutional isomers have different physical properties, determination of the heat of formation would distinguish between the two compounds. (Which would have the greater heat of formation?)

(f)   Enantiomeric resolution would distinguish between the two because (*E*)-4-methyl-2-hexene is chiral, and can be resolved into enantiomers, whereas methylcyclohexane, an achiral compound, cannot be.

7.53   The two isomers of 1,3-cyclohexanediol are enantiomers. Distinguishing between enantiomers requires a technique that has a chiral attribute.

(a)   Boiling point does not distinguish between two enantiomers.

(b)   Optical rotation does have a chiral aspect, and therefore can distinguish between two enantiomers. (See Further Exploration 7.3.)

(c)   Because hexane is an achiral solvent, enantiomers have identical solubilities in hexane.

(d)   Density is not a chiral property, and therefore the two enantiomers have identical densities.

(e)   The solubilities of two enantiomers in principle differ in an enantiomerically pure chiral solvent.

(f)   Dipole moment is not a chiral property, and therefore the two enantiomers have identical dipole moments.

(g)   Because taste buds are chiral, they in principle differentiate between two enantiomers.

7.54   A rapid rotation can occur about the bond between the cyclohexane ring and the ethyl group so that the van der Waals radii of the methyl group and the axial hydrogens do not overlap.

In the conformations *B* and *B′*, the interactions of the axial hydrogens with the hydrogens of the ethyl group are about the same as the corresponding interactions in a methyl–hydrogen 1,3-diaxial interaction. As a result, the ethyl–hydrogen interaction has about the same energy cost as a methyl–hydrogen interaction.

  There is a little more to this situation than meets the eye. Because conformation *A* has relatively high energy, it is effectively "locked out" of the conformational equilibrium. This means that axial ethylcyclohexane has *two* rather than three conformations about the ring–ethyl bond. In other words, axial ethylcyclohexane has *reduced freedom of motion*. This conformational constraint would be expected to show up in its *entropy*, and, in fact, it does. Axial ethylcyclohexane has a lower $\Delta S°$ than axial methylcyclohexane; this lower entropy increases its standard free energy, because $\Delta G° = \Delta H° - T\Delta S°$. However, in equatorial ethylcyclohexane, there are two gauche methyl–CH_2 interactions with the ring that are not present in equatorial methylcyclohexane.

These interactions raise the enthalpy of the equatorial form of ethylcyclohexane relative to methylcyclohexane and raise its free energy. These two changes—decreased entropy in the axial conformation, and increased enthalpy in the equatorial conformation—almost exactly compensate.

7.55   The chair conformations of α-(+)-glucopyranose:

Conformation *A* is more stable and is therefore the predominant conformation in the equilibrium because all the substituents (except one) are equatorial.

7.56   (a)   An achiral tetramethylcyclohexane that undergoes the chair interconversion to give identical molecules:

(b)   An achiral trimethylcyclohexane that undergoes the chair interconversion to give conformational diastereomers:

(c)   A chiral trimethylcyclohexane that undergoes the chair interconversion to give conformational diastereomers:

(d)   A tetramethylcyclohexane that undergoes the chair interconversion to give conformational enantiomers:

conformational enantiomers

7.57 There are four stereoisomers—two enantiomers and two meso compounds. The planes of symmetry of the achiral stereoisomers are shown as dashed lines.

achiral   enantiomers   achiral

7.58 (a) The two chiral stereoisomers are the following enantiomeric pair:

(b) The only candidate structures are those in which the number of axial chlorines equals the number of equatorial chlorines. In addition, the two chlorines within each of the three pairs that have a 1,4-relationship must differ in whether their positions are axial or equatorial. Thus, each of the following stereoisomers has two identical chair conformations. (The numbers refer to corresponding carbons.)

7.59 (a) False. They are diastereomers, and they are not interconvertible without breaking bonds.

(b) False. As noted in (a), they are stereoisomers; they have the same connectivity.

(c) True, as noted in part (a).

(d) True, as noted in part (a).

(e) True, because switching bonds at either of the bridgehead carbons converts one stereoisomer into the other.

(f) True, because the bond switching mentioned in (e) would require breaking a chemical bond.

(g) False, because they are not mirror images (and also because they are diastereomers).

(h) False. They do not interconvert at all.

7.60 A conformational representation of the steroid:

7.61 In order for a compound to exist in a relatively high-energy twist-boat conformation, there must be a good reason for it to avoid the chair. Indeed, compounds *B* and *D* have very severe 1,3-diaxial interactions between a *tert*-butyl group and a methyl group in the chair conformations shown. Compound *D*, a *cis*-decalin derivative, can avoid its 1,3-diaxial interaction by undergoing a chair interconversion. However, compound *B*, a *trans*-decalin derivative, cannot undergo the chair interconversion (see Eq. 7.32, text p. 351, in Sec. 7.6B). Hence, the only way it can avoid the 1,3-diaxial interaction is for the ring containing the *tert*-butyl group to assume a twist-boat conformation. The correct answer, then, is *B*.

7.62 *Cis*-1,3-di-*tert*-butylcyclohexane can exist in a chair conformation (*A*) in which both *tert*-butyl groups are equatorial. However, in either chair conformation of *trans*-1,3-di-*tert*-butylcyclohexane (*B*), there is an axial *tert*-butyl group. The axial *tert*-butyl group can be avoided if compound *B* exists in a twist-boat conformation *C*. in which both *tert*-butyl groups are equatorial. (This conformation is shown as a boat for simplicity.) Evidently the twist-boat conformation is more stable than either chair conformation containing an axial *tert*-butyl group. The interaction of the methyl groups within the *tert*-butyl substituents with the ring hydrogens should be about the same in both conformations *A* and *C*, so that these cancel in the comparison.

7.63   Olean can exist as a pair of non-congruent mirror images, so it is a chiral compound. Additionally, it contains no symmetry elements. Although it may appear that there is a plane of symmetry from the way that it is presented in the problem, you must remember to think about the chair conformations three dimensionally (you may wish to build a model). The compound can also be presented as follows; a clear lack of symmetry is visible this way, and the two enantiomers can be seen as mirror images.

S-(+)-olean   *R*-(–)-olean

(a)   False. It is chiral, because there are no symmetry elements.

(b)   False, as noted in part (a).

(c)   False; although the compound is chiral, it does *not* contain an asymmetric carbon. (Every carbon is connected to two or more identical groups.)

(d)   True; the "spiro" carbon (the carbon common to both rings) is a stereocenter, because exchanging the attachment of any two groups would produce a stereoisomer (specifically, the enantiomer).

(e)   False; although the compound is chiral, it has a stereocenter as noted in part (d).

 Experiments have shown that male fruit flies are only attracted to the (*R*)-(–)-enantiomer and ignore the (*S*)-(+)-enantiomer. The females respond to (*S*)–(+)-enantiomer. This is another example of the effect of chirality in natural systems. The absence of an asymmetric carbon in olean begs the question: How are the *R* and *S* configurations assigned? In a case like this, there is no way to assign group priorities on the basis of atomic number or molecular mass. Special conventions have been developed for cases like this.

7.64   (a)   True. This compound is achiral, as a plane of symmetry exists. Its mirror image is congruent. This compound also contains a stereocenter—two actually. Each of the "spiro" carbons (the carbons common to two rings) are stereocenters (indicated by asterisks in the structure at right), because switching two groups at either one would produce a stereoisomer, specifically a diastereomer of the one shown.

(b)   False; as noted in part (a), the compound contains stereocenters.

(c)   False; the compound is achiral and it contains no asymmetric atoms.

(d)   False; the compound is achiral.

(e)   False; the compound is achiral.

7.65   Within each structure, count the 1,3-diaxial interactions between the methyl group and axial ring hydrogens four carbons away (counting the methyl as one carbon). These are shown with double-headed arrows in the structures of *A* and *B* below. (There are no methyl-hydrogen 1,3-diaxial interactions in compound *C*.)

A   *B*

Each interaction shown adds 3.7 kJ mol⁻¹ to the heat of formation relative to that of *C*. Consequently, *C* has the lowest heat of formation; the heat of formation of *B* is $(2 \times 3.7) = 7.4$ kJ mol⁻¹ (1.8 kcal mol⁻¹) greater; and the heat of formation of *A* is $(3 \times 3.7) = 11.1$ kJ mol⁻¹ (2.7 kcal mol⁻¹) greater than that of *C*.

7.66   (a)   Anti-addition can occur to give two chiral diastereomers, which will be formed in different amounts. Each is a single enantiomer because the starting material is a single enantiomer.

(3R,5R)-3,5-dimethylcyclopentene

chiral diastereomers

(b)   The bromonium ion derived from the reaction of cyclopentene with bromine undergoes opposite-side substitution by water to give the *trans*-bromohydrin, which is chiral. Because the starting materials are achiral, the chiral product is obtained as a racemate.

enantiomers

7.67   The two dibromides are the two diastereomeric *trans*-disubstituted *trans*-decalin derivatives:

Remember that *trans*-decalin derivatives cannot undergo the chair interconversion, so the axial and equatorial bromines cannot interchange.

7.68   Given that addition of bromine at each double bond is anti, two diastereomers of 1,2,4,5-tetrabromocyclohexane can be formed. One is the achiral meso compound, and the other is the racemate.

meso   racemate

Assuming you had samples of the two compounds but didn't know which was which, what experiment could you do that might identify the two compounds? (*Answer:* Carry out an enantiomeric resolution, perhaps by chiral chromatography. The compound that can be separated into optically active components—its two enantiomers—is identified as the racemate.) It turns out that the meso compound is the stereoisomer with the higher melting point. As we show in Chapter 8, of two isomers, the more symmetrical one is typically the one with the higher melting point.

7.69   Compound *A* is *not* consistent with the data because it is meso and cannot be optically active. Compound *B* is consistent with the data because it is chiral and would hydrogenate to give *trans*-1,4-dimethylcyclohexane, an achiral product. Compound *C* is consistent with the data because it is chiral and would hydrogenate to give a mixture of *cis*- and *trans*-1,4-dimethylcyclohexane, both of which are achiral products. Compound *D* is *not* consistent with the data, because, although it is chiral and thus optically active, it would hydrogenate to give a chiral and optically active product, (1S,3S)-1,3-dimethylcyclohexane. Compound *E* is *not* consistent with the data because it is not chiral and therefore could not be optically active.

7.70   Alcohol *A* can be prepared in good yield from 3-hexene (either cis or trans) by either (a) hydroboration–oxidation or (b) oxymercuration–reduction because the alkene is symmetrical.

3-hexene (cis or trans)   *A*

Alcohol *B* cannot be prepared by ether (a) or (b). It has the correct regiochemistry for preparation by hydroboration–oxidation, but the stereochemistry of such an addition would have to be anti to give this alcohol, whereas hydroboration–oxidation is a syn addition.

Alcohol *C* can be prepared by reaction (b). It has the correct regiochemistry, and stereochemistry is not an issue in this case.

1-methylcyclohexene   *C* (racemate)

Alcohol *D* has the correct regiochemistry and stereochemistry to be prepared in a syn addition by reaction (a), hydroboration–oxidation.

1-methylcyclohexene   *D* (racemate)

7.71   If the chirality of structures (a), (b), and (d) is not clear, draw the mirror image in each case and test it for congruence. This process is essential for cases of this nature. (Stereocenters are indicated with a ◊.)

(a) chiral   (b) chiral   (c) achiral (a meso compound)   (d) chiral

7.72   Because this is an amine, both the chair interconversion and amine inversion can occur. The amine inversion interchanges the axial and equatorial positions of the electron pair and the hydrogen on the nitrogen *without* interconverting the chair conformations of the ring. Thus, four chair conformations are in rapid equilibrium:

7.73   (a)   The only stereocenter in the molecule is the nitrogen, which rapidly undergoes inversion, a process that rapidly interconverts enantiomers. Consequently, this compound cannot be resolved into enantiomers at room temperature.

(b)   Because nitrogen inversion interconverts enantiomers, the compound cannot be resolved into enantiomers.

(c)   The bicyclic structure prevents the nitrogens from inverting, and both nitrogens are stereocenters. Consequently, the molecule is chiral and can be resolved into enantiomers.

(d)   The asymmetric ring carbon causes the molecule to be chiral; inversion of the nitrogen interconverts diastereomers. Hence, the compound can be resolved into two enantiomeric sets of rapidly interconverting diastereomers. (This case is much like the one in Problem 6.22, Sec. 6.9, p. 309.)

7.74   In the transition state for inversion of both amines, the nitrogen is sp²-hybridized. The C—N—C angle required by this hybridization is 120°. This angle is much more difficult to achieve within a three-membered ring, which has an optimal C—N—C angle of 60°, than it is in within a five-membered ring, which has an optimal C—N—C angle of 109°. Hence, there is much more angle strain in the transition state for inversion of 1-methylaziridine than there is in the transition state for inversion of 1-methylpyrrolidine. For this reason, the transition state for inversion of the aziridine derivative has a very high energy, and nitrogen inversion in this derivative is very slow. (This is a very unusual situation, since inversion of most amines is very rapid; see text Sec. 6.9B.)

ring angle ~ 60°; sp² hybridization requires 120°; therefore this transition state contains a large amount of angle strain

ring angle ~ 109°; sp² hybridization requires 120°; therefore this transition state contains less angle strain

sp²-hybridized transition states for amine inversion

7.75   (a)   In the chair conformation, for every C—O bond dipole in a given direction, there is another C—O bond dipole of the same magnitude pointing in the opposite direction. Thus, in the diagram at right, the gray dipoles cancel each other, and the black dipoles cancel each other. Because pairs of dipoles cancel, the overall dipole moment is zero.

chair conformation of 1,4-dioxane

(b)   To the extent that the twist-boat conformation is present, it will contribute a nonzero dipole moment, because the C—O bond dipoles do not cancel in this conformation. (We use the simpler boat conformation to illustrate this idea.)

each dipole is the
resultant of two
C—O bond dipoles

resultant dipole vector
for the molecule

boat conformation of 1,4-dioxane

7.76   (a)   Conformation A contains four 1,3-methyl–hydrogen diaxial interactions; the energy cost of these is $(4 \times 3.7) = 14.8$ kJ mol^{-1}. In addition, conformation A contains a *gauche*-butane interaction for an additional 3.7 kJ mol^{-1}, for a total energy cost of $3.7 + 14.8 = 18.5$ kJ mol^{-1}. Conformation B contains two methyl–hydrogen 1,3-diaxial interactions, at an energy cost of $(2 \times 3.7) = 7.4$ kJ mol^{-1}; a *gauche*-butane interaction at 3.7 kJ mol^{-1}; and the methyl–methyl 1,3-diaxial interaction, at an energy cost to be determined. Let the energy cost of a methyl–hydrogen 1,3-diaxial interaction be $\Delta G°(\text{Me–H}) = 3.7$ kJ mol^{-1}, that of a methyl–hydrogen *gauche*-butane interaction be $\Delta G°(\text{gauche}) = 3.7$ kJ mol^{-1}, and let the energy cost of a methyl–methyl 1,3-diaxial interaction be $\Delta G°(\text{Me–Me})$. It is given that the $\Delta G°$ of the overall equilibrium, $\Delta G°(\text{eq})$, equals 8.4 kJ mol^{-1} and that conformation A has lower energy. Then

$$\Delta G°(\text{eq}) = \Delta G°(B) - \Delta G°(A) = 8.4 = 7.4 + 3.7 + \Delta G°\,(\text{Me–Me}) - 18.5$$

with all quantities in kJ mol^{-1}. Solving for $\Delta G°\,(\text{Me–Me})$,

$$\Delta G°\,(\text{Me–Me}) = 15.8 \text{ kJ mol}^{-1}\,(3.8 \text{ kcal mol}^{-1})$$

This calculation shows that a methyl–methyl 1,3-diaxial interaction is more than four times as costly in energy terms as a methyl–hydrogen 1,3-diaxial interaction. This is consistent with the greater size of a methyl group relative to a hydrogen.

(b)   Conformation C has no 1,3-diaxial interactions, whereas conformation D has one methyl–methyl 1,3-diaxial interaction, which, from part (a), costs 15.8 kJ mol^{-1}, and two methyl–hydrogen 1,3-diaxial interactions, which cost $(2 \times 3.7) = 7.4$ kJ mol^{-1}. Thus, conformation D is $(15.8 + 7.4) = 23.2$ kJ mol^{-1} less stable than conformation C. The standard free energy change for the equilibrium is therefore $+23.2$ kJ mol^{-1} (5.54 kcal mol^{-1}).

 In order to measure an equilibrium constant, measurable amounts of both components must be present in the equilibrium. For the interconversion of C and D, the equilibrium constant can't be measured directly because too little of compound D is present. This problem illustrates the "counterpoise" method for making estimates of very unfavorable equilibrium constants. In part (a), the methyl–methyl diaxial interaction on the right is counterbalanced by the two axial methyl groups on the left. This strategy raises the energy of the left side of the reaction and thus leads to measurable amounts of both components. The "cost" of the counterbalance can then be subtracted because its effect is known from the conformational analysis of methylcyclohexane.

7.77   (a)   The approach to this problem is similar to that used in the solution to Problem 7.76. The $\Delta G°$ for this equilibrium, $\Delta G°(AB)$, balances two phenyl–hydrogen 1,3-diaxial interactions, on the right, against two methyl–hydrogen 1,3-diaxial interactions, on the left. Since $\Delta G° > 0$, the equilibrium is unfavorable; evidently a Ph–H diaxial interaction has a greater energy cost than a CH$_3$–H diaxial interaction. This deduction is consistent with a larger size for the Ph group.

(b)   From the analysis in part (a),

$$\Delta G°(AB) = 4.73 = 2\Delta G°(\text{Ph–H}) - 2\Delta G°(\text{CH}_3\text{–H}) = 2\Delta G°(\text{Ph–H}) - 7.4$$

Solving,

$$\Delta G°(\text{Ph–H}) = (4.73 + 7.4)/2 = 6.1 \text{ kJ mol}^{-1}$$

In the chemical equation for this part of the problem, two Ph–H diaxial interactions are balanced against an equatorial Ph. As we have just calculated, the energy cost, and therefore $\Delta G°$ for the equilibrium in this part of the problem is 2×6.1 kJ mol^{-1} = 12.2 kJ mol^{-1}.

7.78   (a)

cis-2-butene   borolane

(b)   The products in part (a) are enantiomers. Since they were formed from achiral reagents, they are formed in equal amounts.

(c)

cis-2-butene   (2R,5R)-2,5-dimethyl-borolane

(d)   The products in part (c) are diastereomers. Diastereomeric transition states with different energies lead to the products, so they were formed at different rates, and thus, in different amounts.

(e)   The organoborane can approach the π bond of cis-2-butene molecule in two distinguishable ways. Approach from one face gives a transition state in which there is a severe van der Waals repulsion between a methyl group of the alkene and a methyl group on the organoborane (structure *A*). Approach from the opposite face gives a transition state in which the van der Waals repulsion is between a *hydrogen* of the alkene and a methyl group of the borane (structure *B*).

greater van der Waals repulsion

A

smaller van der Waals repulsion

B

Given that transition state *B* has less severe van der Waals repulsions, the product borane derived from that transition state is formed more rapidly than the product from transition state *A*.

(f)   When allowed to react with basic H_2O_2, it gives the corresponding alcohol, which has the *S* configuration.

H_2O_2, ^-OH

product from transition state *B*   (*S*)-2-butanol

7.79   (a)   The two diastereomeric products result from addition at the upper (*U*) and (*L*) lower faces of the double bond.

(b)   The methyl group shields the upper face (in steroid terminology, the β face) of the molecule like an umbrella. Therefore, product *L* is the predominant product.

the steric effect of the methyl group forces the reaction to the opposite (lower) face of the double bond

7.80   (a)   The reaction is an anti addition, because the two added groups (CH_3O and F) are trans in the product.
          (b)   We assume that the anti stereochemistry is accounted for by formation of a cyclic ion, which is opened by opposite-side substitution at the carbon with more alkyl substituents. We assume that the CH_3O group adds first because the fluorine is present at the more substituted carbon.

oxonium ion intermediate

(In this equation, the oxonium ion is formed at the upper face of the double bond and the fluoride ion reacts at the opposite (lower) face. Formation of the oxonium ion at the lower face of the double bond is equally probable; reaction of fluoride ion at the upper face would give the enantiomer.)

7.81   Hydration involves a carbocation intermediate. Because a carbocation can react with a nucleophile at either face of its vacant 2p orbital, a mixture of stereoisomeric hydration products *A* and *B* is expected.

from top   *from bottom*

A   *B*

We chose arbitrarily to have the D_3O^+ react at the upper face of the alkene doble bond. It can react with equal probability at the lower face. In that case we would get the enantiomers of the two products shown. Because both modes of addition are equally probable, all of the products shown are racemates—as they must be, because all of the reactants are achiral.

7.82   Because the product contains one Br, it appears that a reaction with the double bond—most reasonably, an addition reaction—has taken place. The product X has the same unsaturation number as the starting material. If the double bond has disappeared, then there must be an additional ring. This deduction, in turn, suggests that a nucleophile other than bromide ion has reacted with the bromonium ion to form a ring, and the only available nucleophile is the —OH group. The following two possibilities for X are consistent with this deduction, in which the bromonium ion is formed at the less crowded upper face of the double bond, and the nucleophilic oxygen is then set up to react at the opposite (lower) side.

A   *B*

If you got to either of these possibilities, you've made a solid analysis. Product A is the correct one. The mechanism starts with the bromonium ion intermediate, formed as shown in Eq. 5.10a, text p. 240, Sec. 5.2A. The OH group reacts as a nucleophile with this ion:

A

Product *B* would be formed by a similar mechanism, with the nucleophilic reaction of the OH group occurring at the other carbon of the bromonium ion.

Product *A* is formed for two reasons. First, as shown in Eq. 5.14 and Fig. 5.1, text pp. 242–243 in Sec. 5.2B, nucleophiles react at the tertiary carbon of a bromonium ion, because it has a high degree of carbocation character. Second, the nucleophilic reaction has to take place so that the nucleophilic oxygen approaches the carbon from the side opposite to the leaving bromine. This creates, in the *transition state*, a 90° O—C—C angle within the six-membered ring, which results in considerable strain. (This point can be verified with a model.)

 We might ask why bromide ion does not react with the bromonium ion to complete the addition of bromine. It is often found that intramolecular reactions that form 5- and 6-membered rings are thousands of times faster than competing intermolecular reactions. This topic is covered in Chapter 12.

7.83 Amide resonance is not important for compound *B* because the resonance structure involves a bridgehead double bond, which violates Bredt's rule. (See Eq. 7.37a–c, text Sec. 7.6C.)

B   *violates Bredt's rule*

7.84 (a) These compounds are noncongruent mirror images and are therefore enantiomers.

(b) We know that in most common cases (for example, *cis*- and *trans*-2-butene), *E,Z* isomers are diastereomers. However, all that is needed to apply the *E,Z* nomenclature is a difference in priorities between the two groups attached to *each* carbon of a double bond. Using compound *A* as an example, the issue, then, is whether the two groups beginning with the carbons *(a)* and *(b)* have different priorities. There is no differences in *connectivity* if we begin at carbons *(a)* and *(b)* and move outwards along the path of highest priority within each group. However, there is a difference at the carbons themselves, and that difference is in their *stereochemical configurations.* Because we did not cover this possibility in Chapter 4, you would have been justified in making up your own priority rule; but Cahn, Ingold and Prelog decided arbitrarily that an *R* configuration receives the higher priority. (See the solution to Problem 6.54d and e, which involves the same issue.) Because carbon *(b)* has the *R* configuration, and it is on the side of the double bond opposite to the higher-priority phenyl (Ph) group, then alkene *A* is the *E* isomer. It follows that alkene *B* is the *Z* isomer. This, then, is an amusing and unusual situation in which *E,Z* isomers are actually enantiomers.

7.85 (a) Two rings in a bridged bicyclic compound can be joined in three stereochemically different ways which might be termed *(out,out)*, *(in,out)*, and *(in,in)*. Assuming hydrogens at both bridgeheads, these ways can be represented schematically as shown at right. (The large dots are the bridgehead carbons.)

out, out   *in, out*   *in, in*

(b)   As these simple diagrams suggest, the *in, out* and *in, in* patterns require that one and two hydrogens, respectively, occupy the region of space *within* a ring. This is sterically impossible for small rings. For small rings, this would actually require such a distortion of the tetrahedral structure that all four bonds to the bridgehead carbon with the *in* configuration would lie on one side of a plane through this carbon. (If the impossibility of achieving this sort of structure is not clear, try to build a model of (*in, out*)-bicyclo[2.2.2]octane.) The larger the rings, the larger the space within the ring, and for very large rings, the "inner" hydrogens can be accommodated easily within the ring and the tetrahedral bonding geometry of the bridgehead carbons can be achieved without significant strain. Hence, the (*in, out*) and (*in, in*) stereoisomers of the 77-carbon bicyclic alkane are stable.

8

Nomenclature and Noncovalent Intermolecular Interactions

CHAPTER 8 SUMMARY

- Alcohols, ethers, alkyl halides, thiols, and sulfides are classified as methyl, primary, secondary, or tertiary according to the number of substituent groups at the alpha (α) carbon.

- Organic compounds are named by both common and substitutive nomenclature. In substitutive nomenclature, the name is based on the principal group and the principal chain. The principal group, specified by priority, is cited as a suffix in the name. Other groups are cited as substituents. Hydroxy (—OH) and mercapto (—SH) groups can be cited as principal groups with the suffixes *ol* and *thiol*, respectively; hydroxy has a higher priority for citation as a principal group. Halogens, alkoxy (—OR) groups, and alkylthio (—SR) groups are always cited as substituents.

- Boiling points depend on the noncovalent attractions between molecules. The types of noncovalent forces that affect boiling point are (1) van der Waals attractions, which include induced dipole–induced dipole attractions, (2) dipole–dipole attractions; and (3) hydrogen bonding. All of these noncovalent attractions are much weaker than covalent bonds.

- The general increase of boiling points (roughly 20–30 °C per carbon) within many homologous series is due to the increasing areas of the molecular surfaces at which molecules can interact by van der Waals attractions.

- Large amounts of branching in hydrocarbon groups reduces their surface area relative to unbranched groups of the same molecular mass; therefore, the boiling points of highly branched molecules tends to be lower than that of their unbranched isomers containing the same functional groups.

- Van der Waals attractions are strongest when interacting molecules are polarizable—that is, when their electron clouds are easily deformed by interactions with external charges. Less polarizable molecules, such as perfluorocarbons, have lower boiling points than more polarizable molecules, such as hydrocarbons, of the same molecular mass and shape.

- Solubilities are a consequence of the noncovalent intermolecular interactions of unlike molecules (solute and solvent) weighed against the noncovalent intermolecular interactions of pure solute molecules and pure solvent molecules. The solubility of a solute in a solvent is a consequence of the free energy of solution, ΔG_s, which is the free energy of the solution itself minus the free energies of pure solvent and pure solute.

- The free energy of solution ΔG_s is composed of two parts: the free energy of interaction, $\Delta G_{interaction}$ and the free energy of mixing, ΔG_{mixing} (Eq. 8.54a). The free energy of mixing originates solely from the entropy of mixing, ΔS_{mixing} by the equation $\Delta G_{mixing} = -T\Delta S_{mixing}$ (Eq. 8.52). The ΔG_{mixing} is always favorable, just as the mixing of blue balls and red balls in a box is favorable; that is, mixing occurs spontaneously.

- Solvents are classified as protic or aprotic, depending on their ability to donate hydrogen bonds; polar or apolar, depending on the magnitude of its dielectric constant; and donor or nondonor, depending on its ability to act as a Lewis base. These categories can and do overlap.

- When the interactions between molecules in the pure substances are similar to those between the solvent and solute in the solution, then the ΔG_s is dominated by the entropy of mixing, and a solution forms. This situation is frequently observed when solute and solvent have the same classification (for example, both protic, or both apolar aprotic). This is the basis of the like-dissolves-like rule of solubility.

■ The tendency of hydrocarbon groups in aqueous solution to associate is called *hydrophobic bonding*. Hydrophobic bonding is driven by the entropy increase of water when the water molecules that solvate hydrocarbon chains (lower-entropy water) are released into solvent water (higher-entropy water). The association of phospholipid molecules to form lipid bilayers and cell membranes is an example of hydrophobic bonding.

■ Within many homologous series, a graph of melting points against the number of carbons in the alkyl chain shows a sawtooth pattern. Highly symmetrical compounds typically have significantly higher melting points than their less symmetrical isomers. Both effects are caused by (a) the greater entropy associated with a crystal containing symmetrical molecules; and (b) by the more efficient crystal packing and, as a consequence, greater van der Waals attractions, in the crystals of the higher-melting compounds.

■ The solubility of a solid is affected by its melting point. Solids with high melting points are typically less soluble than isomers containing the same functional groups with lower melting points. Structural symmetry, because it raises melting point, tends to reduce solubility.

■ The high dielectric constant of a polar solvent contributes to the solubility of an ionic compound by reducing the attractions between oppositely charged ions, as shown by the electrostatic law (Eq. 8.55, Sec. 8.6B). Ionic solubility is also enhanced by donor interactions, charge–dipole attractions, and hydrogen bonding of an ion with the solvent molecules in its solvation shell.

■ Because of the effective solvation of ions, molecules with limited water solubility typically become more soluble when they can be ionized.

■ Crown ethers, cryptands, and other ionophores form complexes with cations by creating artificial solvation shells for them in which the ions are stabilized by ion–dipole and donor interactions. Ionophore antibiotics and ion channels bind cations by the same principle.

 # FURTHER EXPLORATIONS

 ## 8.1   Trouton's Rule

The purpose of this exploration is to understand why the boiling point is a good measure of intermolecular interactions in the liquid state. The basis of this discussion is *Trouton's rule,* which came about from the observations by Frederick Trouton of Trinity College, Dublin, reported in 1884, that the boiling points of different liquids are related to their heats of vaporization in a simple way.

First, let's consider the equilibrium between a pure liquid and its vapor. As in any equilibrium, the free energy difference between the final state (in this case, a vapor) and the initial state (in this case, a liquid) is described by the familiar equation

$$\Delta G = \Delta H - T\Delta S \tag{SG8.1}$$

Let the boiling temperature be T_b (in Kelvin). At this temperature, liquid and vapor are in equilibrium, and therefore $\Delta G = 0$. We can then re-write Eq. SG8.1 as

$$T_b = \frac{\Delta H^{\bullet}_{vap}}{\Delta S^{\bullet}_{vap}} \tag{SG8.2}$$

where the black-dot superscripts (•) mean that these are molar quantities.

At the boiling point, $\Delta H^{\bullet}_{vap}$ is the "latent heat of vaporization," the energy required to convert a liquid to its vapor at its boiling point. If we assume that there are *no* interactions between molecules in the gas phase (that is, the gas is ideal), then $\Delta H^{\bullet}_{vap}$ is a *direct measure* of the energies of the noncovalent interactions in the liquid.

The $\Delta S_{vap}^{\bullet}$ is the entropy difference between gaseous and liquid states. We'll explore the details of entropy later in the chapter, but a simple interpretation in this context is that it measures *freedom of motion*. Our image of a liquid is that the molecules are moving randomly but are packed together rather closely and have a limited degree of motional freedom. (In other words, a molecule can't move very far without colliding with another molecule—or several other molecules.) In a gas, molecules are much farther apart and have a much greater freedom of motion. Therefore, we expect (and find) that $\Delta S_{vap}^{\bullet} > 0$. What is remarkable is that $\Delta S_{vap}^{\bullet}$ is actually *nearly the same* —about +85 J mol^{-1} K^{-1} —for a large number of organic liquids. If this is so, then Eq. SG8.2 becomes

$$T_b \sim \frac{\Delta H_{vap}^{\bullet}}{85}$$

(SG8.3)

This is the "modern version" of Trouton's rule. Because $\Delta H_{vap}^{\bullet}$ is a direct measure of the energies of the noncovalent interactions in the liquid, this proves the assertion that the boiling point (in K) is a quantitative measure of these interactions. Notice that Trouton's rule justifies not only the qualitative notion that we can think about boiling points in terms of intermolecular interactions in the liquid, but also gives us a way to estimate the magnitude of these interactions.

As we might expect, there are some liquids that don't follow Trouton's rule. Water is one. The $\Delta S_{vap}^{\bullet}$ for water is +109.1 J mol^{-1} K^{-1}. This is because water has a "more ordered" (lower-entropy) liquid state than (for example) hydrocarbons. This more ordered liquid state is caused by hydrogen bonding, which constrains the motion of water molecules in the liquid. So, when water is converted into a gas, in which such constraints are gone, more entropy is gained. However, accompanying this larger entropy gain is a larger $\Delta H_{vap}^{\bullet}$. Because of the hydrogen bonds in the liquid, $\Delta H_{vap}^{\bullet}$ is also unusually large because the hydrogen bonds have to be broken to form the gas. In other words, $\Delta H_{vap}^{\bullet}$ and $\Delta S_{vap}^{\bullet}$ tend to compensate in cases like this. Thus, using the boiling point as a crude measure of intermolecular interactions is justified even for liquids that do not follow Trouton's rule exactly.

SOLUTIONS TO PROBLEMS

Solutions to Focused Problems

8.1   (a)   Isobutyl fluoride is a primary alkyl halide.
        (b)   Hexyl iodide is a primary alkyl halide.
        (c)   Cyclopentyl bromide is a secondary alkyl halide.
        (d)   Neopentyl chloride is a primary alkyl halide. Although there is branching in the molecule, it does not occur at the α-carbon.

8.2   (a)

2,2-dichloro-5-methylhexane

   (b)

chlorocyclopropane

   (c)

6-bromo-1-chloro-3-methylcyclohexene

   (d)   I—CH₂—I

methylene iodide

   (e)

1-bromo-1-pentyne

8.3   (a)   3-Bromo-3-chloro-1-methylcyclopropene

(b)   (*Z*)-3-Chloro-2-pentene

(c)   3-Bromo-1,1,1-trichloro-2-fluorobutane

(d)   Chloroform ($HCCl_3$) is the traditional name for trichloromethane.

(e)   Neopentyl bromide [$(CH_3)_3C$—CH_2—Br] is the common name for 1-bromo-2,2-dimethylpropane.

(f)   *trans*-1,3-Dibromocyclobutane

(g)   1-Chloro-3-isopropyl-1-methylcyclohexane

(h)   4-Chloro-2-pentyne

8.4   (a)   (b)   (c)   (d)   (e)

8.5   (a)   1-Butanol

(b)   (*R*)-3-Bromo-1-butanol

(c)   (2*S*,3*E*)-1-Chloro-3-methyl-3-penten-2-ol

(d)   (1*S*,5*R*)-2-Chloro-5-methyl-2-cyclopentenethiol. The 2-refers to the position of the double bond; the position of the —SH group is at the 1-position because it is the principal group; the number 1 is not written because a single principal group on a cyclic system must be at position 1. The stereochemical configurations (*R*, *S*) come first in the name.

(e)   (2*R*,3*R*,4*S*)-3-Butyl-5-hexene-2,4-diol

(f)   2,5-Cyclohexadienol

(g)   1-Mercapto-2-pentanol

(h)   2-Methyl-2-propanethiol

(i)   (*S*)-3-Hexyn-2-thiol

8.6   (a)   (b)   (c)

$CH_3CH_2OCH_2CH_2CH_3$ $(CH_3)_3C$—S—$CH(CH_3)_2$

ethyl propyl ether   ***tert*-butyl isopropyl sulfide**

dicyclohexyl ether

(d)   (e)

H_2C=$CHCH_2$—O—CH_2—⟨⟩   ⟨⟩—O—CH=CH_2

allyl benzyl ether   **phenyl vinyl ether**

(f)   (g)

(2*R*,3*R*)-2,3-dimethyloxirane

5-(ethylthio)-2-methylheptane

8.7   (a)   2-Methoxy-2-methylpropane (commonly called MTBE, for methyl tert-butyl ether)

(b)   2-Ethoxyethanol

(c)   1-(Isobutylthio)-2-methylpropane
(This would typically be known by its common name, diisobutylsulfide.)

(d)   (*E*)-5-Methoxy-3-penten-1-ol

8.8   (a)   The structure of the compound and the correct name is shown at right. The ring is numbered to give the substituent the lower number.

"3-butyl-1,4-dioxane" (incorrect name)
2-butyl-1,4-dioxane (correct name)

(b)   The structure of 2-butoxyethanol is HO—CH$_2$CH$_2$—OCH$_2$CH$_2$CH$_2$CH$_3$.

8.9   In Table 8.1, the bond lengths in a given period are about 4 pm greater in the column containing oxygen and sulfur than in the column containing the halogens. Assuming that this trend holds true for CH$_3$SeH, then the predicted carbon–selenium bond length in this compound is about 198 pm. The prediction for (CH$_3$)$_2$Se can be refined by noting in Fig. 8.1, text Sec.8.3 (p. 395), that the carbon–sulfur bond length in dimethyl sulfide is about 2 pm smaller than that in methanethiol. Assuming a similar trend in the corresponding selenium derivatives, the predicted bond length for (CH$_3$)$_2$Se is 196 pm.

 The point of this problem is that more often than not, molecular properties vary in a regular way throughout the periodic table, and that reasonably good predictions can be made within one group from trends in a nearby group.

8.10   Because of the molecular geometry of *trans*-1,2-dichloroethylene, the C—Cl bonds are oriented in opposite directions, as are the C—H bonds. Consequently, their bond dipoles cancel, and the molecular dipole moment of *trans*-1,2-dichloroethylene is zero. (The cancellation of the C—Cl bond dipoles is shown in the diagram below; the much weaker C—H bond dipoles also cancel for the same reason.) The corresponding bond dipoles of *cis*-1,2-dichloroethylene do not cancel; consequently it has a significant dipole moment. The interaction of the molecular dipole moments of *cis*-1,2-dichloroethylene molecules provides a cohesive force in the liquid state that is not present in the trans stereoisomer (Sec. 8.5B, p. 402.) Because such cohesive forces enhance boiling point, *cis*-1,2-dichloroethylene has the higher boiling point of 60.3°.

Cl   H
  \   / ⟋
   C = C   C—Cl bond dipole vectors
  /   \   cancel; therefore the molecular
H   Cl   dipole moment is zero.

trans-1,2-dichloroethylene
(bp = 47.4 °C)

Cl   Cl
  \   /   the vector sum of the two C—Cl
   C = C   bond dipoles is not zero.
  /   \
H   H   C—Cl bond dipole

cis-1,2-dichloroethylene
(bp = 60.3 °C)

8.11   As in the relationship between pentane and neopentane, shown on text p. 399, octane and 2,2,3,3-tetramethylbutane have the same molecular weights but very different boiling points. Like neopentane, 2,2,3,3-tetramethylbutane is branched, more spherical, and thus has a smaller surface area and weaker van der Waals interactions than octane, which is linear. The energy required to release octane molecules into the gas phase is larger, so octane boils at the higher temperature, 126 °C.

8.12   (a)   Acetaldehyde has a higher boiling point (21 °C) than propene (–47 °C) because there are stronger dipole-dipole interactions between acetaldehyde molecules resulting from the significant C=O bond dipole.

(b)   The curve for boiling points of aldehydes should show up somewhere between the curves for alkanes/ethers/alkyl chlorides and nitriles on Fig. 8.2b. Aldehydes have dipole moments greater than alkanes (and alkenes, which as stated in the problem are about the same as alkanes) so the curve should be higher.

(c)   Because alkyl iodides (iodoalkanes) have much lower boiling points than alkanes of about the same molecular mass, and because alkenes and alkanes with the same branching pattern and molecular mass have about the same boiling points, we would expect the curve for alkyl iodides to fall well below that for alkenes and alkanes—that is, well to the right of the alkyl bromide curve.

8.13 (a) Recall that a bromine contributes about the same molecular mass (80 units) as a pentyl group (73 units), and that alkyl bromides have boiling points *lower* than alkanes of about the same molecular mass; then 2-bromopropane should have a boiling point that is lower than that of an eight-carbon alkane. Because the two alkanes listed have more than eight carbons, their boiling points are correspondingly higher. Because the two alkanes have the same type of branching, their boiling points should be in the same order as their molecular masses. These considerations suggest that the order of increasing boiling points is 2-bromo-propane < 4-ethylheptane < 4-ethyloctane. (The actual boiling points are 59°, 141°, and 168°, respectively.)

(b) A chlorine contributes about the same molecular mass (35 units) as an ethyl group (29 units), and chloroalkanes have about the same boiling points as alkanes of the same molecular mass. Therefore, chloromethane has about the same boiling point as propane, which has a lower boiling point than the five-carbon alkene 1-pentene. The alcohol has the highest boiling point because it has about the same molecular mass as 1-pentene, but can donate and accept hydrogen bonds. Consequently, it has the highest boiling point of all. The order of increasing boiling points is, therefore, chloromethane (–42°) < 1-pentene (30°) < 1-butanol (118°).

8.14 (a) HBr is primarily a hydrogen-bond donor because it is a strong acid; a neutral —Br is a poor acceptor.

(b) The hydrogen in hydrogen fluoride is an excellent hydrogen-bond donor, and the fluorine is an excellent hydrogen-bond acceptor.

(c) Acetone is a hydrogen-bond acceptor; the oxygen can accept hydrogen bonds at its unshared electron pairs.

accepts hydrogen bonds

acetone

(d) *N*-Methylacetamide can serve as both a hydrogen-bond donor and a hydrogen-bond acceptor. The nonbonding electron pair on nitrogen is not basic because of its resonance interaction with the carbonyl group. It is not a very good hydrogen-bond acceptor.

the oxygen can accept hydrogen bonds

N-methylacetamide

the nitrogen can donate its hydrogen to a hydrogen-bond acceptor

(e) Phenol is both a hydrogen-bond donor and a hydrogen-bond acceptor. The O—H can participate in hydrogen-bond donation, and the oxygen can accept hydrogen bonds at its nonbonding electron pairs.

the oxygen can accept hydrogen bonds

phenol

the oxygen can donate its hydrogen to a hydrogen-bond acceptor

(f)   The ethylammonium ion can donate its N—H hydrogens to hydrogen bonds, but it cannot accept hydrogen bonds because it has no nonbonding electron pairs.

$$CH_3CH_2-\overset{\overset{\displaystyle H}{|}}{\underset{\underset{\displaystyle H}{|}}{N^+}}-H----$$ the nitrogen can donate its hydrogens to hydrogen bonds

ethylammonium ion

8.15   The two carboxylic acids are compounds *B* and *C*, and since the problem states that they are solids at room temperature, the two highest melting points belong to them. The greater symmetry of compound *C* molecules allows them to pack the most tightly together in the solid phase and thus possess the higher melting point (168–172 °C). Compound *B*'s melting point is then 143–147 °C. The same analysis can be applied to the two alkenes: compound *A* is more symmetrical, so it has the higher melting point of the two (–74.5 °C); compound *D* melts at –175 °C.

8.16   (a)   First, convert the melting points in °C to kelvins by adding 273.2 °C. The melting point of benzene is 278.7 K and that of toluene is 178.1 K The ratio of melting temperatures is 1.56, whereas the ratio of ΔH_m values is 0.86. In other words, the difference in ΔH_m values is in the wrong direction to account for the difference in melting points. Therefore, ΔS_m must play a significant role in the relative melting points of the two substances.

(b)   From Eq. 8.44b, knowing T_m and ΔH_m for both compounds, we can calculate ΔS_m. This gives ΔS_m for benzene as 20.6 J mol^{-1} K^{-1} and ΔS_m for toluene as 37.3 J mol^{-1} K^{-1}. Now we calculate the ΔS_m difference based on symmetry alone. For this calculation we use Eq. 8.46a, and for which we need the symmetry numbers of benzene and toluene. The symmetry number of benzene is 12. This is derived from the original plus five indistinguishable structures related by successive rotations of 60° about axis *a*, and six indistinguishable structures related by 180° rotations about both axes *b* through *g*.

axis *a*
perpendicular to the page   axes *b*, *c*, and *d*   axes *e*, *f*, and *g*

The symmetry number of toluene is 2. The entropy difference due to symmetry alone between benzene and toluene is

$$\Delta S = 2.3R \log\left(\frac{\sigma_{benzene}}{\sigma_{toluene}}\right) = 2.3R \log(12/2) = 2.3R \log(6) = 14.9 \text{ J mol}^{-1} \text{ K}^{-1}$$

If we add 14.9 J mol^{-1} K^{-1} to 20.6 J mol^{-1} K^{-1} (the ΔS_m for benzene), we obtain 35.5 J mol^{-1} K^{-1}, which is very close to the ΔS_m for toluene. In other words, the difference in the melting points of benzene and toluene is due to a relatively small difference in ΔH_m and a large difference in ΔS_m *almost completely accounted for by the difference in symmetry.*

8.17   (a)   An outcome in which two are heads and two are tails has greater entropy (more possible arrangements) than when one is heads and three are tails. There are six 'two heads two tails' arrangements: HHTT, TTHH, HTHT, THTH, HTTH, THHT. There are only four 'one heads three tails' arrangements: HTTT, THTT, TTHT, TTTH.

(b)   Both of these arrangements have the same number of possible outcomes, so they have equal entropy.

(c)   Use Eq. 8.51 on text p. 415 to solve for the entropy in each situation. The 1 mole situation is solved in the text: +41.9 J K^{-1}. When there are 5 moles of solute, the ΔS_{mixing} is

$$\Delta S_{mixing} = -2.3R(n_1 \log x_1 + n_2 \log x_2) = -2.3R\left[(5.00)\log\left(\frac{5}{60.6}\right) + (55.6)\log\left(\frac{55.6}{60.6}\right)\right]$$

$$= -19.2[-5.40 - 2.08] = +143 \text{ J } K^{-1} mol^{-1}$$

In other words, the ΔS_{mixing} increases as expected for a distribution involving more molecules; there are more ways for the molecules to be dispersed within the mixture.

8.18   First we calculate the number of moles of solvent (hexane) in 1 L. Because the density is 0.659 g mL^{-1}, 1 L of hexane is 659 g and is 659 g/114 g mol^{-1} = 5.78 mol. There are 6.78 mol of both compounds together. The mole fraction of the solute (isotopic hexane) is 1/6.78 = 0.147, and the mole fraction of the solvent (ordinary hexane) is 5.78/6.78 = 0.853. The free energy of mixing is equal to $-T\Delta S_{mixing} = -(298)\Delta S_{mixing}$ (from Eq. 8.52). We then apply Eq. 8.51 to calculate ΔS_{mixing}:

$$\Delta S_{mixing} = -2.3R(n_1 \log x_1 + n_2 \log x_2) = -2.3R[(1.00)\log(0.147) + (5.78)\log(0.853)] = -2.3R(-1.23) \text{ J } K^{-1} mol^{-1}$$

$$= +2.3R(1.23) \text{ J } K^{-1} mol^{-1}$$

With $\Delta G_{mixing} = -T\Delta S_{mixing}$, and with $2.3RT = 5.71$ kJ mol^{-1}, we have

$$\Delta G_{mixing} = -T\Delta S_{mixing} = -2.3RT(1.23) = -5.71(1.23) = -7.02 \text{ kJ } mol^{-1} \ (-1.68 \text{ kcal } mol^{-1})$$

8.19   (a)   2-Methoxyethanol is a polar, protic, donor solvent.

(b)   2,2,4-Trimethylpentane (a major component of gasoline) is an nonpolar, aprotic, nondonor solvent.

(c)   2-Butanone is a polar, aprotic, donor solvent.

8.20   The lower layer is dichloromethane because it has a higher density than water. The long hydrocarbon chain of 1-octanol would both reduce its solubility in water and promote its solubility in dichloromethane. The data on text p. 421 show that 1-hexanol has a very low solubility in water. The solubility of 1-octanol would be much lower still. The 1-octanol would be found primarily in the dichloromethane layer.

8.21   Water can accept a hydrogen bond from the hydrogen of the N—H bond, and can donate a hydrogen bond to the oxygen of the C=O bond. This hydrogen bonding tends to solubilize acetanilide in water. In contrast, the CH_3 group and the phenyl ring cannot form hydrogen bonds with water; consequently, these nonpolar ("greasy") groups, especially the phenyl group, tend to make acetanilide insoluble.

8.22   The lower molecular weight alcohols are the most "water-like"—they can form hydrogen bonds with water molecules and have shorter hydrophobic chains—as evidenced by the miscibility data on text p. 421. Once the chain becomes longer (greater than four carbons), the miscibility decreases with increasing chain length. This decrease in miscibility is associated with a decrease in their entropy of solution. As the carbon chains get longer, the water molecules need to form a larger cavity around the individual molecules, increasing "order" and decreasing entropy, which is unfavorable energetically.

8.23   Terephthalic acid (the 1,4 isomer) has the higher melting point because of its greater symmetry. Because a higher melting point decreases solubility, terephthalic acid is predicted to have the lower solubility. Although both compounds have very small water solubility, the solubility of isophthalic acid is in fact more than 25 times that of terephthalic acid in water, as predicted.

8.24   (a)   Dimethyl sulfoxide has an ion–dipole interaction and a donor interaction with a sodium ion.

(b)   Water can donate hydrogen bonds to the oxygen of DMSO and can have a favorable dipole–dipole interaction with the sulfur.

(c)   Dimethyl sulfoxide also has an ion–dipole interaction with a chloride ion. (In both parts (a) and (c), multiple DMSO molecules cluster around the ions; only one is shown here.)

(a)   (b)   (c)

8.25   (a)   Triethylamine is protonated by the strongly acidic HCl solution to form triethylammonium chloride, an ionic compound that is soluble in water.

triethylamine   **triethylammonium chloride**
(insoluble in water)   (an ionic salt, soluble in water)

(b)   The interactions of trimethylammonium chloride with solvent water:

(c)   Adding a base to the solution, such as sodium hydroxide (NaOH), would deprotonate the triethyl-ammonium ion, converting it back into insoluble triethylamine. Sodium chloride, another water-soluble salt, would be formed in the process.

triethylammonium chloride   **triethylamine**
(an ionic salt, soluble in water)   (insoluble in water)

8.26   (a)   One or both of the $C_{17}H_{35}$ groups could be substituted by any long, unbranched alkyl group with an odd number of carbons; this group can also contain one or more cis double bonds.

phosphatidylserine

(b)   As in part (a), one or both of the $C_{17}H_{35}$ groups could be substituted by any long, unbranched alkyl group with an odd number of carbons; this group can also contain one or more cis double bonds.

phosphatidylcholine
(a lecithin)

8.27   The three hydrogens of the methylammonium ion are perfectly positioned to interact with three of the oxygens of the crown ether by hydrogen bonding. The other three oxygens of the crown ether can stabilize the complex by ion–dipole interactions. In the structure at right, we are viewing the nitrogen as in a Newman projection down the N—C bond, with the methyl group hidden from view.)

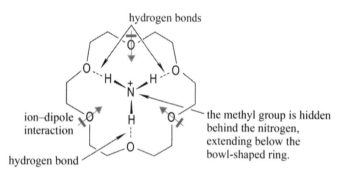

8.28   (a)   (1)   The attraction between ions is by definition an electrostatic attraction.

(2)   The dispersion force (van der Waals attraction) is an attraction between temporary charges on one molecule with temporary charges on another. Although time-dependent, they are nevertheless electrostatic attractions.

(3)   Hydrogen bonding can be viewed as an attraction between the partial positive charge on the hydrogen and the partial negative charge of an acceptor atom, as shown in Display 8.38 on text p. 405.

(4)   A dipole-dipole attraction results from the electrostatic attraction of the positive end of one dipole with the negative end of another, so it is also an electrostatic attraction.

(5)   A charge–dipole attraction is an attraction between a charge and another opposite charge of a dipole—again, fundamentally an electrostatic attraction.

(b)   A dispersion force fluctuates with time. A hydrogen bond probably also has a weak covalent component, and it is an attraction between partial charges. The attractions involving dipoles also involve a partial positive charge in each dipole that is connected to a partial negative charge at the other end of the same dipole.

Solutions to Skills Objectives Problems

8.29 (a) The most important contributor to bond length is the row (period) in the periodic table of the contributing atoms. On this basis, *E* and *F* are candidates for the longest bond. Structure *E* involves a bond from Br to an sp³-hybridized carbon, whereas *F* involves a bond from Br to an sp²-hybridized carbon. Bond length decreases in the order sp³ > sp² > sp. Therefore structure *E* has the longest bond. The bond in *A* is the shortest, because it is an sp²–sp² single bond.

(b) Methanethiol (*A*) should have the smallest bond angle. Bond angles at sulfur are typically just over 90°; the C—S—C angle of dimethyl sulfide should be a little larger than the C—S—H angle of methanethiol to avoid van der Waals (steric) repulsions between the methyl groups. This prediction is verified by the data in Fig. 8.1, Sec. 8.3, text p. 395.

8.30 Compound *A* is a primary alcohol. (Don't be confused by the branching at the β-carbon.) Compound *B* is a tertiary alcohol. Compound *C* contains both primary and secondary alcohols and is also a tertiary alkyl halide. Compound *D* is a secondary alkyl halide.

8.31 (a) Butyl bromide (b) Cyclohexyl alcohol
(c) Methyl phenyl sulfide, or Methyl phenyl thioether

8.32 (a)

isopropyl methyl sulfide

(b)

methyl vinyl ether

(c)

di-*tert*-butyl ether

(d)

allyl fluoride

8.33 (a) The substitutive name of halothane is 2-bromo-2-chloro-1,1,1-trifluoroethane.
(b) The substitutive name of methoxyflurane is 2,2-dichloro-1,1-difluoro-1-methoxyethane.

8.34 The systematic names are (a) (*E*)-2-butene-1-thiol and (b) 3-methyl-1-butanethiol.

8.35 The OH groups are the principal groups, so these dictate the principal chain and its numbering. The name is 6-mercapto-1-methoxy-3-vinyl-2,4-heptanediol.

8.36 (a)

(S)-2-methyl-3-buten-1-ol

(b) $HS—CH_2—C{\equiv}C—CH_2—OH$

4-mercapto-2-butyn-1-ol

8.37 (a) 3-Bromofuran (b) 2-Ethyl-2-isopropyloxirane

8.38 (a)

2-methylthiophene

(b)

meso-2,5-dimethyloxolane

(c)

2-(3-furyl)-1-ethanol

8.39 (a) The order of boiling points is *tert*-butyl alcohol < 2-pentanol < 1-hexanol. (The actual boiling points are 82°, 119°, 158°.) This follows the order of molecular masses. The increased branching of 2-pentanol relative to 1-hexanol, and *tert*-butyl alcohol relative to 2-pentanol, makes the differences between these boiling points even greater than they would be for unbranched alcohols of the same molecular masses.

(b)   The order of boiling points is 1-hexene < 1-chloropentane < 1-hexanol. (The actual boiling points are 63°, 108°, 158°.) Use the fact that the mass of a chlorine (35 units) is about the same as that of an ethyl group (29 units), and that alkyl chlorides have about the same boiling points as alkanes (and alkenes) of the same molecular mass. Consequently, 1-chloropentane has about the same boiling point as 1-heptene, which, in turn, has a higher boiling point than 1-hexene. 1-Hexanol has about the same molecular mass as heptane, and its boiling point is considerably higher because of hydrogen bonding.

(c)   The order of boiling points is propane < diethyl ether < 2-propanethiol < 1,2-propanediol. (The actual boiling points in °C are –42°, 37°, 57–60°, 189°.) Diethyl ether has a higher boiling point than propane because diethyl ether is more polar and because it has a greater molecular mass. 2-Propanethiol (isopropyl mercaptan) has a higher boiling point because it has an even higher molecular mass. 1,2-Propanediol has the highest boiling point because of hydrogen bonding.

(d)   The order of boiling points is cyclobutane < chlorocyclobutane < cyclooctane. Use the facts that the chlorine has about the same mass as an ethyl group, and that alkyl chlorides and alkanes of the same molecular mass have about the same boiling points. Then order the boiling points by molecular mass. (The actual boiling points are 12°, 83°, and 151°, respectively.)

8.40   (a)   The order of boiling points is neopentane < pentane < hexane < octane. The last three follow the trend of increasing boiling point with molecular mass for unbranched alkanes. Neopentane is highly branched and very symmetrical. It therefore has a smaller surface area than pentane and therefore has weaker dispersion interaction between molecules in the liquid state.

(b)   Sodium chloride has the higher melting point because the interionic forces—the ionic bonds—are much stronger than the intermolecular noncovalent interactions in the solid state.

(c)   The *E* isomer has a zero dipole moment because corresponding bond dipoles are oriented in opposite directions. The *Z* isomer has a nonzero dipole moment because the bond dipoles add vectorially. The intermolecular interaction between dipole moments accounts for the higher boiling point of the *Z* isomer.

(Z)-3,4-dichloro-3-hexene   **(E)-3,4-dichloro-3-hexene**
bp = 167–168 °C   bp = 142 °C

(d)   The order of solubility is 1,2-dichloroethane < 3-octanol < 1-butanol. Hydrogen bonding with the solvent water is one of the most important determinants of water solubility, but long alkyl chains (as in 3-octanol) provide significant alkane character which discourages solubility. Alkyl halides are very insoluble in water.

8.41   Acceptor sites are labeled "**A**" and donor sites are labeled "**D**." The nonbonding electron pairs on nitrogen in amides *C* and *E* are not very basic (because they are delocalized into the carbonyl group by resonance) and therefore they aren't classified as acceptors.

8.42 (a) Compound *B* is more symmetrical and therefore it has the higher melting point of 62–65 °C.

(b) Compound *A*, with a symmetry number of 4, is more symmetrical than compound *B*, with a symmetry number of 1. Therefore compound *A* has the higher melting point of 61–62 °C.

(c) We can determine melting points (in kelvins) by dividing ΔH_m by ΔS_m. (Remember to convert the units of ΔS_m to kJ K^{-1} mol^{-1}.) We get the melting point in °C by subtracting 273.16 from the melting point in kelvins. This gives, for the melting point of *A*, 308.8 K or +35.6 °C, which is very close to the observed 35.4 °C measurement. The same procedure for *B* gives the melting point of *B* as 238.6 K or –34.6 °C, which is very close to the –34.5 °C measurement. Notice also the similarities in the relative ΔS_m values of the hydrocarbons pentane and neopentane (Display 8.48 of the text); the difference here has a similar explanation. We expect pivalic acid, with its more compact structure and smaller surface area, to have a lower boiling point than valeric acid, just as neopentane has a lower boiling point than pentane.

8.43 (a) 2,2,2-Trifluoroethanol is polar because it has a dielectric constant $\varepsilon > 15$; this compound is a donor because it has nonbonding electron pairs on oxygen; and the compound is protic because it has a hydrogen bonded to oxygen that can be donated in a hydrogen bond.. *It is a polar, protic, donor solvent.*

(b) 1-Hexene is nonpolar because of its low dielectric constant; it has neither nonbonding electron pairs nor hydrogens capable of hydrogen bonding. *It is a nonpolar, nondonor, aprotic solvent.*

(c) 3-Pentanone is a *polar, aprotic, donor* solvent. (The nonbonding electron pairs on oxygen have donor capability.)

(d) Tetrahydrofuran is an aprotic, apolar, donor solvent.

(e) A solution containing 50 volume % of tetrahydrofuran (THF) and 50 volume % of water, if the two components have equal densities, has many more moles of water than THF. We are not given the densities, but assuming that THF has a lower density than water (as most ethers do), the number of moles of THF is further reduced relative to the number of moles of water. Assuming *equal* numbers of moles of THF and water, and assuming that the dielectric constant is proportional to the number of moles of each, we can calculate that the *minimum* effective dielectric constant should be about (0.5)(7.6) + (0.5)(78) ~ 43. Because the solvent is more than 50 mole percent water, the effective dielectric constant should be greater than this. This calculation places this mixed solvent clearly in the *protic* and *polar* category. Both THF and water have nonbonding electron pairs on their oxygens, so the mixed solvent is also a *donor* solvent.

8.44 (a) Even though retinol contains an OH group, the dominant character of retinol is that of a hydrocarbon. Therefore retinol would be more soluble in hexane than in water.

A *rough* rule of thumb is that water solubility typically requires at least one OH group for every five carbon atoms if there are no other features (such as hydrogen-bond acceptor sites) that promote water solubility.

(b) 4-Ethoxy-1-butanol has both a hydrogen-bond donor site and two hydrogen-bond acceptor sites and only modest hydrocarbon character. In these respects it is much more like ethanol than hexane, and should be substantially more soluble in ethanol.

8.45 Protic, polar, and donor solvents are best at dissolving ions, and formic acid is the only solvent that has all three characteristics.

8.46 (a) For isomers of closely related structure, such as *A* and *B,* we can rely on relative melting points as a guide to solubility; and we can rely on symmetry as a guide to relative melting points. The more symmetric compound *B* should have a higher melting point and should be less soluble in water.

(b) Melting point and symmetry are good guides to solubility, not only in water, but in *all* solvents. Although we expect both *A* and *B* to be considerably more soluble in acetonitrile than they are in water, nevertheless compound *A* should be more soluble than compound *B* (and it is).

(c) You might be tempted to say that compound *B*, with the considerably higher melting point (because it is ionic) should be less soluble. But remember that melting point and solubility correlate only for isomers that differ in symmetry. These two compounds are not isomers, and the effect of ionization on solubility is so profound that compound *B* is much more soluble than compound *A*.

(d) The higher-melting and more symmetrical pivalic acid is less soluble in water than valeric acid.

8.47 (a) Ethanol and 1-propanol each contains a "hydrophilic end" (the OH group) and a "hydrophobic end" (the alkyl chain). The bifunctionality of these compounds allows them to intercalate into cell membranes which are similarly composed of molecules with hydrophilic and hydrophobic ends. However, since the alcohols' alkyl chains are much shorter, they disrupt the packing of the lipids and change the properties of the bilayer, leading to disruption (at high concentrations).

(b) Sugars, despite having six carbon atoms, are surrounded by hydrophilic hydroxy groups. They cannot penetrate the highly "greasy" (hydrophobic) inner layer of a lipid bilayer. Sugars require special receptors to traverse the cell membrane.

(c) Sodium dodecyl sulfate (hypervalent structure at right) has a long un-branched hydrocarbon group and a charged, polar group at one end of the molecule. It resembles a phospholipid and, in fact, it can have significant noncovalent interactions with one or more phospholipid molecules, thereby disrupting a cell membrane. (In Chapter 20 we show that SDS and similar compounds form *micelles*, which can in fact incorporate phospholipids [and other greasy molecules] within their structures.) This is the basis of detergent action.

$$CH_3(CH_2)_{10}CH_2O\overset{\displaystyle O}{\underset{\displaystyle O}{\overset{\|}{\underset{\|}{S}}}}O^- \ Na^+$$

sodium dodecyl sulfate (SDS)

(d) The structure of propofol is largely hydrocarbon-like, which is why it is insoluble in water even though it contains a polar OH group,. However, it is evidently soluble in compounds with a high degree of hydrophobic character, like soybean oil and lecithin. The oil and lecithin together form propofol-containing phospholipid bilayers in water.

(e) The tetrabutylammonium cation is interesting in that it has a net +1 charge, but it is largely composed of hydrophobic alkyl chains. In water, solvation of this ion is difficult because of the large cavity that must be formed within the solvent water molecules. This results in a significant amount of low-entropy water around the hydrocarbon chains (as in Fig. 8.5, p. 424 of the text). DMSO, on the other hand, is a non-hydrogen-bonding solvent, and it has a greater dipole moment (4.0 D compared to 1.9 D for water, from Table 8.2). When the ion is transferred from water to DMSO, the low-entropy water surrounding the butyl groups is released to become part of the higher-entropy solvent water. This accounts for the highly positive ΔS of transfer. Evidently, DMSO does not lose much entropy when it solvates the cation—or, even if it does, it is a much larger molecule than water, and fewer molecules are required to surround the cation.

(f) Remember that solubility of any liquid in water results from the *difference* of the interactions between liquid molecules and the interactions of liquid molecules with water molecules. The lower boiling point of ether shows that the interactions between ether molecules in pure ether are weaker than the interactions between 1-pentanol molecules in pure 1-pentanol. This means that ether molecules have a greater "escaping tendency" than 1-pentanol molecules—whether the "escape" is into the gas phase or into aqueous solution. Weak ether-ether interactions are traded, when ether dissolves in water, for significant ether–water interactions, which no doubt involve accepting hydrogen bonds from water as well as dipole–dipole interactions. Stronger 1-pentanol–1-pentanol interactions, no doubt due to hydrogen bonding and dispersion forces between the alkyl chains, are not adequately compensated by 1-pentanol—water interactions, probably because of a negative entropy change associated with water solvation of the hydrocarbon chains.

Solutions to Integrated Problems

8.48   The answers to all five parts are as follows: (α-Carbons are noted with an asterisk, *.) Although not requested in the problem, the asymmetric carbons in the chiral structures are the α-carbons, except in 2-methyl-1-butanol, in which the asymmetric carbon is the β-carbon.

8.49   The answers to all four parts are as follows: (α-Carbons are noted with an *; common nomenclature is listed first, followed by IUPAC substitutive nomenclature.)

8.50   (a)   The unsaturation number of the ether is 1; because it has no double bonds, it must contain a ring. Two of the many possibilities are

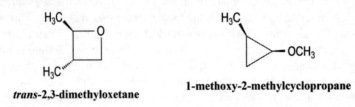

trans-2,3-dimethyloxetane   1-methoxy-2-methylcyclopropane

(b)   The unsaturation number is 2, and both rings and/or multiple bonds are allowed in this case. Two of several possibilities are

(c)   The following glycol, although it has asymmetric carbons, is achiral because it is meso:

cis-4-cyclohexen-1,2-diol

(d)   2,3-Butanediol exists as a meso stereoisomer and two enantiomers.

$$CH_3CH\overset{\overset{\displaystyle OH}{|}}{{}}-\overset{\overset{\displaystyle OH}{|}}{C}HCH_3$$

2,3-butanediol

(e)   1,2-Butanediol exists as a pair of enantiomers:

$$HOCH_2\overset{\overset{\displaystyle OH}{|}}{C}HCH_2CH_3$$

1,2-butanediol

(f)   The six epoxides with the formula C_4H_8O:

cis-2,3-dimethyloxirane
(a meso compound)

trans-2,3-dimethyloxirane
(exists as two enantiomers)

2,2-dimethyloxirane

2-ethyloxirane
(exists as two enantiomers)

8.51   The principal chains are highlighted in each case. In (b), the choice between the two branches containing double bonds is dictated by chain length. The principal chain has more carbons than the 5- or 6-membered ring, and therefore the compound is named as an octane rather than a cyclohexane or cyclopentane derivative. The numbering of substituents then dictates the proper chain numbering in (a); the position of the principal group (OH) dictates chain numbering in (b).

(a)

4-butyl-1-cyclohexyl-3-cyclopentyl-3-methoxyoctane

(b)

(3*R*,4*S*,5*Z*)-3-allyl-4-butyl-3-cyclopentyl-5-octen-1-ol

8.52   In both (a) and (b), hydrogen bonding between molecules of the first compound is an intermolecular attraction that significantly increases the boiling point. (As we show in Chapter 20, acetic acid actually forms discrete hydrogen-bonded dimers; and acetamide, because it has two hydrogens that can be donated to hydrogen bonds, can form even larger hydrogen-bonded aggregates.)

acetic acid
(hydrogen-bonded dimer)

acetamide
(hydrogen-bonded aggregate)

8.53   Isomers in which there is an OH and/or an NH group will provide hydrogen-bonding capabilities that should significantly increase the water solubility. Here are several possibilities; there are many others.

(OH can be a substituent at any one of these carbons)

8.54   Despite the fact that nitromethane cannot form hydrogen bonds, it has a higher boiling point than 2-propanol most likely because nitromethane is a more polar molecule. Apparently, the dipole-dipole interactions of nitromethane are strong enough that its boiling point is higher than a compound that can form hydrogen bonds. Looking up the dipole moments of both molecules would support this conclusion. (In fact, the dipole moment of nitromethane is 3.4 D, and the dipole moment of 2-propanol is less than half, 1.6 D.)

8.55   (a)   Compound A is soluble in hydrocarbon solvents by the "like-dissolves-like" criterion. The alkyl groups, particularly the large cetyl ($C_{16}H_{33}$) group, resemble the hydrocarbon chains of the solvent. In contrast, ammonium chloride would require a hydrogen bond acceptor solvent to dissolve; hydrogen bonds cannot be provided by hydrocarbons.

   (b)   Although compound A is soluble in hydrocarbon solvents, it exists in such solvents as ion pairs and higher aggregates rather than as free ions. (See Fig. 8.8, p. 428 of the text.) The reason is that hydrocarbon solvents have very low dielectric constants ($\varepsilon \approx 2$), whereas a high dielectric constant is required to separate ionic aggregates into free ions. Furthermore, a hydrocarbon offers no solvation by hydrogen bonding to the bromide counter-ion. Hence, this anion remains in proximity to its positive partner.

8.56   In the presence of concentrated acid, dibutyl ether is protonated. The protonated ether is an ionic compound, and ionic compounds are soluble in water:

dibutyl ether
not ionic; not soluble in water

nitric acid

an ionic compound; soluble in water

8.57   (a)   The ammonium ion is solvated by hydrogen bonding with the solvent ethanol; the ion is the donor and the ethanol is the acceptor. Ethanol also stabilizes the dissolved ion by ion–dipole interactions. The chloride ion is also solvated by hydrogen bonding in which ethanol is the donor. (Only one interaction of each type is shown.)

(b)   The sodium ion is stabilized by the dipole of the carbonyl of *N*-methylpyrrolidone. An important resonance structure of *N*-methylpyrrolidone shows that there is an electrostatic (ion-ion) interaction between the chloride and the partial positive charge on nitrogen as well as between the oxygen and the sodium ion.

N-methylpyrrolidone

another resonance form of
N-methylpyrrolidone

8.58   Isopropyl mercaptan is less polar than the other two compounds, and the S—H group, unlike the O—H group of isopropyl alcohol, is not a good donor of hydrogen bonds. (See the solution to Problem 8.64 for experimental evidence for this point.) Therefore, isopropyl mercaptan should have the lowest boiling point, and should have a lower boiling point than isopropyl alcohol. (In fact, the boiling point of isopropyl mercaptan is 57–60 °C, and that of isopropyl alcohol is 83 °C.)

8.59   Hexethal should be (and is) the more potent sedative, because it is more soluble in membranes, and therefore can readily pass through them. It is more soluble because the longer alkyl chain is more like the interior of a membrane than is the shorter ethyl chain. Furthermore, the long alkyl chain makes hexethal *less* soluble in water, because large alkyl groups are not effectively solvated by water.

8.60   Lecithin, a phospholipid, will form phospholipid vesicles in water, as explained on text p. 434–435. These vesicles contain phospholipid bilayers that incorporate molecules of the oil within the "hydrocarbon" part of the bilayer. As with cell membranes, the aqueous sides of the vesicles—the polar head groups—interact well with water. Thus, the vesicles containing the oil intermix with the water and do not form a separate layer.

8.61   (a)   Vitamin C and the vitamin B compounds shown are water-soluble because of their large numbers of hydroxy groups (for hydrogen bonding with water) and other polar attributes (causing dipoles, again making them soluble in water). Vitamin A and D_3 have much more hydrophobic character, making them more likely soluble in fat. Even though they each posess one hydroxy group, the molecules are sufficiently large that their properties are dominated more by their hydrocarbon portions.

(b)   Overdosing on a fat-soluble vitamin would be more dangerous, because it is more difficult to clear the non-water-soluble compounds from the body through natural excretion processes.

8.62   Think of the dissociation reaction of the crown ether–K^+ complex as you would any other equilibrium:

crown ether–K^+ + solvent ⇌ crown ether + K^+ (solvated)

Anything that stabilizes the species on the right side of the equation will increase the dissociation constant of the complex. Because potassium ion, a typical positive metal ion, is solvated more effectively by water than by ether, the dissociation constant for the complex is greater in water.

8.63   The ammonium ion will interact by donating hydrogen bonds to the oxygens of nonactin. (See the solution to Problem 8.27 in this manual, which describes the similar interaction of ammonium ion with a crown ether.) In contrast, the binding of potassium ion by nonactin does *not* involve hydrogen bonds, but rather ion–dipole and donor interactions. It is interesting that both the crown ether and nonactin are selective for binding of the potassium ion, and both also bond the ammonium ion. Evidently, the spatial requirements for both ions is similar. The cavity that fits the potassium ion equally well accommodates a nitrogen and its bound hydrogens.

8.64   (a)   By Le Chatelier's principle, when the concentration of ethanol increases, the concentration of dimer increases.

(b)   Use the usual relationship between $\Delta G°$ and K_{eq}; recall that $2.3RT$ at 298 K is 5.71 kJ mol^{-1}.

$$\Delta G° = -2.30RT(\log K_{eq}) = -5.71(\log 11) = -5.95 \text{ kJ mol}^{-1}$$

(c)   Neglect the volume of ethanol and assume that 1 mole of ethanol is dissolved in 1 L of solution so that the initial concentration of ethanol is 1 M. (This introduces an error of only 2%.) Assume that x M of ethanol react to form dimer. The concentration of ethanol remaining is then $(1.0 - x)$ M, and the concentration of dimer is $x/2$ M. (Remember, it takes two ethanol molecules to form one molecule of dimer.) The equilibrium-constant expression for dimerization becomes

$$K_{eq} = 11 = \frac{[\text{dimer}]}{[\text{ethanol}]} = \frac{x/2}{(1-x)^2}$$

From the quadratic formula, $x = 0.808$ M. Thus, the concentration of dimer is $x/2 = 0.404$ M; that of free ethanol is $(1 - x)$ or 0.192 M. Thus, about 80% of the ethanol is dimerized in a 1 M solution.

(d)   Because dimerization occurs through hydrogen-bond formation, then the greater equilibrium constant for dimerization of ethanol suggests that alcohols have stronger intermolecular hydrogen bonds than thiols.

(e)   The alcohol $CH_3SCH_2CH_2OH$ would have the greater solubility, because it can both donate and accept hydrogen bonds to water. An —SH group is a much poorer hydrogen bond acceptor and hydrogen bond donor than —OH, as the data in part (d) of this problem show.

8.65   A similar "sawtooth" pattern should also be expected for the densities of the solid unbranched alkanes, because their packing is more efficient with even numbered carbons. More efficient packing results in more molecules per unit volume—that is, higher density.

8.66   (a)   Even though they have the same molecular weight, 1-butanol has a higher boiling point because of its larger surface area. *Tert*-butyl alcohol is more spherical and therefore a smaller surface area. This results in weaker attractive van der Waals forces between *tert*-butyl alcohol molecules. (See Study Problem 8.4 on text p. 407.)

(b)   The entropy of solution would be lower (less favorable) for 1-butanol. The longer, tube-like structure of 1-butanol has more surface area per molecule than the more spherical *tert*-butanol. Therefore, more low-entropy (or more highly organized) water molecules are necessary to surround 1-butanol.

(c)   Because *tert*-butyl alcohol has a lower boiling point, its tendency to "escape" from a pure *tert*-butyl alcohol to any other milieu—including an aqueous solution—is greater than the "escaping tendency" of 1-butanol. Moreover, its spherical shape, which results in a smaller surface area, requires a smaller solvating cavity and therefore a less negative $\Delta S_{\text{solution}}$ contribution in water. Therefore, *tert*-butyl alcohol is more soluble in water than 1-butanol; in fact, *tert*-butyl alcohol is miscible with water in all proportions.

8.67   Because $\Delta G° = \Delta H° - T\Delta S°$, and because the free energy of hydrophobic bonding is dominated by its highly positive entropy, as the discussion in Sec. 8.6D shows, increasing the temperature increases the magnitude of the negative entropy term (that is, $-T\Delta S°$ becomes more negative). This change makes $\Delta G°$ more negative. The more negative $\Delta G°$, the larger the equilibrium constant and the more favorable the reaction. In summary, free energy should decrease with increasing temperature and the equilibrium constant should increase.

8.68   (a)   A greater partition coefficient in an octanol/water mixture (P_{ow}) corresponds to more of the drug in the 1-octanol phase. Our understanding of solubility suggests the more hydrophobic of the two drugs would prefer the more hydrophobic 1-octanol phase. Because atenolol contains many more polar bonds and hydrogen-bonding groups, it should be more soluble in water. Propranolol contains fewer polar bonds and hydrogen bonding groups, and it contains an additional aromatic (hydrocarbon) ring that atenolol doesn't have. These features make propranolol more hydrophobic and more soluble in 1-octanol.

(b)   If log P_{ow} = 0.22, then P_{ow} = 1.66. This number is the ratio of the concentration of drug molecules in the 1-octanol phase to that in the water phase. Because there are equal amounts of the two solvents, the P_{ow} is also the ratio of masses of drug in the two phases. If the mass of drug in water is x_w and the mass of drug in 1-octanol is x_{oct}, then x_{oct}/x_w = 1.66. Because $x_{oct} + x_w$ = 1 g, we can solve these two equations simultaneously to obtain x_a = 0.375 g and x_w = 0.625 g.

(c)   We may conclude on the basis of "like dissolves like" that *liquid* atenolol is more soluble in water than *liquid* propranolol because it is more polar that propranolol. However, atenolol's melting point is very high relative to that of propranolol. According to Eq. 8.68, the solubility of a solid is equal to the solubility of the liquid plus the free energy of fusion which increases with the melting point. So, it would not be surprising if the solubilities are close, or if atenolol is even slightly less soluble. (The discussion of the solubilities of nifedipine and 9,10-dihydroanthracene on text p. 426 illustrates a similar situation.) In summary, the partition coefficient of the two drugs, already in the liquid form in the partition experiment, doesn't necessarily indicate relative solubility of the *solids* in water.

(d)   The structures of the conjugate acids:

(e)   At a pH of 7.4, an acid with a pK_a of 9.5 is more than 99% protonated (Eq. 3.67b, text p. 148). The protonated forms are ionic, less soluble in 1-octanol, and more soluble in water. Therefore, for both molecules the partition coefficient would decrease, as the denominator increases relative to the numerator.

(f)   Physiological pH is close to 7.4, so administering the drugs in the protonated form [see part (e)] ensures their water solubility.

8.69   (a)   In compound *A*, the equatorial position of the —OH group is preferred for the usual reason: it avoids 1,3-diaxial interactions with hydrogens on the ring. However, compound *B* prefers to have an axial —OH group because it affords the opportunity for intramolecular hydrogen bonding; this additional bonding is a stabilizing effect, and lowers the energy of the axial conformation below that of the equatorial conformation.

hydrogen bonding stabilizes
the axial confomation

intramolecular hydrogen bonding
is not possible in the
equatorial conformation

(b)   The solution to part (a) shows that intramolecular hydrogen bonding can stabilize conformations that otherwise might be less stable. In this case, intramolecular hydrogen bonding can stabilize the gauche conformations of both stereoisomers. Indeed, in either enantiomer of the racemate, such hydrogen bonding can occur in one conformation in which the large *tert*-butyl groups are anti to each other. However, in the meso stereoisomer, the necessity that the hydroxy groups be gauche in order for hydrogen bonding to occur also means that the large *tert*-butyl groups must also be gauche. The stabilizing effect of intramolecular hydrogen bonding cannot compensate for the magnitude of the resulting van der Waals repulsions between the *tert*-butyl groups.

one of the gauche conformations of
(3*R*,4*R*)-2,2,5,5-tetramethyl-3,4-hexanediol
is stabilized by internal hydrogen bonding, and
the *tert*-butyl groups are anti

severe van der
Waals repulsions

any conformation of
meso-2,2,5,5-tetramethyl-3,4-hexanediol
in which internal hydrogen bonding is possible
is destabilized by van der Waals repulsions between
the *tert*-butyl groups

8.70   (a)   Equilibrium (2) contains more of the conformation with the methyl group in an axial position than equilibrium (1) does; that is, equilibrium (2) lies farther to the right. The reason is that, because C—O bonds are shorter than C—C bonds, the axial methyl group in equilibrium (1) is somewhat closer to the axial hydrogens than the axial methyl group in equilibrium (2). Therefore, the 1,3-diaxial interactions in the axial-methyl conformation of equilibrium (1) are somewhat more severe than those in the axial-methyl conformation of equilibrium (2).

smaller distance;
greater van der Waals repulsions

larger distance;
smaller van der Waals repulsions

axial conformation in equilibrium (1)

axial conformation in equilibrium (2)

(b)   The reasoning is much the same as that in part (a). Because the C—O bonds are shorter in the ether *A* than the corresponding C—C bonds in butane *B*, the methyl groups are brought closer together in the gauche conformation of the ether than they are in the gauche conformation of butane. Consequently, van der Waals repulsions in the gauche conformation of the ether are somewhat greater than they are in the gauche conformation of butane. In contrast, the shorter bonds of the ether should have little effect on the energy of the anti conformation, in which the methyl groups are far apart. Therefore, the greater energy of the gauche conformation of the ether (relative to the anti conformation) causes less of the gauche conformation to be present at equilibrium. Thus, butane contains more gauche conformation at equilibrium.

The Chemistry of Alkyl Halides
Substitution and
Elimination Reactions

CHAPTER 9 SUMMARY

■ Two of the most important types of alkyl halide reactions are nucleophilic substitution and β-elimination.

■ In a nucleophilic substitution reaction of an alkyl halide, a nucleophile replaces a halogen, which leaves as a halide ion. Nucleophilic substitution reactions occur by two mechanisms, called S_N2 and S_N1.

■ Characteristics of the S_N2 reaction:

- The S_N2 reaction occurs in a single step with inversion of stereochemical configuration.

- The S_N2 reaction is characterized by a second-order rate law.

- An S_N2 reaction typically competes with one or more E2 reactions.

- Alkyl substitution in the alkyl halide at the α- or β-carbons retards the S_N2 reaction and, if a good Brønsted base is present, favors competing elimination by the E2 mechanism. The S_N2 reaction is most commonly observed with methyl, primary, and unbranched secondary alkyl halides.

- The S_N2 reaction is especially rapid in polar aprotic solvents, although it can be run in protic solvents in favorable cases.

■ Nucleophilicity—the rate at which a nucleophile reacts in S_N2 reactions—is affected by several factors:

- Base strength: Stronger bases are better nucleophiles in the absence of hydrogen-bonding effects. Correlations with basicity are generally most straightforward when comparing nucleophiles that contain the same nucleophilic atom.

- Solvent: Solvents that can donate hydrogen bonds reduce nucleophilicity. Overall nucleophilicity is a compromise of basicity and hydrogen bonding.

- Polarizability: Nucleophilicity increases with polarizability of the nucleophilic atom. Highly polarizable nucleophiles (such as sulfide and iodide ions) typically contain nucleophilic atoms from higher-numbered periods of the periodic table. Some weak bases are good nucleophiles because they are polarizable.

■ Characteristics of the S_N1 reaction:

- The S_N1 reaction is characterized by a first-order rate law that contains only a factor in alkyl halide concentration.

- The S_N1 reaction occurs mostly in polar protic solvents. A very common example of this reaction is solvolysis, in which the solvent is the nucleophile.

- The S_N1 reaction involves a carbocation intermediate. For this reason, it is accelerated by alkyl substitution at the α-carbon. Consequently, the S_N1 reaction is observed mostly with tertiary and secondary alkyl halides.

- When the alkyl halide is chiral, the products of the S_N1 reaction are mostly racemic, although some products of inverted configuration resulting from the reactions of ion pairs are observed in many cases.

- The S_N1 reactions of some secondary alkyl halides occur with rearrangement when the secondary carbocation intermediate can rearrange to a more stable tertiary carbocation.

■ In a β-elimination, a halide ion and a β proton are removed from an alkyl halide with the formation of an alkene. β-Elimination reactions occur by two mechanisms, called E2 and E1.

■ Characteristics of an E2 reaction:

- An E2 reaction follows a second-order rate law (first order in the base).

- An E2 reaction typically occurs with anti stereochemistry.

- E2 reactions often occur in competition with S_N2 reactions. The E2 reaction is favored both by use of a strong Brønsted base and by α- and β-alkyl substitution in both the alkyl halide and the base. It is the major reaction of tertiary and secondary alkyl halides in the presence of a strong Brønsted base.

- Highly polarizable nucleophiles show a smaller percentage of E2 reaction in the E2–S_N2 competition.

- When an alkyl halide has more than one type of β-hydrogen, more than one alkene product is generally obtained. The alkene with the most alkyl substitution at the double bond is generally the predominant product of an E2 reaction.

■ Characteristics of an E1 reaction:

- The E1 reaction results from an alternative product-determining step of the S_N1 mechanism in which a carbocation intermediate loses a β-proton to form an alkene. The alkene with the greatest number of alkyl substituents on the double bond predominates.

STUDY GUIDE LINKS

9.1 Deducing Mechanisms from Rate Laws

Suppose you are an expert in traffic safety, and you are asked to determine as much as you can about traffic accidents on a bridge. In fact, your job is to come up with a description for a "typical" accident and formulate a strategy to eliminate the accidents. Of course, you would interview eyewitnesses and read police reports to gather your data. But suppose eyewitness accounts are not available to you for some reason. This is the situation in deducing reaction mechanisms. We cannot directly observe transition states (that is, molecular collisions).

Imagine that the traffic data you have available are the number of accidents as a function of the traffic density on the bridge in the two directions. Suppose you find that the number of accidents is proportional to the number of cars traveling east *times* the number of cars traveling west. From this information you can construct a mental picture—a mechanism—for a typical collision. Because the number of accidents in a given time is proportional to the traffic flow in *both* directions, you could reasonably conclude that a typical accident is a head-on collision between a car in one lane and a car in the other. An *unreasonable* picture would be two cars in the same lane having a rear-end collision, because, in this case, the number of accidents in a given time would be proportional to some function of the traffic density in only that lane. The head-on collision is as good a picture of an accident as any other until other data force you to refine it. In fact, it suggests an experiment for reducing the number of accidents: build a wall between the two lanes of traffic. If this "experiment" reduces the number of accidents, it would support your picture of a typical accident. Notice that it would *not prove your picture to be correct;* it would only *show that your picture is not incorrect.* Nothing short of eyewitnesses can prove you correct.

Similarly, a starting point for deducing a chemical mechanism is the rate law. The rate law tells us about "molecular collisions" that lead to a reaction, but does not tell us whether these collisions occur by a head-on mechanism, a rear-end mechanism, or some less direct pathway. The simplest mechanism possible is adopted, and a modification is considered only when required by new data.

9.2   Ring Carbons as Alkyl Substituents

Study Problem 9.2 in text Sec. 9.5G refers to alkyl substituents in alkene *C*. Some students have difficulty in seeing ring carbons as "substituents," even though they have no problem with the idea of substituents in acyclic compounds. The reason for this difficulty is that the "substituents" in a cyclic compound are "tied together" into a ring. For example, "the substituents at the double bond" in compound *C* should be viewed as follows:

(The methyl group is the third substituent on the double bond.)

9.3   Diagnosing Reactivity Patterns in Substitution and Elimination Reactions

The last case study in Study Problem 9.3 (text Sec. 9.7) points out a danger in using a guide such as Table 9.7 in text Sec. 9.7: *you can sometimes expect special cases that are not covered by the table or borderline cases that are hard to predict.* You can't diagnose every possible reaction with a summary table any more than a physician can diagnose a patient with a table of symptoms and diseases! When you encounter borderline cases that are difficult to predict, you should be able to narrow the possibilities of what can happen. The important thing is for you to be able to predict trends and to understand the reasoning behind them. Table 9.7 will prove most useful if you *make the effort to understand each entry in terms of the principles that underlie it,* and then *think* about each problem you are asked to solve in terms of these principles rather than attack it by rote.

A number of professors of organic chemistry, the authors included, find that the section of our organic chemistry courses that deals with substitution and elimination reactions—the section covered by this chapter—is a point at which a number of students have particular difficulty. We have discerned at least two reasons for this problem. First, this is a point at which the number of reactions and ideas have accumulated to the stage that the student who is not carefully organized is overwhelmed. If you are in this situation, you should read (or re-read) Study Guide Link 5.1, "How to Study Organic Reactions," in this manual, and follow the suggestions there. The second reason for difficulty is the fact that all four mechanisms of substitution and elimination—S_N1, S_N2, E1, and E2—can potentially happen simultaneously, and it can be difficult to keep straight exactly when each process can take place. Again, to use an analogy from the medical world, when a physician diagnoses a problem with a patient, they can't focus simply on one system, such as the heart; the physician must understand that the many systems in the body interact, and that there can be many root causes contributing to a given symptom. In some cases, the problems are so complex that they can't be readily solved! Table 9.7 is designed to help you keep all the possibilities in substitution and elimination reactions organized.

A broader philosophical issue is relevant to this discussion. The temptation is for students to believe that there are "recipes" or set formulas for solving every problem. This section on substitutions and eliminations shows that there are *principles* that can guide you into the right "ballpark," but inevitably, uncertainties will arise in some situations. Again: your goal should be not to memorize every conceivable case, but rather to bring to bear your knowledge of the principles so as to narrow the possibilities.

If you were a laboratory scientist trying to predict the outcome of a reaction, you would make your best prediction using the principles that you know, and then you would go into the laboratory and run an experiment to see whether you are correct. A physician facing such uncertainty runs tests to gather as many facts as possible, adopts a hypothesis, and checks it by administering appropriate medication or carrying out other tests. Two people with difficulties in a relationship adopt a hypothesis as to the nature of the problem and, if they are behaving rationally, change their behavior and see whether the problem is solved. For most problems there aren't magic recipes and tables that allow us to solve problems by rote, nor are there simple formulas into which we "plug" data for an answer. Uncertainty is in the very nature of science, and indeed, of life itself. If you think about it, life would be awfully dull without the challenge of such uncertainty. *One reason that organic chemistry is valued as a prerequisite course of study by various disciplines is that a student must, in order to succeed, develop a body of principles that can be applied to situations in which there is inherent uncertainty.* The good news is that if you master this skill in organic chemistry, you'll find it easier to do so in your other endeavors as well.

FURTHER EXPLORATIONS

9.1   Reaction Rates

Because rates are usually functions of concentration, and because concentrations change as a reaction proceeds, you can perhaps appreciate that the rate of a reaction changes over its course. If the rate of a reaction is proportional to the concentrations of one or more reactants, the rate diminishes during a reaction because the concentration of reactants diminishes. (See Problem 9.8 in Sec. 9.3B of the text and its solution.)

Reaction rates, like other rates of change, are more precisely expressed in the language of calculus. Suppose, for example, we have a reaction $S \longrightarrow P$ that follows a rate law that is first order in S. Then the rate can be written

$$\text{rate} = \frac{d[P]}{dt} = -\frac{d[S]}{dt} = k[S] \tag{FE9.1}$$

This is a first-order differential equation that can be solved for $[S]$ as a function of time. If S_0 is the concentration of S at the start of the reaction ($t = 0$), then the solution to this equation is

$$[S] = S_0 e^{-kt} \tag{FE9.2}$$

The concentration of product as a function of time is

$$[P] = S_0(1 - e^{-kt}) \tag{FE9.3}$$

(If you are familiar with differential equations, you can easily verify this solution by differentiating Eq. FE9.2, as follows.

$$d[S]/dt = -kS_0 e^{-kt} \tag{FE9.4}$$

Substituting Eq. FE9.2 into Eq. FE9.1 yields Eq. FE9.4 and thus verifies the solution. You can, in a similar manner, verify by differentiating Eq. FE9.3 that $d[P]/dt = -d[S]/dt$.)

Equation FE9.2 gives the expression for the actual concentration of $[S]$ as a function of time, and shows that it is a decaying exponential. Each rate law has its own characteristic solution for the dependence of concentration on time. Computer-fitting actual concentration versus time data for a reaction of known kinetic order to the appropriate rate law allows the experimentalist to determine the value of the rate constant k. This is how rate constants are usually derived in practice.

9.2   Absolute Rate Theory

The dependence of the first-order rate constant k on the standard free energy of activation can be calculated from Eyring theory, or absolute rate theory. (Henry Eyring [1901–1981] was for many years a professor at the University of Utah, and he also served as President of the American Chemical Society in 1963. Eyring developed absolute rate theory, also called "transition-state theory.") A result of absolute rate theory is that a first-order rate constant is given by

$$k = \frac{\kappa T}{h}\, e^{-\Delta G^{\circ\ddagger}/RT} = (6.21 \times 10^{12}\ \text{sec}^{-1})(e^{-\Delta G^{\circ\ddagger}/RT})$$

(FE9.5)

where κ = the Boltzmann constant (1.381×10^{-23} J K^{-1}), h = the Planck constant (6.626×10^{-34} J s^{-1}), and T is the absolute temperature in kelvins. The evaluation of the $\kappa T/h$ pre-exponential factor in Eq. FE9.5 is at $T = 298$ K, or approximately room temperature. This is the equation used to give the relationship between the standard free energy of activation and rate constant in Table 9.2 in text Sec. 9.3B.

A useful concept use to describe the velocity of a reaction is the **half-life. abbreviated** $t_{1/2.}$

$$t_{1/2} = \ln 2/k = 0.693/k$$

(FE9.6)

In each successive half-life of a reaction, the percent completion of a reaction decreases by half. In other words, a reaction is 50% complete after one half-life; 75% complete after two half-lives; and $100 \times (1.00 - 0.5^n)$% complete after n half-lives. Because the concentration of products is a decaying exponential, a reaction is technically never fully completed; but it is generally accepted that a reaction is considered practically complete after 10 half-lives, after which a reaction is about 99% complete. From Eq. FE9.6, ten half-lives = $(10)(0.693)/k$, or about $7/k$. This is the source of the "time to completion" numbers in Table 9.2.

REACTION REVIEW

I.   REACTIONS OF ALKYL HALIDES

A.   *NUCLEOPHILIC SUBSTITUTION REACTIONS*

1. A nucleophilic substitution reaction, or nucleophilic displacement reaction, is a very general type of reaction:

    a. It is a *substitution* because one group is substituted for (or displaces) another group, which departs.

        i.   The group that is displaced in a nucleophilic substitution reaction is called the leaving group.

        ii. The leaving group in a nucleophilic displacement reaction is the nucleophile in the reverse reaction, and vice-versa.

        iii. The best leaving groups give the weakest bases as products.

    b. It is a *nucleophilic substitution* because the substituting group acts as a nucleophile, or Lewis base.

        i.   Many nucleophiles are anions and others are uncharged.

        ii. The roles of the nucleophile and the leaving group are reversed if the reaction can be run in the reverse direction.

2. Nucleophilic substitution reactions occur by two common mechanisms: the S_N2 mechanism and the S_N1 mechanism.

    a. The S_N2 mechanism occurs in a single step with inversion of stereochemical configuration; it is characterized by a second-order rate law.

$$\text{rate} = k \left[\text{Nuc:}^- \right] \left[\begin{array}{c} \diagdown \\ \diagup \end{array} \text{C} - \text{X} \right]$$

i. An S_N2 reaction is an electron-pair displacement reaction that is conceptually similar to a Brønsted acid–base reaction; the nucleophile is analogous to the Brønsted base and the electrophile is something other than a proton. The leaving group is displaced from the electrophile in an S_N2 reaction, just as a leaving group is displaced from a proton in a Brønsted acid–base reaction.

an S_N2 reaction

a Brønsted acid–base reaction

ii. S_N2 reactions are typically carried out in polar protic solvents, polar aprotic solvents, or mixtures of these. Polar aprotic solvents in many cases accelerate S_N2 reactions because they do not form hydrogen bonds to the nucleophile.

b. The S_N1 mechanism is characterized by a first-order rate law that contains only a term in alkyl halide concentration. The S_N1 reaction can involve significant loss of stereochemical configuration.

$$\text{rate} = k \left[-\overset{|}{\underset{|}{\text{C}}} - \text{X} \right]$$

3. The reaction of an alkyl halide with a protic solvent in which no base or nucleophile has been added is called a *solvolysis*.

   a. The solvolysis reactions of tertiary alkyl halides are fastest in polar, protic donor solvents.

   b. Tertiary and secondary alkyl halides undergo solvolysis reactions by the S_N1 and E1 mechanisms; tertiary alkyl halides are much more reactive.

   c. Rearrangements are observed in S_N1 solvolysis reactions that involve rearrangement-prone carbocations as intermediates.

4. Substitution reactions can be intramolecular—that is, the nucleophile, the electrophile, and the leaving group are part of the same molecule.

5. Nucleophilic substitution reactions can be used to transform alkyl halides into a wide variety of other functional groups. (See Table 9.1 on text p. 452.)

B. *β-ELIMINATION REACTIONS*

1. An elimination reaction is a reaction in which two or more groups are lost from within the same molecule.

   a. In an alkyl halide, the carbon bearing the halogen is often referred to as the α-carbon, and the adjacent carbons are referred to as the β-carbons.

   b. An elimination that involves loss of two groups from adjacent carbons is called a β-elimination.

   c. In its overall outcome, a β-elimination reaction is conceptually the reverse of an addition to an alkene.

2. β-Elimination reactions occur by two mechanisms:

    a. The E2 mechanism, which competes with the S_N2 mechanism, has a second-order rate law and typically occurs with anti stereochemistry.

    anti-elimination

    b. The E1 mechanism is an alternative product-determining step of the S_N1 mechanism in which a carbocation intermediate loses a β-hydrogen to form an alkene.

3. Base-promoted β-elimination reactions in many cases occur by E2 mechanisms; they typically follow a rate law that is second order overall and first order in each reactant.

    a. A Brønsted base reacts at a β-hydrogen of the alkyl halide, not at a carbon atom as in a nucleophilic substitution reaction.

    b. If the reacting alkyl halide has more than one type of β-hydrogen atom, then more than one β-elimination reaction is possible.

    c. Strong bases, such as sodium ethoxide [Na^+ $^-OCH_2CH_3$] and potassium *tert*-butoxide [K^+ $^-OC(CH_3)_3$], promote the β-elimination reactions of alkyl halides.

    d. Often, the conjugate-acid alcohols of these bases are used as solvents, for example, sodium ethoxide in ethanol or potassium *tert*-butoxide in *tert*-butyl alcohol.

🔒 SOLUTIONS TO PROBLEMS

Solutions to Focused Problems

9.1   (a)   The expected nucleophilic substitution product is butyl methyl sulfide, $CH_3CH_2CH_2CH_2$—S—CH_3. Sodium iodide (Na^+ I^-) is a by-product.

    (b)   The product is ethylammonium iodide.

$$CH_3CH_2\overset{+}{N}H_3 \quad I^-$$

ethylammonium iodide

9.2   The net ionic equation for the reaction in Focused Problem 9.1a:

$$^-SCH_2CH_2CH_2CH_3 \ + \ I—CH_3 \longrightarrow CH_3SCH_2CH_2CH_2CH_3 \ + \ I^-$$

9.3 (a) Two elimination products are expected because there are two types of β-hydrogen in the starting alkyl halide.

2-bromo-2,3-dimethylbutane

2,3-dimethyl-1-butene
(from elimination of
a hydrogen of type 1)

2,3-dimethyl-2-butene
(from elimination of
the hydrogen of type 2)

(b) As in part (a), because there are two types of β-hydrogen, two alkenes can be formed:

1-chloro-1-methylcyclohexane

1-methylcyclohexene
(from elimination of a
hydrogen of type 1)

methylenecyclohexane
(from elimination of a
hydrogen of type 2)

9.4 (a) The substitution product can in principle be obtained as two stereoisomers. (Sec. 9.4C of the text shows which of these is formed and why.)

substitution products

elimination products

(b) Methyl iodide, H_3C—I, can form only a substitution product, dimethyl ether, H_3C—O—CH_3. No elimination products are possible because there are no β-hydrogens.

(c) (Bromomethyl)cyclopentane can form one substitution product and one elimination product.

(bromomethyl)cyclopentane

9.5 (a) Because fluoride ion is a stronger base than chloride ion, the equilibrium lies to the left. The analogous acid-base reaction is HF + Cl⁻ ⇌ HCl + F⁻. Using Eq. 3.59b in Sec. 3.6B, the estimated K_{eq} for this reaction roughly 10^{-10}.

(b) Because azide ion is a stronger base than chloride ion, the equilibrium lies to the right. The analogous acid-base reaction is HCl + N_3^- ⇌ HN_3 + Cl⁻, which has a K_{eq} of about 10^{11}.

(c) Because methoxide ion is a much stronger base than chloride ion, the equilibrium lies to the right. The analogous acid-base reaction is HCl + CH_3O^- ⇌ CH_3OH + Cl⁻, which has a K_{eq} of at least 10^{21}.

9.6 (a) The reaction is third order overall; the reaction is first order in alkene and second order in Br_2. The dimensions of the rate constant are such that the overall rate has the dimensions mol L⁻¹ sec⁻¹; thus, the rate constant has the dimensions L² mol⁻² sec⁻¹, or M⁻² sec⁻¹.

(b) The reaction is first order overall, and first order in alkyl halide. The rate constant has the dimensions of sec⁻¹.

9.7   (a)   Apply Eq. 9.19b:

$$\frac{k_A}{k_B} = 10^{(\Delta G_B^{\circ\ddagger} - \Delta G_A^{\circ\ddagger})/2.30RT} = 10^{14/5.71} = 10^{2.45} = 283$$

(b)   Transform Eq. 9.19c to get the difference between the standard free energies of activation.

$$\Delta G_A^{\ddagger} - \Delta G_B^{\ddagger} = 2.3RT \log \frac{k_B}{k_A} = 5.71 \log(450) = 15.2 \text{ kJ mol}^{-1}$$

Therefore, reaction A has the higher $\Delta G^{\circ\ddagger}$ by 15.2 kJ mol^{-1} (3.63 kcal mol^{-1}).

9.8   Because D and E are converted into F, their concentrations decrease with time; therefore, the rate also decreases with time. Because the rate is the *slope* of the curves of [reactants] or [products] with time, the slopes of these curves continually decrease with time, and approach zero at long times. A plot of these concentrations with time is shown.

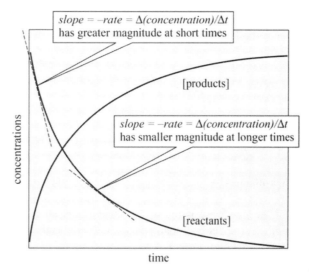

9.9   A mechanism consistent with the rate law is a bimolecular process in which ammonia displaces acetate ion from the proton:

CH$_3$C—Ö—H :NH$_3$ ⟶ [CH$_3$C—Ö$^{\delta-}$---H-----$^{\delta+}$:NH$_3$]‡ ⟶ CH$_3$C—Ö:$^-$ $^+$NH$_4$

acetic acid   **acetate ion**

9.10   (a)   Ethyl bromide and cyanide react as shown:

CH$_3$CH$_2$—Br + :C≡N:$^-$ ⟶ CH$_3$CH$_2$—C≡N: + Br$^-$

ethyl bromide   **propionitrile**
(bromoethane)

Because the S$_N$2 mechanism involves a molecule of alkyl halide and a molecule of nucleophile in a bimolecular reaction, the expected rate law is second order, first order in alkyl halide and first order in cyanide:

rate = k[CH$_3$CH$_2$Br][$^-$CN]

(b)   The simplest (S$_N$2) mechanism involves the nucleophilic reaction of cyanide ion with ethyl bromide and the simultaneous departure of a bromide ion as a leaving group—that is, an electron-pair displacement reaction. In the transition state, there are partial bonds between cyanide and carbon, and between carbon and bromide.

$$\left[\; :N{\equiv}\overset{\delta-}{C}:\text{------}\;\overset{\overset{\displaystyle CH_3}{|}}{CH_2}\text{------}\;\overset{\delta-}{:\!\ddot{B}\ddot{r}\!:}\;\right]^{\ddagger}$$

9.11 Sodium ethoxide reacts orders of magnitude more rapidly with the Brønsted acid than with the alkyl halide. Because this reaction consumes all of the ethoxide, only the HI reacts. The products are ethanol, CH_3CH_2OH, and sodium iodide, $Na^+ I^-$. The methyl iodide is unchanged.

9.12 There is more than enough sodium cyanide to react with both the acid HBr and the alkyl halide. Therefore, the products are hydrogen cyanide (HCN, 0.1 M), "ethyl cyanide" (propionitrile, CH_3CH_2CN, 0.1 M), sodium bromide (NaBr, 0.2 M, half from the reaction with HBr and half from the reaction with ethyl bromide), and unreacted sodium cyanide (0.8 M). However, the HCN and 0.1 M NaBr is formed instantaneously. The rest of the NaBr, as well as the nitrile, are formed much more slowly, because Brønsted bases react much more rapidly with Brønsted acids than with alkyl halides. In other words, Brønsted acid-base reactions are much faster than S_N2 reactions.

9.13 The reaction is an S_N2 reaction with inversion of configuration.

$$CH_3CH_2CH_2\overset{\overset{\displaystyle H}{|}}{\underset{\underset{\displaystyle D}{|}}{C}}\!\!-\!Cl \;+\; K^+ I^- \longrightarrow I\!-\!\overset{\overset{\displaystyle H}{|}}{\underset{\underset{\displaystyle D}{|}}{C}}\!\!-\!CH_2CH_2CH_3 \;+\; K^+ Cl^-$$

(R)-chlorobutane-1-*d*   **(S)-iodobutane-1-*d***

9.14 In an S_N2 reaction, the stronger base is the better nucleophile when the same nucleophilic atom is involved. Because ethoxide is a much stronger base than ethanol, ethoxide is the better nucleophile. In fact, ethanol is such a poor nucleophile that no reaction is observed with ethanol alone.

9.15 (a) The products of the S_N2 reaction between potassium acetate and ethyl iodide:

$$CH_3C\overset{\displaystyle O}{\underset{\displaystyle \ddot{O}-CH_2CH_3}{\diagup\!\!\!\!\diagdown}} \qquad + \qquad K^+ I^-$$

ethyl acetate   **potassium iodide**

(b) Potassium acetate is a better nucleophile in acetone because ethanol is a protic solvent and reduces the nucleophilicity of potassium acetate by hydrogen-bond donation. Consequently, potassium acetate in acetone reacts more rapidly with ethyl iodide than a solution of the same nucleophile in ethanol.

9.16 The discussion of this section shows that elements in higher periods (lower on the periodic table) within a group are better nucleophiles in protic solvents than elements in lower periods of the same group. Thus, triethylphosphine, $Et_3P:$, is a better nucleophile than triethylamine, $Et_3N:$. The products are as follows:

$$H_3C{-}I \;+\; :X(Et)_3 \longrightarrow H_3C{-}\overset{+}{P}(Et)_3 \; I^- \quad or \quad H_3C{-}\overset{+}{N}(Et)_3 \; I^-$$

$$X = P \qquad\qquad\qquad X = N$$

9.17 (a)   (b)   (c)

9.18 The formation of the carbon–anion (carbanion) intermediate, its resonance structure, and the mechanism of its breakdown to product, and are as follows:

resonance structures of the anion

elimination reaction of the anion

The carbanion is more stable than a typical carbanion because it is stabilized by the polar effect of the nearby oxygen, and because it is resonance-stabilized. This stabilization lowers the pK_a for its formation.

➡️ Ionization of the O—H group also occurs, but this does not lead to product.

9.19 (a)   A primary deuterium isotope effect slows the reaction of B relative to C; and the better leaving group accelerates the reaction of A relative to B and C. The reactivity order is therefore $B < C < A$.

(b)   Compound B has a better leaving group. The rates are in the order $A < B$.

9.20 (a)   The hydration of styrene should be (and is) slower in D_2O, because the transferred proton is a deuterium, and a rate-retarding primary deuterium isotope effect operates when the proton transferred in the rate-limiting step is isotopically substituted. Be sure to write this mechanism (see Sec. 4.10C).

The products differ by isotopic substitution. In H_2O/H_3O^+, the hydration product is compound A (at right); in D_2O/D_3O^+, the hydration product is compound B.

$$\underset{A}{\overset{OH}{\underset{|}{PhCHCH_3}}} \qquad \underset{B}{\overset{OD}{\underset{|}{PhCHCH_2D}}}$$

(b)   The hydration rate of the deuterium-substituted styrene should differ very little, if at all, from that of styrene itself, because the deuteriums are not transferred in the rate-limiting step.

9.21 (a)   Analyze one enantiomer of (±)-stilbene dibromide; the result is the same for the other. First, draw sawhorse projection (below) or a line-and-wedge projection. How do we know that it is one of the enantiomers? The two asymmetric carbons have the same configuration—in this case, R. (The two asymmetric carbons would have opposite configurations in the meso stereoisomer.) Also, in the eclipsed conformation shown, with the two hydrogens in identical positions, there is no internal mirror plane. (The meso diastereomer would have an internal mirror plane.) Next, internally rotate about the central carbon–carbon bond until the H and Br that are eliminated are anti (*step 2* below). Finally, draw the alkene that is formed and note its stereochemistry (*step 3* below).

the alkene product has
the Z configuration

(b)   Reasoning identical to that used in part (a) shows that the alkene formed from the meso stereoisomer has the *E* configuration. (We leave it to you to draw the appropriate structures.)

9.22   In a stereospecific reaction with a given stereochemistry—anti-elimination, in this case—a diastereomeric product requires a diastereomeric starting material (either enantiomer). The easiest path to the answer is to convert the starting material in Eq. 9.44a into its diastereomer by the interchange of any two groups at one of the carbons. Either this compound or its enantiomer would give the product of *E* configuration.

9.23   The alkene 1-butene is formed in least amount because it is the alkene with fewest alkyl substituents on the carbons of the double bond.

9.24   (a)   Redraw the structures in a sawhorse or line-and-wedge projection in which the hydrogen or deuterium that is eliminated is anti to the bromine. The stereochemistry of the alkene follows by looking at the relationship of the substituents—gauche or anti. Anti substituents are trans in the product alkene.

The two possible alkenes in reaction (1):   The two possible alkenes in reaction (2):

Because the starting materials are diastereomers, the products must also be diastereomers if the elimination is anti.

(b)   Elimination can also result in a minor product in each reaction through removal of a proton on the other β-carbon. This minor product in reaction (1) is the enantiomer of the one in reaction (2) because of the presence of the asymmetric carbons bearing deuterium.

elimination from this β-carbon — minor product in reaction (1)

elimination from this β-carbon — minor product in reaction (2)

9.25 (a) Let 1-bromo-3-methylbutane react with dimethyl sulfide, preferably in a polar aprotic solvent (although the reaction of a third-period nucleophile would occur at a useful rate in a polar solvent such as ethanol).

1-bromo-3-methylbutane   dimethyl sulfide

(b) In this case, the nucleophile is the conjugate base of ethanethiol, sodium ethanethiolate. Although a polar aprotic solvent could be used, ethanol would probably be the most convenient solvent for solubility reasons. (Because the nucleophile is from the third period, the reaction would occur at a convenient rate in a protic solvent; therefore, experimental convenience determines the choice of solvent.)

1-bromo-3-methylbutane   sodium ethanethiolate

(c) In contrast to part (b), this reaction is an elimination that could be brought about by the reaction of 1-bromo-3-methylbutane with any strong base, such as $K^+ (CH_3)_3CO^-$ (potassium *tert*-butoxide). Use of the highly branched base maximizes elimination and minimizes substitution.

1-bromo-3-methylbutane   potassium *tert*-butoxide

9.26 Analyze the alkyl substitution pattern in each alkyl halide: alkyl substitution at both the α- and β-positions increases the ratio of elimination to substitution. Compound D is secondary and it has two β substituents; compound B is primary and has two β substituents; compound C has one β substituent; and compound A, of course, has no β hydrogens and therefore cannot undergo β-elimination. Therefore, the ratio of E2 elimination to S_N2 substitution decreases in the order $D > B > C > A$.

9.27 (a) Chloride ion is a weaker base than all of the alkoxide bases. Because α-substitution in the alkoxide promotes a greater proportion of elimination, the order is $C > A > B > D$.

(b) Larger, "bulkier" (more sterically crowded) bases tend to favor formation of less substituted alkenes, because access to interior hydrogens is hindered. Base C would probably yield the greatest proportion of alkene Y (see text Eq. 9.49).

9.28 Products A and B result from the Lewis acid–base association reactions of water and ethanol, respectively, with the carbocation intermediate that is formed by ionization of the alkyl halide starting material. Products C and D result from the Lewis acid–base association reactions of water and ethanol, respectively, with the rearranged carbocation intermediate. (Compound D is the product shown in Eq. 9.69.)

[Reaction scheme: ionization of a tertiary alkyl chloride to an unrearranged carbocation + Cl⁻, then rearrangement to a rearranged carbocation + Cl⁻; each carbocation reacts with H₂O and EtOH to give products A, B, C, D.]

$$A \quad B \quad C \quad D$$

$$+ \ R\overset{+}{O}H_2 \ Cl^-$$

Alkenes can form by loss of a β-proton from either of the two carbocation intermediates. (The solvent, indicated below by ROH, acts as the base that removes the proton.)

[Reaction scheme: unrearranged carbocation + Cl⁻ → $(CH_3)_3C-CH=CH_2$ (E) + $R\overset{+}{O}H_2 \ Cl^-$ via ROH]

unrearranged carbocation

[Reaction scheme: rearranged carbocation + Cl⁻ → $(CH_3)_2C=C(CH_3)_2$ (F) + $H_2C=\overset{CH_3}{\underset{}{C}}-CH(CH_3)_2$ (G) + $R\overset{+}{O}H_2 \ Cl^-$ via ROH]

rearranged carbocation

Alkene *F* would be formed in greatest amount because it has the greatest amount of alkyl substitution at the double bond. Alkene *G* would be formed in smaller amount; alkene *E* would be formed in least amount for two reasons: it comes from a carbocation that can rearrange, and it has only one alkyl substituent on the double bond.

9.29   The mechanism for the formation of the rearranged product in Eq. 9.69 (product *D* in the solution to Problem 9.28):

[Reaction scheme showing ionization, rearrangement, and reaction with EtOH, then proton transfer to EtOH to give the ether product + $Et\overset{+}{O}H_2$:Cl̈:⁻]

9.30   The amount of racemization, 66%, corresponds to 33% of the *S* product and 33% of the *R* product. The amount of inversion corresponds to 34% of the *S* product. Therefore, the product mixture consists of 67% of the *S* methyl ether and 33% of the *R* methyl ether.

(33%)   **3-methoxy-3,7-dimethyloctane**   (67%)
(product *A*)

9.31   (a)   An increasing percentage of racemization should occur for a more stable carbocation. Because an increase in carbocation stability increases its lifetime, a greater fraction survives past the ion-pair stage to form a fully solvated, achiral carbocation, as shown in Fig. 9.13.

(b)   A less stable carbocation, by the same reasoning as in part (a), should have a shorter lifetime; a greater fraction will not last past the ion-pair stage, which reacts with the solvent by inversion.

9.32   (a)   Entry 3 of Table 9.7 in the text covers this case; the major product is the substitution product 1-methoxybutane, which is formed by an S_N2 mechanism:

1-methoxybutane

A small amount of the alkene 1-butene will also be formed in an E2 reaction:

1-butene

(b)   2-Bromobutane is a secondary alkyl halide, and potassium tert-butoxide is a strong, highly branched base. Entry 7 of Table 9.7 covers this case. Thus, the E2 reaction is the mechanism that occurs. Two possible alkenes, 1-butene and 2-butene, can form. Either or both can be considered as correct answers. A significant amount of 1-butene is formed because the large base molecule reacts at the least sterically hindered hydrogen.

2-bromobutane   **1-butene**

We leave it to you to show the formation of the 2-butene stereoisomers.

(c)   2-Bromo-1,1-dimethylcyclopentane is a secondary alkyl halide with significant β-alkyl substitution; a protic solvent is used without a strong base. Entry 9 in Table 9.7 covers this case. Rearrangement from the initially formed secondary carbocation *a* to the tertiary carbocation *b* is likely to occur.

a   *b*

Each carbocation can react to give both S_N1 substitution and E1 elimination products. Carbocation *a* gives the substitution product *A* and the elimination product *B*.

Carbocation *b* gives the substitution product *C* and the elimination products *D–F*. (The curved-arrow mechanism for formation of *C* is analogous to the mechanism shown above for the formation of *A*; and the curved-arrow mechanisms for the formation of alkenes *D–F* (by the E1 pathway) are analogous to the mechanism shown above for the formation of *B*.) Products *C*, *D*, and *F* are likely the major products; alkene *E* would be formed in the smallest amount because it is the least substituted alkene.

(d) Bromocyclohexane is a secondary alkyl halide; methanol is a polar, protic solvent; and there is no strong base present. Entry 9 of Table 9.7 covers this situation; both S_N1 product *A* and E1 product *B* are formed.

Solutions to Skills Objectives Problems

9.33 Substitute an H for the alkyl group in each case.

(a)

$$HBr + EtS^- \rightleftharpoons HSEt + Br^-$$

(b)

$$HBr + :\ddot{O}H_2 \rightleftharpoons H-\overset{+}{\ddot{O}}H_2 + Br^-$$

9.34 Use the pK_a values for the acid–base analogy to make this determination, then apply Eqs. 3.59a–b, text p. 144.

(a) HBr, $pK_a \sim -9$; EtSH, $pK_a \sim 11$ $\Delta(\log K) = \Delta pK_a = 11 - (-9) = 20$

(b) HBr, $pK_a \sim -9$; H_3O^+, $pK_a = 0$ $\Delta(\log K) = \Delta pK_a = 0 - (-9) = 9$

The most favorable equilibrium is therefore (a).

9.35 (a) The overall kinetic order is 3.
(b) The reaction is first order in alkene and second order in HBr.

9.36 Transition states A and C are both consistent with the rate law in Problem 9.35 because they both involve two molecules of HBr and one molecule of alkene.

9.37 Because the rate law is zero order in bromine, the rate-limiting step must occur before the entry of Br_2 into the reaction. Therefore, Step (1) is rate-limiting.

9.38 Apply Eq, 9.19b or 9.19c, Sec. 9.3B, text p. 458, as appropriate.

(a) The following calculation shows that the standard free energy of activation of reaction (2) is greater by 7.94 kJ mol^{-1}.

$$\log(\text{relative rate}) = \log\left(\frac{k_1}{k_2}\right) = \log(24.7) = 1.39 = \frac{\Delta G_2^{o\ddagger} - \Delta G_1^{o\ddagger}}{2.3RT}$$

$$\Delta G_2^{o\ddagger} - \Delta G_1^{o\ddagger} = (1.39)(2.3RT) = (1.39)(5.71) = 7.94 \text{ kJ mol}^{-1} \ (1.90 \text{ kcal mol}^{-1})$$

(b) The reaction with the greater standard free energy of activation—reaction (3)—is slower. The following calculation shows that reaction (2) is faster than reaction (3) by a factor of 18,800, or that reaction (3) is slower than reaction (2) by a factor of 1/18,800, or 5.32×10^{-5}.

$$\text{relative rate} = \frac{k_2}{k_3} = 10^{(\Delta G_3^{o\ddagger} - \Delta G_2^{o\ddagger})/2.3RT} = 10^{24.4/5.71} = 10^{4.27} = 18,800$$

9.39 (a)   (b)

$$H_3C-\overset{+}{P}(Ph)_3 \ I^-$$

methyltriphenylphosphonium iodide

$$+ Br^-$$

thiirane

9.40 The potassium ion or sodium ion could be used as counter-ions in any of the following.

(a)

$CH_3CH_2CH_2O^-$ Na^+

sodium propoxide

$+$ CH_3CH_2—I

**iodoethane
(ethyl iodide)**

(b)

Na^+ ^-CN

sodium cyanide

$+$ CH_3CH_2—I

**iodoethane
(ethyl iodide)**

(c)

K^+ ^-O ⌐⌐⌐ O ⌐

potassium 4-methoxybutanolate

$+$ CH_3CH_2—I

**iodoethane
(ethyl iodide)**

(d)

Na^+ ^-S ⌬ ⸗CH_3

**sodium *trans*-3-methylcyclo-
pentanethiolate**

$+$ CH_3CH_2—I

**iodoethane
(ethyl iodide)**

(e)

$(CH_3)_3N:$ $+$ CH_3CH_2—I

**trimethylamine   iodoethane
(ethyl iodide)**

[The stereochemistry in part (d) is retained at the asymmetric ring carbons; the substitution occurs at the α-carbon of ethyl iodide.]

9.41 (a) The order in the S_N2 reaction is $A < B < C < E < D$. The $B < C$ order is determined by Table 9.3 on text page 464, which shows that α substitution slows the rate somewhat more than the same β substitution. The greater rate of D is the result of a better leaving group.

(b) The order in S_N1–E1 reactivity is $C \approx E < D << B << A$.

9.42 In a hydrogen-bonding solvent such as ethanol, reactions of alkyl halides with strong-base nucleophiles are retarded by hydrogen bonding of the solvent to the nucleophile. Therefore, within a column of the periodic table, the element effect on nucleophilicity is that weaker bases are better nucleophiles. Therefore, the thiolate ion (which is a weaker base than the alkoxide ion) will out-compete the alkoxide ion for a limited amount of alkyl halide. As a result, the product of the reaction is ethyl methyl sulfide, CH_3CH_2S—CH_3.

9.43 (a) The S_N1 solvolysis products result from the Lewis acid–base association reaction of the respective solvent molecules on the carbocation intermediate, followed by loss of a proton.

$(CH_3)_3C$—Cl ⇌ $(CH_3)_3C^+$ Cl^-

→ (with $HOCCH_3$, O above) $(CH_3)_3C$—$OCCH_3$ $+$ HCl

***tert*-butyl acetate**

→ (with $HOCH$, O above) $(CH_3)_3C$—OCH $+$ HCl

***tert*-butyl formate**

(b) Both solvents are protic and donor solvents and therefore differ very little in their ability to solvate ions by hydrogen bonding or donor interactions. The major difference between the two solvents is their dielectric constants. The reaction is faster in formic acid because its higher dielectric constant promotes the separation of the chloride-ion leaving group from the carbocation. (This effect is discussed in text Sec. 9.6C.) Because this is the rate-limiting step in the S_N1 reaction, lowering the standard free energy of activation of this step increases the rate.

9.44 (a)

The alkene products predominate, and alkenes *C* are the alkenes formed in greatest amount.

(b)

(c)

(very small amount)

In part (c), the reaction does not affect the stereochemistry at the asymmetric carbons.

9.45 (a) Alkenes from *A* and *B*:

Alkenes from *C* and *D*:

(b) These conditions typically give the most stable alkene—the one with the greatest number of alkyl substituents on the carbons of the double bond. From *A*, alkene *A2* would be formed in greatest amount. Alkene *B1* is the only possibility from *B*. From *C*, alkenes *C2* and *C3* would be formed in greatest amount, with the trans isomer *C2* formed in somewhat larger amount than the cis isomer *C3*. Alkene *D1* is the only possibility from *D*.

(c) With a highly branched base, typically a 1-alkene will be formed in preference to internal alkenes in situations in which this possibility exists; therefore, the 1-alkenes would be *A1*, *C1*, and *D1*. Alkene *B1* is the only possibility.

9.46 (a) The deuterium-substituted cycloalkyl bromide *A* would react more slowly than cycloalkyl bromide *B* because of the primary deuterium kinetic isotope effect on the rate of elimination.

(b) Elimination of H and Br would occur more rapidly than elimination of D and Br because of the primary deuterium kinetic isotope effect on the rate of elimination.

9.47 (a) The alkyl halide undergoes internal rotation so that the hydrogen to be eliminated is anti to the leaving bromine. The stereochemistry of the alkene then follows.

(b) Redraw the structure of the alkyl iodide in a line-and-wedge structure in which the hydrogen that is eliminated and the iodine are anti, the methoxy (OCH$_3$) group and the other hydrogen are connected with dashed wedges, and the phenyl and the methyl group are connected with solid wedges.

9.48 (a) This primary alkyl halide will undergo predominantly the S$_N$2 reaction:

(b) This secondary halide, in the absence of a strong base, and in the presence of a polarizable nucleophile in a polar aprotic solvent, undergoes an S$_N$2 reaction. (It is driven to completion by precipitation of the KBr.)

(c) This is the same alkyl halide as in part (b), but a strong base is present, so an E2 reaction takes place.

(both cis and trans)

9.49 This reaction involves a secondary carbocation intermediate *A* that can rearrange to tertiary carbocation *B*:

carbocation *A*   carbocation *B*

Each carbocation can react with the solvent water to complete the S$_N$1 reaction and give an alcohol.

carbocation *A*

2-methyl-3-pentanol
alcohol *A1*
(both enantiomers)

carbocation *B*

2-methyl-2-pentanol
alcohol *B1*

Each carbocation can lose β-protons to solvent water to complete E1 reactions that give alkenes.

carbocation *A*

2-methyl-2-pentene   **(*E*)-4-methyl-2-pentene**   **(*Z*)-4-methyl-2-pentene**
alkene *A2*   alkene *A3*   alkene *A4*

carbocation *B*

1-methyl-2-pentene   **2-methyl-2-pentene**
alkene *B2*   alkene *B3 = A2*

9.50 (a) The step labeled k_1, ionization of the alkyl halide to give the carbocation, is rate-limiting. The product-determining steps are the subsequent four steps that form the ether and alkene products.

(b) The overall rate is k_1. The rate of formation of each product is proportional to the amount that is formed. The rate of formation of *A* is therefore $(0.595)(1.14 \times 10^{-4}) = 6.78 \times 10^{-5}$ sec^{-1}; the rate of formation of *B* is $(0.235)(1.14 \times 10^{-4}) = 2.68 \times 10^{-5}$ sec^{-1}; the rate of formation of *C* is $(0.156)(1.14 \times 10^{-4}) = 1.78 \times 10^{-5}$ sec^{-1} and the rate of formation of *D* is $(0.014)(1.14 \times 10^{-4}) = 1.60 \times 10^{-6}$ sec^{-1}.

(c) Because more of the substitution product is formed than elimination products combined, substitution is faster.

9.51 The first step in any problem that requires structures is to draw the structures:

$CH_3(CH_2)_4CH_2Br$

1-bromohexane
(1)

3-bromo-3-methyl-pentane (2)

1-bromo-2,2-dimethyl-butane (3)

3-bromo-2-methyl-pentane (4)

2-bromo-3-methyl-pentane (5)

(a) Compound (1) gives the fastest S_N2 reaction with sodium methoxide because it is the only primary alkyl halide with no alkyl branching—that is, only one substituent—at the β-carbon.

(b)   Compound (3) is least reactive to sodium methoxide in methanol because it cannot undergo a β-elimination—it has no β-hydrogens—and the three β-substituents make it virtually unreactive in the S_N2 reaction, much like neopentyl bromide.

(c)   Only compound (1) can give only one alkene in the E2 reaction with potassium *tert*-butoxide.

(d)   Compound (2) will give an E2 but no S_N2 reaction with sodium methoxide in methanol because it is tertiary and the S_N2 reaction is very slow because of steric hindrance to the nucleophilic reaction.

(e)   Compounds (4) and (5) undergo the S_N1 reaction with rearrangement. If compound (3) is forced to react, it will also undergo the S_N1 reaction with rearrangement, but under ordinary conditions it will not react.

(f)   Compound (2) will give the fastest S_N1 reaction because it is the only tertiary alkyl halide.

9.52   The products of the reactions of isopentyl bromide, $(CH_3)_2CHCH_2CH_2Br$, with mechanisms noted: (When the solvent is a by-product, it is not shown.)

(a)
$(CH_3)_2CHCH_2CH_2I$

$+ \; K^+ \; Br^-$

(by S_N2)

(b)
$(CH_3)_2CHCH_2CH_2OH \; + \; (CH_3)_2CHCH_2CH_2OEt$

(by S_N2)

$+ \; (CH_3)_2CHCH{=}CH_2 \; + \; K^+ \; Br^-$

(small amount by E2)

(c)
$(CH_3)_2CHCH{=}CH_2$

$+ \; K^+ \; Br^-$

(by E2)

(d)
$(CH_3)_2CHCH_2CH_2F$

$+ \; Cs^+ \; Br^-$

(by S_N2)

(e)   No reaction, because hydrogen bonds from solvent to fluoride ion are so strong that fluoride ion in a protic solvent cannot act as a nucleophile.

(f)
$(CH_3)_2CHCH_2CH_2OCH_3$

(by S_N2)

$(CH_3)_2CHCH{=}CH_2 \; + \; Na^+ \; Br^-$

(small amount by E2)

9.53   The products of the reactions of 2-bromo-2-methylhexane:

$$CH_3$$
$$CH_3C(CH_2)_3CH_3$$
$$Br$$

2-bromo-2-methylhexane

(a)

$$\underset{OH}{\overset{CH_3}{CH_3C(CH_2)_3CH_3}} \; + \; \underset{OEt}{\overset{CH_3}{CH_3C(CH_2)_3CH_3}} \; + \; \underset{\text{predominant alkene formed}}{\overset{CH_3}{CH_3C{=}CHCH_2CH_2CH_3}} \; + \; \overset{CH_3}{H_2C{=}C(CH_2)_3CH_3}$$

$$H_3O^+ \; + \; Et\overset{+}{O}H_2 \; + \; Br^-$$

(b)

$$\overset{CH_3}{CH_3C{=}CHCH_2CH_2CH_3} \; + \; \overset{CH_3}{H_2C{=}C(CH_2)_3CH_3}$$

$$+ \; Na^+ \; Br^-$$

(c)

$$\underset{OH}{\overset{CH_3}{CH_3C(CH_2)_3CH_3}} \; + \; \underset{I}{\overset{CH_3}{CH_3C(CH_2)_3CH_3}} \; + \; K^+ \; Br^-$$

$$+ \; \text{the alkenes in part (b)} \; + \; H_3O^+ \; Br^-$$

In part (c), the tertiary alkyl iodide product, if it forms at all, will also undergo rapid solvolysis to give the alcohol.

9.54   (a)   The presence of a strong base favors an E2 reaction. Two different protons could be lost, along with a bromide ion. Proton *(a)* is lost by an anti-elimination. To determine the stereochemistry, draw the structure with this proton and the bromine in an anti relationship. (Both line-and-wedge and sawhorse structures are shown.)

(2S,3R)-2-bromo-3-methylpentane

As the equation above shows, the loss of hydrogen *(b)* gives a different alkene. Because this alkene has no stereoisomers, the stereochemistry of the elimination is not relevant. Methanol and $Na^+ Br^-$ are formed as by-products in both eliminations.

(b) The reaction conditions favor a solvolysis reaction, which occurs by an S_N1–E1 mechanism accompanied by rearrangement.

The S_N1 product *A* is derived from the Lewis acid–base association reaction of solvent with the first-formed carbocation. It should be partly racemic, although the exact stereochemical outcome is difficult to predict exactly. It should have largely inverted configuration at carbon-2, because the carbocation will be a backside-solvated ion pair; this carbocation will react faster with the solvent molecule than the bromide ion is replaced by a second solvent molecule. (See Fig. 9.13, text Sec. 9.6D, p. 495.). The stereochemistry of carbon-3 is unaffected. The substitution product *B* is derived from the rearranged tertiary carbocation. Compounds *C* and *D* (*D* is a mixture of *E* and *Z* isomers) are E1 products. The stereochemistry at carbon-3 of alkene *C* is the same as in the starting alkyl halide. Alkene *D* should be the major alkene product because it has the larger number of substituents on the double bond.

Solutions to Integrated Problems

9.55 In all cases, the increase in polarizability is due to an increase in how easily the electron cloud of an atom is distorted, and polarizability increases going down a column on the periodic table (see Sec. 9.4E).

(a) $O < S < Se$ (b)   fluoroform < chloroform < iodoform   (c) $F^- < Cl^- < Br^- < I^-$

9.56 The first thing to do is to draw out the structures.

The order of increasing S_N2 reaction rates is $C < B < E < D < A$. Alkyl halides with three β-substituents (*C*) are virtually unreactive in S_N2 reactions. Secondary alkyl halides with no β-substituents (*B*) react more slowly than primary alkyl halides with two β-substituents (*E*), and the latter react more slowly than unbranched primary alkyl halides (*D*). Methyl halides (*A*) react most rapidly.

9.57   Dimma has attempted the S_N2 reaction of a *tertiary* alkyl halide with a strong base. As Dimma would have realized, had he consulted a superior organic chemistry text such as yours, that these are precisely the conditions that promote the E2 reaction. Consequently, Dimma formed the alkene isobutylene (methylpropene, $(CH_3)_2C{=}CH_2$), which bubbled merrily out of the reaction flask. Ima, on the other hand, allowed a strong base to react with methyl iodide, which cannot undergo elimination; the only alternative is the desired S_N2 reaction.

9.58   Allow the appropriate alkoxide, $(CH_3)_3CCH_2{-}O^- Na^+$, to react with ethyl bromide, $CH_3CH_2{-}Br$. Using sodium ethoxide with neopentyl bromide, $(CH_3)_3CCH_2{-}Br$, won't work; why?

9.59   In base, some of the chlorines are eliminated as chloride ion in E2 reactions. One such reaction is as follows; can you find another? (S_N2 reactions probably do not occur; why?)

9.60   In the transition state of the S_N2 reaction, the carbon at which substitution occurs is sp^2-hybridized; consequently, the ideal bond angles at this carbon are 120°. (See Fig. 9.2 in text Sec. 9.4C, p. 465.) This requirement for a 120° bond angle means that as cyclopropyl iodide is converted into the S_N2 transition state, the bond angle within the cyclopropane ring should increase. The increasing angle causes additional angle strain in the cyclopropane ring that raises the energy of the transition state. Consequently, S_N2 reactions of cyclopropyl halides are unusually slow.

greater angle;
greater strain

Such rate-inhibiting angle strain is present in neither compound *B*, isopropyl iodide, nor its S_N2 transition state. Consequently, the S_N2 reactions of isopropyl iodide are faster than those of cyclopropyl iodide.

9.61   If this compound were to undergo solvolysis, it would have to form carbocation *A*:

As a model will verify, the four asterisked carbons cannot become coplanar because of the constraints of the bicyclic ring system. Yet sp^2 hybridization requires trigonal planar geometry. Remember: *hybridization and geometry are connected.* If a molecule can't achieve the geometry for a given hybridization, then that hybridization will not occur. Because sp^2 hybridization is the lowest-energy hybridization for a carbocation, the inability to achieve this hybridization raises the energy of the carbocation.

But that's not all. Solvation of the carbocation can only occur from one side, because the opposite side of the electron-deficient carbon is blocked by part of the ring system. The absence of effective solvation, then, also raises the energy of this carbocation. Any solvolysis reaction involving such an unstable carbocation is slow.

9.62   (a)   Potassium fluoride, an ionic compound, is not soluble in hydrocarbons such as benzene. Consequently, it will not react in such solvents because its dissolved concentration is essentially zero. When the crown ether is added, the crown ether forms a benzene-soluble complex with the potassium ion, which, by electrical neutrality, includes an accompanying "naked" (that is, essentially unsolvated) fluoride ion. This soluble but unsolvated fluoride ion is a very good nucleophile, and reacts with benzyl bromide to yield benzyl fluoride, $PhCH_2F$, plus "naked" bromide ion, which then exchanges with insoluble potassium fluoride to give more dissolved fluoride and solid KBr. The newly dissolved fluoride can react with more benzyl bromide. The cycle is repeated until one of the reactants is exhausted. To summarize:

crown ether	crown ether · K⁺ F⁻	PhCH₂F	crown ether · K⁺ Br⁻	crown ether · K⁺ F⁻

(soluble)   (soluble)   (soluble)   (soluble)

+   ⇌   +   —PhCH₂Br→   +   ⇌   +

K⁺ F⁻   K⁺ F⁻   K⁺ F⁻   K⁺ Br⁻ + K⁺ F⁻

(insoluble,   (insoluble,   (insoluble,   (insoluble) (insoluble,
in excess)   in excess)   in excess)   in excess)

(b)   The scheme in part (a) shows that complexation of the cation by the crown ether is essential for the reaction to occur, because this is the only way to obtain the dissolved anion. Because [18]-crown-6 does not bind the smaller lithium cation—it is selective for the larger potassium cation—lithium fluoride is not solubilized by the crown ether, and is therefore unreactive whether the crown ether is present or not.

9.63   (a)   The β-hydrogens in the two alkyl bromides are indicated with an asterisk (*).

H*

Br

H* H* H* H*

A   *B*

H*   H

Br

(b)   Alkyl halide *A* can give an alkene from elimination of a hydrogen at each β position—that is, two alkenes (shown below). However, alkyl halide *B* cannot form an alkene by elimination of the bridgehead hydrogen because such an alkene violates Bredt's rule. (See Sec. 7.6C, text p. 352.) Therefore, only one alkene is formed from *B*. Although all of the double bonds in the alkenes from *A* are also bridgehead double bonds, they are not *trans* within a small ring and so are not twisted and can therefore form readily.

H

+

alkenes from *A*

H

H*   H

alkene from *B*

9.64   (a)   The thiosulfate dianion contains two types of nucleophilic atoms: the anionic oxygen and the anionic sulfur. Therefore, alkylation could occur at the oxygen or the sulfur:

O⁻   O⁻

Na⁺   ⁻S—S²⁺—O—CH₃   +   H₃C—S—S²⁺—O⁻ Na⁺

O⁻   O⁻

A   *B*

product of *O*-alkylation   product of *S*-alkylation

(b) In a hydrogen-bonding solvent such as methanol, the more weakly basic atom is the better nucleophile. (The principle is the same as in the solution to Problem 9.42.) Therefore, product *B* is the major one observed.

9.65 (a) The principle to apply is that the right side of the equilibrium is favored by the solvent that better separates and solvates ions. Because ethanol is the more polar solvent—it has the higher dielectric constant—it better separates ions. Because it is a protic, donor solvent, it better solvates ions; consequently, the equilibrium lies farther to the right in ethanol. Furthermore, ethanol's ability to donate hydrogen bonds allows it to provide better solvation for the bromide ion than does diethyl ether.

(b) In this case, the dielectric constants of the two solvents are the same. Consequently, the ability of the solvent to solvate the ions determines the relative position of the equilibrium. Because dimethylacetamide is *aprotic,* it does not solvate anions as well as the aqueous methanol solvent, which has a hydrogen-bonding solvation mechanism available. Consequently, the equilibrium lies farther to the right in aqueous methanol.

9.66 The reaction between methyl iodide and sodium ethoxide is an S_N2 process that has the following rate law:

$$\text{rate} = k[\text{CH}_3\text{I}][\text{CH}_3\text{CH}_2\text{O}^-]$$

This means that the rate of the reaction depends on the concentration of the nucleophile. In the first case, the nucleophile concentration changes during the reaction from 0.1 *M* to 0 *M*; in the second case, the nucleophile concentration changes from 0.5 *M* to 0.4 *M*. In the second case, the reaction is faster because the concentration of the nucleophile is higher at all times during the reaction. However, in either case, there is sufficient nucleophile present to react completely with the alkyl halide. If we wait long enough, the yield of the reaction will be the same in either case.

9.67 In methanol, the iodide ion is an excellent nucleophile but, because of its weak basicity, also a good leaving group. Therefore, the iodide ion reacts rapidly with methyl bromide to give methyl iodide. The methyl iodide then reacts more slowly with the methanol solvent in a slow S_N2 solvolysis reaction.

9.68 Because it is a tertiary alkyl halide without β-hydrogens, trityl chloride can react only by the S_N1 mechanism. A characteristic feature of this mechanism is that the first step is rate-limiting, whereas the Lewis acid–base association reaction of the nucleophile(s) with the carbocation intermediate occurs in the second, product-determining, step. Therefore, the rate of reaction of trityl chloride is *independent of the nucleophile concentration,* and therefore independent of the nucleophile that reacts with the carbocation.

The product-determining steps for the various reactions are different. The principle to apply is that the better nucleophile reacts more rapidly with the carbocation intermediate; therefore, the major product is derived from the better nucleophile.

Thus, in reaction (1), the only nucleophile available is water. The reaction of water with the carbocation and loss of a proton to solvent gives trityl alcohol, Ph_3C—OH, as the solvolysis product. (The solvent acetone can also in principle react with the carbocation, but the reaction product is unstable under the reaction conditions for reasons that we won't discuss here.) In reaction (2), both water and azide ion are in competition for the trityl cation. The basicities of water and azide are determined by the pK_a values of their respective conjugate acids. The conjugate acid of water, H_3O^+, has a pK_a value of 0. The conjugate acid of azide ion, HN_3, has a pK_a value of 4.7. Consequently, azide is the more basic nucleophile, and is evidently the better nucleophile. When, in reaction (2), sodium azide is added to the reaction mixture, trityl chloride ionizes at the same rate, but the carbocation intermediate reacts preferentially with azide ion.

When hydroxide ion is added in reaction (3), the rate of ionization of trityl chloride again remains unchanged; however, because both hydroxide ion and azide are present, both nucleophiles react with the carbocation to give a mixture of trityl alcohol and trityl azide.

9.69   (a)   If retention is the stereochemical scenario, each act of substitution has no effect on the optical activity, and $k°/k_S = 0$. If inversion is the stereochemical scenario, each act of substitution inverts the stereochemical configuration of one alkyl halide molecule, which gives it the opposite sign but the same magnitude of optical rotation. In this case, the optical activities of *two* molecules cancel; so, two molecules' worth of the original optical activity are lost. Therefore, $k°/k_S = 2$. This situation can be envisioned with the following "cartoon:"

In contrast, if equal amounts of retention and inversion occur, then for every event that results in the loss of two units of optical activity, we have another event that results in the loss of no optical activity; on average, $k°/k_S = 1$.

   (b)   The results show that $k°/k_S$ is 1.92 ± 0.18, which is equal to 2.0 within experimental uncertainty. This result establishes inversion as the stereochemical outcome.

The percent experimental uncertainty r of a ratio is determined from the percent uncertainties of the numerator and denominator (n and d, respectively) by the formula $r = \sqrt{n^2 + d^2}$. Using this formula, the percent uncertainty of the ratio is 9.38%, which gives $(0.0938)(1.92) = 0.18$ as the absolute uncertainty.

9.70   Compound A has no conformation in which the β-hydrogens and the bromine are anti, whereas compound B has a conformation (its more stable conformation) in which the β-hydrogens are anti to the bromine. For this reason, compound B should more readily undergo the E2 reaction.

Compound A has no β-H's anti to the Br in either conformation

Compound B has two β-H's anti to the Br in the more stable conformation

9.71 The important keys to solving this problem are (1) that the E2 reaction is fastest when the hydrogen and chlorine are anti; and (2) the substituted cyclohexanes can undergo the chair interconversion rapidly. In order for menthyl chloride to undergo anti-elimination, it must undergo the chair interconversion so that the chlorine is axial:

menthyl chloride
(this conformation cannot undergo *anti*-elimination)

anti to the Cl

2-menthene

The only alkene that can be formed by anti-elimination is 2-menthene. In contrast, neomenthyl chloride has two β-hydrogens, H^a and H^b, that can be lost in anti-eliminations along with the chlorine to give 2-menthene and 3-menthene, respectively:

neomenthyl chloride

elimination of H^a and Cl

2-menthene

elimination of H^b and Cl

3-menthene

More 3-menthene is formed because it is the alkene with the greater number of alkyl substituents at the carbons of its double bond.

9.72 The key to the solution is to assume that E2 reactions are faster when they can occur with anti stereochemistry. In the first compound, two β-hydrogens can be eliminated with the bromine: the hydrogen at the ring junction, and a hydrogen of the methyl group. The hydrogen at the ring junction is eliminated because it gives the alkene with the greater number of alkyl substituents on the double bond.

anti to the Br

In the second compound, the hydrogen at the ring junction cannot be anti to the bromine; therefore, the only anti hydrogen available for elimination is a hydrogen of the methyl group:

syn (gauche) to the Br

anti to the Br

9.73 (a) Product *A* results from anti-elimination by a conventional E2 reaction:

3-cyclohexen-1-ol
(*A*)

(The same sort of elimination reaction gives compound *A* from *cis*-4-chlorocyclohexanol.) Product *B* is formed by an intramolecular nucleophilic substitution reaction of the alkoxide that occurs in the twist-boat conformation (shown here for simplicity as a boat):

+ :Cl⁻   *B*

Product *C* results from an ordinary S_N2 reaction:

C

(b)   The intramolecular substitution that yields the bicyclic compound *B* requires, like all nucleophilic substitutions, approach of the nucleophile to the side opposite to the departing chloride. This can happen only in the boat form of the trans stereoisomer. Because a conformation that permits opposite-side substitution of the intramolecular alkoxide is not available to the cis stereoisomer, it reacts instead by an ordinary S_N2 reaction with the external nucleophile ⁻OH.

> Notice that the formation of product *B* instead of *cis*-1,4-cyclohexanediol from the trans stereoisomer of the starting material means that the *intramolecular* nucleophilic substitution reaction is much faster than the S_N2 reaction of the same compound with hydroxide. Intramolecular reactions are in many cases faster than analogous intermolecular reactions. This point is explored in more detail in Sec. 12.8 of the text.

9.74   The reaction of butylamine with 1-bromobutane is a typical S_N2 reaction. (Bu— = the butyl group = $CH_3CH_2CH_2CH_2$—):

Bu—NH₂   :Br⟶Bu ⟶ Bu—N⁺H₂—Bu   :Br⁻

This mechanism is consistent with a second-order rate law, because the rate law requires one molecule of amine and one molecule of alkyl halide in the transition state.

The second reaction is also a nucleophilic substitution reaction, but, because it is intramolecular (that is, the nucleophile and carbon at which it reacts are part of the same molecule), the reaction is first-order.

⟶   N⁺H₂   :Br⁻

9.75 (a) The reaction occurs largely with retention of configuration and only a small amount of inversion. This result is inconsistent with stereochemical inversion observed with other S_N2 reactions, and it caused a stir for that reason. (By the time this study was done, stereochemical inversion in the S_N2 reaction had become well established.)

(b) The initial formation of *trans-P* occurs by the usual S_N2 mechanism with inversion. The subsequent formation of *cis-P* occurs when iodide ion in solution displaces iodide, again with inversion, from *trans-P* by a second S_N2 reaction with inversion. Starting with either *cis-* or *trans-P* gives the same equilibrium mixture in which *cis-P* is the major component of the equilibrium.

(c) *Cis-P* is favored because both substituents can exist in equatorial arrangements. (See Sec. 7.5B and Focused Problem 7.15 for a discussion of the puckering of cyclobutane rings.) As in substituted cyclohexanes, substituted cyclobutanes are destabilized by 1,3-diaxial interactions.

cis-P
both substituents are equatorial
(more stable)

trans-P
one substituent is axial
(less stable)

9.76 As triethylamine approaches the isopropyl iodide, the alkyl groups of the nucleophile and those of the isopropyl iodide approach so close that van der Waals repulsions occur and the reaction rate is reduced. This steric effect does not occur with quinuclidine because its alkyl groups are in a sense "tied back" and don't "collide" with the methyl groups of isopropyl iodide. Therefore, quinuclidine reacts faster with isopropyl iodine.

Because methyl iodide does not have branching at its α-carbon, there is greatly reduced steric effect with the alkyl groups of triethylamine. Because this rate-retarding effect is reduced, the reactivity of triethylamine is closer to that of quinuclidine.

9.77 Remember that the key to working problems of this type is to find a known structure and work from it. (See point 3 in Study Guide Link 4.3 of this manual.) Compound *B* is an alkene, because it decolorizes Br_2 in an inert solvent. The structure of the ozonolysis product *C* leads to the identity of *B*, shown at right.

The molecular formula of *A*, $C_8H_{13}Br$, is equivalent to that of the alkene *B* plus the elements of HBr. Because the structure of *C* shows that the carbon skeleton must be bicyclic, there can be no additional unsaturation in *A*, a point confirmed by the fact that it does not add bromine. Evidently, compound *A* is an alkyl halide that undergoes a reaction with a strong base to give the alkene *B*; this reaction is then an E2 reaction. The alkyl halide *A*, and its E2 reaction, are shown at right.

9.78 Compound *A* is an alkyl halide with two degrees of unsaturation. Because one unsaturation is accounted for by the cyclohexane ring found in subsequent products, the other must be a double bond. Subsequent elimination reactions of *A* yield two dienes *B* and *C*. The catalytic hydrogenation of these two compounds establishes that compounds *A*–*C* and 1-isopropyl-4-methylcyclohexane have the same carbon skeleton. Because ozonolysis leaves the cyclohexane ring intact, the double bond must be external to the ring; and the identity of product *F* shows that both double bonds are in fact attached to the ring. There is only one possibility, shown at right, for compound *B*.

compound *B*

The following are possibilities for *A*:

A1   *A2*   *A3*   *A4*

Any other possibility for *A* is ruled out both by the structure of *B* and by the fact that compound *A* is achiral. Compound *A1* is ruled out by the fact that it can only give one diene in the E2 reaction; yet compound *A* gives two dienes, compounds *B* and *C*. That compound *A* undergoes S_N1 solvolysis rapidly also rules out compound *A1*, because it is a primary alkyl halide that would not react under S_N1 conditions. Because compound *A* undergoes ozonolysis to give acetone—and because structure *A1* was ruled out—*only structure A2 fits all the data.* The structure of alkene *B* is given above; the structure of alkene *C*, the other E2 reaction product of compound *A* = *A2*, is as follows.

compound *C*

Now that we know compound *A*, we are in a position to derive the structures of compounds *D* and *E*. These are the bromine-addition products of compound *A*. The two products are the two possible diastereomers:

A2 = A

D and *E*

10

Free-Radical Reactions, Main Group Organometallic Compounds, and Carbenes

CHAPTER 10 SUMMARY

- Free radicals are species containing an unpaired electron.

- Carbon free radicals are sp^2-hybridized, and the unpaired electron occupies the unhybridized $2p$ orbital.

- The stability of carbon free radicals is in the order tertiary > secondary > primary, but the effect of alkyl substitution on free-radical stability is considerably less significant than the effect of alkyl substitution on carbocation stability.

- Most reactions involving free radicals as intermediates involve free radical chains. A free-radical chain reaction involves three phases:

  - In the *initiation* phase, radicals are formed from nonradical precursors by the action of free-radical initiators such as peroxides, light, or molecular dissociation.

  - In the *propagation* phase, free radicals undergo reactions that produce new free radicals. Among these are addition to a double or triple bond and atom abstraction. Propagation reactions occur repeatedly in a free-radical chain reaction.

  - In the *termination* phase, radicals are destroyed by reaction with other radicals by radical recombination, β-scission, or other processes. Although these reactions occur throughout a free-radical chain reaction, they become significant generally only when the concentrations of other reactants dwindles.

- In free-radical reactions, electrons are moved individually rather than in pairs.

- Free-radical reactions are described mechanistically by a curved-arrow notation called the *fishhook notation.* In this notation the sources and fates of single electrons are traced with singly-barbed arrows, or fishhooks.

- The ease of formation of free radicals is given by *bond dissociation energies,* which are the enthalpies required to dissociate various types of bonds into free radicals by moving one of the electrons in an electron-pair bond to each atom.

  - Like other reaction enthalpies, bond-dissociation energies can be used in Hess's-law calculations to determine the approximate $\Delta H°$ values of reactions.

- Three types of free-radical chain reactions are described in this chapter: free-radical addition of HBr to alkenes; free-radical polymerization of alkenes; and free-radical substitution reactions of alkanes and halogens.

- In free-radical addition of HBr to alkenes, HBr adds so that the Br goes to the alkene carbon with fewer substituents and the hydrogen goes to the alkene carbon with more alkyl substituents.

  - Free-radical addition of HBr to alkenes and alkynes is initiated by small amounts of peroxides or by other initiators such as AIBN. HI and HCl do not undergo a free-radical addition of this type.

  - The regiospecificity of the peroxide-initiated HBr addition reaction is opposite to the regiospecificity of HBr addition in the absence of peroxides, which does not involve free-radical intermediates.

- In the propagation phase of the reaction, a bromine atom produced in the initiation phase of the reaction reacts at the alkene carbon with fewer substituents; as a result, a carbon radical is formed at the alkene carbon with more substituents. This radical abstracts a hydrogen from an HBr molecule, thereby completing the addition and generating another bromine atom to continue the radical chain. Both free-radical stability and a steric effect are important in the regiospecificity of this reaction.

■ In free-radical polymerization, an initiating radical adds to one alkene carbon to produce a radical at the other alkene carbon, which adds to the double bond of another alkene molecule to produce a new radical. This radical adds to the double bond of yet another alkene to form a new radical, and the process continues so that a large molecule called a *polymer* is formed.

- When asymmetric carbons are formed in a polymer such as poly(propylene), various polymer stereochemistries are possible. In isotactic poly(propylene), the methyl substituents are on the same side of the polymer chain; in syndiotactic poly(propylene), the position of the methyl substituents alternates regularly along the polymer chain; in atactic poly(propylene), the positions of the methyl substituents are random.

■ In free-radical substitution, a bromine or chlorine atom produced from dissociation of the parent halogen molecule in the initiation step substitutes for an alkane hydrogen.

- The halogen atom abstracts a hydrogen from the alkane to form a carbon radical. This radical then reacts with a halogen molecule to form a haloalkane and another halogen atom.

- The relative amounts of the possible haloalkane products is governed by the relative stability of the radical intermediates, by the relative numbers of different hydrogen types in the alkane, and by the reactivity of the halogen.

- More reactive chlorine atoms are less selective than less reactive bromine atoms. The greater selectivity of less reactive reagents is called the *reactivity–selectivity principle.*

■ Internal alkynes can be reduced to trans alkenes with sodium dissolved in liquid ammonia, which is a source of solvated electrons.

- This process is complementary to catalytic hydrogenation with a poisoned catalyst, which gives cis alkenes.

- The mechanism involves a radical-anion intermediate.

■ Organometallic reagents contain a carbon–metal bond.

■ Two of the most important main-group organometallic reagents are Grignard reagents (R—MgX, where R is any carbon group and X is Cl, Br, or I) and organolithium reagents (R—Li).

- Grignard and organolithium reagents have significant carbon-anion (carbanion) character

- An important method for forming both types of reagent is the reaction of an alkyl halide with the metal.

- Ether solvents are critical for the formation of Grignard reagents because they solvate and solubilize the reagents.

■ Grignard and organolithium reagents react rapidly with even weakly acidic compounds such as water and alcohols to give an alkane and the conjugate base ⁻OH or ⁻OR. This is a useful reaction for preparing deuterium-substituted compounds using D_2O as the source of the isotope.

■ The most important reactions of Grignard and organolithium reagents are with carbonyl compounds (discussed in Chapters 19 and 21) and with epoxides (discussed in Chapter 12).

■ 1-Alkynes are the most acidic of the aliphatic hydrocarbon types, with a pK_a of about 25. Their acidity allows the formation of acetylenic organometallic compounds from 1-alkynes.

- Sodium acetylides can be prepared by reactions of 1-alkynes with $NaNH_2$ (sodium amide) in liquid ammonia.

- Acetylenic Grignard and lithium reagents are formed by the acid–base reaction of a 1-alkyne with a Grignard or organolithium reagent.

- Sodium acetylides react with alkyl bromides in S_N2 reactions to form a new carbon–carbon bond, giving internal alkynes from 1-alkynes.

- Carbenes are molecules containing a divalent carbon: a carbon atom with only two covalent bonds, a nonbonding electron pair, and a vacant 2p orbital. Carbenes, like carbocations and free-radicals, are unstable species that are important reactive intermediates in some reactions.

  - Dihalocarbenes ($:CX_2$, X = Br or Cl) can be generated from a haloform ($HCCl_3$ or $HCBr_3$) and a strong base such as K^+ $(CH_3)_3C$—O^- by α-elimination.

- Carbenoids are organometallic reagents that serve as carbene equivalents. The Simmons–Smith reagent is an organozinc reagent approximately represented by I—CH_2—Zn—I that can be prepared from diethylzinc and methylene iodide (diiodomethane, CH_2I_2). This carbenoid reagent serves as a synthetic source of the parent carbene $:CH_2$.

- Both dihalocarbenes and $:CH_2$ from the Simmons–Smith reagent undergo rapid syn additions to alkenes to give cyclopropanes.

FURTHER EXPLORATIONS

10.1   Bond Dissociation Energies and Heats of Reaction

Study Problem 10.4 in Sec. 10.1E, text p. 521, shows how bond dissociation energies can be used to calculate heats of reactions: subtract the bond dissociation energies of the bonds formed from those of the bonds broken. This exploration shows why this process works.

Essentially, this is a Hess's law calculation. This can be illustrated by using bond dissociation energies to calculate the $\Delta H°$ of the following gas-phase reaction:

$$·CH_3 + Cl_2 \longrightarrow Cl—CH_3 + ·Cl \qquad \text{(FE10.1)}$$

The term "bond dissociation energy" will be abbreviated with the acronym "BDE" for convenience. The bond formed is a H_3C—Cl bond, with BDE = 350 kJ mol^{-1}. The bond broken is a Cl—Cl bond, with BDE = 239 kJ mol^{-1}. The $\Delta H°$ for the reaction, then, is 239 – 350 = –111 kJ mol^{-1}.

The reaction in Eq. FE10.1 can be viewed as the sum of the following two reactions and their associated $\Delta H°$ values.

		$\Delta H°$
$·CH_3 + ·Cl \longrightarrow Cl—CH_3$		–BDE of Cl—CH_3
$Cl_2 \longrightarrow 2\ Cl·$		BDE of Cl_2
Sum: $·CH_3 + Cl_2 \longrightarrow Cl—CH_3 + ·Cl$		BDE of Cl_2 – BDE of Cl—CH_3

As this equation shows, the $\Delta H°$ values are the BDEs of the bonds involved. Because the first reaction is simply the bond dissociation of H_3C—Cl written in reverse, its $\Delta H°$ is the negative BDE of H_3C—Cl. Thus, the total $\Delta H°$ is then the BDE of the bond broken less the BDE of the bond formed.

10.2   Mechanism of Formation of Grignard Reagents

Despite being discovered over 100 years ago and studied extensively ever since, the details of the formation of organomagnesium compounds (Grignard reagents) are still debated. It is largely agreed that an alkyl radical is formed on the surface of the metal but it is unclear how "free" this radical is: does it stay adsorbed to the magnesium surface, or diffuse away before it combines with magnesium ions and the halide? A detailed investigation of these theories is beyond the scope of our consideration here. In either scenario, the mechanism of formation of Grignard reagents can be summarized as follows. First, an electron is transferred from the magnesium to the alkyl halide (X = I, Br, or Cl) to form a radical anion. The radical anion then undergoes fragmentation to give an alkyl radical and a halide anion. Finally, the alkyl radical combines with the magnesium ion and halide, which then bonds with the halide, to form the Grignard reagent.

$$R{-}X \ + \ \ddot{M}g \ \longrightarrow \ R{-}X^{\bar{\cdot}} \ + \ Mg^{\dagger}$$
radical anion

$$R{-}X^{\bar{\cdot}} \ \longrightarrow \ R{\cdot} \ + \ X^{-}$$

$$R{\cdot} \ + \ {\cdot}Mg^{+} \ \longrightarrow \ R{-}Mg^{+}$$

$$R{-}Mg^{+} + X^{-} \ \longrightarrow \ R{-}Mg{-}X$$
Grignard reagent

This sequence is *not* a chain reaction. As noted in the text, these reagents are likely part of a complex with solvent molecules (Sec. 10.5B), typically diethyl ether or tetrahydrofuran (THF), which can stabilize the electron-deficient magnesium.

REACTION REVIEW

I.   CONVERSION OF ALKENES AND ALKANES INTO ALKYL HALIDES WITH FREE-RADICAL REACTIONS

A.   *FREE-RADICAL ADDITION OF HYDROGEN BROMIDE TO ALKENES*

1.   Alkenes react with hydrogen bromide in the presence of peroxides so that the bromine adds to the less branched carbon of the double bond. (The "less branched" carbon is the carbon with fewer alkyl substituents.)

a.   This reversal of regioselectivity in HBr addition in the presence of a peroxide is termed the peroxide effect.

b.   This mechanism involves free radicals as reactive intermediates.

2.   The role of peroxides (or other initiators) is to react with a small amount of HBr to generate a bromine atom.

a. A small amount of peroxides undergoes homolysis to peroxy radicals.

$$(CH_3)_3C{-}O{-}O{-}C(CH_3)_3 \longrightarrow 2\,(CH_3)_3C{-}O\bullet$$

di-*tert*-butyl peroxide
(a common free-radical initiator)

b. A peroxy radical removes a hydrogen atom from HBr to give a bromine atom.

$$(CH_3)_3C{-}O\bullet \quad H{-}Br \longrightarrow (CH_3)_3C{-}OH \;+\; \bullet Br$$

c. These are the initiation steps of the reaction.

3. The bromine atom begins a chain reaction in which it reacts with the alkene to give a new radical, which then reacts with HBr to generate another bromine radical.

a. These are the propagation steps of the reaction.

c. Most free-radical reactions are chain reactions: the radical (Br• in the above mechanism) that is consumed in one step is formed again in a subsequent step. It is said to propagate the chain.

4. Free-radical chain reactions end when radical intermediates recombine. Such reactions are called *termination* steps. Because radical intermediates are present in very low concentration, termination reactions typically occur only when other reactants are exhausted.

5. The regioselectivity of HI or HCl additions to alkenes are *not* affected by the presence of peroxides.

B. *FREE-RADICAL HALOGENATION OF ALKANES*

1. Alkanes react with bromine and chlorine in the presence of heat or light in free-radical substitution reactions to give alkyl halides.

$$CH_3{-}CH_3 + Br_2 \xrightarrow[\text{free-radical initiators}]{\text{heat, light, and/or}} CH_3{-}CH_2Br + BrCH_2{-}CH_2Br + CH_3{-}CHBr_2 + H{-}Br$$

(The relative amounts of various bromination products depends on conditions and relative concentrations of hydrocarbons and bromine.)

a. The reaction is initiated when a halogen molecule absorbs energy from heat or light and dissociates homolytically into halogen atoms. (X = halogen)

$$X{-}X \xrightarrow{\text{light or heat}} X\bullet + \bullet X$$

b. The ensuing chain reaction has the following propagation steps.

$$R{-}H \quad \bullet X \longrightarrow R\bullet + H{-}X$$

$$R\bullet \quad X{-}X \longrightarrow R{-}X + \bullet X$$

c. Termination steps result from the recombination of radical species.

2. Abstraction rates of the various types of hydrogen is in the order tertiary > secondary > primary, which follows the relative stabilities of the free-radical intermediates produced.

| isobutane
(large excess) | ***tert*-butyl bromide**
(99.5% of product) | **isobutyl bromide**
(0.5% of product) |

a. The relative numbers of hydrogens of the different types also contributed probabilistically to the relative amounts of products produced.

b. Chlorination is much faster than bromination and is less selective; the relative numbers of hydrogens of each type is more important in determining the relative amounts of various products.

| isobutane
(large excess) | ***tert*-butyl chloride**
(36% of product) | **isobutyl chloride**
(64% of product;
more primary than tertiary
hydrogens in the reactant) |

3. The inverse relationship of selectivity and reactivity is called the *reactivity–selectivity principle*. This principle operates in many types of organic reactions.

4. Free-radical halogenations with chlorine and bromine proceed smoothly, halogenation with fluorine is violent, and halogenation with iodine does not occur.

II. FREE-RADICAL POLYMERIZATION OF ALKENES

1. In the presence of free-radical initiators such as peroxides, many alkenes (monomers) react to form polymers in a reaction called polymerization.

vinyl chloride

poly(vinyl chloride)

a. Polymers are very large molecules composed of repeating units.

b. Polymers such as poly(vinyl chloride) and poly(ethylene) are examples of *chain polymers* (sometimes called *addition polymers*)—polymers in which no atoms of the monomer unit have been lost as a result of the polymerization reaction.

2. Many polymerizations occur by a free-radical mechanism, illustrated here for polyethylene.

III. REDUCTION OF INTERNAL ALKYNES WITH ALKALI METALS IN LIQUID AMMONIA

1. Reaction of an internal alkyne (not a 1-alkyne) with a solution of an alkali metal (usually sodium) in liquid ammonia gives a trans-alkene.

2. This reaction is complementary to catalytic hydrogen of internal alkynes with a poisoned catalyst, which gives cis alkenes. (Sec. 4.10B).

   a. The deep blue solution of alkali metals dissolved in pure liquid ammonia is a source of electrons complexed to ammonia (solvated electrons).

   b. The addition of an electron to the triple bond results in a species (a radical anion) that has both an unpaired electron and a negative charge.

   c. The radical anion is such a strong base that it removes a proton from ammonia to give a vinylic radical (a radical in which the unpaired electron is associated with one carbon of a double bond).

   d. The resulting vinylic radical rapidly undergoes inversion.

   i. The equilibrium between the cis and trans radicals favors the trans radical for steric reasons.

   ii. Because there is much more of the trans radical, the ultimate product of the reaction is the trans alkene.

   e. The vinylic radical accepts an electron to form a vinylic anion.

a trans vinylic radical   a vinylic anion

f.   This anion is more basic than the conjugate base of ammonia; the anion therefore removes a proton from ammonia to complete the addition.

IV.   MAIN-GROUP ORGANOMETALLIC REAGENTS

A.   *FORMATION OF GRIGNARD AND ORGANOLITHIUM REAGENTS*

1.   Reaction of an alkyl or aryl halide with metallic magnesium in an ether solvent, or with lithium in a hydrocarbon solvent, gives an organometallic reagent—a compound with a carbon–metal bond.

an alkyl or aryl halide   a Grignard reagent
(many iodides and
chlorides also work)

an organolithium
reagent

2.   Ether solvents such as anhydrous diethyl ether or tetrahydrofuran are required for Grignard reagent formation because they solvate the magnesium in the Grignard reagent and wash the reagent from the metal surface.

3.   The carbon–metal bond of a Grignard or organolithium reagents is polar and has significant partial negative charge on the carbon. They can be conceptualized as carbon anions (carbanions) partially bonded to metals.

B.   *FORMATION OF ACETYLENIC ORGANOMETALLIC REAGENTS FROM 1-ALKYNES*

1.   Acetylenic Grignard reagents can be formed by the reaction of 1-alkynes with a methyl or ethyl Grignard reagent (CH_3MgBr or CH_3CH_2MgBr).

a 1-alkyne   **ethylmagnesium bromide**   an acetylenic   **ethane**
                                                  Grignard reagent   (a gas)

a.   This is effectively an acid–base reaction in which a stronger acid (the 1-alkyne) reacts with a stronger base (the ethyl anion in the guise of a Grignard reagent) to form a weaker base (the acetylenic anion in the guise of a Grignard reagent) and a weaker acid (ethane).

b.   1-Alkynes are stronger acids ($pK_a \sim 25$) than alkanes ($pK_a \sim 55$–60).

2. Sodium acetylide reagents can be formed from a 1-alkyne with sodium amide in liquid ammonia.

$$R—C{\equiv}C—H \ + \ Na^+ \ :\overset{..}{\underset{..}{N}}H_2 \ \xrightarrow{\text{NH}_3\text{ (liq.)}} \ R—C{\equiv}C\overset{..}{:} \ Na^+ \ + \ :NH_3$$

<table>
<tr><td>a 1-alkyne</td><td>**sodium amide**</td><td>a sodium acetylide
(a sodium 1-alkyn-1-ide)</td></tr>
</table>

This is effectively an acid–base reaction in which a 1-alkyne (a stronger acid) reacts with a stronger base (amide ion) to give a weaker base (an acetylide ion) and ammonia (a weaker acid).

C. *PROTONOLYSIS OF GRIGNARD AND ORGANOLITHIUM REAGENTS*

1. All Grignard or organolithium reagents react vigorously with even relatively weak acids, such as water and alcohols, to give the conjugate-base hydroxide or alkoxide and the conjugate-acid hydrocarbon of the carbanion. This reaction is an example of protonolysis—a reaction with the proton of an acid that breaks chemical bonds.

$$R—MgX \ + \ R'—OH \ \longrightarrow \ R—H \ + \ Mg^{2+} \ + \ X^- \ + \ R'—O^-$$

As this equation illustrates, hydrocarbons can be prepared by protonolysis of a Grignard or organo-lithium reagent.

2. A particularly useful variation of this reaction is the preparation of hydrocarbons labeled with the hydrogen isotopes deuterium (D, or ^{2}H) or tritium (T, or ^{3}H) by reaction of a Grignard reagent with the corresponding isotopically labeled water.

D. *REACTION OF SODIUM ACETYLIDES WITH ALKYL HALIDES AND SULFONATES TO FORM INTERNAL ALKYNES*

1. Acetylenic anions (as sodium acetylides) can be used as nucleophiles in S_N2 reactions to prepare other alkynes.

a. When an alkyl halide or sulfonate is used in this reaction, it must be primary or unbranched secondary.

b. This is another method of carbon–carbon bond formation.

V. FORMATION OF CYCLOPROPANES WITH CARBENES AND CARBENOIDS

A. *FORMATION OF DIHALOCARBENES AND THEIR REACTIONS WITH ALKENES*

1. The compound methylene (:CH$_2$) and its substituted derivatives are called *carbenes*. A dihalocarbene, (a dihalomethylene) is a species with the structure :CX$_2$, where X = halogen.

   a.. Carbenes contain a divalent carbon, which has a nonbonding electron pair and an empty 2p orbital.

   b. Carbenes can react as a nucleophile and an electrophile at the same time.

2. Dihalocarbenes can be formed by an α-elimination reaction of a haloform (a compound with the structure HCX$_3$, where X = halogen) promoted by a strong base.

   a. The haloform first loses a proton to a strong base to form a trihalomethyl anion.

tert-butoxide   a haloform   *tert*-butyl alcohol   a trihalomethyl
                                                                                anion

b.   The trihalomethyl anion loses halide ion to form the dihalocarbene.

a trihalomethyl   a dihalomethylene
      anion

3.   Dihalocarbenes undergo concerted syn-addition reactions with alkenes to form dihalocyclopropanes.

B.   *CARBENOIDS: THE SIMMONS-SMITH REACTION*

1.   The reaction of diethylzinc [(CH$_3$CH$_2$)$_2$Zn] with methylene iodide (diiodomethane, CH$_2$I$_2$) gives an α-iodomethylzinc organometallic compound called the Simmons–Smith reagent.

$$(CH_3CH_2)_2Zn \quad + \quad CH_2I_2 \quad \longrightarrow \quad I-CH_2-Zn-I \quad + \quad CH_3CH_2CH_2CH_3$$

diethylzinc   **methylene**   **Simmons-Smith reagent**
                         iodide

2.   The original preparation of the Simmons–Smith reagent uses a Zn–Cu alloy called a *zinc–copper couple* for this preparation instead of diethylzinc.

a.   The Simmons–Smith reagent can be conceptualized as methylene (:CH$_2$) coordinated to zinc.

b.   Because they show carbenelike reactivity, *α*-halo organometallic compounds are sometimes called carbenoids. A carbenoid is a reagent that is not a free carbene but has carbenelike reactivity.

2.   The reaction of the Simmons–Smith reagent with an alkene results in the concerted syn addition of methylene to the alkene to give a cyclopropane.

Simmons–Smith   benzene
reagent   (solvent)

VI.   MULTISTEP ORGANIC SYNTHESIS

A.   *ASPECTS OF ORGANIC SYNTHESIS*

1.   Functional-group transformation: the transformation of one functional group into another.

a.   It is helpful to construct functional-group transformation charts as reactions are learned.

b.   Examples of these are found in Sec. 10.8, pp. 549–550 of the text.

2.   Carbon–carbon bond formation

3.   Stereochemistry (discussed in Sec. 12.10)

B. *Retrosynthetic Analysis*

1. The compound to be prepared in an organic synthesis is the *target molecule.*
2. Potential synthetic pathways are best developed by starting at the target molecule and working backwards to a reasonable starting material.
3. Make a list of reactions that can form new carbon–carbon bonds and note particularly in your pathway where these reactions might be used.
4. In many cases there can be more than one reasonable synthesis of a target molecule.
5. Examples of retrosynthesis analyses are given in Study Problems 10.5 and 10.6 in Sec. 10.8, of the text.

 # SOLUTIONS TO PROBLEMS

Solutions to Focused Problems

10.1   (a)

$(CH_3)_3C\!-\!C(CH_3)_3 \longrightarrow 2\,(CH_3)_3C\cdot$

(b)

$(CH_3)_2C\!=\!CH_2 \quad \cdot Br \longrightarrow (CH_3)_2\dot{C}\!-\!CH_2\!-\!Br$

(c)

$$\left[H_2C\!=\!CH\!-\!CH\!=\!CH\!-\!\dot{C}H_2 \longleftrightarrow H_2\dot{C}\!-\!CH\!=\!CH\!-\!CH\!=\!CH_2 \right]$$

(d)

$R\!-\!CH_2\!-\!\dot{C}H_2 \longrightarrow R\cdot\ +\ H_2C\!=\!CH_2$

> As part (c) demonstrates, the fishhook notation, like the curved-arrow notation, can be used to derive resonance structures.

10.2   (a)   This reaction is heterolytic, because it requires movement of electron *pairs.*

$:N\!\equiv\!C: \quad CH_2CH_3 \longrightarrow\ :N\!\equiv\!C\!-\!CH_2CH_3\ +\ :\ddot{Br}:^-$
$:\!\ddot{Br}:$

(b)   This reaction is homolytic, because it involves movement of *single* electrons.

$CH_3\dot{C}HOH \longrightarrow CH_3\dot{C}HOH\ +\ H\!-\!\ddot{O}CH_3$
$H\ \cdot\ddot{O}CH_3$

(c)   The derivation of resonance structures for radicals uses the fishhook notation:

$$\left[CH_3\dot{C}H\!-\!\ddot{O}H \longleftrightarrow CH_3CH\!=\!\dot{O}H \right]$$

10.3   The initiation step, shown in the problem, produces a bromine atom. This adds to the π bond in the first propagation step to give the free radical *A.*

$$H_2C{=}CH_2 \quad \xrightarrow{\;\;\bullet Br\;\;} \quad H_2\overset{\bullet}{C}{-}CH_2Br$$

A

In the second propagation step, the radical *A* reacts with Br_2 to give the addition product, 1,2-dibromoethane, and a bromine atom, which can begin the propagation cycle again.

$$Br{-}Br$$

$$H_2\overset{\bullet}{C}{-}CH_2Br \quad \longrightarrow \quad BrCH_2{-}CH_2Br \;\; + \;\; \bullet Br$$

A   **1,2-dibromoethane**

Radical *A* could form the product by reacting with another initiating radical Br·, but this does not happen! The reason is that the initiating radical is present in trace amounts, whereas the reactant Br_2 is present in large amounts. Be sure to read Study Problem 10.3 on text p. 515 if you missed this point.

10.4 (a) The first initiation step is shown in Eq. 10.9, text p. 512. In the second initiation step, the radical produced in the first initiation step abstracts a hydrogen atom from HBr:

$$H{-}Br$$

$$NC{-}\overset{\bullet}{C}(CH_3)_2 \quad \longrightarrow \quad NC{-}\overset{H}{\underset{|}{C}}(CH_3)_2 \;\; + \;\; \bullet Br$$

The bromine atom then adds to the π bond of cyclohexene in the first propagation step (*1*); the resulting radical then reacts with HBr in the second propagation step (*2*) to give the addition product.

(b) Three possible recombination products are *A*, *B*, and *C*, formed by the following recombination reactions. All of the radical reactants are chain-propagating radicals in the reaction mechanism of part (a). The chances of these steps occurring is very small, so these products are typically observed only in very small amounts at the end of the reaction, when the other reactants have been exhausted.

$$2\,Br\bullet \;\longrightarrow\; Br{-}Br$$

A

B   *C*

10.5 (a)

$$H_2C{=}CHCH_2CH_2CH_3 \;\; + \;\; HBr \;\; \xrightarrow{\text{peroxides}} \;\; BrCH_2{-}CH_2CH_2CH_2CH_3$$

1-pentene   **1-bromopentane**

(b) In this case, both of the two possible reaction pathways involve secondary free radicals. Recall, however, from the text that the regioselectivity of free-radical HBr addition is controlled in part by steric factors. Therefore, the Br goes to the less congested side of the double bond.

(E)-4,4-dimethyl-2-pentene   **4-bromo-2,2-dimethylpentane**   **3-bromo-2,2-dimethylpentane**
                                                    (the major product)

10.6   (a)

(b)

(c)

In this mechanism, we show formation of the (Z) stereoisomer of the product, which is actually the observed product. Can you think of one or more reasons why this stereoisomer might be formed?

10.7   (a)   One H—Cl bond (431 kJ mol^{-1}) and one H$_3$C—Cl bond (350 kJ mol^{-1}) are formed; and one H$_3$C—H bond (439 kJ mol^{-1}) and one Cl—Cl bond (239 kJ mol^{-1}) are broken. As shown by Eq. 10.27 on text p. 521, subtract the bond dissociation energies of the bonds formed from those of the bonds broken to obtain the desired estimate for $\Delta H°$.

$$\Delta H° = 239 + 439 - (431 + 350) = -103 \text{ kJ mol}^{-1}$$

(b)   Two Cl—CH$_2$ bonds are formed (355 kJ mol^{-1} each); a Cl—Cl bond (239 kJ mol^{-1}) and a carbon–carbon π bond (≈ 243 kJ mol^{-1}) are broken.

$$\Delta H° = 239 + 243 - 2(355) = -228 \text{ kJ mol}^{-1}$$

10.8   (a)   Calculation of the $\Delta H°$ values involves subtracting the bond dissociation energies of the bonds formed from those of the bonds broken, as in the solution to Focused Problem 10.7. In the second propagation step, a secondary C—H bond is formed, and an H—Br bond is broken.

$$\Delta H° = 368 - 412 = -44 \text{ kJ mol}^{-1}$$

(b)   The corresponding calculation of $\Delta H°$ for HCl is

$$\Delta H° = 431 - 412 = +19 \text{ kJ mol}^{-1}$$

(c)   As noted in the text, endothermic propagation steps are much slower than exothermic ones. Because the second propagation step of the free-radical addition of HCl is endothermic, it is too slow to compete with the very exothermic recombination reactions that effectively shut down the chain mechanism.

10.9   The second propagation step in which Br is abstracted rather than H involves formation of a secondary C—Br bond rather than a secondary C—H bond; the bond broken (H—Br) is the same. Hence, we use the calculation in the solution of Focused Problem 10.8(a), except that the bond dissociation energy of a C—Br bond is substituted for that of a C—H bond. The closest value in Table 5.3 is the bond dissociation energy of $(CH_3)_2CH$—Br, 309 kJ mol^{-1}. The required $\Delta H°$ is therefore

$$\Delta H° = 368 - 309 = 59 \text{ kJ mol}^{-1}$$

Therefore, abstraction of Br is a very *unfavorable* process energetically ($\Delta H° \gg 0$), whereas the abstraction of H is a very favorable process ($\Delta H° < 0$). As is usually the case in free-radical reactions, the relative enthalpies of the processes governs their relative rates. Hence, abstraction of H is *much* more favorable, and therefore is much faster, than abstraction of Br.

10.10   (a)

(±)-hexane-3,4-d_2

(b)

***meso*-hexane-3,4-d_2**

10.11   (a)   If R• is the initiating radical, its reaction with acrylonitrile occurs at the methylene carbon because a resonance-stabilized radical is formed:

Its reaction at the methine (CH) carbon of the double bond gives a radical that is *not* resonance stabilized.

(b)   The structure of the final polymer is the same regardless of the addition regiospecificity as long as the addition occurs with the same regiospecificity at each step.

10.12   (a)

isotactic poly(methyl methacrylate)

(b)

syndiotactic poly(vinyl chloride)

10.13   The free-radical chain mechanism (initiation and propagation steps) for bromination of ethane:

Initiation: Br⎯Br $\xrightarrow{\text{light}}$ 2 Br·

Propagation: Br· + H⎯CH$_2$CH$_3$ ⟶   Br⎯H + ·CH$_2$CH$_3$

ethyl radical

Br⎯Br + ·CH$_2$CH$_3$ ⟶   Br⎯CH$_2$CH$_3$ + Br·

ethyl bromide
(bromoethane)

10.14   In the bromination of ethane, the ethyl radical, CH$_3$ĊH$_2$, is one of the chain-propagating radicals (Focused Problem 10.13). The recombination of two ethyl radicals in a termination step of ethane bromination forms a small amount of butane, as shown at right.

CH$_3$CH$_2$· + ·CH$_2$CH$_3$ ⟶   CH$_3$CH$_2$CH$_2$CH$_3$

butane

10.15   In this reaction, there is only 36% of the tertiary alkyl chloride product, and 64% of the primary chloride. While it would appear that abstraction of a primary hydrogen is favored, remember that there is a 9/1 statistical advantage to abstracting one of these hydrogens relative to the tertiary one. Therefore, we can calculate the relative rate of the two reactions. There is a little more than half, or 0.56 times, as much *tert*-butyl chloride as there is isobutyl chloride (36% ÷ 64%). But on a per hydrogen basis, the tertiary hydrogen is (0.56 ÷ 1/9) = 5 times as reactive as a primary hydrogen. This discrepancy is not as dramatic as in the reaction with bromine (Eq. 10.42, text p. 532), but still favors abstraction of a tertiary hydrogen over a primary one.

10.16   The two major products formed when pentane reacts with bromine in light are 2-bromopentane and 3-bromopentane. These products are formed in a ratio of roughly 2:1, because, although they both come from similarly stable secondary radicals, 2-bromopentane is statistically favored because there are twice as many of the hydrogens at the corresponding positions of pentane as there are at the central carbon. A third possible product, 1-bromopentane (not shown) is only a minor product because it would have to be produced from a much less stable primary radical, and the statistical advantage of the primary hydrogens is only 3:2.

pentane   + Br$_2$ $\xrightarrow{\text{light}}$ **2-bromopentane**   +   **3-bromopentane**

10.17   (a)   The major bromination product is the one that forms from the most stable radical intermediate, in this case, a tertiary radical. Other products would form from less stable secondary or primary radicals, at a slower rate, and in minor amounts.

methylcyclohexane   + Br$_2$ $\xrightarrow{\text{light}}$ **1-bromo-1-methylcyclohexane**

It would be reasonable to ask whether the 10:1 numerical ratio of secondary to tertiary hydrogens might result in a significant proportion of secondary halides. It has been found that the reactivity ratio of secondary to tertiary hydrogens in the free-radical bromination of simple alkanes is 0.00455. If this ratio applies to this case, then there should be about 96% of the substitution product shown in the preceding solution and only about 4% of a mixture of the possible secondary bromides. (There are six possible secondary substitution products, four of which are racemates.)

(b)   The first step in solving a problem like this is to draw the structures of the starting material and the products. There are three different monochlorinated products that can form in this reaction. In 2,2,3-trimethylbutane, there are fifteen primary hydrogens, and only one tertiary, which means the statistical chances of abstracting a primary hydrogen and forming the corresponding monochlorinated product is 15/1. However, as you know from the discussion on selectivity in Sec. 9.10A, tertiary radicals are much more stable than primary ones, and are formed at a faster rate. We're told that the relative reactivity of the tertiary hydrogen under these conditions is 4.5 times greater than any of the primary hydrogens. So, if we multiply 4.5 by 1/15, we get 0.3.

The ratio then of the tertiary alkyl chloride product to the primary alkyl chloride products is 0.3:1. To convert that ratio into percentages, we add $0.3 + 1 = 1.3$, and then divide 0.3 by 1.3 (and multiply by 100%) to get 23.1%, which is the percentage of the tertiary product formed.

The remaining 76.9% must be the two primary products. Since we can assume that there is no chemical preference for any one of the primary hydrogens over any of the others, we assume that there is a 9:6, or 3:2 ratio of these two products to one another, based on the number of hydrogens available that would lead to each product. Thus, there is 30.8% of one of the primary alkyl chlorides, and 46.1% of the other.

2,2,3-trimethylbutane   (23.1%)   (30.8%)   (46.1%)

10.18   (a)

isopropyl bromide   isopropylmagnesium bromide
(a Grignard reagent)

(b)

bromobenzene   phenyllithium
(an organolithium reagent)

(c)   Not all organometallic reagents are prepared from alkyl halides. Here's one from Chapter 5:

10.19   (a)   (b)

$(CH_3)_3C$—Li + $Li^+ Cl^-$

tert-butyllithium

1-cyclohexenylmagnesium bromide

10.20   (a)   The product of the reaction results from protonolysis of the lithium reagent at the carbon of the C—Li bond to give methane, CH_4, and lithium methoxide, $CH_3O^- Li^+$. The curved-arrow notation:

methane

(b)   The products result from protonolysis of the C—Mg bond: isobutane, $(CH_3)_2CHCH_3$, and HOMgCl, which, under the aqueous reaction conditions, is ionized to Mg^{2+}, HO^-, and Cl^-. The curved-arrow notation for the protonolysis:

$$HO—H \quad (CH_3)_2CHCH_2—MgCl \quad \longrightarrow \quad HO: \ MgCl \ + \ (CH_3)_2CHCH_2—H$$

isobutane

10.21   (a) $(CH_3)_2CH—MgBr$ and $CH_3CH_2CH_2—MgBr$ both react with water to give propane.

(b)   The compounds formed in the reactions of the Grignard reagents in part (a) with D_2O are $(CH_3)_2CH—D$ and $CH_3CH_2CH_2—D$, respectively. You should convince yourself of these results by writing a curved-arrow mechanism as shown in the solution to Focused Problem 10.20(b).

10.22   The first reaction is essentially an acid–base reaction, as shown in the foregoing section. The second reaction is a protonolysis of the resulting acetylenic Grignard reagent by D_2O.

CH_4 $C≡C—MgBr$ $C≡C—D$

A   *B*

methane $+ \ Mg^{2+} + \ Br^- + \ ^-OD$

(a gas)   *C*

10.23   (a)   The OH group is ionized under conditions of limited base because its pK_a (~ 16) is lower than that of the acetylenic hydrogen (~ 25).

$$HO \frown\frown C≡CH \ + \ Na^+ \ ^-NH_2 \ \xrightarrow{\text{liquid } NH_3} \ Na^+ \ ^-O \frown\frown C≡CH \ + \ :NH_3$$

$pK_a = 16$ (one equivalent)

(b)   Under conditions of excess base, both the OH and the 1-alkyne are ionized by base to give the di-anion..

$$HO \frown\frown C≡CH \ + \ Na^+ \ ^-NH_2 \ \xrightarrow{\text{liquid } NH_3} \ Na^+ \ ^-O \frown\frown C≡C^- \ + \ 2 \ :NH_3$$

$pK_a = 16$ (excess)

(c)   Because the OH group has been transformed into a methoxy group, which is not acidic, the alkyne remains as the only acid.

$$CH_3O \frown\frown C≡CH \ + \ Na^+ \ ^-NH_2 \ \xrightarrow{\text{liquid } NH_3} \ CH_3O \frown\frown C≡C^- \ + \ :NH_3$$

10.24   (a)   This is an S_N2 reaction of the anion with ethyl iodide to give 2-pentyne, $CH_3C≡CCH_2CH_3$, plus sodium iodide, $Na^+ \ I^-$.

(b)   Sodium acetylide reacts with both alkyl halide groups to give a 1,8-nonadiyne, $HC≡C(CH_2)_5C≡CH$, plus two equivalents of sodium bromide, $Na^+ \ Br^-$.

10.25   Tertiary alkyl halides react with strong bases to give elimination products, not substitution products. Acetylenic anions are strong bases, so student's products were methylpropene and propyne.

$$(CH_3)_3CBr \ + \ Na^+ \ ^-:C≡CCH_3 \ \xrightarrow{\text{an E2 reaction}} \ (CH_3)_2C=CH_2 \ + \ HC≡CCH_3$$

***tert*-butyl bromide**   **methylpropene**   **propyne**

10.26   3,3-Dimethyl-1-butyne is allowed to react **with sodium** amide, and the resulting acetylenic anion with methyl iodide.

$$(CH_3)_3CC{\equiv}CH \xrightarrow[\text{NH}_3\text{ (liq)}]{\text{NaNH}_2} (CH_3)_3CC{\equiv}C{:}^- \ Na^+ \xrightarrow{\text{CH}_3\text{I}} (CH_3)_3CC{\equiv}CCH_3$$

3,3-dimethyl-1-butyne   **4,4-dimethyl-2-pentyne**

10.27   Either alkyl group can be introduced by the reaction with an alkyl halide. The alternative to Eq. 10.65 is to use the acetylenic anion derived from propyne followed by alkylation with 1-bromobutane. The reaction in Eq. 10.65, however, is probably a better alternative because methyl halides are more reactive than butyl halides and because there is no possibility of even a small amount of elimination.

$$CH_3C{\equiv}CH \xrightarrow[\text{NH}_3\text{ (liq)}]{\text{NaNH}_2} CH_3C{\equiv}C{:}^- \ Na^+ \xrightarrow[\text{1-bromobutane}]{\text{CH}_3\text{CH}_2\text{CH}_2\text{CH}_2\text{—Br}} CH_3C{\equiv}CCH_2CH_2CH_2CH_3$$

propyne   **2-heptyne**

10.28   (a)   (b)

cyclohexene   +   HCBr$_3$ **bromoform**

                                      +   PhCH$_2$Br   **benzyl bromide**

2,3-dimethyl-2-butene

10.29   (a)   The product is a meso compound, and it is the only stereoisomer formed.

cyclopentene   +   HCCl$_3$ +   K$^+$ $^-$OC(CH$_3$)$_3$ ⟶   +   HOC(CH$_3$)$_3$ +   K$^+$ Cl$^-$

         (b)   Two diastereomeric products *A* and *B* are formed in different amounts. (Which do you think should predominate? Think about steric effects.)

+   HCCl$_3$ +   K$^+$ $^-$OC(CH$_3$)$_3$ ⟶   HOC(CH$_3$)$_3$ + K$^+$ Cl$^-$

                                                      A   *B*

10.30   (a)

(*Z*)-4-methyl-2-pentene   *cis*-1-isopropyl-2-methyl-cyclopropane (racemic)

Simmons–Smith reagent (I—Zn—CH$_2$I)

         (b)

Simmons–Smith reagent (I—Zn—CH$_2$I)

10.31 (a)

(b)

In part (a), addition occurs on the face of the ring opposite to the angular methyl group for steric reasons.

10.32 (a) In this synthesis we use one equivalent of disiamylborane (Eq. 5.34c, Sec. 5.6A, text p. 254) to hydroborate selectively the 1-alkene, which provides a much smaller steric impediment to the reaction than the internal alkene. It is also possible that hydroboration of the alkene group could be carried out selectively in the presence of the alkyne, and then the alkyne could be completely hydrogenated in one step.

(b)

(c)

10.33 The syntheses in (a) and (b) are identical except for the reduction step. Both begin with formation of an alkyne precursor *A*.

(a)   Catalytic hydrogenation of *A* over a poisoned catalyst gives the cis isomer of the desired alkene:

(b)   Reduction of *A* with sodium in liquid ammonia (the solvated electron reduction) gives the trans isomer.

10.34   Conversion of the alkyne selectively into a cis alkene by catalytic hydrogenation over a poisoned catalyst (such as the Lindlar catalyst) will accomplish the desired conversion.

Solutions to Skills Objectives Problems

10.35   (a)

$CH_3CH_3 + Mg^{2+}$
$+ Br^- + {}^-OH$

(b)

$(CH_3)_4C + Br_2,$
$H—Br$

(c)

(d)

(e)

$(CH_3)_2CH — D$

(f)

(g)

(h)

(i)

(j)

$+ Mg^{2+} + Br^- + {}^-OH$

(k)

(l)

MeMgI, then
$(CH_3)_2CHCH_2CH_2C≡C—MgI + CH_4$ (a gas), then
$(CH_3)_2CHCH_2CH_2C≡C—D$

(m)

(n)

(o)

10.36 (a) The major constitutional isomer formed is the tertiary bromoalkane formed by reaction of bromide ion with the tertiary carbocation formed as a reactive intermediate.

(b) The major isomer formed is the secondary bromoalkane formed by a free-radical chain mechanism involving a tertiary free-radical intermediate.

(c) The major product is the same as in part (a), but formed by a free-radical chain substitution mechanism at the tertiary carbon.

10.37 The fishhook mechanism for the tributyltin hydride mechanism: (Steps are numbered for reference in Problem 38.)

10.38 The initiation steps are (1) and (2), because they are only required once to generate the chain-propagating radical $Bu_3Sn\bullet$. The propagation steps are (3) and (4) because they consume and regenerate the same radicals. Step (5) is one of several possible termination steps.

10.39 (a) Refer to the equation in the hint. There are 12 equivalent C—H bonds in the product and 4 of the same bond type in the starting material, for a net of 8; this makes a contribution $8f$, or $8(423) = 3384$ kJ mol^{-1}. There are 6 equivalent C—C bonds in the product, for a contribution of $6g$; but there is also one bond of this type (not labeled) in the starting material, for a *net* contribution of $5g$, or $5(368) = 1840$ kJ mol^{-1}. In the starting material there is one double bond for a contribution of $1e = 728$ kJ mol^{-1}. There are 4 allylic C—H bonds (bonds on carbons adjacent to double bonds) for a contribution of $4a = 4(372) = 1488$ kJ mol^{-1}. There are 2 vinylic hydrogens *b* for a contribution of $2b = 2(463) = 926$ kJ mol^{-1}. There are 2 vinylic C—C bonds *d* for a contribution of $2d = 2(418) = 836$ kJ mol^{-1}. There are two allylic C—C bonds *c* for a contribution of $2(320) = 640$ kJ mol^{-1}. The H—H *h* bond in the reactant contributes 435 kJ mol^{-1}. Recognizing that

$$\Delta H^\circ_{\text{hydrogenation}} = \text{BDE(bonds broken)} - \text{BDE (bonds formed)},$$

we have

$$\Delta H^\circ_{\text{hydrogenation}} = 728 + 1488 + 926 + 836 + 640 + 435 - 3384 - 1840 = -171 \text{ kJ mol}^{-1} \text{ or } -40.9 \text{ kcal mol}^{-1}$$

(b)   This calculation significantly overestimates the $\Delta H^\circ_{\text{hydrogenation}}$ for cyclohexene.

 One possible reason is that the BDE values are averages for many different compounds, and the values for the compounds in this problem may be slightly different. The calculated value is a small difference between large numbers, so slight differences in several values could cause this discrepancy. Another possible reason is that heats of hydrogenation are determined in solution, whereas BDE numbers are determined in the gas phase. This example shows that calculation of ΔH° values for reactions using BDEs gets us "in the ball park" but is certainly not exact.

(c)   Let the BDE of the bond to a •CH$_2$ be *x*, and the R—CH$_2$ bond be *y*. Then the overall ΔH° of the reaction is approximated by

$$4b + e + x + y - 2g - 4f - x - y = 4b + e - 2g - 4f = 4(463) + 728 - 2(423) - 4(368) = +262 \text{ kJ mol}^{-1} (+62.6 \text{ kcal mol}^{-1})$$

(d)   This calculation suggests that polymerization should *not* go to completion. We've been making the assumption that we can use ΔH° values as approximations for the ΔG° values. (Remember it's the ΔG° values that determine whether a reaction is favorable.) In this case, however, this assumption is not a good one. Why? BDEs are determined at 298 K (25 °C) and 1 atm pressure. However, polymerization involves high temperature and pressure, as shown in the equation cited in the hint. The pressure forces the alkene molecules together and increases their *probability of reaction*, which is what ΔS° measures. In other words, the reaction is favorable because both *T* and ΔS° in the equation $\Delta G^\circ = \Delta H^\circ - T\Delta S^\circ$ are so large (and the $-T\Delta S^\circ$ term so negative) that the ΔH° is not the major component that determines ΔG°.

(e)   The formation of *C* is the first step in a polymerization reaction. In the answer to part (c) we showed that at 25 °C this reaction has an unfavorable ΔH°, and free-radical HBr addition does *not* involve high temperature and pressure that favors polymerization. In contrast, the formation of *B* gains a strong C—H bond (412 kJ mol^{-1}) and breaks a weak H—Br bond (368 kJ mol^{-1}) so that the ΔH° of the overall transformation ($368 - 412 = -44$ kJ mol^{-1}) is highly favorable.

10.40 (a)   This poly(styrene) is atactic because the arrangements of groups along the polymer chain is not stereoregular.

(b)   This (polystyrene) is syndiotactic because the stereochemical arrangement of the Ph substituents alternates regularly along the polymer chain.

10.41 (a)   Three units of poly(propylene carbonate):

poly(propylene carbonate)
(three units)

(b)   Two condensed structures for polyester; either is correct depending on where we decide to make the break. (Others are also possible.)

10.42 (a) The $^+$MgBr can further dissociate to Mg^{2+} and Br^-.

$$CH_3CH_2\text{—}MgBr \longrightarrow CH_3CH_2\text{—}H + \overset{+}{M}gBr \quad \overset{..}{\underset{..}{\text{:}}}\overset{..}{O}H$$

(b) We show the two propagation steps of the free-radical chain reaction:

$$(CH_3)_3C\text{—}CH_2\text{—}H \quad \cdot Br \longrightarrow (CH_3)_3C\text{—}\overset{\cdot}{C}H_2 \quad H\text{—}Br$$

$$(CH_3)_3C\text{—}\overset{\cdot}{C}H_2 \quad Br\text{—}Br \longrightarrow (CH_3)_3C\text{—}CH_2\text{—}Br + \cdot Br$$

(c)

(d)

Initiation:

$$tBuO\text{—}OtBu \longrightarrow 2 \cdot OtBu$$

Propagation:

(e)

Initiation steps:

$$R\text{—}O\text{—}O\text{—}R \longrightarrow 2\ R\text{—}O$$

$$R\text{—}O\cdot \quad H\text{—}Br \longrightarrow R\text{—}O\text{—}H + \cdot Br$$

Propagation steps:

(f)

dibromocarbene

(g)

(h)   We show the Grignard reagent as an ionic compound to stress the acid–base character of the reaction.

(i)

10.43   To undergo a β-elimination a compound must have hydrogens β to the leaving halide. To undergo an α-elimination, the compound must have an α-hydrogen *and no* β-hydrogens—otherwise a β-elimination is preferred. With these stipulations in mind, we conclude that compounds *A* and *C* will undergo a β-elimination, and compound *B* will undergo an α-elimination.

10.44   (a)   The preparation of 1-heptyne from acetylene is shown in Study Problem 10.5 on text p. 547. From there the synthesis proceeds as follows:

1-heptyne   **1-bromoheptane**

(b)   The first step in the following synthesis formed the di-anion of acetylene, which is then allowed to react with two equivalents of bromoethane.

(c)

(d)   The product is shown in an eclipsed conformation for greater ease in showing the stereochemistry, which requires two consecutive syn additions.

‡p.

Deuterium could be used in the first addition, and hydrogen in the second. The Lindlar catalyst, however, must be used in the first addition to stop the hydrogenation at the alkene stage. The (*RS,SR*) designation of stereochemistry is another way of specifying relative configurations in a racemate; this is explained in Further Exploration 7.1 on p. 162 of this manual.

(e)   The preparation of monosodium acetylide (the mono-anion of acetylene) is shown in part (c) of this problem.

5-methyl-1-hexanol

Solutions to Integrated Problems

10.45 (a)   The compound must have a symmetrically placed double bond to form the same HBr addition products in the presence and absence of peroxides. Two possibilities are the following:

cyclopentene   **1,2-dimethylcyclopropene**

(b)   The compound must be a saturated alkane, because its formula is C_nH_{2n+2}. The only compound that gives only two monochlorinated products, one of which is chiral, is 2,3-dimethylbutane. (The asymmetric carbon of the chiral chlorination product is marked with an asterisk,)

2,3-dimethylbutane

(c)   The four compounds must be stereoisomers, and all must be chiral. Additionally, they must contain a ring because they possess one degree of unsaturation and no double bond. Finally, they must contain an acidic proton—one that can be removed by the Grignard reagent to yield ethane, a gas. A compound with two asymmetric carbons could exist as four stereoisomers. The stereoisomeric 2-methyl-1-cyclopropanols fit the bill.

(d)   Because the compound undergoes catalytic hydrogenation to hexane, it must contain an unbranched six-carbon chain. Because it reacts with methyllithium to release a gas, it must be a 1-alkyne. 1-Hexyne fits the data.

1-hexyne

methane
(the gas)

10.46   The gas is propane, formed by the protonolysis of the Grignard reagent with the alcohol.

propane

10.47   (a)   Only the first two products are formed in the absence of peroxides, and only the third is formed in the presence of peroxides. Different products are formed because different mechanisms and reactive intermediates are involved under the different conditions.

(b)   The first product (*B* in the following scheme) results from a carbocation rearrangement, and the second (*A* in the following scheme) from normal regioselective ("Markovnikov") addition.

The third product is the consequence of a free-radical addition mechanism, the propagation steps of which are as follows:

(c)   Peroxide-promoted addition is in competition with normal addition and rearrangement. The normal processes occur at the same rate at which they occur in the absence of peroxides. The fact that *only* the product of peroxide-promoted addition is observed, then, means that this process is much faster than the other, competing, processes.

10.48   (a)   Ozonolysis breaks the polymer at its double bonds:

(b)   Because the *E* and *Z* isomers of alkenes give the same ozonolysis products, gutta-percha is evidently either the all-*E* stereoisomer of natural rubber or a polymer containing both *E* and *Z* alkene units. (In fact, it is the all-*E* stereoisomer.)

gutta-percha

10.49   (a)   Because carbon–carbon bonds are weaker than carbon–hydrogen bonds (see Table 10.2, Sec. 10.1E, text p. 520), the carbon–carbon bond of ethane should break more easily than a C—H bond.

(b) Breaking of a bond to any of the methyl groups (pathway (1) below) gives a methyl radical and a tertiary free radical. Breaking of the central carbon–carbon bond (pathway (2) below) gives two tertiary free radicals. Because process (2) gives two tertiary radicals, it is the one that should occur most readily.

$$
\begin{array}{ccc}
& H_3C-\overset{\overset{\displaystyle CH_3}{|}}{\underset{\underset{\displaystyle CH_3}{|}}{C}}\!\!-\!\!\overset{\overset{\displaystyle CH_3}{|}}{\underset{\underset{\displaystyle CH_3}{|}}{C}}\!\!-\!\!CH_3 & \\
\overset{(1)}{\swarrow} & & \overset{(2)}{\searrow}
\end{array}
$$

Pathway (1): $H_3C-\overset{\overset{CH_3}{|}}{\underset{\underset{CH_3}{|}}{C}}-\overset{CH_3}{\underset{CH_3}{C}}\!\!\cdot\; +\; \cdot CH_3$

Pathway (2): $H_3C-\overset{CH_3}{\underset{CH_3}{C}}\!\!\cdot\quad \cdot\overset{CH_3}{\underset{CH_3}{C}}-CH_3$

(c) Because ethane gives two methyl radicals and 2,2,3,3-tetramethylbutane gives two tertiary radicals, the bond homolysis of 2,2,3,3-tetramethylbutane should occur more readily.

(d) The $\Delta H°$ for the reaction in part (a) is the difference between the heats of formation of products and reactants.

$$\Delta H° = 2\,\Delta H_f°(\cdot C(CH_3)_3) - \Delta H_f°(2,2,3,3\text{-tetramethylbutane})$$
$$= 2(51.5) - (-225.9) = 328.9 \text{ kJ mol}^{-1}\ (78.6 \text{ kcal mol}^{-1})$$

The $\Delta H°$ for the cracking of ethane is

$$\Delta H° = 2\,\Delta H_f°(\cdot CH_3) - \Delta H_f°(\text{ethane})$$
$$= 2(146.6) - (-84.7) = 377.9 \text{ kJ mol}^{-1}\ (90.32 \text{ kcal mol}^{-1})$$

This calculation shows that the cracking of ethane is 49 kJ mol^{-1} (11.7 kcal mol^{-1}) less favorable than the cracking of 2,2,3,3-tetramethylbutane. This demonstrates quantitatively the assertion in part (b), namely, that 2,2,3,3-tetramethylbutane undergoes cracking more readily. The underlying assumption is that the free energies of activation and enthalpies of the cracking reaction are similar.

10.50 (a) In the general reaction, the C—H bond of methane and the X—X bond of the halogen are broken; the C—X bond of the alkyl halide and the H—X bond of the hydrogen halide are formed. The overall energetics of the reactions are as follows, in kJ mol^{-1}:

	Breaking of C—H bond	Breaking of X—X bond	Formation of H_3C—X bond	Formation of H—X bond	Overall energy change ($\Delta H°$)
X = Cl	439	239	−350	−431	−103
X = Br	439	190	−302	−368	−41
X = I	439	149	−241	−297	+50

These calculations show that chlorination and bromination are exothermic, whereas iodination is endothermic. Consequently, iodination does not occur because it is energetically unfavorable.

(b) Because the iodination reaction is energetically unfavorable, its reverse is energetically favorable. Consequently, H—I (from the trace of acid) reacts with H_3C—I as follows:

$$H-I\; +\; H_3C-I\; \longrightarrow\; I-I\; +\; CH_4$$

As this equation shows, iodine is formed in this reaction.

10.51 The two products come from substitution of one or both of the tertiary hydrogens on carbons 2 and 3, because these substitutions involve a tertiary free radical intermediate. Substitution of a methyl hydrogen would involve a less stable primary free radical.

2,3-dimethylbutane

10.52   (a)   Follow the pattern for Steps (3) and (4) in the solution to Problem 10.37.

(b)   One sequence of reactions is to convert the alkyl halide into a Grignard reagent, which is then treated with water:

Another is to carry out an E2 reaction to obtain an alkene (or a mixture of alkenes), which is then subjected to catalytic hydrogenation to give methylcyclohexane.

(c)   The initiation steps of this interesting reaction are the same as for an ordinary reaction of H—SnBu$_3$—homolytic cleavage of AIBN and formation of ·SnBu$_3$. This radical then abstracts a bromine atom from the starting material, as in the usual reaction, to give a free-radical intermediate. This radical, instead of abstracting H from H—SnBu$_3$, adds intramolecularly to the nearby double bond. The resulting radical then abstracts a hydrogen atom from H—SnBu$_3$.

Why should this intramolecular process occur? Intramolecular reactions that form 5-membered rings are typically much faster than the corresponding intermolecular reactions (Sec. 12.8). The trans stereochemistry at the ring junction is simply more stable than the alternative cis stereochemistry.

10.53 (a) As the bond dissociation energies in Table 10.2 of the text show, secondary C—H bonds are weaker than methyl C—H bonds (412 versus 423 kJ mol^{-1}). Therefore, this process is favorable by about 11 kJ mol^{-1}.

(b) The secondary radical becomes one of the chain propagating radicals, thus forming a branch.

10.54 (a) The first step of poly(styrene) polymerization is reaction of an initiating radical with the alkene styrene:

The resulting radical adds to another molecule of styrene to produce a new radical. which adds to another molecule of styrene, and so on.

Neglecting the initiating R group, the propagation process can be represented generally as

(b) Because both "ends" of 1,4-divinylbenzene can be involved in polymer formation, addition of 1,4-divinylbenzene connects, or *crosslink,* polymer chains. Such a crosslink is shown with bolded bonds in the following structure:

Notice that because only a small amount of 1,4-divinylbenzene is used, divinylbenzene does not polymerize with itself.

(c) Crosslinks are introduced into polymers to increase their strength and rigidity.

10.55 (a) A skeletal partial structure of co-polymerized ethylene and 1-butene:

polymerized
1-butene units

(b) The occasional ethyl branches prevent the polymer chains from packing closely, much as branches might prevent the tight stacking of tree limbs. Because "loosely stacked" polymer chains have lass mass per unit volume than "tightly stacked" polymer chains, the copolymer has lower density than poly(ethylene) itself—thus the name, "very low-density polyethylene."

10.56 (a) The methyl radical reacts at the carbon of the double bond with fewer substituents because this mode of reaction gives the more substituted radical, and because the less-substituted carbon is the least sterically hindered site of reaction. (For ethylene, in reaction A, only one product is possible.)

(reaction A) $H_2C{=}CH_2$ + $\cdot CH_3$ $\longrightarrow$ $\cdot CH_2CH_2CH_3$

(reaction B) $(CH_3)_2C{=}CH_2$ + $\cdot CH_3$ $\longrightarrow$ $(CH_3)_2\overset{\cdot}{C}{-}CH_2CH_3$

(reaction C) $(CH_3)_2C{=}CHCH_3$ + $\cdot CH_3$ $\longrightarrow$ $(CH_3)_2\overset{\cdot}{C}{-}CH(CH_3)_2$

(b) When we compare reactions B and A, we see that two methyl groups at the site of the radical accelerate the reaction modestly. This is probably due to the stability of the free radical (tertiary versus primary) and the reflection of this stability in the transition state for its formation (Hammond's postulate). It might be surprising that the effect on rate is so modest. However, if you consider the effect of substituents on alkene stability (Table 4.1, Sec. 4.5B, text p. 198) and the effect of substituents on free-radical stability (Table 10.1, Sec. 10.1D, text p. 518), you can see that these effects are not very different. Remember that relative rates reflect the *difference* in standard free energy of transition states and reactants. Because branching affects both starting material and transition state in the same way, and because the effect is of a similar magnitude in both alkenes and free radicals, the residual effect on the transition state is rather modest.

When we compare reactions B and C, we start with a *more* substituted alkene in C and go to an identically substituted radical in both cases. This should lower the energy of the alkene starting material more than that of the transition state. If so, reaction C should be slower—again, not because the transition state is *higher* in energy, but because the starting material is *lower* in energy. (When you climb two energy mountains, the top of the mountains may be at the same elevation above sea level, but if the base camp is lower for one mountain, then that mountain requires a greater climb.) In addition, a steric effect operates on reaction C that does not operate on reaction B, because the carbon at which reaction C occurs is more substituted. This would tend to raise the energy of the transition state for reaction C and lower the rate, as observed.

10.57 (a) The dissociation of ethanethiol at the C—S bond gives the •SH radical and the ethyl radical. From Table 5.2, text p. 212, we find that the ΔH_f° of the ethyl radical is 121.3 kJ mol^{-1}. Therefore, the bond dissociation energy is

$$\Delta H^\circ = \Delta H_f^\circ (\cdot SH) + \Delta H_f^\circ (\cdot CH_2CH_3) - \Delta H_f^\circ (\text{ethanethiol}) = (143.1 + 121.3) - (-46.2) = 310.6 \text{ kJ mol}^{-1}$$

(b) The reaction of ethanethiol with the *tert*-butoxy radical is

$CH_3CH_2S{-}H$ $\cdot O{-}C(CH_3)_3$ $\longrightarrow$ $CH_3CH_2S\cdot$ + $H{-}O{-}C(CH_3)_3$

ethanethiol   ***tert*-butyl alcohol**

By Eq. 10.27, text p. 521, the ΔH° of this reaction is the bond dissociation energy (BDE) of the ethanethiol S—H minus the BDE of the *tert*-butyl alcohol O—H. In Table 10.2, text p. 520, the closest BDE to that of *tert*-butyl alcohol is the BDE of H—OCH$_3$ (methanol), 438 kJ mol^{-1}. The BDE of H—SCH$_3$ (methanethiol), 366 kJ mol^{-1}, is the closest approximation to that of ethanethiol. The ΔH° of the reaction is then $366 - 438 = -72$ kJ mol^{-1}. Because the reaction is exothermic, it should be a good source of free radicals.

(c)   The first propagation step is (by analogy the first propagation step in Study Problem 10.3, text p. 515), addition of the •SR radical to a double bond, which breaks a carbon–carbon π bond (243 kJ mol^{-1}) and forms a carbon–sulfur bond (from part (a), –307.5 kJ mol^{-1}). The $\Delta H°$ of this step is $243 - 308 = -65$ kJ mol^{-1}. The second propagation step is (by analogy with the second propagation step in Study Problem 10.3) formation of a carbon–hydrogen bond (–423 kJ mol^{-1}) and breaking of a sulfur–hydrogen bond (366 kJ mol^{-1}). The $\Delta H°$ of this step is $366 - 423 = -57$ kJ mol^{-1}. This calculation shows that both propagation steps are exothermic; therefore, both are reasonable propagation steps.

(d)   The initiation step, reaction of the thiol with an alkoxy radical formed by homolysis of a peroxide, is shown in Study Problem 10.3, text p. 515. The radical produced in that step adds to the π bond of the alkene so as to produce the tertiary free radical, and this radical reacts with the thiol to propagate the chain.

Be sure that you did not form the product by a recombination of two radicals; see the discussion in Study Problem 10.3.

10.58   (a)   The H—CN bond is stronger than the O—H bond, so the abstraction of a hydrogen atom from HCN by the *tert*-butoxy radical is endothermic by $528 - 438 = 90$ kJ mol^{-1}. [These numbers are from Table 10.2 with the bond energy of CH$_3$O—H as an approximation for that of (CH$_3$)$_3$CO—H.] The first initiation step, formation of the *tert*-butyloxy radical from a peroxide, is also endothermic. Because *both* initiation steps are highly endothermic, the reaction is not likely to generate a high enough concentration of radicals to initiate a chain reaction.

(b)   This propagation step involves breaking an H—CN bond (528 kJ mol^{-1}) and formation of a secondary carbon–hydrogen bond (–412 kJ mol^{-1}). The $\Delta H°$ of this step is $528 - 412 = +116$ kJ mol^{-1}. This step is highly endothermic and therefore not reasonable as a propagation step.

10.59   (a)   (Be sure to read Study Problem 10.3.) Because the CF$_3$ group has added to the carbon of the double bond with no alkyl branches, the radical •CF$_3$ must be the first one that adds to the alkene. The radical that results from this addition abstracts an iodine atom from CF$_3$I to give the product.

CH$_3$CH$_2$CH$_2$ĊH =CH$_2$ •CF$_3$ $\longrightarrow$ CH$_3$CH$_2$CH$_2$ĊH—CH$_2$—CF$_3$

CH$_3$CH$_2$CH$_2$ĊH—CH$_2$—CF$_3$ I—CF$_3$ $\longrightarrow$ CH$_3$CH$_2$CH$_2$CH—CH$_2$—CF$_3$ +   •CF$_3$

(b)   A •CBr$_3$ radical formed in the initiation steps adds to the double bond; the resulting radical then abstracts a bromine atom from Br$_3$C—Br to propagate the chain.

Br$_3$C• $\longrightarrow$ Br$_3$C   Br—CBr$_3$ $\longrightarrow$ Br$_3$C   Br   +   •CBr$_3$

(c)   The O—Cl bond, a bond between two electronegative atoms, is the weakest bond in the molecule. Homolysis of the O—Cl bond gives a chlorine atom plus an alkoxy radical, which then undergoes a reaction that opens the ring to give a new radical A.

Radical A then reacts with another molecule of the starting material to regenerate an alkoxy radical, which propagates the chain.

10.60   (a)   First, the presence of deuterium in the tertiary-hydrogen position increases the percentage of abstraction of primary hydrogens because the primary deuterium kinetic isotope effect retards the rate of deuterium abstraction. This means that if there is any rearrangement of primary to tertiary free radicals it is more likely to be observed. If the hypothetical rearrangement occurs, we have the following reaction sequence:

This scheme shows that the reaction produces an HCl molecule as well as a *deuterium-substituted tert-*butyl chloride. In the absence of rearrangement, the only pathway to *tert*-butyl chloride is to form DCl along with an *isotopically unsubstituted* alkyl halide. Rearrangement, then, would produce an excess of HCl over deuterium-containing alkyl halide. Because this situation is not observed, rearrangement appears to be ruled out.

(b)   As noted in the first sentence of the previous part, a kinetic deuterium isotope effect would be expected for abstraction of deuterium by a chlorine atom; that is, this deuterium abstraction is slower than hydrogen abstraction. No isotope effect operates in the formation of B. Therefore, the A:B ratio should be reduced over the ratio observed when deuterium is not present.

(c)   Protonolysis of a Grignard reagent in D_2O could provide the required isotopically substituted halide:

Another alternative would be to add DBr to methylpropene in the presence of peroxides, then form the Grignard reagent from the resulting 1-bromo-2-methylpropane-2d, and then subject this Grignard reagent to protonolysis in H_2O.

$$H_3C\!\!-\!\!\underset{H_3C}{\overset{}{C}}\!\!=\!\!CH_2 \xrightarrow[\text{peroxides}]{DBr} H_3C\!\!-\!\!\underset{D}{\overset{CH_3}{C}}\!\!-\!\!CH_2Br \xrightarrow[\text{ether}]{Mg} H_3C\!\!-\!\!\underset{D}{\overset{CH_3}{C}}\!\!-\!\!CH_2MgBr \xrightarrow{H_2O} H_3C\!\!-\!\!\underset{D}{\overset{CH_3}{C}}\!\!-\!\!CH_3$$

10.61   The different boiling points indicate that the three alkyl halides are either constitutional isomers or diastereomers. The outcome of the Grignard protonolysis shows that all of the alkyl halides have the same carbon skeleton, that of 2,4-dimethylpentane. The protonolysis in D_2O confirms the fact that the bromines are bound at different places on the carbon skeleton. The only three possibilities for the alkyl halides are

$$\underset{A}{\underset{\overset{|}{Br}}{CH_3\overset{\overset{CH_3}{|}}{C}CH_2\overset{\overset{CH_3}{|}}{C}HCH_3}} \qquad \underset{B}{BrCH_2\overset{\overset{CH_3}{|}}{C}HCH_2\overset{\overset{CH_3}{|}}{C}HCH_3} \qquad \underset{C}{CH_3\overset{\overset{CH_3}{|}}{C}H\underset{\overset{|}{Br}}{C}H\overset{\overset{CH_3}{|}}{C}HCH_3}$$

Compound *B* is the chiral alkyl halide, and compounds *A* and *C* are the other two. (The absolute configuration of *B* (that is, whether it is *R* or *S*) is not determined by the data.)

The protonolysis products in H_2O and D_2O are

In H_2O:

$$CH_3\overset{\overset{CH_3}{|}}{C}HCH_2\overset{\overset{CH_3}{|}}{C}HCH_3$$
from *A*, *B*, and *C*

In D_2O:

$$\underset{\overset{|}{D}}{CH_3\overset{\overset{CH_3}{|}}{C}CH_2\overset{\overset{CH_3}{|}}{C}HCH_3} \qquad DCH_2\overset{\overset{CH_3}{|}}{C}HCH_2\overset{\overset{CH_3}{|}}{C}HCH_3 \qquad CH_3\overset{\overset{CH_3}{|}}{C}H\underset{\overset{|}{D}}{C}H\overset{\overset{CH_3}{|}}{C}HCH_3$$
from *A* … from *B* … from *C*

10.62   The fact that protonolysis reactions of the corresponding Grignard reagents give the same hydrocarbon indicates that the two compounds have the same carbon skeleton. The conditions of ethanol and no added base are S_N1 conditions. Since compound *A* reacts rapidly to give a solution containing bromide ion, it must be an alkyl halide that readily undergoes an S_N1 reaction, and therefore it is probably a tertiary alkyl bromide. Because the two alkyl halides give the same ether, the product from compound *B* must be formed in a rearrangement. The only tertiary alkyl halide with the formula $C_5H_{11}Br$ is 2-bromo-2-methylbutane, and this is therefore compound *A*:

$$\underset{\overset{|}{Br}}{CH_3\overset{\overset{CH_3}{|}}{C}CH_2CH_3} + CH_3CH_2OH \longrightarrow \underset{\overset{|}{OCH_2CH_3}}{CH_3\overset{\overset{CH_3}{|}}{C}CH_2CH_3} + HBr$$

2-bromo-2-methylbutane
(A)

2-ethoxy-2-methylbutane
(C)

Two possible alkyl halides with the same carbon skeleton as *A* could rearrange in respective S_N1 reactions to give the same carbocation, and hence the same ether product, as *A*; these are labeled *B1* and *B2* below. (An S_N1 reaction of *B1* would be very slow, if it occurred at all.)

B1

B2

However, only *B2* can react in an E2 reaction with sodium ethoxide to give an alkene that furnishes acetone as one of its ozonolysis products:

B2 = B
2-bromo-3-methylbutane

$(CH_3)_2CHCH=CH_2$

does not give acetone
as an ozonolysis product

Consequently, compound *B2* is compound *B*, 2-bromo-3-methylbutane.

10.63 Compound *A* has four degrees of unsaturation, two of which are accounted for by the triple bond of the alkyne group. Because the product of hydrogenation is butylcyclohexane, one degree of unsaturation is accounted for by a 6-membered ring. The triple bond cannot be within the ring (why?), nor can it be at the end of the four-carbon chain because of the absence of a reaction of *A* with ethylmagnesium bromide. Hydrogenation of *A* under conditions that hydrogenate triple bonds to double bonds but leave double bonds unaffected gives an alkene which, on ozonolysis, loses *two* carbons. The fact that a *tricarboxylic* acid is formed on ozonolysis shows that the double bond of *A* is within the ring. It is now possible to write several preliminary structures for *A*:

A1

A2

A3

Compound *A1* is ruled out because it is not chiral; recall that compound *A* is optically active. To distinguish between *A2* and *A3,* imagine the results of hydrogenation over a poisoned catalyst followed by ozonolysis:

A2

chiral; would be
optically active if *A2* were
optically active

$+ CH_3CO_2H$

A3

achiral; would be
optically inactive even if *A3* were
optically active

$+ \ CH_3CO_2H$

The observation of optical activity in the ozonolysis product rules out *A3* and is consistent only with *A2* as the structure of compound *A*.

compound *A* = *A2*
3-(2-butynyl)cyclohexene

10.64 (a) As Display 10.49, Sec. 10.5B, text p. 536, shows, ethers solubilize Grignard reagents by forming Lewis acid–base complexes with them. Because a tertiary amine is also a Lewis base, it can solubilize Grignard reagents by the same type of interaction, shown at right.

(b) A secondary amine *cannot* be used to solubilize a Grignard reagent, because the N—H hydrogen is acidic enough to cause protonolysis of the Grignard reagent:

10.65 Because muscalure is a cis-alkene, we can prepare it by hydrogenation of an alkyne. The required alkyne is unbranched and can be prepared by a series of alkylations involving acetylenic anions and alkyl bromides. One of the required alkyl bromides is 1-bromooctane, which is prepared as follows. (For the preparation of sodium acetylide, $Na^+ \ {}^-C{\equiv}CH$, see the solution to Problem 14.44(c).

As this example shows, the carbon chain of an alkyne can be extended two carbons at a time using the following sequence (referred to below as the *alkyne n + 2 sequence*):

Consequently, 1-bromotridecane, $CH_3(CH_2)_{12}Br$, can be prepared by applying four successive "alkyne *n + 2* sequences" to 1-pentyne (*n* = 2), and finishing with the hydrogenation and HBr-addition steps:

$$CH_3(CH_2)_2C\equiv CH \xrightarrow[\text{sequences}]{\substack{\text{four}\\\text{"alkene } n+2\text{"}}} CH_3(CH_2)_{10}C\equiv CH \xrightarrow[\text{Lindlar catalyst}]{H_2} CH_3(CH_2)_{10}CH=CH_2 \xrightarrow[\text{peroxides}]{HBr} CH_3(CH_2)_{12}Br$$

1-pentyne   **1-bromotridecane**

To complete the synthesis of muscalure, carry out two successive alkylations of acetylenic anions with the alkyl bromides prepared above, and then hydrogenate the resulting product.

$$Na^+ \; \overset{..}{:}C\equiv CH \xrightarrow{CH_3(CH_2)_7Br} CH_3(CH_2)_7C\equiv CH \xrightarrow[\text{NH}_3\text{ (liq.)}]{NaNH_2} CH_3(CH_2)_7C\equiv C\overset{..}{:} \; Na^+ \xrightarrow{Br(CH_2)_{12}CH_3}$$

$$CH_3(CH_2)_7C\equiv C(CH_2)_{12}CH_3 \xrightarrow[\text{Lindlar catalyst}]{H_2}$$

(structure of muscalure: $CH_3(CH_2)_7$ and $(CH_2)_{12}CH_3$ on $C=C$ with H's)

muscalure

10.66   **(a)**   In reaction (1), the "anionic carbon" of ethylmagnesium bromide acts as a base towards the C—H bond of ethynylmagnesium bromide. (The reacting C—Mg bond of the Grignard reagent is shown as ionic to stress its polar character. This is a convenient fiction that helps us to remember that the Grignard reagent is a strong carbon base.)

$$CH_3\overset{..}{C}H_2 \; \overset{+}{M}gBr \qquad H-C\equiv CMgBr \longrightarrow CH_3CH_3 + \overset{..}{:}C\equiv CMgBr$$

                    ethynylmagnesium $\overset{+}{M}gBr$
                       bromide

According to Eq. (2) in the problem, an excess of acetylene reacts with the BrMgC≡CMgBr product of Eq. (1) to regenerate ethynylmagnesium bromide. Thus, the formation of BrMgC≡CMgBr, the undesired by-product, is avoided by "recycling" it in Eq. (2) back to the desired product HC≡CMgBr.

(b)   The "anionic carbon" of one Grignard reagent acts as a base toward the C—H bond of the other.

$$HC\equiv C\overset{..}{:} \; \overset{+}{M}gBr \qquad H-C\equiv CMgBr \rightleftharpoons HC\equiv C-H + \overset{+}{M}gBr \; \overset{..}{:}C\equiv CMgBr$$

(c)   Precipitation of BrMgC≡CMgBr would pull both reactions (1) and (2) to the right, thus depleting the desired product HC≡CMgBr and forming more of the undesired one. Using a solvent that avoids this precipitation avoids these side-reactions.

10.67   **(a)**   The radical anion forms in this case just as it does with internal alkynes. (See Eq. 14.18a in Sec. 10.2, text p. 523.) This radical anion is a strong base, and a 1-alkyne is acidic enough to react with it. (Internal alkynes, of course, lack this acidic hydrogen.) Recall that 1-alkynes have a pK_a of 25, whereas ammonia, the solvent, has a pK_a of about 32. Ammonia is the only acid available in the reaction of internal alkynes, but in the reaction of 1-alkynes, the alkyne itself is the strongest acid present. This is the source of the side reaction.

$$R\overset{.}{C}=\overset{..}{C}H \qquad H-C\equiv CR \longrightarrow R\overset{.}{C}=CH_2 + Na^+ \overset{..}{:}C\equiv CR$$

    radical anion   **vinylic radical**   **acetylenic anion**

The vinylic radical is reduced as in the case of internal alkynes to give a vinylic anion:

$$R\overset{.}{C}=CH_2 + \overset{.}{N}a \longrightarrow R\overset{..}{C}=CH_2 \; Na^+$$

            vinylic radical   **vinylic anion**

Here we have another strong base; the pK_a of a vinylic proton is about 42. This anion therefore consumes yet another molecule of 1-alkyne to give the alkene.

Therefore, *two* acetylenic anions are formed for every one alkene formed; the acetylenic anions do not react further.

(b) Presumably, the side reactions above could be avoided if a proton source were present that is a stronger acid than the acetylide ion. Ammonium sulfate (that is, $^+NH_4$ ion) was used for this purpose; the pK_a of $^+NH_4$ is 9.25. In the mechanism of part (a), simply substitute this acid everywhere a 1–alkyne is acting as an acid. With this ploy, all of the alkyne is reduced.

10.68 (a) This mechanism consists of two successive Brønsted acid–base reactions.

(b) At first sight this looks like a simple substitution of iodide for chloride. However, the involvement of sodium hydroxide suggests another mechanism. Sodium hydroxide acts as a base to form trichloromethyl anion, which then forms dichlorocarbene. (See Eq. 10.7a–b, Sec. 10.7A on text p. 542.) Dichlorocarbene reacts with iodide ion to form a new anion, which is protonated by water. The reaction is not observed in the absence of NaOH because a base is required to form trichloromethyl anion.

(c) Butyllithium acts as a base to promote an α-elimination by forming an anion that decomposes to phenyl-carbene, which then reacts with the alkene to form the cyclopropane.

10.69 (a) The reaction in the problem introduction suggests that carbene *A* should form; if so, it should react with the alkene double bond to form a cyclopropane, which is compound *X*.

(b)   The nitrogen analog of a carbene—a nitrene *B*— is formed as an intermediate, and it reacts with the alkene to form *Y*, the nitrogen analog of a cyclopropane, called an aziridine.

10.70   The OH group is an electron donor, and it can form a Lewis acid–Lewis base adduct with the zinc atom of the Simmons–Smith reagent. This complex necessarily positions the CH_2 group of the reagent over the face of the double bond that is cis to the hydroxy group so that the bonds to the cyclopropane are cis to the OH.

> nonbonding electrons donated
> from oxygen to zinc

> "carbene" carbon is positioned
> over the double bond cis to the OH

It is found generally that cyclopropane formation is generally "directed" cis to allylic oxygen atoms in the Simmons–Smith reaction by this type of interaction. Thus, the stereochemistry of one asymmetric center determines the stereochemistry of the neighboring center.

The Chemistry of Alcohols and Thiols

CHAPTER 11 SUMMARY

- Alcohols ($pK_a \sim$ 15–16) and thiols ($pK_a \sim$ 10) are weak acids. The conjugate bases of alcohols are called alkoxides or alcoholates, and the conjugate bases of thiols are called mercaptides or thiolates.

- Alkoxides are formed by the reaction of alcohols with strong bases such as sodium hydride (NaH) or by reaction with alkali metals. Solutions of thiolates can be formed by the reaction of thiols with NaOH in alcohol solvents.

- The acidity of alcohols in solution is reduced (the pK_a is increased) by branching near the OH group and increased by electron-withdrawing substituents.

- Alcohols and thiols are weak Brønsted bases and react with strong acids to form positively charged conjugate-acid cations that have negative pKa values. Alcohols are considerably more basic than thiols in solution.

- Several reactions of alcohols involve breaking the C—O bond. A unifying principle in all these reactions is that the OH group, itself a very poor leaving group, is converted into a good leaving group.

  - In dehydration and in the reaction with hydrogen halides, the OH group of an alcohol is converted by protonation into a good leaving group. The protonated OH is eliminated as water (in dehydration) to give an alkene, or is displaced by halide (in the reaction with hydrogen halides) to give an alkyl halide.

  - In the reactions of alcohols with $SOCl_2$ and with Ph_3PBr_2, the reagents themselves convert the OH into a good leaving group, which is displaced in a subsequent substitution reaction to form an alkyl halide.

- In the reaction with a sulfonyl chloride, the alcohol is converted into a sulfonate ester, such as a tosylate or a mesylate. The sulfonate group serves as an excellent leaving group comparable to bromide or chloride in substitution or elimination reactions.

- Phosphates and pyrophosphates are in many cases utilized as leaving groups in biology. They are relatively inert in solution but are activated by hydrogen bonding or metal-ion bonding within enzyme active sites.

- In an organic reaction, a compound has been oxidized when the product into which it is converted has a greater (more positive, less negative) oxidation number. A compound has been reduced when the product into which it is converted has a smaller (less positive, more negative) oxidation number. The change in oxidation number from a reactant to the corresponding product is the same as the number of electrons lost or gained in the corresponding half-reaction.

- Alcohols can be oxidized to carbonyl compounds with Cr(VI). Primary alcohols are oxidized to aldehydes (in the absence of water) or carboxylic acids (in the presence of water), secondary alcohols are oxidized to ketones, and tertiary alcohols are not oxidized. Primary alcohols are oxidized to carboxylic acids with alkaline $KMnO_4$ followed by acid.

- Thiols are oxidized at sulfur rather than at the α-carbon. Disulfides and sulfonic acids are two common oxidation products of thiols.

- Sulfur, phosphorus, and other atoms in periods higher than 2 can violate the octet rule by octet expansion. However, hypervalent structures have important octet resonance forms in which there is considerable separation of charge. Sulfonic acid and phosphoric acid derivatives are examples of compounds in which sulfur and phosphorus have expanded octets.

■ Just as two isomeric molecules can be classified as constitutional isomers or stereoisomers by analyzing their connectivities, two groups within a molecule can be classified as constitutionally equivalent or constitutionally nonequivalent. In general, constitutionally nonequivalent groups are chemically distinguishable. Constitutionally equivalent groups are of three types.

- Homotopic groups, which are chemically equivalent under all conditions.

- Enantiotopic groups, which are chemically nonequivalent in reactions with chiral reagents such as enzymes, but are chemically equivalent in reactions with achiral reagents.

- Diastereotopic groups, which are chemically nonequivalent under all conditions.

■ An example of a naturally occurring oxidation is the conversion of ethanol into acetaldehyde by NAD^+; this reaction is catalyzed by the enzyme alcohol dehydrogenase. Because the enzyme and coenzyme are chiral, one of the enantiotopic *a*-hydrogens of ethanol is selectively removed in this reaction.

■ Alcohols play important roles as starting materials or intermediates in organic synthesis. A network diagram of alcohol conversions is given in Sec. 11.12.

■ Methanol and ethanol serve important roles as fuels. Industrial ethanol, once produced by ethylene hydration, is now produced by fermentation of sugars derived from plants such as corn.

STUDY GUIDE LINKS

11.1   More on Half-Reactions

The technique used for balancing the half-reaction in Eq. 11.51a–c of the text, shown in Study Problem 11.3, text p. 588, is relatively straightforward when oxygens are involved. How can half-reactions be written when oxygens are not involved? For example, what is the half-reaction in the following oxidation?

$$H_2C=CH_2 \xrightarrow{Br_2} Br-CH_2-CH_2-Br$$

Because oxygens are balanced with H_2O, the bromines can be balanced with HBr. Taking this approach results in the following half-reaction:

$$2\ HBr + H_2C=CH_2 \longrightarrow Br-CH_2-CH_2-Br + 2e^- + 2H^+$$

The corresponding inorganic half-reaction is

$$2e^- + 2H^+ + Br_2 \longrightarrow 2\ HBr$$

If the two half-reactions are added, and if the "free electrons" and the HBr molecules on each side of the equation are canceled, the sum is the overall bromine addition reaction:

$$2\ \cancel{HBr} + H_2C=CH_2 \longrightarrow Br-CH_2-CH_2-Br + \cancel{2e^-} + \cancel{2H^+}$$

$$\cancel{2e^-} + \cancel{2H^+} + Br_2 \longrightarrow 2\ \cancel{HBr}$$

$$\overline{\qquad H_2C=CH_2 + Br_2 \longrightarrow Br-CH_2-CH_2-Br \qquad}$$

Although the notion of a half-reaction can thus be applied to bromine addition, it isn't very useful because not only "free electrons" but also fictitious HBr molecules have to be used to make it work; HBr is not involved in the bromine addition reaction of alkenes. On the other hand, water and protons *really are* involved in the oxidation of ethanol to acetic acid, so that the idea of a half-reaction is somewhat more realistic.

When you want to determine the number of "electrons lost" and "electrons gained" in an oxidation or reduction reaction, you'll find it much simpler as a rule to calculate the change in oxidation number than to balance a half-reaction. The only time half-reactions are worth the effort in organic chemistry is when you have to balance a complicated oxidation–reduction equation, as demonstrated in Study Problem 11.6 (text p. 593). (One certainly doesn't need half-reactions to balance the bromine-addition reaction above.) Nevertheless, it is important for you to understand the concept of a half-reaction because biochemists frequently discuss redox processes in terms of half-reactions. If you take biochemistry, you'll probably be glad you learned (or re-learned) these concepts here.

 # FURTHER EXPLORATIONS

 ## 11.1 Solvation of Tertiary Alkoxides

The discussion in the text shows that the relative basicities of alkoxides in solution is clearly governed by solvation, because the basicity order in solution (tertiary > secondary > primary) is different from that in the gas phase (tertiary < secondary < primary). How can we visualize this solvation effect?

At first, it was thought that this might be a steric effect; that is, the alkyl branches in a tertiary alkoxide "get in the way" of the solvent that would normally be solvating the alkoxide oxygen. However, this was shown not to be the case in the following way. There is some indication that the $\Delta H°$ of solvation for tertiary alkoxides is actually negative. Steric effects normally are reflected by *increases* in $\Delta H°$. (For example, this is why *cis*-2-butene has a greater $\Delta H_f°$ than *trans*-2-butene.) The unfavorable $\Delta G°$ for solvation, then, resides in a negative $\Delta S°$. (Read about the solubility of hydrocarbons in water in the text, Sec. 8.6D.) The negative $\Delta S°$ means that the solvent is becoming more ordered, less chaotic, when the tertiary alkoxide dissolves in it.

When the tertiary alkoxide dissolves, the solvent surrounding the hydrocarbon groups cannot form hydrogen bonds to the hydrocarbon groups, so they form stronger hydrogen bonds to each other. This releases energy; therefore, the $\Delta H°$ decreases. However, the tighter hydrogen bonds reduce the random motions of the solvent molecules; because randomness decreases, $\Delta S°$ decreases. In effect, the solvent around the alkoxide becomes more like a solid (ice, if the solvent were water). Because of the tightened hydrogen-bond structure of the solvent, the solvent resists disruption by interaction with the alkoxide oxygen, and the alkoxide oxygen is left "naked," that is, unsolvated. This is what we mean when we say that the hydrocarbon groups adversely affect the solvation of the alkoxide ion.

Think of the following analogy. Imagine a large number of little children running around wildly on a playground, more or less at random. A new child (let's call him Josh) enters the playground carrying candy. A few of the children see this and run up to Josh and interact "favorably" with him by accepting some of his candy.

In a second scenario, let's imagine that a teacher is directing a game in which the students have to hold hands tightly and skip around her in a very ordered circle. The teacher has a bag of candy, but the students can't interact with her "candy" because they are holding hands. The "candy" is analogous to the alkoxide; the teacher is analogous to the hydrocarbon groups; and the children are analogous to solvent molecules. Because the students are forced to interact (hold hands) with each other, they can't reach out for the candy.

11.2   Mechanism of Sulfonate Ester Formation

As the text suggests, the formation of sulfonate esters is another nucleophilic substitution reaction. The role of the pyridine is to form a small amount of the conjugate-base alkoxide ion of the alcohol. Removal of a proton also avoids the formation of free HCl and thus avoids strongly acidic conditions, under which some alcohols and tosylates are reactive.

This alkoxide, acting as a nucleophile, displaces chloride ion from the sulfonyl chloride in a nucleophilic substitution reaction to form the sulfonate ester.

There is really nothing new in this mechanism; the only difference is that the sulfur of a sulfonyl chloride instead of the α-carbon of an alkyl halide is the electrophile.

11.3   Symmetry Relationships Among Constitutionally Equivalent Groups

In Sec. 6.1A, you learned that enantiomers are molecules that are noncongruent mirror images. You also learned in Sec. 6.1C that enantiomers lack certain *symmetry elements,* such as internal mirror planes and centers of symmetry. The purpose of this Further Exploration is to show that the different types of constitutionally equivalent groups—homotopic, enantiotopic, and diastereotopic groups—can also be classified by their symmetry relationships.

Homotopic groups are interchanged by a rotation of the molecule that gives an indistinguishable structure. For example, rotation of the dichloromethane (methylene chloride) molecule 180° about an axis (dashed line) bisecting the two C—Cl (or the two C—H) bonds exchanges the hydrogens and the chlorines, yet leaves the molecule looking exactly as it did before the rotations:

Consequently, the hydrogens are homotopic, and the chlorines are also homotopic. (The hydrogens of CH_3Cl are also homotopic, as shown in the text; demonstrate to yourself the rotation of this molecule that interchanges the three hydrogens while leaving the molecule invariant; see Eq. 11.74 on text p. 602.)

Enantiotopic groups cannot be interchanged by such a rotation. For example, a rotation that interchanges the two enantiotopic α-protons of ethanol gives a molecule that looks different:

That is, the CH_3 group is "up" before the rotation and the OH group is "down." However, after the rotation, these groups have changed positions. The second structure is not congruent to the first without turning it over again.

Enantiotopic groups are related as object and mirror image by an internal mirror plane of the molecule. The internal mirror plane of the ethanol molecule contains the CH_3, the OH, and the central carbon, and it bisects the angle between the two C—H bonds at the α-carbon.

The two hydrogens of ethanol are related as object and mirror image by this plane; hence, these two hydrogens are enantiotopic. Notice that two homotopic groups might also be related by an internal mirror plane, but in order to be homotopic, they must *in addition* be related by a rotation of the molecule as described above. Because a chiral molecule by definition cannot contain an internal mirror plane, it follows that a chiral structure *cannot* contain enantiotopic groups (but it can contain homotopic and diastereotopic groups!).

Finally, diastereotopic protons have no symmetry relationship whatsoever.

REACTION REVIEW

I.   ACID–BASE REACTIONS OF ALCOHOLS AND THIOLS

A.   FORMATION OF ALKOXIDES AND MERCAPTIDES

1.   Alkoxides can be formed from alcohols with stronger bases such as sodium hydride, NaH, which is a source of the hydride ion, $H:^-$.

$$R\text{—}\ddot{O}H \ + \ NaH \ \longrightarrow \ R\text{—}\ddot{O}:^- \ + \ Na^+ \ + \ H_2$$

2.   Sodium metal reacts with an alcohol to afford a solution of the corresponding sodium alkoxide; the rate of this reaction depends strongly on the alcohol.

$$2\,R\text{—}\ddot{O}H \ + \ 2\,Na \ \longrightarrow \ 2\,R\text{—}\ddot{O}:^- \ Na^+ \ + \ H_2$$

3. Because thiols are much more acidic than water or alcohols, they, unlike alcohols, can be converted completely into their conjugate-base mercaptide anions by reaction with one equivalent of hydroxide or alkoxide.

$$R-\ddot{S}H \ + \ NaOH \ \xrightarrow{MeOH} \ R-\ddot{\underset{..}{S}}{:}^- \ + \ Na^+ \ + \ H_2O$$

4. Thiols form insoluble mercaptides (sulfides) with many heavy-metal ions, such as Hg^{2+}, Cu^{2+}, and Pb^{2+}.

$$2\,R-\ddot{S}H \ + \ Hg^{2+} \ \xrightarrow{H_2O} \ R\ddot{S}-Hg-\ddot{S}R \ + \ H_3O^+$$

II. SUBSTITUTION AND ELIMINATION REACTIONS OF ALCOHOLS AND THEIR DERIVATIVES

A. SULFONATE ESTERS AND INORGANIC ESTERS OF ALCOHOLS

1. Sulfonate esters are derivatives of sulfonic acids, which are compounds of the form $R-SO_3H$.

    a. A sulfonate ester is a compound in which the acidic hydrogen of a sulfonic acid is replaced by an alkyl or aryl group.

p-toluenesulfonic acid

ethyl p-toluenesulfonate (ethyl tosylate) a sulfonate ester

    b. Sulfonate esters are prepared from alcohols and other sulfonic acid derivatives called sulfonyl chlorides.

p-toluenesulfonyl chloride a sulfonyl chloride

    i. This is a nucleophilic substitution reaction in which the oxygen of the alcohol displaces chloride ion from the tosyl chloride. (See Further Exploration 11.2 in this manual.)

    ii. The pyridine used as the solvent is a base; besides catalyzing the reaction, it also prevents HCl from forming in the reaction.

2. Sulfonate esters have approximately the same reactivities as the corresponding alkyl bromides in substitution and elimination reactions.

    a. Sulfonate anions, like bromide ions, are good leaving groups.

    b. Sulfonate anions are weak bases; they are the conjugate bases of sulfonic acids, which are strong acids.

3. Sulfonate esters prepared from primary and secondary alcohols, like the corresponding alkyl halides, undergo S_N2 reactions in which a sulfonate ion serves as the leaving group.

a nucleophile

a tosylate ester

p-toluenesulfonate
(tosylate) anion
(a weak base)

4. Secondary and tertiary sulfonate esters, like the corresponding alkyl halides, also undergo E2 reactions with strong bases.

a base

a tosylate ester

p-toluenesulfonate
(tosylate) anion
(a weak base)

5. Secondary and tertiary sulfonate esters, like the corresponding alkyl halides, also undergo S_N1–E1 solvolysis reactions in polar protic solvents.

a tosylate ester

p-toluenesulfonate
(tosylate) anion
(a weak base)

carbocation
intermediate

carbocation
intermediate

solvolysis
(S_N1) product

elimination
(E1) product

6. Tertiary sulfonates do not undergo S_N2 reactions.

B.   PREPARATION OF PRIMARY ALKYL HALIDES

1. Primary alkyl chlorides are best prepared from primary alcohols by the thionyl chloride method.

$$CH_3(CH_2)_2CH_2OH \xrightarrow[CH_2Cl_2]{SOCl_2} CH_3(CH_2)_2CH_2Cl$$

2. Primary alkyl bromides are prepared from primary alcohols by the reaction of the alcohol with concentrated HBr or with Ph_3PBr_2.

$$CH_3(CH_2)_2CH_2OH \xrightarrow[\text{Ph}_3\text{PBr}_2]{\text{HBr or}} CH_3(CH_2)_2CH_2Br$$

3. Primary alkyl iodides are prepared from primary alcohols by the reaction of the alcohol with concentrated HI, which is often provided by mixing KI and H_3PO_4.

$$CH_3(CH_2)_2CH_2OH \xrightarrow{\text{KI, H}_3\text{PO}_4} CH_3(CH_2)_2CH_2I$$

4. The sulfonate ester method works well for primary alkyl halides but requires two separate reactions:

   a. Formation of the sulfonate ester

   b. Reaction of the ester with halide ion

$$CH_3(CH_2)_2CH_2OH \xrightarrow[\text{pyridine}]{\text{RSO}_2\text{Cl}} \underset{\text{a sulfonate ester}}{CH_3(CH_2)_2CH_2OSO_2R} \xrightarrow[\substack{\text{polar aprotic}\\\text{solvent}}]{X^-} CH_3(CH_2)_2CH_2X$$

5. Because all these methods have an S_N2 mechanism as their basis, alcohols with a significant amount of β-branching (that is, alkyl substituents at the β-position) will react more slowly or not at all, with one exception. The exception is Ph_3PBr_2, as described by Display 11.45 in text Sec. 11.4D, which will react even with neopentyl alcohols.

C. *PREPARATION OF SECONDARY ALKYL HALIDES*

1. Secondary alkyl chlorides can be prepared from secondary alcohols by the thionyl chloride method if the secondary alcohol has relatively little β-branching, or by the Ph_3PCl_2 method for more hindered alcohols.

$$(CH_3)_2CH-OH \xrightarrow[\text{PPh}_3\text{Cl}_2]{\text{SOCl}_2 \text{ or}} (CH_3)_2CH-Cl$$

2. Secondary alkyl bromides can be prepared from alcohols by the Ph_3PBr_2 method even if the desired alkyl halide product has significant β-branching.

3. Secondary alkyl halides can be prepared from secondary alcohols via the sulfonate ester, which should be treated with the appropriate halide ion in a polar aprotic solvent to avoid rearrangements.

D. *PREPARATION OF TERTIARY ALKYL HALIDES*

1. Tertiary alkyl chlorides can be prepared from the corresponding tertiary alcohols by reaction with HCl under mild conditions.

2. Tertiary alkyl bromides can be prepared from the corresponding tertiary alcohols by reaction with HBr under mild conditions.

E.   PREPARATION OF ALKENES

1. Alkenes can be prepared by the acid-catalyzed dehydration of alcohols.

2. The equilibrium in this reaction is driven towards the alkene by removal of the volatile alkene, removal of water, or both.

3. The relative rates of alcohol dehydration are in the order tertiary > secondary >> primary.

4. Sulfonate esters readily undergo the E2 reaction with alkoxide bases instead of undergoing S_N2 reactions.

5. This reaction is especially useful in the following situations:

    a. When the acidic conditions of alcohol dehydration lead to rearrangements or other side reactions.

    b. For primary alcohols in which dehydration is not an option.

III.   OXIDATION OF ALCOHOLS AND THIOLS

A.   OXIDATION OF ALCOHOLS TO ALDEHYDES AND KETONES

1. Primary and secondary alcohols are oxidized by reagents containing Cr(VI), that is, chromium in the +6 oxidation state, to give carbonyl compounds (compounds containing the carbonyl group, C=O).

    a. Many forms of chromium are commonly used for these oxidations, including chromate ion (CrO_4^{2-}), dichromate ion ($Cr_2O_7^{2-}$), and chromic anhydride or chromium trioxide (CrO_3). Specific reagents include chromic acid (H_2CrO_4) and sodium or potassium dichromate ($Na_2Cr_2O_7$ or $K_2Cr_2O_7$).

    b. The α-carbon atom of an alcohol must bear a hydrogen atom for oxidation to an aldehyde or ketone to occur.

    c. Primary alcohols react with Cr(VI) reagents to give aldehydes, but if water is present, aldehydes are further oxidized to carboxylic acids.

  d. A complex prepared from CrO_3, pyridine, and HCl, called pyridine chlorochromate (PCC), is often used as an anhydrous Cr(VI) preparation for the oxidation of primary alcohols to aldehydes.

  e. The general mechanism for these oxidations involves a hydride transfer to one of the oxygens of the chromium, and oxidation of the alcohol to the ketone (see Eqs. 11.65a–b, text Sec. 11.7A).

B. OXIDATION OF ALCOHOLS TO CARBOXYLIC ACIDS

  1. Primary alcohols can be oxidized to carboxylic acids using aqueous solutions of Cr(VI) such as aqueous potassium dichromate ($K_2Cr_2O_7$) in acid.

  2. Primary alcohols can be oxidized to carboxylic acids with potassium permanganate ($KMnO_4$) in basic solution, followed by acidification.

$CH_3(CH_2)_7CH_2OH$
1-nonanol
a primary alcohol

$KMnO_4/HO^-$ → H_3O^+ →

nonanoic acid
a carboxylic acid

  3. Because $KMnO_4$ reacts with alkene double bonds, Cr(VI) is preferred for the oxidation of alcohols that contain double or triple bonds.

  4. Potassium permanganate is not used for the oxidation of secondary alcohols to ketones because many ketones react further with the alkaline permanganate reagent.

C. OXIDATION OF THIOLS

  1. Oxidation of a thiol takes place not at the carbon, but at the sulfur.

  a. The most commonly occurring oxidation products of thiols are disulfides and sulfonic acids.

R—S—S—R
general structure of
a disulfide

general structure of
a sulfonic acid

  b. The Lewis structures of these derivatives require either separation of formal charge between sulfur and oxygen or expanded octets around sulfur. The structures with octets are the more important resonance structures, but the structures with expanded octets are often written for convenience.

  c. The same oxidation-number formalism used for carbon can be applied to oxidation at sulfur.

$$H_3C-SH$$

$$N_{ox} = -1$$

$$\left[R-\overset{\overset{:\ddot{O}:^-}{\|}}{\underset{:\ddot{O}:^-}{\overset{|}{S}^{2+}}}-\ddot{O}H \longleftrightarrow R-\overset{\overset{:O:}{\|}}{\underset{:O:}{\overset{\|}{S}}}-\ddot{O}H \right]$$

$$N_{ox} = +5$$

2. Sulfonic acids are formed by vigorous oxidation of thiols or disulfides with $KMnO_4$ or nitric acid (HNO_3).

$$CH_3SH \xrightarrow{\ HNO_3\ } CH_3SO_3H$$

   methanethiol　　　　　　　　　　　**methanesulfonic acid**

3. Many thiols spontaneously oxidize to disulfides merely on standing in air (O_2).

4. Thiols can also be converted into disulfides by mild oxidants such as I_2 in base or Br_2 in CCl_4.

   a. These reactions can be viewed as a series of S_N2 reactions in which halogen and sulfur are the electrophiles that react with nucleophilic thiolate anions.

$$H_3C-S-H \xrightarrow[-H_2O]{^-OH} H_3C-S^- \xrightarrow[-I^-]{I_2} H_3C-S-I \xrightarrow[-I^-]{H_3C-S^-} H_3C-S-S-CH_3$$

   methanethiol　　　　　　　　　　　　　　　　　　　　　　　**methyl disulfide**

   b. When thiols and disulfides are present together in the same solution, an equilibrium among them is rapidly established.

$$CH_3SH \ + \ CH_3CH_2S-SCH_2CH_3 \ \rightleftharpoons \ CH_3CH_2SH \ + \ CH_3S-SCH_2CH_3$$

   methanethiol　　　**diethyl disulfide**　　　　　　　**ethanethiol**　　**ethyl methyl disulfide**

 # SOLUTIONS TO PROBLEMS

Solutions to Focused Problems

11.1　(a)　　　　　　　　　(b)　　　　　　　　　(c)　　　　　　　　　　　　(d)

　　Na$^+$ $^-$O —|　　　K$^+$ $^-$O —|<　　　Mg^{2+} (─O~~~)$_2$　　　Li$^+$ $^-$S~~~

　　sodium isopropoxide　**potassium *tert*-butoxide**　　　　　　　　　　　　**lithium 1-butanethiolate**

　　　　　　　　　　　　　　　　magnesium 2,2-dimethyl-1-butanolate

11.2　(a)　Calcium methoxide or calcium methanolate
　　　(b)　Cuprous ethanethiolate [or copper(I) ethanethiolate]

11.3   (a)   The principle is that the polar effect of a chloro substituent enhances acidity; the effect is greater when more chloro substituents are present; and the effect diminishes as the chloro group is separated from the hydroxy group by more carbons. Thus, 3-chloro-1-propanol, $ClCH_2CH_2CH_2$—OH, is least acidic and has the greatest pK_a; 2-chloroethanol, $ClCH_2CH_2$—OH, is more acidic; and 2,2-dichloroethanol, Cl_2CHCH_2—OH, is most acidic and has the lowest pK_a.

(b)   Thiols are more acidic than alcohols, other things being equal (element effect); and a chloro substituent enhances acidity by a polar effect. Thus, ethanol (CH_3CH_2OH) is least acidic and has the greatest pK_a; 2-chloroethanol is more acidic; and 2-chloroethanethiol, $ClCH_2CH_2$—SH, is most acidic and has the lowest pK_a.

(c)   An alkoxy group enhances acidity because the polar effect of the electronegative oxygen. Thus, 1-butanol, $CH_3CH_2CH_2CH_2OH$, is least acidic and has the greatest pK_a; and 2-methoxyethanol, $CH_3OCH_2CH_2OH$, is most acidic and has the lowest pK_a.

11.4   Alcohols are stronger bases than thiols; therefore, the major acidic species is the conjugate acid of 1-propanol, $CH_3CH_2CH_2\overset{+}{\underset{\cdot\cdot}{O}}H_2$.

11.5   (a)   The products are shown below. See the solution to Focused Problem 11.7 for the relative amounts of each.

(b)   The major product, the *E*-alkene, is shown below. The *Z*-alkene is only a minor product (why?).

11.6   The OH group of the alcohol is protonated in a Brønsted acid–base reaction to form the conjugate acid of the alcohol. The oxygen of the alcohol is the Brønsted base. This loses water to form a carbocation in a Lewis acid–base dissociation reaction in which a water molecule is the leaving group. Finally, in a Brønsted acid–base reaction, water acts as a Brønsted base to remove a β-proton from the carbocation, which acts as a Brønsted acid, to give the alkene. The formation of product *A* by removal of proton *(a)* is shown here; the formation of products *B* and *C* occurs in an analogous manner by removal of β-protons *(b)* and *(c)*, respectively.

11.7   The *E* isomers of *A* and *B* will predominate over the *Z* isomers because the *E* isomers are considerably more stable. Because alkenes *A* and *B* have more alkyl substituents at the double bond than *C*, the *E* isomers of *A* and

B will be formed in greater amount than alkene *C*. (Recall that the E1 reaction, of which dehydration is an example, gives the most stable alkene isomers.)

11.8   The carbocation intermediate in the dehydration of 3-ethyl-3-pentanol is derived by protonation of the OH group and loss of water:

11.9   (a)   1-Methylcyclohexanol and either *cis*- or *trans*-2-methylcyclohexanol should give 1-methylcyclohexene as the major product of dehydration, because this is the most stable alkene that could be formed in each case. The tertiary alcohol 1-methylcyclohexanol should react most rapidly because it involves a tertiary carbocation intermediate; dehydrations of the 2-methylcyclohexanols involve a less stable secondary carbocation intermediate (which would probably rearrange to a tertiary carbocation).

(b)   Both 3-methyl-3-pentanol and 3-methyl-2-pentanol should give 3-methyl-2-pentene as the major product. The tertiary alcohol 3-methyl-3-pentanol should dehydrate more rapidly.

11.10   Protonation of the OH group and loss of water as shown in several of the previous solutions, as well as in Eqs. 11.18a–b on p. 570 of the text, gives a secondary carbocation. As the text discussion of Eq. 11.21 suggests, the mechanism involves a rearrangement of the initially formed secondary carbocation to a more stable tertiary carbocation.

initially formed
secondary carbocation

rearranged
tertiary carbocation

Loss of the two possible β-protons gives the two alkene products.

2,3-dimethyl-1-butene
A

2,3-dimethyl-2-butene
B

11.11   (a)

(b)

(c)

Because concentrated HI solutions are not very stable, "concentrated HI" is usually formed by combining potassium iodide (KI) or sodium iodide (NaI) and phosphoric acid.

11.12   This reaction involves a carbocation rearrangement. We show HBr as the acid although, because water is generated as a product, H_3O^+ is the effective acid once the reaction progresses significantly.

11.13   (a)

$CH_3CH_2CH_2$—Br

1-bromopropane

(b)

1,3-diiodopropane

(c)

Br

$(CH_3)_2C$—$CH(CH_3)_2$

2-bromo-2,3-dimethylbutane

(d)   The compound, neopentyl alcohol, is a primary alkyl halide and cannot react by the S_N1 mechanism; and it has too many β-substituents to react by the S_N2 mechanism. Consequently, there is no reaction.

11.14 (a)

isopropyl methanesulfonate
(or isopropyl mesylate)

(b)

methyl *p*-toluenesulfonate
(or methyl tosylate)

(c)

phenyl *p*-toluenesulfonate
(or phenyl tosylate)

(d)

cyclohexyl methanesulfonate
(or cyclohexyl mesylate)

11.15

cyclohexanol

methanesulfonyl
chloride

cyclohexyl mesylate

 Remember that when we outline a synthesis we provide key starting materials and products, but we do not necessarily show by-products and do not necessarily balance the reaction. For example, by-products are missing in the previous solution. This is acceptable as long as the request is for a synthesis rather than for a balanced reaction.

11.16 (a)

$(CH_3)_2CHCH_2CH_2CH_2OH$ $\xrightarrow{\text{TsCl, pyridine}}$ $(CH_3)_2CHCH_2CH_2CH_2OTs$ $\xrightarrow[\text{polar aprotic solvent}]{\text{Na}^+ \text{I}^-}$ $(CH_3)_2CHCH_2CH_2CH_2I$

(b)

—$CH_2CH_2CH_2OH$ $\xrightarrow{\text{NaH}}$ —$CH_2CH_2CH_2O^- \text{Na}^+$ $\xrightarrow{\text{tosyl chloride}}$

+ H_2

—$CH_2CH_2CH_2OTs$ $\xrightarrow{\text{Na}^+ \text{CH}_3\text{S}^-}$ —$CH_2CH_2CH_2SCH_3$

Although a polar aprotic solvent would accelerate the last step, it would probably work in an alcohol solvent. The nucleophile, CH_3S^-, can be generated by allowing the thiol CH_3—SH to react with one equivalent of sodium ethoxide in ethanol.

11.17 (a)   Cyanide ion displaces the tosylate ester formed in the first step.

(b)   The triflate ester formed in the first step is displaced by the fluoride ion. The nucleophilicity of fluoride ion is increased by complexation of the counter-ion K⁺ with [18]-crown-6, and by the use of acetonitrile, a polar aprotic solvent.

11.18   (a)   (b)

CH₃O—P(=O)(OCH₃)—OCH₃

trimethyl phosphate

CH₃CH₂O—P(=O)(OH)—OH

monoethyl phosphate

11.19   (a)   (b)

H₃C—⁺NH₂—CH₃   ⁻OSOCH₃ (with two =O)

dimethylammonium methylsulfate

CH₃OH + Na⁺ ⁻OSOCH₃ + H₃O⁺

methanol

(c)   (d)

CH₃CH₂OCH₃ + Na⁺ ⁻OSOCH₃

ethyl methyl ether

CH₃CH₂CH₂SCH₃ + Na⁺ ⁻OSOCH₃

methyl propyl sulfide

11.20   1-Butanol can be converted into 1-bromobutane by treating it with triphenylphosphine dibromide (Ph₃PBr₂). (See Eq. 11.43–11.44b on text p. 583.) 1-Butanol can also be converted into 1-bromobutane by first converting it into a sulfonate ester such as butyl tosylate, then treating that ester with sodium bromide in a polar aprotic solvent such as DMSO. Finally, and perhaps most simply, 1-butanol can be converted into 1-bromobutane by treatment with concentrated HBr and H₂SO₄.

CH₃CH₂CH₂CH₂OH **1-butanol** → [Ph₃PBr₂; 1) TsCl, pyridine 2) Na⁺ Br⁻, DMSO; concd. HBr, H₂SO₄] → CH₃CH₂CH₂CH₂Br **1-bromobutane**

11.21   (a)   Eq. 11.42b of the text shows that the nucleophilic reaction of the chloride ion on the chlorosulfite ester is an S_N2 reaction; consequently, it should take place with inversion. It follows that the product should be the *R* enantiomer of CH₃CH₂CH₂CHD—Cl.

(b)   Eqs. 11.44a–b (text p. 583) shows that the nucleophilic reaction of the bromide ion on the reactive intermediate occurs by a concerted (S_N2) substitution reaction. Therefore, the reaction should occur with inversion of stereochemistry, and the product would then be (*S*)-2-bromopentane. The S_N2 reaction occurs at an acceptable rate on a secondary carbon in the absence of β substituents. In addition, the reaction is fast because the leaving group is a *very* weak base, and because the reaction is essentially intramolecular.

11.22   We can predict the structure of an *acyl phosphate* by inspecting the structures shown in the text. An acyl chloride, shown in the problem, contains a chloride, so the rest of the compound must be an acyl group. The phosphate group is shown in Eq. 11.46. Combining the two gives us an acyl phosphate.

acyl   phosphate
group   group

an acyl phosphate

Magnesium binding as shown in Eq. 11.48 (or alternatively, hydrogen bonding provided by the enzyme) can stabilize the negative charges on the oxygen. Presumably the bond between the carbon and the oxygen is also stretched, weakening it, making the phosphate group a very good leaving group.

weakened bond
makes C more
electrophilic

11.23 (a) Because the alcohol groups are primary, concentrated HBr/H_2SO_4 will bring about the desired reaction. Treatment with PPh_3Br_2 will also work.

(b) The simplest method for effecting the conversion shown is to treat the alcohol with thionyl chloride and pyridine. Conversion of the alcohol to a sulfonate ester and treatment of the ester with sodium chloride in a polar aprotic solvent would also work, but involves more steps. Use of concentrated HCl is not satisfactory because HCl (unlike HBr) is not reactive enough.

(c) The alcohol reacts with HBr to give a carbocation intermediate that rearranges. However, this is precisely the reaction desired; reaction of the rearranged carbocation with bromide ion gives the desired product. (See Study Problem 11.1 on text p. 572.)

(d) Because a carbocation intermediate, and therefore rearrangements, are a distinct possibility if this alcohol is exposed to acidic reagents such as HBr, the sulfonate ester method should be used. Thus, treat the alcohol with tosyl chloride and pyridine, and treat the resulting tosylate with sodium bromide in a polar aprotic solvent. This type of solvent suppresses carbocation formation. Alternatively, PPh_3Br_2 gives similar results.

11.24 (a) Treatment with HBr/H_2SO_4 would protonate the OH group, converting it into a good leaving group. A secondary carbocation is produced, which would then rearrange via a hydride shift to a more stable tertiary carbocation. The S_N1 reaction would conclude by the bromide adding at the positive center, leaving the rearranged product 2-bromo-2-methylpentane. (See the solution to Focused Problem 11.12 in this manual.)

(b) Treating secondary alcohols with triphenylphosphine dibromide also converts the OH group into a good leaving group. Because the bromide ion is produced very close to the α-carbon, it does not have to find this carbon by random diffusion. Therefore, an S_N2 reaction occurs much faster than an S_N1 reaction. (In Sec. 12.8 we show that *proximity* of the reactants can accelerate reactions by many orders of magnitude.)

3-bromo-2-methylpentane

11.25   (a)   The bromination of methane to give methyl bromide is a two-electron oxidation.

(b)   The conversion of toluene into benzoic acid is a six-electron oxidation.

(c)   The conversion of an alkyl iodide into an alkane is a two-electron reduction.

(d)   The dihydroxylation of an alkene by $KMnO_4$ is a two-electron oxidation.

(e)   The ozonolysis of 2-methyl-2-butene is a six-electron oxidation. Notice that both carbons of the double bond must be considered, even though the molecule is "split in two" as a result of the reaction.

(f)   The addition of HBr to an alkene is neither an oxidation nor a reduction. (One carbon of the alkene is formally oxidized and the other is reduced by the same amount.)

(g)   This is a two-electron reduction; the negative charge contributes –1 to the oxidation number of the product.

11.26   The half-reaction of Problem 11.25, part (b):

Electrons are "lost" in this half-reaction. This observation is consistent with the classification of this reaction in the solution to Problem 11.25(b) as an oxidation. To complete the sentence, "This reaction is a six-electron oxidation."

11.27   (a)   The alkene is reduced, and the H_2 is oxidized.

(b)   The organic compound is reduced, and the $^-AlH_4$ is oxidized.

(c)   The alkene is oxidized, and the Br_2 is reduced.

(d)   The organic compound is oxidized, and oxygen is reduced.

11.28   The "glib" way to work this problem is to balance the changes in oxidation states of the oxidizing agent and the organic compound. The manganese is reduced from Mn(VII) to Mn(IV); this is a three-electron reduction. Toluene undergoes a six-electron oxidation. Therefore, it will take two equivalents of MnO_4^- to oxidize one equivalent of toluene, or two moles of MnO_4^- per mole of toluene, to reconcile the electrons lost and the electrons gained.

Let's now prove that this works by balancing the equations, multiplying the lower one by 2 so that the "free" electrons cancel, and adding the two, canceling H_2O and H^+:

$$2\,H_2O \; + \; PhCH_3 \longrightarrow PhCO_2H \; + \; 6e^- \; + \; 6H^+$$

multiply by 2 ⟶ $\overset{8}{\cancel{4}}H^+ \; + \; \overset{2}{}MnO_4^- \; + \; \overset{6}{\cancel{3}}e^- \longrightarrow \overset{2}{}MnO_2 \; + \; \overset{4}{\cancel{2}}\,H_2O$

$$PhCH_3 \; + \; 2\,MnO_4^- \; + \; 2H^+ \longrightarrow PhCO_2H \; + \; 2\,MnO_2 \; + \; 2\,H_2O$$

As this shows, and as we deduced above, two permanganates are required per mole of toluene for this oxidation.

11.29   (a)   On the assumption that sufficient PCC has been added, both primary alcohols are oxidized:

$$HO—CH_2CH_2CH_2—OH \xrightarrow{PCC} O{=}CHCH_2CH{=}O$$

(b)   Only the primary alcohol is oxidized:

11.30   (a)   The carboxylic acid could be prepared from $(CH_3)_2CHCH_2CH_2CH_2CH_2OH$ (5-methyl-1-hexanol) by oxidation either with aqueous chromic acid [or any other form of aqueous Cr(VI)], or with aqueous potassium permanganate followed by acid.

(b)   This compound (3-pentanone) can be prepared by a PCC oxidation of the corresponding alcohol, 3-pentanol. (Aqueous dichromate could also be used.)

$$\underset{\textbf{3-pentanol}}{CH_3CH_2\overset{\overset{\textstyle OH}{|}}{C}HCH_2CH_3} \xrightarrow{PCC} \underset{\textbf{3-pentanone}}{CH_3CH_2\overset{\overset{\textstyle O}{||}}{C}CH_2CH_3}$$

(c)   The aldehyde can be prepared from 5-methyl-1,5-hexanediol by the PCC oxidation or other variation of nonaqueous oxidation with Cr(VI).

$$\underset{\textbf{5-methyl-1,5-hexanediol}}{CH_3\overset{\overset{\textstyle OH}{|}}{\underset{\underset{\textstyle CH_3}{|}}{C}}CH_2CH_2CH_2CH_2OH} \xrightarrow{PCC} \underset{\textbf{5-methyl-5-hydroxyhexanal}}{CH_3\overset{\overset{\textstyle OH}{|}}{\underset{\underset{\textstyle CH_3}{|}}{C}}CH_2CH_2CH_2CH{=}O}$$

(d)   This aldehyde can be prepared by a PCC oxidation of the corresponding alcohol.

11.31   This oxidation occurs in much the same way that NAD^+ oxidations occur: by transfer of a hydride from the alcohol to the carbocation.[Notice that it is a *hydride* (a hydrogen with two electrons), not a proton, that is transferred.]

11.32   (a)   The two methyl groups *a* and *b* are constitutionally equivalent and diastereotopic. The pair of methyl groups *a* and *c* are constitutionally nonequivalent, as are the pair *b* and *c*.

When investigating the stereochemical relationships among several groups, the appropriate technique is to consider *each pair*. Thus, three relationships must be categorized within three groups (although it is possible that some relationships might be the same.) Here's an amusing little puzzle that uses the same sort of reasoning. Man *A* looks at a picture and comments about the person *B* in the picture, saying, "Brothers and sisters have I none; this man's father is my father's son." What is the relationship between *A* and *B*?

(b) Hydrogens *a* and *b* are constitutionally equivalent and enantiotopic. (The analysis of this case is essentially identical to the analysis of the α-hydrogens of ethanol; see Eq. 11.75, text p. 602.)

(c) The two fluorines are constitutionally equivalent and diastereotopic.

(d) Hydrogens *a* and *b* within a particular chair conformation are constitutionally equivalent and diastereotopic. However, the chair interconversion interchanges the positions of these two hydrogens and makes them completely equivalent, that is, homotopic, *over time.*

(e) Replacing H^a and H^b in turn with a "circled H" shows that these hydrogens are constitutionally equivalent and enantiotopic, as are H^c and H^d. H^a and H^c are also constitutionally equivalent and enantiotopic, as are H^b and H^d. Finally, H^a and H^d are constitutionally equivalent and homotopic, as are H^b and H^c.

(f) Carbon-2 and carbon-4 are constitutionally equivalent and enantiotopic. Hydrogens *a* and *d* are constitutionally equivalent and diastereotopic, as are hydrogens *b* and *c*. Hydrogens *a* and *c* are constitutionally equivalent and enantiotopic, as are hydrogens *b* and *d*. Finally, hydrogens *a* and *b* are constitutionally equivalent and diastereotopic, as are hydrogens *c* and *d*.

These relationships are best seen with substitution tests—one for each pair. For example, the substitution test for carbons 2 and 4 shows that the two molecules formed by circling each carbon in turn are enantiomers; therefore, these carbons are enantiotopic.

11.33 (a) In both, the hydrogen (or deuterium) is delivered from the pro-(*R*) position on NADH (the "up" position when NADH is drawn as it is in the problem) so that it ends up in the pro-(*R*) position of deuterated ethanol. In part (a), the resulting isotopically substituted ethanol is chiral.

(S)-ethanol-1-*d*

(b) Because deuterium is delivered, the α-carbon of the resulting ethanol bears two deuteriums—that is, the product is CH_3CD_2OH—and it therefore has no asymmetric carbon. The molecule is achiral.

11.34 (a) In a thiol, the oxidation number of sulfur is –1; in a sulfonic acid, it is +5; therefore, the change in oxidation number is +6. The oxidation of a thiol to a sulfonic acid is a six-electron oxidation. (This can be verified by writing a balanced half-reaction.)

(b) This is a two-electron oxidation, because a hydrogen (which contributes –1 to the oxidation number of sulfur) is replaced by an OH (which contributes +1).

11.35 (a) Following the methodology in Sec. 11.6A on p. 588, first write the half reaction. Then, balance the oxygens with water, the hydrogens with protons, and finally the positive charges with electrons. The reaction is a two-electron oxidation.

$$Ph_3P: \; + \; H_2O \longrightarrow Ph_3\overset{+}{P}\!\!=\!\!O \; + \; 2H^+ \; + \; 2e^-$$

(b)

11.36 The reaction is a nucleophilic substitution in which one sulfur acts as a nucleophile toward another sulfur, which acts as an electrophile. Because a thiolate is a much better nucleophile than a thiol—thiolates are *much* more basic than neutral thiols—a thiolate serves as the nucleophile in this reaction. A base is required to form the thiolate anion. The reaction is faster in the presence of base because the reaction rate depends on the nucleophile concentration, and base increases the concentration of the nucleophile.

$$CH_3CH_2\ddot{S}-H \quad \ddot{\textrm{:}}\ddot{O}CH_2CH_3 \quad \rightleftharpoons \quad CH_3CH_2\ddot{S}\textrm{:}^- \quad + \quad H-\ddot{O}CH_2CH_3$$

$$CH_3CH_2\ddot{S}\textrm{:}^- \quad \underset{\underset{CH_2CH_2CH_3}{|}}{S}-\ddot{S}CH_2CH_2CH_3 \quad \longrightarrow \quad CH_3CH_2\ddot{S}-SCH_2CH_2CH_3 \quad + \quad {}^-\textrm{:}\ddot{S}CH_2CH_2CH_3$$

11.37 (a) 2-Methyl-3-pentanol can be prepared by hydroboration–oxidation of 2-methyl-2-pentene, which can be prepared by dehydration of 2-methyl-2-pentanol:

$$\underset{\underset{CH_3}{|}}{CH_3\overset{\overset{OH}{|}}{C}CH_2CH_2CH_3} \xrightarrow{H_2SO_4} \underset{\underset{CH_3}{|}}{CH_3C}=CHCH_2CH_3 \xrightarrow[2)\ H_2O_2,\ ^-OH]{1)\ BH_3,\ THF} \underset{\underset{CH_3}{|}}{CH_3\overset{\overset{OH}{|}}{C}HCHCH_2CH_3}$$

2-methyl-2-pentanol   **2-methyl-2-pentene**   **2-methyl-3-pentanol**

(b) The deuterium-containing alkane can be prepared by protonolysis of a Grignard reagent in D_2O; the Grignard reagent can be prepared from an alkyl halide; and the alkyl halide can be prepared from an alcohol.

$$CH_3(CH_2)_4CH_2OH \xrightarrow{HBr,\ H_2SO_4} CH_3(CH_2)_4CH_2Br \xrightarrow[ether]{Mg} CH_3(CH_2)_4CH_2MgBr \xrightarrow{D_2O} CH_3(CH_2)_4CH_2-D$$

(c) The carboxylic acid can be prepared by oxidation of a primary alcohol, which, in turn, can be prepared from the alkene starting material by hydroboration–oxidation.

$$\text{⬡}=CH_2 \xrightarrow[2)\ H_2O_2,\ ^-OH]{1)\ BH_3,\ THF} \text{⬡}-CH_2OH \xrightarrow[2)\ H_3O^+]{1)\ KMnO_4/NaOH} \text{⬡}-CO_2H$$

methylenecyclohexane   **cyclohexylmethanol**   **cyclohexanecarboxylic acid**

(d) The aldehyde can be prepared by oxidation of a primary alcohol; the required primary alcohol can be prepared by hydroboration–oxidation of an alkene; and the required alkene can be prepared by an E2 reaction of a primary alkyl halide using a branched base.

$$\underset{\underset{CH_3}{|}}{CH_3CH_2CH_2CHCH_2Br} \xrightarrow[(CH_3)_3COH]{K^+\ (CH_3)_3CO^-} \underset{\underset{CH_3}{|}}{CH_3CH_2CH_2C}=CH_2 \xrightarrow[2)\ H_2O_2,\ ^-OH]{1)\ BH_3,\ THF}$$

$$\underset{\underset{CH_3}{|}}{CH_3CH_2CH_2CHCH_2OH} \xrightarrow{PCC} \underset{\underset{CH_3}{|}}{CH_3CH_2CH_2CHCH}=O$$

Solutions to Skills Objectives Problems

11.38 (a)

$$CH_3CH_2OH \ + \ Et_3N \ \rightleftharpoons \ CH_3CH_2O^- \ + \ Et_3\overset{+}{N}H$$

       ethanol   **triethylamine**   **ethoxide**   **triethylammonium**

(b) Before reading the solution that follows, try using the process illustrated by Eqs. 3.58–3.59b and Study Problem 3.8 in the text, Sec. 3.6C.

    The pK_a of the acid on the left side of the expression, ethanol, is about 16 (see Display 11.1 in the text). The pK_a of the acid on the right side of the expression is 10.65. Therefore, the weaker acid–base pair is on the left side, and the left side is favored, which means that the equilibrium constant will be less than 1. Taking the difference of the pK_a values, and raising 10 to that power gives an equilibrium constant of about $10^{(10.65 - 16)} = 10^{-5.35}$.

11.39 (a) Lithium 1-butoxide (common) or lithium 1-butanolate (substitutive)
     (b) Copper(II) 3-methyl-1-butanethiolate

11.40 (a)   (b)

$$(CH_3O^-)_2 \ Mg^{2+}$$

11.41 (a) $B < C < A$, or 2-Chloroethanol < 3-chloro-1-propanethiol < 2-chloro-1-propanethiol. Thiols are more acidic than alcohols (element effect). 2-Chloro-1-propanethiol is the more acidic thiol because the electronegative chlorine is closer to the site of negative charge in the conjugate-base thiolate anion (polar effect).

    (b) $C < B < A < D$, or *tert*-butyl alcohol < isopropyl alcohol < propyl alcohol < 1-propanethiol. Alcohols with a greater degree of alkyl substitution near the OH group are less acidic because their conjugate bases are more poorly solvated; and thiols are substantially more acidic than alcohols (element effect).

    (c) $B < A < C$, or $CH_3NHCH_2CH_2CH_2OH < CH_3NHCH_2CH_2OH < (CH_3)_3\overset{+}{N}CH_2CH_2OH$. The last compound is most acidic because it has a full-fledged positive charge that stabilizes a negative charge in the conjugate-base anion. In the other two compounds, the nitrogen stabilizes the conjugate-base anion by its electron-withdrawing polar effect; in the more acidic of these compounds, the nitrogen is closer to the oxygen, which is the site of negative charge in the conjugate-base alkoxide anion (polar effect).

11.42 (a)

$$CH_3CH_2OH \ + \ CH_3CO_2H \ \rightleftharpoons \ CH_3CH_2\overset{+}{O}H_2 \ + \ CH_3CO_2^-$$

    ethanol   **acetic acid**   **protonated**   **acetate**
                                          ethanol

(b) Before reading the solution that follows, try using the process illustrated by Eqs. 3.58–3.59b and Study Problem 3.8 in the text, Sec. 3.6C.

    The pK_a of the acid on the left side of the expression, acetic acid, is 4.76. The pK_a of the acid on the right side of the expression is about –2 (see Display 11.14 in the text). Therefore, the weaker acid–base pair is on the left side, and the left side is favored, which means that the equilibrium constant will be less than 1. Taking the difference of the pK_a values, and raising 10 to that power gives an equilibrium constant of about $10^{((-2) - 4.76)} = 10^{-6.76}$, or about 10^{-7}.

(c) The very small equilibrium constant of about 10^{-7} found in part (b) tells us that there is a very small amount of products (acetate and protonated ethanol) relative to reactants (ethanol and acetic acid) at equilibrium. We know that the concentration of ethanol is about 17 M, and acetic acid is 0.1 M. The concentrations of protonated ethanol and acetate will be equal, which we'll call x. Using Eq. 3.50 in text Sec. 3.6A, we get $10^{-7} = x^2/(0.1 \ M)(17 \ M)$, where x is the concentration of both protonated ethanol and acetate. Solving for *x*, the concentrations of each of the products are 0.0013 *M*, which is negligible relative to the concentrations of either ethanol or acetic acid.

11.43 (a) $C < D < A < B$, or $^-$O—CH$_2$CH$_2$—OH < CH$_3$CH$_2$CH$_2$—OH < CH$_3$O—CH$_2$CH$_2$—OH < CH$_3$O—CH$_2$CH$_2$—$\overset{+}{O}$H$_2$. The protonated alcohol is the most acidic because it has an oxygen with a full positive charge. 2-Methoxyethanol is next most acidic because of the electron-withdrawing polar effect of the oxygen. The anion is least acidic because the negative charge on the oxygen interacts repulsively with a second negative charge formed on ionization of the O—H group:

(b) Relative to the acidity of ethanol, the acidity of the protonated ether is enhanced by the charge effect; and the acidity of the protonated sulfide is enhanced by both the charge effect and the element effect. The order of increasing acidity is $A < C < B$, or

$$CH_3CH_2\overset{..}{\underset{..}{O}}H \; < \; CH_3CH_2\overset{\overset{H}{|}}{\underset{..}{\overset{+}{O}}}CH_2CH_3 \; < \; CH_3\overset{\overset{H}{|}}{\underset{..}{\overset{+}{S}}}CH_3$$

11.44 (a) OTs

(b) OMs

(c) OTf

11.45 (a) CH$_3$CH$_2$CH$_2$CH$_2$Br

(b) CH$_3$CH$_2$CH$_2$CH$_2$$\overset{+}{O}H_2$

HSO$_4^-$

(c) CH$_3$CH$_2$CH$_2$CH=O

(d) CH$_3$CH$_2$CH$_2$CH$_2$O$^-$ Na$^+$

+ H$_2$

(e) CH$_3$CH$_2$CH$_2$CH$_2$OTs

+ ⟨NH⟩$^+$ Cl$^-$

(f) CH$_3$CH$_2$CH$_2$CH$_2$Cl

+ SO$_2$ + ⟨NH⟩$^+$ Cl$^-$

(g) CH$_3$CH$_2$CH$_2$CH$_2$Br

(h) CH$_3$CH$_2$CH$_2$CH$_2$—OSCF$_3$ (with S double-bonded to two O)

+ ⟨NH⟩$^+$ $^-$OSCF$_3$ (with S double-bonded to two O)

11.46 (a) (CH$_3$)$_3$CCl

(b) no reaction

(c) (CH$_3$)$_2$C=CH$_2$

(d) (CH$_3$)$_3$CO$^-$ K$^+$

+ H$_2$

(e) (CH$_3$)$_3$COMs

+ ⟨NH⟩$^+$ Cl$^-$

11.47 (a) S$^-$ Na$^+$

(b) no reaction

(c) SO$_3$H

(d) S—S (diethyl disulfide)

(e)

~~~SO₃H

(f)

~~~S⁻ K⁺

11.48 (a)

CH₃CH₂CH₂CH₂CN

(b)

CH₃CH₂CH₂CH₂F

(c)

(CH₃)₃COCH₃

(d)

H₂C=C(CH₃)₂ + K⁺⁻O₃SCH₃

11.49

~~~CH₃ + H₂O + H₃PO₄   (+ ~~~=CH₂)

major product   minor product

11.50

phosphorous oxybromide

2-bromobutane

11.51

cyclohexanol   **pyridine**

a chlorosulfite ester

+ Cl⁻ HN⁺

pyridinium chloride

sulfur dioxide
(a gas)

11.52 (a)   Neither an oxidation nor a reduction. An internal oxidation–reduction has occurred, but there is no *net* oxidation or reduction.

(b)   This reaction is a two-electron reduction.

(c)   This reaction is a two-electron oxidation.

(d)   This reaction is a two-electron oxidation.

(e)   This reaction is neither an oxidation nor a reduction; it is a simple ionization.

11.53 The glycol is oxidized. This follows from the fact that a bond to carbon is replaced by a bond to oxygen at each carbon of the glycol. The other participant in the reaction, periodate (IO_4^-) must therefore be reduced. Indeed, the ionic product iodate (IO_3^-) contains one less oxygen bound to the iodine; the iodine is converted from I(VII) into I(V).

11.54 (a) The number of electrons involved in the oxidation half-reaction is determined from the oxidation numbers of the carbons that change:

The number of electrons lost is $[(+1) + (+1)] - [0 + 0] = +2$. (This result could also be determined from a balanced half-reaction.)

The half-reaction for periodate is

$$2\,H^+ + 2e^- + IO_4^- \longrightarrow IO_3^- + H_2O$$

Therefore, periodate undergoes a two-electron reduction.

(b) From the balanced equation, shown in the problem, 0.1 mole of periodate is required to oxidize 0.1 mole of the diol.

11.55 The oxidation of a secondary alcohol to a ketone is a 2-electron oxidation. In the process, CrO_3, a form of Cr(VI), is converted into Cr^{3+}, a form of Cr(III); hence the chromium half-reaction is a 3-electron reduction. Therefore, $2/3$ mole of CrO_3 is required to oxidize 1 mole of the alcohol.

The molecular mass of the alcohol is 116; therefore, 10.0 g = 0.0862 mole. Consequently, $(0.667)(0.0862) = 0.0575$ mole of CrO_3 is required for the oxidation. The molecular mass of $CrO_3 = 100$; therefore, 5.75 g of CrO_3 is required for the oxidation.

11.56 First, write a balanced equation using the method of half-reactions:

Because the alcohol oxidation is a four-electron oxidation, and the reduction of nitric acid to NO is a three-electron reduction, it takes four moles of nitric acid to oxidize three moles of the alcohol, that is, $4/3$ mole of nitric acid per mole of alcohol. Therefore, it takes $(4/3)(0.100) = 0.133$ moles of nitric acid to oxidize 0.1 mole of the alcohol.

11.57 In the following answers, chemically equivalent items have the same letter.

(a) In this compound, 1,1,1,2,2-pentafluoroethane, fluorines *a* are constitutionally equivalent and homotopic; fluorines *b* are constitutionally equivalent and enantiotopic; fluorines *a* are constitutionally nonequivalent to fluorines *b*.

(b) Fluorines *a* are constitutionally equivalent and diastereotopic; fluorines *a* are constitutionally non-equivalent to fluorine *b*.

(c) Fluorines *a* are constitutionally equivalent and diastereotopic; fluorines *a* are constitutionally nonequivalent to fluorine *b*.

11.58 (a)

three chemically nonequivalent sets of Hs

(b)

four chemically nonequivalent sets of Hs

(c)

five chemically nonequivalent sets of Hs

(d)

five chemically nonequivalent sets of Hs

11.59 (a)

three chemically nonequivalent sets of Cs

(b)

three chemically nonequivalent Cs

(c)

three chemically nonequivalent sets of Cs

(d)

four chemically nonequivalent sets of Cs

11.60 (a) To introduce a deuterium we need to start with a Grignard reagent, which, in turn, requires an alkyl bromide. Alkyl bromides can be prepared from tertiary alcohols with concentrated HBr, but only after we move the —OH group from carbon-1 to carbon-2 via the alkene.

(b)

(c) Aldehydes are prepared by oxidation of primary alcohols; hydroboration–oxidation provides the required alcohol from the alkene.

from part (b)

(d)

from part (b)

Alternatively, the alkyl bromide could be prepared from the starting alcohol with concd. HBr/H_2SO_4 and then converted in the same last step to the sulfide.

(e) In this question, stereochemistry is important. The product is a racemic mixture of the cis isomers. We can prepare bromides from the corresponding alcohols, in this case the "non-Markovnikov" alcohol. Treatment with triphenylphosphine dibromide inverts the configuration at that carbon, yielding the cis products.

Solutions to Integrated Problems

11.61 (a)

3-ethyl-3-pentanol   **3-ethyl-2-pentene**

(b)

1,2-diphenylethanol   **stilbene**   **benzaldehyde**
                            (mostly trans)

(c)

2-methyl-1-butanol   **3,6-dimethyl-4-octene**

11.62 This is a triester of 1,2,3-propanetriol (glycerol) and nitric acid, known by its traditional name of nitroglycerin. It is derived from glycerol (a triol) and nitric acid:

1,2,3-propanetriol   **nitric acid**   **glyceryl trinitrate**
                                                (nitrogylcerin)

11.63 Protonation of the —OH group, a Lewis acid–base dissociation reaction to form a carbocation, and a Lewis acid–base association reaction of isotopically labeled water with this carbocation are the key steps in this mechanism. (*O represents ^{18}O.)

11.64 (a) This exchange occurs essentially through a series of Brønsted acid–base reactions. As shown below, once the deuterium is incorporated into the solvent, it is significantly diluted, so that its probability of reaction with the alkoxide is very small. In addition, such a reaction is retarded by a significant primary deuterium isotope effect and competes less effectively with the corresponding reaction of water.

becomes diluted
into the solvent H_2O

(b) To prepare $CH_3CH_2CH_2$—OD from $CH_3CH_2CH_2$—OH, use the same reaction with D_2O/NaOD as the solvent.

11.65 (a) We have to use a method that involves one inversion of configuration. Thus, we convert the alcohol to a tosylate; this reaction does not affect the C—O bond. Then we displace the tosylate with inversion using isotopic ⁻OH (that is, ⁻O^{18}—H; the isotopic oxygen is indicated by an asterisk: *O.)

(b) In this case, convert the alcohol into a bromide using a method that involves an inversion of configuration. (Either Ph_3PBr_2 or the two-step alcohol → tosylate → alkyl bromide sequence shown below will work.) Then, in a second inversion step, displace the bromide with ⁻O^{18}—H to provide the alcohol with the desired configuration.

11.66 (a) The sulfonate ester serves as a leaving group in either case:

mesylate   triflate

(b)   The triflate anion is a weaker base than the mesylate anion because the *polar effect* of the fluorines stabilizes the negative charge in the triflate anion and thereby lowers the pK_a of the conjugate sulfonic acid. (See Sec. 3.7E of the text.) The principle to apply is that the better leaving group is the weaker base. This is true because the leaving group is accepting a negative charge and breaking a covalent bond in both the Brønsted acid–base reaction with a base and an electron-pair displacement (S_N2) reaction with a nucleophile. The polar effect of the fluorines should operate in the same way on both processes, because the processes are so similar.

11.67   The essence of this solution is a balanced equation for the oxidation of ethanol by dichromate and conversion of the stoichiometry into a value for percent blood alcohol. Because the dichromate is aqueous, assume that ethanol is oxidized to acetic acid (a four-electron oxidation). A balanced equation for this process is as follows (verify this):

$$16H^+ \; + \; 3\,CH_3CH_2OH \; + \; 2\,Cr_2O_7^{2-} \longrightarrow 3\,CH_3CO_2H \; + \; 11\,H_2O \; + \; 4\,Cr^{3+}$$

This equation shows that $3/2$ mole of ethanol is oxidized for every mole of $Cr_2O_7^{2-}$ ion consumed. Required to solve the problem are the grams of ethanol oxidized and the mL of blood in which that ethanol is contained. The process can be diagrammed as follows:

moles of dichromate $\Rightarrow$ moles of ethanol $\Rightarrow$ grams of ethanol $\left.\begin{array}{l} \\ \\ \end{array}\right\}$ divide and multiply the result by 100

mL of air $\Rightarrow$ mL of blood

The amount of ethanol oxidized is ($3/2$ mol ethanol per mol of dichromate)(0.507×10^{-6} mol of dichromate) = 0.761×10^{-6} mol of ethanol. The molecular mass of ethanol is 46.07 g mol^{-1}; so, the grams of ethanol in the sample of blood is (46.07 g mol^{-1})(0.761×10^{-6} mol) = 3.50×10^{-5}.

The 52.5 mL of air collected is equivalent to 0.0250 mL of blood:

$$mL\ blood = \frac{52.5\ mL\ of\ air}{2100\ mL\ of\ air\ per\ mL\ of\ blood} = 0.0250$$

Finally, the grams of ethanol per mL of blood is (3.50×10^{-5} g of ethanol) ÷ (0.0250 mL of blood) = 1.40×10^{-3}, and, using the formula in the problem, the percent blood alcohol content is this number times 100, or 0.14%. Consequently, Bobbin is legally intoxicated, and Officer Order should make the arrest.

11.68   Because thiols and disulfides equilibrate to give a mixture of all possible thiols and disulfides (see Eq. 11.89a–b, text p. 609), Stench has evidently obtained a complex mixture of di-2-octyl disulfide, dibutyl disulfide, and butyl 2-octyl disulfide. Although the desired product is *statistically* favored (it is twice as likely to be formed as the other disulfides, other things being equal), it is not likely to be more than 50% of the reaction mixture.

11.69 (a) First, convert the poor leaving group (the hydroxy group) into a good leaving group (the tosylate group). Then displace the leaving group with bromine to give the product.

(b) Treatment with HI protonates the OH, turning it into a good leaving group ($-\overset{+}{O}H_2$). Iodide then displaces the H_2O by an S_N2 mechanism, which gives the following product.

(c) For the structure and mode of action of dimethyl sulfate, see Eq. 11.40 on text p. 581.

$$(CH_3)_2CH-SH + CH_3O^- \longrightarrow (CH_3)_2CH-S^- \xrightarrow{\text{dimethyl sulfate}} (CH_3)_2CH-SCH_3$$
$$+ CH_3OH$$

(d) The hydroxy group is converted into a good leaving group, $-OS(O)Cl$, which is displaced by chloride as shown by the mechanism on Eq. 11.42b, text p. 582, to give the product below.

$$\underset{CH_3CHCH_2CH_2Ph}{\overset{Cl}{|}}$$

(e) Similar to the reaction of triphenylphosphine dibromide (Eqs. 11.44a–b, text p. 583) the reaction of alcohols with triphenylphosphine dichloride proceeds by an S_N2 mechanism with inversion of configuration.

(f) Treatment of the alcohol with strong acid protonates the hydroxy group making it a good leaving group. Once the water leaves, there are two different but equivalent protons adjacent to the positively charged carbon. Because there is an equal opportunity to remove either one leading to one of two enantiomers, a racemic mixture of E1 elimination products is generated.

(g) The first step occurs by an S_N1 or S_N2 mechanism to give the alkyl bromide, which then undergoes an E2 elimination to give the alkene. There are two different but equivalent protons adjacent to the bromine leaving group. Because there is an equal opportunity to remove either one leading to one of two enantiomers, a racemic mixture of E2 elimination products is generated.

(mixture of cis
and trans)

enantiomers

(racemic mixture)

(h) An equilibrium is established between the various sulfides and thiol, as in the solution to Problem 11.68.

$$(CH_3)_2CHCH_2CH_2—S—S—CH_2CH_3 \;+\; (CH_3)_2CHCH_2CH_2—S—S—CH_2CH_2CH(CH_3)_2 \;+$$

$$CH_3CH_2—S—S—CH_2CH_3 \;+\; (CH_3)_2CHCH_2CH_2—SH \;+\; CH_3CH_2—SH$$

11.70 (a) The primary deuterium kinetic isotope effect of 6.6 suggests that the step of the mechanism shown in Eq. 11.65b, text p. 596, is rate-limiting, because the isotopically substituted hydride is transferred in this step. (Review the discussion of the *deuterium kinetic isotope effect,* Sec. 9.5D.)

(b) Because of the primary kinetic isotope effect in part (a), the deuterium is transferred more slowly than the hydrogen (about seven times more slowly). Thus, more of the aldehyde containing deuterium is formed. Because PCC is an achiral reagent, it cannot differentiate between the two enantiotopic α-hydrogens of ethanol; each is removed with equal frequency. Consequently, the isotope effect causes isotopic discrimination *regardless* of the stereochemical positions of the hydrogen and deuterium.

(c) The enzyme alcohol dehydrogenase, which is a chiral catalyst, distinguishes between the α-hydrogens of ethanol so completely that it removes *only* the pro-(R) hydrogen whether this hydrogen is isotopically substituted or not. Thus, when this hydrogen is substituted by deuterium, the deuterium is removed, and, as a result, the aldehyde contains no deuterium. Although there may be an isotope effect associated with removal of this hydrogen, it is not reflected in the relative amounts of H and D removed from a given enantiomer because the stereochemical discrimination (which may be a factor of 1000 or more) is a much larger effect than the primary deuterium isotope effect (which could be a factor only as large as 7 or so).

11.71 The two branches of citrate at the central carbon are *enantiotopic.* Hence, they are chemically distinguishable to a chiral catalyst such as an enzyme. Evidently, the difference is such that the dehydration occurs into the unlabeled branch, as shown in the reaction scheme.

One difference in H_2SO_4 solution is that the carboxylate groups are not ionized, but this is not the key difference. The point of the problem is that an achiral laboratory reagent will not make the distinction between enantiotopic groups. Therefore, equal amounts of dehydration should occur into each branch, and there is no reason to expect exclusively the Z stereochemistry observed in the product of the enzyme-catalyzed reaction.

+ the (E)-stereoisomers

11.72 First, draw the 2S,3R stereoisomer of the product so that the stereochemistry of the addition can be deduced.

(2S,3R)-3-deuteriomalate

Only a malate stereoisomer with the 2*S* configuration will dehydrate; if the enzyme is stereospecific in one direction, it must be stereospecific in the other. (See Eq. 7.44, text Sec. 7.7A.) The pro-*R* hydrogen at carbon-3 comes from the solvent; if the reaction had been run in H$_2$O, this would be a hydrogen. This is an anti addition; we leave it to you to confirm this point. (See Problem 7.47(a), text p. 377.)

(a) Because this stereoisomer has the 2*S* configuration, it will dehydrate. Although the pro-*R* hydrogen at carbon-3 will be lost to solvent, the stereochemistry at this position is irrelevant because there is no isotopic label. The deuterium in the starting material is retained, because this deuterium is not part of the reaction. The product is shown at right.

(E)-2-deuteriofumarate

What if the reaction, when it reverses back to malate, gives the product with the OH group on the carbon *without* the deuterium? This is possible, because deuterium and hydrogen differ in size by only a tiny amount, and the enzyme will not distinguish them by size alone. This reaction, because of the stereospecificity of the enzyme, must also form the 2*S* enantiomer, and it must be formed in an anti addition. This malate product, then, must be the diastereomer of the 2*S*,3*R* stereoisomer we started with, as shown at right. To **(2S,3S)-3-deuteriomalate** the extent that any malate remains at equilibrium, it will ultimately be a mixture of the 2*S*,3*S* and 2*S*,3*R* diastereomers. Now, what if this stereoisomer then reverses back to fumarate? Will the deuterium be washed out? It will not be, and we leave it to you to apply the principle of microscopic reversibility to understand why.

(b) The fumarate stereoisomer obtained in this reaction is the same as in part (a). The same comments apply to the dynamic reversal of the reaction; the starting material, to the extent that it remains at equilibrium, will be a mixture of the 2*S*,3*S* and 2*S*,3*R* diastereomers, and the deuterium will not wash out.

(c) Because this malate stereoisomer has the 2*R* configuration, and because the enzyme is specific for 2*S*-malate, this isomer will not react.

11.73 (a) Following the steps in Sec. 11.6A, we can balance the equation. First, we don't need to add any water because no oxygens are involved. Next, we add one H$^+$ to the right side to replace the one that was lost from serotonin. Finally, we add two electrons to the right side to cancel the plus charges. Thus, two electrons were lost, or, we can say this reaction is a two-electron oxidation.

(b) The way that the serotonin structure is shown, the hydrogen pointing away is the pro-*R* hydrogen (see Sec. 11.9B).

Because the enzyme is specific and can specifically remove the pro-*R* hydrogen instead of the pro-*S* hydrogen, different products will result depending on whether compound *A* or *B* is fed to the enzyme.

(c)   Compound *A* reacts faster, due to the deuterium isotope effect, discussed in Sec. 9.5D. When the pro-*R* hydrogen is H (as in *A*) it is abstracted faster than when it is D (as in *B*).

11.74   The reactivity data and the molecular formula of *A* indicate that compound *A* is an alkene with one double bond. The identity of compound *D* follows from the oxidation of 3-hexanol; it can only be 3-hexanone (see following equation). It is given that 3-hexanone is an ozonolysis product of alkene *A* (along with H—CO₂H (formic acid), not shown in the following equation). Since alkene *A* has seven carbons and one double bond, and 3-hexanone has six carbons, the carbon of alkene *A* not accounted for by 3-hexanone must be part of a $=CH_2$ group. Therefore, the identity of alkene *A* is established as 2-ethyl-1-pentene. The identities of compounds *B* and *C* follow from the reactions of *A*.

11.75   Their reactions with NaH, their formulas, and the fact that they can be resolved into enantiomers show that compounds *A* and *B* are chiral isomeric alcohols. Their eventual conversion into methylcyclohexane shows that they have the same carbon skeleton as methylcyclohexane. When *optically active A* and *B* are converted into tosylates and subjected to the E2 reaction with potassium *tert*-butoxide they give *optically active,* and therefore *chiral*, alkenes. The only possible chiral alkenes with the carbon skeleton of methylcyclohexane are 4-methyl-cyclohexene and 3-methylcyclohexene, which are compounds *C* and *D*; the data do not determine which is which. To summarize:

4-methylcyclohexene   **3-methylcyclohexene**   **methylcyclohexane**

compounds *C* and *D*
(the data do not determine which is which)

Because compounds *A* and *B* are chiral, their tosylates must also be chiral. The two chiral tosylates that would both give a mixture of compounds *C* and *D* are the tosylates of the stereoisomeric 3-methylcyclohexanols. Because alcohols *A* and *B* have different melting points, they must be the *diastereomeric* 3-methylcyclohexanols. (Both stereoisomers of 4-methylcyclohexanol are ruled out because both are achiral.) The individual identities of these alcohols follow from an analysis of the E2 reactions of their tosylates. An anti elimination can take place only when the tosylate group is in an axial position. In the tosylate of *cis*-3-methylcyclohexanol, the tosylate group can only assume the axial position required for E2 elimination when the molecule is in a very unstable conformation in which the methyl group is also axial:

In the tosylate of *trans*-3-methylcyclohexanol, the tosylate group can assume the axial position in a conformation in which the methyl group is equatorial. Furthermore, the presence of one axial group in both chair conformations tends to make their energies more nearly equal than in the tosylate of compound *A*, in which the diaxial conformation is *much* less stable.

The compound containing the greater amount of the conformation in which the tosylate group is in the axial position—the tosylate of *B*—undergoes elimination more rapidly. Consequently, compound *A* is *cis*-3-methylcyclohexanol, and compound *B* is *trans*-3-methylcyclohexanol.

cis-3-methylcyclohexanol
(compound *A*)

trans-3-methylcyclohexanol
(compound *B*)

11.76 Unlike HBr, HCN is a weak acid ($pK_a = 9.4$), and it provides very little cyanide ion because it is essentially un-ionized. In addition, it is not acidic enough to protonate the alcohol to any reasonable extent, because the pK_a of a protonated alcohol is –2. Because the OH group of the alcohol is not protonated, it is not a good leaving group. With a nucleophile concentration of nearly zero and a poor leaving group, a substitution reaction cannot take place.

Addition of H_2SO_4 to the reaction mixture would indeed solve the leaving group problem, because the H_2SO_4 would protonate the OH group. However, added protons suppress the ionization of HCN even more than it is suppressed normally. Without a reasonable concentration of nucleophile, the reaction is doomed. And, unless he is working in a very efficient fume hood, Bluelip is doomed, too, because addition of H_2SO_4 to a solution of cyanide ion produces lethal HCN gas.

11.77   Ethanol is partially protonated under the acidic conditions. Water is displaced from a protonated ethanol molecule by another ethanol molecule:

HÖCH₂CH₃

CH₃CH₂—ÖH₂ ⟶ CH₃CH₂—Ö⁺—CH₂CH₃ ⇌ CH₃CH₂—Ö—CH₂CH₃ + H₂ÖCH₂CH₃

protonated ethanol
(the conjugate acid
of ethanol)

H   HÖCH₂CH₃

+ ÖH₂

Ethanol is a weak base, and it is therefore not a very potent nucleophile. This is why the reaction requires high temperature.

The outcome of the reaction is dependent on the reaction conditions. The reaction is run in a sealed pressure vessel, because the temperature is well above the boiling points of the reactants, products, and solvent. If the vessel were not sealed, ethylene would be formed by alcohol dehydration and, as the lowest-boiling component, would escape from the vessel. In a sealed vessel, any ethylene that is formed remains in solution and can undergo acid-catalyzed addition of ethanol.

11.78   (a)   The oxidation of the alcohol RCH_2OH to the aldehyde $RCH{=}O$ is a two-electron oxidation. The reduction of DMSO to the sulfide is a two-electron reduction. Although the carbon of carbon monoxide has undergone a net one-electron reduction, the carbon of CO_2 has undergone a compensating net one-electron oxidation.

(b)   The oxidizing agent is the sulfur atom in DMSO; it is reduced (because it oxidized the alcohol carbon).

(c)

RCH₂—Ö—H ← :NEt₃ ⇌ RCH₂—Ö:⁻   H—⁺NEt₃

RCH₂—Ö:⁻ (CH₃)₂S⁺—Ö—C—C—Cl ⟶
                              A
        :O: :O:

RCH₂—Ö—S⁺(CH₃)₂ Cl⁻ + Ö=C=Ö + C + Et₃⁺NH Cl⁻
              B   :O:

(d)   To solve this problem, first write out the reaction that occurs with deuterium-labeled alcohol:

D
|
R—C—O—S⁺(CH₃)₂   ⟶   R—C—D   +   H₃C—S—CH₂D
| ‖
D   O

Because deuterium ends up incorporated into one of the methyl groups on sulfur, one of those hydrogens must have been removed at some point. Use triethylamine to remove a proton, generating an intermediate called a *sulfur ylide*. The sulfur ylide then can react *intramolecularly* to remove a deuterium from the nearby carbon, oxidizing that carbon, and eliminating dimethyl sulfide containing deuterium. The sulfur ultimately accepts electrons, so it is oxidized, as noted in the solution to part (b).

D   CH₂—H
|
R—C   :NEt₃ ⟶
| O S⁺
D   CH₃

D   CH₂   sulfur accepts electrons
|
R—C   + H⁺NEt₃ ⟶ products
| O S⁺
D   CH₃

a sulfur ylide

11.79   (a)

(b)

(c)

11.80   (a)   The carbocation rearrangement in this solution, in which a tertiary carbocation is converted into a secondary carbocation, is driven by relief of ring strain.

the protonated alcohol

A
+ :ÖH₂

B

The product shown above is the major product (74%). But this mechanism begs a question: Why isn't the major product one that forms from carbocation *B* rearranging again by a methyl shift to a tertiary carbocation? [What would the product(s) be?] The answer may be that the empty 2p orbital of this carbocation is not aligned optimally with the ring–methyl bonds for rearrangement to occur. (For rearrangement to occur, overlap must occur between the bonding orbitals and the empty orbital.)

this angle should be 0° for optimum overlap

empty 2p orbital

ring bonds {

CH₃

H

CH₃

If the rate of this rearrangement of carbocation *B* is decreased, then the large excess of bromide ion (from ionization of HBr) completely consumes carbocation *B* before it can rearrange. The alkyl bromide, which is very insoluble in the partially aqueous solvent, forms a separate layer and is removed from the reaction. Undoubtedly, in a solvolytic medium such as ethanol, in which the alkyl halide would be soluble, rearrangement of carbocation *B* would be observed.

Carbocation *A* has no problem with orbital alignment because there is more or less free (or less constrained) rotation about the bonds to the cationic carbon. The rearrangement is evidently much faster than the capture of this ion by bromide ion.

(b)   The carbocation rearrangement in this mechanism also involves a ring expansion. In this case, a tertiary carbocation is converted into another tertiary carbocation. The reaction occurs because a more stable six-membered ring forms from the slightly less stable five-membered ring. Other products are likely observed, but the question only asks us to provide a mechanism for the one shown.

the protonated alcohol   + :ÖH₂   H   :ÖH₂   + H—ÖH₂

(c)   The product contains eight carbons; therefore, it involves the reaction of two methylpropene carbon skeletons, joined as follows:

this carbon has an additional H

this carbon has an additional H

The first step in this process is the reaction of one alkene molecule with HF to give a *tert*-butyl cation *A*:

$(CH_3)_3C^+$:F̈:⁻

A

(The fluoride counterion is omitted in subsequent steps.) Carbocation *A* then reacts in a Lewis acid–base association reaction with a second molecule of alkene to give carbocation *B*:

$(CH_3)_3\overset{+}{C}$ $H_2C{=}C(CH_3)_2$ $\longrightarrow$ $(CH_3)_3C{-}CH_2{-}\overset{+}{C}(CH_3)_2$

B

Finally, carbocation *B* reacts with isobutane by removing a *hydride* (a hydrogen with two electrons). This is exactly like a carbocation rearrangement by hydride shift, except that the reaction occurs across two different molecules:

The product is accompanied by a *tert*-butyl cation *A*, which is the cation produced in the initial protonation step. This cation can thus react with another molecule of alkene to give another carbocation *B*, and so on. Notice that if a *tert*-butyl cation *A* reacts with a molecule of isobutane by a hydride shift, another *tert*-butyl cation *A* and another isobutane are formed, that is, such a "side reaction" (which undoubtedly occurs) has no net effect. Finally, the hydride-shift mechanism accounts for the results with deuterium.

This reaction is actually a *chain reaction*, in which the *tert*-butyl cation propagates the chain. Although most chain reactions are *free-radical* chain reactions, this is an interesting example of a *cationic* chain reaction.

(d) This is an addition to the alkene that is conceptually similar to hydration or HBr addition. Trifluoromethanesulfonic acid (triflic acid), a strong acid, protonates the alkene double bond to give a carbocation, which then undergoes a Lewis acid–base association reaction to give the product.

11.81 (a) In a resonance structure, both carbon and phosphorus possess eight electrons, either as nonbonding pairs or shared.

(b) Carbanions are very strong bases because they are the conjugate bases of very weak acids. As the resonance structures in part (a) show, this compound has significant negative charge on carbon, and therefore significant carbocation character.

12

The Chemistry of Ethers, Epoxides, Glycols, and Sulfides

CHAPTER 12 SUMMARY

■ Ethers are weak bases with conjugate-acid pK_a values of –2 to –3. Sulfides are much weaker bases with conjugate-acid pK_a values of –6 to –7. Because ethers are Lewis bases, they donate electrons to Lewis acids such as BF_3 and Grignard reagents.

■ Ethers can be synthesized by the Williamson ether synthesis (an S_N2 reaction), by alkoxymercuration–reduction, or by the dehydration of alcohols or the related acid-catalyzed addition of alcohols to alkenes.

■ Epoxides can be synthesized by the syn-oxidation of alkenes with peroxycarboxylic acids or by the cyclization of halohydrins. Epoxides of allylic alcohols can be prepared in high enantiomeric purity by asymmetric epoxidation using a reagent consisting of titanium isopropoxide, *tert*-butyl hydroperoxide, and either (+)- or (–)-diethyl tartrate.

■ Ordinary ethers are relatively unreactive compounds. The major reaction of ethers is cleavage, which occurs under strongly acidic conditions but not under typical basic conditions. The acid-catalyzed cleavage of tertiary ethers occurs more readily than the cleavage of primary or methyl ethers because tertiary carbocation intermediates can be formed.

■ Because of their ring strain, epoxides undergo ring-opening reactions with ease. For example, epoxides react with water to give glycols, ethylene oxide reacts with Grignard reagents to give primary alcohols, and other epoxides react with lithium organocuprate reagents to give secondary or tertiary alcohols. In acidic solution, a protonated epoxide preferentially reacts with nucleophiles at

the tertiary carbon. Bases react with an epoxide at the carbon with fewer alkyl substituents.

■ Ring-opening reactions of epoxides in either acid or base occur with inversion of configuration at carbon stereocenters.

■ Glycols (1,2-diols) can be prepared from alkenes by treatment with OsO_4 followed by hydrolysis, by variations of this reaction in which a catalytic amount of OsO_4 is used along with other oxidants, or by treatment with aqueous alkaline $KMnO_4$. This stereospecific reaction results in a net syn-addition of two hydroxy groups to the alkene double bond.

■ Glycols can be oxidatively cleaved into two carbonyl compounds (aldehydes or ketones) by treating them with periodic acid.

■ Oxonium and sulfonium salts react with nucleophiles in substitution and elimination reactions; oxonium salts are more reactive than sulfonium salts. S-Adenosyl-methionine (SAM) is a sulfonium salt used in nature as a methylating agent.

■ A number of intramolecular reactions are faster than their intermolecular counterparts because intramolecular reactions have larger (less negative) $\Delta S^{\circ\ddagger}$ values. The rate acceleration of an intramolecular reaction is characterized by the proximity effect, or effective molarity, which is the ratio of rate constants for the intramolecular and corresponding intermolecular reactions. Proximity effects due to $\Delta S^{\circ\ddagger}$ differences can be as high as $10^8\ M$.

■ The catalytic power of enzymes resides in the fact that they form enzyme–substrate complexes; because the reactions that take place within these complexes are intramolecular, they are strongly accelerated.

■ Intramolecular nucleophilic substitution reactions are most common when the reaction forms six-membered and smaller rings.

■ Alkyl halides with β-alkylthio substituents react intramolecularly by forming three-membered sulfonium ions, which are rapidly opened with nucleophiles. If the α-carbon of the alkyl halide is a stereocenter, such reactions involve two consecutive inversions of configuration, which then results in overall retention of configuration.

■ Ethers, on standing in air, slowly form explosive peroxides, and diethyl ether is particularly flammable. Both oxidation and combustion reactions represent safety hazards of ethers. Ethers are largely inert toward other oxidation reactions.

■ Sulfides are readily oxidized to sulfoxides and sulfones.

■ The three fundamental operations of organic synthesis are (1) functional-group transformation, (2) control of stereochemistry, and (3) carbon–carbon bond formation.

STUDY GUIDE LINKS

12.1 Learning New Reactions from Earlier Reactions

Quite often the text points out the close relationship between two reactions—a new one, and one you have already studied. Such connections will help you to understand new reactions in terms of what you already know and will substantially reduce the amount of *really new* material that you must master. Alkoxymercuration–reduction is such a case; this reaction is closely related to oxymercuration–reduction. In order to make the connection between the two reactions, it almost goes without saying that you have to know what oxymercuration–reduction is! If you did not know the meaning of this term when you read this section, did you go back and find out?

The question at the end of Study Problem 12.1 on text p. 626—why *tert*-butyl bromide and sodium methoxide will not work in a Williamson ether synthesis—is another effort to get you to think about earlier material, in this case the reactivities of alkyl halides in the presence of a strong base. Did you answer this question? (If you can't answer it, review Sec 9.5G as suggested in the problem.)

Remember that *continued review* is one of the keys to successful study in organic chemistry. You must try constantly to remain active rather than passive when you study. Answering questions such as the ones posed in the text will assist you. If you can't answer one of these questions, write it down and get help at the next available opportunity!

12.2 Common Intermediates from Different Starting Materials

An important thing to notice about this section is that the same reactive intermediate—in this case, a carbocation—can sometimes be generated in several different ways. For example, consider the many different ways that a *tert*-butyl cation can be formed as a reactive intermediate:

Whether a carbocation is formed by the protonation of an alkene, by the dehydration of an alcohol, or by the solvolysis of an alkyl halide, the same product is obtained if a large excess of the same nucleophile is present in each case. Thus, if methanol is used as the solvent, *tert*-butyl methyl ether is the product regardless of the origin of the carbocation. On the other hand, if concentrated HBr is used, *tert*-butyl bromide is the product.

It is natural at first to think of hydrogen halide addition to alkenes, alcohol dehydration, alkene hydration, and solvolysis of tertiary alkyl halides as different reactions. What you now should be able to see is that these reactions are all linked through a common intermediate, the carbocation. The product that is obtained depends on the conditions used.

FURTHER EXPLORATIONS

12.1   Mechanism of OsO₄ Addition

Even though the OsO₄ addition is a concerted reaction—a reaction that occurs in a single step—let's break the reaction into fictitious individual steps so that we can see the relationship of this reaction to other electrophilic additions we have studied.

First, let's think about the electronic character of OsO₄. Remember that osmium in a +8 oxidation state is very electronegative. When a nucleophile reacts at any of the oxygens, electrons are transferred onto the osmium.

In this sense, osmium serves as a "leaving group" in the nucleophilic substitution reaction on oxygen—except that it remains tethered to that oxygen by its one remaining bond. The mechanism of OsO₄ addition can now be viewed as a close analogy to the mechanism of ozone addition to alkenes (Further Exploration 5.2). The electrophilic oxygen accepts the pair of π electrons from the alkene, and a carbocation is generated from the alkene. We can write a resonance structure for the osmium species in which an electron pair (and negative charge) on osmium is delocalized onto an oxygen. This oxygen undergoes a Lewis acid–base association to complete the addition.

an osmate ester

Compare this to the "fictitious" ozonolysis mechanism, in which the central oxygen serves as an "electron sink" and the terminal, negatively-charged, oxygen reacts with the carbocation:

Even though these mechanisms are concerted, thinking of them as two-step processes, as shown above, helps to understand why these reactions take place, and helps to see the connection between these reactions and other electrophilic addition reactions that involve carbocations explicitly.

REACTION REVIEW

I.   SYNTHESIS OF ETHERS, EPOXIDES, GLYCOLS, AND SULFIDES

A.   *WILLIAMSON ETHER SYNTHESIS*

1. The Williamson ether synthesis is an S_N2 reaction in which the conjugate base of an alcohol (or thiol) reacts as a nucleophile with an unhindered alkyl halide or sulfonate ester to form an ether (or sulfide).

$(CH_3)_3C-O^- Na^+$ + H_3C-OTs ⟶ $(CH_3)_3C-O-CH_3$ + $TsO^- Na^+$

sodium 2-methyl-2-propanolate   **methyl tosylate**   **2-methoxy-2-methylpropane (or *tert*-butyl methyl ether)**   **sodium tosylate**

$Ph-S^- Na^+$ + CH_3CH_2Br ⟶ $Ph-S-CH_2CH_3$ + $Na^+ Br^-$

sodium thiophenolate   **bromoethane**   **(ethylthio)benzene (or ethyl phenyl sulfide)**   **sodium bromide**

a. Methyl halides, primary alkyl halides, or the corresponding sulfonate esters can be used in a Williamson ether synthesis.

b. Tertiary and many secondary alkyl halides cannot be used in a Williamson ether synthesis because the E2 (elimination) reaction occurs instead.

$H_3C-O^- Na^+$ + $(CH_3)_3C-I$ ⟶ H_3C-O-H + $Na^+ I^-$ + $(CH_3)_2C=CH_2$

sodium methanolate   ***tert*-butyl iodide (or 2-iodo-2-methylpropane)**   **methanol**   **sodium iodide**   **methylpropene**

2. In principle, two different Williamson ether syntheses are possible for any ether with two different alkyl groups.

a. The preferred synthesis is usually the one that involves the alkyl halide with the greater S_N2 reactivity.

b. Tertiary and secondary alkyl groups should be derived from the alkoxide.

B.   ALKOXYMERCURATION–REDUCTION OF ALKENES

1. An alkoxymercuration–reduction reaction occurs when oxymercuration–reduction is carried out in an alcohol solvent instead of water.

$$(CH_3)_2C=CH_2 \xrightarrow[\text{(CH}_3)_2\text{CHOH}]{\text{Hg(OAc)}_2} \begin{matrix}(CH_3)_2CHO\\ |\\ (CH_3)_2C-CH_2\\ |\\ HgOAc\end{matrix} \xrightarrow{\text{NaBH}_4} \begin{matrix}(CH_3)_2CHO\\ |\\ (CH_3)_2C-CH_2\\ |\\ H\end{matrix}$$

2. The mechanism of the alkoxymercuration reaction is completely analogous to the mechanism of oxymercuration, except that an alcohol instead of water is the nucleophile that reacts as a nucleophile with the mercurinium ion intermediate.

3. The alkoxymercuration–reduction reaction can be used to prepare ethers that cannot be prepared by a Williamson ether synthesis.

C.   ETHERS FROM ALCOHOL DEHYDRATION AND ALKENE ADDITION

1. Two molecules of an alcohol can undergo dehydration to give an ether (a reaction used primarily in industry).

$$2\ CH_3CH_2OH \xrightarrow[140°]{\text{H}_2\text{SO}_4} CH_3CH_2-O-CH_2CH_3$$
ethanol   **diethyl ether**

a. This is generally restricted to the preparation of symmetrical ethers derived from primary alcohols.

b. Secondary and tertiary alcohols cannot be used because they undergo dehydration to alkenes.

c. The formation of ethers from primary alcohols is an S_N2 reaction in which one alcohol displaces water from another molecule of protonated alcohol.

protonated alcohol

2. Tertiary alcohols can be converted into unsymmetrical ethers by treating them with dilute solutions of strong acids in an alcohol solvent.

$$(CH_3)_3C-OH + CH_3OH \xrightarrow{\text{H}_2\text{SO}_4\ \text{(trace)}} (CH_3)_3C-O-CH_3 + H_2O$$
***tert*-butyl alcohol   methanol**   ***tert*-butyl methyl ether**
                  (solvent;
                  in excess)

a. The formation of an ether from a tertiary alcohol and a primary alcohol solvent is an S_N1 solvolysis reaction in which the protonated tertiary alcohol loses water to give a relatively stable carbocation that is attacked by solvent to give the product.

conjugate acid   + $H_2\ddot{O}:$ conjugate acid   + $H_2\overset{+}{\ddot{O}}-CH_3$
of the tertiary alcohol   of the product

b.   One of the alcohol starting materials must readily lose water after protonation to form a relatively stable carbocation; typically this is a tertiary alcohol.

c.   The alcohol that is used in excess must be one that either cannot lose water after protonation to give a carbocation or should form a carbocation much less readily than the tertiary alcohol. Typically, this is a primary alcohol.

d.   Any alkene that does form (by an E1 pathway) is not removed but is reprotonated by the acid to give back a carbocation, which eventually reacts with the alcohol solvent.

e.   Ether formation from tertiary alcohols and the dehydration of tertiary alcohols are alternative branches of a common mechanism. (See Study Guide Link 12.2 and Sec 11.2 in the text.)

3.   Treatment of an alkene with a large excess of alcohol in the presence of an acid catalyst gives an ether, provided that a relatively stable carbocation intermediate is involved.

1-methylcyclohexene   **1-ethoxy-1-methylcyclohexane**

D.   *OXIDATION OF ALKENES WITH PEROXYCARBOXYLIC ACIDS*

1.   One of the best laboratory preparations of epoxides involves the direct oxidation of alkenes with peroxycarboxylic acids.

cyclopentene   a peroxycarboxylic acid   **cyclopentene oxide**   a carboxylic acid
(1,2-epoxycyclopentane)

a.   The oxidizing agent, a peroxycarboxylic acid, is a carboxylic acid that contains an —O—O—H (hydroperoxy) group rather than an —OH (hydroxy) group.

b.   The general terms *peroxyacid* or *peracid*, denoting any acid containing a hydroperoxy group, are sometimes used instead of peroxycarboxylic acid.

c.   Many peroxycarboxylic acids are unstable and are formed just prior to use by mixing a carboxylic acid with hydrogen peroxide.

2.   The formation of an epoxide from an alkene and a peroxycarboxylic acid is a concerted addition reaction. (See Sec. 12.3A in the text for a further discussion of this mechanism.)

alkene   peroxycarboxylic   epoxide   carboxylic
acid   acid

a.   This mechanism is very similar to that for the formation of a bromonium ion in bromine addition to alkenes.

b.   The formation of epoxides with peroxycarboxylic acids is a stereoselective reaction; it takes place with complete retention of the alkene stereochemistry. (See example under point 1.)

E.   *CYCLIZATION OF HALOHYDRINS*

1.   Epoxides can be synthesized by the treatment of halohydrins with base.

a. This reaction is an intramolecular variation of the Williamson ether synthesis in which the alcohol and the alkyl halide are part of the same molecule.

b. The alkoxide anion, formed reversibly by reaction of the alcohol with NaOH, displaces halide ion from the neighboring carbon.

2. Like other S_N2 reactions, this reaction takes place by opposite-side substitution of the nucleophilic oxygen anion at the halide-bearing carbon.

a. The nucleophilic oxygen anion and the leaving halide assume an anti relationship in the transition state of the reaction.

b. Halohydrins derived from cyclic compounds must be able to assume the required anti relationship through a conformational change if epoxide formation is to succeed.

forms an epoxide in basic solution

does not form an epoxide in basic solution

F.   ASYMMETRIC EPOXIDATION

1. Allylic alcohols react with *tert*-butyl hydroperoxide in the presence of titanium(IV) isopropoxide catalyst to form an epoxide at the double bond of the allylic alcohol functionality.

an allylic alcohol   ***tert*-butyl hydroperoxide**   an epoxide   ***tert*-butyl alcohol**

a. Only double bonds with an allylic —OH group form an epoxide; other double bonds usually do not react.

b. In the presence of $(2R,3R)$-(+)-diethyl tartrate, (+)-DET, the epoxide formed is a preponderance of only one enantiomer; in the presence of $(2S,3S)$-(−)-diethyl tartrate, (−)-DET, the epoxide formed is a preponderance of the other enantiomer.

"O" adds from below

"O" adds from above

G.   PREPARATION OF GLYCOLS

1. Some glycols can be prepared by the acid-catalyzed reaction of water with epoxides. (See also Section II.C.4.)

an epoxide   a glycol

2. Alkenes can be oxidized to glycols with OsO_4.

a. The osmium in OsO_4 is in a high oxidation state, Os(VIII), and thus attracts electrons.

   i. This electron-attracting ability results in a concerted cycloaddition reaction between OsO_4 and an alkene that reduces the Os(VIII) to Os(VI) and gives an intermediate called an osmate ester.

   an osmate ester

   ii. A glycol is formed when the cyclic osmate ester is treated with water.

   an osmate ester   a glycol

b. A mild reducing agent such as sodium bisulfite ($NaHSO_3$) is added in many cases to convert the osmium-containing by-products into reduced forms of osmium that are easily removed by filtration.

c. Osmium and its compounds are very toxic and are quite expensive. By including an amine oxide such as *N*-methylmorpholine-*N*-oxide (NMMO) or trimethylamine-*N*-oxide (TMAO) in the reaction mixture, only catalytic amounts of OsO_4 are required, because the amine oxide converts the osmium by-products back into OsO_4.

an alkene   (catalytic amount)   water/*tert*-butyl alcohol (solvent) pyridine   an osmate ester   a glycol   Os(VI)

Os(VI)   an amine oxide   catalyst regenerated   an amine

d. The formation of glycols from alkenes is a stereospecific syn-addition. The five-membered osmate ester ring is easily formed when two oxygens of OsO_4 are added to the same face of the double bond by a concerted mechanism.

osmate ester   *cis*-**1,2-cyclohexanediol**

e. Hydrolysis of epoxides and the OsO_4 oxidation of alkenes are complementary reactions because they provide glycols with different stereochemistry.

3. Glycol formation from alkenes can also be carried out with potassium permanganate ($KMnO_4$), usually under aqueous alkaline conditions.

cis-**1,2-cyclohexanediol**

a. This reaction is also a stereospecific syn-addition, and its mechanism is probably similar to that of OsO_4 addition.

  i. Visually, when oxidation occurs, the brilliant purple color of the permanganate ion is replaced by a murky brown precipitate of magnesium dioxide (MnO_2).

  ii. This color change can be used as a test (called the Baeyer test) for functional groups that can be oxidized by $KMnO_4$.

b. A problem with the use of $KMnO_4$ is that yields are low in many cases because over-oxidation occurs.

II. REACTIONS OF ETHERS, EPOXIDES, GLYCOLS, AND SULFIDES

A. BASICITY OF ETHERS AND SULFIDES

1. Ethers are relatively unreactive compounds; however, the oxygen of an ether can be protonated under strongly acidic conditions.

$pK_a = -2$ to -3

2. Thiols are less basic than ethers, but react similarly in very strong acid.

$pK_a = -5$ to -7

3. Ethers can also act as Lewis bases with electron deficient Lewis acids.

R—O—R' + H—B—H → an ether / a Lewis base + borane a Lewis acid

B. CLEAVAGE OF ETHERS

1. The C—O bonds of ethers react under acidic conditions (HI or HBr) to give alcohols and alkyl halides in a reaction called ether cleavage.

R—O—R' + H—I → R—OH + R'—I → (in some cases) R—I + H_2O

a. Ethers containing only methyl and/or primary alkyl groups require strong acid and relatively harsh conditions to bring about ether cleavage.

b. Secondary and tertiary ethers cleave more readily than primary or methyl ethers because the S_N1 mechanism intervenes and carbocation intermediates can be formed.

c. The alcohol formed in the cleavage of an ether can go on to react with HI or HBr to give a second molecule of alkyl halide.

2. The cleavage mechanism of ethers containing only methyl and/or primary alkyl groups involves the following steps:

a. Protonation of the ether oxygen.

b. A substitution reaction of the protonated ether by the iodide or bromide ion (a good nucleophile) in an S_N2 reaction to form an alkyl halide, which liberates an alcohol as a leaving group.

$CH_3(CH_2)_4CH_2$—O—CH_3 ⇌ (HBr / heat) $CH_3(CH_2)_4CH_2$—O⁺—CH_3 → $CH_3(CH_2)_4CH_2$—OH + H_3C—Br

1-methoxyhexane   **1-hexanol**   **bromomethane**

$CH_3(CH_2)_4CH_2$—OH → (HBr / heat) $CH_3(CH_2)_4CH_2$—Br

1-hexanol   **1-bromohexane**

3. The mechanism of cleavage in ethers containing tertiary alkyl groups (which occurs under milder conditions) involves the following steps:

a. Protonation of the ether oxygen.

b. Formation of a carbocation by loss of the alcohol leaving group.

c. A Lewis acid–base association reaction of the carbocation with the halide ion.

4. Because the S_N1 reaction is faster than competing S_N2 processes, none of the primary alkyl halide is formed.

H_3C—O—$C(CH_3)_3$ ⇌ (HI) H_3C—O⁺—$C(CH_3)_3$ → H_3C—OH + $(CH_3)_3C^+$:I⁻ → $(CH_3)_3C$—I

***tert*-butyl methyl ether**   **methyl alcohol**   a tertiary carbocation   ***tert*-butyl iodide**

5. Secondary alcohols can react by both mechanisms.

C.   REACTIONS OF EPOXIDES WITH ACID OR BASE

1. Because of their ring strain, epoxides undergo ring-opening reactions with ease; the opening of an epoxide relieves the strain of the three-membered ring.

2. The ring opening of epoxides by bases involves opposite-side substitution by the nucleophile at the epoxide carbon.

    a. A reaction of this type is essentially an S_N2 reaction in which the epoxide oxygen serves as the leaving group.

    b. Under basic or neutral conditions, nucleophiles typically react with unsymmetrical epoxides at the carbon with fewer alkyl substituents.

      i.   This regioselectivity is expected from the effect of branching on the rates of S_N2 reactions.

      ii. Alkyl substitution at the epoxide carbon slows the rate of nucleophilic substitution, so a reaction at an unbranched carbon is faster and leads to the observed product.

    c. Inversion of configuration is observed if attack occurs at a stereocenter.

3. The regioselectivity of the ring-opening reaction is different under acidic conditions.

    a. The nucleophile reacts at the more branched carbon of the epoxide.

    b. If neither carbon is tertiary, a mixture of products is formed in most cases.

      i.   Bonds to tertiary carbon atoms are weaker than bonds to primary carbon atoms and the protonated oxygen is a good leaving group.

      ii. The regioselectivity of acid-catalyzed epoxide ring opening is very similar to the nucleophilic reactions of water and alcohols with bromonium ions.

4. Acid-catalyzed epoxide hydrolysis is generally a useful way to prepare glycols.

an epoxide   a glycol

    a. This reaction occurs by opposite-side substitution of water on the protonated epoxide.

    b. This reaction, when combined with epoxide formation from an alkene, gives glycols with different stereochemistry from that obtained from the OsO_4 reaction with alkenes. (Compare with the reaction in Sec. I.G.2.d.)

(±)-*trans*-1,2-cyclohexanediol

5. Base-catalyzed hydrolysis of epoxides also gives glycols, although, in some cases, polymerization occurs as a side reaction under the basic conditions.

D.   REACTION OF ETHYLENE OXIDE WITH GRIGNARD REAGENTS

1. Grignard reagents react with ethylene oxide to give, after a protonation step, primary alcohols.

    a. The carbon of the C—Mg bond of the Grignard reagent has carbanion character and can react with the epoxide as a nucleophile.

    b. The magnesium of the Grignard reagent is a Lewis acid and coordinates with the epoxide oxygen.

    c. Coordination of an oxygen to a Lewis acid makes the oxygen a better leaving group.

    d. After the Grignard reagent has reacted, the alkoxide is converted into the alcohol product in a separate step by the addition of water or dilute acid.

2. The reaction of a Grignard reagent with ethylene oxide provides a method for the formation of carbon–carbon bonds.

3. The reaction of Grignard reagents with epoxides other than ethylene oxide is not very useful because it gives mixtures of products caused by rearrangements and other side reactions.

E.   OXIDATIVE CLEAVAGE OF GLYCOLS

1. The carbon–carbon bond between the OH groups of a vicinal glycol can be cleaved with periodic acid to give two carbonyl compounds.

    a. The cleavage of glycols with periodic acid takes place through a cyclic periodate ester intermediate that forms when the glycol displaces two —OH groups from H_5IO_6.

    b. The cyclic ester spontaneously breaks down by a cyclic flow of electrons in which the iodine accepts an electron pair.

a cyclic periodate ester;
iodine (VII)

iodine (V)

+ 2 H_2O

2. A glycol that cannot form a cyclic ester intermediate is not cleaved by periodic acid.

3. The periodate cleavage reaction has been used as a test for glycols as well as for synthesis.

F.   OXIDATION OF ETHERS AND SULFIDES

1. Except for peroxide formation (which occurs when ethers are allowed to stand in air) and combustion, ethers are relatively inert towards oxidizing conditions.

2. Sulfides oxidize at sulfur rather than at carbon when they react with many common oxidizing agents; sulfides are readily oxidized to sulfoxides and sulfones.

    a. Sulfoxides and sulfones can be prepared by the direct oxidation of sulfides with one and two equivalents, respectively, of hydrogen peroxide, H_2O_2.

a sulfide

a sulfoxide

a sulfone

b. Other common oxidizing agents, such as $KMnO_4$, HNO_3, and peroxyacids, readily oxidize sulfides.

3. Nonionic Lewis structures for sulfoxides and sulfones cannot be written without violating the octet rule. Although these structures are convenient and often used, the octet structures are more important.

a sulfoxide

a sulfone

G. REACTIONS OF OXONIUM AND SULFONIUM SALTS

1. If the acidic hydrogen of a protonated ether is replaced with an alkyl group, the resulting compound is called an oxonium salt; the sulfur analog of an oxonium salt is a sulfonium salt.

$(CH_3)_3O^+ \ ^-BF_4$

trimethyloxonium tetrafluoroborate

$(C_2H_5)_3S^+ \ ^-BF_4$

triethylsulfonium tetrafluoroborate

2. Oxonium and sulfonium salts react with nucleophiles in S_N2 reactions.

$$R-OH \ + \ (CH_3)_3O^+ \ ^-BF_4 \ \longrightarrow \ R-O-CH_3 \ + \ H_3C-O-CH_3 \ + \ HBF_4$$

an alcohol   **trimethyloxonium tetrafluoroborate** (an oxonium salt)   a methyl ether   **dimethyl ether**   **tetrafluoroboric acid**

a. Oxonium salts are among the most reactive alkylating agents known (they react very rapidly with most nucleophiles). They are usually isolated as $^-BF_4$ salts because the $^-BF_4$ counterion is not nucleophilic. These salts must be stored in the absence of moisture.

b. Sulfonium salts are considerably less reactive than the corresponding alkyl chlorides in S_N2 reactions and therefore are handled more easily than oxonium salts.

3. The important biological sulfonium salt *S*-adenosylmethionine (SAM) is a methylating agent for biological nucleophiles.

SOLUTIONS TO PROBLEMS

Solutions to Focused Problems

12.1   Compound *B* is the weakest acid; there are only protons attached to carbon atoms and none of them are very acidic. The strength of the sulfur-hydrogen bond is much less than the oxygen-hydrogen bond, so compound *A* is a stronger acid than compound *D*. The greater electronegativity of oxygen relative to nitrogen makes compound *D* a stronger acid than compound *C*. (See text Sec. 3.7B for a detailed description of the element effect on acidity.) Thus, the compounds arranged in order of increasing acidity are:

B	*C*	*D*	*A*
weakest acid			strongest acid

12.2   (a)   In this reaction, an alkoxide, formed by reaction of the alcohol with sodium, is alkylated by methyl iodide to give isopropyl methyl ether:

$$(CH_3)_2CHOH \xrightarrow{Na} (CH_3)_2CHO^- Na^+ \xrightarrow{CH_3I} (CH_3)_2CHOCH_3 + Na^+ I^-$$
$$+ H_2$$

isopropyl methyl ether
(2-methoxypropane)

(b)   A thiolate ion, formed by reaction of NaOH with the thiol, is alkylated by allyl chloride to give $H_2C=CH-CH_2-S-CH_3$ (allyl methyl sulfide) and $Na^+ Cl^-$. NaOH is a strong enough base to form the thiolate anion, but would not be strong enough to form the conjugate base of an alcohol.

(c)   Alkoxide bases react with tertiary alkyl halides to give elimination products rather than substitution products. (See Sec. 9.5G.)

$$CH_3O^- Na^+ + (C_2H_5)_3C-Br \xrightarrow{CH_3OH} CH_3CH_2-\overset{\overset{\displaystyle CH_3CH_2}{|}}{C}=CHCH_3 + CH_3OH + Na^+ Br^-$$

(d)   Neopentyl halides and sulfonate esters do not undergo S_N2 reactions at room temperature; furthermore, they do not undergo β-elimination because there are no β-hydrogens. Thus, no reaction occurs.

12.3   (a)

cyclohexyl–$CH_2CH_2OH \xrightarrow{NaH}$ cyclohexyl–$CH_2CH_2O^- Na^+ \xrightarrow{CH_3CH_2CH_2I}$ cyclohexyl–$CH_2CH_2OCH_2CH_2CH_3$

A second Williamson ether synthesis, the reaction of sodium propoxide with (2-bromoethyl)cyclohexane, is in principle possible; however, because the alkyl halide has more alkyl branching this reaction would give, in addition to the desired product, a significant amount of alkene resulting from elimination.

(b)

$$(CH_3)_2CHSH \xrightarrow[EtOH]{Na^+ EtO^-} (CH_3)_2CHS^- Na^+ \xrightarrow{CH_3I} (CH_3)_2CHSCH_3$$

The alternative synthesis, reaction of the methanethiolate ion with isopropyl bromide, is less desirable because secondary alkyl halides react more slowly than methyl halides and give some elimination products.

(c)   Di-*tert*-butyl ether cannot be prepared by a Williamson ether synthesis, because its preparation would require an S_N2 reaction of *tert*-butoxide with a *tert*-butyl halide. With this combination of reagents, an E2 reaction would occur instead; methylpropene and *tert*-butyl alcohol would be the only organic products formed. (See the solution to part (c) of Focused Problem 12.2.)

12.4   (a)   Let $-R = -CH_2CH_2CH_2CH_3$. The mechanism is essentially identical to that shown in Eq. 5.20-21 on text p. 246, except that the nucleophile is isopropyl alcohol rather than water.

The nucleophilic reaction of the alcohol with the mercurinium ion intermediate occurs at the carbon with the greater number of alkyl substituents because this carbon–mercury bond is weaker (that is, easier to break) than the bond to primary carbon; alternatively, this carbon has the greater degree of carbocation character.

(b)   Synthesis of the ether in Eq. 12.10b would require a secondary alkyl halide and an alkoxide with α-substituents. A significant, if not predominant, amount of alkene by-product would be obtained by the E2 mechanism.

12.5   The product results from alkoxymercuration of the double bond followed by reduction (by $NaBH_4$) to remove the mercury. The net effect is overall addition of the elements of ethanol to the double bond with the ethoxy group on the carbon with the greater number of alkyl substituents.

$(CH_3)_2CHCHCH_3$
|
OCH_2CH_3 **2-ethoxy-3-methylbutane**

12.6   Because each carbon of the alkene double bond has one alkene substituent, there is no strong preference for the reaction of methanol at either carbon of the resulting mercurinium ion. Consequently, two constitutional isomers of the product ether are formed. (The same result would be obtained regardless of the stereochemistry of the alkene starting material.)

OCH_3
|
$CH_3CH_2CHCH_2CH_3$ + $CH_3CH_2CH_2CHCH_3$
                                                 |
                                          OCH_3

3-methoxypentane   **2-methoxypentane**

12.7   (a)   Cyclohexene is subjected to oxymercuration in cyclohexanol, and the product is then reduced with $NaBH_4$:

dicyclohexyl ether

(b)   Isobutylene (methylpropene) is subjected to alkoxymercuration in isobutyl alcohol, and the resulting organomercury compound is reduced with $NaBH_4$.

$$(CH_3)_2C{=}CH_2 \; + \; HOCH_2CH(CH_3)_2 \; \xrightarrow[\text{2) NaBH}_4,\text{ NaOH}]{\text{1) Hg(OAc)}_2} \; (CH_3)_3C{-}OCH_2CH(CH_3)_2$$

isobutylene   **isobutyl alcohol**   **_tert_-butyl isobutyl ether**

12.8   If we start with two primary alcohols, ROH and R'OH, we would expect them to have similar basicities and similar nucleophilicities. Each alcohol could react with each protonated alcohol. Consequently, three possible products would be formed: R—O—R, R'—O—R, and R'—O—R'. None of the alcohols would be formed in very high yield, and separation of the products could be quite laborious.

The reason we can let a *tertiary* alcohol react with a primary alcohol to give an unsymmetrical ether is that the tertiary alcohol forms a carbocation in acidic solution *much* faster than either it or the primary alcohol react by the S_N2 mechanism; and, once the carbocation is formed, it is rapidly consumed by its Lewis acid–base association reaction with the large excess of primary alcohol that is present.

12.9   The carbocation formed from the tertiary alcohol reacts with ethanol to give the following ether:

12.10   Use the discussion in Study Guide Link 12.2: What alkene would give the same carbocation as the starting alcohol in Focused Problem 12.9?

carbocation intermediate

12.11   (a)   Because this ether is symmetrical, and because both alkyl groups are primary, alcohol dehydration is the appropriate method. This ether can be prepared by treating $ClCH_2CH_2OH$ (2-chloroethanol) with H_2SO_4 at high temperature.

(b)   Because this is an ether with one tertiary alkyl group and one methyl group, it can be prepared by dehydration of the tertiary alcohol in the presence of methanol, or by acid-catalyzed addition of methanol to either of two possible alkenes.

2-methyl-2-butanol   **2-methyl-1-butene**   **2-methyl-2-butene**   **2-methoxy-2-methylbutane**

(c)   Alkoxymercuration–reduction of isobutylene using 1-butanol, $CH_3CH_2CH_2CH_2OH$, as the solvent in the alkoxymercuration step is preferred. (Acid-catalyzed addition of 1-butanol to methylpropene might also work.)

$$CH_3(CH_2)_3OH \; + \; (CH_3)_2C{=}CH_2 \; \xrightarrow[\text{2) NaBH}_4,\text{ NaOH}]{\text{1) Hg(OAc)}_2} \; (CH_3)_3C{-}O(CH_2)_3CH_3$$

1-butanol   **butyl _tert_-butyl ether**
(solvent)

(d)   Dibutyl ether, $CH_3CH_2CH_2CH_2OCH_2CH_2CH_2CH_3$, is a symmetrical ether that can be prepared by the acid-catalyzed dehydration of 1-butanol.

12.12   (a)   (b)   (c)   (d)

In parts (c) and (d), the stereochemistry of the product and the net syn-addition dictate the stereochemistry of the alkene starting materials.

12.13   (a)   (b)

CH₃CH₂'''''C—C''''''H
         H   CH₂CH₃

(the racemate)

12.14   The key to solving this problem is to realize that, in order for epoxide formation to occur, the oxygen of the conjugate-base alkoxide must be anti to the bromine in the transition state so that it can execute an opposite-side substitution. Draw each of the stereoisomeric alkoxides in such a conformation and evaluate the relative energies of the resulting transition states.

In the transition state for the reaction of the alkoxide conjugate base of *A* there are two significant gauche interactions (brackets).

(2R,3S) diastereomer (*A*)
(or its enantiomer)

+ Br⁻

the transition-state conformation
is the most stable conformation

The transition-state conformation is very close to the most favorable conformation of *A*. The transition-state conformation of *B*, however, has three significant gauche interactions, but its more stable conformation has two (brackets).

(2R,3R) diastereomer (*B*)
(or its enantiomer)

+ Br⁻

transition-state
conformation

most stable
conformation

Consequently, isomer *B* must gain additional energy associated with a methyl–methyl gauche interaction in order for it to react. This represents an additional component to the activation free energy of *B* that is not present in the reaction of *A*. Consequently, the reaction of *A* is faster.

12.15 As in the solution to Focused Problem 12.14, the alkoxide oxygen and the halogen leaving group must be anti in the transition state. This anti relationship is possible in a chair conformation only in the reaction of the trans stereoisomer (see Eq. 12.27 on text p. 633). In the cis stereoisomer, such an anti relationship is not possible. Because the cis stereoisomer cannot achieve the appropriate transition-state conformation for epoxide formation, it is unreactive.

trans-2-chlorocyclohexanol
(conjugate-base alkoxide)
an anti conformation is accessible

cis-2-chlorocyclohexanol
(conjugate-base alkoxide)
an anti conformation is not accessible

12.16 **(a)** In the protonated ether (see Eq. 12.30a, text p. 634, for a typical protonation reaction) the iodide ion can react as a nucleophile at either the carbon of the methyl group or the α-carbon of the butyl group. Because this is an S_N2 reaction, and because S_N2 reactions of methyl compounds are faster than those of butyl compounds (why?), the nucleophilic reaction occurs at the methyl carbon to give methyl iodide and 1-butanol.

(b) Because alcohols react with HI to give alkyl iodides, the initially formed 1-butanol would be expected to give 1-iodobutane eventually.

(c) The discussion in Sec. 12.4 of the text indicates that the cleavage reactions of tertiary ethers, which are S_N1 reactions, occur under milder conditions than the S_N2 cleavage reactions of methyl and primary ethers. Hence, cleavage of the protonated ether occurs so as to give the *tert*-butyl cation and methanol. (See Eqs. 12.31a-b on text pp. 634-5 for the mechanism of a similar reaction.) The Lewis acid–base association reaction of iodide with the *tert*-butyl cation gives *tert*-butyl iodide.

(d) Tertiary ethers and sulfides cleave by an S_N1 mechanism. The rate-limiting step in S_N1 reactions is formation of the carbocation by Lewis acid–base dissociation. The rate of this dissociation reaction is proportional to the concentration of the reacting species—in this case, the conjugate acid of the ether or sulfide. Because sulfides are much less basic than ethers (Sec. 12.1, text p. 624), the concentration of the reacting species is much smaller in the case of sulfides. Consequently, sulfides are less reactive than ethers toward acid-promoted cleavage.

12.17 **(a)** Diisopropyl ether reacts initially to give isopropyl alcohol and isopropyl iodide by either an S_N1 or S_N2 mechanism, or by both mechanisms operating simultaneously. Once the alcohol is formed, it reacts with concentrated HI to give isopropyl iodide as well. Therefore, the product is isopropyl iodide.

diisopropyl ether

(b) This tertiary ether reacts rapidly by a carbocation (S_N1) mechanism to give, initially, the tertiary iodide and ethanol. The ethanol subsequently reacts more slowly to give ethyl iodide.

2-ethoxy-2,3-dimethylbutane

12.18   Under basic or neutral conditions under which the epoxide is not protonated, the nucleophiles react at the carbon of the CH$_2$ group. The configuration of the asymmetric carbon is not affected because no bonds to this carbon are broken.

(a)

(b)

12.19   The strategy in this problem is to let the —OH group originate from the epoxide oxygen. In the starting material, this oxygen must be attached to the same carbon as the —OH group in the product as well as to an adjacent carbon; the nucleophile becomes attached to the adjacent carbon. To summarize:

(a)

(b)

The strategy outlined above suggests another possibility:

    If you came up with this idea, you are reasoning correctly. However, every good idea has to be tempered by practical reality. In this case, the strategy will not work because the most common source of nucleophilic hydride, LiAlH$_4$ (lithium aluminum hydride), also reacts with the cyano (CN) group.

12.20   (a)   The nucleophilic reaction of methanol on the protonated epoxide occurs with inversion to give (3S,4R)-4-methoxy-3-hexanol:

(3S,4R)-4-methoxy-3-hexanol

(You should verify that the nucleophilic reaction of methanol at the other asymmetric carbon of the protonated epoxide gives the same result.)

(b) The enantiomer of the epoxide in part (a) gives the enantiomer of the product formed in part (a), (3R,4S)-4-methoxy-3-hexanol. (Verify this by writing the mechanism.)

12.21 (a) Under acidic conditions, the nucleophilic reaction of water occurs at the carbon of the epoxide ring with the methyl substituents (on the protonated epoxide); consequently, the OH group at that carbon is enriched in ^{18}O. In basic solution, the reaction of hydroxide occurs at the CH_2 carbon of the epoxide, and the OH group at that carbon is enriched in ^{18}O. ($^*O = {}^{18}O$.)

(b) Inversion of configuration would occur at the carbon of the epoxide (or protonated epoxide) at which the nucleophilic reaction occurs, giving enantiomeric products. (The nucleophilic oxygen is labeled with an asterisk.)

12.22 (a)

3-methyl-1-pentanol

(b) When an epoxide is opened by an organometallic reagent, the OH group of the product alcohol is on the carbon adjacent to the carbon that is the electrophilic target of the "carbon anion" from the organometallic reagent. This strategy reveals two choices for possible reagent combinations, with the alkyl groups that could be provided by the complex organocuprate shown in boxes:

OH
|
CH_3⎯CH_2⎯C⎯CH_2⎯$CH_2CH_2CH_3$
|
⇩ CH_3CH_2 ⇩

3-ethyl-3-heptanol

$(CH_3)_2Cu(CN)Li_2$ + H_2C⎯C⎯CH_2CH_3 $\xrightarrow[]{H_3O^+ \; | \; H_3O^+}$ $(CH_3CH_2)_2C$⎯CH_2 + $(CH_3CH_2CH_2)_2Cu(CN)Li_2$
|
$CH_2CH_2CH_2CH_3$

12.23 (a)

⎯CH_2CH_2OH

(b)

H   OH
\ |
H_3C ⎯ C ⎯ C ⎯⎯ H
/   \
Ph CH_3

(racemate)

(c)

CH_3CH_2C≡CCH_2CH_2OH

The racemate is formed in (b) because the phenyl (Ph) group can react at either asymmetric carbon with equal probability. Because the epoxide is a meso compound, and is therefore achiral, chiral products must be formed as racemates.

12.24 (a)

1-methylcyclopentene   (racemate)

(b)

trans-2-butene   **(±)-2,3-butanediol**
(the racemate)

12.25 (a)   The alkene required is 4-ethoxy-1-butene, $CH_3CH_2OCH_2CH_2CH$=CH_2.

(b)

methylenecyclobutane

(c)

$CH_3CH_2CH_2$ $CH_2CH_2CH_3$
\   /
C=C
/   \
H   H

cis-4-octene

(d)

$CH_3CH_2CH_2$ H
\   /
C=C
/   \
H $CH_2CH_2CH_3$

trans-4-octene

In part (c), the syn-addition of two identical groups—in this case, OH groups—to a symmetrical cis-diene gives a meso diol. Alternatively, in part (d), the syn-addition of two identical OH groups to a symmetrical *trans*-diene gives a racemic diol. Be sure to analyze these additions in detail if these points are not clear. (See Sec. 7.8E for a similar type of analysis.)

12.26 This mechanism is much like the one shown for OsO_4 in Eq. 12.57, text p. 646. A cyclic electron flow that gives either of the resonance structures of the cyclic intermediate is correct.

 For a list of errors in this manual, please visit https://web.ics.purdue.edu/~loudonm/teaching/

12.27 (a)

(c)

(b)   The products are phenylacetaldehyde, $PhCH_2CH{=}O$, and formaldehyde, $O{=}CH_2$.

12.28 (a)

(b)

12.29 Because iodide ion is a good nucleophile and the trimethyloxonium ion is an excellent alkylating agent, alkylation of the iodide ion occurs to give dimethyl ether and methyl iodide.

12.30 In each case, a nucleophile reacts with a methyl group of the oxonium salt, expelling dimethyl ether as a leaving group.

(a)

(b)

$$(CH_3)_3\overset{+}{S} \quad \bar{B}F_4 \; + \; (CH_3)_2\overset{..}{O}$$

12.31 An acid-base reaction must precede every methylation to reveal a nonbonding electron pair on the nitrogen to make them available for an S_N2 reaction with the methyl group on SAM.

12.32   The intramolecular product tetrahydrofuran results from an internal nucleophilic substitution reaction of the alkoxide on the alkyl halide. The intermolecular product 1,4-butanediol results from the S_N2 reaction of hydroxide ion with the alkyl halide.

4-bromo-1-butanol

tetrahydrofuran
(the intramolecular
substitution product)

1,4-butanediol
(the intermolecular
substitution product)

To form the cyclic product tetrahydrofuran, –OH must react with the alcohol to ionize it; the conjugate base anion of the alcohol then cyclizes in an internal substitution reaction. In contrast, to form 1,4-butanediol, –OH must react with the alkyl halide in an S_N2 reaction. Ionization reactions are much faster than S_N2 reactions; and once the ionization has occurred, the cyclization, because it is intramolecular, is much faster than the S_N2 reaction between hydroxide and the alkyl halide. So, tetrahydrofuran would be the major product.

Another reasonable intermolecular possibility you may have considered is the following:

4-(4-bromobutoxy)-1-butanol

To form this product, ionized 4-bromobutanol must react with the alkyl halide "end" of another molecule of 4-bromobutanol in an S_N2 reaction. Because such a reaction is bimolecular, the intramolecular reaction of the same ion to give tetrahydrofuran is much faster.

12.33   Because the $\Delta H^{\circ\ddagger}$ of the two reactions is the same, the difference in $\Delta G^{\circ\ddagger}$ is the negative of the difference in the $T\Delta S^{\circ\ddagger}$:

$$\Delta G_B^{\circ\ddagger} - \Delta G_A^{\circ\ddagger} = T(0.180 - 0.030) = 44.7 \text{ kJ mol}^{-1}$$

Reaction B has the higher $\Delta G^{\circ\ddagger}$. To convert this different into a relative rate, apply Eq. 9.19a in Sec. 9.3B on text p. 458:

$$\frac{\text{rate}_A}{\text{rate}_B} = 10^{(\Delta G_B^{\circ\ddagger} - \Delta G_A^{\circ\ddagger})/2.30RT} = 6.7 \times 10^7$$

In other words, reaction A is about 70 million times faster than reaction B.

12.34   (a)   The formation of product A "freezes" fewer bond rotations, and therefore has the less negative entropy of activation. This would make the formation of product A faster, except for the fact that it has more strain—that is, a greater enthalpy—than product B. The product that is formed more rapidly depends on the balance between the entropy and enthalpy of activation.

(b)   The first reaction is a bimolecular reaction, whereas the second is an intramolecular reaction. Forming the transition state of the first reaction requires "freezing" the translation of one molecule relative to the other, whereas forming the transition state of the second reaction requires no freezing of translations. Because the bond changes in the two reactions are about the same, we suspect that the second reaction will be much faster because its $\Delta S^{\circ\ddagger}$ is less negative (or more positive).

12.35   (a)   The S_N2 mechanism with ‾OH as the nucleophile:

(b)   The intramolecular reaction mechanism, in the rate-limiting step, results in an unstable intermediate called an α-lactone:

an α-lactone

The α-lactone reacts rapidly with ⁻OH in another S_N2 reaction:

(c)   The proximity effect is

$$\frac{k_1}{k_2} = \frac{1.2 \times 10^{-4}\ s^{-1}}{6.4 \times 10^{-4}\ M^{-1}s^{-1}} = 0.19\,M$$

where k_1 is the first-order rate constant, and k_2 is the second-order rate constant. This is a rather low effective molarity.

(d)   The intramolecular rate law is

$$rate_1 = k_1[S]$$

where S is the starting material; and the second-order rate law for the reaction of hydroxide with the alkyl halide is

$$rate_2 = k_2[S][^-OH]$$

From the ratio of the two rate constants calculated in part (c), the two rates are the same when $[^-OH] = 0.19\ M$.

(e)   The ratio of the two rates, $rate_2/rate_1$, in 1 M NaOH is $1.0/0.19 = 5.3$. Therefore, the second-order reaction accounts for about 84% of the product in 1 M NaOH.

(f)   The carbonyl carbon ideally has bond angles of 120°. When it is forced into a 3-membered ring, its orbitals have to be "bent" to accommodate an interatomic angle closer to 60°. In an epoxide, a saturated carbon atom has a preferred bond angle of 109.5°. Consequently, there is more angle strain in the α-lactone; this accounts for its greater instability. The instability of the α-lactone intermediate is the reason that the intramolecular reaction has a rather small effective molarity despite its entropic advantage.

12.36   An analysis identical to the one shown in Study Problem 12.6, text pp. 658-659, results in the following episulfonium ion, which is the diastereomer of the one shown in Eq. 12.77b.

In this case, however, the episulfonium ion is meso and therefore achiral. The nucleophilic reactions of water at C-2 and C-3 give enantiomers, which must be formed in identical amounts.

12.37   The intramolecular mechanism shown in the solution to Focused Problem 12.35(b) indicates that two inversions of configuration, which is equivalent to net retention of configuration, occur at the same carbon in reaching the final product. Because this mechanism predominates at low NaOH concentration, its stereochemical consequences—retention of configuration—are manifest under the same conditions. As the NaOH concentration is increased, the intermolecular conventional S_N2 mechanism takes over and becomes the predominant one at high NaOH concentration. In this mechanism, inversion of configuration is expected and is observed.

12.38   If intramolecular participation occurs, an intermediate episulfonium ion is formed. In this ion, the two carbons in the three-membered ring are equivalent and indistinguishable except for the label. The reaction with water is equally probable at each of these carbons. Consequently, 50% of the label ends up at each carbon.

If, however, water reacts as a nucleophile directly at the labeled carbon to displace the chloride ion in a conventional S_N2 reaction, an episulfonium ion would not be involved, and the label would not scramble. That is, product *A* would form exclusively.

12.39   The two reactions proceed through a common episulfonium ion intermediate that results from an intramolecular substitution of the protonated OH group by the sulfur.

Evidently, the nucleophilic reaction of the chloride ion with this intermediate takes place at the carbon with the methyl substituent. Because the same intermediate is formed in both reactions, the product is the same.

A separate question is why the chloride ion should react only at the more branched carbon. The cyclic intermediate is much like a protonated epoxide (or a bromonium ion or mercurinium ion); therefore, the nucleophilic reaction of chloride ion occurs at the more branched carbon for the same reason it does in a protonated epoxide. (See discussion of Eq. 12.38b on text p. 639.).

12.40   Cyclophosphamide, a nitrogen mustard, crosslinks two DNA strands by reacting with guanine residues on each strand. Using the same abbreviations as Eq. 12.81 in the text, the structure is shown here:

crosslinked DNA strands

12.41 (a)

a sulfoxide

(b)

a sulfone

12.42 (a) Lengthening a carbon chain by two carbons is required; what better way than the reaction of a Grignard reagent with ethylene oxide?

methylpropene isobutyl alcohol isobutyl bromide

An alternate synthesis of isobutyl bromide is to treat methylpropene with HBr in the presence of peroxides.

(b) The sulfone must be obtained by oxidation of dibutyl sulfide; and the sulfide is obtained from the S_N2 reaction of the thiolate conjugate base of 1-butanethiol with 1-bromobutane.

dibutyl sulfone

(c) The approach is similar to that used in Study Problem 12.8 on text p. 668, except that a carbon–carbon bond is formed in this case; a complex organocuprate is used for this step.

(±)-*trans*-1-ethoxy-
2-methylcyclopentane

12.43 (a)

(b)

(c)

12.44 (a)

(b)

12.45 (a) A substituent that is cis to the —CH₂OH group points back into the catalyst, and the resulting steric repulsions can either reduce the binding affinity of the compound for the catalyst, or, more likely, push the alkene group out so that it is not optimally aligned with the peroxy —OH group for epoxidation.

(b) The same phenomenon would be observed with (–)-DET, because the catalyst is enantiomeric to the one formed from (+)-DET and the alkene is not chiral; so, the same energetics should apply.

Solutions to Skills Objectives Problems

12.46

conjugate acid
of dipropyl sulfide

conjugate acid
of dipropyl ether

pK_a −6 to −7   −2 to −3

12.47 Compound C is the weakest acid because all protons are attached to carbon atoms and are not very acidic. The charge effect tells us that compounds B and D are more acidic than A and E. The SH bond is weaker than the OH bond, so compound A is more acidic than compound E, and compound B is more acidic than compound D. (See Sec. 3.7A-B for a detailed description of the charge effect and element effect on acidity trends.) The compounds arranged in order of increasing acidity are:

12.48 (a) This is a Williamson ether synthesis by the S_N2 mechanism:

(b) An alcohol dehydration converts a primary alcohol into a symmetrical ether. The acid is used to protonate the OH group of one alcohol. A neutral alcohol acts as the nucleophile.

$C_{10}H_{22}O$

(c) This is an alkoxymercuration-reduction that gives the following ether:

(d) A bromohydrin is formed in the first step, and then the alcohol is deprotonated by base. An intramolecular S_N2 reaction results in the formation of an epoxide. Only one enantiomer is shown here, but the product would be formed as a racemic mixture.

(e) This is an epoxidation with a peroxy acid, leading to retention of alkene configuration. The epoxide is then protonated by acid, and opened by water in the second step.

(f) Tertiary alcohols are converted into unsymmetrical ethers when treated with dilute acid in the presence of excess alcohol, usually as the solvent. The tertiary alcohol here is protonated by the acid and loses water in a dissociation reaction.

conjugate acid
of the tertiary alcohol

conjugate acid
of the product

(g)   This reaction is a variation of the Williamson ether synthesis to synthesize a sulfide.

(h)   An asymmetric (Sharpless) epoxidation converts the following allylic alcohol into the enantiopure epoxide shown.

(i)   One equivalent of hydrogen peroxide will oxidize the sulfide to a sulfoxide.

(j)   The Z-alkene will be oxidized to the glycol with OsO_4 in a stereospecific syn-addition, retaining the configuration of the alkene in the glycol product. The product will be formed as a racemic mixture.

R = —$CH_2(CH_2)_5CH_2CO_2H$
R′= —$CH_2(CH_2)_6CH_3$

12.49   (a)   (b)   (c)

dibutyl sulfoxide   **dibutyl sulfone**   ***cis*-2,3-diethyloxirane**

(d)

CH₃CH₂   OH

H—C—C····CH₂CH₃

H₃C   H

(±)-(3*R*,4*R**)-4-methyl-3-hexanol

(e)

CH₃CH₂   OH

H—C—C····CH₂CH₃

HO   H

(±)-(3*R*,4*R**)-3,4-hexanediol

(f)

CH₃CH₂CH=O

(2 equivalents)

The asterisks in the name indicate relative rather than absolute configurations. In the case of parts (d) and (e), the (3*R**,4*R**) notation means that the two carbons have the same configuration; if one carbon is *R*, the other is also *R*; but if one is *S*, the other is also *S*. This notation tells us which diastereomeric pair we are dealing with. (See also Further Exploration 7.1 on p. 162 of this manual for a discussion of the same notation in another context.)

(g)

CH₃CH₂   OCH₃

H—C—C····CH₂CH₃

H₃C   H

(±)-(3*R**,4*R**)-3-methoxy-
4-methylhexane

(h)

OCH₂CH₃

CH₃CH₂CCH₂CH₂CH₃

CH₃

(±)-3-ethoxy-
3-methylhexane

(i)

OCH₃

CH₃CH₂CCH₂CH₂CH₃

CH₃

(±)-3-methoxy-
3-methylhexane

(j)   All of the compounds in this sequence have the *R* configuration at the asymmetric carbon. (The configuration of the asymmetric carbon is not shown explicitly because it is not affected by the sequence.) In the last step, the alkoxide produced from the Grignard reaction is not protonated with H_3O^+, as is usual, but rather is alkylated directly with CH_3I.

$$BrCH_2CH_2CHCH_2CH_3 \xrightarrow[\text{ether}]{Mg} BrMgCH_2CH_2CHCH_2CH_3 \xrightarrow{\triangle}$$

                    | |

                   CH₃   CH₃

(*R*)-1-bromo-3-methylpentane

$$BrMg^+ \ ^-OCH_2CH_2CH_2CH_2CHCH_2CH_3 \xrightarrow{H_3C-I} CH_3OCH_2CH_2CH_2CH_2CHCH_2CH_3$$

                            | |

                           CH₃   CH₃

(*R*)-1-methoxy-5-methylheptane

12.50   It is assumed that the starting material is racemic. As an additional exercise, you should consider the stereochemical course of the reactions if the starting material is a single enantiomer.

O

H₃C—△

CH₃CH₂

2-ethyl-2-methyloxirane

(a) and (b)

OH

CH₃CH₂CCH₂OH

CH₃

(c)

OH

CH₃CH₂CCH₂OCH₃

CH₃

(d)

OCH₃

CH₃CH₂CCH₂OH

CH₃

(e)

OH

CH₃CH₂CCH₂CH₃

CH₃

(f)

$$\text{CH}_3\text{CH}_2\overset{\overset{\displaystyle Br}{|}}{\underset{\underset{\displaystyle CH_3}{|}}{C}}\text{CH}_2\text{OCH}_3$$

(g)

$$\text{CH}_3\text{CH}_2\overset{\overset{\displaystyle Br}{|}}{\underset{\underset{\displaystyle CH_3}{|}}{C}}\text{CH}_2\text{OH}$$

(h)

$$\text{CH}_3\text{CH}_2\overset{\overset{\displaystyle OCH_3}{|}}{\underset{\underset{\displaystyle CH_3}{|}}{C}}\text{CH}_2\text{OCH}_3$$

(i)

$$\text{CH}_3\text{CH}_2\overset{\overset{\displaystyle OCH_3}{|}}{\underset{\underset{\displaystyle CH_3}{|}}{C}}\text{CH}_2\text{OCH}_2\text{CH}_3$$

(j)

$$\text{CH}_3\text{CH}_2\overset{\overset{\displaystyle O}{\|}}{C}\text{CH}_3 \ + \ \overset{\overset{\displaystyle O}{\|}}{C}\text{H}_2$$

(k)

$$\text{CH}_3\text{CH}_2\overset{\overset{\displaystyle MgBr}{|}}{\underset{\underset{\displaystyle CH_3}{|}}{C}}\text{CH}_2\text{OCH}_3$$

(l)

$$\text{CH}_3\text{CH}_2\overset{\overset{\displaystyle CH_2CH_2OH}{|}}{\underset{\underset{\displaystyle CH_3}{|}}{C}}\text{CH}_2\text{OCH}_3$$

12.51 (a) The ether cleavage here follows an S_N1 mechanism, as both alkyl branches are able to make tertiary or secondary carbocations. The first step is protonation of the oxygen by the acid.

(b) This primary ether is cleaved by the S_N2 mechanism, as primary carbocations will not be formed. The first step is protonation of the oxygen by the acid.

(c) Under basic conditions, HCN is first deprotonated by base, a very fast reaction. Then the epoxide ring is opened by the cyanide nucleophile attacking the less substituted carbon of the epoxide. The alkoxide ion is protonated by HCN.

(d) Under acidic conditions, the epoxide ring is protonated and then opened by attack on the more branched carbon of the epoxide by methanol. Another methanol molecule will act as a base to afford the final product.

(e) A Grignard reagent reacts with ethylene oxide to extend the carbon chain by two carbons. The alkoxide ion is protonated during aqueous workup.

(f)   The higher-order organocuprate reagent opens the epoxide on the side containing fewer alkyl substituents. The alkoxide is protonated during acidic workup to afford the tertiary alcohol.

(g)   Methoxide deprotonates the alcohol of epoxide *A* very rapidly, and the resulting anion reacts intra-molecularly to give an isomeric epoxide *B*. The two epoxides are in a relatively rapid equilibrium because the forward and reverse reaction are both intramolecular. Methoxide can open either epoxide, but the reaction at the unsubstituted carbon of epoxide *B* is fastest; all other points of ring opening have at least one substituent. As a result, glycol *C* is formed preferentially. Cleavage of glycol *C* gives the products shown.

(h)   The carboxylate anion is alkylated via an S_N2 reaction with the trimethyloxonium ion.

(i)   The cis-glycol is cleaved via oxidation with periodic acid.

(j) After methoxide deprotonates the alcohol, and intramolecular S_N2 reaction displaces the bromide leaving group.

12.52 (a)

(b)

3-methyl-1-pentene

(c)

The organometallic reagent will probably(see Sec. 5.6A, remove the OH proton of epoxide *A* to give a lithium alkoxide. The assumption of this synthesis is that a second equivalent of the organometallic reagent could then open the epoxide. The lithium from the alkoxide formed in the reaction with the OH group might even activate the opening of the epoxide:

(d) As the hint suggests, the using a sterically hindered borane reagent such as disiamylborane (Sec. 5.7, text p. 259) will cause the boron to bond to the carbon of the alkyne farther from the *tert*-butyl group in order to minimize steric interference. It is likely that even diborane itself would be sensitive enough to the steric effect of the *tert*-butyl group that it could be used.

$$(CH_3)_3C-C\equiv C-H \xrightarrow[\substack{2) \triangle}]{\text{1) NaNH}_2/\text{NH}_3 \text{ (liq)}} (CH_3)_3C-C\equiv C-CH_2CH_2O^- \xrightarrow{CH_3CH_2I}$$

3,3-dimethyl-1-butyne

$$(CH_3)_3C-C\equiv C-CH_2CH_2OEt \xrightarrow[\text{2) NaOH/H}_2O_2]{\text{1) disiamylborane}} (CH_3)_3C-CH_2\overset{O}{\overset{\|}{C}}-CH_2CH_2OEt$$

(e)

$$H_2C=CHCH_2Cl \xrightarrow[\text{EtOH}]{\text{NaOEt}} H_2C=CHCH_2OEt \xrightarrow[\text{OsO}_4\text{ (cat.)}]{\text{NMMO, H}_2O} HO\diagdown\diagup\overset{OH}{\diagup}\diagdown OEt$$

Solutions to Integrated Problems

12.53 (a) Two of the many nine-carbon ethers that cannot be prepared by the Williamson ether synthesis:

$$H_3C-\overset{\overset{\displaystyle CH_3}{|}}{\underset{\underset{\displaystyle CH_3}{|}}{C}}-O-\overset{\overset{\displaystyle CH_3}{|}}{\underset{\underset{\displaystyle CH_3}{|}}{C}}-CH_2CH_3 \qquad H_3C-\overset{\overset{\displaystyle CH_3}{|}}{\underset{\underset{\displaystyle CH_3}{|}}{C}}-O-CH_2-\overset{\overset{\displaystyle CH_3}{|}}{\underset{\underset{\displaystyle CH_3}{|}}{C}}-CH_3$$

2-methyl-2-*tert*-butoxybutane   ***tert*-butyl neopentyl ether**

The Williamson ether synthesis involves an S_N2 reaction between an alkoxide ion and an alkyl halide or sulfonate ester. Thus, one of the alkyl groups of an ether to be prepared by the Williamson synthesis must be derived from an alkyl halide that will undergo the S_N2 reaction. The foregoing ethers contain only tertiary and neopentyl alkyl groups, and neither tertiary nor neopentyl alkyl halides undergo the S_N2 reaction.

(b) Any ether containing only methyl or primary alkyl groups is a correct answer. Two of several possible examples are $CH_3O(CH_2)_7CH_3$ (methyl octyl ether, or 1-methoxyoctane) and $CH_3(CH_2)_3O(CH_2)_4CH_3$ (butyl pentyl ether, or 1-butoxypentane).

(c) Because the alkyl halides formed are in the same molecule, the ether is cyclic:

tetrahydrofuran

(d) Dipropyl ether $(CH_3CH_2CH_2)_2O$, would give propyl bromide as the only alkyl halide.

(e) Each of the following alkenes would give diastereomeric glycols as a result of the different syntheses. The $KMnO_4$ treatment would give *meso*-2,3-dimethyl-2,3-butanediol from *cis*-2-butene and would give (±)-2,3-dimethyl-2,3-butanediol from *trans*-2-butene; the epoxidation–hydration sequence would give the opposite results with the two alkenes. The $KMnO_4$ treatment would give *cis*-1,2-cyclobutanediol from cyclobutene, and the epoxidation–hydration sequence would give *trans*-1,2-butanediol.

***cis*-2-butene**   ***trans*-2-butene**   **cyclobutene**

(f)   1-Butene, $H_2C{=}CHCH_2CH_3$, gives racemic 1,2-butanediol with either reagent.

(g)   Since the alkene has an unsaturation number = 3 and forms di-epoxides, it must also have a ring.

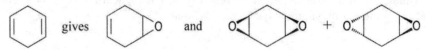

1,3-Cyclohexadiene is not a correct answer because it gives two stereoisomeric mono-epoxides and three stereoisomeric di-epoxides.

(h)   A correct answer must be an alkene that gives a glycol that cannot exist as diastereomers. The alkene 1-hexene ($H_2C{=}CHCH_2CH_2CH_2CH_3$) is one such compound.

12.54   Because an alkoxide ion is much less basic than a carbanion, the epoxide reacts most rapidly.

12.55   The second reaction, opening of the epoxide, should occur most rapidly, because the epoxide is considerably more strained than tetrahydrofuran. It is the ring strain that causes epoxide opening to be so fast.

12.56   Epoxides add HCl to give neutral chloro alcohols. This reaction is so fast and so favorable that HCl can be quantitatively (that is, completely) removed by adding 2-methyloxirane to a reaction mixture.

$$\underset{H_3C}{\overset{O}{\triangle}} + HCl \longrightarrow H_3C{-}\underset{\underset{Cl}{|}}{CH}{-}CH_2OH$$

12.57   (a)   Sodium ethoxide reacts with water to give ethanol and sodium hydroxide. Ethoxide is a stronger base than hydroxide and water is present in excess because it is the solvent. The equilibrium therefore favors sodium hydroxide. Consequently, the alcohol $(CH_3)_2CHCH_2CH_2OH$ rather than the ether will be formed as the major substitution product.

$$Na^+ \; C_2H_5O^- + H_2O \;\rightleftharpoons\; Na^+ \; {}^-OH + C_2H_5OH$$

$$\underset{\substack{\text{solvent;}\\\text{in large excess}}}{} \qquad \underset{\substack{\text{reacts with}\\\text{the alkyl halide}}}{\phantom{Na^+ \; {}^-OH + C_2H_5OH}}$$

Another problem is that the alkyl halide is insoluble in water. This insolubility makes any reaction that does occur very slow because the concentration of alkyl halide is limited to the very small amount that will dissolve. Changing the solvent to ethanol would solve all of these problems.

(b)   The alkyl halide from which the Grignard reagent is formed is also an alcohol. Because alcohols react rapidly with Grignard reagents in Brønsted acid–base reactions, the Grignard reagent will be destroyed. (See Eq. 10.56, text p. 538.)

12.58   (a)   2-Pentanol cannot be prepared by the reaction of a Grignard reagent with ethylene oxide, because this method can only be used for the preparation of certain primary alcohols. Alternatively, using a substituted epoxide and the higher-order organocuprate $Et_2Cu(CN)Li_2$ (text p. 643) will afford the desired alcohol after protonolysis.

$$(CH_3CH_2)_2Cu(CN)Li_2 \xrightarrow[\text{2) } H_3O^+]{\text{1) } \overset{O}{\underset{}{\triangle}}\!{-}CH_3} CH_3CH_2{-}CH_2\underset{\underset{OH}{|}}{C}HCH_3$$

2-pentanol

(b)   1-Pentanol, a primary alcohol, can be prepared by the reaction of propylmagnesium bromide with ethylene oxide followed by protonolysis.

$$CH_3CH_2CH_2-MgBr \xrightarrow[\text{2) } H_3O^+]{\text{1) }} CH_3CH_2CH_2-CH_2CH_2-OH$$

1-pentanol

12.59 (a) Reaction of ethylene ($H_2C{=}CH_2$) with OsO_4 in the presence of NMMO gives ethylene glycol. This compound ($HO-CH_2CH_2-OH$) is achiral and thus cannot be resolved into enantiomers.

(b) The reaction of OsO_4 with any alkene is a syn-addition. The reaction with *cis*-2-butene gives *meso*-2,3-butanediol. Because meso compounds are achiral, this compound cannot be resolved into enantiomers.

meso-**2,3-butanediol**

(c) In contrast, the syn-addition of two OH groups to *trans*-2-butene gives (±)-2,3-butanediol, that is, the racemate. Any racemate can, in principle, be resolved into enantiomers.

(±)-**2,3-butanediol**

(d) The syn-addition of two OH groups to *cis*-2-pentene gives (±)-(2R*,2S*)-2,3-pentanediol—that is, the racemate of 2,3-pentanediol in which the two asymmetric carbons have opposite configurations. (See the solutions to Problems 12.49(d)–(e) for an explanation of this notation.) Because this compound is a racemate, it can be resolved into enantiomers.

(±)-**(2R*,3S*)-2,3-pentanediol**
(the 2S,3R enantiomer is shown)

This outcome is different from that of the reaction in (b) because *cis*-2-butene and 2,3-butanediol have symmetry not present in *cis*-2-pentene and 2,3-pentanediol—that is, no meso stereoisomer is possible in 2,3-pentanediol.

12.60 Because the hydroxide ion reacts with an epoxide at the CH_2 group, the configuration at the asymmetric carbon is unaffected. Consequently, (+)-2-methyloxirane has the *R* configuration.

(R)-(+)-2-methyloxirane   **(R)-(−)-1,2-propanediol**

12.61 In this reaction the nucleophile (water) reacts at the protonated epoxide at the more substituted carbon with inversion of stereochemical configuration.

(S)-2-ethyl-2-methyloxirane   **(R)-2-methyl-1,2-butanediol**

this oxygen
comes from water

12.62 The formula of the product indicates one degree of unsaturation. Because the double bond undergoes addition with mercuric acetate, this unsaturation cannot be due to a double bond and therefore is accounted for by a ring. As the hint and the molecular formula of the product suggest, this is an alkoxymercuration reaction in which the OH group from within the same molecule serves as the nucleophile:

12.63 As a result of this reaction, carbon-2 becomes asymmetric, and diastereomers are formed corresponding to the two possible configurations of this carbon. The product with the S configuration at carbon-2 (compound B) is optically active; however, the other product (compound A) is a meso compound, and hence, is optically inactive.

(3S,4S)-4-methoxy-3-methyl-1-pentene

compound A
(meso; therefore achiral
and optically inactive)

compound B
(optically active)

12.64 Synthesis (1) will not accomplish the desired objective, because the —OH group of the alcohol would be lost as water. (See the mechanism in Eqs. 12.15–12.16, text p. 629.) In contrast, synthesis (2) would result in complete incorporation of the isotopic oxygen into the ether. Because this is a conventional Williamson ether synthesis, the carbon–oxygen bond to the ring is not broken. (See Eq. 12.7, text p. 626.) Consequently, you should advise Flaskclamper to use the second synthesis.

12.65 The structures of the compounds without stereochemical detail are as follows. Compounds (2) and (3) are diastereomers; compound (2) is the meso diastereomer. Although we do not know which enantiomer of compound (3) is the (+)-enantiomer, all we need to know to solve the problem is that it is one or the other.

2,4-dimethoxy-1,3-butanediol
compound (1)

1,4-dimethoxy-2,3-butanediol
compound (2) = the meso diastereomer;
compound (3) = a chiral diastereomer

3,4-dimethoxy-1,3-butanediol
compound (4)

Because compound (2) is meso, and no meso compound can be optically active, compound (2) = D. All of the other compounds are enantiomerically pure chiral compounds and are therefore optically active. Diastereomers (2) and (3) would give the same products of periodic acid oxidation; hence, compound (3) = C. Because compound (1) is not a vicinal glycol, it would not react with periodic acid; hence, compound (1) = A. Therefore, compound (4) = B; and, indeed, it would react to give periodate cleavage products that are different from those obtained from C (= compound 3) or D (= compound 2).

12.66 (a) A bromohydrin is formed in the first step, and this is oxidized to an α-bromo ketone by the CrO_3.

an α-bromo ketone

(b) The epoxide is opened to a glycol by the acidic aqueous reaction conditions, and the glycol is cleaved as it is formed by the periodic acid.

(c) The azide nucleophile opens the epoxide with inversion of configuration to give the racemate of the trans azido alcohol.

(d) Sulfide ion displaces one chloride to give a thiolate ion, which then reacts in an internal nucleophilic substitution to give the cyclic sulfide tetrahydrothiophene.

tetrahydrothiophene

(e) A tosylate of one OH group is formed. (Introduction of a second tosylate is much slower because of the resulting van der Waals repulsions caused by the introduction of a second large tosylate group.) The resulting hydroxy tosylate reacts like a bromohydrin. (Remember, you can think of the tosylate group as a "large halide.") Thus, in base, the OH group ionizes to an alkoxide, which undergoes an internal nucleophilic substitution reaction with the tosylate expelled as a leaving group to give an epoxide. This is very much like epoxide formation from a bromohydrin.

meso-**2,3-butanediol**

Because the starting diol is achiral (it is meso) and the reagents are achiral, all chiral compounds in this reaction scheme are racemates.

12.67 (a) The desired compound can be prepared by a ring-opening reaction of (2R,3R)-2,3-dimethyloxirane with acidic methanol, because epoxide ring-opening occurs with inversion of configuration:

(2R,3S)-3-methoxy-2-butanol

(Be sure to convince yourself that the nucleophilic reaction of methanol at either carbon of the epoxide ring gives the same stereoisomer of the product.)

(b)   Carbon-3, the asymmetric carbon, has the *S* configuration, which is the same configuration that carbon-3 has in the product of part (a). Consequently, oxidation of the alcohol product of part (a) gives the desired compound:

(2R,3S)-3-methoxy-2-butanol
from part (a)

(c)   If the OH group can be converted into an ethyl ether without changing the configuration of either carbon, the problem will be solved. The Williamson ether synthesis is a good way to accomplish this objective, because it does not break a carbon–oxygen bond.

(2R,3S)-3-methoxy-2-butanol
from part (a)

(d)   Because the carbon bearing the ethoxy group is the one that must be inverted, use the ring-opening reaction in part (a), except substitute ethanol for methanol. Then carry out the Williamson ether synthesis used in part (c) with methyl iodide instead of ethyl iodide.

(2R,3S)-3-ethoxy-2-butanol

Don't be confused by the fact that the carbon numbering changes because of numbering conventions. Thus, carbon-2 in the alcohol becomes carbon-3 in the product, and vice versa.

12.68   In both parts the mCPBA reacts at the side of the alkene π-bond that involves the less severe van der Waals repulsions.

(a)   One of the methyl groups must be axial, and therefore it creates a 1,3-diaxial interaction with the incoming reagent on one face of the ring. Therefore, the reaction occurs at the other face.

A
(major product)

B

(b)   The face of the ring on the same side of the methyl group is blocked, so the reaction occurs at the opposite face.

A
(major product)

B

12.69   (a)   The OsO_4 (in the presence of water) effects glycol formation from the alkene starting material. This produces Os(VI), which is re-oxidized to Os(VIII) by periodic acid. One equivalent of the periodic acid effects this oxidation, continually recycling the osmium until the alkene is fully utilized; this role of periodic acid is the same as that of NMMO in that method. The second equivalent of the periodic acid then brings about cleavage of the glycol. To summarize:

(b)

12.70   (a)   The curved-arrow mechanism for salt formation: (The two reactions occur separately.)

(b)   Counting the nonbonding electron pair, each sulfur in the product bears four different groups. Consequently, the sulfurs are stereocenters. Because the sulfurs do not undergo rapid inversion (text Sec. 6.9B, p. 309), stereoisomeric sulfonium salts that differ in configuration at sulfur can be isolated. Because each salt contains two sulfur stereocenters, the two salts are diastereomers. One diastereomer is the meso, or (*R,S*), stereoisomer; the other is the racemate, a 1:1 mixture of the (*R,R*) and (*S,S*) stereoisomers.

12.71   Because sulfur in a sulfoxide does not undergo inversion, it is an asymmetric atom. Because there are two asymmetric carbons in methionine (asterisks in the structure below), methionine sulfoxide can exist as diastereomers. (See the solution to Problem 12.73.)

methionine sulfoxide

12.72   When butyl methyl sulfoxide is treated with excess hydrogen peroxide, the sulfur is further oxidized to the sulfone. The sulfoxide sulfur is an asymmetric atom, thus butyl methyl sulfoxide can be separated into two optically active enantiomers. However, the sulfur in a sulfone is not asymmetric, so the optical activity is lost when the sulfoxide is oxidized further.

12.73   If sulfur does invert over a long period of time, its configuration goes from S to R. Because there are other asymmetric carbons in SAM which retain their configuration, this new compound is a diastereomer of naturally occurring SAM.

sulfur has the R configuration
($R^2 > R^1 > CH_3$. Electrons are lowest priority.
See Fig. 12.1 in the text for the structure of SAM.)

12.74   The anion that results from the ring-opening reaction with hydroxide reacts as a nucleophile in another epoxide ring-opening reaction; and so on.

12.75   (a)   The role of the $^+NH_4\,Cl^-$ is to protonate the nitrogen of the aziridine. Without this protonation, the aziridine nitrogen is too basic to serve as a good leaving group.

Like the ring opening of epoxides, the ring opening of aziridines involves opposite-side substitution and therefore gives the trans stereoisomer of the product.

   (b)   Cleavage of an epoxide generates an alkoxide leaving group (conjugate-acid $pK_a \approx 15$). Although this is quite basic, it is much less basic (by about 17 pK_a units) than the amide (——NH) leaving group in an aziridine (conjugate-acid $pK_a \approx 32$). Normally, alkoxides do not make very good leaving groups, because they are too basic; but, as explained in Sec. 12.5A in the text, epoxides are opened because of the great ring strain of the three-membered ring. The leaving group in an aziridine is too poor a base to be opened without protonation, despite its ring strain.

   (c)   An HCl solution at 0.01 M has a pH = 2. This is acidic enough to protonate the azide ion, thus eliminating its nucleophilicity. The ammonium ion is not acidic enough to protonate the azide ion significantly.

12.76   (a)   Because the large *tert*-butyl group forces the molecule into the vast predominance of a chair conformation in which this group is equatorial, the two —OH groups in this conformation of compound A are in a trans-diaxial arrangement. The cyclic periodate ester intermediate cannot be formed from two hydroxy groups in this arrangement without introducing significant ring strain. Because the required intermediate cannot form, the oxidation of compound *A* cannot occur. In contrast, compound *B* can be oxidized with periodic acid because the two —OH groups are trans-diequatorial.

<table>
<tr><td style="text-align:center;">the cyclic periodate
ester cannot form

A</td><td style="text-align:center;">the cyclic periodate
ester can be formed

B</td></tr>
</table>

(b)   Compound *B* is locked into a conformation in which the two hydroxy groups are trans-diaxial; consequently, this compound cannot be oxidized for the same reason that compound *A* cannot be oxidized in part (a).

<table>
<tr><td style="text-align:center;">*A*
can form a periodate ester</td><td style="text-align:center;">*B*
cannot form a periodate ester</td></tr>
</table>

12.77   Reaction (1) is an S_N2 reaction in which oppositely charged species react. This reaction should be promoted by conditions that increase the interaction between charges, that is, by a smaller dielectric constant. Therefore, reaction (1) is faster in ethanol ($\varepsilon = 32$) than it is in water ($\varepsilon = 80$). (Solvent dielectric constants are found in Table 8.2 on text p. 418.) Reaction (2) is an S_N1 reaction in which the rate-limiting step involves dissociation of an alkyl halide to a carbocation and a chloride ion; this reaction is promoted by conditions that foster separation of charged groups, that is, by a greater dielectric constant. Consequently, reaction (2) is faster in water. Reaction (3) is an S_N1 reaction, but, because the starting sulfonium ion is positively charged, formation of the carbocation intermediate does not involve any change in the number of charges. Consequently, the rate of this reaction is not affected very much by a change of solvent.

This problem utilizes the hypothetical case of pure water as one of the solvent options. Recall that alkyl halides are rather insoluble in water, so a co-solvent such as acetone is typically added to solubilize them. However, typical acetone–water mixtures have a much higher dielectric constant than ethanol. The answer to the problem is still valid if acetone–water is used as a solvent instead of pure water.

12.78   The principles to use in solving this problem are that (*a*) the *tert*-butyl group, because it is so large, must assume the equatorial position; and (*b*) the —OH and —Cl groups must be able to achieve an anti relationship in order for opposite-side substitution (and thus epoxide formation) to occur. The required anti relationship is impossible for compound (3) because the —OH and —Cl groups are cis and are thus fixed in either chair conformation at a dihedral angle of about 60°. Therefore, compound (3) is *A*. In the chair conformation in which the *tert*-butyl group is equatorial, the —OH and —Cl groups in compound (1) are trans-diaxial, and are ideally set up for the formation of an epoxide. Consequently, compound (1) is *B*, and the epoxide formed from it is *D*.

<table>
<tr><td style="text-align:center;">compound (1) = *B*</td><td style="text-align:center;">$\xrightarrow[\text{CH}_3\text{OH}]{\text{KOH}}$</td><td style="text-align:center;">epoxide *D*</td></tr>
</table>

Finally, consider compound (2): the —OH and —Cl groups are trans, but they are diequatorial. The only way that they can achieve an anti relationship is either for the ring to undergo a chair interconversion, or for it to assume a twist-boat conformation. In either case, the conformation required for epoxide formation has a very high energy, and very little of it will be present. Because the transition-state conformation has very high energy, epoxide formation from compound (2), although possible, is very slow. Therefore, compound (2) = *C* and the epoxide formed from it is *E*. Epoxides *D* and *E* are stereoisomers, as required by the problem.

12.79 As illustrated by Eq. 12.27 on text p. 633, and as discussed in the solution to the previous problem, the —OH and —Br groups must be able to assume a trans-diaxial arrangement in the transition state for opposite-side substitution to occur.

Immediately we rule out compound *D*, as the —OH and —Br groups are cis, and it will not form an epoxide.

The —OH and —Br groups in compounds *A*–*C* are trans. Because all of the compounds shown are trans-decalin derivatives, they cannot undergo the chair interconversion. Therefore, the —OH and —Br groups could be trans-diequatorial or trans-diaxial. Examine each one in turn.

Compound *A* must assume a twist-boat conformation in order to undergo epoxide formation by opposite-side substitution. Because this conformation, and therefore the transition state of the reaction, has a very high energy, epoxide formation is likely to be very slow.

The remaining two bromohydrins *B* and *C* have trans-diaxial arrangements and react readily:

12.80 (a)   The sulfur can serve as a nucleophile in an intramolecular nucleophilic substitution reaction. The product is a cyclic sulfonium salt.

(b)   The assumption is that the reaction of water with the starting sulfide occurs at about the same rate as it does with 1-chlorohexane. If so, then the relative rate of the intramolecular reaction above and the competing intermolecular S_N2 reaction of water with the same compound is 21, as given in the problem. A calculation very similar to the one in Study Problem 12.5, text p. 657, shows that $k_1/k_2 = (21)(20\ M) = 420\ M$. This is the proximity effect.

(c)   The proximity effect for this reaction is smaller than the one referenced in the problem because there is a higher entropy cost to form a 5-membered ring than there is to form a 3-membered ring. Forming the 5-membered ring "freezes" 4 internal rotations around single bonds (Eq. 12.73, text p.655), decreasing the rate of the reaction in comparison to forming the 3-membered ring. Lowering the rate of the reaction will decrease the proximity effect, as calculated here.

12.81   The transformation of compound *A* to compound *B* occurs with *retention of configuration*. A retention of configuration can be the result of *two successive inversions of configuration*. This fact and fact (3) suggest that neighboring-group participation is occurring. Indeed, a neighboring-group mechanism involves a meso and therefore *achiral* cyclic sulfonium-ion intermediate *D*, formed by an intramolecular substitution with inversion of configuration:

Because intermediate *D* is achiral, any chiral products that result from it must be formed as racemates (Sec. 7.7A); this accounts for fact (2). Reactions of intermediate *D* with ethanol at the two carbons of the three-membered ring give respectively the two enantiomers of product *B*. (The formation of one enantiomer is shown below; you should show formation of the other.) This substitution reaction occurs with inversion of configuration; coupled with the first inversion that takes place in the formation of *D*, this accounts for the overall retention noted in fact (1).

Because reactions involving neighboring-group participation are typically accelerated by orders of magnitude, we have accounted for fact (3): the much greater rate of compound *A*.

12.82 (a)   The substitution with retention of configuration suggests a neighboring-group mechanism involving two substitutions involving a boat-like intermediate *X*.

This intermediate is unstable because it incorporates two fused four-membered rings and a boat cyclohexane; it is highly strained, and thus opens readily.

(b)   The substitution with retention of configuration suggests a neighboring-group mechanism involving a bicyclic episulfonium ion *Y*.

12.83   (a)   Because the product is meso, it must have opposite configurations at the two asymmetric carbons—in other words, it has the 2*S*,3*R* configuration. (In a meso compound, this is the same as the 2*R*,3*S* configuration.) Substitution of the OH by a Br must occur, therefore, with retention of configuration, and such a stereochemical result suggests that neighboring-group participation takes place. The protonated OH group departs with concomitant opposite-side substitution by the bromine within the same molecule to form a bromonium ion. The bromonium ion then reacts with bromide ion (from the ionized HBr in solution) with inversion to give the *meso*-2,3-dibromobutane.

The center of symmetry shows that the product is meso. Alternatively, you can form an eclipsed conformation to see an internal mirror plane.

Reaction of the bromide ion at the other carbon of the bromonium ion is equally probable. We leave it to you to show that this reaction gives the same result.

(b)   The similar reaction of the 2*S*,3*S* stereoisomer gives a meso bromonium ion, which is achiral. Reactions of bromide ion at the two carbons give respectively (2*S*,3*S*)- and (2*R*,3*R*)-2,3-dibromobutane in equal amounts—that is, the racemate.

(2S,3S)

(2R,3R)   (2S,3S)
(A)   (B)

12.84 (a) In this reaction, the initially formed bromonium ion reacts intramolecularly with the oxygen of the OH group. (The mechanism for the formation of a bromonium ion is shown in Eq. 5.10a, text p. 240.)

(b) The alkoxide, formed by ionization of the alcohol OH group, opens the epoxide intramolecularly. The hydroxide catalyst is regenerated by protonation of the resulting alkoxide ion.

formed by the reaction
of hydroxide with the
alcohol

(c) This mechanism is very similar to the mechanism for the intramolecular substitution involved in the formation of product *B* in Problem 9.73 on text p. 505. The thiolate anion, formed by ionization of the thiol (not shown), displaces the tosylate group within a twist-boat conformation (shown here for simplicity as a boat).

+ ⁻:ŌTs

(d) The epoxide chemistry in this chapter has focused on ring-opening substitution reactions; however, you have learned that elimination competes with substitution. This reaction is an E2-like elimination reaction in which either of the two distinguishable β-hydrogens is removed and the epoxide oxygen serves as a leaving group. The mechanism for the formation of the major alkene product is shown below; the minor product is formed by an essentially identical mechanism involving the other β-hydrogen. Substitution (that is, nucleophilic ring-opening) evidently does not occur because of the van der Waals repulsions that would result in the transition state of such a reaction between the alkyl branches of the base and those of the epoxide.

(e) The carbocation intermediate, which is formed by protonation of the double bond, undergoes an intramolecular Lewis acid–base association reaction with the sulfur nucleophile:

(f)

(g) The key to this mechanism is to notice that inversion of configuration has occurred at carbon-2, and that cyanide ion has reacted at carbon-1. The formation of another epoxide by neighboring-group participation with inversion of configuration at C-2 is followed by reaction of the cyanide ion as a nucleophile at C-1, which is the less substituted carbon of the epoxide. (The rearrangement of one epoxide to another by the intramolecular reaction is called the *Payne rearrangement*. For another example, see the solution to Problem 12.51g.)

(h) This appears to be a relatively rare instance of neighboring-group participation involving a four-membered ring. The ring is opened by the nucleophilic reaction of methanol at the more substituted carbon.

(i)   The *pinacol rearrangement* involves a methyl group shift to afford a resonance-stabilized carbocation intermediate. As the problem suggests, start the mechanism by protonating an OH group and losing water to form a tertiary carbocation.

12.85 (a)   An episulfonium ion (analogous to a bromonium ion) X is formed at one double bond; reaction with the liberated chloride ion results in a net anti addition.

The sulfur in this product then serves as the electrophile for the reaction at the other double bond, which involves another episulfonium ion Y. (Because of the pucker of the cyclooctene ring, as a model will show, the distance from the double bond to the sulfur is not as far as it might seem.)

(b)   The substitution with retention of stereochemical configuration suggests that neighboring-group participation has taken place. The first substitution involves the formation of episulfonium ion Y from part (a) with loss of chloride ion under the rather extreme conditions. This ion is opened by the azide ion.

The second substitution occurs by an essentially identical mechanism.

12.86 (a) In the racemic mixture, only the enantiomer with the correct stereochemistry, in this case the S configuration, will undergo epoxidation. The other enantiomer will remain unreacted.

the S enantiomer
forms the epoxide

the R enantiomer
remains unreacted

(b) The (+)-DET Sharpless asymmetric epoxidation reagent reacted with the pro-S branch of the achiral allylic alcohol. Prochirality is discussed in the text Sec. 11.9B.

(c) The enantiomeric epoxidation reagent reacts with the pro-R branch of the molecule.

12.87 Compound A is an octene because it undergoes typical alkene reactions and it gives octane on catalytic hydrogenation. Compounds C and D are stereoisomeric glycols, and glycol D is achiral. The only octene isomer that would give an achiral glycol is one of the stereoisomers of 4-octene. *Cis*-4-octene reacts with mCPBA to give an *achiral* (meso) epoxide B, which, upon ring opening in aqueous acid, gives racemic 4,5-octanediol. The same alkene reacts in a syn-addition with OsO4 to give *meso*-4,5-octanediol, an achiral compound. Therefore, compound A is cis-4-octene. To summarize:

Introduction to Spectroscopy
Infrared Spectroscopy and
Mass Spectrometry

CHAPTER 13 SUMMARY

- Spectroscopy deals with the interaction of matter and electromagnetic radiation. Electromagnetic radiation is characterized by its energy E, wavelength λ, and frequency ν, which are interrelated by $E = h\nu = hc/\lambda$, where h is Planck's constant and c is the velocity of light.

- Infrared spectroscopy deals with the absorption of infrared radiation by molecular vibrations. An infrared spectrum is a plot of the percentage of infrared radiation transmitted through a sample as a function of the wavenumber or wavelength of the radiation.

- The frequency of an absorption in the infrared spectrum is equal to the frequency of the bond vibration involved in the absorption.

- The wavenumber or frequency of an absorption is greater for vibrations involving stronger bonds and smaller atomic masses (Eqs. 13.12 and 13.14, text pp. 693-694). The smaller of two atomic masses involved in a bond vibration has the greater effect on the frequency of the vibration.

- The intensity of an absorption increases with the number of absorbing groups in the sample and the size of the dipole moment change that occurs in the molecule when the vibration occurs. Absorptions that result in no dipole moment change are infrared-inactive.

- The infrared spectrum provides information about the functional groups present in a molecule. The =C—H stretching and bending absorptions and the C=C stretching absorption are very useful for the identification of alkenes. The ≡C—H and C≡C stretching absorptions are diagnostic for terminal alkynes. The O—H stretching absorption is diagnostic for alcohols. The C=O stretching absorption

(discussed in detail in Secs. 19.4A, 20.3A, and 21.4A) is a very important diagnostic absorption of carbonyl compounds.

- In electron-ionization (EI) mass spectrometry, a molecule loses an electron to form the molecular ion, a radical cation, which in most cases decomposes to fragment ions. The relative abundances of the fragment ions are recorded as a function of their mass-to-charge ratios m/z, which, for most ions, equal their masses. Both molecular masses and partial structures can be derived from the masses of these ionic fragments.

- Associated with each peak in a mass spectrum are other peaks at higher mass that arise from the presence of isotopes at their natural abundance. Such isotopic peaks are particularly useful for diagnosing the presence of elements that consist of more than one isotope with high natural abundance, such as chlorine and bromine.

- Ionic fragments are of two types: even-electron ions, which contain no unpaired electrons; and odd-electron ions, which contain an unpaired electron. Even-electron ions have odd masses if they contains the elements C, H, O, and halogen; they have even masses if they contain an odd number of nitrogens. Odd-electron ions have even masses unless they contain an odd number of nitrogen atoms, in which case they have odd masses.

- In chemical-ionization (CI) mass spectrometry, molecules are ionized by protonation in the gas phase. Because this is a much gentler ionization technique than EI, a CI mass spectrum typically contains a greater proportion of molecular ion (as its conjugate acid) than the EI spectrum of the same compound.

375

FURTHER EXPLORATIONS

13.1   The Vibrating Bond in Quantum Theory

In the text we've used Hooke's law to describe bond vibrations. If the vibrating bond were like a classical vibrating spring, its energy (and therefore its frequency) could be increased by any arbitrary amount. However, vibrating bonds are subject to quantum theory, which requires that bond vibrations must have discrete energy levels. The energy levels of a vibrating bond are given by

$$E_n = (n + \tfrac{1}{2})h\nu_0$$

(SG13.1)

where n is a *vibrational quantum number,* and ν_0 is the fundamental vibrational frequency. The quantum number can assume the integral values 0, 1, 2, … The lowest energy level, E_0, is obtained by substituting $n = 0$ into this equation:

$$E_0 = \tfrac{1}{2} h\nu_0$$

(SG13.2)

This is called the *zero-point energy.* The wavenumber of this lowest-energy vibration is given by the Hooke's law expression shown in Eq. 13.12 on p. 693 of the text, and the frequency ν_0 is obtained by multiplying the wavenumber by the velocity of light c.

There is a certain length of a spring (or bond), r_m, at which the potential energy of a spring is a minimum. In terms of simple springs, this is the length of the "relaxed" spring—the length at which the spring is neither stretched nor compressed. Hooke's law assumes that the potential energy stored in a spring (bond) is proportional to the square of the amount by which the spring is stretched or compressed from this distance:

$$E_{\text{potential}} = \tfrac{1}{2}k(r - r_m)^2 = \tfrac{1}{2}k\Delta r^2$$

(SG13.3)

An oscillating spring gathers potential energy when it is stretched or compressed, and this potential energy is released as kinetic energy as the spring returns to r_m. The conservation of energy requires that the sum of the potential and kinetic energies of the spring remains constant (barring frictional losses). The potential energy function in Eq. SG13.3 describes a parabola, and this is not a bad approximation for the potential energy of a vibrating bond for the first few energy levels. Such a parabola is shown in **Fig. SG13.1**. Superimposed on this parabola are the discrete energy levels required by quantum theory given by Eq. SG13.1. The total energy of the vibrating bond—potential + kinetic—in a particular energy level is given by the position of the energy-level line.

This brief discussion of quantum theory can be used to justify two points made in the text. The first point is that the frequency of light that brings about energy absorption must equal the fundamental frequency of the bond vibration ν_0. For a bond to absorb energy from light, it must absorb an amount of energy exactly equal to $E_1 - E_0$; that is, the energy of the vibration must "jump" to the next energy level. From Eq. SG13.1, the energy difference between E_1 and E_0 is

$$E_1 - E_0 = \Delta E_{1,0} = \tfrac{3}{2}h\nu_0 - \tfrac{1}{2}h\nu_0 = h\nu_0$$

(SG13.4)

This equation shows that promotion of the bond to energy level E_1 from E_0 requires light with energy $= \Delta E_{1,0} = h\nu_0$ and thus a frequency ν_0. But this is just the fundamental frequency of the bond vibration, as asserted.

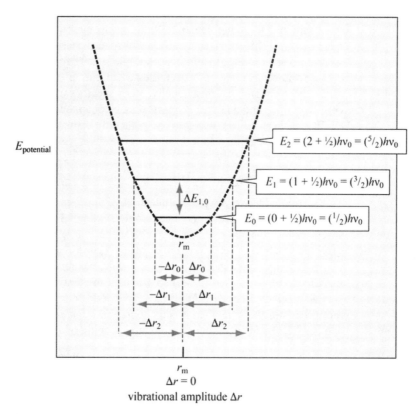

$E_{\text{potential}}$

$E_2 = (2 + \frac{1}{2})h\nu_0 = (^5/_2)h\nu_0$

$E_1 = (1 + \frac{1}{2})h\nu_0 = (^3/_2)h\nu_0$

$\Delta E_{1,0}$

$E_0 = (0 + \frac{1}{2})h\nu_0 = (^1/_2)h\nu_0$

r_{m}

$-\Delta r_0$ Δr_0

$-\Delta r_1$ Δr_1

$-\Delta r_2$ Δr_2

r_{m}
$\Delta r = 0$
vibrational amplitude Δr

Figure SG13.1   A plot of potential energy versus vibrational amplitude for a Hooke's law potential. The potential energy curve is the heavy dashed line. The vibrational quantum levels are superimposed on this curve. (Only the first three levels are shown.) $\Delta E_{1,0}$ is the energy required for excitation of a vibration in the lowest energy level, E_0, to the next level, E_1.

The second point is that the amplitude of the vibration increases when light is absorbed. Compare the points on the potential-energy curve (dashed line) in Fig. SG13.1 for the E_0 and E_1 energy levels. You can see that the vibrational amplitude—the magnitude of the stretching and compression indicated by the Δr values—increases as the energy of the vibration is increased from E_0 to E_1. The points at which an energy line meets the parabola essentially define the limits of the oscillation at that energy, and these points become further apart with increasing energy. The mechanical analogy is that to increase the magnitude of the oscillations in a spring, we have to energize it (usually by stretching it).

13.2   FT-IR Spectroscopy

This section will give you a better idea how an FT-IR spectrometer works. In an FT-IR spectrometer, radiation to be passed through the sample is split and sent over two paths simultaneously by a system consisting of two mirrors, one moving and one fixed (a Michelson interferometer). The difference between the lengths of the two light paths at various times is precisely calibrated with a laser. The superimposed beams of light from the two paths are passed through the sample. The interference patterns, or interferograms, that result can be thought of as superimposed plots of absorption versus time from both light beams in which all wavelengths in the spectrum contribute simultaneously. The interferograms are stored in a small computer and analyzed by *Fourier transformation*. Fourier transformation is a mathematical technique for decomposing time-based wave motions into their contributing frequency components. (For example, the unique sound of a bell as it dies out over time might be analyzed in terms of the different contributing pitches by the same technique.) Fourier analysis converts the interferograms into a plot of absorption versus wavenumber—the IR spectrum. The entire process takes just a few seconds.

In addition to saving time, FT-IR has much better wavenumber resolution than conventional IR spectroscopy. (This means that two closely-spaced peaks would appear as separate peaks rather than as a single broad peak.) In addition, multiple spectra from the same sample can be digitally added; because "noise" is random, it decreases when multiple spectra are summed, and the signal due to the IR spectrum grows. Consequently, many spectra can be accumulated from the same sample in a very short time, and thus strong spectra can be obtained from very dilute samples. The principles of FT-IR have been known for many years, but only in relatively recent times has the technique become practical because of the availability of inexpensive, high-powered computers.

13.3   The Mass Spectrometer

This section will give you a more quantitative understanding of how a mass spectrometer operates. The magnetic-sector instrument diagrammed in Fig. 13.22, text p. 718, is the simplest type of mass spectrometer; it is still used for the mass spectrometry of small molecules. When an ion in such a mass spectrometer is accelerated to a voltage V, its kinetic energy is $mv^2/2$, where m is its mass, and v is its velocity. (Note that in this section, italic v is for *velocity*, which is different than ν, which is the Greek letter nu, used in SGL13.1 and in the text, for frequency. These characters look very similar in print.) Because all ions have, to a good approximation, the same velocity v as a result of the accelerating voltage, then it is the *ionic mass* that gives one ion a different kinetic energy from another. The basic equation describing the bending of an ion by a magnetic field **B** is

$$m/z = k\,\frac{\mathbf{B}^2 r^2}{V}$$

(SG13.5)

where r is the radius of the circular path over which the ion is bent, z is the charge on the ion, and k is a proportionality constant. This equation shows that the paths of ions of a larger mass are bent over a circle of greater radius. Similarly, for an ion of a given mass, increasing the magnetic field **B** decreases the radius of its path.

In the magnetic-sector mass spectrometer, a collector is positioned to sense only the ions traversing a path of a particular radius r. By increasing the magnetic field **B** in a regular way, ions of progressively increasing mass are focused onto the collector. The relative intensity of the ion beam of a given mass is recorded as an ion current. Thus a plot is obtained of ion current versus m/z—the mass spectrum.

Other types of ion sorting are used in modern spectrometers. In FT-ICR (Fourier Transform-Ion Cyclotron Resonance) mass spectrometry, ions are trapped in a chamber by a magnetic field, and the trapped ions move in closed, circular loops perpendicular to the magnetic field. The angular frequency of this motion is inversely proportional to the m/z of the ion. An analysis of the angular frequencies is tantamount to an analysis of the masses present. FT-ICR mass spectrometry gives very precise values of m/z.

In TOF (time-of-flight) mass spectrometers, ions are accelerated down a tube by an electric field. The time that it takes an ion to traverse the tube is proportional to the square root of its m/z. The measurement of the times at which ions reach the detector at the end of a flight tube provides essentially a measurement of the ion masses.

 # SOLUTIONS TO PROBLEMS

Solutions to Focused Problems

13.1   (a)   Apply Eq. 13.1.

$$\nu = \frac{3.00 \times 10^8 \, \text{m s}^{-1}}{9 \times 10^{-6} \, \text{m}} = 3.33 \times 10^{13} \, \text{s}^{-1}$$

(b) Apply Eq. 13.1 and include the conversion factor 10^{-10} m Å^{-1}.

$$\nu = \frac{3.00 \times 10^8 \, \text{m s}^{-1}}{(4800 \, \text{Å})(10^{-10} \, \text{m Å}^{-1})} = 6.25 \times 10^{14} \, \text{s}^{-1}$$

13.2 (a) Multiply the frequency obtained in the solution to Focused Problem 13.1(a) times Planck's constant:

$$E = h\nu = (3.99 \times 10^{-13} \, \text{kJ s mol}^{-1})(3.33 \times 10^{13} \, \text{s}^{-1}) = 13.3 \, \text{kJ mol}^{-1}$$

(b) Use the same procedure as in part (a) of this problem:

$$E = h\nu = (3.99 \times 10^{-13} \, \text{kJ s mol}^{-1})(6.25 \times 10^{14} \, \text{s}^{-1}) = 249 \, \text{kJ mol}^{-1}$$

13.3 (a) The energy of X-rays is greater than that of any visible light, including blue light. (In fact, the energy is so much greater that prolonged exposure to X-rays is harmful.)

(b) The energy of radar (that is, microwaves) is less than that of any visible light, including red light.

If the energy of microwaves is much less than that of red light, why is it that microwave energy rapidly cooks food, whereas red light does not? The reason is that microwaves are absorbed by molecules in food, and the absorbed energy is rapidly converted into heat. In contrast, red light is reflected or transmitted, but it is mostly not absorbed, and is therefore not converted into heat.

13.4 (a) Apply Eq. 13.7 on text p. 687:

$$\tilde{\nu} = \text{wavenumber} = (1 \times 10^4 \, \mu\text{m cm}^{-1}) \div 6.0 \, \mu\text{m} = 1667 \, \text{cm}^{-1}$$

(b) Apply rearranged Eq. 13.7 from text p. 687:

$$\lambda = \text{wavelength} = (1 \times 10^4 \, \mu\text{m cm}^{-1}) \div 1720 \, \text{cm}^{-1} = 5.81 \, \mu\text{m}$$

13.5 The peak at 1380 cm^{-1} has a percent transmittance (%T) of about 41. By Eq. 13.9b, its T is 0.41. Using Eq. 13.9c, solve for absorbance, A, which is the negative log of T, or 0.39.

13.6 From Eq. 13.8a on text p. 687 we know that $\nu = c\tilde{\nu}$. Convert the frequency into wavenumber and then consult Fig. 13.4 on text p. 686.

$$\tilde{\nu} = \frac{\nu}{c} = \frac{4.41 \times 10^{13} \, \text{s}^{-1}}{3.00 \times 10^{10} \, \text{cm s}^{-1}} = 1470 \, \text{cm}^{-1}$$

The strong peak in the center of the spectrum between 1400 and 1500 cm^{-1} is due to C—H bending vibrations in nonane.

13.7 Using Eq. 13.8a on text p. 687, convert the wavenumber to a frequency, which is the "times per second" unit.

$$\nu = c\tilde{\nu} = (3.00 \times 10^{10} \, \text{cm s}^{-1})(2143 \, \text{cm}^{-1}) = 6.43 \times 10^{13} \, \text{s}^{-1}$$

13.8 According to the text, bond strength is the major determining factor of IR absorption frequency as we go across a row of the periodic table. Table 10.2 in Sec. 10.1E is essentially a table of bond strengths. From this table, we find that bond strengths are in the order

$$\text{H—F} > \text{H—OH} > \text{H—NH}_2 > \text{H—CH}_3$$

Because the strongest bonds vibrate with the highest frequencies, the order of stretching frequencies is the same; that is, H—F has the highest frequency, and H—CH₃ the lowest.

13.9 Take the ratio of two equations like Eq. 13.13a in Study Problem 13.1, one for the C—H bond, and the other for the C—D bond. Everything cancels except the wavenumbers and the square roots of the masses.

$$\frac{\tilde{v}_D}{\tilde{v}_H} = \sqrt{\frac{m_H}{m_D}}$$

or

$$\tilde{v}_D = \tilde{v}_H \sqrt{\frac{m_H}{m_D}} = (3090 \text{ cm}^{-1}) \sqrt{\frac{1}{2}} = 2185 \text{ cm}^{-1}$$

In fact, C—D vibrations appear in the IR at lower energy than the corresponding C—H vibrations (to the right in conventional IR spectra).

13.10 (a) Active. The double bond is located asymmetrically; therefore, even a small dipole moment will be affected by the stretching vibration.

(b) Inactive. The dipole moment of this molecule is zero, and it remains zero when the C=C bond is stretched.

(c) Inactive. The zero dipole moment of cyclohexane is unaffected by this symmetrical vibration.

(d) Inactive. The zero dipole moment of this alkyne is not changed by stretching the triple bond.

(e) Active. Because the nitro group is bent, stretching the two N—O bonds symmetrically will increase these bond dipoles and will also increase the resultant dipole of the molecule.

(f) Active. The dipole moment of the molecule is due mostly to the bond dipole of the C—Cl bond, which is the only bond in the molecule with significant polarity. Increasing the length of this bond dipole will increase the dipole moment of the molecule.

13.11 The carbon skeleton of the hydrogenation product defines the following alkenes as possibilities.

(1)	(2)	(3)	(4)	(5)
2-methyl-1-pentene	2-methyl-2-pentene	(E)-4-methyl-2-pentene	(Z)-4-methyl-2-pentene	4-methyl-1-pentene

The strong 912 and 994 cm⁻¹ C—H bending absorptions are very close to the standard values of 910 and 990 cm⁻¹, which are typical of —CH=CH₂ groups. Therefore compound *A* is *(5)*. The weak or absent C=C stretching absorption in the 1660–1670 cm⁻¹ region suggests that the candidates for *B*, *C*, and *E* are compounds *(2)*, *(3)*, and *(4)*. The strong 967 cm⁻¹ C—H bending absorption is the definitive absorption for trans alkenes. Therefore, compound *E* is *(3)*. Compound *C* must be the cis alkene *(4)*, from the C—H bending absorption at 714 cm⁻¹, because the C—H bending absorption for cis alkenes occurs at lower wavenumber than the C—H bending absorption of other alkene types. The 1650 cm⁻¹ C=C stretching absorption as well as the 885 cm⁻¹ C—H bending absorption point to *(1)* as the structure of compound *D*. This leaves only compound *B* unassigned, and it must therefore have structure *(2)*. Its C—H bending absorption is consistent with this assignment.

13.12 The moderately strong C=C stretching absorption at about 1650 cm⁻¹ and the C—H bending vibration at 890 cm⁻¹ suggest that (a) is the spectrum of 2-methyl-1-hexene. The C—H bending absorption at about 970 cm⁻¹ and the virtual absence of a C=C stretching absorption in spectrum (b) confirms that this is the spectrum of *trans*-2-heptene.

13.13 The absence of a C=C stretching absorption near 1650 cm⁻¹ and an O—H stretching absorption near 3300 cm⁻¹, and the presence of a strong absorption near 1100 cm⁻¹ (C—O stretching absorption), show that this is the spectrum of the ether.

13.14 (a) The two C—O bonds of an ether can undergo both symmetrical and unsymmetrical stretching vibrations; each of these normal vibrational modes has an associated infrared absorption. (These vibrations are described on the first two lines of Fig. 13.8 on text p. 695.)

   (b) For an ether with two of the same alkyl groups attached, the two unsymmetrical stretches would cancel each other. The general structure would be R^1—O—R^2 where $R^1 = R^2$, for example, CH_3OCH_3.

13.15 Think of molecular association by hydrogen bonding as a reaction like any other and apply Le Chatelier's principle.

$$2\,RO-H \;\rightleftharpoons\; RO-H\cdots:\ddot{O}R$$

unassociated
alcohols

$\mid$
H

hydrogen-bonded
alcohols

Increasing the concentration of alcohol drives this equilibrium to the right, and decreasing the concentration of the alcohol drives this equilibrium to the left. At sufficiently low concentrations of alcohol, the spectrum is that of the unassociated alcohol. The O—H bond in a hydrogen-bonded O—H absorbs at a different frequency than the unassociated O—H.

The broadness of the absorption for a hydrogen-bonded O—H is due to the presence of hydrogen bonds of many different strengths. Not only does the absorption change position when the sample is diluted, but it also becomes much sharper.

13.16 Table 13.3 on text p. 709 shows that Si has the following isotopic abundances: ^{28}Si, 92.21; ^{29}Si, 4.67; and ^{30}Si, 3.10. These abundances result in the following relative intensities for the $m/z = 74$ and 75 peaks:

$m/z = 74$: relative intensity $(4.67)/(92.21) = 0.0506$, or 5.06%
$m/z = 75$: relative intensity $(3.10)/(92.21) = 0.0336$, or 3.36%

To this must be added the contribution due to ^{13}C. The molecular mass of $(CH_3)_4Si$ is 88; therefore, the base peak at $m/z = 73$ results from loss of 15 mass units (CH_3). Therefore, the $m/z = 74$ peak will have a contribution from three isotopic carbons equal to $3(1.11)/(92.21) = 0.0361$ (3.61%). Add this contribution to that calculated above to get the final value for the $m/z = 74$ peak:

$m/z = 74$: relative intensity $= 5.06\% + 3.61\% = 8.67\%$

The contribution of ^{13}C to the $m/z = 75$ peak is negligible. Therefore, the $m/z = 74$ peak has a relative intensity of 8.67%, and the $m/z = 75$ peak has a relative intensity of 3.36%.

13.17 Assume the molecular ion (base) peak at $m/z = 50$ has an abundance of 100%. This peak is due to molecules that contain ^{12}C, ^{1}H, and ^{35}Cl. The M + 1 peak is due to molecules that contain either ^{13}C, ^{1}H, and ^{35}Cl or ^{12}C, ^{2}H, and ^{35}Cl. The intensity of the M + 1 peak due to ^{13}C *relative to the base peak* is $(0.0111/0.989) = 1.12\%$, and the relative intensity due to ^{2}H is $3(0.00015)/(0.99985) = 0.0004$, or 0.04%. Therefore, the M + 1 peak at $m/z = 51$ is due almost entirely to molecules that contain ^{13}C, ^{1}H, and ^{35}Cl, but we'll include the contribution of ^{2}H for completeness. Because there is one carbon, the intensity of the $m/z = 51$ peak is 1.16% (1.12% + 0.04%). The M + 2 peak at $m/z = 52$ is due almost entirely to ^{12}C, ^{1}H, and ^{37}Cl; according to Table 13.3, the ratio of this peak to the base peak should be $0.2423/0.7577 = 0.320$, or 32%. (This peak has a contribution from molecules that contain ^{13}C, ^{2}H, and ^{35}C equal to $3 \times (0.0111/0.989)(0.00015/0.99985) = 0.00017\%$ that can be ignored). Finally, there is a peak at M + 3, or $m/z = 53$, which is due to molecules that contain either ^{13}C, ^{1}H, and ^{37}Cl or ^{12}C, ^{2}H, and ^{37}Cl. The contribution of ^{2}H is only 0.0004. Thus, the relative abundance from the *simultaneous* presence of ^{13}C and ^{37}Cl is $(0.0112)(0.320) = 0.00358$, or 0.36%. This peak is almost negligible.

13.18 (a) An $m/z = 71$ corresponds to the formula C_5H_{11}. The only rational fragmentation that could give this formula is between the two CH_2 carbons:

(b)   The mechanism of this fragmentation:

$$m/z = 71$$

(c)   A rearrangement to a more stable tertiary carbocation occurs.

$$m/z = 71$$

Actually, this rearrangement is probably concerted with the fragmentation shown in part (b) so that formation of a primary carbocation is avoided.

13.19   When the molecule contains only C, H, and O, odd-electron ions have even mass, and even-electron ions have odd mass. Therefore, (a) and (c) are even-electron ions, and (b) is an odd-electron ion.

13.20   (a)   Because even-electron ions of compounds containing only C, H, and Cl have odd mass, the ion with $m/z = 71$ is the even-electron ion and the ion with $m/z = 70$ is the odd-electron ion.

(b)   HCl can be lost to give the odd-electron ion.

(c)   The molecular mass of 2-chloropentane is 106 (for the ^{35}Cl isotope). The $m/z = 71$ peak is due to loss of a chlorine atom from the molecular ion by an give the 2-pentyl cation. (You'll learn in Sec. 13.6D that this is called *inductive cleavage*. It's the mass-spectrometric analog of an S_N1 reaction!)

2-pentyl cation

The $m/z = 70$ peak is due to hydrogen-atom transfer within the molecular ion followed by loss of HCl in a process analogous to that shown for loss of water in Eq. 13.26a on text p. 712:

$$m/z = 70$$

13.21   (a)   An m/z of 45 indicates a loss of 43 mass units, which corresponds to an isopropyl group. An α-cleavage mechanism could give such a loss:

molecular ion $m/z = 45$

(b)   Inductive cleavage at either side of the oxygen would give a primary carbocation. With di-*sec*-butyl ether, in contrast, a secondary carbocation is formed by inductive cleavage. Because of the greater stability of secondary carbocations, more inductive cleavage occurs in the fragmentation of di-*sec*-butyl ether.

(c)   The major difference would be a large abundance of molecular ion and fewer low-mass fragments.

13.22   (a)   A fragment ion with $m/z = 101$ corresponds to a loss of 30 mass units, which corresponds to two methyl groups—that is, a molecule of ethane:

CI molecular ion
$m/z = 131$

$m/z = 101$

(b)   The same process occurs, except that 16 mass units—that is, methane—is eliminated.

CI molecular ion
$m/z = 131$

$m/z = 115$

(c)   As long as we're eliminating groups on the α-carbon, we might as well also eliminate the hydrogen:

CI molecular ion
$m/z = 131$

$m/z = 73$

This elimination is a little different from the other two in that the hydrogen on the positively charged oxygen is not eliminated; instead, one of the alkyl groups is eliminated. A β-hydrogen could also be eliminated along with the hydrogen on the oxygen to give a loss of H_2. Can you see any evidence for this in the mass spectrum of Fig. 13.21b?

Solutions to Skills Objectives Problems

13.23   (a)   Using Eq. 13.8a in Sec. 13.2A, convert the wavenumber to a frequency, which is the "times per second" unit.

$$\nu = c\,\tilde{\nu} = (3 \times 10^{10}\ \text{cm s}^{-1})(1714\ \text{cm}^{-1}) = 5.14 \times 10^{13}\ \text{s}^{-1}$$

(b)   Use Eq. 13.7 in Sec. 13.2A, rearranging and solving for λ.

$$\lambda = (10^4\ \mu\text{m cm}^{-1})/\tilde{\nu} = (10^4\ \mu\text{m cm}^{-1})/(1714\ \text{cm}^{-1}) = 5.83\ \mu\text{m}$$

(c)   Use Eq. 13.8b in Sec. 13.2A.

$$E = hc\,\tilde{\nu} = (3.99 \times 10^{-13}\ \text{kJ s mol}^{-1})(3 \times 10^{10}\ \text{cm s}^{-1})(1714\ \text{cm}^{-1}) = 20.5\ \text{kJ mol}^{-1}$$

13.24   Use Eq. 13.9c on text p. 688. The initial %*T* of 10% is equal to an absorbance of 1, and after dilution, the sample now has a %*T* of 50% which is equal to an absorbance of 0.3. So, the sample was diluted by a factor of 0.3.

13.25   Use an approach similar to that of Study Problem 13.1 in the text on p. 693. We are given $\tilde{\nu}$ for a carbon–carbon double bond, which we'll call $\tilde{\nu}_2$, and we're asked to estimate $\tilde{\nu}_1$, the stretching frequency of a carbon–carbon single bond. Let the force constant for a double bond be κ_2, and that of a single bond be κ_1. Now take the ratio of $\tilde{\nu}_2$ and $\tilde{\nu}_1$ as given by Eq. 13.12.

$$\frac{\tilde{\nu}_2}{\tilde{\nu}_1} = \sqrt{\frac{\kappa_2}{\kappa_1}}$$

All of the mass terms cancel because they are the same in both cases—the mass of a carbon atom.

The problem states that the ratio κ_2/κ_1 is 2.16, of which the square root is 1.47. With $\tilde{\nu}_2 = 1650$ cm^{-1}, we can solve for $\tilde{\nu}_1 = (1650$ cm$^{-1})/1.47 = 1122$ cm^{-1}.

13.26   Take the ratio of two equations 13.12 as in Focused Problem 13.9, one for the O—H bond, and the other for the O—D bond. Everything cancels except the wavenumbers and the square roots of the masses.

$$\frac{\tilde{\nu}_D}{\tilde{\nu}_H} = \sqrt{\frac{m_H}{m_D}}$$

or

$$\tilde{\nu}_D = \tilde{\nu}_H \sqrt{\frac{m_H}{m_D}} = (3600 \text{ cm}^{-1})\sqrt{\frac{1}{2}} = 2546 \text{ cm}^{-1}$$

13.27   (a)   The vibrational modes of water must transform the molecule in a way that makes the molecule distinguishable from the original. The three modes that meet this criterion are

O—H symmetrical stretch

O—H unsymmetrical stretch

O—H symmetrical bend (scissor)

(b)   The classifications are shown in the diagram in part (a).

(c)   Stretching vibrations typically occur at higher frequency than bending vibrations. Therefore, the 1595 cm^{-1} vibration is the bending vibration; the other two vibrations are the stretching vibrations.

13.28   The two stretching vibrations of the nitro group correspond to the symmetrical and unsymmetrical stretching modes.

13.29   (a)   Inactive. The zero dipole moment of acetylene is not changed by stretching the triple bond.

(b)   Active. The stretching of the triple bond will increase and decrease the dipole moment of this molecule.

(c)   Active. Because ammonia is trigonal pyramidal, stretching the three N—H bonds symmetrically will increase these bond dipoles and will also increase the resultant dipole of the molecule.

(d)   Active. This molecule has a dipole moment so stretching of the C=C bond would change the dipole moment.

(e)   Inactive. This molecule has a zero dipole moment that remains zero when the C=C bond is stretched.

13.30   (a)   This absorption is not given. Carbon-carbon triple bonds show up between 2100-2200 cm^{-1}.

(b)   *G*   (c) *A*   (d) *I*   (e) *H*   (f) *B*   (g) *C*

13.31 Spectrum (1) has no peaks characteristic of alkenes or alcohols, but it does have a strong peak near 1100 cm^{-1} that is a C—O stretching absorption. This is the spectrum of compound (d), dipropyl ether.

Spectrum (2) shows the typical absorptions of a —CH=CH$_2$ group: the double C—H bending absorption at 910 and 990 cm^{-1}, and the strong C=C stretching absorption at 1640 cm^{-1}. This is the spectrum of compound (a), 1,5-hexadiene.

Spectrum (3) has a C≡C stretching absorption around 2100 cm^{-1}, and a C≡C—H stretching absorption at 3300 cm^{-1}, characteristic of terminal alkynes. This is the spectrum of compound (h), 1-hexyne.

Spectrum (4) is virtually featureless; the only strong absorptions are the C—H stretching absorptions in the 2800–3000 cm^{-1} region. This is the spectrum of compound (f), cyclohexane.

Spectrum (5) shows both O—H stretching and C=C stretching absorptions, and is the spectrum of compound (c), 1-hexen-3-ol. The double peaks at 910 and 990 cm^{-1} confirm the presence of a —CH=CH$_2$ group. Compound (g), 3-hexanol, is ruled out because it should not have alkene absorptions in its IR spectrum.

Spectrum (6) has C=C stretching absorption at 1675 cm^{-1} and shows the typical high-wavenumber shoulder at about 3050 cm^{-1} for the alkene C—H stretching absorption. The position of the C=C stretching absorption along with the C—H bending absorption near 800 cm^{-1} indicates an alkene with three alkyl substituents on its double bond. This is the spectrum of compound (b), 1-methylcyclopentene. The definite absence of a C—H bending absorption near 970 cm^{-1} rules out compound (e), *trans*-4-octene.

13.32 (a) Ions *A* and *C* could be initially formed species in an electron-ionization mass spectrometry experiment, because they are radical cations. Ion *C* would be formed instead of ion *B* because nonbonding electrons are more easily removed than bonding σ electrons. Ion *D* is a cation, but not a radical cation and therefore is not initially formed in EI mass spectrometry,.

(b)

13.33 (a) Use Eq. 13.20b in Sec. 13.6b to determine the relative abundance of ions that have one ^{13}C atom in place of ^{12}C atoms in this five-carbon ion.

$$\text{relative abundance} = (5\ \text{carbons})(0.0111) = 0.0555$$

Now multiply the intensity of the molecular ion peak (55%) by the relative abundance (0.0555) to get an intensity for the M + 1 peak of 3.05%.

(b) First determine the most likely fragmentation that occurred. The parent compound has a molecular weight of 84, and the base peak has a molecular weight of 69. The difference is 15, which corresponds to a methyl group (—CH$_3$). If a methyl group was lost, the base peak has only four carbons. Use Eq. 13.20b again to determine the relative abundance of ions that have one ^{13}C atom in place of ^{12}C atoms in this four-carbon ion.

$$\text{relative abundance} = (4 \text{ carbons})(0.0111) = 0.0444$$

Now multiply the intensity of the base peak (100%, by definition) by the relative abundance (0.0444) to get an intensity for the M + 1 peak of 4.44%.

13.34   The three peaks correspond to dichloromethane molecules that have, respectively, two atoms of ^{35}Cl, one atom each of ^{35}Cl and ^{37}Cl, and two atoms of ^{37}Cl. The probability of any combination of two isotopes is the product of their separate probabilities. Make a table of the possibilities using the relative abundances from Table 13.3, text p. 709:

	Cl #1	Cl #2	
$m/z = 84$	35	35	relative probability = $(0.7577)^2 = 0.574$
$m/z = 86 \Big\}$	35	37	relative probability = $(0.7577)(0.2423) = 0.184$
	37	35	relative probability = $(0.2423)(0.7577) = 0.184$
			total relative probability 0.368
$m/z = 88$	37	37	relative probability = $(0.2423)^2 = 0.059$

Taking the peak at $m/z = 84$ as 100%, the ratios of the peaks are 100%, 64.1%, and 10.3%, respectively.

The middle entry of the table above shows that two *different* isotopes can occur within the same compound in two ways. An analogy is the combinations that can be rolled with a pair of six-sided dice. There is only one way to roll a "2" (a "1" on each of the dice), but there are two ways to roll a "3" (a "1" on one of the dice and a "2" on the other, and vice-versa).

13.35   Hint (2) says that the molecular ion is formed by ejection of an oxygen electron:

The fragment that forms from loss of a methyl group (see solution to Problem 13.33b) has $m/z = 69$, an odd mass, and therefore the ion must be an even-electron ion. Therefore, the methyl group must fragment as a methyl radical so that one electron is left behind, and the oxygen-containing fragment has an even number of electrons:

13.36   (a)   The parent compound has molecular mass = 115. The fragment with $m/z = 72$ corresponds to a loss of 43 mass units, which is accounted for by loss of a propyl radical by an α-cleavage mechanism.

molecular ion $m/z = 72$

(b)   Loss of a propyl radical from the molecular ion of 3-methyl-3-hexanol (molecular mass = 116) by an α-cleavage mechanism similar to that in part (a) gives a fragment with $m/z = 73$.

$$CH_3CH_2-\overset{\overset{\displaystyle +\ddot{O}H}{|}}{\underset{\underset{\displaystyle CH_3}{|}}{C}}-CH_2CH_2CH_3 \longrightarrow CH_3CH_2-\overset{\overset{\displaystyle +\ddot{O}H}{\|}}{\underset{\underset{\displaystyle CH_3}{|}}{C}} \ + \ \cdot CH_2CH_2CH_3$$

molecular ion $m/z = 73$

(c)   1-Pentanol has molecular mass = 88; therefore, a fragment with $m/z = 70$, which must be an odd-electron ion, could result from the loss of H_2O (18 mass units.) A hydrogen-transfer mechanism followed by loss of H_2O, exactly as shown for 1-heptanol in Eq. 13.26a on text p. 712, accounts for this ion.

$$CH_3(CH_2)_2\overset{|}{\underset{}{C}}H-\overset{+\cdot\ddot{O}H}{CH_2} \longrightarrow CH_3(CH_2)_2\dot{C}H-\overset{H-\overset{+}{\ddot{O}}H}{CH_2} \longrightarrow CH_3(CH_2)_2\dot{C}H-\overset{+}{C}H_2 \ + \ H-\ddot{\underset{..}{O}}H$$

molecular ion of $m/z = 70$
1-pentanol

(d)   Fragmentation of neopentane at any one of its four carbon–carbon bonds gives a methyl radical and the *tert*-butyl cation, which has the correct mass:

$$H_3C-C(CH_3)_3 \xrightarrow{-e^-} H_3C \overset{+}{\underset{j}{}} C(CH_3)_3 \longrightarrow H_3C\cdot \ + \ {}^+C(CH_3)_3$$

neopentane   ***tert*-butyl cation**
                                                                $m/z = 57$

13.37   (a)   The molecular mass of 1-methoxybutane is 88. It can lose a propyl group (43 mass units) by α-cleavage to give a peak at $m/z = 45$:

$$CH_3CH_2\dot{C}H_2-CH_2-\overset{+}{\underset{..}{O}}CH_3 \longrightarrow CH_3CH_2\dot{C}H_2 \ + \ CH_2=\overset{+}{\underset{}{O}}CH_3$$

molecular ion of 1-methoxybutane $m/z = 45$

It can also lose methanol following a hydrogen transfer. [This mechanism is analogous to the one by which water is lost from primary alcohols, shown in Eq. 13.26a, text p. 712., and the solution to Problem 13.36(c).]

$$CH_3CH_2\overset{|}{\underset{}{C}}H-\overset{+\cdot\ddot{O}CH_3}{CH_2} \longrightarrow CH_3CH_2\dot{C}H-\overset{H-\overset{+}{\ddot{O}}CH_3}{CH_2} \longrightarrow CH_3CH_2\dot{C}H-\overset{+}{C}H_2 \ + \ H-\ddot{\underset{..}{O}}CH_3$$

molecular ion of $m/z = 56$
1-methoxybutane

(b)   2-Methoxybutane, molecular mass = 88, can lose either a methyl group (15 mass units) or an ethyl group (29 mass units) by α-cleavage. Such losses would give rise to peaks at $m/z = 73$ and $m/z = 59$, respectively. The ethyl group is lost preferentially because of the relative stability of the ethyl versus the methyl radical.

$$CH_3\dot{C}H_2-\overset{\overset{\displaystyle +\cdot\ddot{O}CH_3}{|}}{CH}-CH_3 \longrightarrow CH_3\dot{C}H_2 \ + \ \overset{\overset{\displaystyle +\ddot{O}CH_3}{\|}}{CH}-CH_3$$

molecular ion of $m/z = 59$
2-methoxybutane

13.38   (a)   This fragment has an odd mass (odd number of hydrogens), so it is an even-electron fragment (see Sec. 13.6C in the text).

(b) This fragment has an even mass, and it contains only C, H, and O, so it is an odd-electron fragment.

13.39 The presence of an odd number of nitrogens in a molecule containing, as the other atoms, any combination of C, H, O, and halogen reverses the odd-electron/even-electron mass correlations in Sec. 13.6C because such a molecule must have an odd molecular mass. (Molecules containing no nitrogen or an even number of nitrogens have even molecular masses.)

(a) The molecular ion of any molecule, because it is formed by ejection of a single electron, must be an odd-electron ion.

(b) A fragment of even mass containing one nitrogen is an even-electron ion. (The ion in the solution to Problem 13.36(a) is an example.)

(c) A fragment ion of odd mass containing a single nitrogen must be an odd-electron ion.

13.40

$$Bu_3N: \; + \; CH_5^+ \longrightarrow \; Bu_3\overset{+}{N}H \; + \; CH_4$$

molecular mass = 186

13.41

$$H_3C \; + \; CH_5^+ \longrightarrow \left[\; \longleftrightarrow \; \right] + \; CH_4$$

Solutions to Integrated Problems

13.42 The principle is that compounds that have different physical properties have different IR spectra.

(a) Because 3-pentanol and racemic 2-pentanol are constitutional isomers, they have different IR spectra.

(b) Because enantiomers have identical physical properties, the IR spectra of (R)- and (S)-2-pentanol are identical.

(c) This one is a little tricky! The two chair conformations of cyclohexanol are diastereomers and, because diastereomers have different physical properties, the two conformations also have different IR spectra. The problem is *observing* the two conformations independently. Any sample of cyclohexanol at room temperature is a mixture of these diastereomeric conformations, and the IR spectrum of cyclohexanol is therefore a spectrum of the mixture, with the spectrum of each conformation present in proportion to its concentration.

> Even though the two chair conformations of cyclohexanol are rapidly interconverting, the IR absorption process is "fast enough" to record the spectra of the two diastereomeric conformations independently. Therefore, the spectrum observed is a superposition of the spectra of the two compounds.

13.43 (a) Catalytically hydrogenate the alkene, and confirm the reaction by observing loss of the C=C stretching absorption at 1660–1675 cm^{-1}.

$$\text{1-methylcyclohexene} \; + \; H_2 \xrightarrow{\text{Pt/C catalyst}} \text{methylcyclohexane}$$

1-methylcyclohexene   **methylcyclohexane**

(b)   Carry out a Williamson ether synthesis. Confirm the reaction by observing loss of the O—H stretch of the alcohol in the 3200–3400 cm^{-1} region of the IR spectrum, and by intensification of the C—O stretching absorption.

$$CH_3(CH_2)_4CH_2OH \xrightarrow{NaH} CH_3(CH_2)_4CH_2O^- \; Na^+ \xrightarrow{CH_3I} CH_3(CH_2)_4CH_2OCH_3$$

1-hexanol   **1-methoxyhexane**
                                                                                              (hexyl methyl ether)

13.44   Only spectrum (2) has the C=C stretching absorption near 1640 cm^{-1} and the high-wavenumber shoulder at about 3050 cm^{-1} for the alkene C—H stretching absorption; this is therefore the spectrum of compound *A,* and spectrum (1) is the spectrum of compound *B.*

13.45   (a)   Because stronger bonds absorb at higher wavenumber (text Sec. 13.3A), the order of increasing C—H bond strength is alkane < alkene < alkyne. This deduction can be verified by consulting the relevant C—H bond dissociation energies in Table 10.2, text p. 520; these bond energies are alkane (423 kJ mol^{-1}), alkene (463 kJ mol^{-1}), and alkyne (558 kJ mol^{-1}).

(b)   [Note that your textbook *may have* a typo in this question that was corrected in the second printing. The bond dissociation energy of the ≡C—H bond is 558 kJ mol^{-1}. This triple bond in your text may be shown as a double bond. If so, redo this problem using the alkyne stretching frequency, as this is what the solution below shows.]
From Eq. 13.12 (text p. 693) and the associated arguments, the stretching frequencies are proportional to the square roots of the force constants κ. The force constants, in turn, are roughly proportional to the bond strengths. We then have

$$\frac{\tilde{\nu}_{alkyne\; C—H}}{\tilde{\nu}_{alkane\; C—H}} = \frac{3300}{2850} = 1.157 = \sqrt{\frac{558 \text{ kJ mol}^{-1}}{x \text{ kJ mol}^{-1}}}$$

Solving,

$$x = (558 \text{ kJ mol}^{-1})/(1.157)^2 = 417 \text{ kJ mol}^{-1}$$

The value in Table 10.2 is 423 kJ mol^{-1}. This calculation thus gives the correct answer within 5% of the difference between the two bond energies.

13.46   The solution to Problem 13.45 noted the correlation between bond dissociation energies and IR absorption frequencies. Triple bonds are clearly stronger than double bonds, which are stronger than single bonds. The question is which of the two double bonds is stronger. Table 10.2 on text p. 520 shows that the bond dissociation energy of a C=C double bond is 728 kJ mol^{-1}, and that of a C=O double bond is 749 kJ mol^{-1}. On this basis, the C=O bond is predicted to absorb at higher frequency. (This is realized in practice; typical C=O absorptions occur in the 1710 cm^{-1} region, whereas C=C absorptions occur in the 1650 cm^{-1} region.) In summary, then, the order of increasing bond strengths and increasing IR absorption frequencies is:

$$C—C < C=C < C=O < C≡C$$

13.47   (a)   The bond dipole of the S—H bond is much less than that of the O—H bond because the electronegativities of sulfur and hydrogen differ less than do the electronegativities of oxygen and hydrogen (Table 1.15, text p. 56). As a result, the S—H vibration is less active in the infrared, and the corresponding absorption is less intense. The S—H absorption occurs at lower frequency mostly because the S—H bond is much weaker than the O—H bond; compare the bond dissociation energies for H—SH and H—OH in Table 10.2, text p. 520.

(b)   The difference between the masses of sulfur and oxygen has little effect on the vibration frequency because the mass of the smaller atom, hydrogen, is the major determinant of the vibration frequency, as shown by

Eq. 13.14 on text p. 694. Therefore, the difference in bond strengths is the major cause of the different absorption frequencies.

(c) Compound B has a typical strong, broad O—H stretching absorption in the 3200–3400 cm^{-1} region, and is therefore the alcohol. Compound A, then, is the thiol. Note the considerably weaker S—H stretching absorption in the spectrum of compound A at about 2530 cm^{-1}, as suggested by part (a) of the problem.

13.48 (a) Both compounds have the same absorptions, except that the absorptions of compound A in Fig. P13.48 are displaced to lower frequency, an observation that implies a higher mass for the absorbing group; see the discussion of the mass effect associated with Eq. 13.14 on text p. 694. In particular, a peak at 3000 cm^{-1} in compound B, undoubtedly a C—H stretching absorption, is displaced to 2240 cm^{-1} in compound A. Therefore, compound B is CHCl$_3$.

Eq. 13.14 on text p. 694 gives the quantitative basis of the mass effect. If the force constants of C—H and C—D bonds are nearly the same (and they are), then the ratio of stretching frequencies of these two bonds should be equal to the square root of the ratios of the mass terms. Eq. 13.14 predicts a ratio of $\sqrt{2} = 1.414$; the more exact equation, Eq. 13.12 (text p. 693), predicts a ratio of 1.36. The actual ratio, (3000/2240) = 1.34, is very close to this prediction.

(b) The two compounds could be distinguished by mass spectrometry by the masses of their parent ions. Both CHCl$_3$ and CDCl$_3$ have four molecular ions. (Why four? See Table 13.3, text p. 709.) Each molecular ion peak of CDCl$_3$ lies at one unit higher mass than the corresponding molecular ion peak of CHCl$_3$.

13.49 The reaction with NaH followed by CH$_3$I is a Williamson ether synthesis to form a methyl ether.

$$R\text{—}OH \xrightarrow{\;NaH\;} \xrightarrow{\;CH_3I\;} R\text{—}OCH_3$$

mass = 102   mass = 116

A B

Whatever the mass of the starting alcohol A, the methylation adds 14 mass units. Because the CI mass spectrum of B has $m/z = 117$, ether B itself has $m/z = 116$, and alcohol A then has a mass of $116 - 14 = 102$. The $m/z = 87$ peak in the mass spectrum of A represents a loss of 15 mass units, which is a methyl group. The $m/z = 73$ peak represents a loss of 29 mass units, which is an ethyl group. The tertiary alcohol C—OH group has a mass of 29. The mass of 102 can be accounted for if the carbon of this alcohol group bears two ethyl groups (58 mass units) and one methyl group (15 mass units). Compound A is therefore 3-methyl-3-pentanol, and its fragmentation with loss of an ethyl group occurs by α-cleavage:

:ÖH $\overset{+}{\cdot}$ÖH $\overset{+}{\cdot}$ÖH

CH$_3$CH$_2$—C—CH$_3$ $\xrightarrow{-e^-}$ CH$_3$CH$_2$—C—CH$_3$ ⟶ CH$_3$CH$_2$—C—CH$_3$ + CH$_3$ĊH$_2$

CH$_3$CH$_2$ CH$_3$CH$_2$ $m/z = 73$

mass = 102   molecular ion

A

3-methyl-3-pentanol

(We leave it to you to show the formation of the $m/z = 87$ fragment.)

13.50 First ask what products are expected from the reaction. Two constitutional isomers, both with molecular mass = 88, are anticipated:

H$_3$C   H   1) BH$_3$/THF   OH   OH

C=C $\xrightarrow{\text{2) H}_2\text{O}_2/\,^-\text{OH}}$ CH$_3$CHCH$_2$CH$_2$CH$_3$ + CH$_3$CH$_2$CHCH$_2$CH$_3$

H   CH$_2$CH$_3$ **2-pentanol**   **3-pentanol**

trans-2-pentene

Spectrum (b) is consistent with 3-pentanol; the base peak at $m/z = 59$ corresponds to loss of CH_3CH_2—, and there are two ethyl groups in this compound that could be lost by α-cleavage. Spectrum (a) has a base peak at $m/z = 45$ that corresponds to loss of 43 units (a propyl group). 2-Pentanol has a propyl branch that could be lost as a radical by α-cleavage.

When unknown compounds come from a chemical reaction, use what you know about the reaction as a starting point for postulating structures.

13.51 (a) A mass of 28 units could be accounted for by a molecule of ethylene, $H_2C{=}CH_2$.

(b) Loss of 18 mass units could indicate loss of H_2O. (See the solution to Problem 13.36(c) for an example of such a loss.)

(c) The data indicate the presence of chlorine, which has a mass of 35; loss of 36 mass units, then, could correspond to loss of HCl. (See the solution to Problem 13.20(c) for an example of such a process.)

13.52 The three peaks correspond to dibromomethane molecules that have, respectively, two atoms of ^{79}Br, one atom each of ^{79}Br and ^{81}Br, and two atoms of ^{81}Br. The probability of any combination of two isotopes is the product of their separate probabilities. Follow the procedure used in the solution to Problem 13.34, except that the relative abundances are those of the bromine isotopes.

	Br #1	Br #2	
$m/z = 172$	79	79	relative probability = $(0.5069)^2 = 0.257$
$m/z = 174\}$	79	81	relative probability = $(0.5069)(0.4931) = 0.250$
	81	79	relative probability = $(0.5069)(0.4931) = 0.250$
			total relative probability 0.500
$m/z = 176$	81	81	relative probability = $(0.4931)^2 = 0.243$

The probabilities, and therefore the relative abundances, of the molecular ions in dibromomethane are in the ratio 0.257:0.500:0.243, or 1.00:1.95:0.95, or about 1:2:1, as asserted in the problem.

13.53 (a) The molecular mass of ethyl bromide (CH_3CH_2Br) is 108 (for ^{79}Br) and 110 (for ^{81}Br). Consequently, the ion with $m/z = 110$ is the molecular ion of the molecule containing the heavier isotope. It is formed by ejection of an electron from one of the bromine unshared pairs:

$$CH_3CH_2{-}\ddot{\underset{\cdot\cdot}{Br}}{:} \xrightarrow{-e^-} CH_3CH_2{-}\overset{+}{\underset{\cdot\cdot}{Br}}{:}$$
$$m/z = 110,\ 108$$

(b) The ion with $m/z = 108$ is the molecular ion containing the lighter bromine isotope, and it is formed by the process shown in part (a).

(c) The fragment at $m/z = 81$ corresponds to the formation of a bromine cation and an ethyl radical from the compound containing the heavier bromine isotope:

$$CH_3\overset{\frown}{C}H_2{-}\overset{+}{\underset{\cdot\cdot}{Br}}{:} \longrightarrow CH_3\dot{C}H_2 + {}^+\ddot{Br}{:}$$
$$m/z = 81,\ 79$$

(d) The ion with $m/z = 79$ is the bromine cation formed from the molecular ion containing the lighter bromine isotope, and it is formed by the mechanism shown in part (c).

(e) The fragment at $m/z = 29$ is the ethyl cation formed by inductive cleavage at the carbon-bromine bond:

$$CH_3CH_2\overset{\frown}{{-}}\overset{+}{\underset{\cdot\cdot}{Br}}{:} \longrightarrow CH_3\overset{+}{C}H_2 + :\dot{Br}{:}$$
$$m/z = 29$$

(f)   The ion with $m/z = 28$ is an odd-electron ion formed from the molecular ion by hydrogen transfer followed by loss of HBr:

$$H_2\dot{C}—CH_2 \longrightarrow H_2\dot{C}—\overset{+}{C}H_2 + H—\ddot{B}r:$$
$$m/z = 28$$

(g)   The ion at $m/z = 27$ could arise by elimination of a hydrogen atom by α-cleavage from the ion with $m/z = 28$ formed in part (f).

$$H_2\dot{C}—\overset{+}{C}H \longrightarrow H_2C=\overset{+}{C}H + H\cdot$$
$$m/z = 28 \qquad\qquad m/z = 27$$

13.54 (a)   Sigma (σ) electrons are held closer to the nucleus than π electrons; by the electrostatic law, attraction between particles of opposite charge (such as an electron and a nucleus) is greater when the particles are closer. Therefore, there is a greater energy of attraction of the nucleus for σ electrons. This means that more energy must be used to remove a σ electron; therefore, a π electron is removed more easily. An equivalent explanation is that the bond dissociation energy of a typical π bond is considerably less than that of a σ bond (see Table 10.2, text p. 520). It should take less energy to eject an electron from a weaker bond—a π bond—than it does from a stronger bond—a σ bond.

(b)   The structure of the ion formed by loss of a π electron from 1-heptene:

$$CH_3(CH_2)_4CH=CH_2 \xrightarrow{-e^-} \left[CH_3(CH_2)_4\dot{C}H—\overset{+}{C}H_2 \longleftrightarrow CH_3(CH_2)_4\overset{+}{C}H—\dot{C}H_2 \right]$$

1-heptene

 Because the unpaired electron is in a π molecular orbital and is therefore shared between the two carbons, we can put the unshared electron (and therefore the positive charge) on either carbon.

(c)

$$CH_3CH_2CH_2\dot{C}H_2—CH_2—\dot{C}H—\overset{+}{C}H_2 \longrightarrow CH_3CH_2CH_2\dot{C}H_2 + CH_2=CH—\overset{+}{C}H_2$$

1-butyl radical   **allyl cation**

Nuclear Magnetic Resonance Spectroscopy

CHAPTER 14 SUMMARY

- The NMR spectrum records the absorption of energy by nuclei from a radiofrequency (rf) source in the presence of a magnetic field. Absorption results in "spin-flipping"—that is, promotion of nuclear spins to a higher energy level.

- Only nuclei with spin give NMR spectra; protons (^1H) and carbon-13 (^{13}C) are two of the nuclei that have spin $\pm\frac{1}{2}$.

- The four elements of a proton NMR spectrum are the number of resonances (also called lines or signals), which correspond to the number of chemically nonequivalent sets of protons; the chemical shift, which provides information on the chemical environment of the observed protons; the integral, which indicates the relative number of protons being observed; and the splitting, which gives information about the number of protons (or other spin-active nuclei) on adjacent atoms.

- Proton and carbon-13 chemical shifts can be estimated using Fig. 14.4 and Fig. 14.22, respectively.

- Chemically nonequivalent nuclei in principle have different chemical shifts. Constitutionally nonequivalent nuclei and diastereotopic nuclei are chemically nonequivalent.

- The $n + 1$ rule determines the splitting observed in many spectra. When a nucleus is split by more than one chemically nonequivalent set of nuclei, the observed splitting is multiplicative; that is, it is the result of successive applications, in any order, of the splittings caused by each set.

- Splitting by nuclei of spin $\pm\frac{1}{2}$ such as the proton, results in splitting patterns in which the individual lines ideally have the relative intensities shown in

Table 14.2. In practice, splitting patterns show leaning: they deviate from these intensities, and the deviation is greater as the difference in the chemical shifts of the mutually split protons is smaller.

- The magnitude of the splitting between two nuclei is given by the coupling constant J. For splitting between vicinal protons, the magnitude of J varies with the dihedral angle between the bonds to the coupled protons according to the Karplus relationship (Fig. 14.9, p. 754). For example, the splitting between anti protons is much greater than that between gauche protons. This relationship can be used to determine molecular conformations in some cases. When the dihedral angle between coupled protons varies rapidly because of conformational interconversion, the observed coupling constants are weighted averages.

- Splitting more complex than that predicted by the $n + 1$ rule is observed when the resonances of two coupled protons (in Hz) differ in chemical shift by an amount that is comparable to their mutual coupling constant. Because the chemical shift in Hz increases in proportion to the operating frequency, but coupling constants do not, many compounds that give non-first-order spectra at a lower magnetic field give first-order spectra at a higher manetic field.

- Deuterium resonances are not observed in a proton NMR spectrum. Shaking the solution of an alcohol with D_2O (the "D_2O shake") brings about the rapid exchange of deuterium for the hydrogen of OH and of other rapidly exchangeable protons. This results in the elimination of the OH resonance and elimination of splitting of nearby coupled protons.

■ A time-averaged NMR spectrum is observed for species involved in rapid equilibria. It is possible to observe absorptions for the individual species by decreasing the rates of the processes involved in the equilibria (for example, by lowering the temperature).

■ The NMR of carbon nuclei in a compound can be observed as a ^{13}C NMR spectrum, despite the low natural abundance of this isotope.

■ In a proton-decoupled ^{13}C NMR spectrum, the carbon–proton couplings are removed; each chemically nonequivalent set of carbons appears as a single line. The number of protons attached to each carbon can be determined using the DEPT technique.

■ Magnetic resonance imaging (MRI) is a safe, noninvasive imaging technique used in medicine. MRI images rely on differences in the proton relaxation times of water molecules in different physiological environments. Two relaxation times are observed: the longitudinal relaxation time T_1, which measures the rate of return of z magnetization to equilibrium; and the transverse relaxation time T_2, which measures the rate of randomization of nuclear precessional phase angles (*xy* magnetization). Coordination compounds of the paramagnetic ion Gd^{3+} significantly reduce T_1 values; these compounds are widely used as contrast agents in MRI.

STUDY GUIDE LINKS

14.1 Approaches to Problem Solving

The approach given in Sec. 14.3F in the text for solving NMR structure problems is not necessarily the only approach. For example, you might take step 5 first. The point is that *you should be systematic and extract all the information you can from each piece of evidence in the spectrum before starting to write complete structures.*

A good analogy (suggested to the authors by a reviewer) is the various approaches to working a jigsaw puzzle. Some people start by working around the borders, and others by looking for unique patterns or colors in the picture on the box in which the puzzle is packaged. The important point is that fitting together pieces of the puzzle at random is not likely to yield a solution in a reasonable amount of time; *your approach must be systematic.*

The dangers of using "recipes" for problem solving were discussed in Study Guide Link 9.3 in this manual. Remember that exceptions to any recipe will occur, and you will be best equipped to deal with such situations by developing your own approach and understanding the basis of each step. Do not let any "recipe" inhibit you from trying your own style of problem solving. Ultimately the style that works for you is the right one. The way to become proficient in developing your own style of problem solving is to practice. As you develop your problem-solving techniques, you are doing more than learning about NMR (or other aspects of organic chemistry); you are developing a life skill.

14.2 More NMR Problem-Solving Hints

When you solve structure problems with NMR (or any sort of spectroscopy), it is very important that you be *intellectually honest* with yourself. Do *not* force a structure to fit the data. That is, write down each observation from the spectrum and then write down what it tells you. This point can be illustrated with the jigsaw-puzzle analogy in Study Guide Link 14.1. Anyone who has worked a jigsaw puzzle has encountered pieces that look as if they should go together, but do not quite fit; the temptation to force-fit the pieces is often great—but force-fitting never solves the puzzle. When you propose a structure, be able to defend it on the basis of *evidence that you can cite.* If you must guess a structure, then you haven't really solved the problem. If you are stymied by a structural problem, write down *all possible structures* that seem to fit the data. Ask yourself carefully whether each of these structures fits *all* of the information available: chemical shift, integration, and splitting. Sometimes the act of writing down and analyzing the information in this way will lead to a solution. If you must consult the solution without having completely solved the problem, then at least you have defined the points of uncertainty. If, after looking at the solution, you still don't see why your structures are ruled out, seek assistance! You can sometimes learn more from a situation like this, as frustrating as it is, than you can from a correctly worked problem!

Now to some specific tips. The authors have observed a tendency among some students to assume that protons that are physically adjacent in a structure must also have absorptions that are adjacent in their NMR spectra. Don't be caught in this trap. Although this may be true in specific cases, in general it is not something that you can rely on.

The last hint is to remind you about common splitting patterns in NMR spectra that appear over and over again. These patterns are so easy to recognize that they *immediately* suggest the presence of certain groups. Let's consider a few of these.

A three-proton triplet at high field along with a two-proton quartet at lower field [**Fig. SG14.1(a)**] is the earmark of an ethyl group in a structure CH_3CH_2—X, in which the group X has no coupled protons. Ethyl esters and ethyl ethers show this type of pattern. The chemical shift of the quartet provides information about the nature of the group X.

A one-proton septet or multiplet at low field along with a six-proton doublet at high field [**Fig. SG14.1(b)**] is characteristic of an isopropyl group. Sometimes the one-proton pattern, because of its complexity, is buried among other absorptions, or is otherwise difficult to distinguish. In such cases, the six-proton doublet alone can be used to diagnose an isopropyl group. Likewise, a three-proton doublet at relatively high field is characteristic of the methyl group in the following partial structure:

$$\overset{\diagdown}{\underset{\diagup}{C}}H—CH_3 \longleftarrow \text{a 3-proton doublet}$$

Figure SG14.1   Two splitting patterns commonly observed in proton NMR spectra. In this diagram, absorption peaks are represented as lines.

A three-proton singlet almost always suggests a methyl group. For example, a three-proton singlet in the δ 3–3.5 region in many cases corresponds to a CH_3O— group. A nine-proton singlet corresponds to a *tert*-butyl group. Be on the lookout for these common patterns when you solve unknowns with NMR.

FURTHER EXPLORATIONS

14.1 Quantitative Estimation of Chemical Shifts

This section provides some tools for making more quantitative estimates of chemical shifts than Fig. 14.4 in the text. The basis of all these methods is that the influence of surrounding groups on the chemical shift of a set of protons is largely additive.

The chemical shifts of compounds H_3C—G, where G is a substituent group, can be read directly from the second column of **Table SG14.1** on p. 398. For example, the chemical shift of CH_3I is δ 2.2. (Tables are found at the end of this section.)

Table SG14.1 also provides the data that can be used for calculating the chemical shifts of methylene groups in a compound G^1—CH_2—G^2, where G^1 and G^2 are attached groups. To estimate the chemical shift, we simply add the *shift contributions* of the two groups to 0.2. That is,

$$x\delta\,(G^1\!\!-\!\!CH_2\!\!-\!\!G^2) = 0.2 + \sigma(G^1) + \sigma(G^2) \qquad \text{(SG14.1)}$$

where $\sigma(G^1)$ and $\sigma(G^2)$ are the shift contributions of the two groups. These shift contributions are given in the last column of Table SG14.1.

For example, suppose you want to calculate the chemical shift of the methylene protons of methylene chloride (CH_2Cl_2). In this example, $G^1 = G^2 = Cl$. Applying Eq. SG14.1, with $G^1 = G^2 = 2.5$, we have δ (Cl—CH_2—Cl) = 0.2 + 2.5 + 2.5 = 5.2. The actual chemical shift of methylene chloride is δ 5.3. (This method gives results that are generally accurate to within a few tenths of a ppm.)

Here are some other examples:

$H_3COCH_2C\!\equiv\!N$: 0.2 + 2.4 + 1.7 = δ 4.3; the observed shift is δ 4.25.
$H_3COCH_2OCH_3$: 0.2 + 2.4 + 2.4 = δ 5.0. (Figure 14.1 shows that the actual shift is δ 4.6.)
H_3COCH_2Cl: 0.2 + 2.5 + 2.4 = δ 5.1; the observed shift is δ 5.5.

When using this method, the effect of a β-halogen, β-ether, or β-OH group can be taken into account by adding another 0.5 ppm. For example, let's estimate the chemical shift of the boldfaced protons in the following compound:

$$Br\!-\!\textbf{CH}_2\!-\!CH_2\!-\!Cl$$

In this example, the chlorine is β to the protons of interest. Using Eq. SG14.1, we have

$$\delta = 0.2 + 0.6 + 2.3 + 0.5 = 3.6$$

where the last 0.5 is the contribution for the β-halogen. (The observed shift is δ 3.3.)

A review of this approach can be found in two articles in the *Journal of Chemical Education,* which you should be able to find online or in your college library:

E. C. Friedrich and K. G. Runkle, *J. Chem. Educ.,* **1984,** *61,* 830–832; ibid., **1986,** *63,* 127–129.

The second of these articles extends the method discussed in the text to the chemical shifts of methine protons.

Another method for estimating chemical shifts makes use of a very useful tabulation of chemical-shift data that was assembled by Tom Curphey (who became a professor in the Dartmouth School of Medicine) while a graduate student at Harvard in 1961, and subsequently modified by Harry Morrison (now Professor Emeritus of Chemistry and Dean Emeritus of Science at Purdue University). The Curphey-Morrison data are given in **Table SG14.2** on p. 399. (Tables follow at the end of this section.)

The use of this table can be illustrated by a calculation of the chemical shifts of the various protons of the following compound.

$$\overset{a}{H_3C}-\overset{\overset{\displaystyle Cl}{\displaystyle |}}{\underset{b}{CH}}-\overset{c}{CH_2}-\overset{d}{CH_2}-Cl$$

Protons a are methyl protons; therefore begin the calculation with a base shift of δ 0.9. To this, add δ 0.63 for a chlorine β to a methyl group (first line, last column of the chlorine entry). The predicted chemical shift of protons a is δ 1.53. (The observed shift is δ 1.60.)

Proton b is a methine proton; begin with a base shift of δ 1.55. To this add δ 2.55 for an α-chlorine contribution (third line for chlorine under "Alpha shift") for a predicted chemical shift of δ 4.10. (The observed shift is δ 4.27.)

Protons c are methylene protons with *two* β-chlorines. The predicted shift is δ 1.20 + δ 0.53 + δ 0.53 = δ 2.26. (The observed shift is δ 2.15.)

Try to predict the chemical shift of the protons d. (The observed shift is δ 3.72.)

The comparisons given above between calculated and observed shifts illustrate the point that the chemical shifts calculated by this (or any other) method are estimates; do not rely on them for accuracy to more than a few tenths of a part per million.

The Curphey-Morrison table gives explicitly the contributions for β shifts as well as the shifts for methine protons. Its advantage over the Table SG14.1 is that it can be used to calculate the shifts of methine protons. However, it does not predict very well the shifts of CH_2 or CH protons bound to two or more electronegative groups. (For example, the predicted shift of CH_2Cl_2 is δ 5.8; the method above using Eq. SG14.1 predicts δ 5.2; the observed shift is δ 5.3.)

14.2   Fourier-Transform NMR

In FT-NMR, the sample is pulsed with a very intense, short pulse of rf radiation that causes all the nuclear spins to absorb energy simultaneously. The decay of absorption is then monitored for a second or two. During this time, all the nuclear spins in the sample return to their normal equilibrium distribution of +1/2 and –1/2 spins and, in doing so, emit radiofrequency energy, each at its characteristic frequency. An analogy is the ringing of a bell. When we "pulse" (strike) the bell, all frequencies (pitches) in the bell are activated simultaneously. As the bell rings, the pitches add together to give the characteristic sound of the bell, which dies out over time. An actual graph of this rf energy decay for dimethoxymethane (the compound whose spectrum is shown in Fig. 14.1 in Sec. 14.1 of the text) is shown in **Fig. SG14.2** on p. 400. Remember, this is the total rf emitted from all the different proton frequencies simultaneously. You can see the die-out over time, as well as the periodic pulsing of the rf, in the same sense that you can hear pulsations in the sound of a bell when it rings. (In fact, this trace is sometimes called a "ringdown.") A mathematical technique called *Fourier transformation* can be used to translate the ringing of the bell over time into a plot of sound intensity versus pitch (frequency). Such a plot would be a spectrum showing the different pitches of which the sound is composed and their relative intensities. The same technique can be applied to the decaying NMR signal over time to convert it into a plot of absorption intensity versus frequency (or field), which is the conventional NMR spectrum. Thus, Fourier transformation of the data in Fig. SG14.2 provides the actual NMR spectrum of dimethoxymethane shown in Fig. 14.1 in the text. This mathematical process, once very laborious, is performed in a fraction of a second on a desktop computer.

Table SG14.1	Group Contributions for Estimating Chemical Shifts of CH_2 Protons Using Eq. SG14.1	
Functional group G	**δ for CH_3—G**	**σ(G) for —CH_2—G**
—H	0.2	0.0
—CR_3 or —CR_2— (R = H, alkyl)	0.9	0.6
—F	4.3	3.6
—Cl	3.0	2.5
—Br	2.7	2.3
—I	2.2	1.8
—CR=CR_2 (R = H, alkyl)	1.8	1.3
—C≡CR (R = H, alkyl)	2.0	1.4
—OH	3.5	2.6
—OR (R = alkyl)	3.3	2.4
—OR (R = aryl)	3.7	2.9
—SH, —SR	2.4	1.6
(Ph, phenyl)	2.3	1.8
—C(=O)—R	2.1 (R = alkyl) 2.6 (R = H)	1.5
—C(=O)—OR (R = alkyl, H)	2.1	1.5
—O—C(=O)—OR	3.6 (R = alkyl) 3.8 (R = aryl)	3.0
—C(=O)—NR_2 (R = alkyl, H)	2.0	1.5
R—N—C(=O)—R (R = alkyl, H)	2.8	—
—NR_2 (R = alkyl, H)	2.2	1.6
—N—R (R = aryl)	2.9	—
—C≡N	2.0	1.7

Table SG14.2	The Curphey–Morrison Table for Estimating Chemical Shifts*

Effect of a Substituent G on the Chemical Shifts of H^α and H^β

$$G-\underset{H^\alpha}{\overset{|}{C}}-\underset{H^\beta}{\overset{|}{C}}-$$

Base Values of Shift Positions: $-CH_3$ δ 0.90 $-CH_2-$ δ 1.20 $-\overset{|}{CH}-$ δ 1.55

Functional group G	Type of hydrogen	Alpha shift	Beta shift
—Cl	CH_3	2.43	0.63
	CH_2	2.30	0.53
	CH	2.55	0.03
—Br	CH_3	1.80	0.83
	CH_2	2.18	0.60
	CH	2.68	0.25
—I	CH_3	1.28	1.23
	CH_2	1.95	0.58
	CH	2.75	0.00
—Aryl	CH_3	1.40	0.35
	CH_2	1.45	0.53
	CH	1.33	0.22
—CH=O and —CR=O	CH_3	1.23	0.25
	CH_2	1.10	0.30
	CH	0.95	—
—CO_2H and —CO_2R	CH_3	1.23	0.25
	CH_2	1.05	0.30
	CH	1.05	—
—C≡C—	CH_3	0.78	0.05
	CH_2	0.75	0.10
	CH	1.25	0.00
—OH and —OR	CH_3	2.45	0.35
	CH_2	2.30	0.15
	CH	2.20	—
—O—Aryl	CH_3	2.95	0.40
	CH_2	3.00	0.45
	CH	3.30	—
—O—CO_2R	CH_3	2.88	0.38
	CH_2	2.98	0.43
	CH	3.45	0.36

(Table continues)

Functional group G	Type of hydrogen	Alpha shift	Beta shift
—NR$_3$	CH$_3$ CH$_2$ CH	1.25 1.40 1.35	0.13 0.13 0.00
—C≡N	CH$_3$ CH$_2$ CH	1.10 1.10 1.05	0.45 0.40 0.47
—NO$_2$	CH$_3$ CH$_2$ CH	3.50 3.15 3.05	0.65 0.85 —
—C≡C—	CH$_3$ CH$_2$ CH	0.90 0.80 0.35	0.15 0.05 —

*Reprinted by permission from the Ph. D. dissertation of T. J. Curphey, Harvard University, 1961, as modified by T. J. Curphey and H. A. Morrison.

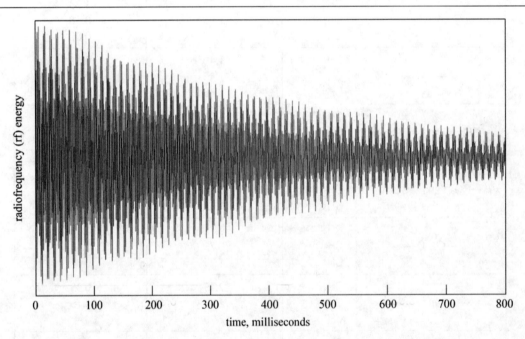

Figure SG14.2 An NMR "ringdown." This is the emission of all frequencies as a plot of radiofrequency (rf) energy versus time following the pulse of dimethoxymethane with a short, powerful burst of rf energy. Fourier transformation of these data give the spectrum in Fig. 14.1 in Sec. 14.1 of the text.

SOLUTIONS TO PROBLEMS

Solutions to Focused Problems

14.1 (a) Use Eq. 14.6 in text Sec. 14.2:

$$\nu_p = \frac{\gamma_H}{2\pi}\mathbf{B}_p = \frac{26{,}753 \text{ rad G}^{-1}\text{ s}^{-1}}{2\pi \text{ rad}}(117{,}430 \text{ G}) = 500 \times 10^6\,\text{s}^{-1} = 500 \text{ MHz}$$

A 500 megahertz NMR spectrometer would be required.

 Whenever π enters into a formula, its units are radians (rad). The gauss unit is abbreviated G. Remember that hertz (Hz) is another way of representing the unit reciprocal seconds (s^{-1}).

(b) Use the same equation and solve for $\mathbf{B}_p$:

$$\nu_p = 900 \times 10^6 \text{ Hz} = \frac{\gamma_H}{2\pi}\mathbf{B}_p = \frac{26{,}753 \text{ rad G}^{-1}\text{ s}^{-1}}{2\pi \text{ rad}}\mathbf{B}_p$$

Solving,

$$\mathbf{B}_p = 211{,}373 \text{ G}$$

 Because ν_p and $\mathbf{B}_p$ are proportional, we could also have used the result of the last part and multiplied the value of $\mathbf{B}_p$ in the last part by the ratio of frequencies (= 900 MHz/500 MHz).

14.2 (a) The reduction in shielding at protons a is the field equivalent of, the chemical shift *difference*, 1005 Hz. Using Eq. 14.6 in text Sec. 14.2 for both protons a and the TMS protons, and then subtracting the two, we obtain

$$\nu_a - \nu_{TMS} = 1005 \text{ Hz} = \frac{\gamma_H}{2\pi}(\mathbf{B}_a - \mathbf{B}_{TMS}) = \frac{26{,}753 \text{ rad G}^{-1}\text{ s}^{-1}}{2\pi \text{ rad}}(\mathbf{B}_a - \mathbf{B}_{TMS})$$

Solving for $\mathbf{B}_a - \mathbf{B}_{TMS}$,

$$\mathbf{B}_a - \mathbf{B}_{TMS} = 0.24 \text{ G}$$

With an applied field of 70,500 G (given in the problem), this is 3.4×10^{-6} of the applied field $\mathbf{B}_0$, which is exactly the chemical shift in parts per million. From this example, we see that we could have calculated the chemical shift in gauss by multiplying the applied field (70,500 G) by the chemical shift in ppm and including the factor 10^{-6}. The point of this problem is to show how small chemical shifts are as a percentage of the applied field. This is why the units are "parts per million."

(b) Following the procedure used in part (a),

$$\mathbf{B}_b - \mathbf{B}_{TMS} = (10^{-6})(4.56 \text{ ppm})(\mathbf{B}_0) = 0.32 \text{ G}$$

14.3 The greater the chemical shift, the *less shielded* are the protons. Therefore, the protons at δ 5.5 are least shielded), and those at δ 1.3 are the most shielded.

14.4 (a) Using Eq. 14.9, we can calculate that the operating frequency v_0 at 21,100 G is 89.8 MHz. Applying Eq. 14.1 for the chemical shift of the methyl protons—protons a—of dimethoxymethane (3.35 ppm), we obtain Δv = the chemical shift in Hz = (3.35 ppm)(89.8 MHz) = 301 Hz.

(b) Following the procedure in part (a), we calculate that the operating frequency v_0 at 141,000 G is 600 MHz. Applying Eq. 14.1 for the chemical shift of the methyl protons of dimethoxymethane (3.35 ppm), we obtain Δv = the chemical shift in Hz = (3.35 ppm)(600 MHz) = 2011 Hz.

14.5 (a) Subtract the two chemical shifts in ppm and apply Eq. 14.1. That is,

$$\text{chemical shift difference} = \delta_2 - \delta_1 = \frac{\Delta v}{v_0} = \frac{45}{60} = 0.75 \text{ ppm}$$

(b) Using the procedure from part (a) gives

$$\text{chemical shift difference} = 45/300 = 0.15 \text{ ppm}$$

14.6 Parts (a)–(c) are answered in the text discussion that follows the problem (p. 737).

(d) Because Si is more electropositive (less electronegative) than any of the other atoms in the table, hydrogens near the Si are *more shielded*. Because chemical shift decreases with increased shielding, this means that $(CH_3)_4Si$ has a smaller chemical shift. A derivative $(CH_3)_xM$, in which M is an element more electropositive than Si, should have a negative chemical shift. $(CH_3)_2Mg$ and $(CH_3)_4Sn$ are two of several possible correct answers.

14.7 (a) The order is $C < B < A$. The protons of methylene chloride (dichloromethane, CH_2Cl_2) have the greatest chemical shift, because chlorine is more electronegative than iodine. The chemical shift of methylene iodide is greater than that of methyl iodide because two iodines have a greater chemical shift contribution than one.

(b) The order is $B < A < D < C$. The protons in compound C (1,1,2,2-tetrachloroethane) have the greatest chemical shift of compounds A–C because they are adjacent to two chlorines. In comparing C and D, compound C has the greater chemical shift because of the β chlorines in C, as well as the fact that the proton in C is a methine proton, whereas the protons in D are methylene protons. In comparing A to B, the protons in A are methylene protons that are α to a chlorine, whereas in B the protons are methyl protons that are β to chlorine. Finally, in comparing A with D, the protons in D are α to two chlorines, whereas the protons in A are α to only one chlorine. The effect of an α-chlorine is much greater than that of a β-chlorine, because chemical shift falls off with the distance of the electronegative group from the proton.

(c) The order is $B < C < D < A$. This follows the electronegativity order of the central atom; the protons in the compound containing the least electronegative element have the smallest chemical shift.

14.8 (a) Because protons H^a and H^b are diastereotopic, their chemical shifts are different.

(b) Because protons H^a and H^b are constitutionally nonequivalent, their chemical shifts are different.

(c) Because protons H^a and H^b are diastereotopic, their chemical shifts are different. Because the chair interconversion switches the positions of these two protons, their chemical shifts are the same when averaged out over time. (Sec. 14.8 addresses this point.)

(d) Because protons H^a and H^b are enantiotopic, their chemical shifts are identical.

14.9 The question asks essentially how many chemically nonequivalent sets of protons there are in each case.

(a) This compound has four chemically nonequivalent sets of protons; so, its NMR spectrum consists of four resonances. Only the methyl groups of the *tert*-butyl group are equivalent.

(b) This compound has two chemically nonequivalent sets of protons; hence, its NMR spectrum consists of two resonances.

(c) Three resonances are observed. Notice that the two methyl groups are diastereotopic and therefore chemically nonequivalent.

14.10 (a) This is the structure given in Problem 14.9(b).

$$\delta\ 1.08 \quad \overset{\overset{\displaystyle Cl}{|}}{(CH_3)_3C-C(CH_3)_2}$$
$$\delta\ 1.59$$

Another structure consistent with the data is shown to the right. This is a satisfactory answer at this point in the chapter. However, as you will learn in Sec. 14.4 of the text, the spectrum of this compound will consist of a number of lines that result from a phenomenon called *splitting*, which occurs when protons are on *adjacent carbons*.

$$CH_3CH_2-\overset{\overset{\displaystyle Cl}{|}}{\underset{\underset{\displaystyle CH_2CH_3}{|}}{C}}-CH_2CH_3$$

(b)

$$\delta\ 6.55 \quad \overset{\overset{\displaystyle CH_3}{|}}{Cl_2CH-\underset{\underset{\displaystyle CH_2Cl}{|}}{C}-CH_3} \quad \overset{\displaystyle \diagdown \delta\ 1.99}{}$$
$$\delta\ 4.31$$

14.11   The spectrum will contain three resonances, which arise from the protons labeled *a, b,* and *c*. Protons *a*, integrating for 9*H*, occur near δ 3.2. Protons *b*, integration for 3*H*, will have nearly the same chemical shift. Because these protons are chemically nonequivalent to protons *a*, their chemical shift should be different, but only slightly. Protons *c*, integrating for 6*H*, should have a somewhat greater chemical shift than protons *a* or *b*—probably about δ 3.5—because methylene protons have a slightly greater chemical shift than methyl protons in similar chemical environments.

$$\overset{\displaystyle b}{\overset{\displaystyle OCH_3}{|}}$$
$$\overset{a}{CH_3}O-\overset{c}{CH_2}-\overset{}{\underset{\underset{\underset{\displaystyle c}{\displaystyle CH_2-OCH_3}}{|}}{C}}-\overset{c}{CH_2}-\overset{a}{OCH_3}$$

14.12   (a)   This is a quantitative analysis problem that can be solved by assigning the resonances and using their integrals. The resonance at δ 1.8 is that of *tert*-butyl bromide, and the resonance at δ 2.2 is that of methyl iodide. (These assignments follow from Table 14.1 and Fig. 14.4 in text Sec. 14.3C.) The ratio of methyl iodide to *tert*-butyl bromide is 15:1. This follows from the fact that there are three times as many protons in the *tert*-butyl bromide molecule as in the methyl iodide molecule; thus, for a given concentration, the *tert*-butyl bromide gives a resonance that is three times as intense as that of methyl iodide.

$$\frac{\text{5-proton resonance for } CH_3I}{\text{3 protons per molecule}} \times \frac{\text{9 protons per molecule}}{\text{1-proton resonance for } (CH_3)_3Br} = \frac{\text{15 molecules of } CH_3I}{\text{1 molecule of } (CH_3)_3Br}$$

The mole percent of CH_3I is the number of moles of CH_3I (15) divided by the total moles of both compounds (16) expressed as a percent. Therefore, the mole percent of CH_3I is (15/16) × 100, or 94 mole percent; the mole percent of $(CH_3)_3CBr$ is (1/16) × 100, or 6 mole percent.

(b)   The $(CH_3)_3CBr$ impurity in CH_3I is more easily detected, because a given mole fraction of $(CH_3)_3CBr$ gives a resonance that is three times as strong as the resonance for the same amount of CH_3I, as the solution to part (a) of this problem demonstrated.

14.13   In the following discussions, the integral is not mentioned. It corresponds to the number of protons under observation in each case.

(a)   The resonance of the CH_3 protons should be a doublet (one neighboring proton) in the alkyl region, which is δ 0.7–1.7. Because of the β-chlorines, this doublet will probably be at the higher end of this range. The —$CHCl_2$ proton is a quartet at considerably greater chemical shift. Fig. 14.4 (text Sec. 14.3C) puts δ 4.2 as the limit on the chemical shift of a CHCl proton; because there are two chlorines, the resonance would be at higher shift. (The shift of CH_2Cl_2 in Table 14.1 is δ 5.3; a methine proton would be shifted even more.)

(b)   The protons of the $ClCH_2$— group would be a doublet at δ 4.2 or higher. The protons of the —$CHCl_2$ group would be a triplet at considerably greater chemical shift. This chemical shift would be similar to, and probably somewhat greater than, the chemical shift of the —$CHCl_2$ protons in part (a). (The greater shift would be caused by the β-chlorine.)

(c) The methyl groups would be a large doublet in the δ 0.7–1.7 range, probably near the middle. (The β-halogen would increase the shift, but methyl protons are usually near the low end of the ranges given in Fig. 14.4, text Sec. 14.3C.) The methine proton would be a septet near δ 3.0.

(d) This compound, 1,2-dimethoxyethane, should have two resonances. The methylene resonances for the boldfaced protons $CH_3OCH_2CH_2OCH_3$ are equivalent and should be a singlet in the δ 3.5 region. The methyl groups should be one large singlet near δ 3.2.

(e) In this compound (oxetane), the methylenes closest to the oxygen are a triplet near δ 3.5, and the central —CH_2— should be a quintet (a five-line pattern) near δ 1.8.

(f) The spectrum of this compound consists of two singlets. The —CH_2— resonance occurs at a chemical shift greater than δ 4.2 (because of the β-chlorine), and the methyl groups should appear in the δ 0.7–1.7 region, probably towards the higher end because of the chlorines.

 More precise estimates of chemical shift can be made using the methods discussed in Further Exploration 14.1 in this manual.

14.14 There are four possibilities for the spin of three neighboring equivalent protons b; these are shown in the following table. There are three ways in which two spins can be the same. This table shows that the resonance for protons a would be split by the three protons b into a quartet whose lines are in the intensity ratio 1:3:3:1.

	b protons											
	1	2	3	1	2	3	1	2	3	1	2	3
	all (+)			two (+), one (−)			one (+), two (−)			all (−)		
spin combinations	+½	+½	+½	+½ +½ +½ −½	+½ −½ +½	−½ −½ +½	+½ −½ −½	−½ +½ −½	−½ −½ +½	−½	−½	−½

14.15 (a) This structure is ruled out by several facts. First, there are *eight* nonequivalent sets of protons in this candidate; notice that the two CH_3O— groups on the left, as well as the CH_2 protons, are diastereotopic, and therefore nonequivalent. The actual NMR spectrum shows five sets of protons. Also, this candidate should have three methoxy singlets unless two of them coincidentally overlap. Furthermore, the methyl group on the right should be a doublet, whereas it is actually a singlet.

(b) This candidate would also be expected to have five nonequivalent sets of protons. (Can you identify them?) However, this would also not show two triplets. Assuming the diastereotopic CH_2 protons have different chemical shifts, the would appear as two doublets.

14.16 (a)
δ 1.03   δ 3.40
$H_3C—CH_2—CH_2—Br$
δ 1.88

(b)
δ 3.98   δ 5.87
$ClCH_2—CHCl_2$

(c)
CH_2Br
|
$BrCH_2CCH_2Br$
|
CH_2Br

14.17 The molecular formula indicates no degrees of unsaturation; this rules out rings or double bonds. The IR spectrum rules out an alcohol. All of the oxygens, then, are associated with ether groups. The 9H singlet at δ 3.2 requires three equivalent CH_3O groups. The unsplit 3H resonance at δ 1.2 indicates a methyl group on a quaternary carbon. The 6H triplet at δ 3.4 suggests three equivalent CH_2 groups attached to an oxygen. These CH_2 groups, from their splitting, are also adjacent to three other CH_2 groups, which must be the six protons at δ 1.4, and these are also attached to a quaternary carbon. because they are split only by the CH_2 groups. A structure with this number of carbons and only four resonances must involve a high degree of symmetry. A structure that fits all the data is shown at right at the bottom of the previous page.

δ 1.2
CH_3 δ 3.2
$CH_3OCH_2CH_2—C—CH_2CH_2OCH_3$
$CH_2CH_2OCH_3$
δ 1.4   δ 3.4

14.18   (a)   The reaction is the peroxide-promoted HBr addition to the alkene (Sec. 10.1 in the text):

$$BrCH_2CH{=}CH_2 \ + \ H{-}Br \ \xrightarrow{\text{peroxides}} \ BrCH_2CH_2CH_2Br$$

3-bromo-1-propene   **1,3-dibromopropane**
                                                  compound A

(b)   The chemical shifts and splittings are indeed consistent with this structure. The resonance of the central methylene group is predicted to be a quintet in the δ 0.7–1.7 range (Fig. 14.4); the β-bromines will probably bring this to the high end of this range or beyond; and the resonance for the two methylenes α to the bromines is predicted to be a triplet near δ 3. The splittings agree exactly, and the observed chemical shifts are acceptably close.

14.19   (a)   The resonance for protons H^d will be a singlet. The resonances for protons H^a, H^b, and H^c will all be triplets at their respective chemical shifts since the coupling constants for protons $a - c$ are all equal, the splitting will follow the $n + 1$ rule.

(b)   The resonance for protons H^d will be a singlet as in part (a). The resonance for proton H^a will be split into a triplet by the two H^b protons, by whatever the coupling constant between those two protons is. The resonance for protons H^c will also be split into a triplet by the two H^b protons, but the triplet will have different spacing between its peaks as the result of the different coupling constant between protons b and c. The resonance for the H^b protons will be split into a doublet of doublets instead of a triplet as in part (a) as a result of the unequal coupling constants.

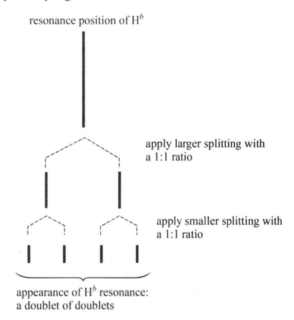

resonance position of H^b

apply larger splitting with a 1:1 ratio

apply smaller splitting with a 1:1 ratio

appearance of H^b resonance:
a doublet of doublets

Note that in the above solution, the larger of the two coupling constants was applied first, followed by the smaller. It is arbitrary which gets applied first; doing the reverse (smaller followed by larger) will result in the same splitting pattern. Prove it to yourself!

14.20   The resonances for protons H^c and H^a will be doublets at their respective chemical shifts, and the resonance of proton H^b will be split into a quartet ($J = 6.6$ Hz) by the three protons H^c, and each line of this quartet will be split into a doublet ($J = 3.6$ Hz) by the single proton H^a. The resonance for H^b can be analyzed with a splitting diagram:

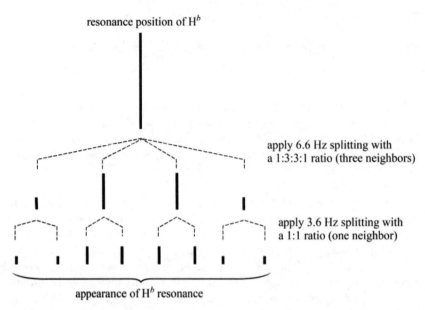

The complete spectrum is sketched here:

14.21 (a) Because protons H^b and H^c are diastereotopic, they are chemically nonequivalent; therefore, counting the methyl group, there are four chemically nonequivalent sets of protons. The methyl group is split into a doublet by H^a. The resonance for proton H^a is split into a quartet by the methyl protons, and each line of this quartet is split into a triplet by H^b and H^c, to give a "quartet of triplets," or twelve lines, for proton H^a. (The triplet is the result of two overlapping doublets with $J_{ab} = J_{ac}$; see Fig. 14.11 in text Sec. 14.5A for a related case.) Because H^b and H^c are diastereotopic, they absorb at different chemical shifts. Proton H^b is split by proton H^a into a doublet, and each line of this doublet is split into another doublet by H^c. Therefore, proton H^b is a doublet of doublets, or four equally intense lines. The same is true of H^c—its resonance will be four equally intense lines at a slightly different chemical shift from the resonance of H^b. Several of these patterns may overlap to create a very complex-looking spectrum.

(b) The difference from part (a) is that, in the quartet resonance for proton H^a, each line of this quartet is split into a *doublet* by H^b and into *another doublet* by H^c, to give a "quartet of doublet of doublets," or sixteen lines. Coincidental overlaps could reduce this number.

14.22 (a) The integration of the δ 3.42 resonance and its splitting show that compound *A* has the partial structure —CH_2CH_2Br. The δ 0.91 resonance indicates the two methyl groups within an isopropyl group. This means that compound *A* can only be

$$(CH_3)_2CHCH_2CH_2—Br$$

1-bromo-3-methylbutane
(isoamyl bromide; compound *A*)

Compound *B*, from the chemical shifts, has no hydrogens α to the bromine. Therefore, this is a tertiary alkyl halide. We see two methyl singlets and an ethyl quartet, which determine the structure.

$$CH_3CH_2—\underset{\underset{CH_3}{|}}{\overset{\overset{CH_3}{|}}{C}}—Br$$

2-bromo-2-methylbutane

(b)   Running the spectrum of compound *A* at higher field could simplify the complexity of the δ 1.7-1.8 resonance. This region contains the combined resonances for the following protons:

δ 1.7–1.8
$$(CH_3)_2\overbrace{CH—CH_2}—CH_2—Br$$

The difference in chemical shift between these protons, in Hz, would increase (double), but the coupling constants would remain unchanged. Thus, the likelihood that Eq. 14.22 (text Sec. 14.5B) is valid would be increased. In the event that all resonances are first-order, and all coupling constants are nearly equal, the —CH₂— group would become a quartet and the methine CH of the isopropyl group would remain complex (21 lines even if first-order, barring coincidental overlaps).

14.23   The δ 1.2–1.5 resonance corresponds to the following protons in 1-chlorohexane:

δ 1.2–1.5
$$H_3C—\overbrace{CH_2—CH_2—CH_2—CH_2}—CH_2—Cl$$

1-chlorohexane

The complexity, and thus the inability to assign an exact chemical shift, arises because of complex splitting of the protons of interest. To solve this problem, we would prepare the deuterium-substituted analogs, in which this splitting is not present.

$$D_3C—CH_2^a—CD_2—CH_2^c—CH_2^d—CH_2—Cl \qquad H_3C—CD_2—CH_2^b—CD_2—CH_2—CH_2—Cl$$

compound *A*   compound *B*

In the NMR spectrum of compound *A*, protons *a* would be a singlet, and protons *c* would be a simple triplet. (These are well separated from protons *d*; see Fig. 14.13, text Sec. 14.5B.) In the NMR spectrum of compound *B*, protons *b* would be a singlet.

14.24   (a)   Assuming that in the absence of D₂O, the sample is very dry, the spectrum of 3-methyl-2-buten-1-ol would change as follows as the result of a D₂O shake:

3-methyl-2-buten-1-ol

(b)   Assuming that in the absence of D₂O, the sample is very dry, the spectrum of 3,3-dimethyl-2-butanol would change as follows as the result of a D₂O shake:

quartet of
doublets   doublet   quartet   no resonance in ^1H NMR
↓   ↓   ↓   ↓

$(CH_3)_3C$—CH—OH $\xrightarrow{D_2O \text{ shake}}$ $(CH_3)_3C$—CH—OD
        | |
        CH_3 CH_3

(A "quartet of doublets" can also be referred to as a "doublet of quartets.")

14.25 The two protons in the δ 4.6–4.8 region are vinylic protons, and they are not chemically equivalent. Their very small splitting is consistent only with geminal (that is, H_2C=) protons. Thus, we have the partial structure

The resonance at δ 1.8 integrates for 3 protons, and it is in the allylic region. This means that there is an allylic methyl group. The resonance at δ 1.03 integrates for nine protons. Because it is a singlet, this can only be a *tert*-butyl group. This completes the structure:

δ 1.3 (3*H*)
δ 4.6–4.8 (2*H*)

2,2,3-trimethyl-1-butene

14.26 (a) Nothing but ethyl groups and one oxygen! This can only be diethyl ether, CH_3CH_2—O—CH_2CH_3.

(b) The large coupling constants for both sets of protons indicate splitting by fluorine. The fact that they are triplets suggest that they are adjacent to the carbon bearing both fluorines, but not on carbons adjacent to each other; otherwise they would split each other as well. The δ 3.63 protons are α to the chlorine. These considerations demand the following structure:

δ 1.75   δ 3.63
         F
         |
H_3C—C—CH_2—Cl
         |
         F

1-chloro-2,2-difluoropropane

Remember that the fluorines are not observed in proton NMR. What would the ^{19}F NMR spectrum look like? *Answer:* a quartet of triplets (or triplet of quartets) with two splittings, coupling constants $J = 13$ Hz and 17.5 Hz, respectively.

14.27 The spectrum of chloroethane would resemble that of bromoethane (Fig. 14.6, text Sec. 14.4). It would consist of a typical ethyl pattern—a triplet (for the CH_3) and a quartet (for the CH_2), except that the quartet for the CH_2 in the chloroethane spectrum would be at a somewhat greater chemical shift than it is in bromoethane. In fluoro-ethane, the resonance of the methyl group would be a *triplet of doublets*. (Splitting by the CH_2 group gives three lines; each of these are split into two by the fluorine for a total of six.) The resonance of the CH_2 group would be a *doublet of quartets*. (Splitting by the fluorine gives a widely spaced doublet; each line of the doublet is split into a quartet by the CH_3 group.) The chemical shift of the CH_2 resonance in fluoroethane would be expected to be somewhat greater than that of the CH_2 group in chloroethane.

14.28 (a) The unknown is *tert*-butyl alcohol, $(CH_3)_3C$—OH. Notice the absence of an α-proton.

(b)   The exchangeable proton suggests the presence of an alcohol. The resonance at δ 5.40 is a vinylic proton adjacent to two other protons, and the resonance at δ 4.10 arises from two protons adjacent to both the oxygen and the vinyl group. (Their chemical shift is too great to be caused by either the oxygen or the vinyl group alone.) Evidently the sample is wet enough that the splitting between the —OH proton and the α-proton is not observed; this situation is very common. (Contrast this with the solution to Problem 14.24a, which involves the same compound.) 3-Methyl-2-buten-1-ol, $(CH_3)_2C$=$CHCH_2OH$, is the structure consistent with the data.

14.29   At room temperature, the three conformations of 1-bromo-1,1,2-trichloroethane are in rapid equilibrium. Therefore, the NMR spectrum of this compound is a singlet whose chemical shift reflects the environment of the proton averaged over time across all conformations. Any changes observed at low temperature would be related to the fact that the interconversion between the different conformations is slowed, and thus certain conformations can be individually observed at the lower temperature. There are three such conformations, two of which are enantiomers:

enantiomeric conformations
have identical NMR spectra

In conformation A, protons H^a and H^b are equivalent and have the same chemical shift. In either one of the enantiomeric conformations, H^a and H^b are chemically nonequivalent because they are diastereotopic. In either of these conformations, the resonance of H^a is split by H^b and is therefore a doublet; likewise, the resonance of H^a is split by H^b and is also a doublet. Consequently, the spectrum of conformation B should consist of two doublets, that is, four lines of equal intensity, assuming no overlap. Because conformations B and C are enantiomeric, their spectra will be identical. Therefore, the total spectrum should consist of one line for conformation A, and two doublets for conformations B and C, for a total of five lines at low temperature, assuming no overlaps. The relative integrals depend on the amount of form A relative to the amounts of forms B and C; the latter two forms will be present in equal amounts.

14.30   The resonance for the methyl group at room temperature is a singlet. When the temperature is lowered, the resonance of this methyl group should consist of two singlets, one for the conformation of 1-chloro-1-methylcyclohexane in which the methyl group is axial, and one for the conformation in which the methyl group is equatorial. The relative integrals of the two singlets will be proportional to the relative amounts of the two conformations. Because chlorine and methyl are about the same size, there should be about equal amounts of the two conformations.

1-chloro-1-methylcyclohexane

14.31 Because of its symmetry, 4-heptanol has only four chemically nonequivalent sets of carbons, and therefore its spectrum consists of four lines, and is consequently spectrum 1. All of the carbons of 3-heptanol are chemically nonequivalent; therefore, the carbon NMR spectrum of this compound consists of seven lines, and is therefore spectrum 2. These structures show equivalent carbons with the same numbers.

$$CH_3CH_2CH_2CHCH_2CH_2CH_3$$
$$\quad 1 \quad 2 \quad 3 \quad 4 \quad 3 \quad 2 \quad 1$$

4-heptanol

$$CH_3CH_2CHCH_2CH_2CH_2CH_3$$
$$\quad 1 \quad 2 \quad 3 \quad 4 \quad 5 \quad 6 \quad 7$$

3-heptanol

14.32 The carbon NMR spectrum of 1,1-dichlorocyclohexane should consist of four lines, whereas that of *cis*-1,2-dimethylcyclohexane should consist of three lines. (Equivalent carbons have the same numbers in these structures.)

1,1-dichlorocyclohexane

cis-1,2-dichlorocyclohexane

 Because the chair interconversion is rapid, we can use the planar structures of cyclohexanes in resolving questions of symmetry. Can you deduce how the carbon NMR spectra of these compounds would differ—if at all—at very low temperature?

14.33 (a) This compound has five nonequivalent sets of carbons and therefore should give five lines, not three as observed.

(b) This compound should have a carbon NMR spectrum with three lines, as observed. Furthermore, the attached hydrogen ratio is also consistent with this structure. What is *not* consistent, however, is the chemical shift information. The δ 112.9 resonance suggests a carbon bound to *more than one* oxygen. In this structure, no carbon is bound to more than one oxygen, whereas in the correct structure, the methine (CH) carbon is bound to three oxygens.

14.34 (a) First, draw the structures!

trans-1-chloro-2-butene

2-chloro-1-butene

Two reasons that *trans*-1-chloro-2-butene (*A*) is not consistent with the data are that this alkene would have a spectrum in which the integral indicates a total of *two,* not three, vinylic protons, and *two,* not one, protons on the carbon α to the chlorine.

　　Among the reasons that 2-chloro-1-butene (*B*) does not fit the data are that it offers no explanation for the δ 4.5 absorption, and it has only two vinylic protons, whereas the observed NMR spectrum indicates three.

 When considering possible structures for consistency with NMR data, some beginning students try to wrestle with splitting data before they have considered other aspects of the data that are more straightforward and at least as powerful. Notice in the foregoing solution that the alternative structures can be clearly ruled out on the basis of relative integral or number of absorptions.

(b) The assignments for the vinylic protons are determined by both relative integrals and chemical shifts. Terminal vinylic protons always occur, other things being equal, at a smaller chemical shift than internal vinylic protons, because alkyl substitution results in an increased chemical shift. Secondly, the integral for the δ 5.0–5.3 resonances is twice that of the δ 5.9–6.1 resonances, and there are twice as many terminal vinylic protons as internal ones. Finally, the δ 5.0–5.3 resonances show splittings consistent with *both* a cis and a trans relationship to other vinylic protons, and the only way this can occur is if there are two such protons—one cis to the remaining vinylic proton and the other trans.

And this brings us to the issue of detailed assignment. Trans splittings are larger than cis splittings. Because the vinylic proton at lower chemical shift (about δ 5.1) in the δ 5.0–5.3 set has the smaller splitting, it is the one that has the cis relationship to the internal vinylic proton. The vinylic proton at δ 5.3 has the trans relationship. This assignment also makes sense because the vinylic proton with the larger chemical shift is closer to the chlorine.

split by both vinylic Hs and by adjacent alkyl H (8 lines in theory; 2 lines overlap; therefore, 7 are observed)

(c) (It is helpful to use graph paper for this question.) The proton signal at δ 5.9–6.1 is split by the three adjacent protons: the alkyl hydrogen H^a, $J = 6.6$ Hz, the cis vinyl proton H^c, $J = 10.1$ Hz, and the trans vinyl proton H^t, $J = 17.1$ Hz. The splittings can be applied successively in any order; we have chosen to apply them in decreasing order (but the same pattern results regardless). See that with this combination of coupling constants, the center two lines are close enough to overlap and appear as one line in the spectrum. So, while there are "really" 8 lines, only 7 lines appear.

proton signal at δ 5.9-6.1

apply 17.1 Hz splitting from H^t

apply 10.1 Hz splitting from H^c

apply 5.3 Hz splitting from H^a

eight lines, but middle two appear as one, for seven visible peaks

14.35 Compound A has a different formula from that given in the study problem, so we can rule it out immediately. Beyond that, it would be easy to believe that the methyl groups of A could account for the twelve-proton singlet at δ 1.22, and the two OH groups for the two-proton exchangeable resonance at δ 1.96; but this leaves no hydrogens to account for the δ 1.57 resonance. In other words, structure A has two nonequivalent sets of hydrogens, whereas the NMR spectrum indicates at least three. Structure A also has two nonequivalent sets of carbons, whereas the carbon NMR spectrum indicates three.

Structure B is also ruled out by its molecular formula, which is $C_8H_{16}O_2$. But the NMR spectrum is also not consistent with structure B, which should have two methyl singlets, each integrating for 6H, and one methyl singlet integrating for 3H, along with one exchangeable OH proton. Although it would have the required three resonances, the integration ratio (6:3:1) is different from that observed (12:4:2 or 6:2:1). This difference is well within the ability of the spectrometer to differentiate. But the carbon NMR is even more definitive. Structure B has five nonequivalent sets of carbons, whereas the carbon NMR spectrum indicates three sets. Even if the methyl resonances overlapped accidentally, which is unlikely, the attached-hydrogen analysis for B would predict no carbon with two attached protons.

Solutions to Skills Objectives Problems

14.36 The four pieces of information are
1.   The **number of individual resonances**, or peaks (not counting splitting), that are determined by how many different sets of nonequivalent protons are present;
2.   The **chemical shift**, which is used to determine what functional groups are near the proton(s) under observation;

3.   The **integral**, which is used to determine how many protons are being observed relative to the total number in the molecule;

4.   The **splitting** patterns, which are used to determine connectivity—that is, how many protons are on atoms adjacent to the carbon that bears the proton being observed.

14.37   (a)   NMR spectroscopy requires that the sample be situated in a magnetic field before absorption of electromagnetic radiation can occur. The field establishes the energy difference between nuclear spins.

(b)   Some of the protons change from a lower-energy spin to a higher-energy spin.

(c)   The circulating electrons within the molecule generate their own magnetic fields which shield the protons somewhat from the full externally applied magnetic field.

(d)   See Sec. 14.7A and Fig. 14.15. At a vinylic proton, the magnetic field induced by the circulating π electrons of the double bond reinforces the applied magnetic field of the NMR instrument. As a result, vinylic protons have NMR absorptions at relatively large chemical shifts.

(e)   See Sec. 14.7C and Fig. 14.19. At an acetylenic proton, the magnetic field induced by the circulating π electrons of the triple bond opposes the applied magnetic field of the NMR instrument. As a result, acetylenic protons have NMR absorptions at relatively small chemical shifts.

(f)   The coupling constant of vicinal protons varies as the angle between them changes, according to the Karplus relationship (Eq. 14.16, text Sec. 14.5A). You can use this formula to calculate the coupling constant (J) for each set of protons, or use the Karplus curve plotted in Fig. 14.9 in Sec. 14.5A to estimate the coupling constant for each dihedral angle (θ). By either method, the dihedral angle of 20° gives J_{ab} of about 9 Hz, and the dihedral angle of 80° gives J_{bc} of about 1.5 Hz. So, $J_{bc} < J_{ab}$.

(g)   The chemical shift in ppm (δ) does not change with operating frequency ν_0 because it is *defined* as the proportionality constant that relates the chemical shift in Hz, $\Delta\nu$, to the operating frequency ν_0. (See Eq. 14.1 in text Sec. 14.1.) Because $\Delta\nu$ and ν_0 are proportional, δ for a given resonance is the same at any operating frequency.

(h)   The coupling constant J is independent of the applied field (and also the operating frequency).

(i)   The chemical shift difference *in frequency units* (Hz) between the two protons must be considerably greater than the coupling constant between these protons (in the same units). This relationship is embodied in Eq. 14.22 in text Sec. 14.5B. More simply, the splitting patterns of the two coupled protons should not overlap.

(j)   As noted in Sec. 14.9A, the operating frequency for carbon NMR is about one quarter that of proton NMR at the same field strength. This is a consequence of the smaller magnetogyric ratio of carbon, which is 25% of that for protons (Eq. 14.34, text p. 774, and Table 14.4, text p. 775). For a spectrometer operating at 500 MHz for proton NMR, the operating frequency for carbon NMR would be about 125 MHz.

(k)   NMR and MRI both use high magnetic fields and radiofrequency radiation. Unlike structural NMR, which involves chemical shifts and splitting between coupled protons, MRI monitors the relaxation rates of protons in water to produce images of soft tissues in medical applications. See text Sec. 14.12 for additional details.

14.38   (a)   Use Eq. 14.10 in text Sec. 14.3B to convert the chemical shifts (δ) of H^a and H^b into frequencies ($\Delta\nu$) at a ν_0 of 300 MHz. Solving for ν_a gives 1035 Hz and solving for ν_b gives 1080 Hz. So, the chemical shift difference of H^a and H^b is 45 Hz. (Alternatively, you can take the difference of 3.60 − 3.45 = 0.15 ppm, and use the same equation to solve for $\Delta\nu$ to get the same answer.)

(b)   See Sec. 14.4A in the text. Statement (3) is true: because coupling between protons is mutual, then $J_{ba} = J_{ab} = 6.0$ Hz.

(c)   The splitting pattern for protons H^b will be a triplet. The table of spin combinations is identical to the one shown in Display 14.14, text Sec. 14.4B. The triplet will have the same general shape as the one shown in Fig. 14.8 (three lines, with the middle one twice as tall as the other two).

(d)   The splitting pattern of H^a will also be a triplet.

(e)   In general, obtaining a spectrum on a higher-field NMR spectrometer (in this case, 300-MHz versus 60-MHz) yields a more well-resolved spectrum. "Well-resolved" means that there is generally less overlap between signals.

    In this specific case, we have already determined in parts (a) through (d) that H^a and H^b produce triplets at chemical shifts of δ 3.45 and δ 3.60, respectively. On a 300-MHz spectrometer, these peaks appear at frequencies of 1035 Hz and 1080 Hz, respectively. If you rework part (a) using ν_0 of 60 MHz, you see that the peaks would appear at 207 Hz and 216 Hz, respectively, on a 60-MHz NMR spectrum. Ignoring the splitting of each of these signals for a moment, we can plot both peaks on a spectrum using both scales.

Here we can see how chemical shift in ppm is independent of the operating frequency of the NMR spectrometer. While the absorption frequencies, $\Delta\nu$, of each proton differ depending on the instrument, the chemical shift in ppm (δ) is the same regardless.

It's not until we consider splitting that the differences between the spectra arise. Each signal will be split into a triplet as determined in parts (c) and (d). If we take each absorption and sequentially add and subtract the coupling constant, we get a triplet pattern where the middle line appears at the absorption frequency, and the outer two lines are given by $\Delta\nu + J$ and $\Delta\nu - J$. Therefore, on the 300-MHz spectrum, the two triplets appear like this:

The two triplets are well *resolved*, or separated, from each other.

Constructing the triplet pattern on the 60-MHz NMR still gives two triplets, but the relative spacing between the peaks of each triplet is much larger, and they cross, or overlap.

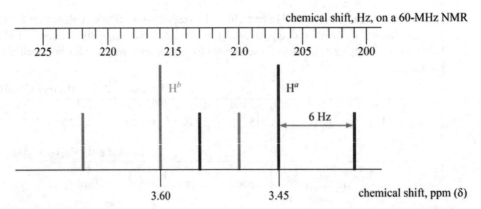

The reason the peaks in the triplets are spaced so much further apart is because the ratio of the size of the coupling constant is much larger relative to the separation of the individual resonances. Therefore, on the 60-MHz spectrometer, we are less likely to see first order splitting. We see that the triplets are *less resolved* relative to the spectrum obtained at 300-MHz.

14.39 (a) The NMR spectrum of a very dry sample would show splitting between protons 1 and 2. Therefore, the resonance of proton 1 should be a triplet; the resonance of protons 2 should be either a quartet (if $J_{12} = J_{23}$) or a doublet of triplets (if $J_{12} \neq J_{23}$), where the J's are coupling constants. In a wet sample, the splitting associated with proton 1 is obliterated. In that case, the resonance of proton 1 should be a singlet; the resonance of protons 2, which is split only by protons 3, should be a triplet. The resonance of protons 3 should be a triplet in both the wet and the dry sample.

$$\overset{3}{Cl}—\overset{}{CH_2}—\overset{2}{CH_2}—\overset{1}{OH}$$

2-chloroethanol

(b) Following a D$_2$O shake, the resonance of proton 1 should disappear; and the resonance of protons 2 should be the same as it is in the wet sample—that is, a triplet.

14.40 Both stereoisomers are based on a *trans*-decalin-like structure. You could determine which of the stereoisomers is which by inspecting the indicated protons. In compound A, the protons are anti to one another, and therefore would have a coupling constant between 8–15 Hz (see Sec. 14.5A and Fig. 14.9). In compound B, they are gauche to one other and would have a coupling constant between 2–5 Hz.

Note that because of the nearby sp^2-hybridized carbon (the one that is double-bonded to the oxygen), the two protons that we've focused on may not be exactly in gauche or anti orientations relative to each other. However, our analysis holds: we can inspect these two signals and expect a larger coupling constant between the protons in compound A because the angle between them is greater.

14.41 For each of these problems, you need to be systematic. Because the number of lines in a splitting pattern is one more than the number of nearby nonequivalent protons, you might try noting the number of signals seen, then applying an "$n - 1$ rule" to each signal. For example, if a sextet is given ($n = 6$), then you know that there are five ($6 - 1$) nearby nonequivalent protons (these, of course, would need to be on more than one attached carbon because one carbon can't have that many protons). After you've established $n - 1$ for each signal, then experiment with possible structures, using X and Y to fill out carbon's octet when necessary.

(a) CH_3CH_2—X

(b) $(CH_3)_3C$—X

(c) $(CH_3)_2CH$—X

(d)   A "triplet, quintet" pattern means one set of protons is split by 2 protons, and another set is split by 4 protons. Because a carbon cannot have 4 protons attached and also be bonded to another carbon, split this group of 4 into two groups of 2. Then we can propose the following structure: X—CH_2—CH_2—CH_2—X. The middle set of hydrogens gives the quintet, and the first and third sets give the same triplet because they are equivalent.

(e) CH_3CHX_2

(f)   Use the answer to part (d) as a starting point. Here there are two triplets, so while the general structure is similar, we can differentiate between the first and third CH_2 sets by adding X to one and Y to the other to give: X—CH_2—CH_2—CH_2—Y.

In theory, if X ≠ Y then the coupling constants of the first and third CH_2 groups may be different. This would cause the middle CH_2 quintet to possibly become a multiplet (a triplet of triplets). However, the problem says to assume that the splitting patterns all have the same coupling constants, so we can assume that the middle CH_2 signal remains a quintet.

(g)   X—CH_2—CHY—CH_3

(h)   X—CH_2—CH_2—CH_3

(i) CH_3—X

14.42   (a)   The compound is neopentane (2,2-dimethylpropane), $(CH_3)_4C$.

(b)   The compound is cyclopentane. The chemical-shift difference between (a) and (b) is due to largely to the typical difference between methylene and methyl protons.

(c)   The compound is 1,1-dimethoxyethane, $(CH_3O)_2CH$—CH_3.

(d)   The D_2O exchange result indicates the presence of an alcohol. Given that the splitting indicates an ethyl group, the integral indicates three equivalent ethyl groups. Because no resonance occurs in the δ 3–4 region, there is no α proton, that is, no **CH**—O—. Thus, the alcohol is a tertiary alcohol. The compound is

$$CH_3CH_2$$
$$|$$
$$CH_3CH_2-C-OH$$
$$|$$
$$CH_3CH_2$$

3-ethyl-3-pentanol

(e)   The spectrum shows indicates vinylic protons with no cis or trans splittings. So, a partial structure is

$$H_2C{=}C\Big\langle$$

The resonances at δ 2.0 and δ 1.8 are nearly singlets and integrate for 2 and 3 protons, respectively. These protons, from their chemical shifts, must be allylic. We now have the partial structure

$$H_2C{=}C\Big\langle{}^{CH_2-}_{CH_3}$$

The very small splittings of these groups are evidently allylic splittings (Table 14.3, text Sec. 14.7A). The remaining resonance is a nine-proton singlet, which can only be a *tert*-butyl group. This completes the structure, which is consistent with the result of catalytic hydrogenation.

2,4,4-trimethyl-1-pentene   **2,2,4-trimethylpentane**

(f) The compound has one degree of unsaturation and a *tert*-butyl group. This, plus the requirement for two chlorines and a partial structure CH_2—CH required by the splitting leaves only the following possibility:

$$(CH_3)_3C — CH_2 — CH = CCl_2$$

1,1-dichloro-4,4-dimethyl-1-pentene

(g) The two protons are evidently equivalent (a single chemical shift), and, from the coupling constants, they must be close to the fluorines, but not on the same carbon as the fluorines. The compound is $BrCH_2CF_2Br$, 1,2-dibromo-1,1-difluoroethane.

Why not the following compound?

1,2-dibromo-1,2-difluoroethane

The proton NMR spectrum of this compound would be a doublet of doublets. (Why?) In addition, with *two* halogens on the same carbon, the chemical shift would be much greater. Furthermore, one of the H—F coupling constants would be considerably larger because of the presence of a fluorine on the same carbon.

(h) The compound is F_3C—CH_2—I (1,1,1-trifluoro-2-iodoethane). The splitting of the protons is caused by the fluorines.

(i) Because there are sixteen protons, the integrals account for 2, 12, and 2 protons, respectively. The compound is $(CH_3O)_2CHCH_2CH(OCH_3)_2$ (1,1,3,3-tetramethoxypropane), better known to chemists as malonaldehyde dimethyl acetal. What makes this problem tricky—and the whole point of the problem—is that not all of the equivalent protons of a given type are on the same carbon. For example, the two CH protons are equivalent and therefore have exactly the same chemical shift; they are split into a triplet by the adjacent CH_2 protons. Likewise, the CH_2 protons are split by the two adjacent protons into a triplet.

(j) This compound has two degrees of saturation, which could be two double bonds, a ring and a double bond, a triple bond, or other combinations. The evolution of a gas (ethane) upon reaction with EtMgBr suggests that there is an acidic proton present. That proton cannot be attached to an O, however, as it is also given that there is no evidence of an —OH in the IR spectrum. So, this compound must contain an internal O, as in an ether (R—O—R). The 3H singlet suggests an isolated methyl group, and its chemical shift is consistent with a CH_3O—. So, with the three remaining carbons and two degrees of unsaturation to still account for, the other half of the ether could reasonably be —CH_2—C≡CH. The splitting is due to longer-range coupling between the alkyne and CH_2 protons as described by Display 14.26 in text Sec. 14.7B. The terminal alkyne's acidic proton reacts with the Grignard reagent to evolve ethane. Therefore, the structure is $CH_3OCH_2C≡CH$.

14.43 In a proton-decoupled carbon NMR spectrum, the carbon–proton couplings are removed; each chemically non-equivalent set of carbons appears as a single line. The number of protons attached to each carbon can be determined using the DEPT technique.

14.44 Because all carbons are chemically nonequivalent, 4-methyl-1-penten-3-ol should have six carbon resonances in its carbon NMR spectrum. (Notice that the carbons of methyl groups 5 and 6 are diastereotopic.)

4-methyl-1-penten-3-ol

14.45 (a) To solve this problem, draw the structures and use the carbon chemical-shift chart (Fig. 14.22 in text Sec. 14.9B). Compound A is eliminated because it has eight chemically nonequivalent carbons and there are only four carbon signals. Compounds $B, C, D,$ and E all have four sets of nonequivalent carbons. The peak at δ 80.5 suggests a carbon in a carbon-carbon triple bond, so compounds B and D can be ruled out. We can next rule out compound C because the two carbons in its carbon-carbon triple bond are nonequivalent; it would give two distinct signals in the alkyne region, but we only see one. So, we have left only compound E, which fits the data with one set of chemically equivalent alkyne carbons and three alkyl carbons in the low chemical shift range.

(b) Compounds A and B have four sets of nonequivalent carbons and compound C has three. Because the given spectrum indicates four nonequivalent sets of carbons we can rule out compound C. We now have to decide between compounds A and B. An important difference in these structures is that only A contains a carbon connected to *two* oxygens. From the chemical-shift chart (Fig. 14.22 in text Sec. 14.9B), we see that a typical carbon in the environment —C—O— has a chemical shift between δ 40 and δ 80. The maximum chemical shift in the spectrum, however, is nearly δ 100, and this would be consistent with a compound in which a carbon is bound to two oxygens. Therefore, the compound is A. To further rule out B, note that the spectrum contains no methyl (3-proton) carbon in the ether region.

14.46 (a) At room temperature, the three conformations of 2,2,3,3-tetrachlorobutane are in rapid equilibrium. Therefore, the NMR spectrum of this compound is a singlet whose chemical shift reflects the environment of the six equivalent protons averaged over time across all conformations. Any changes observed at low temperature would be related to the fact that the interconversion between the different conformations is slowed, and thus certain conformations can be individually observed at the lower temperature. There are three such conformations, two of which are enantiomers:

enantiomeric conformations
have identical NMR spectra

In conformation A, all six protons are equivalent, and their resonance should be a singlet. In either one of the enantiomeric conformations, all six protons are still chemically equivalent to each other, but they are now chemically nonequivalent to the protons in conformation A. So, conformations B and C also give a singlet, but it will have a different chemical shift than the one given by A.

(b) Because conformations B and C are enantiomeric, their spectra will be identical. Therefore, the total spectrum should consist of one line for conformation A, and one line for conformations B and C, for a total of two lines at low temperature. The relative integrals depend on the amount of form A relative to the amounts of forms B and C; the latter two forms will be present in equal amounts. In other words, $[A] \neq [B] + [C]$, so the intensity of the A peak will be different than the $B+C$ peak.

14.47 Number the carbons and view the Newman projection about the C1–C2 bond, with carbon-2 nearer the observer:

$$\overset{3}{C}H_3 - \overset{2}{C}H_2 - \overset{1}{C}H_2 - Br$$

enantiomeric conformations have
identical ^{13}C NMR spectra

Assuming the spectrum is proton-decoupled, we would expect to see three resonances—one for each of the carbons—at room temperature. On cooling the sample, we would see two sets of three resonances. One set is due to the two enantiomeric conformations, and the other set is due to the remaining conformation. The largest chemical-shift change should be in the resonance for carbon-3, because its proximity to the bromine changes between the conformations. Smaller changes would be anticipated for carbons 1 and 2.

14.48 (a) The D_2O exchange indicates an alcohol, which must be tertiary, because of the absence of a resonance in the δ 3–4 region. The ^{13}C NMR spectrum indicates that there are four nonequivalent sets of carbons. The doublet-septet pattern indicates an isopropyl group, and the 6-proton singlet indicates two methyl groups. The compound, with assigned ^{13}C NMR chemical shifts, is

2,3-dimethyl-2-butanol

(b) The IR spectrum suggests a 1-alkyne. In the NMR spectrum, the δ 3.31 singlet indicates a methoxy group, and the six-proton δ 2.41 singlet suggests two equivalent propargylic methyl groups bound to a quaternary carbon. The compound is 3-methoxy-3-methyl-1-butyne.

3-methoxy-3-methyl-1-butyne

(c) The IR absorbance at 3200-3600 cm^{-1} suggests an —OH group is present, and there is a strong absorbance at 1090 cm^{-1} which suggests C—O single bonds, present either in alcohols or ethers. The NMR shows a resonance that disappears with a D_2O shake, confirming that an —OH group is present. Additionally from this spectrum, a six-hydrogen singlet suggests two methyl groups attached to a carbon that does not possess any protons. The other three-hydrogen singlet suggests another isolated methyl group, perhaps as part of an ether. With three of the six carbons and nine of the 14 hydrogens accounted for as belonging to methyl groups, you could start trying various structures to see what fits the rest of the data. The compound below fits the spectral data given:

δ 3.4 (triplet that disappears after D₂O shake)   δ 3.8 (quartet that becomes a triplet after D₂O shake)

δ 3.4 (triplet)

The quartet at δ 3.8 is split by both the —OH proton and the adjacent CH₂ protons. After the D₂O shake, the —OH becomes an —OD and the signal at δ 3.8 is the only split by the CH₂ protons, into a triplet.

(d)   The IR data suggest a 1-alkyne. The presence of a methoxy group is indicated by the three-proton singlet at δ 3.79. There is one additional degree of unsaturation; because three carbons are already accounted for, the unsaturation must be in the remaining two. Consequently, there is a double bond. Cis stereochemistry is suggested by the small 6-Hz coupling constant between the vinylic protons. (See Table 14.3 in text Sec. 14.7A.) The splitting suggests that one of the vinylic protons (δ 4.52) is also coupled to an acetylenic proton. The chemical shift of the other vinylic proton (δ 6.38) suggests that it is α to the methoxy oxygen. All of these data conspire to give (Z)-1-methoxy-1-buten-3-yne.

(Z)-1-methoxy-1-buten-3-yne

(e)   First, by calculating the unsaturation number (Sec. 4.3), it can be determined that the compound has either a double bond or a ring. The sharp absorbance on the IR spectrum at 1658 cm⁻¹ suggests a carbon-carbon double bond (as do the absorbances at 3085 and 875 cm⁻¹, which correspond to protons attached to carbons of carbon-carbon double bonds). Both the broad stretch in the IR at 3200-3600 cm⁻¹ and the signal in the NMR spectrum that disappears with a D₂O shake confirm an —OH group. Further inspection of the NMR spectrum shows a 3H singlet that tells us there is an isolated CH₃ group. The signals at δ 4.8 and 4.9 correspond to protons attached to a carbon that is part of a double bond. At this point, we have enough information to start trying various structures to see what fits the rest of the data. The only alkene alcohol that fits the data is:

δ 4.8, 4.9 (singlets)

Here is a related structure that is close but doesn't quite fit the data. Why not?

(f)   The IR data suggest that the compound is both a 1-alkyne and an alcohol. The δ 4.6 resonance for 1H indicates that the alcohol is secondary. Its splitting also shows that this α-hydrogen is adjacent to a methyl group, and its chemical shift shows that other functionality, probably the C≡CH group, is adjacent. The dry/wet contrast also indicates that this is the resonance of an alcohol α-proton. A small splitting of thè α-hydrogen by the acetylenic hydrogen would account for the secondary splitting of each quartet line into a doublet. (Recall that vinylic, and by extension, acetylenic, splitting can occur over four bonds.) The spectrum is consistent with only 3-butyn-2-ol.

OH
|
HC≡CCHCH₃

3-butyn-2-ol

(g)   The mass spectrum indicates that the compound contains a single bromine (double molecular ion). The NMR spectrum indicates an ethoxy group (CH_3CH_2O—); this comes from the typical triplet–quartet pattern in which the —CH_2— group is at a typical α-oxygen chemical shift and shows splitting with *only* the CH_3. The spectrum indicates two vinylic hydrogens that split each other with a very small splitting, which, however, is larger than a geminal splitting. The IR spectrum shows no evidence for a trans alkene, nor does it show the characteristic absorption for a =CH_2 group. Therefore, the vinylic hydrogens are cis. (The IR absorption at 694 cm^{-1} could be the C—H bending absorption for a cis alkene.) The ethoxy group counts for 29 mass units; the ^{79}Br for 79 mass units; two vinylic CH groups would complete the mass of 150. The structure can be only (*Z*)-1-bromo-2-ethoxyethene.

(*Z*)-1-bromo-2-ethoxyethene

Solutions to Integrated Problems

14.49   (a)   The spectrum of cyclohexane is a singlet. The spectrum of its constitutional isomer *trans*-2-hexene has many resonances, including a vinylic resonance in the δ 5–6 region that is not present in cyclohexane.

(b)   Only 1-hexene will have a complex vinylic proton absorption that integrates for 25% of the total absorption (that is, 3 protons); the vinylic proton absorption of *trans*-3-hexene will consist of a triplet integrating for 17% of the total absorption (that is, 2 protons).

(c)   The resonance at greatest chemical shift in the NMR spectrum of 1,1-dichlorohexane would be a one-proton triplet; this resonance would be at a chemical shift of at least δ 5.9. The resonance at greatest chemical shift in the NMR spectrum of 1,6-dichlorohexane would be a four-proton triplet at about δ 3.3. The resonances at highest chemical shift in the spectrum of 1,2-dichlorohexane would be a complex pattern in the δ 3.3–4.3 region. How complex depends on whether the diastereotopic protons at carbon-1 have different chemical shifts and whether they have different coupling constants with the methine proton at carbon-2.

1-proton triplet present at highest chemical shift—at least δ 5.9
↓
$Cl_2CHCH_2CH_2CH_2CH_2CH_3$
1,1-dichlorohexane

4-proton triplet present at about δ 3.3
$ClCH_2CH_2CH_2CH_2CH_2CH_2Cl$
1,6-dichlorohexane

3-proton multiplet in the δ 3.3–4.3 region
$ClCH_2CHCH_2CH_2CH_2CH_3$
|
Cl
1,2-dichlorohexane

(d)   The spectrum of *tert*-butyl methyl ether consists of two singlets; that of isopropyl methyl ether contains a singlet for the methoxy group, but a more complex doublet-septet pattern for the isopropyl group.

9*H* singlet at δ 1.2   3*H* singlet at δ 3.2
$(CH_3)_3C$—OCH_3
***tert*-butyl methyl ether**

1*H* septet at δ 3.7
6*H* doublet at δ 1.2   3*H* singlet at δ 3.2
$(CH_3)_2CH$—OCH_3
isopropyl methyl ether

(e) The first compound, $Cl_3CCH_2CH_2CHF_2$, has the absorption at greatest chemical shift and the more complex spectrum; furthermore, this high-shift absorption will show not only H–H splitting, but also the typically large H–F splitting, and will thus appear as a triplet of triplets—that is, a triplet in which each individual line is a more closely spaced triplet. Only the second compound, $CH_3CH_2CCl_2CClF_2$, will show a three-proton triplet at about δ 1 as part of a typical ethyl triplet-quartet pattern.

14.50 (a) The carbon spectrum of cyclohexane would be a single line, as all six carbons are chemically equivalent. *Trans*-2-hexene would show six lines.

(b) *Trans*-3-hexene has three nonequivalent groups of carbons, and thus three resonances, while all six carbons in 1-hexene are nonequivalent, giving six resonances.

(c) The spectrum of 1,6-dicholorohexane would differ from the other two because it has only three unique carbons, and would give three resonances, while the other two each have six. The spectrum of 1,1-dichlorohexane would contain the carbon resonance with the highest chemical shift resulting from the heavily deshielded 1-carbon with two chlorines attached. The spectrum of 1,2-dicholorohexane would have two resonances in the 20–60 ppm range, indicating carbons 1 and 2 with one chlorine attached to each, but no signal as high as that one discussed previously in 1,1-dichlorohexane.

14.51 (a) (b) (c) (d) (e)

cyclohexane

2,3-dimethyl-2-butene

di-*tert*-butyl ether

2,2,4,4-tetramethylpentane

1,1,2,2,tetramethyl-cyclopropane

14.52 (a) The NMR spectrum of the starting material has two sets of resonances, one of which is in the vinylic-proton region (δ 4.5–5.0). The other would be a singlet in the allylic methyl range, perhaps around δ 1.8. The NMR of the product is a singlet near δ 1.7. To follow the reaction, look for the disappearance of the vinylic proton resonance and one singlet and the appearance of the singlet arising from the product.

(b) The NMR spectrum of the starting material is a singlet at δ 1.7; see the solution to Problem 14.51(b). The NMR spectrum of the product should contain an isopropyl doublet-septet pattern at about δ 0.9 and δ 2, respectively, as well as a six-proton methyl singlet near δ 1.4. To follow the reaction, look for the disappearance of the singlet at δ 1.7 and the appearance of a doublet around δ 0.9.

14.53 The most obvious difference is that the spectrum of 1-methylcyclohexene should show *one* vinylic proton and a three-proton singlet (neglecting any allylic splitting); whereas the spectrum for 3-methylcyclohexene should show *two* vinylic protons and a three-proton doublet.

1-methylcyclohexene

3-methylcyclohexene

14.54 (a) All methyl groups are homotopic; all methine carbons are homotopic; and all methylene carbons are homotopic. (Prove this by a substitution test if this isn't clear.) Each carbon of one type is constitutionally nonequivalent to all carbons of other types. Consequently, the carbon NMR spectrum of this compound should consist of three resonances.

(b)   Because the chair interconversion is rapid, the relationships of the carbons in these compounds can be decided from the planar projections shown. The compound shown in this part should have six ^{13}C NMR absorptions because there are six chemically nonequivalent sets of carbons (unless there are coincidental overlaps between some absorptions). In the structure of this compound shown at right, carbons with the same number are enantiotopic, and therefore chemically equivalent. (Notice the plane of symmetry.) The following pairs of carbons are diastereotopic, and therefore chemically nonequivalent: 3 and 4; 5 and 6; 1 and 2. All other pairs are constitutionally nonequivalent.

14.55   (a)   The NMR spectrum of the first compound, $(CH_3)_2CH—Cl$ (isopropyl chloride), should consist of the doublet-septet pattern characteristic of isopropyl groups, with the doublet at about δ 1.2 and the septet at about δ 3.7. The NMR spectrum of the deuterium-substituted analog, $(CH_3)_2CD—Cl$, should consist of a singlet at essentially the same chemical shift as the doublet in the first compound. (The splitting between H and D nuclei on adjacent carbons is nearly zero.)

(b)   The NMR spectrum of the compound containing deuterium should consist of two triplets of equal intensity. (The coupling constant between H and D nuclei on adjacent carbons is nearly zero.) The NMR spectrum of the other compound, as shown in Fig. 14.7 in text Sec. 14.4A, consists of a triplet at low field and a quintet at higher field; the triplet should have twice the integrated intensity of the quintet. Each set of resonances in one spectrum should have the same chemical shift as the corresponding set of resonances in the other.

(c)   First analyze the third compound. The resonance of the methyl group will be a three-proton singlet. The two methylene hydrogens are diastereotopic and are therefore chemically nonequivalent. Therefore, they should have different (if only slightly different) chemical shifts, and each will split the other into a doublet. Thus, the resonances for these two protons should consist of a closely spaced doublet of doublets.

The two diastereomers should have different NMR spectra. To begin with, neglect the splitting by deuterium. (Deuterium splitting, which is considered in Problem 14.66(c), can be eliminated by an instrumental technique.) Each spectrum should consist of a three-proton singlet for the methyl group and a one-proton singlet for the methylene hydrogen. Although the methyl singlets will likely have identical chemical shifts in the two compounds, the chemical shifts of the other hydrogens will differ by essentially the same amount as the chemical shifts of the two methylene hydrogens in the third compound. The resonance of the methylene hydrogen in the 1S,2R diastereomer will have the same chemical shift as the resonance of the pro-S hydrogen in the third compound; and the resonance of the methylene hydrogen in the 1R,2R diastereomer will have the same chemical shift as the resonance of the pro-R hydrogen in the third compound. In each compound, both hydrogens will be singlets, neglecting any splitting by deuterium.

In these spectra, the H^a hydrogen is arbitrarily assigned the greater chemical shift, but the chemical shifts of these two diastereotopic hydrogens H^a and H^b might be reversed.

14.56   Compound A is 2,3-dimethyl-2-butene, $(CH_3)_2C{=}C(CH_3)_2$. [See Problem 14.51(b).] Remember that symmetrical alkenes don't typically show C=C stretching vibrations in IR (see text Display 13.15 and surrounding discussion). That it reacts with Br_2 confirms that it contains a double bond. The two peaks in the carbon NMR help to distinguish it from the very similar proton NMR data for compound D (see below).

The IR spectrum of compound B indicates a $H_2C=C$ group. The NMR spectrum shows a typical isopropyl pattern: a septet/doublet pattern (δ 2.2, δ 1.0, respectively); from its chemical shift, the isopropyl CH group is allylic. The isopropyl group must be attached to a carbon that bears no protons, because the CH of the isopropyl group is split only by the adjacent methyl groups. The presence of two vinylic hydrogens (δ 4.6–4.7) with very weak splitting is in agreement with the geminal vinylic hydrogens suggested by the IR spectrum, and the three equivalent hydrogens remaining probably correspond to a methyl group. Compound B is 2,3-dimethly-1-butene.

$(CH_3)_2CH—C=CH_2$
|
CH_3

2,3-dimethyl-1-butene
(compound B)

Compound C shows a nine-proton singlet characteristic of a *tert*-butyl group. The remaining three protons are vinylic. The IR spectrum shows $C=C$ stretching and $C—H$ bending absorptions that confirm the presence of a $—CH=CH_2$ group. Although it is not necessary to interpret the splitting to define the structure, it is very similar to that in Figs. 14.16–14.17 in text Sec. 14.7A. Compound C is therefore $(CH_3)_3C—CH=CH_2$, 3,3-dimethyl-1-butene.

That compound D is not an alkene follows from both the chemical data and the NMR absorption. A six-carbon hydrocarbon that is a singlet in the NMR, that has one degree of unsaturation, and that is not an alkene can be only cyclohexane. [You may have seen this one before: Problem 14.51(a).] The single peak in carbon NMR helps to distinguish this compound from compound A.

14.57 (a) Protons a and b are diastereotopic protons and thus give different chemical shifts as discussed in text Sec. 14.3D. You can verify for yourself that they are diastereotopic by performing the substitution test (text Sec. 11.9A) and comparing the virtual products.

(b) Proton a is split both by proton b and the proton on the adjacent carbon. Because the two coupling constants are different, application of the $n + 1$ rule becomes more complex here; a doublet of doublets is observed. [See the solution to Problem 14.19(b) in this manual for a similar situation.]

14.58 The spectrum shows at least four nonequivalent sets of protons in the ratio 1:1:4:6. Let's see how the possible structures stack up against this analysis.

cis-3-hexene (A)   (Z)-1-ethoxy-1-butene (B)   2-ethyl-1-butene (C)

$Cl_2CH—CH(OCH_2CH_3)_2$

1,1-dichloro-2,2-diethoxyethane (D)

$Cl—CH—CH—Cl$
| |
CH_3CH_2O OCH_2CH_3

1,2-dichloro-1,2-diethoxyethane (E)

Compounds A–C are ruled out immediately. The vinylic hydrogens in A would be a single triplet. In compound B, there would be two methyl triplets and a more complex splitting for the two vinylic hydrogens. In compound C, the vinylic hydrogens would be a singlet (neglecting allylic splitting). The two methine hydrogens of D would definitely give a pair of doublets at fairly high chemical shift, as observed. The two methine hydrogens of E are completely equivalent and would give a singlet as the absorption at greatest chemical shift. Therefore, compound E is ruled out, and compound D is the remaining possibility.

As to the complexity of the δ 3.7 resonance, the two CH_2 hydrogens in either ethyl group on both D and E are *diastereotopic,* and therefore, chemically nonequivalent. For example, in the case of D,

H^a and H^b are diastereotopic; H^c and H^d are also diastereotopic. On the other hand, H^a is enantiotopic to H^c, and H^b is enantiotopic to H^d. (You should verify these statements by a substitution test.) Therefore, H^a and H^c are chemically equivalent, as are H^b and H^d. But H^a and H^b are chemically *nonequivalent*, as are H^c and H^d. Consequently, there are *five* nonequivalent sets of hydrogens.

It is perhaps not surprising that the chemical shifts of the diastereotopic protons are not very different, because their chemical environments are very similar. This is very often the case with diastereotopic protons; they're different, but not by that much.

Let's now consider the splitting patterns of these diastereotopic protons. We've already said that H^a and H^b occur at slightly different chemical shifts. This would account for two absorptions. Each of these is split into a doublet by the other to give four lines. Each of these is split into a quartet by the neighboring methyl group. This gives $4 \times 4 = 16$ lines. Because H^c is chemically equivalent to H^a, and H^d is chemically equivalent to H^b, the analysis is complete. You can actually see all sixteen lines for these diastereotopic hydrogens in the spectrum.

14.59   (a)   Protons *a* and *b* are diastereotopic and thus give different chemical shifts as discussed in text Sec. 14.3D. You can verify for yourself that they are diastereotopic by performing the substitution test (text Sec. 11.9A) and comparing the virtual products. They are each split into a doublet by proton *c*.

   (b)   The signal for proton *c* is split with $J = 7.0$ Hz into a quartet by protons *a*, and that quartet is split again with $J = 7.0$ Hz, into a septet by protons *b*.

Finally, each peak in that septet is split into a doublet by 4.3 Hz by proton *d*.

appearance of H^c resonance:
a septet of doublets

The signal for proton *c* appears on a spectrum as a "septet of doublets", though one would most likely just refer to it as a "multiplet".

14.60   The sample of 2,5-hexanediol is a mixture of diastereomers, the racemate and the meso compound. Each of these stereoisomers has a carbon NMR spectrum that consists of three resonances, but the two spectra are different, if only slightly so.

(±)-2,5-hexanediol
(the 2*R*,5*R* enantiomer is shown)

meso-2,5-hexanediol

14.61 The IR spectrum and the D_2O shake results indicate that compound A is an alcohol, and the absence of a resonance in the δ 3–4 region indicates that the alcohol is tertiary. Treatment of compound A with H_2SO_4 yields a compound B, which, from its δ 5.1 resonance in its NMR spectrum, is evidently an alkene. Evidently, the reaction with H_2SO_4 is a dehydration. If so, the molecular mass of 84 for compound B means that compound A has the molecular mass of compound B plus 18 (the molecular mass of H_2O). Hence, the molecular mass of A is $84 + 18 = 102$.

Now let's turn to the integrals for compound A. The integrals (from high to low shift) are in the approximate ratio 1:4:6:3. Let's adopt the hypothesis that the O—H proton accounts for $1H$. If so, then compound A contains 14 hydrogens, and its formula is $C_6H_{14}O$, which gives the molecular mass of 102 hypothesized above. Why couldn't the alcohol have a formula $C_7H_{14}O$? Because this would give the wrong molecular mass.

In the NMR of compound A, the singlet at δ 1.2 integrates for $6H$. This can only be two methyl groups. The triplet just below δ 1.0 integrates for $3H$, and, from its splitting, corresponds to a CH_3CH_2— group. However, the methylene protons in this group must be split by other protons because they are not a simple quartet. The following structure would account for the complex splitting at δ 1.4, as well as the other facts:

complex pattern at δ 1.3–1.5

$$CH_3CH_2CH_2\underset{\underset{CH_3}{|}}{\overset{\overset{OH}{|}}{C}}CH_3 \xrightarrow{H_2SO_4} CH_3CH_2CH{=}C\begin{smallmatrix}CH_3\\ \\CH_3\end{smallmatrix} + H_2O$$

2-methyl-2-pentanol
(compound A)

2-methyl-2-pentene
(compound B)

The dehydration would then give compound B shown above. In the NMR of this compound we expect a single vinylic proton split into a triplet, and this is what is observed at δ 5.1. (Some additional, very small, allylic splitting is also observed.) The NMR spectrum of compound B is rationalized as follows:

δ 5.1 (triplet with allylic splitting)

δ 1.2 (triplet)

$$CH_3CH_2CH{=}C\begin{smallmatrix}CH_3\\ \\CH_3\end{smallmatrix}$$

diastereotopic methyls; two singlets at δ 1.6–1.7

δ 2.0 (quintet)

14.62 The formula, the IR spectrum, and the D_2O shake indicate that compound A is a five-carbon alcohol with one degree of unsaturation. This could be a ring or a double bond, but the NMR spectrum clearly indicates two vinylic protons near δ 5.5, a fact that suggests a double bond as the source of unsaturation. Moreover, the IR shows weak C=C stretch and a strong absorption at 965 cm^{-1}, which indicates a *trans*-alkene. The ^{13}C NMR resonances at δ 125.5 and δ 135.5, each with one attached hydrogen, support the diagnosis of an internal alkene. The alcohol is secondary because there is one proton in the α-oxygen region (δ 4.2) and one attached hydrogen for the corresponding carbon at δ 68.8; the latter proton is shifted a little more than normal for this type of proton; therefore, this proton is also allylic. Furthermore, this α-proton is adjacent to four other protons. Two methyl groups are adjacent to single hydrogens. One of these methyl groups is allylic (δ 1.7) and the other is not (δ 1.2). To summarize, we have the following structural elements, which can only be assembled to give *trans*-3-buten-2-ol as the structure. The chirality of the proposed structure is consistent with the fact that compound A can be resolved into enantiomers.

trans-3-buten-2-ol

14.63 The two doublets centered at δ 6.2 are assigned to the diene protons 1 and 2. These protons are split by each other, but not significantly by any other protons. (Coupling constants depend on dihedral angle, and the large splitting evidently suggests an anti relationship about the single bond resulting from an *s*-trans conformation of the diene unit.) The two resonances centered at δ 4.9 are assigned to the geminal alkene protons 3 and 4. These two protons are diastereotopic, and are therefore chemically nonequivalent. Their splitting, although not expanded, is evidently the small splitting typical of geminal alkene protons (Table 14.3, text Sec. 14.7A). The multiplet just below δ 4 is assigned to the α-alcohol methine proton 6. We are given no information about a D₂O shake, so the proton 5 is not identified; it is probably within the complex pattern between δ 1 and δ 3. Within the inset on the right, the doublet integrating for three protons probably corresponds to a methyl group. The only CH₃—CH group is the one with methyl protons 8. In the inset, the doublet of doublets corresponds to six protons, which corresponds to the two diastereotopic methyl groups 9 and 10, each split by the adjacent methine hydrogen and having slightly different chemical shifts. The resonance at δ 0.5 is due to an unsplit methyl group and must correspond to protons 7.

vitamin D₃

14.64 The unsaturation number = 1. There are no alkene absorptions in the IR; therefore, we postulate the presence of a ring. The presence of an alcohol is indicated by the IR data, and the triplet at δ 3.5 must be an OH proton because it is lost as the result of a D₂O shake. Furthermore, it must be part of a —CH₂OH group because of its triplet splitting. The protons at δ 3.7 must be the α-protons, because their splitting collapses after a D₂O shake; furthermore, these protons are attached to a quaternary carbon, because they are not split further. This quaternary carbon must also bear a methyl group, because the 3*H* resonance at δ 1.3 is unsplit. We therefore have the following partial structure:

We are missing two carbons and an oxygen; recall that there must be a ring. The two doublets between δ 4.4 and δ 4.6 must be two CH₂ groups, and the remaining oxygen must complete a four-membered ring. Consider the following two structures:

In structure *B*, the two CH₂ groups would be at least triplets, not doublets (and their splitting would likely be more complex, because the two protons on each carbon are diastereotopic). Furthermore, the chemical shifts of both

CH_2 sets indicate that they are both α to an oxygen; in structure B, only one CH_2 meets this criterion. Therefore, structure B is ruled out, leaving structure A as the only reasonable candidate.

The splitting of the CH_2 protons in A is rationalized as follows.

The two H^a protons are enantiotopic and therefore chemically equivalent; the same is true for the two H^b protons. Protons H^a and H^b split each other, and each is split into a doublet by its neighbor. Presumably H^a is the proton at greatest chemical shift because of its proximity to the OH group.

14.65 (a) From Fig. 14.9, text Sec. 14.5A, you know anti protons have large coupling constants and gauche protons have smaller coupling constants. When H^a is in the axial position (conformation E), it is split by adjacent protons that are anti and gauche. These protons would split H^a by different coupling constants. When H^a is in the equatorial position (conformation A), it is split by adjacent protons that are both gauche. These protons would split H^a by coupling constants that are about equal. Inspection of Fig. P14.65 shows a resonance that is split into a triplet of triplets, or, a splitting pattern that would result from splitting by protons with different coupling constants, one large and one small. (This situation is similar to the one described by Fig. 14.10, text Sec. 14.5A.) So, the resonance shown in P14.65 belongs to proton H^a in the axial position, or conformation E.

(b) The other resonance, not shown, would come from H^a in the equatorial position (conformation A), and the adjacent splitting protons are both gauche. As described in part (a), the coupling constants coming from both would be small, and probably about equal. Thus, you might predict a quintet for this resonance, with small coupling constants between 3–4 Hz.

(c) One conformation is present in 3.39 times greater concentration than the other. The equilibrium constant, K_{eq}, is the ratio of the two conformations, 3.39:1, or 3.39. Using this value, the temperature (convert Celsius to kelvins) of 193 K, and the gas constant of 8.314×10^{-3} kJ mol^{-1} K^{-1}, we can solve for $\Delta G°$.

$$\Delta G° = -2.3RT \log K_{eq}$$
$$= -2.3(8.314 \times 10^{-3} \text{ kJ mol}^{-1} \text{ K}^{-1})(193 \text{ K}) \log 3.39$$
$$= -1.96 \text{ kJ mol}^{-1}$$

You know from previous discussions of the conformations of monosubstituted cyclohexanes (text Sec. 7.3) that the larger substituent prefers to be in the equatorial position to minimize 1,3-diaxial interactions. So, the smaller resonance (not shown, relative integral = 1) corresponds to H^a in the axial position, and thus the iodine in the equatorial position, conformation E. So, conformation E is present at a 3.39 times greater conformation that A.

14.66 (a) The nitrogen can take on three values of spin with equal probability: +1, 0, and –1. Therefore, the *protons* near a nitrogen of each spin state are shifted accordingly (or not shifted, in the case of spin = 0.) The *proton* NMR spectrum of $^+NH_4$ should (and does) consist of three lines of equal intensity.

(b) The reason the ^{13}C NMR spectrum of $CDCl_3$ is a 1:1:1 triplet is the same reason that the proton NMR spectrum of $^+NH_4$ is a 1:1:1 triplet [see part (a)]: deuterium, like nitrogen, can have spins of +1, 0, and –1, and deuterium splits ^{13}C resonances just as it splits proton signals.

(c) The compound with one deuterium, H_2CDI, would show a triplet proton resonance with lines that have a 1:1:1 intensity ratio, because the deuterium can have three different values of spin (+1, 0, and –1). Because these are equally probable, they give rise to lines in the proton splitting pattern of equal intensity. The compound D_3CI has no proton NMR spectrum. For HCD_2I, the two deuteriums have the following combinations of spin values:

Total spin = +2: +1, +1 (one way)
Total spin = +1: 0, +1 and +1, 0 (two ways)
Total spin = 0: 0, 0; +1, –1; and –1, +1 (three ways)
Total spin = –1: 0, –1 and –1, 0 (two ways)
Total spin = –2: –1, –1 (one way)

Each value of the total deuterium spin makes a different contribution to the position of the proton resonance; thus, there are five lines, with relative intensities proportional to relative probabilities of deuterium spin (that is, to the number of different ways that each spin can be achieved): 1:2:3:2:1.

Mass spectrometry could be used to easily determine the identity of each sample, as the samples each differ in molecular mass. IR could also be used: only D_3CI will show a C—D stretching absorption and no C—H stretching absorption in the IR spectrum.

14.67   The ^{17}O NMR resonance of ^{17}O-water is split into a triplet by the attached two protons when a small amount of water is dissolved in CCl_4. The splitting pattern suggests that, in the presence of acid, the ^{17}O resonance is split by *three* protons. Evidently, this splitting is due to the hydronium ion, H_3O^+.

14.68   Note that the ^{13}C resonances from *unenriched* compounds will be much weaker than those from the isotopically enriched compounds and can be ignored. The proton-decoupled ^{13}C NMR spectrum will be a composite spectrum that consists of the spectra of the individual species present. These species, and their relative abundances, will be as follows: (*C = ^{13}C):

*CH_3—*CH_2—Br relative probability = (0.5)(0.5) = 0.25; two doublets
CH_3—*CH_2—Br relative probability = (0.5)(0.5) = 0.25; one singlet
*CH_3—CH_2—Br relative probability = (0.5)(0.5) = 0.25; one singlet
CH_3—CH_2—Br relative probability = (0.5)(0.5) = 0.25; no ^{13}C spectrum

Each resonance will show an apparent "triplet" pattern: the line from the singly-enriched species will be in the center of a doublet for the doubly-enriched species. The singly-enriched species are singlets because there is so little ^{13}C resulting from the natural abundance of ^{13}C at each carbon that it can be ignored. Each line of the doublet will have half the total intensity of the doublet, and therefore half the intensity of the singlet.

14.69   (a)   The resonance of the unpaired electron is split into four lines by the three neighboring protons according to the same $n + 1$ rule that applies to splitting between nuclei.

(b)   Use Eq. 14.9 in text Sec. 14.3B with $\gamma_n = \gamma_{electron} = 17.6 \times 10^6$ radians gauss^{-1} sec^{-1}, and $\mathbf{B_0} = 3400$ gauss.

$$v_{electron} = \frac{\gamma_{electron}}{2\pi} \mathbf{B_0} = \frac{17.6 \times 10^6 \text{ rad G}^{-1} \text{ s}^{-1}}{2\pi \text{ rad}} (3400 \text{ G}) = 9.5 \times 10^9 \text{ s}^{-1} \approx 10^{10} \text{ Hz}$$

Fig. 13.2 on text Sec. 13.1A shows that a frequency of 10^{10} Hz is in the microwave region of the electromagnetic spectrum. Indeed, ESR instruments employ microwave radiation to detect the magnetic resonance of electrons.

14.70   Shielding decreases chemical shift. Because a "naked" proton is completely *unshielded* by electrons, its chemical shift should be very large, and it should feel the full effect of the applied field. Because the range of proton chemical shifts within organic compounds extends to about 11–12 ppm (Fig. 14.4 in text Sec. 14.3C), we expect the chemical shift of a "naked" proton to be even greater than this. Therefore, the first answer, $\delta > 8$, is the correct one.

Dienes and Aromaticity

CHAPTER 15 SUMMARY

- Molecules containing conjugated double bonds have additional stability, relative to unconjugated isomers, that can be attributed to the continuous overlap of their carbon 2p orbitals to form π molecular orbitals.

- The delocalization energy, or resonance energy, of a conjugated π-electron system containing *j* double bonds is the difference between its π-electron energy and the π-electron energy of *j* ethylenes. The most stable conformation of 1,3-butadiene and other conjugated dienes is the *s*-trans conformation, which is an anti conformation about the central single bond of the diene unit.

- A cumulene is a compound with one or more sp-hybridized carbon atoms that are part of two double bonds. An allene is a cumulene with two cumulated double bonds. Adjacent π bonds in a cumulene are mutually perpendicular; appropriately substituted allenes are chiral.

- Heats of formation for isomers are generally in the following order: conjugated dienes < ordinary dienes < alkynes < cumulenes.

- Compounds with conjugated double or triple bonds have UV–visible absorptions at $\lambda_{max} > 200$ nm.

- Each conjugated double or triple bond in a molecule contributes 30–50 nm to its λ_{max}. When a compound contains many conjugated double or triple bonds, it absorbs visible light and appears colored.

- The intensity of the UV–vis absorption of a compound is proportional to its concentration (Beer's law). The constant of proportionality ε, called the molar extinction coefficient, is the intrinsic intensity of an absorption.

- Some molecules that absorb energy from UV or visible radiation can lose some of that energy by emitting light (fluorescence) at greater wavelength. Many fluorescent molecules have rigid chromophores with extended π-electron systems. Fluorescence is a very sensitive analytical method that has important applications in biology, such as the green fluorescent protein (GFP).

- The Diels–Alder reaction is a pericyclic reaction that involves the cycloaddition of a conjugated diene and a dienophile (usually an alkene). When the diene is cyclic, bicyclic products are produced.

- The diene assumes an *s*-cis conformation in the transition state of the Diels–Alder reaction; dienes that are locked into *s*-trans conformations are unreactive.

- Each component of the Diels–Alder reaction undergoes a syn-addition to the other. In many cases the endo mode of addition is kinetically favored over the exo mode.

- Conjugated dienes react with hydrogen halides to give mixtures of 1,2- and 1,4-addition (conjugate addition) products. Such a mixture of products is accounted for by the formation of a resonance-stabilized allylic carbocation intermediate, which can react with halide ion at either of two positively charged carbons.

- When the products of a reaction do not come to equilibrium under the reaction conditions, the reaction is said to be kinetically controlled. A kinetically controlled reaction can give a mixture of products that is substantially different from the mixture obtained if the products are allowed to come to equilibrium. The predominance of the 1,2-addition product in the reaction of hydrogen halides with conjugated alkenes is an example of kinetic control. If the products of a reaction come to equilibrium under the reaction conditions, the reaction is said to be thermodynamically controlled.

■ The molecular-orbital basis of resonance stabilization is that the filled π molecular orbitals of lowest energy in a conjugated molecule, ion, or radical extend across the single bonds that connect the conjugated double bonds. These low-energy MOs make a significant negative contribution to the energy of the molecule.

■ Benzene is the prototype of a class of compounds, including some ions, that have a special stability called aromaticity. All aromatic compounds contain $4n + 2$ π electrons in a continuous, planar, cyclic array.

■ The basis of the $4n + 2$ rule is that bonding MOs are completely filled in molecules with $4n + 2$ π elec-

trons, and that the bonding MOs have very low energy.

■ Compounds that contain $4n$ π electrons in a continuous, planar, cyclic array are antiaromatic and are especially unstable.

■ Aromatic rings can have noncovalent attractions with other aromatic rings. Two configurations are commonly observed: a parallel-offset stacking interaction and an edge-to-face interaction. Aromatic rings can also have noncovalent attractions with positive ions. In this type of attraction, called a pi–cation interaction, a positive ion is situated at the face of the ring where it can interact favorably with the ring π electrons. These interactions are very important in some protein structures and in the binding of substrates and inhibitors to some enzymes and receptors.

STUDY GUIDE LINKS

 ### 15.1   A Terminology Review

The purpose of this Study Guide Link is to review and to distinguish between the following terms used to describe addition reactions:

1.   cycloaddition reactions
2.   concerted reactions
3.   pericyclic reactions

These terms in some cases may seem synonymous, but they are in fact different, and have very precise meanings. A cycloaddition is an addition that results in ring formation. Ozone addition and the Diels–Alder reaction are two examples of cycloadditions. The term cycloaddition succinctly describes the outcome of a reaction, but says nothing about the mechanism of the reaction. The terms concerted and pericyclic, in contrast, are descriptions of mechanism. Reactions that occur by mechanisms involving a single step—that is, mechanisms involving no reactive intermediates—are referred to as concerted reactions. Concerted reactions that occur by a cyclic flow of electrons are called pericyclic reactions (pericyclic = "around the circle"). All pericyclic reactions are concerted, but some concerted reactions are not pericyclic. Thus, hydroboration of alkenes (Sec. 5.6A of the text) is a concerted pericyclic reaction (although it is not a cycloaddition because a ring is not formed). The S_N2 reaction (Sec. 9.4 of the text) is concerted, but it is not pericyclic because the flow of electrons in the mechanism is not cyclic. The Diels–Alder reaction is a concerted cycloaddition, and it is also a pericyclic reaction. (Chapter 28 is devoted to a more detailed study of pericyclic reactions.)

FURTHER EXPLORATIONS

 ## 15.1   More on UV Spectroscopy

Why are peaks in UV spectra so broad? We are led to ask this question because the molecular orbital theory of UV spectroscopy discussed in the text implies that a $\pi \rightarrow \pi^*$ transition should occur at a very precise value of the energy.

The answer lies in the rate at which electronic transitions take place. It turns out that the motions of electrons are thousands of times faster than the motions of nuclei. As you learned in Chapter 13, the nuclei in molecules are vibrating constantly. When energy is absorbed from UV light by a molecule, this act of absorption "catches" the molecule in a variety of vibrational states. That is, some bonds are stretched, some are compressed, and all possible combinations of these states for all possible bonds can occur. Because the electronic transition occurs so much faster than bond vibrations, the "molecule" that absorbs UV energy is actually a complex mixture of molecules in various vibrational states. These states differ in energy, as do the excited states into which they are converted. These energy differences are reflected in a broadening of the energy at which the UV transition is observed. In gas-phase UV spectra, one can in many cases see very narrow peaks associated with the different vibrational states. (Spectroscopists call these peaks "vibrational fine structure.") Interaction of the absorbing molecules with solvent further blurs these vibrational energies, so that the absorption observed in solution is something of an "envelope" of all of these states. Thus, UV absorptions in solution are typically very broad.

 ## 15.2   The π Molecular Orbitals of Benzene

Only the benzene bonding π molecular orbital of lowest energy is shown in Fig. 15.16b on text p. 842. Benzene has five other π molecular orbitals, and they are shown in **Fig. SG15.1** on the following page.

As you can see, there are three bonding and three antibonding orbitals—six 2p orbitals, six MOs. Two of the *bonding* MOs—π_2 and π_3—have the same energy. Orbitals of the same energy are said to be *degenerate*. Because these MOs have the same energy, they have the same number of nodes (one). Two of the *antibonding* MOs—π_4^* and π_5^*—are also degenerate. Because there are six π electrons, there are just enough electrons to fill the three bonding MOs. All three of these MOs, because they are bonding, are lower in energy than an isolated 2p orbital, whose energy is indicated by the dashed line. Therefore, benzene is stabilized by molecular-orbital formation.

In Sec. 15.7D, you will learn to calculate the energies of these MOs. When you do, you will find that the energy of π_1 is *particularly* low. This filled, low-energy MO, which encircles the entire molecule with π-electron density above and below the ring, is particularly important for the stability of benzene. This is the MO that is shown in Fig. 15.16b of the text. (Undoubtedly, the fact that the bond angles in benzene are exactly right for sp²-hybridized carbons contributes to the molecule's stability as well.)

Figure SG15.1 The π molecular orbitals of benzene superimposed on a ball-and-stick model. Wave peaks are black, and wave troughs are gray. The $E = 0$ dashed line is the energy of an electron in an isolated 2p orbital.

REACTION REVIEW

I. ADDITION OF HYDROGEN HALIDES TO CONJUGATED DIENES

A. *1,2- AND 1,4-ADDITIONS*

1. Conjugated dienes react with hydrogen halides to give two types of **addition product**.

   a. The major product results from 1,2-addition. In a 1,2-addition, the hydrogen and halogen add to adjacent carbons.

   b. The minor product results from 1,4-addition, or conjugate addition. In a 1,4-addition, the hydrogen and halogen add to carbons that have a 1,4-relationship.

1,2-addition product   1,4-addition product

2. The 1,2-addition mechanism is analogous to the addition of HX with an ordinary alkene.

3. The 1,4-addition, or conjugate-addition, mechanism involves the following steps:

   a. Protonation of a double bond to give an allylic carbocation in which the positive charge is not localized, but is instead shared by two different carbons.

   b. Formation of two constitutional isomers by a Lewis acid–base association reaction of the halide ion with the carbocation at either of its electron-deficient carbons.

1,2-addition product   1,4-addition product

4. The product distribution in hydrogen-halide addition to a conjugated diene is determined by the relative rates of the product-determining steps.

   a. Formation of the 1,2-addition product is faster but reversible.

   b. Formation of the 1,4-addition product is slower but virtually irreversible.

II. DIENE POLYMERS

A.   *FORMATION OF POLYMERS FROM CONJUGATED DIENES*

1. The polymerization of 1,3-butadiene gives poly(butadiene), an important type of synthetic rubber used in the manufacture of tires.

1,3-butadiene   **poly(butadiene)**

   a. Poly(butadiene) is referred to as a diene polymer because it comes from polymerization of a diene monomer.

   b. Poly(butadiene) has only one double bond per unit, because one double bond is lost through the addition that takes place in the polymerization process.

2. Polymerization of dienes is a free-radical reaction. Although the polymerization product is largely the result of 1,4-addition, a small amount of 1,2-addition can occur as well.

3. Styrene–butadiene rubber, SBR, is an example of a copolymer, a polymer produced by the simultaneous polymerization of two or more monomers, in this case, styrene ($PhCH=CH_2$) and 1,3-butadiene.

III.   THE DIELS–ALDER REACTION

A.   *INTRODUCTION*

1.   Many conjugated dienes undergo addition reactions, called Diels–Alder reactions, with certain alkenes or alkynes.

   a.   The conjugated diene component of this reaction is referred to simply as the diene.

   b.   The alkene component of this reaction is referred to as the dienophile.

dienophile ⟶ $H_2C{=}CH{-}CO_2CH_3$

$+$

diene ⟶

CO_2CH_3

2.   Some of the dienophiles that react most readily in the Diels–Alder reaction bear substituent groups such as esters ($-CO_2R$), nitriles ($-CN$), or certain other unsaturated, electronegative groups that are conjugated with the alkene double bond. (Some alkynes also serve as dienophiles.)

3.   When the diene is cyclic, bicyclic products are obtained in the Diels–Alder reaction.

$+$ CO_2CH_3 ⟶ CO_2CH_3 $\equiv$ CO_2CH_3

4.   The Diels–Alder reaction is

   a.   a cycloaddition reaction (an addition reaction that results in the formation of a ring).

   b.   a pericyclic reaction (a concerted reaction that involves a cyclic flow of electrons).

$H_2C{=}CH{-}CO_2CH_3$ ⟶ CO_2CH_3

   c.   a 1,4-addition, or conjugate addition, reaction (an addition that occurs across the outer carbons of the diene).

   d.   a stereoselective syn-addition to both diene and dienophile.

B.   *EFFECT OF DIENE CONFORMATION ON THE DIELS-ALDER REACTION*

1.   In the Diels–Alder transition state, the diene component is in the *s*-cis conformation.

   a.   Dienes "locked" into *s*-trans conformations are unreactive in Diels–Alder reactions.

   b.   Dienes "locked" into *s*-cis conformations are unusually reactive and in many cases much more reactive than corresponding noncyclic dienes.

1,3-cyclopentadiene

"locked" into the *s*-cis diene
conformation by its structure

more reactive in Diels–Alder reactions

(2E,4E)-2,4-hexadiene

prefers the *s*-trans conformation

less reactive in Diels–Alder reactions

2.   The Diels–Alder reactions of a *cis*-diene are much slower than the corresponding reactions of its *trans*-diene stereoisomer, which does not have the destabilizing repulsion in its *s*-cis conformation.

cis-diene

less reactive in Diels–Alder reactions

van der Waals repulsion

trans-diene

no methyl–methyl repulsion

more reactive in Diels–Alder reactions

    a.   This van der Waals repulsion destabilizes the *s*-cis conformation of the diene.

    b.   This van der Waals repulsion destabilizes the transition states for the Diels–Alder reactions of the *s*-cis conformation of the diene.

C.   STEREOCHEMISTRY OF THE DIELS-ALDER REACTION

   1.   When a diene and dienophile react in a Diels–Alder reaction, they approach each other in parallel planes.

     a.   This type of approach allows the π-electron clouds of the two components to overlap and form the bonds of the product.

     b.   Each component adds to the other at one face.

       i.   The diene undergoes a syn-addition to the dienophile.

      ii.   The dienophile undergoes a 1,4-syn-addition to the diene.

     c.   Groups that are cis in the alkene starting material are also cis in the product.

     d.   Syn-addition to the diene is apparent if the terminal carbons of the diene unit are stereocenters.

   2.   A syn-addition requires the following structural features in the Diels–Alder product:

     a.   The two inner substituents of the diene always have a cis relationship.

     b.   The two outer substituents of the diene also always have a cis relationship.

     c.   Tn inner substituent on one carbon of the diene is always trans to an outer substituent on the other.

     d.   Cis substituents in the dienophile are cis in the product; trans substituents in the dienophile are trans in the product.

   3.   For a given diene and dienophile, two diastereomeric syn-addition products are possible and are abbreviated with the terms endo and exo.

     a.   In the endo product, the dienophile substituents are cis to the outer diene substituents.

endo addition → product with labels:
- CH₃ ← outer diene substituent ⎫
- CO₂CH₃ ← dienophile substituent ⎬ cis
- CO₂CH₃ ← dienophile substituent ⎫
- CH₃ ← outer diene substituent ⎬ cis

b. In the exo product, the dienophile substituents are trans to the outer diene substituents.

exo addition → product with labels:
- CH₃ ← outer diene substituent ⎫
- CO₂CH₃ ← dienophile substituent ⎬ trans
- CO₂CH₃ ← dienophile substituent ⎫
- CH₃ ← outer diene substituent ⎬ trans

c. In many cases, the endo mode of addition is favored over the exo mode, particularly when cyclic dienes are used.

SOLUTIONS TO PROBLEMS

Solutions to Focused Problems

15.1 (a) 1,3,5-Hexatriene (either cis or trans) has six MOs; three are bonding, and three are antibonding. These MOs are shown in **Fig. SG15.2** on the following page.

(b) Each MO of successively higher energy has an additional planar node.

(c) The nodes are shown in Fig. SG15.2. The nodes are symmetrically arranged about the center of the molecule. (The nodes of all MOs are shown.)

15.2 The delocalization energy is the energy of each MO times the number of electrons in that MO, minus the same energy for three ethylenes:

$$\text{delocalization energy} = 2(1.80\beta) + 2(1.25\beta) + 2(0.44\beta) - (3)(2)(1.00\beta) = 0.98\beta$$

15.3 First, draw the structures:

$H_2C{=}CH$, H — $C{=}C$ — H, CH_3

(E)-1,3-pentadiene

$H_2C{=}CHCH_2CH{=}CH_2$

1,4-pentadiene

$H_2C{=}CH$, H — $C{=}C$ — H, CH_2CH_3

(E)-1,3-hexadiene

$H_2C{=}CHCH_2$, H — $C{=}C$ — H, CH_3

(E)-1,4-hexadiene

(Solution continues on the following page below Fig. SG15.2)

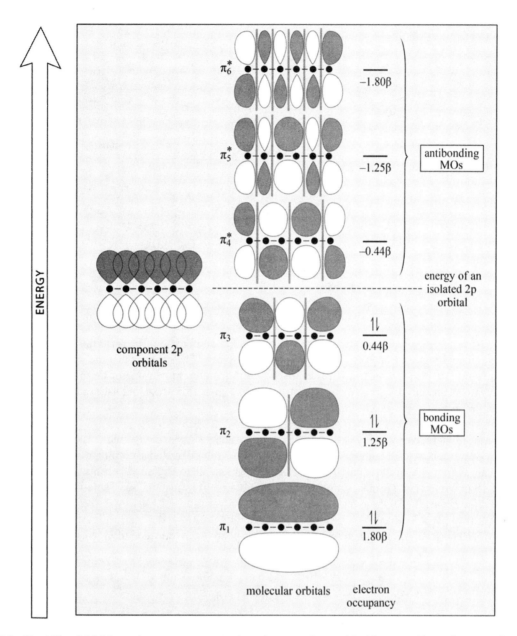

Figure SG15.2   The MOs of 1,3,5-hexatriene, to accompany the solution to Focused Problem 15.1. The carbons are shown as black spheres. All MOs are projected onto the plane of the page. The nodes are the gray vertical lines.

(*E*)-1,3-pentadiene has a double bond with one substituent and a double bond with two substituents (counting the carbon of other double bond), whereas 1,4-pentadiene has two double bonds with a single substituent. Therefore, the energy difference between these two pentadiene isomers reflects not only the effect of conjugation but also the effect of additional double-bond substituents. Both effects stabilize the conjugated isomer. The two hexadiene isomers contain identical numbers of substituents; therefore, the energy difference between these two isomers reflects only the effect of conjugation. Because the comparison of pentadiene isomers contains two effects that operate in the same direction, the energy difference between these isomers is expected to be greater, as observed.

15.4   The conformations of (2E,4E)-2,4-hexadiene:

s-cis   s-trans

The conformations of (2E,4Z)-2,4-hexadiene:

greater
van der Waals
repulsions

s-cis   s-trans

(2E,4E)-2,4-hexadiene has the greater proportion of s-cis conformation because the van der Waals repulsions between inner groups are smaller in the (2E,4E)-stereoisomer. The reason is that the inner groups in the (2E,4E)-stereoisomer are both hydrogens, whereas, in the (2E,4Z)-stereoisomer, one of these groups is a methyl group, which is considerably larger than hydrogen. It follows then that (2E,4E)-2,4-hexadiene would also have the greater proportion of the gauche conformation.

15.5   The order of increasing stability is $C < A < B$. As a cumulated diene, compound C is the least stable, and would have the largest heat of formation. The conjugated diynes can be ranked according to the relative energies of the terminal and internal alkynes; alkynes, like alkenes, are stabilized by alkyl substitution at the carbons of the π bond. Thus, compound B is most stable, with the lowest heat of formation. See Table 15.1 on text p. 801 for some of the relative energies of these functional groups.

15.6   (a)   The energy of light is given by Eq. 13.3 on text p. 684 with the value of h, in kJ s mol^{-1}, given in Eq. 13.5 on text p. 684.

$$E = \frac{hc}{\lambda} = \frac{(3.99 \times 10^{-13} \text{ kJ s mol}^{-1})(3.00 \times 10^{8} \text{ m s}^{-1})}{450 \times 10^{-9} \text{ m}} = 266 \text{ kJ mol}^{-1}$$

(The corresponding value in kcal mol^{-1} is 63.6.)

   (b)   The calculation is identical to that in part (a) with different numbers:

$$E = \frac{hc}{\lambda} = \frac{(3.99 \times 10^{-13} \text{ kJ s mol}^{-1})(3.00 \times 10^{8} \text{ m s}^{-1})}{250 \times 10^{-9} \text{ m}} = 479 \text{ kJ mol}^{-1}$$

(The corresponding value in kcal mol^{-1} is 114.)

15.7   (a)   Use Eq. 15.7 on text p. 808. An absorbance $A = 1.0$ means that $(I_0/I) = 10$; that is, one-tenth, or 10%, of the incident radiation is transmitted. When $A = 0$, all of the incident radiation is transmitted.

   (b)   Use Eq. 15.7 on text p. 808 with $(I_0/I) = 2$. Thus, $A = \log (2) = 0.30$.

15.8   The piece with greater absorbance transmits less of the incident radiation. Therefore, the thick piece of glass has greater absorbance.

15.9   (a)   Apply Beer's law (Eq. 15.8 on text p. 809), with $\varepsilon = 10,750$ absorbance units cm^{-1} L mol^{-1}, $l = 1.0$ cm, and, from Fig. 15.5 on text p. 808, $A = 0.800$ absorbance units. The concentration of isoprene is therefore

$$\frac{0.800 \text{ absorbance units}}{(10,750 \text{ absorbance units cm}^{-1} \text{ L mol}^{-1})(1 \text{ cm})} = 7.44 \times 10^{-5} \text{ mol L}^{-1}$$

(b)   The absorbance of the isoprene sample in Fig. 15.5 on text p. 808 at 235 nm is 0.225. With the concentration determined from part (a), Beer's law gives $A = 0.225$ absorbance units $= (\varepsilon)(7.44 \times 10^{-5} \text{ mol L}^{-1})(1 \text{ cm})$, or $\varepsilon = 3.02 \times 10^3$ absorbance units L mol^{-1} cm^{-1}. Another way to determine the extinction coefficient at a different wavelength is based on the fact that the ratio of absorbances at different wavelengths equals the ratio of the extinction coefficients. Therefore, the extinction coefficient at 235 nm is

$$\varepsilon_{235} = \varepsilon_{225} \frac{A_{235}}{A_{225}} = (10,750 \text{ absorbance units L mol}^{-1} \text{ cm}^{-1}) \frac{0.225 \text{ absorbance units}}{0.800 \text{ absorbance units}}$$

or $\varepsilon_{235} = 3.02 \times 10^3 = 3023$ absorbance units L mol^{-1} cm^{-1}.

15.10 (a)   The two alkyl substituents contribute +10 nm to the base λ_{max} of 217 for a predicted λ_{max} value of 227 nm.
       (b)   The —CH$_2$— groups of the ring attached to the double bond are alkyl substituents. (If this point is not clear, see Study Guide Link 9.2 on p. 219 of this manual.) For three alkyl substituents the predicted λ_{max} is $(217 + 15)$ nm $= 232$ nm. (This calculation assumes that the molecule exists predominantly in the *s*-trans conformation.)

15.11 At pH $= 9$, fluorescein has the greater quantum yield, because the peak height of fluorescence is nearly as intense as the absorption. At pH $= 5$, the fluorescence is a smaller fraction of the absorbance.

15.12 Compound *C* is likely to emit yellow fluorescence. Of all the compounds given, it is the most highly conjugated and is the most rigid because of its fused aromatic ring system.

15.13 (a)   (b)

15.14 (a)   An analysis like that employed in Study Problem 15.1 (text pp. 821-822) suggests two possibilities.

Pair *A* is preferred because, in many cases, the most reactive dienophiles are those with conjugated electronegative substituents. But if your answer was pair *B*, you have analyzed the problem correctly.
       (b)   There are two possibilities:

Pair *A* is the more practical alternative because of the availability of starting materials, and especially because dienophiles with electron-withdrawing substituents are more reactive. But if your answer was pair *B*, you have analyzed the problem correctly.

15.15 (a) With 1,3-butadiene as the diene and ethylene as the dienophile, the product would be cyclohexene.

1,3-butadiene   **ethylene**   **cyclohexene**

(b) Cyclohexene, the product from the reaction of ethylene and 1,3-butadiene in part (a), is not observed because 1,3-butadiene reacts with itself to give 4-vinylcyclohexene:

4-vinylcyclohexene

(c) Using an excess of ethylene (many more equivalents relative to 1,3-butadiene) could favor the formation of cyclohexane. The chances of 1,3-butadiene colliding with a molecule of ethylene would become much higher.

15.16 (a) The dienophile can be oriented in two different ways relative to the diene.

(b) As in part (a), two possible orientations of the diene and dienophile lead to the following two possible constitutional isomers:

and

15.17 The Diels–Alder reaction requires that the diene assume an *s*-cis conformation in the transition state. In this conformation, the (2*E*,4*Z*)-diene has a methyl–hydrogen van der Waals repulsion that is absent in the (2*E*,4*E*)-diene. (This point was also used in the solution to Focused Problem 15.4.) This repulsion raises the energy of the Diels–Alder transition state of this diene, and therefore reduces the rate of its reaction. Since the (2*E*,4*Z*)-diene reacts much more slowly than the (2*E*,4*E*)-diene, it follows that the (2*E*,4*E*)-diene is consumed in the reaction, and the (2*E*,4*Z*)-diene remains unreacted.

smaller
van der Waals repulsions

greater
van der Waals repulsions

s-cis conformation of
the 2*E*,4*E* diene

s-cis conformation of
the 2*E*,4*Z* diene

15.18   The triene contains two diene units with one double bond common to both. The dienophile reacts with the diene unit that is locked in an *s*-cis conformation.

tetracyanoethylene
(TCNE)

15.19   (a)   The two acetoxy groups are cis in the product because they are both outer substituents in the *s*-cis conformation. (The racemate is formed; only one enantiomer is shown below.)

(b)   The two products, both isolated as racemic mixtures, correspond to the two possibilities in Eq. 15.32 on text p. 827. They result from addition of the diene at either of the two faces of the alkene (or the alkene at either of the two faces of the diene).

endo stereoisomers   exo stereoisomers

Notice that both stereoisomers result from syn-addition. In other words, there are two possible modes of syn-addition.

(c)   The "inner" groups of the diene are "tied together" as the —CH₂— group of the five-membered ring.

CH₂ group of the diene

CO₂CH₃

15.20 (a)

(b)

(c)

(d)

(that is, the reaction of two
1,3-cyclopentadiene molecules)

15.21 (a) As shown by Eq. 15.32 on text p. 827, the endo diastereomer is the one in which the dienophile substituents (which, in this case, are the arms of the five-membered ring) are cis to the outer substituents when the diene is drawn in the *s*-cis conformation.

(b) Here are the diagrams. Notice that the product is formed initially in a boat-like conformation that subsequently changes to its more stable conformation. Be sure to make models if you have difficulty seeing these relationships.

endo product

exo product

15.22 (a) The carbocation intermediate in the addition of HCl has two sites of electron deficiency, either of which can undergo a Lewis acid–base association reaction with the chloride ion.

$H_2C=CH-CH=CH_2$
1,3-butadiene

$$\left[H_2C=CH-\overset{+}{C}H-CH_3 \longleftrightarrow H_2\overset{+}{C}-CH=CH-CH_3 \right]$$

(A) :Cl:$^-$ (B)

:Cl:
|
$H_2C=CH-CH-CH_3$ + $H_2C-CH=CH-CH_3$
          A |
                                   :Cl:
                                    B

(b)   The products are a mixture of constitutionally isomeric ethers because the carbocation intermediate in the S$_N$1 reaction has two sites of electron deficiency that can react by a Lewis acid–base association reaction with the nucleophile ethanol.

3-ethoxy-3-methyl-
cyclohexene

3-ethoxy-1-methyl-
cyclohexene

It is common (and useful) to show the derivation of different products by showing the separate "reactions" of different resonance structures. Be sure you remember, however, that the resonance structures are different structures used to depict *one* species. Thus, in the structures above, the carbocation is *one* species that reacts in *two* different ways, *not* a mixture of different species that undergo separate reactions.

15.23   (a)   An allylic carbocation intermediate, formed by a dissociation reaction, undergoes a Lewis acid–base association reaction with bromide ion at the two sites of electron deficiency.

$$CH_3CH=CH-CH_2-\overset{+}{\overset{|}{O}}H_2 \longrightarrow \left[CH_3CH=CH-\overset{+}{C}H_2 \longleftrightarrow CH_3\overset{+}{C}H-CH=CH_2 \right] + \overset{..}{O}H_2$$

:Br:$^-$

(A) :Br:$^-$ (B)

$CH_3CH=CH-CH_2-\overset{..}{B}r:$ + $CH_3CH-CH=CH_2$
          A |
                                             :Br:
                                              B

(b)   This is an S_N1 reaction; the two products result respectively from the Lewis acid–base association reaction of water at the two electron-deficient carbons of the carbocation intermediate. (Acetone is a co-solvent that helps to dissolve the alkyl chloride in the aqueous solution.)

15.24   (a)   Hammond's Postulate allows us to conclude that the more stable reactive intermediate is formed most rapidly. Therefore, carbon 4 is the carbon most likely to be protonated first, affording a tertiary allylic carbocation. The 1,2- and 1,4-addition products are shown here:

(*E*)-3-methyl-1,3-pentadiene

1,2-addition product   1,4-addition product

(b)   The 1,4-addition product will predominate at equilibrium as it is the more stable of the two alkene products. The alkene formed from 1,4-addition has three alkyl branches on the double bond, whereas the product of 1,2-addition is a terminal alkene with only one alkyl branch.

15.25   (a)   The two electrons that make up the pi bond are shown explicitly here to help you track their movement.

(b)   The incorporation of the dienes in a 1,4- or 1,2-addition, along with the incorporation of styrene could, in theory, occur in any order, so your structure may look different than the one provided here.

(c)   Before each bond-forming reaction, the radical is able to relax into the most stable isomer. The less stable cis-isomer will rotate along one of its bonds to become the more stable *trans*-isomer, as shown below. Once a reaction happens with the next diene, the stereochemistry of the *trans*-double bond is set.

15.26   As the problem suggests, the MOs for the allyl anion are the same as those for the allyl cation, shown in Fig. 15.14 on text p. 826; the two ions differ only in electron occupancy.

(a)   The allyl anion has four π electrons; two occupy π_1, and two occupy π_2.

(b)   The negative charge is caused by the fourth electron, which occupies π_2. Because π_2 has a node through the central carbon, the negative charge is delocalized onto the terminal carbons.

(c)   The resonance structures of the allyl anion give the same conclusion; the nonbonding electron pair (and thus the negative charge) is delocalized to the two terminal carbons.

allyl anion

15.27   (a)   The allyl radical has three π electrons. Two paired electrons occupy π_1, and an unpaired electron occupies π_2. Because π_2 has a node through the central carbon, the unpaired electron is delocalized onto only the terminal carbons.

(b)   The resonance structures of the allyl radical give the same conclusion; the unpaired electron is delocalized to the two terminal carbons.

allyl radical

15.28   The bicyclic amide *A* is not stabilized by resonance, because its resonance would be geometrically impossible. The orbital interaction required for resonance interaction requires an impossible twisting of the molecule. (See discussion in Sec. 7.6C, p. 353 in the text.) To put it another way, the hypothetical resonance structure violates Bredt's rule. In contrast, amide *B* is stabilized by resonance, as the orbitals involved in the resonance are not on bridgehead atoms.

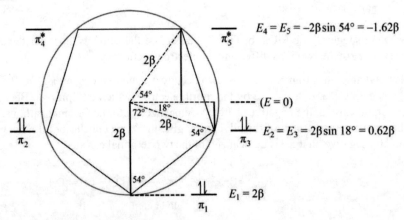

15.29   Follow the procedure in Study Problem 15.2 on text p. 844.

Step 1: Inscribe a regular pentagon within a circle of radius 2β with one vertex in the vertical position. (This is shown in the construction in the following step.)

Step 2: Place an energy level at each vertex. This gives five energy levels for the five MOs. Notice that two of the antibonding MOs, π_4^* and π_5^*, are degenerate (have the same energy), and two of the bonding MOs, π_2 and π_3, are also degenerate.

Step 3: Calculate the energies as multiples of β. Remember that the radius of the circle, and therefore E_1, is 2β. (Remember also that β is a negative number.) Some useful trigonometry is that the internal angle in a pentagon is 108°; the bisected internal angle is 54°; and, from the fact that the sum of the three angles of a triangle must be 180°, the angle between the lines that connect successive vertices to the center must be 72°. These angles are labeled in the construction.

Step 4: Add the π electrons to the MOs. This anion contains 6 π electrons distributed in three bonding MOs: two electrons in π_1, the MO at 2.0β, and two electrons in each of the two (degenerate) MOs, π_2 and π_3, at 0.62β. The antibonding MOs π_4^* and π_5^* at −1.62β are unoccupied.

15.30   (a)   Such a compound contains 10 MOs; 5 are bonding, and 5 are antibonding. One of the bonding π MOs lies at 2β; one of the antibonding π* MOs lies at −2β. The four other bonding MOs are degenerate pairs—that is, $\pi_2 = \pi_3$, and $\pi_4 = \pi_5$—at energies < 0, and the four other antibonding MOs also lie in degenerate pairs at energies > 0.
   (b)   The compound contains 10 π electrons
   (c)   Because 10 is a 4n + 2 number, there are just enough π electrons to fully occupy the 5 bonding MOs.

15.31   Given that furan is aromatic, it can have 4n + 2 π electrons (six electrons) if one of the electron pairs on oxygen is in a 2p orbital that is part of the π-electron system of the ring. The other electron pair on oxygen occupies an sp^2 orbital whose axis is in the plane of the ring. (The electronic structure of furan is drawn in Fig. 26.2 on text p. 1429.)

15.32   An aromatic compound cannot have a single unpaired electron as part of its π-electron system, because the number of π electrons required for aromaticity, $4n + 2$, must be an even number; a single unpaired electron would result in an odd number of electrons.

However, a free radical could certainly be aromatic if the unpaired electron were not part of the π-electron system. An example is the phenyl radical, which could be formed conceptually by abstraction of a hydrogen atom from benzene:

phenyl radical

15.33   (a)   Thiophene is aromatic. One electron pair on sulfur is part of the aromatic π-electron system; the other is not (as in the oxygen of furan; see the solution to Focused Problem 15.31.)

(b)   This ion is not aromatic because it is not a continuous cycle of 2p orbitals. The carbon at the bottom has no p orbitals; it is a saturated carbon atom.

(c)   This ion is aromatic. Each double bond contributes two electrons to the π-electron system. The cationic carbon bears an empty 2p orbital that contributes no electrons but is part of the π-electron system. One electron pair on the oxygen is part of the π-electron system, and the other is not, as in furan (Focused Problem 15.31).

(d)   This ion contains $4n$, not $4n + 2$ π electrons, and is therefore not aromatic.

15.34   Because compounds A and B are antiaromatic, they distort as much as possible to *avoid* overlap of the π orbitals involved in the two double bonds. (See the discussion of cyclobutadiene Sec. 15.7E in the text.) The single bonds are much longer than the double bonds, and the two types of carbon–carbon bonds are not equivalent. Because the deuteriums bridge a double bond in A and a single bond in B, and because the two types of carbon–carbon bonds are different, A and B are different compounds and are therefore in principle capable of independent existence. In contrast, all carbon–carbon bonds in benzene are equivalent—each is half single bond, half double bond. Consequently, all molecules with deuteriums on adjacent positions are identical and indistinguishable. To summarize:

15.35   Compound (d) contains $4n$ π electrons and is, in principle, antiaromatic.

15.36   The positively-charged end of the acetylcholine ion would orient itself inside the barrel of the molecule, much like a ring sliding on a finger. The four faces of the benzene rings would point towards the positive charge, similar to the interaction shown in Fig. 15.20a on p. 855.

Solutions to Skills Objectives Problems

15.37   Compounds A and D are conjugated alkenes. Compounds C and E are cumulated alkenes.

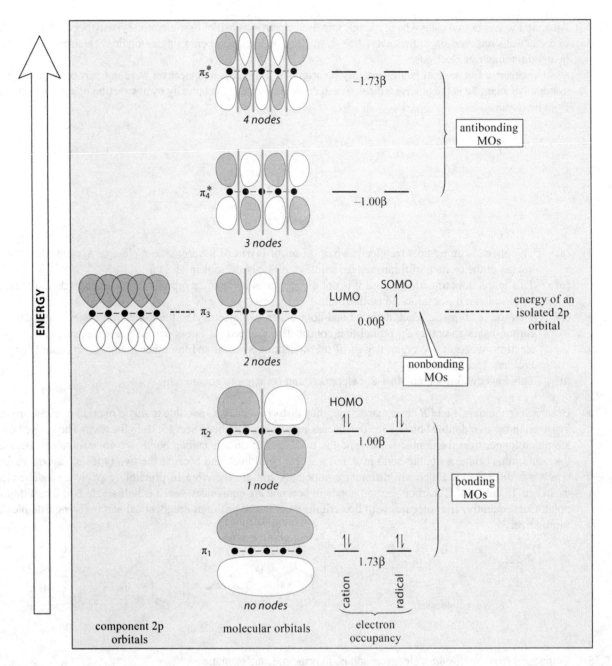

Figure SG15.3 MO diagram to accompany the answers to Problems 15.38 and 15.39. The MOs of the 2,4-pentadien-1-yl cation (15.38) and 2,4-pentadien-1-yl radical (15.39) are depicted in parallel columns. The carbons are shown as black spheres. All MOs are projected onto the plane of the page. The nodes are the gray vertical lines.

15.38 (a) The 2,4-pentadien-1-yl cation is depicted using three resonance structures:

2,4-pentadien-1-yl cation

(b) The molecular orbital energy diagram of the cation is depicted in **Fig SG15.3**, including the energies and electron occupancy of each orbital.

(c)   The nodes and orbital projections are depicted on the left side of Fig SG15.3.

(d)   The classification of the MOs as bonding, antibonding, or nonbonding are indicated in white boxes on the right side of the figure.

(e)   The resonance energy (delocalization energy) is the energy of each MO times the number of electrons in that MO, minus the same energy for two ethylene molecules. Remember that β is a negative number.

$$\text{delocalization energy} = 2(1.73\beta) + 2(1.00\beta) - (2)(2)(1.00\beta) = 1.46\beta$$

(f)   The MO diagram predicts the same distribution of positive charge as the resonance structures in part (a). The LUMO is π_3, which has two nodes running through carbons 2 and 4, indicating that the positive charge will only be shared by carbons 1, 3, and 5.

15.39   (a)   The 2,4-pentadien-1-yl radical is depicted using three resonance structures:

2,4-pentadien-1-yl radical

(b)   Because of the similarities of the MOs with the cation in the previous problem, the MO diagram of the radical is on the same figure, Fig SG15.3.

(c)   The unpaired electron doesn't affect the delocalization energy because it resides in an MO with $\beta = 0$.

$$\text{delocalization energy} = 2(1.73\beta) + 2(1.00\beta) + 1(0.00\beta) - (2)(2)(1.00\beta) = 1.46\beta$$

(d)   The MO diagram predicts the same distribution of unpaired-electron character as the resonance structures in part (a). The unpaired electron occupies π_3, which has two nodes running through carbons 2 and 4, indicating that the lone electron will only be distributed to carbons 1, 3, and 5.

15.40   Remember that increasing heat of formations correspond to decreasing stabilities.

(a)   *Heats of formation*: $A < B < D \ll C$. *Reasons*: Compound A is more stable than compound B because a molecule with conjugated double bonds is more stable than an isomer with isolated double bonds. Compound B is more stable than compound D because trans alkenes are more stable than cis alkenes. Compound C is least stable because an allene is the least stable type of diene.

(b)   *Heats of formation*: $C < A < B$. *Reasons*: Compound C is most stable because it is aromatic. Compound A is more stable than compound B because conjugated alkenes are more stable than isomeric alkynes.

(c)   *Heats of formation*: $C < B \ll A$. *Reasons*: Compound C is the most stable because the allylic carbocation can be drawn as three resonance structures, whereas compound B can only be drawn as two. Compound A is the least stable; even though it appears as a doubly allylic secondary carbocation, it is an antiaromatic ring system with 4 π electrons.

(d)   *Heats of formation*: $B < C < A$. *Reasons*: Compound B is most stable because it is aromatic. Compound A is least stable because molecules with cumulated double bonds are less stable than isomeric molecules with conjugated double bonds.

15.41   (a)   When UV light is absorbed, the electronic transition occurs between the HOMO and the LUMO, as shown at the top of the following page.

(b)   The UV absorption of (E)-1,3,5-hexatriene is reported as 268nm in Table 15.2 (text p. 812) with a HOMO-LUMO gap of 0.88β (Figure 15.7, text p. 811). The UV absorption of 3-methylene-1,4-pentadiene requires an energy of $2(0.517\beta) = 1.034\beta$. Because electronic excitation requires more energy, 3-methylene-1,4-pentadiene will absorb a shorter wavelength of light. (In fact, its λ_{max} is 232 nm.)

3-methylene-1,4-pentadiene

15.42   First, use Beer's law to determine the concentration of the diene:

$$A = 0.356 \text{ absorbance units} = \varepsilon cl = (10{,}750 \text{ absorbance units L mol}^{-1} \text{ cm}^{-1})(c)(1.0 \text{ cm})$$

from which we calculate $c = 3.31 \times 10^{-5}$ mol L^{-1}.

Since the entire sample was diluted to one liter, the entire sample contains 3.31×10^{-5} mole of isoprene. (Any hydrogenation product lacks conjugated double bonds and does not have UV absorption.) The amount of isoprene present is

$$(1.00 \text{ L})(3.31 \times 10^{-5} \text{ mol L}^{-1})(68.12 \text{ g mol}^{-1}) = 2.25 \times 10^{-3} \text{ g} = 2.25 \text{ mg}$$

Because the mass of the entire sample is 75 mg, the mass percent of isoprene is $(2.25 \text{ mg}/75.0 \text{ mg})(100) = 3.00\%$. That is, 3% by mass of the isolated material is unreacted isoprene.

15.43   The principles are that conjugation increases the λ_{max} by 40–50 nm per double bond; each alkyl substituent increases λ_{max} by 5 nm; and an *s*-cis conformation of the diene unit increases the λ_{max} and decreases the absorption intensity relative to that of an analogous *s*-trans diene.

(a)   1,4-Cyclohexadiene is not conjugated and therefore has no UV absorption in the useful region of the spectrum; 1,3-cyclohexadiene, an *s*-cis diene, has absorption at 256 nm (see text page 813). The compound with the UV spectrum is therefore the 1,3-diene.

1,4-hexadiene   **1,3-hexadiene**
has no UV spectrum   conjugated; has a UV spectrum

(b)   The first compound has three alkyl substituents on conjugated double bonds; the second compound has two. Therefore, the λ_{max} of the first compound should be about 5 nm greater than the λ_{max} of the second.

(c)   The second compound has four alkyl substituents on conjugated double bonds; the first compound has two. (The CH_2 group of the ring counts as a substituent on both double bonds.) Therefore, the λ_{max} of the second compound should be about 10 nm greater than the λ_{max} of the first.

(d)   The first compound has three conjugated double bonds, whereas the second compound has two; one double bond is not conjugated with the other two. Therefore, the first compound should have a considerably greater λ_{max} than the second.

(e)   Although the first compound is shown in the *s*-trans conformation, it doesn't exist in that conformation because of severe van der Waals repulsions between the *tert*-butyl groups. It can avoid these repulsions by internally rotating into its *s*-cis conformation. Although the *s*-cis conformation has a repulsion between the inner hydrogen and a *tert*-butyl group, this is much less severe than the repulsions between two *tert*-butyl groups in the *s*-trans conformation. Nevertheless, it is likely that the *s*-cis conformation will be distorted from planarity to relieve this repulsion. This will significantly lower the intensity of its UV absorption.

s-trans conformation   *s*-cis conformation

The *s*-trans conformation of the second diene, however, is more stable than the *s*-cis conformation; therefore, the second diene exists mostly in its *s*-trans conformation. Because *s*-cis dienes have considerably greater λ_{max} values than related *s*-trans dienes, the spectrum of the first diene should have a greater λ_{max} value, but a much lower intensity, than the spectrum of the second diene.

15.44　(a)　The fluorescence wavelength is *greater than* the absorption wavelength.

(b)　Fluorescence involves *different* electronic energy levels than absorption.

(c)　The quantum yield of fluorescence is the ratio of *excited states that return to the ground state compared to the excited states that do not return to the ground state.*

(d)　Fluorescence involves *different* vibrational energy levels than absorption.

(e)　Fluorescence spectroscopy is a more sensitive technique than absorption spectroscopy because *it is measured by the absolute intensity of the light emitted (compared against total darkness), rather than a comparison of light emitted against a significant background light.*

15.45　(a)　Use a method like the one in the solution to Focused Problem 15.21b:

(b)　The Diels-Alder reaction affords both exo and endo products of two constitutional isomers, all four of which are isolated as racemates.

CH₂   O₂N—CH   ...   NO₂   +   NO₂

CH₂

CH

CH   **nitroethylene**

CH₃   CH₃   CH₃

(E)-1,3-pentadiene   endo product   exo product

CH₂   CH₂   +

CH   CH

CH   O₂N   NO₂   NO₂

CH₃   **nitroethylene**   CH₃   CH₃

(E)-1,3-pentadiene   endo product   exo product

15.46   (a)   (b)

H   H—C—CO₂Me

H   +   C   +

MeO₂C—C

H   H

15.47   (a)   The increasing order of Diels-Alder reactivity with a dienophile is $A \ll C < B$. Diene A is locked in an s-trans conformation, and will not react with a dienophile in a Diels-Alder mechanism. In contrast, diene B is locked in an s-cis conformation, and will react rapidly with a dienophile. Diene C is free to rotate and will react with a dienophile when the s-cis conformation is achieved.

       (b)   The increasing order of Diels-Alder reactivity with a dienophile is $B < A < C$. Diene C is locked in an s-cis conformation, and will react rapidly with a dienophile. Steric interactions must be considered in the analysis of dienes A and B. Referring back to the answer for Focused Problem 15.4, the all-trans diene (diene A) will minimize steric interactions in the s-cis conformation, whereas the diene with a cis alkene (diene B) will have greater steric interactions making the required s-cis conformation higher energy and therefore slower to form.

15.48   The analysis in Eqs. 15.27a–b on text p. 824 shows that the s-cis conformations of dienes with cis double bonds are destabilized by van der Waals repulsions, whereas the s-cis conformations of dienes with trans double bonds do not suffer the same repulsions.

H   H   H   H

C—C   C—C

H—C   C—H   R—C   C—R

R   R   H   H

van der Waals   *s*-cis conformation
repulsions   of an all-trans diene

s-cis conformation
of an all-cis diene

These van der Waals repulsions also destabilize the transition states of Diels–Alder reactions, which require s-cis conformations of the reacting dienes. Because the diene in the problem undergoes the Diels–Alder reaction under mild conditions, it is probably the all-trans diene.

15.49   The answer to this problem lies in the s-cis conformations of the three dienes, which are as follows:

4-methyl-1,3-pentadiene
s-cis conformation

identical van der Waals
repulsions

(Z)-1,3-pentadiene
s-cis conformation

(E)-1,3-pentadiene
s-cis conformation

The reactivities of these three dienes in the Diels–Alder reactions correlates nicely with the accessibility of their *s*-cis conformations. (Remember, the transition state of the Diels–Alder reaction involves the *s*-cis conformation of the diene.) The van der Waals repulsions shown above destabilize the *s*-cis conformations of both 4-methyl-1,3-pentadiene and (Z)-1,3-pentadiene to the same extent because the interacting groups are identical. Therefore, their reactivity is reduced by about the same amount relative to that of (E)-1,3-pentadiene, whose *s*-cis conformation has no such repulsions.

15.50   (a)   First, draw the structures of the two carbocations that would be produced by the alternative modes of protonation:

(b)   Although both carbocations are allylic, carbocation *A* has a tertiary carbocation resonance structure, shown above, while neither resonance form of carbocation *B* is tertiary, Therefore, carbocation *A*, from protonation at carbon-1, is the more stable carbocation.

(b)   Because carbocation *A* is the more stable ion, it is the carbocation intermediate involved in HBr addition, and the products are thus derived from it. The two products come from the Lewis acid–base association reaction of bromide ion with the (electron-deficient) carbons that bear positive charge, which we deduce from the resonance structures of the ion.

resonance structures of carbocation *A*   1,2-addition product   1,4-addition product

(c)   The major kinetically controlled product is the 1,2-addition product. This product results from the collapse of the initially formed ion pair at the nearest site of positive charge. (See Eq. 15.42 on text p. 832.)

15.51   (a)   Protonation at carbon-1 of *trans*-1,3,5-hexatriene gives the carbocation with the greatest number of resonance structures. (Verify this point by examining the carbocations obtained from protonation at other carbons.)

The three possible products are derived from the Lewis acid–base association reaction of Br⁻ at the positively charged carbons.

(b)   The kinetically controlled product is the 1,2-addition product, labeled compound *A* in part (a). The reason is given in the text; the 1,2-product results from collapse of the initially formed ion pair at the nearest site of positive charge. (See Eq. 15.42 on text p. 832.)

15.52   As a first initiation step, homolytic cleavage of peroxides is followed by hydrogen abstraction from HBr by the peroxy radical to form a bromine atom. (This step is shown in Sec. 10.1C, Eq. 10.10.) The bromine atom then reacts with the diene in one of the following two ways, propagating the addition of HBr to the diene. The 1,4-addition is the major product because the more stable alkene is formed in the reaction.

1,3-butadiene   1,4-addition   major product   + Br·

1,3-butadiene   1,2-addition   minor product   + Br·

15.53   (a)   The transformations involved in the reaction between 1,3-cyclopentadiene and maleic anhydride form two different products under different conditions:

(b)   Because the initially formed product distribution is markedly different from the equilibrium distribution, the reaction is kinetically controlled.

15.54   (a)   Given that the addition occurs to the triple bond, the addition should occur regiospecifically to give the product in which the chlorine ends up on the internal carbon of the triple bond.

(b)   To deduce the structure of neoprene, follow the pattern for 1,4-addition polymerization shown in the solution to Focused Problem 15.25 in this manual.

15.55   (a)   This species contains six π electrons: two from the double bond, two from the anion, and two from the oxygen. Because $6 = 4n + 2$ for $n = 1$, this species is aromatic.

(b)   This radical has three π electrons: two from the double bond and one from the unpaired electron. It is therefore not aromatic.

(c)   This molecule contains six π electrons: two from each double bond and one from the nitrogen attached to hydrogen. That nitrogen is sp^2-hybridized in order to put its unshared electrons in a 2p orbital to achieve the 6 π electrons needed for aromaticity. It is, indeed, an aromatic compound.

(d)   Each double and triple bond contributes a pair of π electrons for a total of 18; therefore, this compound is likely to be aromatic. Notice that the second pair of π electrons in a triple bond is in a π orbital that is in the plane of the ring; therefore, only the electrons in the π orbital that is perpendicular to the plane of the ring contribute to the aromatic π-electron system.

(e)   This anion has ten π electrons: two from each of the four double bonds and two from the anionic carbon. It is likely to be aromatic.

15.56   The species with $4n + 2$ π electrons are likely to be planar because the planar species are aromatic. Thus, tropylium ion and cyclooctatetraenyl dianion are planar species. Oxepin, with eight π electrons, avoids planarity and, like cyclooctatetraene, thus avoids antiaromaticity.

15.57   Borazole is very stable because it is aromatic; each nitrogen contributes two π electrons and each boron contributes zero electrons, for a total of six π electrons in the aromatic system. The resemblance of borazole to benzene is more obvious from its other two resonance structures:

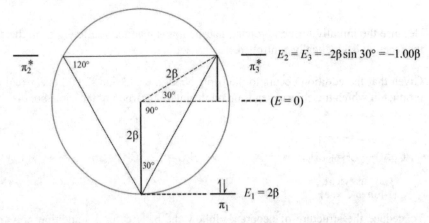

15.58   The Frost circle construction for the cyclopropenyl cation is as follows. See the answer to Focused Problem 15.29 for more information on the correct construction of a Frost circle.

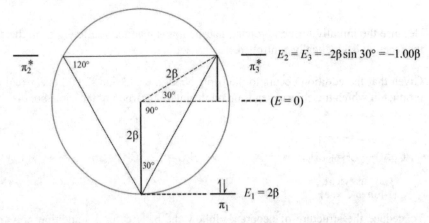

$$E_2 = E_3 = -2\beta \sin 30° = -1.00\beta$$

$$(E = 0)$$

$$E_1 = 2\beta$$

This cation contains two π electrons, which both occupy the bonding MO π_1. From Fig. 15.14, text p. 836, the π-electron energy of the allyl cation is $2(1.41\beta) = 2.82\beta$. The energy of the cyclopropenyl cation is 4.0β. (Remember, β is a negative number.) The extra stabilization, 1.18β, is the contribution of aromaticity. (This does not take into account the destabilizing contribution of ring strain.)

15.59   (a)   The $H_3\overset{+}{N}$— group of the lysine side chain would be oriented towards the face of the aromatic ring of the phenylalanine residue, similar to the interaction shown in Fig. 15.20(a) on p. 855.

(b)   The most energetically favorable interaction between two phenylalanine residues in a protein structure would be an edge-to-face interaction, as shown in Eq. 15.75 on p.854.

Solutions to Integrated Problems

15.60   First, draw the structure of the starting material:

$$H_3C \quad \quad H$$
$$C=C$$
$$H \quad \quad CH=CH_2$$

***trans*-1,3-pentadiene**

(a)

1,2-addition products

(cis and trans)
1,4-addition product

(b)   Within each pair, the first product is derived from the more stable carbocation intermediate (why?), and is therefore likely to be formed in greater amount. Notice that one of the 1,2-addition products is identical to one of the 1,4-addition products—that is, this compound can be formed by either process.

CH₃CH=CHCHCH₃ + CH₃CH₂CHCH=CH₂ + CH₃CHCH=CHCH₃ + CH₃CH₂CH=CHCH₂Br
(with Br substituents)

(cis and trans)

1,2-addition products

(cis and trans)
1,4-addition products

(c) $CH_3CH_2CH_2CH_2CH_3$

(d)   Hydration follows the same pattern as HBr addition; notice that, as in part (b), one of the 1,2-addition products is identical to one of the 1,4-addition products.

(cis and trans)

1,2-addition products

(cis and trans)
1,4-addition products

(e)   No reaction occurs.

(f)   A Diels–Alder reaction takes place, and the two possible adducts are formed. (Typically, the endo adduct predominates.)

endo   exo

15.61   (a)   Since the perpendicular relationship is necessary for chirality of a cumulene, allenes, as well as other cumulenes with an even number of cumulated double bonds, can be chiral. Indeed, 2,3-heptadiene is a chiral allene, and therefore exists as a pair of enantiomers that can, in principle, be separated by an enantiomeric resolution.

enantiomeric 2,3-heptadienes

(b)   The answers to both (a) and (b) follow from the geometry of cumulenes. Each additional cumulated double bond results in a 90° twist of one end of a molecule with respect to the other. So, in alkenes, the atoms connected to the double bond are in the same plane; in allenes, the atoms connected to the ends of the cumulated double bonds are in perpendicular planes; in cumulenes with three contiguous double bonds, the atoms attached to the terminal double bonds are in the same plane, as they are in alkenes. These can exist as E and Z stereoisomers, which are diastereomers:

Z isomer   *E* isomer

the diastereomeric 2,3,4-heptatrienes

15.62   1,3-Cyclohexadiene gives 3-bromocyclohexene by either 1,2- or 1,4-addition of HBr.

1,3-cyclohexadiene   **3-bromocyclohexene**

15.63   Because the bromine atom is added to the diene first, sterics will play a role in the site of its addition. The large bromine atom will add to carbon 1, forming an allylic radical intermediate. The unpaired electron is allylic, thus propagation can proceed with the abstraction of the hydrogen atom in either a 1,4- or 1,2-addition of HBr to the diene.

4-methyl-1,3-pentadiene

4-methyl-1,3-pentadiene

15.64   Severe van der Waals repulsions between the inner hydrogens (shown in the following structure) force the compound out of planarity. These van der Waals repulsions are so great in the planar conformation that not even aromaticity can compensate for the resulting destabilization.

15.65   The color of β-carotene is due to its chromophore of extensively conjugated double bonds. Catalytic hydrogenation would result in addition of hydrogen to these double bonds; therefore, catalytic hydrogenation of a β-carotene sample would convert it from a red-orange conjugated alkene into a colorless alkane.

15.66   Compounds *B* and *C* are highly conjugated, and would thus absorb in the UV-A and UV-B range and could be used as sunscreens. Compound *A* contains only a single benzene ring and would not be an effective sunscreen because it would absorb at a smaller wavelength.

15.67   In order to observe fluorescence, a compound must first absorb light to excite electrons to a higher energy state. Without sunlight, there wouldn't be sufficient energy available to excite those electrons.

15.68   The energy difference between the two waves can be calculated using Eq. 15.10. The 10^9 factor at the end of the calculation converts the difference in reciprocal wavelengths from nm to m.

$$\Delta E = \frac{hc}{\lambda_1} - \frac{hc}{\lambda_2} = hc\left(\frac{1}{\lambda_1} - \frac{1}{\lambda_2}\right) = (3.99 \times 10^{-13}\ \text{kJ mol}^{-1}\ \text{sec})(3.00 \times 10^8\ \text{m sec}^{-1})\left(\frac{1}{488} - \frac{1}{510}\right)\text{nm}^{-1}(10^9\ \text{nm m}^{-1})$$

$$= (12.0 \times 10^{-5}\ \text{kJ mol}^{-1})(8.84 \times 10^{-5}\ \text{nm}^{-1})(10^9\ \text{nm m}^{-1}) = 10.6\ \text{kJ mol}^{-1}$$

15.69   (a)   Other resonance structures are possible, derived from movement of just one or two curved arrows at a time. Those shown represent the most important structures because they demonstrate varying degrees of aromaticity.

| right ring is aromatic | left ring is aromatic | all rings are aromatic |

(b)   The explanation for part (a) applies to this part as well.

15.70   The carbon skeleton of compound *A*, from the hydrogenation data, is related to that of hexane; that is, it has an unbranched chain of six carbons. The two degrees of unsaturation indicate that the compound is either a diene or an alkyne. The IR data suggests that the compound is not an alkyne; the partial hydrogenation data and the observation of optical activity indicate that the compound is an allene. The only possible chiral allene is 2,3-hexadiene, $H_3C—CH{=}C{=}CH—CH_2CH_3$. Indeed, partial hydrogenation of this diene would give a mixture of *cis*-2-hexene and *cis*-3-hexene.

15.71   In order for a diene to react in a Diels–Alder reaction, it must be able to assume an *s*-cis conformation in the transition state.

van der Waals repulsions are
much more severe when R = *tert*-butyl
than when R = H

When R = *tert*-butyl, the van der Waals repulsions in the *s*-cis conformation are much more severe than when R = methyl. Consequently, when R = *tert*-butyl, the transition state of any Diels–Alder reaction is destabilized by these van der Waals repulsions to a greater extent than when R = methyl. Evidently, the destabilization is so great that the Diels–Alder reaction with maleic anhydride does not occur at all when R = *tert*-butyl.

15.72 The product of such a Diels–Alder reaction would be most interesting:

This is a cyclic triene containing three cumulated double bonds within the ring. Such an adduct could never form, because the requirement for linear geometry in the cumulated double bonds of the product would result in a compound with immense angle strain. A related explanation is that the 2p orbitals of maleic anhydride and those of the alkyne must overlap as shown in Figure 15.13 on text p. 824. Because the alkyne is constrained to linear geometry, the transition state for such a reaction would either be very strained (if the alkyne units were bent) or would contain extremely poor electronic overlap (if the alkyne units remained linear).

15.73 (a) The dissociation constant, K_d, can be calculated by using a form of Eq. 3.61a on p. 145. The dissociation constant for *N*-methylquinolinium ion with molecule *A* is

$$K_d = 10^{-\left(\frac{\Delta G°}{2.3RT}\right)} = 10^{-\left(\frac{28.9 \text{ kJ mol}^{-1}}{5.71 \text{ kJ mol}^{-1}}\right)} = 2.96 \times 10^{-6}$$

The dissociation constant for 4-methylquinoline with molecule *A* is 5.66×10^{-5} (or 56.6×10^{-6}, for comparative purposes). This is a larger dissociation constant, meaning that at equilibrium there is a larger proportion of the dissociated species (molecule *A* and 4-methylquinoline) than there is of the complexed pair, relative to *N*-methylquinolinium ion and molecule *A* and its complexed pair. Restated, *N*-methylquinolinium ion binds more strongly to molecule *A* than *N*-methylquinoline does.

(b) Since *N*-methylquinolium ion is positively charged, it is attracted to the aromatic rings of the barrel-shaped molecule *A* more strongly than neutral 4-methylquinoline.

(c) In addition to the pi-cation interaction described in part (b), there are also electrostatic attractions between the negatively charged groups in molecule *B* and the positively charged *N*-methylquinolium ion.

15.74 (a) The 1,2- and 1,4-addition products formed in the reaction of 1,3-pentadiene and HCl are identical if we assume that all double bonds retain their *E* stereochemistry.

H₂C=CH—CH=CH—CH₃   H₂C=CH—CH=CH—CH₃

H   Cl   (1,2-addition)   H   Cl   (1,4-addition)

H₃C—CH—CH=CH—CH₃   H₃C—CH=CH—CH—CH₃
         | |
         Cl   Cl

identical

(b)   There can be no preference for 1,2- versus 1,4-addition on the basis of the product stability, because, as shown in part (a), the products are the same. The problem is that we can't tell one mode of addition from the other! The use of D—Cl, however, solves this problem. The use of deuterium allows us to distinguish between the hydrogen that has added and the ones that were originally in the diene.

H₂C=CH—CH=CH—CH₃   H₂C=CH—CH=CH—CH₃

D   Cl   (1,2-addition)   D   Cl   (1,4-addition)

H₂C—CH—CH=CH—CH₃   H₂C—CH=CH—CH—CH₃
    | | | |
    D   Cl   D   Cl

The two products are different. The isotope has a negligible effect on product stability.

(c)   According to the text, the product of 1,2-addition is preferred, and it is shown in the solution to part (b). In fact, this result was observed experimentally. The actual ratio is about 70:30 in favor of 1,2-addition. (How would we tell one product from the other?) This experiment, reported in 1979 by J. E. Nordlander of Case Western Reserve University, established clearly that product stability has nothing to do with the kinetic preference. The arguments in the text are the only reasonable alternative.

15.75 (a)   Unless the reaction is allowed to run for a very long time, this is a kinetically controlled reaction, because the ratio of the two products is different from the ratio at equilibrium.

(b)   Compound C has a double bond with four substituents; compound B has a double bond with three substituents. Because substitution at the double bond enhances alkene stability, C is more stable and therefore would be favored at equilibrium.

(c)   The reaction occurs through the following carbocation intermediate by a typical E1 process (see Sec. 11.2).

abstraction of either proton by ⁻OTs gives product B

Ph

H   CH₃

H   H   abstraction of only this proton by ⁻OTs gives product C

The formation of the product involves loss of a proton from this intermediate. Alkene B can form in two ways, namely, by abstraction of either one of two protons. Alkene C can form in only one way. Therefore, the formation of alkene B is favored on statistical grounds. It is interesting that, in Fig. P15.75, the ratio of products at zero time is indeed about 2:1.

15.76 (a) Product *A* is the kinetic product because it is formed more rapidly.

(b) The relative stabilities of the two products can be deduced from their chair conformations. The 1,3-diaxial interactions of the axial bromines with the axial methyl group (on the top) and the axial hydrogen (on the bottom) cause compound *A* to be less stable. (See the conformations of *A* and *B* in the next part of the solution.)

(c) Two possible bromonium ions can be formed, which we'll call *X* and *Y*.

The carbons are numbered for reference. We'll come back later to the issue of which is preferred.

Here's the key point: The nucleophilic reaction of Br⁻ with the bromonium ion *must result in a product conformation in which the two Br atoms are anti!* This is a consequence of the geometry of opposite-side substitution. Such a conformation can be reached by the reaction of Br⁻ at either carbon-3 or carbon-4 in each of the bromonium ions. We now analyze each of these reaction modes in turn. Opposite-side substitution of Br⁻ at carbon-3 of *X* gives a *boat conformation of compound B*. This conformation can then relax to *B* itself.

As the nucleophilic reaction occurs, carbon-3 swings down and the three groups bound to it swing up—inversion of configuration. If you have trouble seeing why this reaction should generate a boat, look at the reverse reaction. Start with compound *B* and imagine formation of a bromonium ion from *B* by the reverse reaction. To form this ion, the two bromines must achieve an anti conformation—that is, a dihedral angle of 180°. (Just run this backwards to see this.) Only by forming a boat can backside substitution occur in the reverse direction; therefore, it's also the only way backside substitution can occur in the forward direction. The forward and reverse of a reaction must occur by the same mechanism! (Principle of microscopic reversibility; text Sec. 4.10C, p. 228.)

The nucleophilic reaction of Br⁻ at carbon-4 of ion *X*, however, gives the required anti arrangement of the two bromines in a *chair conformation*:

This gives compound *A* directly. Therefore, bromonium ion *X* is faced with two non-optimal choices: it can open to give a boat, or it can open to give the diaxial form *B*. Because boats have a very high energy, the latter option is the least energetic alternative and is therefore the one observed.

Now let's consider ion *Y* in the same way. The analysis is the same, except that the reaction of Br⁻ at carbon-4 gives a boat conformation of *B*, and reaction of Br⁻ at carbon-3 gives *A*. Here's the reaction at carbon-4 to give a boat:

Y → another boat bonformation of *B*

(We leave it to you to show the reaction of Br⁻ at carbon-3, which gives isomer *A* in the chair conformation.) Therefore, ion *Y* is faced with the same two nonoptimal choices as ion *X*, and the result is the same. In other words, the less stable product is formed more rapidly. The more stable product is formed slowly because of the high-energy boat intermediate required.

Thus, the relative stability of the two ions *X* and *Y* is irrelevant. *Y* might be a little more stable because it avoids the 1,3-diaxial interaction with the methyl group. But the outcome is the same no matter which ion forms. In other words, the outcome is determined by the relative rates of the two product-determining steps in the reaction of either bromonium ion.

(d)   Either bromonium ion *X* or bromonium ion *Y* can form from compound *A*; this process is merely the reverse of the formation of the dibromides from the bromonium ion.

A → *X*

A → *Y*

We've shown previously how bromide can react with either of these bromonium ions to give the more stable product via boat conformations. Although these reactions occur slowly, once they occur, they form the more stable product, thus providing a route to bring the two compounds into equilibrium.

15.77   Assume that *A* and *B* are 1,2- and 1,4-addition products.

$$H_2C{=}CH{-}CH{=}CH_2 \xrightarrow{Cl_2} H_2C{-}CH{-}CH{=}CH_2 + ClCH_2{-}CH{=}CH{-}CH_2{-}Cl$$

(with Cl and Cl on the first two carbons)

1,2-addition product   1,4-addition product

Which is *A* and which is *B*? And what is the stereochemistry of the 1,4-addition product? To answer these questions, note that further addition of Cl₂ to each gives a 1,2,3,4-tetrachlorobutane. Assume that addition of Cl₂ is an anti-addition. Because the 1,4-addition product has two stereocenters, anti-addition of Cl₂ will give the meso tetrachloride if the alkene is trans, and will give the racemic tetrachloride if the alkene is cis. (If this point is not clear, review text Sec. 7.8C.) Because compound *B* gives a single tetrachloride *C* with meso stereochemistry, it follows, then, that the 1,4-addition product is compound *B*, and that it has trans stereochemistry. The 1,2-addition product, then, is compound *A*. Addition of Cl_2 to compound *A* gives both diastereomers of the tetrachloride. To summarize:

3,4-dichloro-1-butene
compound *A*

***meso*-1,2,3,4-tetrachlorobutane**
compound *C*

(±)-1,2,3,4-tetrachlorobutane
compound *D*

compound *B*

15.78 Reaction of KH with cyclopentadiene forms the potassium salt of the cyclopentadienyl anion (structure *X* below) plus dihydrogen, H_2. In this anion, all carbons, except for the isotope, are equivalent by resonance. The isotope makes no detectable difference in the relative importance of the resonance contributors—that is, except for the position of the isotope, all resonance contributors are identical. Therefore, protonation of this anion by H_2O occurs at each carbon with equal (20%) probability. Protonation at carbons 2 and 5 gives the same product; and protonation at carbons 3 and 4 gives the same product. Therefore, each of these products is obtained in (2 × 20%) = 40% yield.

from proton at—   C-1   C-2 and C-5   C-3 and C-4

15.79 The conjugate-base anions of both compounds are aromatic. However, pyrrole itself is aromatic, whereas 1,3-cyclopentadiene is not. Therefore, upon ionization, 1,3-cyclopentadiene gains aromaticity, whereas pyrrole does not. An ammonium ion is typically 23 pK_a units more acidic (32 versus 55) than an alkane; the gain of aromaticity in the ionization of the hydrocarbon makes it 2 pK_a units more acidic than the amine. The net difference of 25 pK_a units corresponds to $(2.3RT)(\Delta pK_a) = 5.71 \times 25 = 143$ kJ mol^{-1} for the gain of aromaticity. (See Eq. 3.61b, text p. 145.) This energetic advantage of aromaticity is remarkably similar to the resonance energy of benzene (Sec. 15.7C of the text).

15.80 (a) The conjugate-base enolate ion is stabilized by the polar (electron-withdrawing) effect of the nearby carbonyl bond dipole as well as by resonance:

Recall (Sec. 3.6, text p. 140) that stabilization of a conjugate base increases the acidity of its conjugate acid.

(b) While the ionization of the molecule at either position would benefit from the polar effect of the nearby carbonyl group, only the conjugate-base enolate ion resulting from the ionization of hydrogen H_b would be enhanced by the resonance effect. The enolate ion resulting from the ionization of H_a is not stabilized by resonance, because resonance is geometrically impossible. The orbital interaction required for resonance interaction requires an impossible twisting of the molecule. (See discussion on pp.837-8 in the text.) To put it another way, the resonance structure violates Bredt's rule:

the conjugate-base anion
from the loss of proton H^a

Therefore, the ionization of H_b would have the lower pK_a; that is, H_b is more acidic.

15.81 The carbocation intermediate in the S_N1 solvolysis of compound *A* is resonance-stabilized, but the carbocation intermediate in the solvolysis of compound *B* is not. Because the relative energies of transition states reflect the relative energies of the corresponding carbocation intermediates (Hammond's postulate), the solvolysis of *A*, which gives the more stable carbocation, is much faster.

carbocation intermediate in the
solvolysis of compound *A*

carbocation intermediate in the
solvolysis of compound *B*

If compound *C* gives a doubly allylic carbocation, why should it be solvolytically inert? The answer is that the cation has a continuous cycle of π electrons. Because this cation contains $4n$ ($n = 1$) π electrons, it is antiaromatic and therefore very unstable. In fact, it is so unstable that it does not form.

antiaromatic
and unstable

15.82 This compound behaves like a salt because it *is* a salt. Tropylium bromide ionizes readily to bromide ion and the tropylium cation, which, because it has a continuous cycle of six ($4n + 2$, $n = 1$) π electrons, is aromatic, very stable, and very easily formed.

tropillium bromide

tropillium cation;
aromatic and stable

15.83 The hybridization of the nitrogen in azepine is sp³, whereas the hybridizations of the nitrogen atoms in both pyrrole and pyridine are sp². Pyridine's nitrogen is sp² hybridized as it participates in a pi bond with an adjacent carbon atom; its unshared pair of electrons is in the plane of the ring. Pyrrole's nitrogen is sp²-hybridized in order for the unshared pair of electrons on nitrogen to occupy a 2p orbital, contributing to the $4n + 2$ π-electron system to make the ring aromatic. The azepine ring remains sp³ hybridized to avoid having $4n$ π electrons and the resulting antiaromaticity.

15.84 (a) The triple bond hydrogenates, but the phenyl rings resist hydrogenation because they have aromatic character. The product of syn-addition is the cis alkene.

cis-stilbene
(*cis*-1,2-diphenylethylene)

(b) This is an ordinary Diels–Alder reaction; because the diene is cyclic, a bicyclic product is formed. Two diastereomers are possible. (Each is formed as the racemate, only one enantiomer of which is shown below.)

endo   exo

(c) This is a Diels–Alder reaction with the alkyne as the dienophile. The phenyl rings are cis because, in the s-cis conformation of the diene, they are both "outer" groups. (See Eqs. 15.29a–b, text pp. 825–6.)

(d) The conjugated-diene part of the triene undergoes a Diels–Alder reaction with a double bond of the benzoquinone. Because there are two double bonds in the quinone, a second Diels–Alder reaction also occurs. Two constitutional isomers result from the second addition because the diene can add in two different orientations.

product of first
Diels–Alder reaction

Moreover, each of these constitutional isomers can exist as two diastereomers:

and

(Which of these are chiral and can therefore exist as enantiomers, and which are meso compounds?)

(e)   The maleic anhydride reacts with the diene unit that is locked into an *s*-cis conformation. The diastereomer of the product formed results from addition to the face of the steroid opposite to the angular methyl groups, which blocks the β-face of the molecule. (See Sec. 7.6D of the text.) In this mode of addition, van der Waals repulsions between maleic anhydride and the angular methyl groups are avoided.

(f)   Because the product has the same number of carbons as the starting material, the reaction is intramolecular. In fact, it is an intramolecular Diels–Alder reaction.

 The bridgehead double bond in this product does not violate Bredt's rule (Sec. 7.6C) because it is part of a large ring (9 carbons).

(g)   The product is nickelocene, the nickel analog of ferrocene. (See Eq. 15.68, text p. 849.)

nickelocene

15.85   Although mycomycin has no asymmetric carbons, it does contain stereocenters: the outer carbons of the allene unit. Evidently, mycomycin is one enantiomer of this chiral allene. The chirality of certain allenes is discussed in Sec. 15.1C.)

15.86   Apply the reverse of the mechanism shown to the Diels–Alder adduct provided to give 1,3-diphenylisobenzofuran and a dienophile. Analysis of the product leads to the conclusion that the dienophile must be cyclopentyne.

cyclopentyne   **1,3-diphenylisobenzofuran**

Cyclopentyne is formed by a β-elimination of $MgBr_2$ from the Grignard reagent formed from 1,2-dibromocyclopentene. Cyclopentyne is particularly unstable because it contains a triple bond in a small (five-membered) ring; the bond angles of the sp-hybridized carbons deviate severely from 180°. It helps to understand the elimination by thinking of the Grignard reagent as a carbanion:

cyclopentyne
(reacts with 1,3-diphenyl-
isobenzofuran)

15.87   Follow the procedure used in solving Problem 15.86: mentally reverse the Diels–Alder reaction so that maleic anhydride is obtained; the other fragment obtained is compound *X*, which thus has the following structure:

reverse
Diels–Alder

**maleic
anhydride**

s-cis conformation

s-trans conformation

compound *X*

The formation of *X* results from ionization of the starting chloride to a tertiary allylic carbocation *A* followed by ring opening of *A* to a secondary carbocation *B*, which is captured by the nucleophile water. (Why should the tertiary allylic carbocation *A* rearrange to a secondary carbocation *B*?)

15.88   The structure of spiropentadiene is shown below; it undergoes a Diels–Alder reaction with two molar equivalents of 1,3-cyclopentadiene. Spiropentadiene is unstable because of its great ring strain. Although cyclopropane rings are retained in the product, they contain less ring strain than cyclopropene rings. (Why?)

spiropentadiene

15.89   (a)   This reaction involves a rearrangement of carbocation *X1* to a more stable carbocation *X2* followed by loss of a β-proton from the rearranged ion. (Only the part of the molecule involved in the reaction is shown.)

(b)   This is a reverse Diels–Alder reaction followed by another Diels–Alder reaction.

1,3-cyclopentadiene

maleic anhydride

maleic anhydride   (endo product)

(c)   The product *X* is a Diels–Alder adduct between α-phellandrene and the alkyne:

α-phellandrene

X

(Because the absolute configuration of α-phellandrene is not specified, it could be either of the two enantiomers, or the racemate, and the same is true of the Diels–Alder adduct *X*.)

Compound *X* undergoes a reverse Diels–Alder in a different manner to lose 3-methyl-1-butene and the aromatic compound diethyl 4-methylphthalate. The driving force for this reaction is, first, the formation of a volatile product (the alkene), and, second, the aromatic stability of the ultimate product.

X

3-methyl-1-butene

diethyl 4-methylphthalate

15.90   Remember that benzene rings have resonance structures in which the double bonds are completely delocalized. Let's draw [4]phenylene using the other resonance structures for the benzene rings:

cyclobutadiene

cyclobutadiene

cyclobutadiene

This structure shows that, in effect, that [4]phenylene incorporates three cyclobutadiene rings. Because these are antiaromatic, they are very unstable. What this means is that all of the benzene rings have no resonance stabilization, because when their π bonds are delocalized, cyclobutadienes result.

Hydrogenation solves this problem. When the central benzene ring is hydrogenated, the π bonds in the remaining benzene rings can be delocalized. (Cyclobutenes, unlike cyclobutadienes, are not continuous cycles of p orbitals and are *not* antiaromatic.) In other words, hydrogenation of the central benzene ring is especially favorable because it is accompanied by restoration of the aromaticity of all three outer benzene rings.

16 | The Chemistry of Benzene and Its Derivatives

CHAPTER 16 SUMMARY

- Benzene derivatives are distinguished spectroscopically by the NMR absorptions of their ring protons, which occur at greater chemical shift than the absorptions of vinylic protons. The unusual chemical shifts of aromatic protons are caused by the ring-current effect. The ^{13}C NMR absorptions of ring carbons are observed in about the same part of the spectrum as the absorptions of the vinylic carbons of alkenes.

- The most characteristic reaction of aromatic compounds is electrophilic aromatic substitution. In this type of reaction, an electrophile reacts with the π electrons of a benzene ring to form a resonance-stabilized carbocation. Loss of a proton from this ion gives a new aromatic compound.

- Examples of electrophilic aromatic substitution reactions discussed in this chapter are halogenation, used to prepare halobenzenes; nitration, used to prepare nitrobenzene derivatives; sulfonation, used to prepare benzenesulfonic acid derivatives; Friedel–Crafts alkylation, used to prepare alkylbenzenes; and Friedel–Crafts acylation, used to prepare aryl ketones.

- Electrophilic iodination of tyrosine, which occurs in the biosynthesis of thyroid hormones, is an example of electrophilic substitution in biology.

- Derivatives containing substituted benzene rings can undergo further substitution either at the ortho and para positions or at the meta position, depending on the ring substituent.

- Benzene rings with alkyl substituents or substituent groups that delocalize positive charge by resonance typically undergo substitution at the ortho and para positions; these substituent groups are called ortho, para-directing groups.

- Benzene rings with electronegative substituents that cannot stabilize carbocations or delocalize positive charge by resonance typically undergo substitution at the meta position. These substituents are called meta-directing groups.

- Whether a substituted benzene undergoes substitution more rapidly or more slowly than benzene itself is determined by the balance of resonance and polar effects of the substituents. Monosubstituted benzene rings containing an ortho, para-directing group other than halogen react more rapidly in electrophilic aromatic substitution than benzene itself. Monosubstituted benzene rings containing a halogen substituent or any meta-directing group react more slowly in electrophilic aromatic substitution than benzene itself. The effects of multiple substituents are roughly additive.

- The activating and directing effects of substituent groups must be taken into account when planning a synthesis.

- Alkene double bonds can generally be hydrogenated without affecting the benzene ring. Benzene derivatives, however, can be hydrogenated under relatively harsh conditions to cyclohexane derivatives.

- Polycyclic aromatic hydrocarbons are metabolized in phase I metabolism to epoxides and diol-epoxides, which can, in some cases, alkylate DNA. Benzo[a]pyrene is a potent environmental carcinogen that is metabolized this way.

STUDY GUIDE LINKS

16.1   NMR of Para-Substituted Benzene Derivatives

Be sure to realize that the NMR spectrum of a para-disubstituted benzene derivative shows the "two leaning doublet" pattern only when the two para substituents are different. When the substituents are the same, all ring protons are equivalent, and the resonance for the ring protons is a singlet.

The splitting pattern for para-disubstituted benzene derivatives is a little more subtle than might first appear, and can be rationalized as follows. Each ring proton has an ortho relationship to one nonequivalent proton and a para relationship to the other nonequivalent proton. Consider, for example, *p*-bromoethylbenzene, the structure of which is shown below, and the NMR spectrum of which is shown in Fig. 16.3 on text p. 875. In this structure, H^a is ortho to one H^b and para to the other H^b. Actually, both ortho and para splittings are present, but the para splittings are typically very small and not readily discernible in the spectrum. The major splitting observed is the one between adjacent protons—about 8 Hz in Fig. 16.3, a value within the range for ortho splitting given in Table 16.1 on text p. 874. Thus, to a useful approximation, the resonance of each proton is split into a doublet by its neighboring ortho proton. Since there are two chemically different types of protons—H^a and H^b—the spectrum consists of two such doublets.

16.2   Different Sources of the Same Reactive Intermediate

The example of Eqs. 16.21a–b on text p. 807–808 is a further illustration of an idea that has been discussed in an earlier chapter (see Study Guide Link 12.2): *Common Intermediates from Different Starting Materials.*

In the Friedel–Crafts alkylation, the reactive intermediate is a carbocation. In this section of the text, you've learned that carbocation intermediates can be generated from alkyl halides and Lewis acids. But you've also learned that carbocations are formed when alkenes are subjected to strongly acidic conditions:

The carbocation intermediates formed in such reactions can serve as electrophiles in electrophilic substitution just as carbocations generated in other ways can. Similarly, we have learned that secondary and tertiary alcohols react with strong acids to generate carbocation intermediates (Sec. 11.2). These carbocations are transformed into alkene dehydration products if the reaction conditions are designed to remove the alkene product from the reaction mixture. But if the alkene is not removed, and an aromatic hydrocarbon is present, these carbocation intermediates from alcohol dehydration can also serve as electrophiles in aromatic substitution. Problem 16.15(b) on text p. 887 is an illustration of this idea in practice.

16.3   Reaction Conditions and Reaction Rate

What do we mean by "harsh" or "mild" reaction conditions? In electrophilic aromatic substitution, harsh conditions include high temperature, strong Lewis acids, high concentrations of reagents, and/or the use of reagents that generate high concentrations of electrophiles. Mild conditions include lower temperature, weaker Lewis acids (or no Lewis acids), and lower concentrations of reagents.

Why is it that such conditions affect reaction rate? Remember, to say that an aromatic compound is highly activated means that it *reacts relatively rapidly* under a given set of conditions. If it is deactivated, it *reacts much more slowly* under the same conditions. Recall (Sec. 4.8B) that reaction rates increase with increasing temperature. Recall also, from your study of rate laws (Sec. 9.3A), that many reaction rates increase with increasing reagent concentration. So, raising the temperature or the reagent concentration, or using a reagent that generates a high concentration of an electrophile, increases the reaction rate. Such strategies are necessary in order to get a highly *deactivated* (unreactive) compound to react at a convenient rate. However, use of harsh reaction conditions on a highly *activated* (reactive) compound in many cases leads to a greater degree of substitution than desired (or to other side reactions); therefore, for highly activated compounds, milder conditions are used, such as weaker Lewis acids (or no Lewis acids), lower temperatures, or lower concentrations of reagents.

The examples in the text illustrate in a practical sense what is meant by "harsh" or "mild" conditions. It is *not* important to learn the exact conditions for each reaction. It *is* important for you to understand *why* certain compounds are more activated toward substitution than others, and to understand conceptually that the success of reaction can depend on the rational choice of reaction conditions.

REACTION REVIEW

I.   ELECTROPHILIC AROMATIC SUBSTITUTION REACTIONS OF BENZENE

A.   HALOGENATION OF BENZENE

1. Benzene reacts with bromine in the presence of a Lewis acid catalyst (FeBr₃) to yield a product in which one bromine is substituted for a ring hydrogen.

$$Br_2 \ + \ \text{(benzene)} \ \xrightarrow{FeBr_3} \ \text{(bromobenzene)}\!-\!Br \ + \ HBr$$

2. An analogous chlorination reaction using Cl₂ and FeCl₃ gives chlorobenzene.

3. Halogenation of benzene differs from the halogenation of alkenes by the type of product obtained and the reaction conditions.

    a.   Alkenes give addition products and react spontaneously under mild conditions.

    b.   Benzene gives substitution products and requires a Lewis acid as well as relatively severe conditions.

4.   The mechanism of the bromination of benzene consists of the following steps:

    a.   The first step is formation of a complex between Br_2 and the Lewis acid $FeBr_3$.

        i.   This complexation makes one of the bromines a much better leaving group.

        ii.   This complex of Br_2 and $FeBr_3$ reacts as if it contained the electron-deficient species $:Br:^+$.

    b.   In the second step, this complex reacts with the π electrons of the benzene ring.

resonance-stabilized
carbocation intermediate

        i.   This step disrupts the aromatic stabilization of the benzene ring (thus the need for the harsh conditions).

        ii.   The carbocation intermediate forms only because it is resonance stabilized.

    c.   The reaction is completed when a bromide ion acts as a base to remove the ring proton and give the products; by losing a β-proton, the carbocation can form a stable aromatic compound.

B.   NITRATION OF BENZENE

1.   Benzene reacts with concentrated nitric acid, usually in the presence of a sulfuric acid catalyst, to form nitrobenzene.

2.   In this reaction, called nitration, the nitro group, $-NO_2$, is introduced into the benzene ring by electrophilic substitution; the electrophile is the nitronium ion, $^+NO_2$.

$$H_2SO_4 \; + \; HNO_3 \longrightarrow \; ^+NO_2 \; HSO_4^- \; + \; H_2O$$

C.   SULFONATION OF BENZENE

1.   Benzene reacts with a solution of sulfur trioxide in H_2SO_4 called fuming sulfuric acid to form benzenesulfonic acid.

2.   In this reaction, called sulfonation, the sulfonic acid group, $-SO_3H$, is introduced into the benzene ring by electrophilic substitution; the electrophile is the neutral compound sulfur trioxide, SO_3.

3. Sulfonic acids such as benzenesulfonic acid are rather strong acids.
4. Sulfonation, unlike many electrophilic aromatic substitution reactions, is reversible; the sulfonic acid group is replaced by a hydrogen when sulfonic acids are heated with steam.

(steam)

D. FRIEDEL–CRAFTS ALKYLATION OF BENZENE

1. The reaction of an alkyl halide with benzene in the presence of a Lewis acid gives an alkylbenzene; this type of reaction is called Friedel–Crafts alkylation.

2. The electrophile in a Friedel–Crafts alkylation reaction is formed by complexation of the Lewis acid $AlCl_3$ with the halogen of an alkyl halide.

   a. Either the alkyl halide–Lewis acid complex, or the carbocation derived from it, can serve as the electrophile in a Friedel–Crafts reaction.

   b. Rearrangements of alkyl groups are observed in some Friedel–Crafts alkylations if the carbocation intermediate is prone to rearrangement.

3. A catalytic amount (much less than one equivalent) of the $AlCl_3$ catalyst can be used in this reaction.
4. The alkylbenzene products are more reactive than benzene itself.

   a. The product can undergo further alkylation, and mixtures of products alkylated to different extents are observed.

   b. A monoalkylation product can be obtained in good yield if a large excess of the starting material is used; this strategy is practical only if the starting material is cheap, and if it can be readily separated from the product.

5. Alkenes or alcohols with an acid catalyst can also be used as the alkylating agents in Friedel–Crafts alkylation reactions.

6. Friedel–Crafts alkylation is generally not useful on compounds that are more deactivated than benzene itself.

E. *FRIEDEL–CRAFTS ACYLATION OF BENZENE*

1. When benzene reacts with an acid chloride in the presence of a Lewis acid catalyst such as aluminum trichloride, $AlCl_3$, a ketone is formed.

    a. This reaction is an example of a Friedel–Crafts acylation.
    b. An acyl group, typically derived from an acid chloride, is introduced into an aromatic ring in the presence of a Lewis acid catalyst.

2. The electrophile in the Friedel–Crafts acylation reaction is a carbocation called an acylium ion.
    a. This ion is formed when the acid chloride reacts with the Lewis acid $AlCl_3$.

    b. The substitution occurs when the benzene π electrons react with the acylium ion, and a proton is lost from the resulting carbocation intermediate.

3. The ketone product of the Friedel–Crafts acylation reacts with the Lewis acid catalyst to form a complex that is catalytically inactive; this has two consequences:
    a. Slightly more than one equivalent of the catalyst must be used:
        i.   One equivalent reacts with the product.
        ii. An additional catalytic amount ensures the presence of catalyst throughout the reaction.
    b. The complex must be destroyed before the ketone product can be isolated; this is usually accomplished by pouring the reaction mixture into ice water.

4. Because the ketone products of acylation are much less reactive than the benzene starting material, acylation occurs only once.

5. Friedel–Crafts acylation does not occur on a benzene ring substituted solely with one or more meta-directing groups. Thus, for example, nitrobenzene does not undergo Friedel–Crafts acylation.

6. The Friedel–Crafts acylation occurs intramolecularly when acylation results in the formation of a five- or six-membered ring.

a. The intramolecular process is much faster than reaction with the acylium ion on the phenyl ring of another molecule.

b. This type of reaction can only occur at an adjacent ortho position.

7. The Friedel–Crafts acylation reaction is important for two reasons.

a. It is an excellent method for the synthesis of aromatic ketones.

b. It is another method for the formation of carbon-carbon bonds (see a partial list on text page 889 or a complete list in Appendix VI, text page A-13).

II. ADDITION REACTIONS OF BENZENE DERIVATIVES

A. HYDROGENATION OF BENZENE DERIVATIVES

1. Because of its aromatic stability, the benzene ring is resistant to conditions used to hydrogenate ordinary double bonds.

2. Aromatic rings can be hydrogenated under extreme conditions of temperature and/or pressure.

3. Catalytic hydrogenation of benzene derivatives gives the corresponding cyclohexanes, and cannot be stopped at the cyclohexadiene or cyclohexene stages.

4. Hydrogenation of the first double bond of benzene is an endothermic (thermodynamically unfavorable) reaction and requires energy (heat or pressure) to take place.

5. Once hydrogenation of the first double bond has taken place, the aromatic character of the benzene ring is lost; consequently, hydrogenation of the remaining double bonds under the harsh reaction conditions is instantaneous.

TABLE SG16.1	Electrophilic Aromatic Substitution Summary	
Electrophile	**Reaction Name**	**Product**
Cl—$\overset{+}{C}$l ---- $\bar{F}eCl_3$	chlorination	chlorobenzene
Br—$\overset{+}{B}$r ---- $\bar{F}eBr_3$	bromination	bromobenzene
SO_3 in H_2SO_4	sulfonation	benzenesulfonic acid
$^+NO_2$	nitration	nitrobenzene
R—C≡$\overset{+}{\ddot{O}}$ $^-AlCl_4$	Friedel–Crafts acylation	an aryl ketone
R—Cl ---- $AlCl_3$ or R^+ $^-AlCl_4$	Friedel–Crafts alkylation	an alkylbenzene

SOLUTIONS TO PROBLEMS

Solutions to Focused Problems

16.1   (a)   *m*-Chloroethylbenzene or 1-chloro-3-ethylbenzene
(b)   *o*-Diethylbenzene or 1,2-diethylbenzene
(c)   *p*-Nitrostyrene or 1-nitro-4-vinylbenzene
(d)   2,4-Dichlorophenol
(e)   2-Bromo-1-chloro-5-fluoro-4-iodo-3-nitrobenzene. Because there is no substituent eligible to be cited as a principal group, the order of alphabetical citation (chlorine before fluorine) determines numbering.
(f)   Benzylbenzene or (phenylmethyl)benzene (also commonly called diphenylmethane)

16.2   (a)

***p*-chlorophenol**

(b)

***m*-nitrotoluene**
3-nitrotoluene

(c)

3,4-dichlorotoluene

(d)

1-bromo-2-propylbenzene

(e)

methyl phenyl ether
(anisole)

(f)

benzyl methyl ether,
(methoxymethyl)benzene

(g)

***p*-xylene**

(h)

***o*-cresol**
(2-methylphenol)

(i)

2,4,6-trichloroanisole

16.3   Add about 25 °C per carbon relative to benzene (80 °C):

(a)   ethylbenzene: 130 °C (actual: 136 °C)   (c)   *p*-xylene: 130 °C (actual: 138 °C)
(b)   propylbenzene: 155 °C (actual: 159 °C)

16.4 The aromatic compound has NMR absorptions with greater chemical shift in each case because of the ring current (Fig. 16.2, text p. 873).

(a) Because thiophene is aromatic, its NMR absorptions occur at greater chemical shift than those of divinyl sulfide.

(b) The protons of benzene have considerably greater chemical shift because benzene is aromatic and 1,4-cyclohexadiene is not.

16.5 (a) This hydrocarbon is a continuous cycle of 18 π electrons. Because all atoms are trigonal planar and all bond angles are 120°, this compound can be planar. It meets the $4n + 2$ criterion for aromaticity for $n = 4$.

(b) The ring current brings about the expected increase in the local field at protons *outside* the ring, which are *deshielded*; but, from Fig. 16.2 on text p. 873, it should have exactly the opposite effect on the protons *inside* the ring. These protons should experience a smaller local field—that is, they are *shielded* and have an unusually small chemical shift.. The resonance at δ 9.28 is that of the outer protons, and the resonance at δ (–2.99) is that of the inner protons. The relative integral is in agreement with this assignment.

16.6 (a) Mesitylene should give an NMR spectrum containing only two singlets in the integral ratio 3:1 for the methyl groups and ring protons, respectively. *p*-Ethyltoluene should contain a methyl triplet at smallest chemical shift (around δ 1.0), a methyl singlet at the next smallest chemical shift (around δ 2.3), and a methylene quartet at somewhat greater chemical shift. There are two possibilities for its ring protons. The most general one is a typical para-substitution pattern (apparent pair of doublets) for the ring protons in the δ 6.5–7.5 region. However, because the methyl and ethyl groups are so similar in their electronic characteristics, another possibility is that the nonequivalence of the ring protons is so slight as not to be apparent. In fact, this is the case; the ring protons are a singlet.

mesitylene
two singlets in its NMR spectrum

p-ethyltoluene
a methyl singlet, a methyl triplet, a methylene quartet,
and aromatic absorption in its NMR spectrum

(b) Among other features, the NMR spectrum of 1-bromo-4-ethylbenzene has a typical ethyl quartet/triplet splitting pattern and a typical para-substitution pattern for the ring protons, as shown in Fig. 16.3, text p. 875. The spectrum of (2-bromoethyl)benzene, in contrast, should show a pair of triplets for the methylene protons and a complex pattern for the ring protons. If this isn't enough to distinguish the two compounds, the integral of the ring protons relative to the integral of the remaining protons is different in the two compounds

1-bromo-4-ethylbenzene

(2-bromoethyl)benzene

(c) These two NMR spectra will be quite different. The NMR of 1-bromo-4-methylbenzene will consist only of a methyl singlet (around δ 2.3) and two apparent doublets in the aromatic region (around δ 6.5–7.5). The NMR of *trans*-1-bromo-4-methylcyclohexane will be quite complicated with 12 aliphatic protons and complex splitting of the ring protons. Additionally, the methyl group will be split into a doublet.

trans-1-bromo-4-methylcyclohexane

1-bromo-4-methylbenzene

16.7   (a)   The signal for an exchangeable proton at δ 3.76 indicates that the compound contains an —OH group. It also contains aromatic absorptions with a splitting pattern that suggests para substitution. The splitting patterns of the δ 1.27 and δ 4.60 resonances indicate the partial structure H₃C—CH, and the chemical shift of the CH is consistent with its being α to both the ring and the OH group. The compound is shown at right on the previous page.

1-(4-methylphenyl)ethanol

   (b)   The IR spectrum indicates the presence of an OH group, and the chemical shift of the broad NMR resonance (δ 6.0) suggests that this could be a phenol. The splitting patterns of the δ 1.17 and δ 2.58 resonances show that the compound also contains an ethyl group, and the splitting pattern of the ring protons shows that the compound is a para-disubstituted benzene derivative. The compound is *p*-ethylphenol.

p-ethylphenol
(4-ethylphenol)

16.8   The following two methyl ethers with the formula C₇H₆OCl₂ would show five lines in their proton-decoupled ¹³C NMR spectra.

1,3-dichloro-2-methoxybenzene   **1,3-dichloro-5-methoxybenzene**

16.9   Count the resonances. Mesitylene has three resonances; isopropylbenzene has six.

isopropylbenzene

mesitylene

16.10   (a)   The two benzene rings are conjugated in compound *A*; because there are more π bonds in conjugation, the UV absorption occurs at longer wavelength than it does in ethylbenzene.

   (b)   Compound *B* is like two connected molecules of compound *C* in the same sense that compound *A* is like two connected ethylbenzene molecules. It might seem that the absorption of compound *B* should occur at greater wavelength than that of compound *C* for the reason given in part (a). That this is not the case means that the two rings in compound *B* are not conjugated. They are not conjugated because the rings are not coplanar; and they are not coplanar because the ortho methyl groups on the two rings have severe van der Waals repulsions that are relieved when the molecule adopts a nonplanar conformation:

van der Waals repulsions between methyl groups

van der Waals repulsions between methyl groups
planar conformation

van der Waals repulsions are relieved

van der Waals repulsions are relieved

nonplanar conformation

16.11 Because styrene has a double bond in conjugation with the ring and ethylbenzene does not, styrene has a greater λ_{max} in its UV spectrum.

16.12 Apply the steps shown in Eqs. 16.16a–b on text pp. 879-880 to the para position of bromobenzene.

16.13 The protonation of SO_3 by H_2SO_4:

The mechanism of sulfonation with the conjugate acid of SO_3 as the electrophile:

16.14 Apply the sulfonation mechanism shown in Eq. 16.21 on text p. 883 to the para position of toluene. Any of the resonance structures can be deprotonated by the base to regenerate aromaticity in the ring. Other resonance structures, not shown, exist with the negative charge delocalized on to each of the oxygens attached to the sulfur.

p-toluenesulfonic acid
(*p*-TsOH)

16.15   (a)   As discussed in Study Guide Link 16.2, carbocations generated in a variety of ways can be used as electrophiles in Friedel–Crafts reactions. In this case, the carbocation produced by the protonation of cyclohexene is the electrophile.

(b)   The same product is obtained because the same carbocation electrophile is involved. It is generated from cyclohexanol by protonation and loss of water. The carbocation then reacts with benzene as in part (a).

the same carbocation
electrophile involved
in part (a)

16.16   The product is *tert*-butylbenzene. The role of the Lewis acid BF_3 is to promote the ionization of HF. The mechanism of the reaction is as follows:

methylpropene

tert-butylbenzene

16.17   The hint, an unsaturation number of 5, and the fact that the product has the same number of carbons as the starting material suggest an intramolecular Friedel–Crafts alkylation to form a new ring.

> Remember that intramolecular reactions involving the formation of five- and six-membered rings are particularly rapid. (Text Sec. 12.8)

16.18 (a)

isobutyrophenone

(b)

benzophenone

16.19 The two possible Friedel–Crafts reactions:

16.20 That the product contains the same number of carbons as the starting material suggests an intramolecular Friedel–Crafts reaction. The unsaturation number of 6 corroborates this deduction. The mechanism below begins with the acylium ion. (For the formation of this ion from the acid chloride, see Eq. 16.31 on text p. 887).

an acylium
ion

A

16.21 (a) Table 16.2 indicates that the methoxy group is an ortho, para-directing group. (The product is mostly the para isomer.)

p-methoxyacetophenone   *o*-methoxyacetophenone

(b)   Table 16.2 indicates that alkyl groups are ortho, para-directing groups, and the ethyl group is a typical alkyl group:

1-ethyl-4-methylbenzene
(*p*-ethyltoluene)

1-ethyl-2-methylbenzene
(*o*-ethyltoluene)

16.22   Let E^+ be a general electrophile. The four resonance structures of the carbocation intermediate that results from reaction of E^+ at the position ortho to the methoxy group of anisole are as follows:

16.23   (a)   Because substitution occurs para to the phenyl group, the phenyl group is evidently an ortho, para-directing substituent.

(b)   The following resonance structures show that the electrons of the phenyl substituent can be used to stabilize the carbocation intermediate when substitution occurs at the para position. The electrons of the phenyl substituent cannot be delocalized in this way when substitution occurs at the meta position.

16.24   Notice that, except for alkyl and aryl groups, all ortho, para-directing groups have electron pairs on atoms adjacent to the benzene ring.

(a)   This substituent, like the methoxy group, is an ortho, para-directing group, a point that is confirmed by the fourth entry of Table 16.2 on text p. 891.

p-bromoacetanilide, or
N-(4-bromophenyl)acetamide

o-bromoacetanilide, or
N-(2-bromophenyl)acetamide

(b)   The three large carbon-fluorine bond dipoles result in substantial partial positive charge on the carbon of the CF_3 group; consequently, this is a meta-directing group.

Br
—CF$_3$

1-bromo-3-trifluoromethylbenzene

16.25 (a) The reaction-free energy profiles for electrophilic substitution of benzene, chlorobenzene at the para position, and chlorobenzene at the meta position are shown in **Fig. SG16.1**. Notice that chlorobenzene is less reactive (that is, has a greater standard free energy of activation) than benzene because chlorine is a deactivating substituent. Notice also that para-substitution reactions on chlorobenzene are faster than meta-substitution reactions because chlorine is an ortho, para-directing group.

   (b) The reaction-free energy profiles for electrophilic substitution of benzene, nitrobenzene at the meta position, and nitrobenzene at the para position are shown in **Fig**. **SG16.2** (following page). Notice that nitrobenzene is less reactive than benzene because the nitro group is a deactivating substituent. Notice also that meta-substitution reactions on nitrobenzene are faster than para-substitution reactions because the nitro group is a meta-directing group.

16.26 Although both the CH$_3$O— and the CH$_3$S— groups are activating, ortho, para-directing groups, the orbitals on sulfur containing the nonbonding electron pairs, like those on chlorine, are derived from quantum level 3. Consequently, the overlap of sulfur orbitals with the carbon orbitals of the ring is poorer than the overlap of the oxygen orbitals. (See Fig. 16.7 on text page 899; the orbitals on sulfur are analogous to those on chlorine in the figure.) In other words, the electron-donating resonance interaction of the sulfur is weaker than that of the oxygen. Although the electron-withdrawing (and therefore rate-retarding) polar effect of oxygen is much greater than that of sulfur, the resonance interaction of oxygen is so much more powerful that it is the dominant effect.

16.27 Bromination of *N,N*-dimethylaniline is faster because nitrogen has nonbonding electron pair that can stabilize the carbocation intermediate by resonance. As in the case of oxygen, the electron-withdrawing polar effect of nitrogen is much less important than its electron-donating resonance effect.

Figure SG16.1   Reaction-free energy profiles to accompany the solution to Problem 16.25(a).

Figure SG16.2   Reaction-free energy profiles to accompany the solution to Problem 16.25(b).

16.28   In this reaction, iodide ion, I^-, is oxidized by hydrogen peroxide, H_2O_2. Every oxidation is accompanied by a reduction, so the hydrogen peroxide is reduced to water. The balanced half reactions are shown below. The reactions were balanced according to the method shown in Sec. 11.6A.

$$2H_2O + I^- \longrightarrow I{-}OH + H_3O^+ + 2e^-$$
$$2H_3O^+ + 2e^- + H_2O_2 \longrightarrow 4H_2O$$
$$\overline{H_3O^+ + I^- + H_2O_2 \longrightarrow 2H_2O + I{-}OH}\quad \text{(Text Eq. 16.46)}$$

16.29   Tyrosine (A) would undergo iodination more rapidly. The aromatic ring of B is deactivated because the carbocation intermediate is destabilized by the electron-withdrawing polar effect of the fluorines.

carbocation intermediate is destabilized
by nearby positve charges

16.30   (a)   Each substituent is an ortho, para-directing group. Two products satisfy the directing effects of both groups.

2-bromo-4-methyl-   +   **4-bromo-2-methyl-**
benzenesulfonic acid   **benzenesulfonic acid**

(b)   As in part (a), each substituent is an ortho, para-directing group. Two products satisfy the directing effects of both groups.

**2-bromo-1-nitro-
4-iodobenzene**   **4-bromo-1-nitro-
2-iodobenzene**

16.31   (a)   The order of increasingly harsh reaction conditions is *m*-xylene < benzene < *p*-dichlorobenzene. The reason is that the methyl groups of *m*-xylene are activating groups and the chlorines of *p*-dichlorobenzene are deactivating groups.

(b)   The order is anisole < toluene < chlorobenzene. Chlorobenzene requires the harshest conditions because chlorine is a deactivating group. Anisole requires the mildest conditions because the methoxy group is more activating than the methyl group of toluene. (See Table 16.2 on text p. 891.)

16.32   Because both groups are meta-directing groups, it might seem that either could be introduced first. However, the Friedel–Crafts acylation cannot be carried out on nitrobenzene because, as discussed on text p. 905, the nitro group is too deactivating. Consequently, the acetyl group must be introduced first.

benzene   **acetophenone**   **m-nitroacetophenone**

16.33   (a)   Hydrogenate cyclohexylbenzene, which, in turn, is prepared as shown in Eqs. 16.29a–b on text pp. 886–887. Note that chlorocyclohexane and AlCl$_3$, or cyclohexanol and either H$_2$SO$_4$ or H$_3$PO$_4$, may be used as the source of the electrophile instead of cyclohexene; see Problem 16.15(b), text p. 887 and Study Guide Link 16.2.

cyclohexylbenzene   **cyclohexylcyclohexane**

(b)   Hydrogenate *tert*-butylbenzene, which, in turn, is prepared by Friedel–Crafts alkylation as shown in Eq. 16.26 on text p. 886 or by the reaction shown in the solution to Problem 16.16.

tert-butylbenzene   **tert-butylcyclohexane**

16.34   The nucleophilic nitrogen adds to the less sterically-hindered side of the epoxide, and from the top face, as drawn. A proton transfer (likely mediated by water) follows, yielding the neutral, covalently-modified DNA molecule.

16.35 Propene is protonated by sulfuric acid to produce the electrophile, a secondary carbocation. The electrophilic aromatic substitution proceeds from there, generating a resonance-stabilized carbocation intermediate before losing a proton to regain its aromaticity.

Solutions to Skills Objectives Problems

16.36 (a) 1,2,4,5-Tetrachlorobenzene
 (b) *p*-Ethylanisole or 4-ethylanisole
 (c) Phenylacetylene, phenylethyne, ethynylbenzene
 (d) 3-Bromo-5-nitrophenol

16.37 (a)

2-chlororesorcinol

(b)

***p*-nitroaniline**

(c) $H_2C{=}CH$—⟨benzene ring⟩—$CH{=}CH_2$

1,4-divinylbenzene

(d) HS—⟨benzene ring⟩—CH_3

4-methylbenzenethiol

16.38   Only compound *A* should have a proton NMR spectrum containing one six-proton singlet in the alkyl region (about δ 1.5). Only compound *B* should have a proton NMR spectrum that contains two closely spaced singlets in the integral ratio 2:1 in the benzylic proton region of the spectrum (about δ 2.3). Only compound *C* should have a proton NMR spectrum that contains two well-separated singlets in the integral ratio 3:1, the larger in the benzylic proton region and the smaller at about δ 4.5. All three compounds are also readily distinguished by the relative integrals of their aromatic proton absorptions.

16.39   Ethylbenzene and styrene each have 6 resonances in their carbon NMR spectra. However, all of the carbons in styrene occur at very high chemical shift (in the alkene and aromatic region), whereas two of the ethylbenzene resonances are in the benzyl and alkyl region at much smaller chemical shift. Xylene, because of its symmetry, has only three resonances in its carbon NMR spectrum.

16.40   (a)   The IR spectrum indicates the presence of an aromatic ring. The singlet at δ 3.72 suggests a methoxy group, and the apparent pair of doublets indicates a para-disubstituted benzene ring. The M + 2 peak in the mass spectrum with about one-third the intensity of the M = 142 peak suggests the presence of chlorine. A chlorine, a —OCH_3 group, and a para-disubstituted benzene ring fully account for the mass of 142. The compound *p*-chloroanisole (1-chloro-4-methoxybenzene) is consistent with all the data.

Cl—⟨benzene ring⟩—OCH_3

p-**chloroanisole**
(1-**chloro-4-methoxybenzene**)

When interpreting mass spectra, don't forget that you must use exact masses for isotopes. Thus, mass M = 142 corresponds to the compound containing ^{35}Cl, and the mass of the M + 2 peak corresponds to the compound containing ^{37}Cl.

(b)   The singlet at δ 3.8 suggests a methyl ether. The doublet of doublets centered at δ 7 suggest a para-disubstituted benzene ring. The complex pattern centered at δ 6.1 corresponds to two vinylic hydrogens, and the UV spectrum suggests a vinylic group conjugated with the benzene ring. The chemical shift of the δ 1.9 resonance suggests a vinylic methyl group, and its splitting of 6.6 Hz shows that this methyl group is coupled to one of the vinylic hydrogens. The 15.7 Hz splitting within the vinylic proton resonance shows that these protons are trans; the 965 cm^{-1} IR absorption confirms this deduction. All of these data conspire to suggest that compound *B* is (*E*)-4-methoxy-1-(1-propenyl)benzene, known more commonly as *trans*-*p*-methoxy-β-methylstyrene.

CH_3O—⟨benzene ring⟩—$C{=}C$ ... H, CH_3, H

(*E*)-**4-methoxy-1-(1-propenyl)benzene**
(*trans*-*p*-**methoxy-β-methylstyrene**)

16.41   (a)

NO₂ / CH₂CH₃ (1-ethyl-4-nitrobenzene) + NO₂ / CH₂CH₃ (1-ethyl-2-nitrobenzene)

1-ethyl-4-
nitrobenzene

1-ethyl-2-
nitrobenzene

(b)

SO₃H / CH₂CH₃ (4-ethylbenzenesulfonic acid) + SO₃H / CH₂CH₃ (2-ethylbenzenesulfonic acid)

4-ethylben-
zenesulfonic acid

2-ethylben-
zenesulfonic acid

(c)

O=C—CH₂CH₃ / CH₂CH₃ (p-ethylpropiophenone) + CH₂CH₃ / C(=O)CH₂CH₃ (o-ethylpropiophenone)

p-ethylpro-
piophenone

o-ethylpro-
piophenone

(d)

CH₃ / CH₂CH₃ (p-ethyltoluene) + CH₃ / CH₂CH₃ (o-ethyltoluene)

p-ethyltoluene   *o*-ethyltoluene

(e)

Br / CH₂CH₃ (1-bromo-4-ethylbenzene) + Br / CH₂CH₃ (1-bromo-2-ethylbenzene)

1-bromo-4-
ethylbenzene

1-bromo-2-
ethylbenzene

In part (d), ethylbenzene must be present in large excess for the monosubstitution products to be obtained. If ethylbenzene is not present in excess, products of dimethylation, trimethylation, etc., will also be formed; see Eq. 16.27 on text p. 886 and the discussion on text p. 905.

16.42   (a)

Cl / NO₂ (1-chloro-3-nitrobenzene)

1-chloro-3-nitrobenzene

(b)

NO₂ / NO₂ (1,3-dinitrobenzene or m-dinitrobenzene)

1,3-dinitrobenzene or
***m*-dinitrobenzene**

(c)   No electrophilic aromatic substitution reaction takes place. Friedel–Crafts acylation does not take place on any benzene derivative less reactive than the halobenzenes. In fact, nitrobenzene can be used as an inert solvent for Friedel–Crafts acylation.

16.43   (a)   Each alkyl halide group reacts with benzene in a separate Friedel–Crafts alkylation reaction.

(b)   The formula indicates that successive electrophilic aromatic substitution reactions have occurred; the product is triphenylmethane, Ph_3CH.

(c)   The reaction is an intramolecular Friedel–Crafts alkylation that forms a ring. Note that the oxygen activates the substitution.

[structure of chroman derivative with O and CH₃]

CH₃

(d)   As in the reaction mechanism shown in Eqs. 16.47a–b, hypoiodous acid will install an iodine ortho to an OH group on a benzene ring.

Since there is a large excess of I—OH, the other ortho position may be iodinated as well.

(e)   The cyclohexyl group, an alkyl group, directs nitration to the ortho and para positions of the benzene ring.

(mostly)

(f)   The nitro group is directed by both substituents to the position ortho to the methoxy group, and the bromine in the second reaction is directed to the other position ortho to the methoxy group.

3-bromo-4-methoxy-5-nitrobenzenesulfonic acid

16.44   (a)   Electrophilic aromatic substitution follows a three-step pattern. The first step involves the generation of an electrophile, E. The π electrons of the aromatic ring react with the electrophile, displacing a leaving group, and generating a non-aromatic resonance-stabilized carbocation with a tetrahedral, sp^3 carbon at the site of substitution. The leaving group acts as a base to remove a proton from the sp^3 carbon to regenerate a newly-substituted aromatic ring.

(b)   This reaction begins by generating an electrophile after the protonated alcohol dissociates to form a tertiary carbocation.

The mechanism then proceeds following the model above. The phosphoric acid catalyst is regenerated as it deprotonates the arenium ion to regenerate its aromaticity.

16.45 Because carbon-4 is para to the methoxy group, the carbocation intermediate involved in nitration at this carbon can be stabilized by electron delocalization from the oxygen, as it is in nitration at the para position of anisole. (See the colored structure in Eq. 16.37 on text p. 892.) Therefore, the methoxy group activates substitution at carbon-4, and nitration at carbon-4 of 1-methoxynaphthalene is faster than nitration of naphthalene.

When nitration occurs at carbon-5, the unshared electrons of the oxygen can be used to stabilize the carbocation intermediate by resonance; consequently, nitration at carbon-5 of 1-methoxynaphthalene is faster than nitration of naphthalene itself.

The carbocation intermediate from nitration at carbon-5:

Nitration at carbon-6 is analogous to nitration at a meta position of anisole; see Eq. 16.38 on text p. 892. In this case, the oxygen electrons cannot be delocalized to stabilize the carbocation. The only effect of the methoxy group in this case is its rate-retarding polar effect. Therefore, nitration at carbon-6 of 1-methoxynaphthalene is slower than nitration of naphthalene.

The carbocation intermediate from acylation at carbon-6:

This sample of three resonance structures illustrates that the positive charge cannot be delocalized to the carbon atom adjacent to the oxygen, thus the oxygen cannot participate in the delocalization of charge.

16.46   Friedel–Crafts acylation (or any other electrophilic substitution reaction) at carbon-2 of furan gives a carbocation intermediate with more resonance structures than acylation at carbon-3. Consequently, the carbocation intermediate from acylation at carbon-2 is more stable, and, by Hammond's postulate, acylation at carbon-2 is faster.

The carbocation intermediate from acylation at carbon-2:

three resonance structures

The carbocation intermediate from acylation at carbon-3:

two resonance structures

16.47   (a)   The boron–oxygen bond dipoles on benzeneboronic acid place substantial positive charge on boron. The boronic acid group thus has the characteristics of a meta-directing group, and the principal mononitration product is *m*-nitrobenzeneboronic acid (at right).

(b)   The carbocation intermediate involved in meta substitution has a greater separation between the positive charge of the carbocation and the positive charge of the substituent than does the carbocation intermediate involved in ortho, para substitution. Therefore, the $-\overset{+}{N}(CH_3)_3$ group is a meta-directing group, and the product is the meta-nitro derivative.

(c)   One benzene ring serves as an ortho, para-directing substituent on the other. (The directing effect of the phenyl substituent is explained in the solution to Focused Problem 16.23.) The product is the *p*-nitro derivative (plus perhaps a small amount of the *o*-nitro isomer).

(d)   Both the phenyl and the methoxy groups direct substitution to the positions marked with an asterisk (*). Nitration occurs at the three asterisked positions; the products are as follows:

16.48   Compounds C and D undergo bromination at the ortho and para positions; compound A undergoes bromination at the meta position (refer to the answer for Problem 47b); and the position of substitution in compound B depends on the balance of the stabilizing effect of the alkyl group and the destabilizing effect of the positive charge. (In fact, this compound brominates in the ortho and para positions.)

16.49   (a)   The boron on benzeneboronic acid (Problem 16.47a) has an empty 2p orbital and is not capable of stabilizing a carbocation by resonance. It would be difficult to introduce electron deficiency into a compound that is already electron deficient; therefore, this compound undergoes substitution much more slowly than benzene itself. The positive charge on the nitrogen in Problem 16.47b destabilizes the carbocation intermediate formed during electrophilic aromatic substitution and thus slows the reaction relative to benzene. Because a phenyl substituent can stabilize the carbocation intermediate formed when biphenyl undergoes nitration (Problem 16.47c), it is an activating group, and biphenyl is therefore more reactive than benzene. The ring containing the methoxy group in Problem 16.47d is activated toward substitution, thus reacting faster than benzene, because the carbocation intermediate can be stabilized by both the phenyl and the unshared pair of electrons on oxygen.

   (b)   Compound A from Problem 16.48 is less reactive in nitration, as explained in the solution to Problem 16.49a, and is undoubtedly less reactive in electrophilic substitution in general. Compound C is activated toward substitution because the substituent has nonbonding electron pair that can be used to stabilize the intermediate carbocation by resonance; therefore, compound C should be more reactive than benzene. Compounds B and D have alkyl substituents, which stabilize carbocations. However, the alkyl group of compound B contains a positively charged group that would interact unfavorably with a carbocation, likely offsetting the stabilizing effect of the alkyl carbon. Consequently, compound B is probably neither an activating nor a deactivating group. Compound D reacts faster than benzene itself.

16.50   (a)   In mesitylene, the three methyl groups all direct nitration to the same position; that is, they activate the same position of the ring. In 1,2,4-trimethylbenzene, only two of the three methyl groups activate the same ring position. Both trimethylbenzenes are more activated than toluene. Therefore, the order is

toluene < 1,2,4-trimethylbenzene < mesitylene

   (b)   The reactivity order follows from the relative activating effects of the substituents. (See the last column of Table 16.2 on text p. 891.)

nitrobenzene < chlorobenzene < benzene

   (c)   In p-chloroanisole, the chlorine and methoxy group direct to different positions; that is, the chlorine strongly deactivates the positions that the methoxy group activate, and the methoxy group deactivates the positions that are least deactivated by the chlorine. In the meta isomer, the directing effects of the two substituents do not conflict. In anisole, the deactivating chloro group is absent altogether. The order is:

p-chloroanisole < m-chloroanisole < anisole

   (d)   The reactivity order follows from the relative activating effects of the substituents. (See the last column of Table 16.2 on text p. 891.)

p-bromoacetophenone < acetophenone < p-methoxyacetophenone

16.51   The two possibilities are the acylation of anisole by benzoyl chloride (pair A), or the acylation of benzene by p-methoxybenzoyl chloride (pair B). Because the methoxy group activates electrophilic substitution at the ortho and para positions, the Friedel–Crafts reaction of pair A should occur under the milder conditions.

benzoyl chloride   anisole   p-methoxybenzoyl chloride   benzene

A   B

(Both reactions would work, however; and the use of pair B avoids the possibility of ortho substitution, although, with the Friedel–Crafts reaction, this does not generally occur to a great extent.)

16.52   In each synthesis that involves substitution on a benzene derivative that contains an ortho, para-directing group, only the product resulting from para substitution is shown.

(a)

benzene $\xrightarrow{CH_3Cl,\ AlCl_3}$ toluene $\xrightarrow{HNO_3,\ H_2SO_4}$ p-nitrotoluene
(excess)

(b)

benzene $\xrightarrow[\text{heat}]{Br_2\ (\text{excess}),\ FeBr_3}$ p-dibromobenzene

(c)

benzene $\xrightarrow{Cl_2,\ FeCl_3}$ chlorobenzene $\xrightarrow[\text{2) } H_2O,\ H_3O^+]{\text{1) } ClCCH_3,\ AlCl_3\ (1.1\ \text{equiv})}$ p-chloroacetophenone

(d)

benzene $\xrightarrow{HNO_3,\ H_2SO_4}$ nitrobenzene $\xrightarrow{H_2SO_4,\ SO_3}$ m-nitrobenzenesulfonic acid

(e)

chlorobenzene $\xrightarrow{HNO_3,\ H_2SO_4}$ p-chloronitrobenzene
prepared in part (c)

(f)

nitrobenzene $\xrightarrow{HNO_3\ (\text{fuming}),\ H_2SO_4}$ $\xrightarrow[\text{H}_2\text{SO}_4\ (\text{fuming})]{HNO_3\ (\text{fuming}),}$ 1,3,5-trinitrobenzene
prepared in part (d)

(g)

H3C—⟨benzene ring⟩—NO2 $\xrightarrow{\text{Br}_2 \text{ (excess), FeBr}_3}$ H3C—⟨benzene ring with Br, Br⟩—NO2

p-nitrotoluene
prepared in part (a)

2,6-dibromo-4-nitrotoluene

(h)

⟨toluene ring with CH3, NO2⟩ $\xrightarrow[\text{heat}]{\text{Br}_2 \text{ (excess), FeBr}_3}$ ⟨ring with CH3, NO2, Br, Br⟩

o-nitrotoluene
by-product of the
synthesis in part (a)

2,4-dibromo-6-nitrotoluene

The nitration of toluene actually gives more ortho isomer than para isomer of toluene, and the two nitrotoluene isomers are readily separated by fractional distillation. (See text Sec. 16.5A, subsection: The Ortho, Para Ratio.) o-Nitrotoluene is a good starting material for a number of ortho-substituted benzene derivatives.

(i)

⟨benzene⟩ $\xrightarrow{\text{C}_2\text{H}_5\text{Cl, AlCl}_3}$ ⟨benzene⟩—CH2CH3 $\xrightarrow[\text{2) H}_3\text{O}^+, \text{H}_2\text{O}]{\text{1) CH}_3\overset{\text{O}}{\text{CCl}}, \text{AlCl}_3}$ CH3C(=O)—⟨benzene⟩—CH2CH3 $\xrightarrow{\text{HNO}_3, \text{H}_2\text{SO}_4}$

benzene
(excess)

ethylbenzene

p-ethylacetophenone

CH3C(=O)—⟨benzene ring with NO2⟩—CH2CH3

4-ethyl-3-nitroacetophenone

(j)

⟨benzene⟩ + (CH3)3CCl $\xrightarrow{\text{AlCl}_3}$ (CH3)3C—⟨benzene⟩—C(CH3)3 $\xrightarrow[\substack{\text{high pressure}\\\text{heat}}]{\text{H}_2, \text{Ni}}$ (CH3)3C—⟨cyclohexane⟩—C(CH3)3

benzene

**2-chloro-
2-methylpropane
(tert-butyl chloride)
(excess)**

1,4-di-tert-butylbenzene

cis-1,4-di-tert-butylcyclohexane

(k)

⟨benzene⟩ + ⟨cyclopentene⟩ $\xrightarrow{\text{H}_2\text{SO}_4}$ ⟨benzene⟩—⟨cyclopentane⟩

benzene   **cyclopentene**
(excess)

cyclopentylbenzene

Cyclopentanol can be substituted for cyclopentene in this synthesis, or chlorocyclopentane and AlCl3 catalyst can be used instead of cyclopentene and H2SO4. Note that each of these possible starting materials serves as a source of the same carbocation, the cyclopentyl cation.

16.53   Any compound must have an unsaturation number of 6 and must contain a benzene ring.

H_2C=CH—⬡—CH=CH_2 HC≡C—⬡—CH_2CH_3

p-divinylbenzene   **1-ethyl-4-ethynylbenzene**

HC≡C—⬡—CH=CH_2 / H

structures such as this without a benzene ring
would hydrogenate to *cis*-1,4-diethylcyclohexane

Structures such as the one on the right that do not contain a benzene ring do not meet the criterion, because they would undergo hydrogenation of the ring double bonds.

Solutions to Integrated Problems

16.54   A compound containing a benzene ring must have at least six carbon atoms and four degrees of unsaturation. Compound *A*, with three degrees of unsaturation, cannot contain a benzene ring; compound *C* has only five carbons and therefore cannot contain a benzene ring. Compounds *B* and *D* have more than six carbons and have five and four degrees of unsaturation, respectively, and therefore could contain a benzene ring.

16.55   (a)   Assume that the carbon-chlorine bond dipoles dominate the dipole moment. In the following diagram, the small gray arrows are the C—Cl bond dipoles and the darker arrows are the resultant molecular dipole moments.

This diagram shows that the order of increasing dipole moment is *p*-dichlorobenzene ($\mu = 0$) < *m*-dichlorobenzene < *o*-dichlorobenzene.

(b)   Boiling points are increased by increasing attractions between molecules. To the extent that molecular attractions are dominated by attractions between molecular dipole moments, the order of increasing boiling point is predicted to be the same as the order of increasing dipole moment. (In fact, the boiling point of the ortho isomer (180 °C) is 6–7 °C higher than the boiling points of the meta and para isomers, which are within one degree of each other.)

(c)   Melting points are increased by intermolecular interactions in the solid structure. These interactions are stronger between more symmetrical molecules. The para isomer has much greater symmetry than either the ortho or meta isomer. The more symmetrical para isomer has a much higher melting point than the other two isomers. In fact, the para isomer is the only isomer that is a solid at room temperature (melting point = 53.5 °C). The larger dipole moment, and therefore stronger intermolecular forces, of the ortho isomer gives it a slightly higher melting point (–17 °C) than the meta isomer (–22 °C), though they are quite similar, and both are much lower than that of the para isomer.

16.56   (a)   The cyclopentadienyl anion is aromatic; its five resonance structures show that all carbons (and hydrogens) are chemically equivalent. Therefore, the proton NMR spectrum of the ion consists of one singlet.

hybrid structure

(b)   Fig. 16.2 on text p. 873 shows that aromatic protons located in the plane of the ring and outside of the ring experience an augmented local field and thus a greater chemical shift. However, in the region above and below the ring, the induced field has the opposite direction, and consequently protons located in this region experience a reduced local field and thus a smaller chemical shift. Such is the case with the methyl group in the problem; the local field at this group is so small that its resonance occurs at nearly 1.7 ppm smaller chemical shift than that of TMS.

16.57   The electron density of the unshared electron pairs on the oxygen of the methoxy group is delocalized to the ortho and para positions on the ring, but not to the meta position:

Thus, there is more electron density at the ortho proton H^a than at the meta proton H^b. Since greater electron density results in a reduced local field $\mathbf{B_0}$ (see Sec. 14.3A), the chemical shift of proton H^a is smaller.

16.58   First draw the resonance structure of compound B that has an electronic octet around phosphorus:

This is an important resonance structure because of its adherence to the octet rule. With this structure in mind, consider then the carbocation intermediate in nitration:

For an oxygen unshared pair to exert a stabilizing resonance effect on the carbocation, a destabilizing interaction must be simultaneously introduced as a result of the adjacent positive charges on oxygen and phosphorus. Therefore, this resonance interaction is relatively unimportant—or much less important than in compound A. Because the resonance interaction of its electron pairs is weak or nonexistent, the oxygen acts more like an electronegative substituent without unshared pairs. Such a substituent (like the nitrogen in a nitro group) should be a deactivating, meta-directing group, as observed.

Phenyl acetate has a dipole resonance structure for the C=O group as well, but such a structure does not have an octet on carbon. Furthermore, it has no other oxygens bound to the carbon to exert an electron-withdrawing polar effect. Therefore, the resonance structure that involves stabilization of the carbocation intermediate by the oxygen is relatively more important—so much so, that the para position is activated.

16.59 The unsaturation number of compounds *A* and *B* is 5; therefore, both compounds could be aromatic. Compound *C* is a hydrocarbon; because hydrogenation of *C* gives a compound (compound *D*) with the molecular formula C_9H_{10}, then the formula of compound *C* could be C_9H_8. If so, then the reaction of compounds *A* and *B* with acid is a dehydration. If compounds *A* and *B* are alcohols, their —OH groups must be on adjacent carbons in order for them to dehydrate to the same alkene. Compound *D*, the hydrogenation product, has an unsaturation number of 5; nitration confirms that it and, by deduction, the other compounds, are aromatic. Since compound *D* can contain no ordinary double bonds (they would have been hydrogenated), it must contain a ring in addition to the aromatic ring. The structures of the compounds that meet all these criteria are as follows:

Note that compound *D*, because of its symmetry, would give only two nitration products:

The nitration results rule out isomers such as the following:

would give four
mononitration products;
therefore, cannot be *D*.

If the compound at left
cannot be *D*,
this cannot be *A*.

16.60 The λ_{max} of styrene is due to the conjugation of the alkene π bond with the p orbitals of the benzene ring. (Only one ring p orbital is shown below). Both π bonds of phenylacetylene cannot be simultaneously conjugated with the 2p orbitals of the benzene ring. (The nonoverlapping p orbitals are shown in gray below.) Therefore, the λ_{max} of phenylacetylene is also due to the conjugation of a single π bond with the benzene ring. Consequently, the structural features of both compounds responsible for their UV spectra are identical.

ring p orbitals
(only one shown)

these p orbitals
do not overlap
with the benzene
ring p orbitals

styrene

phenylacetylene

16.61 For this product to form, benzene must react with the electrophile shown in the following equation:

The electrophile, in turn, is formed from benzenesulfonic acid by the following mechanism:

16.62   Run the electrophilic aromatic substitution reaction in reverse.

There are several variations on this mechanism. For example, SO_3H could be lost from the carbocation intermediate to give protonated SO_3 (that is, $^+SO_3H$; see Problem 16.13 on text p. 883) which could react with water to give SO_3 (sulfur trioxide) and H_3O^+. Sulfur trioxide reacts vigorously with water to give H_2SO_4 (sulfuric acid). The important aspect of the mechanism is the protonation of the ring and loss of a species which would serve as an electrophile in the reverse reaction.

16.63   (a)   Drawing all of the resonance structures formed when naphthalene substitutes at either carbon-1 or carbon-2 illustrates that the carbocation formed from substitution at carbon-1 is more stable, with more total resonance structures, including more *important* resonance structures, with intact benzene rings.

Substitution at carbon-1:

Substitution at carbon-2:

intact benzene   intact benzene

(b) Here we are faced with a kinetic versus thermodynamic argument. Substitution at carbon-1 is preferred kinetically because of the carbocation stability illustrated in part (a). However, substitution at carbon-2 is preferred thermodynamically in order to avoid the unfavorable steric interactions between the sulfonic acid group and the hydrogen atom on carbon-8. The carbocation stability outweighs the steric interaction in determining the rate, because the electrophile approaches the naphthene from below (or above) the plane, rather than *in* the plane. This carbocation, according to Hammond's postulate, resembles the transition state for the reaction, so there is a lower energy barrier to substituting at carbon-1. However, once aromaticity is restored, the sulfonic acid group moves *into* the plane of the ring and the steric interaction with the hydrogen on carbon-8 raises the energy level of the final product. Given enough time, or enough heat, the sulfonation will reverse to unsubstituted naphthalene, and eventually occur at carbon-2, despite the larger kinetic energy barrier. The more stable product forms in this case, and is less likely to revert back to naphthalene, and naphthalene-2-sulfonic acid accumulates as the major product at higher temperature.

16.64 (a) The product has an unsaturation number of 6, which means that, relative to the starting material, the product has an additional ring or double bond. The fact that it does not react with Br_2 suggests that a new ring has been formed. In fact, formation of a carbocation by protonation of the alcohol and loss of water leads to an electrophilic aromatic substitution reaction that forms a new six-membered ring. The mechanism shown here does not consider stereochemistry.

(b) Because the new ring formation could occur with either a cis or a trans ring fusion, two diastereomeric products are formed in this reaction. Trans-fused rings are typically more stable and favored in product-formation, but in this situation, the tendency to form the trans-fused ring, *A*, is even greater. The cis-fused ring, *B*, would have unfavorable 1,3-diaxial interactions between the bridgehead hydrogen and the methyl group, and would therefore be a *very* minor product.

A

B

16.65   (a) and (b)

Generation of the electrophile: The electrophile is the carbocation generated by protonation of the alcohol oxygen and loss of water.

$$(CH_3)_3C \qquad \ddot{O}H_2 \qquad \longrightarrow \qquad (CH_3)_3C \qquad \overset{+}{C}(CH_3)_2 \quad + \quad :\ddot{O}H_2$$

A Lewis acid–base association reaction of the benzene π electrons with the electrophile to generate another carbocation:

Loss of a β-proton to the Brønsted base H₂O to form the new aromatic compound:

$$+ \quad H - \overset{+}{\ddot{O}}H_2$$

A

(c)   The aromatic ring of compound *A* has three alkyl substituents. Two of them—the *tert*-butyl group and one of the ring bonds—direct substitution by their electronic effects to the positions indicated by the asterisk:

1) CH₃CCl, AlCl₃
2) H₂O

Electrophilic substitution might have occurred at either of these positions to give either or both of the compounds shown in the foregoing equation. Both of these positions, however, are ortho to the very large *tert*-butyl group, and one of them is ortho to *two* highly branched groups. For steric reasons substitution cannot occur at these very congested positions. Therefore, it occurs at the remaining ring position.

Note that ring positions meta to alkyl substituents are not deactivated; they are simply less activated than positions that are ortho and para to alkyl substituents. Furthermore, this remaining ring position is activated by one alkyl substituent.

16.66 Protonation of the double bond of styrene by the acid gives a carbocation *A*, which serves as an electrophile to react with the double bond of a second molecule of styrene to give carbocation *B*, which loses a proton to give alkene *X*.

Carbocation *A* reacts with the double bond of styrene rather than the ring because the ring, being aromatic, is more resistant to electrophilic substitution than the "ordinary" double bond of styrene.

Carbocation *B*, formed by re-protonation of *X*, undergoes an internal (intramolecular) electrophilic aromatic substitution. This reaction is accelerated over an ordinary electrophilic aromatic substitution because it is intramolecular (proximity effect; Sec. 12.8).

Y and *Z*
(diastereomers;
each is a racemate)

Carbocation *B* is chiral, but because it is formed from achiral starting materials, it is the racemate. When each stereoisomer of the racemate reacts, a new stereocenter is formed. Both enantiomers of *B* give both possible configurations of the stereocenter. Therefore, two diastereomers (*Y* and *Z*, which are cis–trans isomers) are formed, each of which is a racemate.

16.67   The formula of compound B would be useful in solving this problem. Consider the integral in the NMR spectrum of compound B. The entire spectrum integrates for ten hydrogens, and we can hypothesize that no carbons are lost on the reaction with sodium ethoxide. Therefore, the formula of compound B is C_9H_{10}. Evidently, compound A loses the elements of HBr when it reacts with a base—an elimination reaction—and adds one molar equivalent of H_2 on catalytic hydrogenation. Therefore, compound B is an alkene formed by an E2 reaction from compound A. The NMR of compound B shows that it has five aromatic hydrogens; therefore, the remaining unsaturation of compound B is due to a monosubstituted benzene ring. The resonances for the vinylic hydrogens in the NMR spectrum centered at δ 5.1 integrate for two protons. These two protons are chemically nonequivalent, and they show no large splittings typical of a cis- or trans-alkene. Therefore, the vinylic protons are in the partial structure $H_2C{=}$. The chemical shift and integration of the δ 2.1 resonance suggests a vinylic methyl group. The methyl group cannot be benzylic because this would require one fewer aromatic hydrogen and one more vinylic hydrogen. The structure of B is determined (at right) and the structure of C, its hydrogenation product, is also shown.

B
isopropenylbenzene
(α-methylstyrene)

C
isopropylbenzene

Two possible alkyl halides can be considered for A. The chirality of A allows us to decide between them.

A
(2-bromo-1-methylethyl)benzene

16.68   (a)   *p*-Dibromobenzene can give only one mononitro derivative; therefore, it must be compound A. *o*-Dibromobenzene can give two mononitro derivatives, and is therefore compound B. *m*-Dibromobenzene is compound C.

(b)   Answer this question by deciding in how many different ways a single nitro group may be substituted for a hydrogen in each isomer. Because isomer separation methods (such as crystallization) based on conventional physical properties were used to differentiate isomers, enantiomeric differences between products were not evident; that is, only constitutional isomers or diastereomers are considered to be different compounds.

A   *B*   *C*

(c) The following two products derived from nitration of compound *C* were probably formed in smallest amount. In the formation of compound *C1*, the nitro group and the two bromines are involved in severe van der Waals repulsions; and the formation of compound *C2* satisfies the directing effect of neither bromine substituent.

C1   *C2*

(d) The three dibromobenzene isomers would be differentiated most readily by the number of resonances in their proton-decoupled CMR spectra. Compound *A*, the para isomer, has two resonances; compound *B*, the ortho isomer, has three resonances; and compound *C*, the meta isomer, has four resonances. The number of resonances in each case corresponds to the number of chemically nonequivalent sets of carbons.

16.69 (a) Comparison of the formula of the product to that of naphthalene shows that one molar equivalent of the acylating agent has been introduced. Since there is no chlorine in the product, the reaction must involve a double acylation of the naphthalene ring by both ends of the acid chloride. The only way that this can occur with the formation of rings of reasonable size is for the acylation to occur across ortho positions of one benzene ring, or across the peri positions, which are the two positions on either side of the ring junction. These three possibilities account for the three products:

(b) Ferrocene, like other aromatic compounds, undergoes a Friedel–Crafts acylation reactions.

16.70 The balance between protonation on oxygen and protonation on a carbon of the ring depends on the relative basicities of an aromatic "double bond" and the oxygen of an aromatic ether. The fact that anisole protonates on oxygen shows that an ether oxygen is more basic than the aromatic π-electron system:

conjugate acid of anisole

That 1,3,5-trimethoxybenzene protonates on carbon, then, means that the resulting conjugate-acid cation is stabilized sufficiently that the aromatic carbons have enhanced basicity. The stability of this cation is due to resonance stabilization involving all three oxygens:

16.71 (a) Furan derivatives, like 1,3-cyclopentadiene derivatives, are in effect conjugated dienes locked into s-cis conformations. Such dienes are reactive in Diels–Alder reactions. The product *A* results from such a reaction. (The endo stereoisomer is shown, but the product could be the exo stereoisomer or a mixture of both.)

compound *A*

(b) Evidently, the acidic conditions promote loss of the bridging ether oxygen as water. The mechanistic steps involved are protonation, carbocation formation, loss of a proton, and then repetition of a similar sequence. The driving force for this reaction is the formation of the aromatic ring.

16.72 (a) Furan is an aromatic compound. Because the oxygen atom is sp²-hybridized, furan has a cyclic arrangement of p orbitals, with a total of 6 π electrons, a solution to the $4n + 2$ requirement for aromaticity, when $n = 1$.

(b) The second hydrogenation reaction releases more energy because that reaction does not need to destroy the aromaticity of the ring. Some of the energy released in the first hydrogenation is used to overcome the loss of the stability due to the aromaticity, so the resulting value is much lower than the energy released once the aromaticity is already lost.

(c) Utilizing Hess's Law, summing the two reactions and the corresponding energies provides the following reaction with an overall energy equal to -159.6 kJ mol^{-1} (-38.2 kcal mol^{-1}).

$$\Delta H° = -159.6 \text{ kJ mol}^{-1} (-38.2 \text{ kcal mol}^{-1})$$

(d) Using the value from the second reaction as the energy released during the hydrogenation of an "ordinary" oxygen-substituted double bond, the energy of reducing two of these bonds would be -214 kJ mol^{-1} (-51.2 kcal mol^{-1}). Comparing this value to the energy calculated in part (c), the empirical resonance energy of furan can be calculated to be -54.4 kJ mol^{-1} (-13.0 kcal mol^{-1}).

16.73 In this reaction a *tert*-butyl cation is lost rather than a proton from the carbocation intermediate. The electrophile, a nitronium ion $^+NO_2$, is generated by the mechanism shown in Eqs. 16.19a–d on text pp. 882–883.

16.74 At the higher temperature, the two ortho methyl groups (labeled *a* in the problem) are chemically equivalent on the NMR time scale because rotation of the isopropyl group is rapid; the resonance of these groups occurs at δ 2.25. However, at the lower temperature, rotation of the isopropyl group is slow, and, as a result, the two ortho methyl groups are no longer chemically equivalent. Therefore, the resonances of the ring methyl groups occur at different chemical shifts at the lower temperature. Evidently, one ortho methyl group is wedged between the two isopropyl methyls, and the other is in the same plane as the isopropyl C—H group, as shown by the structure in the problem.

16.75 (a) The two terminal rings of hexahelicene cannot lie in the same plane because, if they did, their hydrogens would have severe van der Waals repulsions. Consequently, the molecule is somewhat bent out of plane so that one of the terminal rings lies over the other. This causes the molecule to trace one turn of a helix, which is a chiral object. This optical rotation, by the way, is huge!

(b) Normally the two rings of a biphenyl derivative lie in the same plane to maximize conjugation, that is, to maximize overlap of their π-electron systems. In this case, however, if the rings were to lie in the same plane, the very large sulfonic acid (—SO₃H) group on one ring would have severe van der Waals repulsions with an ortho hydrogen of the other ring. Consequently, the molecule adopts the conformation shown in the problem, in which the planes of the two rings are perpendicular. However, this conformation is chiral and is thus capable of showing optical activity. Evidently, rotation about the central carbon–carbon bond is slow enough that the molecule can be resolved into isolable enantiomeric conformations! When the molecule is heated, the internal rotation "reaction" (like all reactions) is accelerated, and the optically active molecule racemizes. This situation is also discussed in Further Exploration 6.2 on page 140 of this manual; a similar situation is the subject of the solution to Problem 16.10(b) earlier in this chapter.

17

Allylic and Benzylic Reactivity

CHAPTER 17 SUMMARY

■ Functional groups at allylic and benzylic positions are in many cases unusually reactive.

■ The addition of hydrogen halides to dienes, and the solvolysis of allylic and benzylic alkyl halides are reactions that involve allylic or benzylic carbocations. The acceleration of reactions that involve these carbocations can be attributed to the resonance stabilization of the carbocations.

■ A mixture of isomeric products is typically obtained from a reaction involving an unsymmetrical allylic carbocation as a reactive intermediate because charge is shared by more than one carbon in the ion, and because a nucleophile can react at each charged carbon. A reaction involving a benzylic carbocation, in contrast, typically gives only the product derived from the reaction of a nucleophile at the benzylic position to avoid disrupting the aromaticity of the ring.

■ Free-radical halogenation is selective for allylic and benzylic hydrogens because the allylic or benzylic free-radical intermediates that are involved are resonance-stabilized.

■ *N*-Bromosuccinimide (NBS) in CCl_4 solution is used to carry out allylic and benzylic bromination reactions. Benzylic bromination can also be carried out with bromine and light.

■ Because the unpaired electron in allylic radicals is shared on different carbons, some reactions that involve allylic radicals give more than one product.

■ The substitution of a hydrogen in a C—H bond by —OH is an important reaction of phase I metabolism promoted by cytochrome P450. This reaction occurs by a radical mechanism and is especially prevalent at allylic and benzylic C—H bonds,

although ordinary C—H bonds sometimes undergo oxidation as well.

■ Allylic and benzylic anions are stabilized by resonance and by the polar effect of double bonds.

■ Allylic Grignard and organolithium reagents undergo rapid allylic rearrangements. The transition state for this rearrangement is stabilized because it has the character of a resonance-stabilized allylic carbanion.

■ A β-elimination reaction involving allylic or benzylic β-hydrogens is accelerated because the anionic character in the transition state is stabilized in the same manner as an allylic or benzylic anion, and because the developing double bond in the transition state is conjugated.

■ S_N2 reactions at allylic and benzylic positions are accelerated because their transition states are stabilized by the overlap of 2p orbitals.

■ Allylic and benzylic alcohols can be oxidized selectively by manganese dioxide. The selectivity of the reaction is due to the formation of an allylic or benzylic radical in the reaction mechanism.

■ In aromatic compounds, alkyl side chains that contain benzylic hydrogens can be oxidized to carboxylic acid groups under vigorous conditions.

■ Terpenes, or isoprenoids, are natural products with carbon skeletons characterized by repetition of the five-carbon unit with isoprene connectivity.

■ Terpenes are synthesized in nature by enzyme-catalyzed processes involving the reaction of allylic carbocations with double bonds. The biosynthetic precursor of all terpenes is isopentenyl pyrophosphate (IPP).

■ Isopentenyl pyrophosphate is the biosynthetic precursor of steroids. Squalene, a triterpene containing a head-to-head linkage, is converted into an epoxide, which, after protonation, undergoes a concerted cyclization that can be viewed as a series of electrophilic additions of carbocations to double bonds. The carbocation product of the cyclization undergoes a series of rearrangements and loss of a proton to give lanosterol, in which the typical steroid ring system is established. Lanosterol is a biosynthetic precursor of cholesterol.

 # STUDY GUIDE LINKS

 ## 17.1. Synthetic Equivalence

The benzylic oxidation provides an excellent way to introduce a carboxylic acid group into a benzene ring. (None of the electrophilic aromatic substitution reactions in Chapter 16 provide a way to introduce this group directly.) Suppose, for example, we wish to prepare *p*-nitrobenzoic acid from benzene. The desired acid can be formed by benzylic oxidation of any *p*-alkyl nitrobenzene with benzylic hydrogens—say, *p*-ethylnitrobenzene:

$$\text{p-ethylnitrobenzene} \xrightarrow[\text{2) } H_3O^+]{\text{1) KMnO}_4,\ ^-\text{OH}} \text{p-nitrobenzoic acid}$$

p-ethylnitrobenzene   **p-nitrobenzoic acid**

The *p*-ethylnitrobenzene comes from the nitration of ethylbenzene:

$$\text{ethylbenzene} \xrightarrow{\text{HNO}_3,\ H_2SO_4} \text{p-ethylnitrobenzene}$$

ethylbenzene   **p-ethylnitrobenzene**

And the ethylbenzene comes from Friedel–Crafts alkylation of benzene (Eq. 16.28, text p. 886).

If we wanted to prepare *m*-nitrobenzoic acid, we would first oxidize ethylbenzene to benzoic acid, and then nitrate the benzoic acid, thus taking advantage of the meta-directing effect of the carboxylic acid group:

$$\text{ethylbenzene} \xrightarrow[\text{2) } H_3O^+]{\text{1) KMnO}_4,\ ^-\text{OH}} \text{benzoic acid} \xrightarrow{\text{HNO}_3,\ H_2SO_4} \text{m-nitrobenzoic acid}$$

ethylbenzene   **benzoic acid**   **m-nitrobenzoic acid**

In these schemes, an alkyl group—specifically, the ethyl group—has been utilized as the *synthetic equivalent* of a carboxylic acid group. That is, the ethyl group is something that is easily converted into a carboxylic acid group. (Some people use the word *synthon* to mean the same thing as synthetic equivalent.) A person who is skilled in synthesis tends to see various functional groups in terms of their synthetic equivalents. For example, when such a person sees a carboxylic acid group on a benzene ring, an alkyl group pops into mind. For an aldehyde, a primary alcohol comes to mind. (What would you think of as a synthetic equivalent of a *cis*-alkene?) Of course, a group may have more than one synthetic equivalent, and the appropriate synthetic equivalency will generally depend on the exact situation at hand.

By coupling the Friedel–Crafts alkylation with benzylic oxidation, a connection has been made that was not explicitly discussed in the text. As you improve in your mastery of organic chemistry, this is the sort of connection that you should begin to make on your own. A good student will begin to think of compounds "out of context," that is, in ways not directly related to the text material. For example, a student will see a structure and ask, "How could I synthesize this?" even though its synthesis might not be the context in which the structure is encountered. Professional chemists tend to think this way. A structure or a reaction presented in one context will trigger the imagination to think of a wholly new context in which it might be useful. This is one way that new ideas are born! Awareness of this "intellectual triggering" process is one of the best reasons that scientists study the professional literature. And this works not only in chemistry; it works in fields as diverse as physics and medicine.

FURTHER EXPLORATIONS

17.1   Addition versus Substitution with Bromine

The purpose of this Further Exploration is to explain the effect of low bromine concentration in promoting free-radical substitution. The explanation of this effect centers on the rate laws for addition and substitution. The rate law for ionic addition (Sec. 5.2A) at low bromine concentrations is

$$\text{rate of ionic addition} = k_{\text{addition}}[\text{alkene}][\text{Br}_2]$$

Although it is not discussed in the text, addition of Br_2 to an alkene can also take place by a free-radical chain mechanism. At low Br_2 concentrations, the rate of free-radical addition follows the rate law

$$\text{rate of free-radical addition} = k_{\text{fr-addition}}[\text{alkene}][\text{Br}_2]^{3/2}$$

The rather strange-looking $^3/_2$-order in bromine, roughly speaking, is due to the involvement of a bromine atom ("half of a bromine molecule") produced in the initiation step together with the involvement of a bromine molecule in the rate-limiting propagation step—one and one-half bromine molecules total.

Finally, the rate of free-radical substitution at low Br_2 concentrations has the rate law

$$\text{rate of free-radical substitution} = k_{\text{substitution}}[\text{alkene}][\text{Br}_2]^{1/2}$$

The $^1/_2$-order dependence in bromine is due to the involvement of a single bromine atom produced in the initiation step.

A comparison of these rate laws reveals that the substitution reaction has the least dependence on bromine concentration. Therefore, the rate of free-radical substitution not only increases less rapidly with increasing bromine concentration than the other reactions, but also decreases less rapidly with decreasing bromine concentration. At very low bromine concentration, then, free-radical substitution has the greater rate, and is therefore the observed reaction.

The success of *N*-bromosuccinimide (NBS) in CCl_4 as a reagent for substitution of allylic and benzylic hydrogens by bromine is due to the fact that it provides an experimentally convenient way to maintain a low bromine concentration. It wasn't planned that way; NBS in CCl_4 for many years was known as a good reagent for carrying out allylic and benzylic brominations with no competing additions. Careful research eventually developed the rationale presented here and in the text.

REACTION REVIEW

I.   ALLYLIC AND BENZYLIC SUBSTITUTION REACTIONS

A.   *ALLYLIC AND BENZYLIC BROMINATION WITH NBS*

1.   *N*-Bromosuccinimide (NBS) can be used to bring about substitution of allylic and benzylic hydrogens with bromine.

N-bromosuccinimide
(NBS)

2.   The mechanism of allylic or benzylic bromination with NBS entails the following steps:

a.   The initiation step is the formation of a bromine atom by homolytic cleavage of the N—Br bond in NBS itself.

b.   The ensuing substitution reaction has three propagation steps:

i.   The bromine atom abstracts an allylic or a benzylic hydrogen.

ii.   The HBr thus formed reacts with the NBS in the second propagation step (by an ionic mechanism) to produce a Br_2 molecule.

iii.   The last propagation step is the reaction of this bromine molecule with the radical formed in the first propagation step.

c. The Br_2 concentration remains low because it can be generated no faster than an HBr molecule and an allylic radical are generated.

d. The low solubility of NBS in CCl_4 and the consequent low $[Br_2]$ is crucial to the success of allylic bromination with NBS. (See Further Exploration 17.1.)

3. Because the unpaired electron of an unsymmetrical allylic radical is shared by two chemically nonequivalent carbons, such a radical can react to give two constitutionally isomeric products.

B. BENZYLIC BROMINATION

1. Benzylic hydrogens can be substituted with Br_2 in the presence of light.

2. This is a free-radical chain reaction that involves the following propagation steps:

a. Abstraction of a benzylic hydrogen by a bromine atom in preference to a nonbenzylic hydrogen.

  i. This propagation step gives rise to the selectivity for substitution of the benzylic hydrogen.

  ii. Because of resonance, the benzylic radical that is formed has greater stability than a nonbenzylic radical.

b. Reaction of the benzylic radical with a molecule of bromine to generate a molecule of product along with another bromine atom, which can then react with another benzylic hydrogen.

3. Allylic bromination also occurs under these conditions but is generally not used because of the competition with bromine addition to the double bond; NBS bromination is the preferred method for carrying out allylic bromination.

II. BENZYLIC OXIDATION REACTIONS

A. ALLYLIC AND BENZYLIC ALCOHOLS

1. Allylic and benzylic alcohols are oxidized selectively by a suspension of activated manganese(IV) dioxide, MnO_2, because they react much more rapidly than "ordinary" alcohols. (See Eqs. 17.38a–c, text p. 933 for the mechanism.)

a.   Primary allylic alcohols and primary benzylic alcohols are oxidized to aldehydes.

a primary
allyl alcohol   an aldehyde   a primary
benzyl alcohol   an aldehyde

b.   Secondary allylic alcohols and secondary benzylic alcohols are oxidized to ketones.

a secondary
allyl alcohol   a ketone   a secondary
benzyl alcohol   a ketone

c.   "Activated MnO_2" is obtained by the oxidation–reduction reaction of potassium permanganate, $KMnO_4$, with an Mn^{2+} salt such as $MnSO_4$ under alkaline or acidic conditions, followed by thorough drying.

$$2\, H_2O \;+\; 3\, MnSO_4 \;+\; 2\, KMnO_4 \;\xrightarrow{\;H_2O\;}\; 5\, MnO_2 \;+\; 2\, H_2SO_4 \;+\; K_2SO_4$$

**manganese
dioxide**

B.   OXIDATION AT THE ALLYLIC OR BENZYLIC POSITION BY CYTOCHROME P450

1.   The enzyme cytochrome P450 (CyP450) can catalyze the replacement of allylic or benzylic hydrogens with hydroxy groups. (Less commonly, other hydrogens can also be replaced.) It is likely that a radical mechanism is operating (see Sec. 17.5B).

C.   SIDE-CHAIN OXIDATION OF ALKYLBENZENES

1.   Treatment of alkylbenzene derivatives with strong oxidizing agents under vigorous conditions converts the alkyl side chain into a carboxylic acid group.

a.   The benzene ring is left intact.

b.   The alkyl side chain, regardless of length, is converted into a carboxylic acid group at the benzylic position.

c. The conditions for this side-chain oxidation are generally vigorous: heat, high concentrations of oxidant, and/or long reaction times.

d. Common oxidants are $Na_2Cr_2O_7$ (sodium dichromate), CrO_3 (chromium trioxide), $KMnO_4$ (potassium permanganate), or O_2 and special catalysts.

2. Oxidation of alkyl side chains requires the presence of a benzylic hydrogen; benzene derivatives with no benzylic hydrogens are resistant to side-chain oxidation.

III.   BIOSYNTHESIS OF TERPENES AND STEROIDS

A.   BIOSYNTHESIS OF TERPENES

1. The biosynthesis of terpenes starts with isopentenyl pyrophosphate (IPP), which is interconverted biologically to dimethylallyl pyrophosphate (DMAP).

2. Formation of the carbon skeletons of terpenes typically involves the electrophilic addition of a carbocation, derived from the S_N1-type ionization of an allylic pyrophosphate, to the double bond of a second molecule.

held in proximity by their noncovalent binding
to the enzyme prenyl transferase

B.   BIOSYNTHESIS OF STEROIDS

1. The biological construction of the steroid ring system begins with the biosynthesis of lanosterol

a. The triterpene squalene is formed from IPP and DMAP, with one head-to-head bond formed by a mechanism not discussed in the text.

b. Squalene is converted into squalene epoxide by an enzyme-catalyzed reaction with flavin hydroperoxide. (See Eq. 17.57, text p. 947.)

head-to-head bond
between two
sesquiterpene units

flavin hydroperoxide
squalene oxidase

squalene

squalene epoxide

    c   Squalene epoxide undergoes enzyme catalyzed epoxide ring opening and cyclization to lanosterol.

        i.   The cyclization involves the concerted formation of four rings to give an enzyme-bound carbocation.

        ii.   The carbocation is converted into lanosterol by four successive stereospecific carbocation rearrangements followed by loss of a proton to give a double bond. (For mechanistic details, see text pp. 948–949.)

enzyme catalyst

squalene epoxide

lanosterol

2.   Lanosterol, a steroid, is converted into cholesterol in 19 additional enzyme-catalyzed steps.

SOLUTIONS TO PROBLEMS

Solutions to Focused Problems

17.1   The allylic carbons are indicated with an asterisk (*).

(a)

(b)

17.2 The benzylic carbons are indicated with an asterisk (*).

(a)

(b)

17.3 (a) The order of increasing S_N1 reactivity is $(2) < (1) < (3)$. The rates depend on the stabilities of the respective carbocation intermediates. One resonance structure for the carbocation intermediate in the solvolysis of compound (3) is a tertiary carbocation; the resonance structures of the carbocation intermediates derived from compounds (1) and (2) are secondary carbocations.

carbocation intermediate in the solvolysis of compound (3)

Consequently, relative to the starting alkyl halide, the carbocation intermediate derived from compound (3) is more stable than the intermediates in the solvolysis reactions of the other compounds. The intermediate in the solvolysis of compound (2) is least stable because of the electron-withdrawing polar effect of the oxygen. An oxygen in the meta position cannot be involved in resonance; consequently, only its rate-reducing polar effect operates.

Note that the rates of S_N1 solvolysis reactions parallel the rates of electrophilic aromatic substitution reactions. Just as a methoxy group activates an aromatic substitution reaction at an ortho or para position, it also activates a reaction involving a benzylic cation at an ortho or para position. Just as a methoxy group deactivates aromatic substitution at a meta position, it also deactivates a reaction involving a benzylic cation at a meta position.

(b) The reactivity order is $(2) < (3) < (1)$. The S_N1 reaction of compound (2) is slowest because the polar effect of the *meta*-chloro substituent destabilizes the intermediate carbocation. The reaction of compound (3) is faster because the resonance effect of the *para*-chloro group partially offsets its polar effect.

carbocation intermediate in the solvolysis of compound (3)

Compound (1) reacts most rapidly because the carbocation intermediate is not destabilized by the deactivating polar effect of a chloro substituent, which outweighs its activating resonance effect.

17.4 The alkyl halide is the one that reacts to give the same carbocation intermediate, that is, a carbocation with the same resonance structures:

$$(CH_3)_2\underset{|}{C}-CH=CH_2 \xrightarrow{-Cl^-} \left[(CH_3)_2\overset{+}{C}-CH=CH_2 \longleftrightarrow (CH_3)_2C=CH-\overset{+}{C}H_2 \right]$$
$$Cl$$

the same carbocation intermediate
that is involved in text Eq. 17.9

17.5 The carbocation formed when trityl chloride ionizes, the *trityl cation* (Ph_3C^+), is stabilized by delocalization of electrons from *all three* phenyl rings. This carbocation has more resonance structures than the carbocations formed from the other alkyl halides in the table, and is thus so stable that the transition state leading to its formation also has very low energy; consequently, it is formed very rapidly.

trityl chloride
(triphenylmethyl chloride)

trityl cation
(triphenylmethyl cation)

17.6   The number of products depends on (1) whether all of the benzylic or allylic positions are equivalent, and (2) whether the resonance structures of the free-radical intermediate are identical.

(a)   All allylic positions of cyclohexene are chemically equivalent, and the two resonance structures are identical. Therefore, only one allylic bromination product is possible.

cyclohexene
(starting material)

free-radical intermediate
(identical resonance structures)

3-bromocyclohexene
product

(b)   The two allylic positions are nonequivalent, but only one has allylic hydrogens; therefore, only one can react. The two resonance structures of the resulting free radical are nonequivalent; therefore, two different products can be formed.

3,3-dimethylcyclohexene
(starting material)

free-radical intermediate

6-bromo-3,3-
dimethylcyclohexene

3-bromo-4,4-
dimethylcyclohexene

products

(c)   *Trans*-2-pentene (middle structure in the figure below) has two nonequivalent allylic positions. In each of the allylic radicals that results from hydrogen abstraction at the two positions, the unpaired electron is delocalized to two different carbons. (See Study Problem 17.1.) The abstraction of H^a leads to two products, but abstraction of H^b leads to one product, because the resonance structures of the intermediate free radical are identical (assuming that the double bond retains its trans stereochemistry).

free-radical intermediate
from abstraction of H^a

free-radical intermediate
from abstraction of H^b
(equivalent resonance structures)

(E)-1-bromo-2-pentene   **3-bromo-1-pentene**
products from abstraction of H^a

(E)-4-bromo-2-pentene
product from abstraction of H^b

(d)   A benzylic hydrogen is abstracted from the methyl group, and a resonance-stabilized benzylic free-radical intermediate is obtained. The *tert*-butyl group does not have any benzylic hydrogens, and abstraction of a hydrogen from one of the three methyls of the *tert*-butyl group would not result in a resonance-stabilized free-radical intermediate.

4-*tert*-butyltoluene   free-radical intermediate   **1-bromomethyl-4-*tert*-butylbenzene**

(e)   A benzylic hydrogen is abstracted from the isopropyl group rather than a hydrogen of the two methyl groups because a more stable benzylic free-radical intermediate is obtained.

1-isopropyl-4-nitrobenzene   free-radical intermediate   **1-(1-bromo-1-methylethyl)-4-nitrobenzene**

17.7   (a)   The initially formed Grignard reagent undergoes a rapid allylic rearrangement, as depicted in the middle two structures below. Each Grignard reagent in the equilibrium can react with D$_2$O.

allyl Grignard reagent in equilibrium

(b)   Because the two Grignard reagents in rapid equilibrium are identical, only one product is obtained:

identical

identical

17.8   (a)   Although this alkyl halide has two chemically nonequivalent types of β-hydrogens, an allylic hydrogen is more acidic than a nonallylic one; therefore, the product is the conjugated diene derived from elimination of the allylic hydrogen and the bromine.

H ⟵ This β-proton is abstracted by the base because it is allylic.

Br

$K^+ \; ^-OC(CH_3)_3$

1,3-cyclohexadiene

(b)   The benzylic proton is abstracted; β-elimination gives a vinylic ether.

CH_3O
C=CH₂

(1-methoxyvinyl)benzene
(α-methoxystyrene)

17.9   (a)   Reaction with concentrated HBr involves an allylic carbocation intermediate; consequently, a mixture of products, *A* and *B*, could be formed:

H₃C   H
C=C
H   CH₂—ÖH

→H—ÖH₂   ⇌   ÖH₂ +

H₃C   H
C=C
H   CH₂—ÖH₂ ⇌

***trans*-2-buten-1-ol**

ÖH₂ +

[H₃C   H
C=C
H   CH₂
+

⟷

H₃C   H
⁺C—C
H   CH₂   ]

→

H₃C   H
C=C
H   CH₂Br:
A

+   H₃C—CH—CH=CH₂
:Br:
B

:Br:

(b)   Reaction of the tosylate with NaBr in acetone, in contrast, involves an S_N2 reaction; consequently, there is no reactive intermediate, and only product *A* is formed.

trans-2-buten-1-ol

17.10 (a) (b) (c) (d)

17.11 (a) (b) (c)

17.12 PCC oxidizes all primary alcohols to aldehydes and all secondary alcohols to ketones; MnO_2 oxidizes only allylic or benzylic alcohols (primary alcohols to aldehydes and secondary alcohols to ketones).

(a) Both methods give the same product.

(b) MnO_2 does not oxidize the non-allylic alcohol.

(c) There is no reaction in either case; tertiary alcohols are not oxidized under either set of conditions.

(d) Only the allylic alcohol is oxidized by MnO_2. Both alcohols are oxidized by PCC.

17.13 Oxidation by CyP450 occurs at the benzylic position.

(a)

(b)

17.14 (a) The benzylic alcohol is oxidized to benzoic acid. The nitro group is not affected.

p-nitrobenzyl alcohol   **p-nitrobenzoic acid**

(b) The butyl group is oxidized because it has benzylic hydrogens; the *tert*-butyl group is not affected.

1-butyl-4-*tert*-butylbenzene   **4-*tert*-butylbenzoic acid**

17.15 (a) The oxidation product shows that compound *A* is an ortho-disubstituted dialkylbenzene. To be consistent with this analysis and with the formula compound *A* is *o*-xylene.

(b) Because one carbon is lost as a result of the oxidation, and because the benzene ring accounts for all four degrees of unsaturation, compound *B* must be ethylbenzene.

ethylbenzene   **benzoic acid**

17.16 (a) Vitamin A is a diterpene because it contains four isoprene skeletons, which are shown as heavy bonds.

vitamin A

(b) Caryophyllene is a sesquiterpene because it contains three isoprene skeletons, which are shown as heavy bonds.

caryophyllene

17.17 (a) A biosynthetic mechanism for limonene is as follows: (B: = a base.)

limonene

(b) The ionization of geranyl pyrophosphate is followed by reaction with the pyrophosphate anion on the other electron-deficient carbon of the resonance-stabilized carbocation; rotation about a single bond is followed by ionization of pyrophosphate to give the desired carbocation.

geranyl pyrophosphate

17.18 (a)

C D

(b)   Intermediates A, B, and D are tertiary carbocations, and intermediate C is a secondary carbocation. If intermediate D is formed as part of a stepwise process, as shown in part (a), it would require that carbocation C exist as a discrete intermediate. However, it's energetically unfavorable that a secondary carbocation would form from tertiary carbocation B.

(c)   Because of the energetically unfavorable conversion of a tertiary carbocation into a secondary carbocation, a concerted mechanism is more likely, where all of the bonds involved break and form simultaneously.

17.19   As referenced in Focused Problem 17.18c, it is likely the following mechanism showing the formation of lanosterol is actually a concerted process; however, the steps are illustrated discretely here to more clearly show the rearrangements leading to the final elimination step.

base in the enzyme active site

lanosterol

Solutions to Skills Objectives Problems

17.20   (a)   Compound *A* has the lower heat of formation because the *p*-methoxy group stabilizes the benzylic carbocation through resonance (see Eq. 17.8 on text p. 920). (The resonance structure at right is a particularly important one.) The meta methoxy group in Compound *B* cannot contribute through resonance, and only destabilizes the carbocation through the polar effect.

(b)   Compound *B* is more stable because it can delocalize the unpaired electron via resonance onto the para nitro group. This stabilization is not possible in Compound *A*.

(c)   Compound *B* has a lower heat of formation because it has more resonance important resonance contributors than Compound *A*.

resonance of compound *A*

resonance of compound *B*

17.21   (a)   The solution to this problem requires an analysis of the relative stabilities of the carbocation intermediates involved in the S_N1 reactions of the two compounds. The carbocation intermediate in the solvolysis of compound *A* is resonance-stabilized:

The carbocation intermediate involved in the solvolysis of compound *B* is not resonance-stabilized, and in fact is somewhat destabilized by the electron-withdrawing polar effect of the oxygen. The greater stability of the carbocation derived from compound *A* results in a greater solvolysis rate.

(b)   Compound *A* should react much more rapidly because the carbocation intermediate (shown below) has three important resonance contributors. In particular, an unshared pair of electrons on the oxygen can be delocalized in this cation; in the solvolysis of the other compound, the unshared pairs on oxygen have no resonance interaction with the positive charge in the carbocation.

$$\left[CH_3\ddot{O}-CH\!\!=\!\!CH\!-\!\!\overset{+}{C}H_2 \longleftrightarrow CH_3\ddot{O}-\overset{+}{C}H-CH\!\!=\!\!CH_2 \longleftrightarrow CH_3\overset{+}{O}\!\!=\!\!CH-CH\!\!=\!\!CH_2 \right] \quad :\!\ddot{C}l\!:^{\!-}$$

carbocation intermediate involved in the solvolysis of *A*

By Hammond's postulate, the reaction involving the more stable carbocation intermediate is faster.

17.22   Alkyl halides with allylic or benzylic β-hydrogens undergo more rapid E2 reactions because the allylic or benzylic hydrogens are more acidic than ordinary hydrogens (See Sec. 17.3B in the text.) Because compound *A* has allylic β-hydrogens and compound *B* does not, the E2 reaction of compound *A* is faster, and this reaction proceeds in the following manner:

In the presence of excess base this reaction does not stop here but would undoubtedly give the following anion as a product. Can you see why? Can you give a curved-arrow mechanism for this transformation?

17.23   Compound *B* undergoes free-radical bromination faster than Compound *A* because a resonance-stabilized tertiary benzylic free-radical intermediate forms in the mechanism.

17.24   The benzylic Grignard reagent has more ionic character than a typical Grignard reagent because the benzyl anion is resonance-stabilized.

The benzyl Grignard reagent has significant ionic character.

For that reason, it is likely to have enhanced nucleophilic character. As the benzylic Grignard reagent forms, it rapidly reacts as a nucleophile in an S_N2 reaction with benzyl chloride that is also present in the reaction mixture. This reaction forms bibenzyl.

bibenzyl
[(2-phenylethyl)benzene]

It is possible to minimize this reaction with special experimental arrangements that reduce the contact of the benzylmagnesium chloride with benzyl chloride.

17.25   The structure of the starting material is:

trans-2-butene

(a)

H_3C, H $C=C$ H CH_2Br $+$ $CH_3CHCH=CH_2$ (Br)
3-bromo-1-butene

(E)-1-bromo-2-butene

(b)

H_3C, H $C=C$ H CH_2OH $+$ $CH_3CHCH=CH_2$ (OH)
3-buten-2-ol

(E)-2-buten-1-ol

(c)

H_3C, H $C=C$ H CH_2MgBr $\rightleftharpoons$ $CH_3CHCH=CH_2$ (MgBr)

(d)

H_3C, H $C=C$ H CH_2D $+$ $CH_3CHCH=CH_2$ (D)

17.26 (a) This is an S_N1 reaction, and water acts as a nucleophile at each of the two electron-deficient carbons of the resonance-stabilized carbocation. In product A the double bond is conjugated with the benzene ring, whereas in product B the double bond is not conjugated. Therefore product A would absorb at a greater wavelength in the UV spectrum than product B.

(structures of A and B)

A $\qquad$ $+$ $\qquad$ B

(b) Two allylic Grignard reagents are in equilibrium, and, each reacts with D_2O.

$H_2C=CH-CH=CH-CH_2MgBr \rightleftharpoons H_2C=CH-CH-CH=CH_2$ (MgBr)

$\downarrow D_2O$ $\qquad\qquad\qquad\qquad\qquad \downarrow D_2O$

$H_2C=CH-CH=CH-CH_2D \quad + \quad H_2C=CH-CH-CH=CH_2$ (D)

(c) This is an S_N2 reaction of the ethanethiolate anion; because there is no reactive intermediate, only one product is formed:

$CH_3CH_2SCH_2$, H $C=C$ H $CH(CH_3)_2$

(E)-1-ethylthio-4-methyl-2-pentene

(d) In the absence of a good nucleophile, and in a protic solvent, the allylic bromide undergoes a slow S_N1 ionization, and ethanol reacts as a nucleophile at the two electron-deficient carbons of the resonance-stabilized carbocation to give two constitutionally isomeric ethyl ethers.

$CH_3CH_2O-CH_2-CH=CH-CH(CH_3)_2 \quad + \quad CH_2=CH-CH-CH(CH_3)_2$ (OCH$_2$CH$_3$)

(e) Two allylic Grignard reagents are in equilibrium, and each reacts with ethylene oxide:

$$H_2C=CH-CH=CH-CH_2MgBr \rightleftharpoons H_2C=CH-\underset{\underset{MgBr}{|}}{CH}-CH=CH_2$$

1) (epoxide)
2) H₃O⁺   1) (epoxide)
2) H₃O⁺

$$H_2C=CH-CH=CH-CH_2CH_2CH_2OH \;+\; H_2C=CH-\underset{\underset{CH_2CH_2OH}{|}}{CH}-CH=CH_2$$

(f) The bicyclic alkene has 4 equivalent allylic positions; free radical bromination at any one of them affords the indicated product.

It is possible that a second product is formed in this reaction; can you deduce what it is? (Think about resonance structures of the intermediate free radical.) Can you think of one or more reasons why the product in the problem should be the predominant one? ([*Hints:* (1) Think statistically. (2) Think about double-bond stability.]

(g) The dibromide formed in the bromine-addition reaction undergoes two successive E2 reactions to give naphthalene, an aromatic compound. Both eliminations involve benzylic protons, and the final elimination is driven by the formation of an aromatic product.

(structures showing dihydronaphthalene → Br₂ → dibromide → KOH/ethanol → naphthalene)

naphthalene

(h) All benzylic carbons—even the ones in the ring—are oxidized to carboxylic acid groups.

(structure with CO₂H, CO₂H, CO₂H groups)

**1,2,3-benzenetricarboxylic acid
(hemimellitic acid)**

(i) Only allylic alcohols are oxidized under these conditions, so both diastereomers below will afford the given product.

(two cyclohexene structures with OH and HO groups, "or" between them)

17.27 Remember that once a benzylic hydrogen (or in this case, deuterium) is removed by CyP450, resonance can put the unpaired electron on either of two carbons. The deuteriums present in the alkene serve to "label" the allylic carbons and differentiate them from the alkene carbons, so that the two products, which would be the same in the absence of the deuterium label, are distinguishable.

17.28 CyP450 can convert alkenes into epoxides through the following variation of the "rebound" hydroxylation mechanism.

CyP450

17.29 Compounds (a), (d), (e), and (f) are terpenes. Safrole, compound (b) is not a terpene. Modhephene is believed to be formed from a terpene by a skeletal rearrangement, but its terpene origin is not obvious from its structure. In the following terpene structures, the isoprene skeletons are shown with heavy bonds. Compound (f) has a head-to-head junction between the two isoprene skeletons. The others have more common head-to-tail junctions.

(a)

ipsdienol

(d)

β-thujone

(e)

periplanone B

(f)

meta-cymene

17.30 (a) Geranyl pyrophosphate is converted into farnesyl pyrophosphate by a mechanism exactly analogous to the one shown in Eqs. 17.52–53 on text p. 944. Then farnesyl pyrophosphate hydrolyzes to farnesol; see text Eq. 17.54.

geranyl pyrophosphate

farnesyl pyrophosphate

H_2O (see Eq. 17.54)

farnesol

(b)   To see the new bond connections that have to be made, draw α-pinene in a planar projection. The projection on the left shows the molecule as it is in the text; turning this projection clockwise 120° in the plane of the page yields the second projection; and turning this projection over yields the third projection, which can be related more easily to the carbocation in Focused Problem 17.17a.

α-pinene

Here is the mechanism:

17.31 (a) Picric acid protonates the epoxide oxygen, making the more-branched carbon more electrophilic (See text Sec. 12.5B).

(b) The Lewis acid BF_3 undergoes a Lewis acid–base addition reaction with the oxygen of the epoxide, The resulting positive charge on oxygen makes oxygen a good leaving group, so the epoxide opens readily. However, following the instructions, we use general acids and bases AH and A^- in the mechanism.

Solutions to Integrated Problems

17.32 (a)

(b)

(c)   In the first set of reactions, note that allyl chloride is a particularly reactive alkylating agent.

$CH_3(CH_2)_3C \equiv CH$ $\xrightarrow[\text{2) ClCH}_2\text{CH}=\text{CH}_2]{\text{1) NaNH}_2}$ $CH_3(CH_2)_3C \equiv CCH_2CH=CH_2$ $\xrightarrow{\text{H}_2\text{, Lindlar catalyst}}$

1-hexyne

(Z)-1,4-nonadiene

(d)

(e)

cumene   **m-nitrobenzoic acid**

(f)

cumene   **p-nitrobenzoic acid**

Compare the solutions to parts (e) and (f) and notice how reversing the sequence of the oxidation and nitration steps brings the directing effects of different substituents into play during nitration.

(g)

1,2-dimethoxy-4-methylbenzene

3,4-dimethoxybenzaldehyde

(h)   An inversion of stereochemistry requires the use of an S_N2 reaction.

17.33   In reaction (1), the acidic conditions promote protonation of the alcohol and ionization to a carbocation, which reacts in a Lewis acid–base association reaction with ethanol at the two nonequivalent electron-deficient carbons. (The mechanism, which is discussed in Sec. 12.2C in the text, is outlined below; fill in the curved-arrow notation.)

In reaction (2), the carbon-oxygen bond is never broken; consequently, no allylic rearrangement can occur. This is a Williamson ether synthesis that involves formation of the alkoxide anion, which, in turn, is alkylated by ethyl iodide. (See Sec. 12.2A in the text; Problem 12.64 on text p. 677 illustrates the same idea.)

17.34   (a)   The allylic-rearrangement product B [(E)-1-bromo-2-butene] is shown in part (b). It can be formed by the ionization to a carbocation and a bromide ion followed by reaction with the bromide ion on the other electron-deficient carbon.

(b)   The curved-arrow notation is shown in the following scheme.

3-bromo-1-butene
(compound A)

(E)-1-bromo-2-butene
(compound B)

(c)   Compound B, the rearrangement product, is favored at equilibrium because it has the double bond with the greater number of alkyl branches.

17.35   (a)   Zoapatanol is a diterpene; the four isoprene units are shown with heavy bonds.

(b)   The allylic alcohol in zoapatanol is selectively oxidized to an aldehyde by MnO_2.

17.36   First, examine each structure and determine if it is optically active or not. Then, since MnO_2 oxidizes allylic alcohols and benzylic alcohols selectively, identify the structures containing one or more allylic alcohol groups. Then draw the corresponding structures that would result from oxidation by MnO_2.

(a)   Structures *B* and *E* are optically active compounds and are oxidized by MnO_2 to optically inactive compounds.

(b)   Structure *C* is an optically active compound that is oxidized by MnO_2 to an optically active compound.

(c)   Structure *D* is an optically inactive compound that is oxidized to an optically inactive compound.

(d)   Structure *A* is not an allylic alcohol and therefore is not oxidized by MnO_2.

17.37   The compounds that give the most stable carbocation intermediates are the ones that undergo the most rapid solvolysis. This problem deals with the effect of substituent on the stability of the carbocation intermediate. The key is to analyze the balance of resonance and polar substituent effects just as you would for electrophilic aromatic substitution. The order of increasing reactivity is (4) < (1) < (3) < (2). Compound (2) reacts most rapidly because the carbocation intermediate is stabilized by the *electron-donating resonance effect* of the *para*-methoxy substituent:

As in electrophilic substitution, the resonance effect of the *p*-methoxy group strongly outweighs its electron-withdrawing polar effect (see text Sec. 16.5B). In compound (3), there is a similar resonance effect; however, the polar effects of halogen substituents outweigh their resonance effects. Consequently, compound (3) reacts more slowly. The nitro group exerts no resonance effect in the carbocation intermediates derived from compounds (1) and (4); the question is then whether its polar effect is stronger from the meta or para position. As in electrophilic aromatic substitution, a *para*-nitro group destabilizes a carbocation intermediate more than a *meta*-nitro group because, in a *para*-nitro carbocation, positive charge is on adjacent atoms:

positive charge is on
adjacent atoms

In the *meta*-nitro carbocation, positive charge does not reside on adjacent atoms. Consequently, the *meta*-nitro carbocation is more stable (or perhaps better described as *less unstable*) than the *para*-nitro carbocation, and *m*-nitro-*tert*-cumyl chloride solvolyzes more rapidly than *p*-nitro-*tert*-cumyl chloride.

17.38   Acid-catalyzed dehydration involves a carbocation intermediate; see Sec. 11.2 of the text. The problem, then, really asks, "Which compound gives the most stable carbocation intermediate relative to the starting alcohol?" Compounds *B* and *C* both involve a *p*-methoxy substituent; the carbocation-stabilizing effect of this substituent in a similar situation is shown in Eq. 17.8, text p. 920, and Problem 17.37 above. In the dehydration of *B*, the carbocation intermediate, a tertiary carbocation, is more stable than the carbocation in the dehydration of *C*, a secondary carbocation. The *p*-nitro carbocation is the least stable because the nitro group is an electron withdrawing, carbocation-destabilizing group, just as it is in electrophilic aromatic substitution (and for the same reasons; see the discussion on text p. 894–895). Therefore, the order of increasing reactivity is *A* < *C* <.

17.39   The question is, essentially, "Which —OH group, when protonated and lost as water (Sec. 11.2A), gives the more stable carbocation?" In the image below, loss of the OH on the left gives a carbocation that is both tertiary and doubly benzylic. Loss of the OH on the right gives a carbocation that is secondary and singly benzylic. The choice is clear; the —OH group on the left is lost.

this —OH group is lost because
the more stable carbocation is formed

terfenadine

dehydration product

It is conceivable that, under strongly acidic conditions, both alcohols could dehydrate; the would give the following diene as the dehydration product.

17.40 The greater λ_{max} value indicates the formation of a species with more extensive conjugation. Protonation of the hydroxy group and loss of water gives the benzyl cation, which includes an additional p orbital on the benzylic carbon, in conjugation with the aromatic ring.

benzyl cation
$\lambda_{max} = 442$ nm

Reaction of the benzyl cation with NaOH gives back the starting benzyl alcohol in a Lewis acid-base association reaction.

17.41   (a)   Both the conjugate-base anion of 1,4-pentadiene and the conjugate-base anion of 1,3-cyclopentadiene are doubly allylic and resonance-stabilized. However, the conjugate-base anion of 1,3-cyclopentadiene is also aromatic. (See text pp. 847–848 for a discussion of this case.) Consequently, much less energy is required for the ionization of 1,3-cyclopentadiene, and its pK_a is therefore much lower. (The pK_a difference between these two compounds is estimated to be 10–15 units.)

(b)   The carbocation intermediate in the solvolysis of 3-bromo-1,4-pentadiene is doubly allylic and therefore resonance-stabilized. Consequently, formation of this ion is relatively fast. Although the carbocation intermediate that results from the ionization of 5-bromo-1,3-cyclopentadiene appears to be doubly allylic and resonance-stabilized, it is in fact a cyclic system of $4n$ π electrons, and is therefore antiaromatic and particularly unstable. Accordingly, this ion does not form. (Problem 15.81 on text p. 866 also deals with this case.)

17.42 This reaction is brought about by a series of acid-base equilibria. The equilibrium lies to the right because the product is more stable than the starting material for two reasons: (1) because the double bonds in the products are conjugated with the benzene ring and the double bond in the starting material is not; and (2) because the double bond in the product has more alkyl branches than the double bond in the starting material. Recall that alkyl substitution at double bonds is a stabilizing effect (text Sec. 4.5B).

double bond is *not* conjugated and has one branch

double bond *is* conjugated and has two branches

The cis and trans alkenes undoubtedly come to equilibrium under these conditions. The trans alkene predominates because of the unfavorable van der Waals interaction between the phenyl ring and the methyl group in the cis alkene.

17.43 The equilibrium lies to the right because the double bond has four alkyl substituents whereas, in the starting material, it has three. Recall that alkyl substitution at double bonds is a stabilizing effect (Sec. 4.5B in the text). The mechanism involves the protonation of the double bond to give the benzylic cation, followed by loss of a proton to give the product.

17.44 (a) Formation of a resonance-stabilized propargylic anion (that is, an anion that is "allylic" to a triple bond) is followed by protonation at the internal carbon. Formation of another resonance-stabilized anion followed by protonation gives the 1-alkyne, which is the most acidic species in the reaction mixture. Consequently, its conjugate-base anion is formed irreversibly, and all equilibria are thus pulled toward this anion. The net result is the migration of the triple bond to the end of the carbon chain, in what is referred to as the "acetylene zipper" reaction.

 Addition of water to the reaction mixture gives the 1-alkyne. Therefore, the reaction shown here followed by addition of water results in the migration of a triple bond from an interior position to the 1-position of a carbon chain provided there are no branches. It has been shown that the triple bond migrates with equal frequency to both ends of the chain.

You might ask why a 1-alkyne is more acidic than a propargylic hydrogen when the conjugate-base anion of the latter is resonance-stabilized, and an acetylenic anion is not. This is a reflection of the effect of hybridization on acidity discussed in Sec. 3.7C of the text. A doubly allylic hydrogen is more acidic than an acetylenic hydrogen, but a singly allylic hydrogen is not.

(b) The question is whether the triple bond migrates to the end of the carbon chain nearer to the methyl branch or to the end of the chain farther from the methyl branch. Once we consider the mechanism shown in part (a), the answer becomes clear. The migration of the triple bond must occur away from the methyl branch, because the mechanism of the reaction requires a stepwise migration of the triple bond, and a triple bond cannot form at a carbon that bears a branch because a carbon can have no more than four bonds.

triple bond cannot form
between these two carbons

$CH_3CH_2CH - C \equiv C - CH_2CH_2CH_3$ $\xrightarrow[\text{[see part (a)]}]{\text{B:}^- \atop \text{zipper reaction}}$ $CH_3CH_2CHCH_2CH_2CH_2 - C \equiv C\overset{..}{:}{}^-$

          | |
CH_3 CH_3

3-methyl-4-octyne

17.45 The fact that benzoic acid is obtained by chromic acid oxidation shows that all compounds contain a monosubstituted benzene ring. The NBS reaction is a benzylic bromination, and the alcohol produced by solvolysis of the resulting bromide must be tertiary, because it cannot be oxidized with CrO_3 and pyridine. The structures of compound *A*, *B*, and *C* are therefore as follows:

A   *B*   *C*

17.46 Compound *A* has an unsaturation number of 5. Because it ultimately affords phthalic acid, it is an ortho-disubstituted benzene derivative. The enantiomeric resolution and oxidation results show that compound *C* is a chiral secondary alcohol, and the α-carbon of this alcohol is the asymmetric carbon, because conversion to a ketone destroys its chirality. Compound *C*, in turn, arises from solvolysis of *B*, a product of either allylic or benzylic bromination. Because *A* contains a benzene ring, the formation of compound *B* is probably a benzylic bromination. A by-product of the formation of compound *C* is compound *D*, presumably an alkene, because it can be hydrogenated back to compound *A*. The only way to accommodate all of these data is for the additional degree of unsaturation in compound *A* to be due to a ring. Structures of all compounds that fit these data are as follows:

A   *B*   *C*   *D*   *E*

The following structure for compound *A* is ruled out because it would give a tertiary alkyl halide, which, in turn, would give a tertiary alcohol on solvolysis; a tertiary alcohol cannot be oxidized to a ketone.

17.47 (a) Propargylic Grignard reagents, like allylic Grignard reagents, are an equilibrium mixture of two constitutional isomers. Each reacts with H_2O.

(b) Protonation of the triple bond gives a resonance-stabilized vinylic cation which reacts intramolecularly with the oxygen of the OH group. (H—OTs = *p*-toluenesulfonic acid; see text p. 577 for its structure.)

(c) A hydrogen on the central carbon is more acidic than an acetylenic hydrogen because the conjugate-base anion resulting from removal of the central hydrogen is both allylic and propargylic, and is therefore doubly resonance-stabilized. The conjugate-base anion is alkylated by allyl bromide.

a resonance-stabilized anion

In the resonance structures of the anion intermediate, the negative charge is delocalized to two other carbons. (Draw these structures.) While the mechanism above shows why the indicated product is reasonable, it does not explain why products derived from the other possible resonance structures are not observed. Can you think of a reason?

(d) Protonation of the alcohol and loss of water give an allylic carbocation that can react with ethanol at either of two electron-deficient carbons to give a mixture of two constitutionally isomeric ethyl ethers. The following mechanism begins with the protonated alcohol.

(e)   The formation of a cyclopropane suggests the intermediacy of the following carbene:

$$(CH_3)_2C = C = C:$$

This carbene is formed by loss of chloride ion from an acetylenic ion. The carbene then adds to cyclohexene to give the cyclopropane. (See Eq. 10.70, text p. 543.)

17.48 (a)   This is a case of *neighboring group participation,* which was discussed in Sec. 12.8A, p. 653, of the text. This cannot be a simple S_N1 ionization, because there is no obvious reason that such an ionization should be significantly accelerated. The source of the acceleration is neighboring-group participation: the π electrons of the double bond are acting as an intramolecular nucleophile.

Compound *B* has no double bond; and the rate of its S_N1 reaction gives us an idea of what we can expect for an ordinary S_N1 reaction without such neighboring-group participation.

(b)   The retention of stereochemistry suggests two successive displacements with inversion of stereochemistry in each. (Sec. 12.8C, text p. 658) The opposite-side displacement of chloride by the double bond is followed by opposite-side reaction of the acetate nucleophile to re-form the double bond.

An orbital picture of the intermediate shows that this is a two-electron-three-center bond. One side of the carbon is blocked by the involvement of the π electrons of the double bond. The electrons of the double bond overlap with the 2p orbital of the carbon. This orbital being occupied, it can no longer accept electrons. The incoming nucleophile reacts at the lowest-energy antibonding molecular orbital (LUMO) of the cation. This interaction must occur at the side opposite the double bond (Fig. 9.3, p. 463 of the text).

17.49   Text Section 7.8C shows that anti stereochemistry is one of the major pieces of evidence that a bromonium ion is involved in bromine addition to alkenes; similar conclusions hold for chlorine addition. The text also shows that a mechanism involving carbocation intermediates predicts mixed syn- and anti-addition (Eq. 7.62, text p. 369).

This problem deals with the *competition* between the two mechanisms: one involving a cyclic *chloronium ion* intermediate and the other involving a carbocation intermediate. The loss of stereoselectivity is evidence for the involvement of a carbocation intermediate. When the carbocation intermediate is resonance-stabilized (last two entries in the table), it is stable enough to compete energetically with the chloronium ion as a reactive intermediate. When R = *p*-methoxyphenyl, the carbocation intermediate is further resonance-stabilized by the para substituent. (See, for example, Eq. 17.8 on text p. 920.) In this case, the carbocation intermediate is stable enough, and the carbocation mechanism thus important enough, that nearly complete loss of stereoselectivity is observed.

17.50   Compound *A* is a benzylic bromination product, and it appears that compounds *B* and *C* are S_N1 products, each derived respectively from reaction of one of the nucleophiles present with a carbocation intermediate derived from *A*. Recall that S_N1 reactions are generally accompanied by E1 reactions (Sec. 9.6 in the text); therefore, it is reasonable to suppose that MPTP is an alkene that results from the elimination reaction. The molecular formula of MPTP, $C_{12}H_{15}N$, indicates an unsaturation number of 6. Because five unsaturations are accounted for by the two rings in the starting material, the formula is consistent with the hypothesis that MPTP is an alkene.

17.51   From the structure of the Diels–Alder product *C*, compound *B* must be the conjugated diene 2-chloro-1,3-butadiene. Because compound *B* is an allylic rearrangement product of compound *A*, compound *A* must be the allene 4-chloro-1,2-butadiene. (Note that 1-chloro-1,3-butadiene would not give the same Diels-Alder product.)

17.52 (a)

(b)   In an S_N1-like mechanism, the pyrophosphate leaves, producing a resonance-stabilized allylic carbocation that is attacked by water and deprotonated to yield linalool.

geranyl pyrophosphate   **allylic carbocation**

linalool

(c)   There are two potential places where the epoxide could form: at either the trisubstituted alkene or at the monosubstituted alkene. Inspecting compound *B* shows that the monosubstituted double bond still exists intact, so the epoxide must have formed at the trisubstituted alkene (compound *A*). Compound *A* then undergoes an intramolecular cyclization, and proton transfers yield product *B*.

17.53 (a)   A deuterium-labeled compound, such as the following one, could be used to establish that a reaction takes place with stereochemical inversion. If the product has the opposite configuration as the starting material, then it the mechanism proceeds with inversion—that is, by an S_N2 mechanism.

geranyl phosphate   geraniol

(b)   As noted in Sec. 11.9B, alcohol dehydrogenase catalyzes the removal of the pro-*R* hydrogen of ethanol. It can be deduced that the geraniol produced in part (a) would also have its pro-*R* hydrogen removed by the enzyme. Therefore, if the geraniol product in part (a) is oxidized by the enzyme, the aldehyde produced would contain deuterium.

geraniol

(Alternatively, if you chose to begin with the *S*-deuterium-labeled geranyl phosphate in part (a), then the product geraniol would be *R*, and the product of oxidation by alcohol dehydrogenase would not contain deuterium.)

17.54   (a)

geranyl pyrophosphate   SAM

2-methylgeranyl pyrophosphate

(b)   When trying to solve problems like these, especially with structures containing rings, it's always a good idea to number carbons when trying to figure out the nature of the transformation that occurs. (In this case, the numbers are arbitrary and have nothing to do with nomenclature.) Because we need to take this linear molecule and form one containing two rings, we should begin with an intramolecular reaction. The –OPP group is a good leaving group, so use one of the alkene double bonds as a nucleophile. Here, we're joining C-2 with C-7, as numbered below, forming a six-membered ring.

The intermediate formed as a result of this intramolecular reaction leaves a positive charge behind so that a tertiary carbocation forms at C-1. Because a second ring must be formed, a bond forms between C-5 and C-1. This leaves a carbocation at C-6, which can then react with water and be deprotonated to yield the final product.

2-methylisoborneol

17.55   First analyze the relationship of the isoprene skeletons.

eudesmol

Then use steps like the ones shown in Eqs. 17.52–17.54, text p. 944, to assemble the parts from IPP and DMAP. Start with farnesyl pyrophosphate, the biosynthesis of which is shown in the solution to Problem 17.30(a), text pp. 952-953. Note that B: = a base. (We neglect stereochemistry.)

farnesyl pyrophosphate
biosynthesis is in the
solution to Problem 17.30(a)

eudesmol

A different arrangement of isoprene units in eudesmol can also be envisioned:

eudesmol
(alternate arrangement
of isoprene units)

A biosynthetic scheme based on this arrangement would be an equally correct answer. An isotope-labeling experiment would be required to distinguish between the two arrangements.

17.56   Compound *A* can be protonated at the "upper" double bond to give a tertiary carbocation *X*, which serves as the electrophile in a Lewis acid–base association reaction with the "lower" double bond to give a carbocation *Y*. Loss of acetone gives intermediate *B*.

The conversion of a tertiary carbocation X to a secondary carbocation Y is driven by the replacement of a π bond by a σ bond, and ultimately by the loss of the very stable molecule acetone.

Compound B protonates to give carbocation Z, which rearranges to a tertiary carbocation. This carbocation undergoes a Lewis acid–base association with water to give geosmin after loss of a proton.

17.57 (a) CyP450 oxidizes compounds at their benzylic positions (see Sec. 17.5B).

(b) The positive charge on the nitrogen makes it a good leaving group for the ring-opening reaction, driven by the formation of a carbonyl group.

17.58 (a) The triphenylmethyl anion (trityl anion) is stabilized by resonance interaction with all three benzene rings:

(b)   The resonance interaction shown in part (a) is optimized if the three rings are coplanar; when the rings are coplanar, optimal overlap occurs between the 2p orbitals of the central carbon and those of the rings. However, when the rings are coplanar, there are significant van der Waals repulsions between hydrogens on different rings:

severe van der Waals repulsions

These van der Waals repulsions force the rings in the triphenylmethyl anion to twist significantly away from coplanarity, with a resulting cost in orbital overlap. (The molecule resembles a molecular "propeller.") In other words, in the triphenylmethyl anion, conjugation is not as effective as it would be if the rings were all coplanar.

In fluoradene, the hydrogens that are the source of these repulsions are replaced by bonds that constrain the rings to coplanarity. The resulting increase in orbital overlap brings about a concomitant increase in stabilization of the anion. As a result, the pK_a of the conjugate acid is lowered significantly relative to that of triphenylmethane. The conjugate-base anion of fluoradene is really the "ultimate benzylic anion."

17.59   (a)   The stability of the trityl radical is due to delocalization of the unpaired electron into all three benzene rings.

many other structures

(b)   Hexaphenylethane could be formed by the recombination of two trityl radicals:

hexaphenylethane

(c)   The dimer of the trityl radical is formed by the following mechanism:

Hexaphenylethane is destabilized by van der Waals repulsions between gauche pairs of phenyl groups.

severe van der Waals repulsions

This destabilization is evidently so great, and the drive for recombination so powerful, that aromatic stability in one of the benzene rings can be sacrificed in forming the dimer shown at the beginning of part (c).

17.60   (a)   Fact (1) indicates that the transition state involves a molecule of Et_2NH and a molecule of alkyl halide. Fact (2) indicates that the indicated starting materials undergo the reaction shown; had this not been established, one could have postulated that one of the starting alkyl halides is converted into the other by the reaction conditions, and that the observed product could have originated from only one of the two starting compounds. Fact (3) establishes that the observed product is the actual product of the reaction, and is not derived from a subsequent reaction of its allylic isomer.

Now to the mechanisms. In Reaction (2) an allylic rearrangement has occurred. Such allylic rearrangements generally suggest the involvement of carbocation intermediates, which have two resonance forms, and can react as. nucleophile at two different carbons. However, the observation of second-order kinetics rules out an S_N1 reaction. Therefore, the reaction is evidently a direct substitution at the allylic carbon, thus:

(b)   This concerted allylic substitution mechanism has been termed the S_N2' mechanism. The other isomer reacts by the conventional S_N2 mechanism, in which the nucleophile reacts at the carbon bearing the halide. The large size of the nucleophile dictates that its nucleophilic reaction occurs at the unbranched carbon to avoid van der Waals repulsions. In one case, this is the allylic carbon; in the other, it is the α-carbon.

17.61   (a)   The acid protonates the methylene (CH_2) group to form the resonance-stabilized carbocation A.

resonance-stabilized carbocation A

(b)   At low temperature, the methyl groups fall into four distinct chemical environments. The integral ratio of 2:2:2:1 corresponds to a ratio of 6:6:6:3 hydrogen atoms. Because of the symmetry of the structure, the protons on either side of the molecule are equivalent. The four different NMR resonances are assigned on the resonance hybrid structure shown here:

δ 1.54

H₃C CH₃

δ 2.36 ⟶ H₃C

CH₃

δ 2.63 ⟶ H₃C

CH₃

CH₃ ⟵ δ 2.82

resonance hybrid of carbocation *A*

(c)   As the temperature is raised, the rate of the rearrangement increases so that all 7 methyl groups become equivalent. Therefore, the four separate lines coalesce into one broad line in the NMR spectrum. (See Sec. 14.8 of the text.)

"methyl walk" continues

The Chemistry of Aryl Halides, Vinylic Halides, and Phenols

18

CHAPTER 18 SUMMARY

- Neither aryl halides, vinylic halides, nor phenols undergo S_N2 reactions because the sp hybridization of the S_N2 transition state has relatively high energy and because the opposite-side substitution reaction with nucleophiles is sterically blocked.

- Neither aryl halides, vinylic halides, nor phenols undergo S_N1 reactions because of the instability of the carbocation intermediates that would be involved in such reactions.

- The aryl–oxygen bond is unreactive in both S_N1 and S_N2 reactions.

- Vinylic halides with β-hydrogens undergo β-elimination reactions in base under vigorous conditions to give alkynes.

- Aryl halides substituted with ortho or para electron-attracting groups, particularly nitro groups, undergo nucleophilic aromatic substitution. In this type of reaction, the nucleophile reacts at the ring carbon to form a resonance-stabilized anion (Meisenheimer complex), from which the halide leaving group is then expelled.

- The important concepts for electron counting in transition-metal complexes are

  - *Oxidation state*: The number of X-type ligands plus the charge.

  - *The d^n configuration*: The number of unshared electrons on a metal within a complex, equal to the electron count of the neutral atom minus the oxidation state.

  - *The electron count*: The electron count of the neutral atom plus the number of X-type ligands plus twice the number of L-type ligands, minus the charge.

- Some fundamental processes in transition-metal catalysis are

  - *Ligand association and dissociation*. A succession of dissociation–association steps gives ligand substitution.

  - *Oxidative addition*.

  - *Reductive elimination*. (Oxidative addition and reductive elimination are conceptually the reverse of each other.)

  - *Ligand insertion*.

  - *β-Elimination*. (1,2-Ligand insertion and β-elimination are conceptually the reverse of each other.)

- Several transition-metal-catalyzed reactions can bring about the coupling of vinylic and aryl groups. These are catalyzed by Pd(0), which is typically present as $Pd(PPh_3)_4$ or a related derivative. $Pd(OAc)_2$ can also be used; it is reduced in the reaction mixture to form Pd(0).

  - The Heck reaction brings about the coupling of an alkene with an aryl bromide or iodide. The reaction is typically used to form aryl-substituted alkenes.

  - The Suzuki reaction brings about the coupling of a vinylic or aryl bromide or iodide with an aryl or vinylic boronic acid or borane. The reaction is used to form biaryls, aryl-substituted alkenes, and conjugated alkenes.

  - The Stille reaction is the coupling of an organic group derived from an organostannane reagent with the aryl group of a phenol in which the phenol group has been converted into a triflate group. The Stille reaction can be used to form both alkyl- and aryl-substituted benzenes.

The key mechanistic steps in all of these reactions are an oxidative addition of the aryl halide or triflate to the metal and a reductive elimination of the coupling product.

■ In alkene metathesis, two alkenes can be transformed into other alkenes by breaking the carbon–carbon double bond and scrambling the resulting fragments. The ruthenium-based Grubbs catalysts G1 and G2 can be used to bring about this transformation. A series of ruthenium metallacycles are key intermediates in alkene metathesis. Formation of cyclic alkenes is one of the most important applications of this reaction.

■ Phenols are considerably more acidic than alcohols because phenoxide ions, the conjugate bases of phenols, are stabilized both by resonance and by the electron-attracting polar effect of the aromatic ring. Phenols containing substituent groups that stabilize negative charge by resonance or polar effects (or both) are even more acidic.

■ Phenols can be oxidized to quinones. Two classes of quinones are *ortho*-quinones and *para*-quinones. *Para*-quinones are typically more stable.

■ A hydroquinone (a 1,4-benzenediol) derivative can undergo reversible one-electron oxidation to a semiquinone, and another reversible one-electron oxidation to a quinone. Coenzyme Q is an important biological quinone that undergoes reversible conversion into its hydroquinone, QH_2, as part of electron transport in biology, a process used to fuel energy generation in the cell. Vitamin K is another biologically important hydroquinone that plays a crucial role in blood clotting.

■ Phenols are free-radical inhibitors because abstraction of a hydrogen by a radical gives a resonance-stabilized phenoxy radical. Vitamin E serves as a biological radical scavenger in living systems.

■ Some electrophilic aromatic substitution reactions of phenols show unusual effects attributable to the OH group. For example, phenol brominates three times in bromine water because the OH group can ionize. Phenol is rather unreactive in Friedel–Crafts acylation reactions because the OH group reacts with the $AlCl_3$ catalyst.

STUDY GUIDE LINKS

18.1   Contrast of Aromatic Substitution Reactions

The name "aromatic substitution" may suggest some connection to the reactions you studied in Chapter 16, such as nitration, halogenation, sulfonation, and so on. Other than the fact that a group is replaced on an aromatic ring, *nucleophilic aromatic substitution and electrophilic aromatic substitution have little in common*. In *electrophilic* aromatic substitution, an *electrophile*, or Lewis acid, reacts with the benzene π electrons, and a *proton* is lost from the benzene ring at the site of substitution. In *nucleophilic* aromatic substitution, a *nucleophile*, or Lewis base, reacts at a carbon of the benzene ring and a *halide* is lost from the benzene ring at the site of substitution..

Electrophilic aromatic substitution involves introduction of *electron deficiency* (and positive charge) into the benzene ring. Nitro groups and other electron-attracting groups *destabilize* this charge and thus slow the reaction. Nucleophilic aromatic substitution involves introduction of *more electrons* (and negative charge) into the benzene ring. Nitro groups and other electron-attracting groups *stabilize* this charge and thus accelerate this type of reaction.

Notice that *substitution*, like *addition*, is a type of reaction that can occur by a variety of mechanisms. The word *substitution* by itself conveys no information about mechanism. A substitution can occur by a concerted (S_N2) mechanism, an ionization (S_N1) mechanism, a two-step nucleophilic mechanism (as in nucleophilic aromatic substitution), a two-step electrophilic mechanism (as in electrophilic aromatic substitution), a free-radical mechanism, or (as shown in Secs. 18.5–18.6 of the text) other mechanisms. The mechanism that operates depends on the reactants and the reaction conditions. Students sometimes forget this point and assume that the word *substitution* means "S_N2 reaction," because that is the first substitution reaction they often learn.

18.2 The Cumene Hydroperoxide Rearrangement

The formation of phenol and acetone (Eq. 18.114 on text p. 1012) is an acid-catalyzed rearrangement of cumene hydroperoxide. This Study Guide Link discusses the mechanism of this reaction.

In the first step of the mechanism, an oxygen of the hydroperoxide is protonated:

We'll now treat the rearrangement as if it were a stepwise reaction. Loss of water gives an *electron-deficient oxygen*, the oxygen analog of a carbocation. Rearrangement occurs to give a much more stable resonance-stabilized carbocation. (The electron-deficient oxygen structure is so unstable that loss of water and the rearrangement undoubtedly take place in a concerted manner.)

a resonance-stabilized carbocation

This rearrangement is very similar to the rearrangement that occurs in the oxidation of organoboranes, the mechanism of which was discussed in Further Exploration 5.1 in this manual.

The Lewis acid–base association reaction of water with the carbocation gives a type of alcohol known as a hemiacetal.

a hemiacetal

You should now be able to complete the acid-catalyzed conversion of the hemiacetal into phenol and acetone. Start by protonating the oxygen of the PhO— group. What has to happen next to form phenol? Finish the mechanism to give acetone.

FURTHER EXPLORATIONS

18.1 The Fries Rearrangement

There is more to the Friedel–Crafts reactions of phenols than meets the eye. It turns out that phenols react with acid chlorides to give **phenyl esters**; this is a reaction you'll study in Chapter 21.

phenol   a phenyl ester

(Acid chlorides are electrophiles in this reaction just as they are in the Friedel–Crafts reaction. The phenol oxygen, like the aromatic ring, is a source of electrons because of its unshared pairs.) In the presence of AlCl₃—the catalyst used in Friedel–Crafts reactions—phenyl esters rearrange to ketones:

This reaction, known as the *Fries rearrangement*, can be clearly demonstrated because phenyl esters, which can be prepared as pure compounds in the absence of AlCl₃, can be shown to rearrange when treated with AlCl₃. The products of the Fries rearrangement, as you can see, are exactly what would be expected from a direct Friedel–Crafts reaction. Therefore, the "Friedel–Crafts reaction" of a phenol almost surely involves ester formation followed by the Fries rearrangement, or a *combination* of the Fries rearrangement and a direct Friedel–Crafts reaction. In any event, the *end result* is as if a Friedel-Crafts reaction had taken place directly.

It is interesting that the "Friedel–Crafts reactions" of phenols in many cases give significant amounts of *ortho*-substituted products. These can be separated from the para isomers by application of the principles of hydrogen bonding. The ortho isomers contain *intramolecular* hydrogen bonds:

Such intramolecular hydrogen bonding is not possible for the isomeric *para*-substituted phenol. Because the hydrogen-bonding capability of the phenolic —OH group is satisfied within the *ortho*-substituted molecule, this —OH group forms hydrogen bonds less effectively to acceptor atoms of *other molecules*—for example, solvents or chromatography supports such as silica gel—than the —OH group of the *para*-substituted isomer. This is another manifestation of the *proximity effect* (Sec. 12.8)—in this case, an effect that enhances an intramolecular chemical equilibrium relative to an analogous intermolecular equilibrium. Therefore, the *ortho*-substituted isomer is generally much less soluble in solvents that accept hydrogen bonds, and moves through a silica gel chromatography column more rapidly, than the *para*-substituted isomer. Thus, the great difference in hydrogen-bonding properties of the two isomers makes separation a relatively simple matter. Exactly the same principles can be utilized in designing separations of *o*- and *p*-nitrophenol.

REACTION REVIEW

I. REACTIONS OF ARYL AND VINYL HALIDES

A. ELIMINATION REACTIONS OF VINYLIC HALIDES

1. Base-promoted β-elimination reactions of vinylic halides do occur and can be useful in the synthesis of alkynes.

$$Ph-CH=C{Br \atop Ph} \xrightarrow[(CH_3)_3COH]{(CH_3)_3CO^-K^+} Ph-C\equiv C-Ph \ + \ K^+Br^- \ + \ (CH_3)_3COH$$

B. NUCLEOPHILIC AROMATIC SUBSTITUTION

1. Aryl halides that have one or more nitro groups ortho or para to the halogen undergo nucleophilic substitution reactions under relatively mild conditions.

2. The mechanism involves a resonance-stabilized anionic intermediate called a Meisenheimer complex. (See discussion of mechanism and reactivity of a Meisenheimer complex on pages 967-968 of the text.)

3. The reaction is faster when there are more nitro groups ortho and para to the halogen leaving group.

   a. Ortho and para nitro groups accelerate the reaction because the rate-limiting transition state resembles the Meisenheimer complex, and the nitro groups stabilize this complex by resonance.

   b. Other substituents that can provide resonance stabilization to the Meisenheimer complex can also activate nucleophilic aromatic substitution.

4. The effect of the halogen on the rate is quite different from that in the S_N1 or S_N2 reaction of alkyl halides.

$$Ar-F \gg Ar-Cl \sim Ar-Br \sim Ar-I$$

   a. Electronegative halogens further stabilize the Meisenheimer complex.

   b. The effect is due to the fact that the halide ion is not lost in the rate-determining step, which is the reaction of the nucleophile with the aryl halide.

C. THE HECK REACTION

1. In the Heck reaction, an alkene is coupled to an aryl bromide, aryl iodide, or aryl triflate under the influence of a Pd(0) catalyst.

   a. The actual catalytically active species is PdL_2, formed by two ligand dissociations.

b. The PdL$_2$ generated enters into the catalytic cycle, which consists of the following steps:

i.   The aryl halide undergoes oxidative addition.

ii. A ligand substitution occurs when one of the catalyst ligands is replaced by the alkene.

iii. A 1,2-insertion of the alkene occurs, in which the aryl group migrates to carbon-2 of the alkene.

iv. The β-elimination of the newly formed aryl substituted alkene is followed by ligand association.

v.   The aryl-substituted alkene undergoes ligand dissociation.

vi. A ligand undergoes association with the catalyst.

vii. A reductive elimination of hydrogen halide occurs, which is neutralized by the basic triethylamine present.

$$HBr + N(CH_2CH_3)_3 \longrightarrow Br^- \ HN^+(CH_2CH_3)_3$$

c. Pd(II) catalysts can be used with aryl iodides; it is thought that the Pd(II) is reduced to Pd(0) during the reaction.

2. With cyclic alkenes, β-elimination occurs by loss of the hydride at the β-carbon to which the aryl group was not added. This is required because the elimination, like the ligand insertion that precedes it, is a syn process.

3. When the Heck reaction is applied to unsymmetrically substituted alkenes, two products are in principle possible, because insertion might occur at either of the alkene carbons.

a. When R = phenyl, CO_2R (ester), CN, or other relatively electronegative group, the aryl halide tends to react at the unsubstituted carbon to give the E (trans) stereoisomer.

b. When R = alkyl, mixtures of products are often observed.

4. The Heck reaction is another example of a reaction that can be used to form carbon–carbon bonds.

D. *THE SUZUKI COUPLING*

1. A Suzuki coupling (also known as a Suzuki reaction or a Suzuki–Miyaura coupling reaction) joins an aryl or vinylic boronic acid to an aryl or vinylic iodide or bromide.

a. The coupling is catalyzed by a Pd(0) catalyst ($Pd(PPh_3)_4$ or $Pd(OAc)_2/PPh_3$, as in the Heck reaction).

b. The reaction occurs in the presence of a base, typically aqueous NaOH or Na_2CO_3.

c. The reaction may be used to prepare the following compound types:

i. Biaryls—compounds in which two aryl rings are connected by a σ bond.

ii. Aryl-substituted alkenes.

iii. Conjugated alkenes.

> d.   Coupling occurs with retention of alkene stereochemistry.
> e.   Suzuki coupling avoids issues of regiochemistry that can sometimes occur with the Heck reaction.

2.   The necessary boronic acids may be purchased commercially or prepared in two ways:

> a.   The reaction of a Grignard or organolithium reagent with trimethyl borate, followed by aqueous acid (see Eq. 18.56, text p. 986).
> b.   The hydroboration of 1-alkynes with catecholborane or disiamylborane (see Sec. 5.7, text p. 258).

E.   *ALKENE METATHESIS*

1.   In an alkene metathesis, the groups at each end of the double bonds are interchanged.

2.   A number of transition-metal catalysts have been developed for alkene metathesis. Some are based on tungsten and molybdenum, which tend to be air-sensitive and less tolerant to other functional groups. The most common laboratory catalysts are based on ruthenium, which can be easily handled in the presence of alcohols, phenols, and other groups. The Grubbs G1 and G2 catalysts (Fig. 18.6, text p. 989) are examples.

3.   In many cases, the catalysts bring some or all of the possible alkenes into equilibrium. For an alkene metathesis reaction of two unsymmetrical alkenes, ten alkenes (not counting stereoisomers) can be formed. The production of the desired product can be increased by one or more of the following strategies:

> a.   Using an excess of one of the reactants.
> b.   Utilizing Le Chatelier's principle (e.g., where the by-product is a gas and bubbles out of the reaction mixture).
> c.   Exploiting the fact that different alkenes undergo metathesis at greatly different rates. Alkene metathesis is very sensitive to the steric environment of the alkene double bonds.

4.   One of the most important applications of alkene metathesis is for closing rings, even medium- and large-sized rings.

II.   OTHER EXAMPLES OF TRANSITION-METAL CATALYZED REACTIONS

A.   *ZIEGLER–NATTA CATALYST*

1.   One of the most important transition-metal catalysts in commerce is a catalyst formed from $TiCl_3$ and $(CH_3CH_2)_2AlCl$, called the Ziegler–Natta catalyst.

2. This catalyst brings about the polymerization of ethylene and alkenes at 25 °C and 1 atm pressure to give high-density poly(ethylene).

$$n\ H_2C=CH_2 \xrightarrow[\text{(CH}_3\text{CH}_2)_2\text{AlCl}]{\text{TiCl}_3} \left(\!CH_2CH_2\!\right)_n$$

3. One possible mechanism involves a two-step iterative process:
   a. First, a ligand association occurs.

   b. Then the complex undergoes a 1,2-ligand insertion.

B. HYDROFORMYLATION

1. Hydroformylation is another commercially important process that involves a transition-metal catalyst, in this case, tetracarbonylhydridocobalt(I) catalyst, sometimes called the oxo process. (See Problem 18.29 on text p. 994.)

C. CATALYTIC HYDROGENATION

1. An important transition-metal catalyzed reaction is the homogeneous catalytic hydrogenation of akenes using a soluble rhodium(I) catalyst called Wilkinson's catalyst, $ClRh(PPh_3)_3$. (See Problem 18.15, text p. 982.)

2. Catalytic hydrogenation is an extremely important reaction that occurs over carbon-supported transition metals such as Ni, Pd, and Pt. Although the mechanism of catalytic hydrogenation is not definitively known, it probably involves intermediates on the catalyst surface similar to those described in this chapter.

III. REACTIONS OF PHENOLS AND PHENOLATES

A. IONIZATION OF PHENOLS

1. Phenols are weakly acidic; the pK_a of phenol itself is 9.95.

2. Phenols are acidic enough to be converted completely into their conjugate-base phenolate ions (phenoxide ions) by one equivalent of NaOH or NaOEt.

phenol
$pK_a = 9.95$

B.   USE OF PHENOXIDES AS NUCLEOPHILES

1. Phenoxides can be used as nucleophiles.
2. Aryl ethers can be prepared by the reaction of a phenoxide anion and an alkyl halide (another example of the Williamson ether synthesis).

C.   SUBSTITUTION AT THE ARYL–OXYGEN BOND: THE STILLE REACTION

1. Pd(0) catalysts can catalyze substitution at the aryl–oxygen bond if the phenolic OH group is converted into a trifluoromethanesulfonate ester, which is a very good leaving group.

   a. Aryl trifluoromethanesulfonate esters are prepared from phenols and trifluoromethanesulfonic anhydride (triflic anhydride).

   b. Trifluoromethanesulfonate esters are nicknamed triflates; the triflate group is often abbreviated —OTf.

   c. The triflate group is one of the best leaving groups because it is an extremely weak base.

2. Aryl triflates react readily with organotin derivatives in the presence of Pd(0) catalysts to give coupling products.

   a. The organotin compounds used in the Stille reaction are either commercially available or readily prepared from Grignard Reagents and commercially available trialkyltin chlorides.

   b. In general, vinylic groups and aryl groups are transferred from the tin more rapidly than alkyl groups.

   c. If a tetraalkylstannane is used, alkyl groups can also be transferred.

3. The mechanism of the Stille reaction probably involves the following steps:

   a. The aryl triflate undergoes oxidative addition to the 14-electron $Pd(PPh_3)_2$.

   b. Excess chloride ion (as LiCl) performs a ligand substitution to the resulting unstable complex.

   c. A second ligand substitution occurs, in which the aryl or alkyl groups on the organotin compound (which resembles a carbanion and is nucleophilic) substitutes for the chloride on the Pd.

   d. A reductive elimination occurs to give the product and regenerate the catalyst.

4. The Stille reaction can be carried out with aryl halides instead of aryl triflates.
5. The transferred alkyl group does not rearrange in the Stille reaction.
6. This is another method that can be used to form carbon–carbon bonds.

D. *ELECTROPHILIC AROMATIC SUBSTITUTION REACTIONS OF PHENOLS*

1. Phenols undergo electrophilic aromatic substitution reactions.
2. Phenols react rapidly with bromine.

   a. Phenol can be halogenated once under mild conditions that do not affect benzene itself.

   b. Phenol reacts with Br_2 in H_2O (bromine water) to give 2,4,6-tribromophenol.

   i. Bromine reacts with water to give protonated hypobromous acid, a more potent electrophile than bromine itself.

   ii. In aqueous solutions near neutrality, phenol partially ionizes to its conjugate-base phenoxide anion; this anion is very reactive and brominates instantly, thereby pulling the phenol–phenolate equilibrium to the right.

   iii. Phenoxide ion is much more reactive than phenol because the reactive intermediate is not a carbocation, but is instead a more stable neutral molecule.

(also ionizes)

c. In strongly acidic solution, in which formation of the phenolate anion is suppressed, bromination can be stopped at the 2,4-dibromophenol stage.

3. Phenol is very reactive in other electrophilic substitution reactions.

a. Phenol can be nitrated once under mild conditions.

b. Direct nitration is not the preferred method for synthesis of di- and trinitrophenol, because the concentrated HNO_3 required for multiple nitrations is an oxidizing agent. Instead, nucleophilic aromatic substitution is used. (See text p. 1006 for a discussion of mechanism.)

c. The great reactivity of phenol in electrophilic aromatic substitution does not extend to the Friedel–Crafts acylation reaction.

   i. Friedel–Crafts acylation of phenol occurs slowly, but may be carried out successfully at elevated temperatures; the ring is acylated only once.

   ii. Phenols are less reactive because they react rapidly with the $AlCl_3$ catalyst. (See also Further Exploration 18.1.)

E. OXIDATION OF PHENOLS TO QUINONES

1. *p*-Hydroxyphenols (hydroquinones), *o*-hydroxyphenols (catechols), and phenols with an unsubstituted position ortho or para to the hydroxy group are oxidized to quinones.

HO—⟨benzene ring⟩—OH $\xrightarrow[\text{HOAc}]{\text{CrO}_3}$ O=⟨cyclohexadiene ring⟩=O

hydroquinone   ***p*-benzoquinone**

2. The oxidation of hydroquinone and its derivatives to the corresponding *p*-benzoquinone can also be carried out reversibly in an electrochemical cell.

F.   *INDUSTRIAL PREPARATION AND USE OF PHENOL*

1. Phenol and acetone are prepared from a single starting material, cumene, which in turn comes from benzene and propene, two compounds obtained from petroleum.

2. The production of phenol and acetone is a two-step process.

    a. Cumene undergoes an autoxidation with molecular oxygen to form cumene hydroperoxide.

⟨benzene ring⟩—C(CH₃)(CH₃)—H $\xrightarrow{\text{O}_2}$ ⟨benzene ring⟩—C(CH₃)(CH₃)—O—O—H

      cumene   **cumene hydroperoxide**

    b. The cumene hydroperoxide undergoes an acid-catalyzed rearrangement that yields both acetone and phenol. (See Study Guide Link 18.2.)

⟨benzene ring⟩—C(CH₃)(CH₃)—O—O—H $\xrightarrow{\text{H}_3\text{O}^+}$ ⟨benzene ring⟩—OH + O=C(CH₃)(CH₃)

      cumene hydroperoxide   **phenol**   **acetone**

SOLUTIONS TO PROBLEMS

Solutions to Focused Problems

18.1   (a)   *p*-Bromotoluene reacts most slowly (that is, not at all) because it is an aryl halide; aryl halides do not undergo S$_N$2 reactions. Benzyl bromide is a benzylic halide and therefore reacts most rapidly in an S$_N$2 reaction (see text Sec. 17.4, text p. 931).

Br—⟨benzene ring⟩—CH₃   <<   ⟨benzene ring⟩—CH₂CH₂CH₂Br   <<   ⟨benzene ring⟩—CH₂Br

     ***p*-bromotoluene**   **(3-bromopropyl)benzene**   **benzyl bromide**

   (b)   1-Bromocyclohexene, a vinylic halide, does not react by the S$_N$2 mechanism; 1-(bromomethyl)cyclohexene, an allylic halide, reacts most rapidly. (See text Sec. 17.4, text p. 931.)

1-bromocyclohexene bromocyclohexane 1-(bromomethyl)cyclohexene

18.2 Assume that E2 reactions of vinylic halides follow the same trends as E2 reactions of alkyl halides. Therefore, because an anti-elimination is faster than the corresponding syn-elimination (Sec. 9.5E, text pp. 477-479), the reaction of compound *B* is faster than that of compound *A*; and the reaction of compound *A* is faster than that of compound *C* because bromide is a better leaving group than chloride. Because eliminations involving benzylic β-hydrogens are faster than other eliminations, compounds *A–C*, in which the β-hydrogens are benzylic, react more rapidly than compound *D*, in which the β-hydrogens are not benzylic. Furthermore, the elimination reaction of compound *D* gives a triple bond that is conjugated with only one benzene ring (and thus has only one phenyl branch; see structures that follow), whereas all of the others give triple bonds that are conjugated with two benzene rings (and thus have two phenyl branches). Therefore, the desired order is $D < C < A < B$.

The product of the elimination reactions of compounds *A–C* is diphenylacetylene, Ph—C≡C—Ph; the product of the elimination reaction of compound *D* is phenylacetylene, Ph—C≡C—H.

18.3 (a) The order of increasing S_N1–E1 reactivity is $A \ll B < C$. The benzylic halide *C* reacts fastest because its ionization gives a resonance-stabilized benzylic carbocation; and the vinylic halide *A* reacts slowest because vinylic halides are nearly inert in S_N1 reactions for the reasons given in Sec. 18.3 of the text.

(b) The reactivity order is $B \ll C < A$. The reaction of compound *B* is slowest because vinylic halides are virtually inert in S_N1 reactions; and the reaction of compound *A* is fastest because its ionization gives a resonance-stabilized allylic carbocation.

18.4 (a) The nitrogen of the amine acts as a nucleophile in a nucleophilic aromatic substitution reaction.

$$O_2N—\langle\;\rangle—\overset{+}{N}H_2Et\;\;Cl^- \;\;\rightleftarrows\;\; O_2N—\langle\;\rangle—\ddot{N}HEt \;+\; HCl$$

with NO$_2$ substituent

The nitro groups reduce the basicity of the product to the point that it is deprotonated under the reaction conditions. (Can you give a reason why?) Because you have not yet studied the basicity of aromatic amines, however, either answer shown in equilibrium above should be considered correct.

(b) The product results from nucleophilic aromatic substitution by the thiolate group:

$$\langle\;\rangle—\ddot{\underset{\cdot\cdot}{S}}—Bu \;+\; F^-$$

with NO$_2$ substituent

(c) No reaction, because the compound contains no substituent that can activate nucleophilic aromatic substitution by stabilizing the anionic intermediate (the Meisenheimer complex).

18.5 (a) The second compound, *p*-fluoronitrobenzene, reacts most rapidly because only in the reaction of this compound is the intermediate Meisenheimer complex stabilized by resonance interaction of an unshared electron pair with the nitro substituent.

(b)   The first compound should react more rapidly because the Meisenheimer complex is stabilized by the resonance interaction of both substituents; the complex in the case of the second compound is stabilized by resonance involving only a nitro substituent.

18.6   The bond to one allylic carbon is an X-type bond. That is, if we break the metal–carbon bond, a carbon anion is formed. The double bond is an L-type ligand. Breaking its coordination gives a neutral alkene.

L-type interaction

X-type interaction

As with the case of the cyclopentadienyl anion discussed on text pp. 971-972, this is a bookkeeping fiction; the L- and X-type character is actually spread out over the entire allyl system:

It also happens that allyl can bond to a metal in some cases as an X-type ligand; that is, the π electrons in some cases aren't involved in coordination at all. (See Footnote ** in Table 18.1, text p. 972.) However, this situation is not relevant to the present problem, as the problem asks only about the LX-type coordination.

18.7   (a)   For each electron pair moved out of a metal–oxygen bond onto an attached oxygen, the oxygen takes on a negative charge. Therefore, by definition, each bond to oxygen counts as one X-type interaction. The number of X-type bonds to Mn is 7; and, because the overall charge is on oxygen, the charge on the metal, Q_M is 0. (See Eq. 18.28, text p. 973.) Therefore, the oxidation state is +7.

 The following is an equivalent structure for permanganate in which the negative charge is on Mn:

$$
\begin{array}{c}
O \\
\parallel \\
O = \overset{\displaystyle \quad}{Mn} = O \\
\parallel \\
O
\end{array}
$$

In this case the oxidation state is +8 − 1 = +7; obviously, since resonance structures are different representations of the same compound, we get the same result from either structure.

(b)   The PPh_3 ligands are L-type ligands; therefore, there are no X-type ligands, and, because the charge on Pd is 0, the oxidation state of Pd is 0.

> You might be wondering about the prefix *tetrakis* in the name of this complex. The prefixes *bis*, *tris*, and *tetrakis* are used as numerical prefixes instead of *di*, *tri*, and *tetra* when the group that is enumerated itself contains multiple substituents. Thus, the ligand triphenylphosphine has three phenyl groups on the phosphorus (thus the prefix *tri* in the name of this ligand). There are four triphenylphosphine ligands—thus the prefix *tetrakis*.

(c)   The Cp ligand (Table 18.1) can be viewed as an L_2X ligand for bookkeeping purposes. Therefore, with two X-type bonds and no charge on the iron, the oxidation state of Fe is +2.

18.8   With one X-type ligand (Cl) and zero charge, oxidation state of rhodium (Rh) in the starting material is +1. In the product, two new X-type bonds have been added; hence, the oxidation state is +3. The overall reaction is an oxidation. (This is an example of an oxidative addition, an important type of reaction that is considered later in this section.)

18.9   (a)   Consulting Fig. 18.3, text p. 970, we see that neutral tungsten (W) has 6 valence electrons. Add two electrons for the charge. The ligands are L-type ligands. Therefore, this is a d^8 complex. Alternatively, the oxidation state of tungsten is, by Eq. 18.28, text p. 973, 0 − 2 = −2. Apply Eq. 18.30, text p. 974: n = valence electron count in neutral W (6) − the oxidation state (−2) = 6 − (−2) = 8.

(b)   Pd has ten valence electrons in the neutral atom. There are no charges and no X-type ligands in the complex, so this is a d^{10} complex. Using Eq. 18.30, text p. 974, with an oxidation state of 0, affords the same answer.

(c)   This is the same complex as in Problem 18.8. In the solution to that problem the oxidation state of rhodium (Rh) was calculated to be +3. Fig. 18.3 shows that neutral Rh has 9 valence electrons. Therefore, application of Eq. 18.30, text p. 974, shows that this is a d^6 complex. Alternately, starting with 9 electrons and removing 3 for the X-type ligands gives 9 − 3 = 6 for the value of n.

18.10   (a)   For Rh, start with the number of electrons in the neutral atom (9); subtract charges on the metal (0); add the number of X-type ligands (3); and add twice the number of L-type ligands ($2 \times 3 = 6$) to get $9 + 3 + 6 = 18$ electrons. Alternatively, because it is known that $n = 6$ from the solution to Problem 18.9(c), Eq. 18.31 on p. 974 can be applied to obtain the electron count. This is an 18-electron complex.

(b)   For Tc, start with the number of electrons in the neutral atom (7); subtract charges on the metal (1); add the number of X-type ligands (0); and add twice the number of L-type ligands ($2 \times 6 = 12$) to get $7 − 1 + 12 = 18$ electrons.

18.11   Neutral iron (Fe) has 8 electrons. Because CO is an L-type ligand, it is counted twice in the electron count. Simply solve for x in $8 + 2x = 18$ and obtain $x = 5$. $Fe(CO)_5$, or pentacarbonyliron(0), is in fact a stable complex that can be purchased commercially.

18.12   Vanadium (V) in $V(CO)_5$ has an electron count of $5 + (2 \times 6) = 17$. If the tendency of metals is to have a total of 18 electrons, then V should have a significant driving force to obtain an extra electron, that is, to be reduced. Indeed, the total electron count in $[V(CO)_6]^-$ is 18, and this anion is relatively stable.

18.13   (a)   Cyanide (⁻CN) is an X-type ligand. If we strip the four cyanide ligands from the Zn (zinc), a Zn^{2+} ion remains. From Fig. 18.3 on text p. 970, Zn has 12 valence electrons, so Zn^{2+} has 10. This is exactly the number needed to fill all of the 3d orbitals with two electrons each. Four valence orbitals—the 4s and the 4p orbitals—are left empty, and these are hybridized to form more directed orbitals. (This situation is exactly like carbon hybridization in methane, except that we are using orbitals from period 4.) Hybridizing one 4s and three 4p orbitals gives four sp^3 hybrid orbitals, which, as we know from methane, are directed to the corners of a regular tetrahedron. Each of these empty orbitals accepts a pair of electrons from a cyanide ligand. Thus, the $[Zn(CN)_4]^{2-}$ complex is tetrahedral (the gray dashed lines show the tetrahedron; the black lines are bonds).

Unhybridized Zn^{+2} (d^{10}):

(b)   Triphenylphosphine (PPh_3) is an L-type ligand. If we strip the four PPh_3 ligands from the Pd (palladium), a Pd (0) atom remains. From Fig. 18.3 on text p. 970, Pd has 10 valence electrons. This is exactly the number needed to fill all of the 4d orbitals with two electrons each. Four valence orbitals—the 5s and the 5p orbitals—are left empty, and these are hybridized to form more directed orbitals. (This situation is exactly like carbon hybridization in methane, except that we are using orbitals from period 5.) Hybridizing one 5s and three 5p orbitals gives four sp^3 hybrid orbitals, which, as we know from methane, are directed to the corners of a regular tetrahedron. Each of these empty orbitals accepts a pair of electrons from a PPh_3 ligand. Thus, the $Pd(PPh_3)_4$ complex is tetrahedral (the gray dashed lines show the tetrahedron; the black lines are bonds).

Unhybridized Pd (0) (d^{10}):

18.14   The student has neglected to understand that when an X-type ligand reacts as a nucleophile with a metal to displace an L-type ligand, the charge changes by one unit, because electrons from X-type ligands are assumed to be divided between the metal and the ligand. In other words, charge is not balanced in this equation. If we put a negative charge on the Pd, as required by this reaction, the oxidation state of Pd is unchanged.

18.15   (a)   *Oxidative addition of H₂ to the catalyst:*

oxidation state: +1   oxidation state: +3
electron count: 16e⁻   electron count: 18e⁻

Ligand substitution of one PPh₃ by the alkene:

oxidation state: +3
electron count: 18e⁻

oxidation state: +3
electron count: 18e⁻

1,2-Insertion of the alkene into a Rh—H bond and re-addition of the previously expelled PPh₃ ligand:

oxidation state: +3
electron count: 18e⁻

Note: empty orbital
on Rh because H departed
with its two electrons

oxidation state: +3
electron count: 16e⁻

oxidation state: +3
electron count: 18e⁻

Reductive elimination of the alkene product to regenerate the catalyst:

oxidation state: +3
electron count: 18e⁻

oxidation state: +1
electron count: 16e⁻

(b)   Regardless of whether H₂ or D₂ is used, the 1,2-insertion is a syn-addition:

And the reductive elimination is also syn:

Because we started with a symmetrical cis-alkene, the product is the meso stereoisomer. This is the same stereochemical outcome expected when catalytic hydrogenation over a Pd/C catalyst is used. As discussed in the text (Sec 7.8E p. 371), catalytic hydrogenation is also a syn-addition.

18.16   The steps in Eq. 18.49b, text p. 983, of the text are numbered for reference.

Step 1:
    Fundamental process: oxidative addition
    Oxidation state of Pd starting catalyst (PdL$_2$): 0
    Electron count of Pd in the starting catalyst: 14e$^-$. (Note that neutral Pd is a 10-electron atom.)
    Oxidation state of Pd in the product: +2
    Electron count of Pd in the product: 16e$^-$

Step 2: (From here on, the reactant has the same properties as the product of the previous step.)
    Fundamental process: ligand substitution
    Oxidation state of Pd in the product: +2
    Electron count of Pd in the product: 16e$^-$

Step 3:
    Fundamental process: 1,2-ligand insertion
    Oxidation state of Pd in the product: +2
    Electron count of Pd in the product: 14e$^-$

Step 4:
    Fundamental process: β-elimination
    Oxidation state of Pd in the product: +2
    Electron count of Pd in the product: 16e$^-$

Step 5:
    Fundamental process: ligand dissociation
    Oxidation state of Pd in the product: +2
    Electron count of Pd in the product: 14e$^-$

Step 6:
    Fundamental process: ligand association
    Oxidation state of Pd in the product: +2
    Electron count of Pd in the product: 16e$^-$

(Steps 5 and 6 together result in a ligand substitution.)

Step 7:
    Fundamental process: reductive elimination
    Oxidation state of Pd in the product: 0
    Electron count of Pd in the product: 14e$^-$

(The catalyst is regenerated in Step 7.)

18.17   The two products result from coupling at the two chemically nonequivalent carbons of the alkene:

iodobenzene   propene   (1-propenyl)benzene   isopropenylbenzene
                          (from coupling at C-1)   (from coupling at C-2)

That is, in step 3 of the mechanism described in Eq. 18.49b, text p. 983, and in the previous solution, 1,2-ligand insertion can occur to either carbon of the double bond. Because these two carbons are nonequivalent, two products are obtained.

18.18 Either aryl substituent could originate from the aryl halide or from the alkene. Remember, if the alkene substituent is aryl (as in these cases), substitution occurs mainly at the less branched carbon.

18.19 The environment around the Pd following the initial oxidative addition is rather congested with the large PPh$_3$ ligands, the aryl group, and the halide. Ligand replacement by the alkene is more favorable for alkenes containing fewer bulky substituents. As a consequence of the Heck reaction itself, the product is always more branched than the starting material, and therefore, the starting alkene will "win" the competition for ligand replacement over the product alkene more often than not. Assuming that the subsequent rate constants for substitution are not very different for the two alkenes, the higher concentration of the complex between the catalyst and the starting alkene leads to a greater reaction rate for this alkene. (Remember, rate = k[complex], where k = the rate constant. Given equal k values, the complex present in higher concentration "wins" the rate competition.)

18.20 This is essentially like the cyclohexene case given in Eq. 18.50 (text p. 983) and subsequent discussion.

18.21 As illustrated in Eq. 18.55, text p. 986, Suzuki coupling occurs with retention of the alkene stereochemistry.

(a)   (b)   (c)

18.22 The starting materials for a Suzuki coupling can be determined by mentally "splitting" the aryl–aryl bond to form an aryl bromide with one part and an aryl boronic acid with the other part.

Preparing the two boronic acids:

Synthesis Sequence #1:

Synthesis Sequence #2:

18.23   Start by breaking the compound at the bond between the aryl ring and the alkene; then place a bromine atom (Br) on one of the cleaved fragments and a boronic acid [B(OH)$_2$] or catecholborane (see Eq. 18.57 text p. 986) on the other:

18.24   This reaction involves three sequences of similar steps: a Brønsted acid–base reaction with H$_3$O$^+$; a Lewis acid–base dissociation of methanol from boron; and a Lewis acid–base association of water (from solvent) with boron. Proton transfers are mediated by solvent, even when they're not shown explicitly.

18.25   First, mentally break the two alkenes at the double bond to produce four halves: *A*, bearing the R^1 group; *B* bearing the R^2 group; *C* bearing the R^3 group; and *D* bearing the R^4 group. Now systematically combine one half with another to form trans alkenes—don't forget that one half can combine with another half of the same kind and that *A* + *B* will give the same product as *B* + *A*.

(Solution continues on the following page)

$$A = A \ + \ A = B \ + \ A = C \ + \ A = D \ +$$
$$B = B \ + \ B = C \ + \ B = D \ +$$
$$C = C \ + \ C = D \ +$$
$$D = D$$

18.26 (a)

(b)

(c)

18.27 (a) (The symbol ⇨ means "implies as starting material.)

citronellol
(oil of roses)

(b)

(c)

18.28

18.29   A mechanistic outline:

The classifications of the various steps are:
(1) Substitution of one L-type ligand for another
(2) 1,2-Ligand insertion
(3) Association with L-type ligand
(4) 1,1-Ligand insertion
(5) Oxidative addition
(6) Reductive elimination
(7) Association with L-type ligand

18.30   (a)   2,4-Dinitrophenol is more acidic because its conjugate-base anion has more important resonance structures. In the 2,4-dinitrophenoxide anion, both nitro substituents are involved in resonance stabilization; in 2,5-dinitrophenoxide anion, only the 2-nitro group is involved in resonance stabilization. (The 5-nitro group is not involved for the reasons shown in the discussion of Eq. 18.76 on text p. 996.) The actual pK_a values are

2,5-dinitrophenol
$pK_a = 5.22$

2,4-dinitrophenol
$pK_a = 4.11$

(b)   *Meta*-chlorophenol is more acidic because the conjugate-base anion, *m*-chlorophenoxide, is stabilized by the electron-withdrawing polar effect of the chloro substituent. The actual pK_a values are

phenol
$pK_a = 9.95$

m-chlorophenol
$pK_a = 9.02$

(c)   The *para*-isomer of the conjugate-base phenoxide anion has more important resonance structures. The key resonance structure is the one below, in which the substituent group participates in the resonance stabilization. (The actual pK_a values are 7.7 and 8.0 for the *para*- and *meta*-isomers, respectively.)

$$\left[\text{resonance structures} \longleftrightarrow \right]$$

18.31   (a)

4-nitrophenol
(*p*-nitrophenol)

4-nitroanisole
(*p*-nitroanisole)

(b)

Ph—OH $\xrightarrow{\text{NaOH}}$ Ph—O⁻ Na⁺ $\xrightarrow[\text{2) } H_3O^+]{\text{1)}}$ Ph—OCH₂CH₂OH

phenol

2-phenoxyethanol

18.32   The solubility of most phenols in base is due to their −1 charge when ionized. Their conjugate bases are ionic compounds, and ionic compounds have greater solubility than nonionic ones. The compound shown in the problem is cationic by virtue of its positively charged nitrogen; ionization of the OH group converts it into a neutral compound. Neutral species are less soluble in water than compounds with a net charge.

Compounds that contain both a positive and a negative charge, such as the conjugate base of the compound shown in the problem, are called *zwitterions* ("double-ions"), or inner salts. These types of compounds are discussed in Chapter 27.

18.33   (a)   (b)

9,10-phenanthroquinone
(an *o*-quinone)

1,4-phenanthroquinone
(a *p*-quinone)

18.34   (a)   A *para*-quinone is formed as in Eq. 18.80, text p. 998; the nitro group is unaffected.

(b)   The *para*-quinone is formed because it is more stable than the alternative *ortho*-quinone.

18.35 (a) Let the long alkyl chain in vitamin E (structure on text p. 1002) be abbreviated as —alkyl. The most stable radical is formed by abstraction of the phenolic hydrogen atom:

Another possible resonance structure that can derived by the fishhook notation is structure *B*:

The oxygen with the unpaired electron in this structure violates the octet rule—it has nine electrons. In molecular orbital terms, this structure would require the occupancy of an antibonding orbital and thus would reduce the stability of the radical. Therefore, this resonance structure is not important.

(b) As in part (a), the phenolic hydrogen atom is abstracted because a resonance-stabilized radical is formed.

18.36 (a)

(b)

18.37 (a)

(b)   You can use the second resonance structure shown in part (a) to react with the radical R•.

18.38   The long, hydrophobic hydrocarbon chain of ubiquinone (Eq. 18.89, text p. 1001) anchors it in the cellular hydrophobic membrane of the mitochondria organelle (see Sec. 8.7B).

18.39 (a)

(b) Unlike the phenolic groups of urushiol, the methyl ether groups would not be oxidized to a quinone and would thus not react with a thiol group of a protein to form the "foreign" protein sensed by the immune system. An allergic skin response like that provoked by urushiol would probably not occur.

18.40 (a)   (b)   (c)

18.41 The electrophile is the *tert*-butyl cation, which is formed by protonation of the alcohol and the Lewis acid–base dissociation of water.

tert-**butyl cation**

18.42 (a) Neopentyl phenyl ether does not readily cleave with hot concentrated HBr. Such a reaction would require either an S_N1 or S_N2 reaction of bromide ion at a phenyl–oxygen bond, which does not occur, or an S_N1 or S_N2 reaction at the primary carbon of a neopentyl group. The S_N1 reaction does not occur because it would require the formation of a primary carbocation. (Such a reaction could be forced with very long reaction times and heat, and would occur with rearrangement of the neopentyl group.) The transition state of an S_N2 reaction at the primary carbon of a neopentyl group is so overwhelmed by van der Waals repulsions that such a reaction does not take place; see Fig. 9.4(b) on text p. 465.

In contrast, benzyl methyl ether can cleave by both S_N1 and S_N2 mechanisms at the benzylic carbon, or by an S_N2 mechanism at the methyl carbon. Here is the cleavage by an S_N2 mechanism involving the nucleophilic reaction of Br⁻ at the methyl carbon of the protonated ether:

PhCH₂—O⁺—CH₃ ⟶ PhCH₂—ÖH + :Br—CH₃

benzyl methyl ether   **benzyl alcohol   methyl bromide**
(after protonation by HBr)

And here is the S_N2 mechanism at the benzylic carbon (Sec. 17.4 in the text):

PhCH₂—O⁺—CH₃ ⟶ PhCH₂—Br: + HÖ—CH₃

benzyl methyl ether   **benzyl bromide   methyl alcohol**
(after protonation by HBr)

An S_N1 mechanism at the benzylic carbon is also possible that parallels the mechanism shown in part (a): protonation of the ether, formation of a benzylic cation, and a Lewis acid–base reaction of Br⁻ with this carbocation. The products are the same as those derived from the S_N2 reaction immediately previous.
p-Methoxytoluene can cleave at the methyl–oxygen bond by an S_N2 mechanism.

H₃C—⟨ ⟩—O⁺—CH₃ ⟶ H₃C—⟨ ⟩—Ö—H + :Br—CH₃

p-methoxytoluene   **p-cresol**   **methyl bromide**
(after protonation by HBr)

(b) Diphenyl ether does not cleave with hot, concentrated HBr, because such a cleavage would require either an S_N1 reaction or an S_N2 reaction at a phenyl–oxygen bond; as this section of the text shows, such reactions do not occur. In contrast, *tert*-butyl phenyl ether cleaves by an S_N1 mechanism involving protonation of the oxygen and loss of phenol to form a *tert*-butyl cation, which undergoes a Lewis acid–base association reaction with bromide ion to form *tert*-butyl bromide.

Ph—Ö—C(CH₃)₃ ⟶ Ph—O⁺—C(CH₃)₃ —S_N1→ Ph—ÖH + ⁺C(CH₃)₃ ⟶ :Br—C(CH₃)₃

tert-butyl phenyl ether   **phenol**   **tert-butyl bromide**

18.43   The product is phenylacetylene, Ph—C≡CH, along with the usual by-products, Li⁺ ⁻OTf and ClSi(CH₃)₃.

18.44   The triflate derivative of *p*-nitrophenol (prepared from *p*-nitrophenol and triflic anhydride; Eq. 18.107, text p. 1009) and the Z stereoisomer of the appropriate trimethylstannyl derivative would be required.

O₂N—⟨ ⟩—OTf   +   (CH₃)₃Sn—C=C(—H)—C(=O)—OCH₂CH₃

p-nitrophenyl triflate

18.45 The alcohol A can form a relatively stable tertiary benzylic carbocation under the acidic conditions of the reaction. This serves as a Friedel–Crafts alkylating agent for phenol. (Note the similarity to the mechanism shown in the solution to Problem 18.42, text p. 1008.)

Solutions to Skills Objectives Problems

18.46 (a) Bromobenzene B reacts most slowly (that is, not at all) because it is an aryl halide; aryl halides do not undergo S_N2 reactions. Benzyl bromide C is a benzylic halide and therefore reacts most rapidly in an S_N1 reaction (see text Sec. 17.1, p. 931). Therefore, the order of increasing rate is: $B \ll A < C$.

(b) Compound A reacts most slowly because it is a vinylic halide; vinylic halides do not undergo S_N2 reactions. Compound C is an allylic halide, and therefore reacts most rapidly in an S_N2 reaction (see text Sec. 17.4, p. 931). The order of increasing rate in an S_N2 reaction is: $A \ll B < C$.

18.47 (a) Bromobenzene B reacts most slowly because it is an aryl halide; aryl halides do not undergo S_N1 reactions. Benzyl bromide C is a benzylic halide and therefore reacts most rapidly in an S_N2 reaction (see text Sec. 17.4, pp. 918-922). The order of increasing rate is: $B \ll A < C$.

(b) Compound C is an allylic halide, and therefore reacts most rapidly in an S_N1 reaction by forming a resonance-stabilized carbocation intermediate (see text Sec. 17.1, pp. 918-922). Compound A is a vinylic halide and compound B is a primary alkyl halide, neither of which form the carbocation intermediate required in order to undergo S_N1 reactions (see text Sec 18.3 and Sec 9.8C, respectively). The order of increasing rate in an S_N1 reaction is: $A \sim B \ll C$.

18.48 In order for an aryl halide to undergo the nucleophilic aromatic substitution reaction, the reactive intermediate (Meisenheimer complex) must be stabilized via resonance. The *para*-nitro groups of compounds A and C will be able to stabilize the reactive intermediate, so they are both more reactive than compound B. Fluoride is a better leaving group than chloride in nucleophilic aromatic substitution reactions (see text Sec. 18.4), so the order of increasing reactivity is $B < A < C$.

18.49 (a) In the structure of heme, the iron (Fe) atom has 2 X-type ligand interactions and 3 L-type ligand interactions.

(1) The charge on the metal, Q_M is 0. Using Eq. 18.28, the oxidation state of Fe is calculated to be +2.

(2) Iron has 8 valence electrons in its neutral state (Fig. 18.3). Therefore, using Eq. 18.29, the Fe atom in the heme complex is a d^6 metal.

(3) Using Eq. 18.33, the electron count for the heme complex is $16e^-$.

(b) Metals react with ligands to achieve the more favorable electron count of 18. Oxygen binding to the Fe in heme is a favorable process because, as an L-type ligand, oxygen gives the Fe atom 2 additional electrons to increase its electron count to 18.

18.50 (a) Phenol is most acidic because its conjugate-base anion is stabilized by both the polar and resonance effects of the phenyl group. The conjugate-base anion of benzyl alcohol is stabilized by the polar effect of the phenyl group. The conjugate-base anion of cyclohexanol has none of these stabilizing contributions. The acidity order is cyclohexanol   benzyl alcohol   phenol

◄————————————— pK_a —————————————
————————————— acidity ————————————►

(b)   The *p*-nitro substituent stabilizes the *p*-nitrophenolate ion by both its resonance effect (see text Eq. 18.75 on p. 995) and its polar effect; the *p*-chloro substituent stabilizes *p*-chlorophenolate ion only by its polar effect. The third compound is nitric acid, a strong acid. Its conjugate-base anion, the nitrate ion, is stabilized both by the polar effect of the nitro group, which is much closer to the anionic oxygen than it is in the *p*-nitrophenolate ion, and by a resonance effect:

nitrate ion

Here is the acidity order:

p-chlorophenol   *p*-nitrophenol   nitric acid

◄————————————— pK_a —————————————
————————————— acidity ————————————►

(c)   Mercaptans are more acidic than phenols; and aryl mercaptans (thiophenols) are more acidic than alkyl mercaptans for the same reason that phenols are more acidic than alcohols. The acidity order is

cyclohexanol   cyclohexyl mercaptan   benzenethiol

◄————————————— pK_a —————————————
————————————— acidity ————————————►

(d)   4-Nitrophenol is more acidic than phenol (see text p. 995), and benzenethiols are more acidic than phenols (element effect.) The acidity order is

phenol   4-nitrophenol   4-nitrobenzenethiol

◄————————————— pK_a —————————————
————————————— acidity ————————————►

(e)   The *para*-aryl substituents all stabilize the respective conjugate-base alkoxide ions by both a resonance and a polar effect. The nitro group of the *meta*-nitrophenyl substituent exerts an additional stabilizing polar effect; and the nitro group of the *para*-nitrophenyl substituent stabilizes the conjugate-base phenoxide ion by both a polar effect and a resonance effect:

$$ C \quad < \quad A \quad < \quad D \quad < \quad B $$

$$ \longleftarrow \text{p}K_a \longrightarrow $$

$$ \longleftarrow \text{acidity} \longrightarrow $$

18.51 The steps are numbered for reference. The classifications of the various steps are:

(*1*) oxidative addition
(*2*) ligand dissociation
(*3*) ligand insertion
(*4*) β-elimination
(*5*) reductive elimination

18.52 The Simmons–Smith reagent is formed in the 2-step process shown here:

18.53 In (a), fluoride ion is displaced in a nucleophilic aromatic substitution reaction by ethanethiolate ion. In (b), the product is formed by a nucleophilic aromatic substitution reaction in which chloride is displaced by hydroxide ion. The product is ionized by $^-$OH present in the reaction mixture; addition of acid (H_3O^+) forms the neutral phenol. In (c), the potassium carbonate ionizes the phenol. The resulting phenolate ion is methylated by methyl iodide to form the methyl ether.

(a)

$+ \ F^-$

(b)

(c)

(d)   No reaction occurs under the mild conditions.

Part (e) is a Suzuki coupling reaction in which pyridine derivatives are used instead of benzene derivatives. Part (f) is another Suzuki coupling. In part (g), bromine is introduced at the ring position that is activated by two hydroxy groups.

(e)

(f)

$+ \; Na^+ \; {}^-OTf + B(OH)_3$

(g)

Part (h) is a Stille reaction; the product results from a coupling between the stannane reagent and the triflate derivative of 1-naphthol. Part (i) is a Heck reaction between a *vinylic* halide and an alkene. Vinylic halides react in many cases like aryl halides in Pd-mediated coupling reactions. The coupling with cyclic alkenes results in coupling at an allylic position, as in Eq. 18.50 on text p. 983. Part (j) is an oxidation to a quinone; the oxidation occurs within the same ring.

(h)

**1-naphthol
(α-naphthol)**

(i)

(j)

1,4-naphthoquinone

(k)   The methanesulfonate derivative of 2,4-dinitrophenol undergoes a nucleophilic substitution reaction in which the methanesulfonate group is displaced by methoxide.

2,4-dinitrophenol   **1-methoxy-2,4-dinitrobenzene**

In part (l), nucleophilic aromatic substitution occurs at the position that is ortho and para to the nitro groups, because this involves the more stable anionic intermediate (Meisenheimer complex). In part (m), oxidation occurs to give the quinone. In part (n), the methyl ethers are cleaved but the phenyl ethers are not. [See the solution to Problem 18.62(d).]

(l)

(m)

m-chlorophenol

**2-chloro-2,5-cyclohexadien-
1,4-dione**

(n)

$+ \; 2\,CH_3Br$

Part (o) is an alkene metathesis in which the volatile alkene ethylene is eliminated. In part (p), the vinylic halide undergoes an E2 elimination to produce an alkyne. Part (q) is another alkene metathesis in which the reaction is driven by elimination of ethylene.

(o)

(E)-2-heptene-1-ol

+ H_2C=CH_2 **ethylene**
(gas)

(p)

$(CH_3)_3C$—C≡C—OCH_3

(q)

2,5-dimethyl-2-hexene

+ H_2C=CH_2 **ethylene**
(gas)

18.54   (a)

4-hydroxy-3-methoxycinnamic acid
(ferulic acid)

(b)   No. The phenolic group is meta to the alkene and carboxylic acid groups and the radical that is formed by hydrogen-atom abstraction cannot delocalize to the alkene and carboxylic acid groups.

3-hydroxy-4-methoxycinnamic acid

single electron skips over this carbon

Solutions to Integrated Problems

18.55   (a)   No reaction
(b)   No reaction

(c)

(d)

+ ClMgI

(e)

(f)

(g)

+   (CH₃)₃SnCl

+   Li⁺ ⁻OTf

(h)

(i)

+ Na⁺ ⁻I + B(OH)₃

(j)

18.56   (a)

(b)

(c)   Both benzylic bromination and ring bromination take place; see Eq. 18.98, text p. 1005.

(d)

(e)

(f)

(g)

(h)

OSO$_2$CF$_3$

CH$_3$

(i)

CH$_3$

CH$_3$

(j)

H

H$_3$C

C=C

CH$_3$

H

18.57 (a) The intramolecular alkene metathesis forms a 12-membered cyclic alkene. Recall that 8- or larger membered cyclic alkenes can accommodate a trans double bond (see Sec. 7.6C in the text). Thus compounds X and Y are cis/trans isomers.

····CH$_3$
H
+
H ``CH$_3$
+ H$_2$C=CH$_2$

compounds X and Y

Since the only difference between compound X or compound Y is the stereochemistry at the alkene double bond, catalytic hydrogenation of either compound X or compound Y produces compound Z.

····CH$_3$
H

compound Z

(b) Alkene metathesis produces the more stable trans alkene A; asymmetric epoxidation (see Sec. 12.11 in the text) produces the epoxide C (see Eq. 12.88a, text p. 670, for an equivalent reaction).

H   CH$_2$OH

C=C

(CH$_3$)$_3$C   H

A

H   CH$_2$OH

(CH$_3$)$_3$C····C—C····H

O

C

18.58 The enol A is more acidic (by about five pK_a units) because its conjugate-base anion is stabilized by resonance as shown below, whereas the conjugate-base anion of compound B is not resonance-stabilized.

:Ö:⁻

H$_3$C—C=CH$_2$ ⟷   H$_3$C—C—C̈H$_2$

:O:

conjugate base of compound A

18.59   (a)

(b)   The resonance structures below show that the negative charge is delocalized to the α-carbon as well as over the two oxygens, much like a carboxylate ion, so it has a carboxylic acid-like pK_a.

(c)   The conjugate base form of Warfarin predominates at pH = 7.4. When the pH of a solution is more than two units greater than the pK_a of a particular proton, we can say that the proton is >99% dissociated, as stated by point 3 in Sec. 3.6E, text p. 149. Additionally, you can solve for the ratio of the dissociated to associated forms by using the Henderson-Hasselbalch equation, Eq. 3.65d on text p. 147.

18.60   Hydrogenation of phenol *A* yields the corresponding cyclohexanol. Compound *B* reacts readily with HBr to give the benzylic bromide; the corresponding Grignard reagent undergoes protonolysis to give *p*-xylene. The ether group in compound *C* is cleaved by HBr to give bromoethane, CH_3CH_2Br (which is volatile under the high-temperature reaction conditions), and *m*-cresol.

(a)   (b)   (c)

3,5-dimethylphenol
(compound *A*)

p-methylbenzyl alcohol
(compound *B*)

m-ethoxytoluene
(compound *C*)

m-cresol

+ CH_3CH_2Br

**ethyl bromide
(bromoethane)**

18.61   (a)   Cyclohexanol is ionized only to a very small extent in aqueous NaOH solution; phenol ionizes almost completely. (The pK_a of cyclohexanol is about 17; the pK_a of phenol is about 10.)

(b)   Both compounds react rapidly with NaH to give H_2 and their conjugate bases: sodium cyclohexanolate in the case of cyclohexanol, and sodium phenolate (sodium phenoxide) in the case of phenol.

sodium cyclohexanolate   sodium phenolate

(c)   Both compounds are converted into their triflate derivatives.

cyclohexyl triflate   phenyl triflate

$$-O-\overset{\overset{\displaystyle O}{\|}}{\underset{\underset{\displaystyle O}{\|}}{S}}-CF_3 \;=\; -OTf$$

trifluoromethanesulfonate
(triflate group)

(d)   Only cyclohexanol reacts with concentrated HBr. Because the C—O bond in phenol cannot be broken by an S_N1 or S_N2 reaction, phenol is inert except for a small degree of protonation of the phenol oxygen. The product in the case of cyclohexanol is bromocyclohexane.

—Br   **bromocyclohexane**

(e)   The phenol is brominated on the ring to give the following products; the alcohol is unaffected.

2-bromophenol   **4-bromophenol**

(f)   Cyclohexanol is oxidized to cyclohexanone, whereas phenol is oxidized to *p*-benzoquinone.

cyclohexanone   ***p*-benzoquinone**

(g)   Cyclohexanol undergoes dehydration to cyclohexene; phenol undergoes sulfonation, an electrophilic aromatic substitution reaction.

cyclohexene   **2-hydroxy-**   **4-hydroxy-**
                      benzenesulfonic   **benzenesulfonic**
                      acid   **acid**

18.62   (a)   The second compound reacts in a nucleophilic aromatic substitution reaction. The first compound does not react in this way because it is not activated toward nucleophilic aromatic substitution; the nitro group is not attached to the ring.

(reaction scheme: 2-nitro-4-fluorotoluene + RO⁻K⁺ (R = H or CH₂CH₃) → 2-nitro-4-alkoxytoluene + K⁺ F⁻)

 In the interest of total accuracy we point out that the first compound actually does react, but in a different way. A hydrogen of the —CH₂— group is acidic enough to ionize in alcoholic KOH; can you see why? But it does not liberate F⁻.

(b)   The last compound, diphenyl ether, cannot be prepared by a Williamson ether synthesis, because such a synthesis would require an S_N2 reaction to occur at the carbon of a benzene ring. Notice that the first ether, tetrahydrofuran, can be prepared by an intramolecular Williamson synthesis from 4-bromo-1-butanol, Br—(CH₂)₄—OH. The second ether, ethoxybenzene, can be prepared by a reaction of sodium phenolate, PhO⁻ Na⁺, with ethyl iodide.

(c)   The first compound undergoes solvolysis by an S_N1 mechanism to give a mixture of an alcohol, an ether, an alkene, and HBr (thus the acidic solution). The second compound is inert, because aryl halides do not undergo solvolysis under neutral conditions.

(reaction scheme: 2-phenyl-2-bromopropane + ROH (R = H or CH₂CH₃) → 2-phenyl-2-alkoxypropane + α-methylstyrene + HBr)

(d)   Diphenyl ether will not cleave with HI because such a cleavage would require an S_N1 or S_N2 reaction at the carbon of a benzene ring, and such reactions do not occur. Phenyl cyclohexyl ether, however, would cleave to phenol and iodocyclohexane:

(reaction scheme: cyclohexyl phenyl ether + HI → phenol + iodocyclohexane; diphenyl ether (inert to HI))

cyclohexyl phenyl ether   **diphenyl ether**
(inert to HI)

18.63   The phenol group is unaffected in both compounds (except for a small amount of protonation of the phenol oxygen); the benzyl alcohol and ether groups react.

(structures: 3-(bromomethyl)phenol; and benzene-1,3-diol + CH₃Br)

18.64   Sodium ethoxide converts the thiol completely into its conjugate-base thiolate ion.

(a)   The thiolate ion is alkylated. (Remember from Sec. 17.4 of the text, that allyl bromide is a particularly reactive alkylating agent.)

(reaction scheme: 3,5-dimethylthiophenolate sodium salt + BrCH₂CH=CH₂ (allyl bromide) → 3,5-dimethylphenyl allyl sulfide (SCH₂CH=CH₂) + Na⁺ Br⁻)

product of thiol ionization   **allyl bromide**

(b) No further reaction of the thiolate occurs because bromobenzene, an aryl halide, is inert to nucleophilic substitution and elimination reactions. (Sec. 18.1 of the text.)

18.65 (a) The reaction of phenol as a base with the acid H_2SO_4:

(b) First, consider polar effects. The phenyl group has an electron-withdrawing polar effect; this destabilizes the positively-charged conjugate acid of phenol. Such an effect is absent in the conjugate acid of the alcohol cyclohexanol because its ring has no double bonds. Therefore, the polar effect reduces the basicity of the phenol relative to the alcohol. Second, consider the resonance effect of the benzene ring. The source of this effect is the delocalization of an oxygen nonbonding electron pair into the ring:

resonance structures of phenol

Such an effect is absent in the conjugate acid of the phenol because it would place two positive charges on the oxygen, an electronegative atom.

conjugate acid of phenol

Therefore, resonance stabilizes phenol relative to its conjugate acid and therefore reduces its basicity. (Another way to think of this situation is to imagine that delocalization of the oxygen nonbonding electron pair makes it "less available" for protonation.) Such a resonance effect is not present in the alcohol. The conclusion is that a phenol is much less basic than an alcohol because of both the polar effect and the resonance effect of the ring. Therefore, the alcohol is the stronger base.

18.66 Because NaOH can ionize a phenol, particularly a relatively acidic phenol such as *p*-nitrophenol, which has a pK_a of 7.21, the spectrum *B* (and the corresponding yellow color) is due to the conjugate base of the phenol. The UV absorption reflects conjugation of the extra unshared pair with the ring and its delocalization into the *p*-nitro group, as shown in Eq. 18.75, text p. 995. Addition of acid lowers the pH so that the un-ionized phenol is regenerated, and the UV spectrum of the phenoxide disappears. To summarize:

18.67 It is the un-ionized form of vanillin that has the typical odor. In NaOH solution, the phenol group of vanillin ionizes to its conjugate-base phenoxide ion; because vanillin is no longer present, and because ionic compounds such as the conjugate-base phenoxide are not volatile, the odor disappears. (A compound has to be volatile—that is, it must have a significant vapor pressure—to enter the gas phase and thus reach the nostrils.) Acidification of the solution brings about protonation of the phenoxide and regeneration of vanillin and, therefore, the characteristic odor.

neutral compound:
volatile, has vanilla odor

ionic compound:
nonvolatile, has no odor

18.68 Two principles involved in the solution to this problem are, first, that phenoxide ions are more soluble in water than are their conjugate-acid phenols; and second, that an acid is ionized when the pH of the solution is well above the pK_a of the acid. The second principle can be demonstrated in two ways. Either use the rule of thumb that when the pH is more than two units greater than the pK_a of the acid the acid is >99% dissociated (point 3 on text p. 149), or use Eq. 3.67a (or Eq. 3.67b) to find the percent of the acid in the ionized form (or un-ionized form) at each given pH. These formulas can be solved to find precisely the ratio of $[A^-]/[HA]$.

Condition (1) corresponds to a pH of about 1; neither phenol is ionized (much less than 1%), and therefore neither is soluble in this solution. Under condition (2), some fraction of 2,4-dinitrophenol is ionized (about 44% by Eq. 3.67a), but under condition (3), which is a pH value almost three units higher than the pK_a of the phenol, 2,4-dinitrophenol is >99% ionized. Because this pH value is more than three units below the pK_a of p-cresol, this phenol is not ionized at pH = 7. Therefore, 2,4-dinitrophenol should be relatively soluble at this pH because it is ionized, and p-cresol remains insoluble in the aqueous buffer and can be recovered from the ether layer. This is the best condition to separate the two phenols by extraction. Condition (4) corresponds to a pH of about 13; both phenols are ionized at this pH, and both should be relatively soluble (as their conjugate-base phenoxides) in the aqueous layer.

18.69 (a) Because the S_N2 reaction requires approach of the nucleophile from the backside of the C—Br bond, and because this would require approach of the nucleophile along the axis of the triple bond, the S_N2 reaction is impossible.

(b) 1-Haloalkynes cannot undergo an S_N1 reaction. Such a reaction would require that a carbocation be formed at an sp-hybridized carbon. The positive nucleus of a carbon with sp hybridization is very exposed because the two π bonds hold the electrons relatively far from the carbon nucleus. This means that the electrons in the C—Br bond are held very tightly; this bond is very strong, so the C—Br bond resists ionization. Moreover, there is no mechanism by which the π electrons on the sp-hybridized carbon can be used to stabilize the carbocation; they cannot overlap with the empty sp orbital.

18.70

18.71   (a)   The oxidation state is the number of valence electrons in the neutral atom (9) less the charge (+1) less the number of bonds to X-type ligands (2). (The nitrogens that have three bonds in addition to the metal–nitrogen bond are L-type ligands; the other nitrogen is an X-type ligand. The second X-type ligand is the cyano (—CN) group.) The oxidation state of Co is +3.

(b)   The number n is the valence electrons in the neutral atom (9) less the oxidation state (+3). Therefore, Co in this complex is a d^6 metal.

(c)   The electron count is the number of electrons in the neutral atom (9) less the charge (+1) plus the number of X-type ligands (2) plus twice the number of L-type ligands (4). The electron count is therefore $18e^-$. This complex thus follows the eighteen-electron rule.

18.72   (a)   The number n is the valence electrons in the neutral atom (8) less the number of X-type ligands (5: two nitrogens, sulfur, and count oxygen twice because it is doubly bonded). Therefore, Fe in this complex is a d^3 metal.

(b)   The electron count is n (3) plus twice the number of all ligands (7). The electron count is therefore $17e^-$.

18.73   The π electrons of phenol react with the electrophile to form a carbocation intermediate. (Formation of the electrophile is shown in Eq. 18.99b, text p. 1005.) After a proton is lost from the hydroxy group, the carbocation intermediate becomes the neutral compound that precipitates.

18.74   Solvolysis of the benzylic bromide results in a carbocation intermediate that can lose a proton to give the bracketed intermediate, an example of a quinone methide, which reacts with hydroxide ion as shown in the following equation. (See also Problem 18.90, text p. 1022, for the formation of a similar type of intermediate.)

18.75   (a)

(b)

(c)

(d)

(e)

prepared in part (a)   **2-chloro-4,6-dinitrophenol**

(f)   See Eq. 18.105, text p. 1007. Other sources of the *tert*-butyl cation electrophile could also be used, such as methylpropene and acid. *Tert*-butyl chloride and AlCl$_3$ would probably not work well because of the tendency of phenols to form complexes with AlCl$_3$.

p-cresol   **2,6-di-*tert*-butyl-4-methylphenol
(BHT)**

(g)

(h)

fluorobenzene

1-nitro-4-(2-phenylethoxy)benzene

(i)   First, prepare bromobenzene from benzene.

bromobenzene

Next, prepare 1-bromo-3-nitrobenzene from benzene.

1-bromo-3-nitrobenzene

Then carry out a Heck reaction of either aryl halide with ethylene. We'll use 1-bromo-3-nitrobenzene.

1-bromo-3-nitrobenzene   **m-nitrostyrene**

Finally, carry out another Heck reaction using *m*-nitrostyrene and bromobenzene (using somewhat harsher conditions).

m-nitrostyrene   **bromobenzene**

(j)   First, prepare the Grignard reagent from the aryl bromide, and use it to prepare the corresponding aryl boronic acid. Couple the resulting aryl boronic acid with (*E*)-1-bromo-2-phenylethene in the presence of a Pd(0) catalyst and aqueous base (Suzuki coupling).

Alternatively, couple the aryl bromide with (*E*)-1-catecholboranyl-2-phenylethene in the presence of a Pd(0) catalyst, sodium ethoxide, and a non-aqueous solvent (Suzuki coupling)—water would hydrolyze the catecholboranyl reagent.

(k)   First, nitrate bromobenzene to produce 1-bromo-4-nitrobenzene.

bromobenzene   **p-bromonitrobenzene**

Then carry out a Heck reaction with ethylene.

$$O_2N-C_6H_4-Br + H_2C=CH_2 \xrightarrow[\text{(CH}_3\text{CH}_2)_3\text{N}]{\text{Pd(PPh}_3)_4,} O_2N-C_6H_4-CH=CH_2$$

p-bromonitrobenzene   **ethylene**   **p-nitrostyrene**

Finally, react *p*-nitrostyrene with cis-2-buten-1,4-diol in an alkene metathesis.

p-nitrostyrene   **cis-2-buten-1,4-diol**

(l)   First, prepare the Grignard reagent from the alkyne, and use it to prepare the corresponding stannane.

$$PhC\equiv C-H + CH_3MgBr \longrightarrow PhC\equiv C-MgBr + CH_4$$

$$PhC\equiv C-MgBr + ClSn(CH_3)_3 \longrightarrow PhC\equiv C-Sn(CH_3)_3 + ClMgBr$$

Next, prepare the triflate of the phenol.

4-methoxyphenol

Finally, couple the triflate and the stannane in a Stille reaction.

(m) The target compound can be obtained by a Suzuki coupling. First, prepare a vinylic boronic acid derivative by reacting 1-hexyne with catecholborane (see Eq. 18.57, text p. 986).

1-hexyne

catecholborane

Next, prepare the aryl halide. Brominate resorcinol under mild conditions to produce a monobrominated product. Then carry out a Williamson ether synthesis with an excess of ethyl iodide.

Finally, couple the vinylic boronic acid derivative with the aryl halide in a Suzuki reaction.

Another route to this compound involves alkene metathesis between 1-hexene and 2,4-diethoxystyrene. Show how each of these compounds can be synthesized.

(n) Start with (2-cyclohexenyl)benzene, which is prepared from iodobenzene and cyclohexene by a Heck reaction as shown in Eq. 18.50, text p. 983.

(2-cyclohexenyl)benzene

It is reasonable to suppose that epoxidation will occur at the face of the ring opposite to that occupied by the bulky phenyl group. (As noted in the problem, the product is a racemate because the reagents are achiral.)

18.76   (a)   The simplest way to visualize the polymer resulting from the ROMP reaction is to break the alkene at the C=C bond and then draw the result with a set of parentheses. Finally, mentally break the section between the fragment ends, and then place them on either side of a C=C bond.

(b)

18.77   (a)   In the Diels–Alder reaction of 1,3-cyclopentadiene to yield endo-dicyclopentadiene, one of the 1,2-cyclopentadiene molecules as the "diene" and the other acts as the "dienophile."

endo-**dicyclopentadiene**

(b)   The ROMP reaction using *endo*-dicyclopentadiene produces a cyclopentane ring bearing four alkene "ends" that can recombine is many different ways producing a 3-dimensional network. Consider the difference in strength between a thread and a piece of cloth (2-dimensional lattice). A 3-dimensional lattice is magnitudes stronger and more rigid than a 1- or 2-dimensional structure.

a 3-dimensional polymer

This structure is simplified, because each double bond is connected to a double bond at *any* of the possible positions in another monomer.

18.78   An E2 elimination is most rapid when it occurs with anti stereochemistry. In the first reaction, anti-elimination leads to the observed product. In the second reaction, formation of the alkyne requires a slower syn-elimination; consequently, another process can compete, namely, elimination of a methyl hydrogen and the bromine to form the allene. In the first reaction, elimination of the methyl hydrogen to form the allene is a slower process because allenes are not so stable as alkynes. The transition state for elimination is destabilized by its allene-like character. (The methyl hydrogens are approximately as acidic as the vinylic hydrogens because they are allylic.) It is also possible that the allene is the only product formed in the second reaction and that the alkyne is formed by a base-catalyzed isomerization of the allene:

18.79 Because vinylic halides do not undergo S_N2 or S_N1 reactions, this substitution must occur by an unusual mechanism. The key to this solution is to recognize that the leaving-group effect and the effect of a *para*-nitro substituent are much the same as they are in nucleophilic aromatic substitution. This reaction, in fact, occurs by a mechanism that is much like the mechanism of nucleophilic aromatic substitution, except that it takes place at a vinylic carbon. The *para*-nitro group accelerates the reaction by affording additional resonance stabilization of the anionic intermediate *A*.

anionic intermediate *A*

anion *A*

18.80 The spectrum is consistent with the formation of an anionic intermediate—a stable Meisenheimer complex. (The dotted lines symbolize resonance delocalization of the negative charge.) If you draw out the resonance structures for this ion, you will see that charge is delocalized to the carbons bearing the nitro groups, and it can also be delocalized into the nitro groups. However, charge is *not* delocalized to the other carbons. Therefore, the protons on these carbons do not show the smaller chemical shift that would be expected if there were high electron density on these carbons.

$\delta\,3.3$ $\delta\,6.3$
CH_3O H
O_2N NO_2
$\delta\,8.7$ H H $\delta\,8.7$
NO_2

18.81 The reaction that forms compound *X* is a hydroalumination, and it is completely analogous to hydroboration. The structure of *X* is

compound *X*

The second reaction is very much like a Stille coupling. We expect the carbon–aluminum bond to behave much like other carbon–metal bonds, for example, the carbon–tin bond in a stannane. And the vinylic iodide group is much like a triflate group in the sense that it is a good leaving group. The structure of the coupling product *Y* is compound *Y*

The by-product of the reaction is the dialkyliodoalane, I—Al(*i*Bu)$_2$.

18.82   If PPh$_3$ reacts with a chlorine, an additional electron pair can be moved to the Pd.

18.83   (a)   This is a nucleophilic aromatic substitution reaction. Notice that the cyano group (:N≡C—) can stabilize the intermediate *A* by resonance. Be sure to draw those structures.

A

(b)   Protonation of methylpropene gives the *tert*-butyl cation, which serves as the electrophile in an electrophilic aromatic substitution reaction.

tert-butyl cation

tert-butyl cation
(redrawn)

18.84   (a)   The actual catalytic species is probably Pd(PPh$_3$)$_2$, as shown by Eq. 18.49a on text p. 983. The likely steps include an oxidative addition, followed by a ligand substitution, and finally a reductive elimination to yield the product.

Pd(0)
an 18e⁻ complex

Pd(0)
a 14e⁻ complex

(b)

Grignard reagent

18.85   (a)

(b)   As noted on text pp. 469-470, ^{18}F-containing compounds are useful in positron emission tomography, or PET.

(c)   The compounds must be prepared and used quickly because the radioactive isotopes have a short half-life and their activity decreases sharply over time.

18.86   The large bulky 9-BBN converts the alkene to an organoborane (see Section 5.6A in the text) in which the boron adds regioselectively to the terminal carbon. Subsequent reaction with the vinylic bromide in the presence of a palladium catalyst and aqueous sodium carbonate couples to alkyl borane and the vinylic bromide. In this reaction, a new stereocenter (*) is formed. Neglecting the stereochemistry at this center for the moment, the result of the reaction is as follows:

Note that the cis-1,5-cyclooctanediol is produced when 9-BBN is hydrolyzed; 9-BBN is formed by reacting borane with 1,5-cyclooctadiene.

Now to the stereochemistry of the new stereocenter. There can be no doubt that the 9-BBN reacts with the face of the double bond at the side opposite the angular methyl group. However, there are two possible conformations about the single bond between the ring and the vinylic carbon at which such a reaction is possible; each conformation gives a different stereoisomer.

conformation A

conformation B

The product from conformation B has much more steric interference with the angular methyl group because of the large group attached to boron. Therefore, the product is derived from conformation A. It has R stereochemistry because introduction of the OH group changes priorities at the stereocenter.

S stereoisomer

R stereoisomer

H_2O_2, ^-OH

18.87 Because the glycol starting material is also a chlorohydrin, and because the phenoxide salt is a base, epoxide formation takes place rapidly, accompanied by the formation of NaCl. (See Sec. 12.3B in the text.) It is the epoxide that actually alkylates the phenol more slowly. The hydroxide ion produced in the last step maintains the phenol in its ionized form.

fast

$^-\ddot{O}H$ (from last step)

slow

mephenesin

18.88 The only way to accommodate these results is for the labeled carbon (carbon with *) and an adjacent carbon to become equivalent at some point. The hint suggests a β-elimination to give an alkyne called *benzyne*.

benzyne

Because benzyne is highly strained (why?), it undergoes an unusual reaction: a nucleophilic reaction of the $^-NH_2$ at a carbon of the triple bond to give a carbanion, which is protonated to give the aniline product. Because benzyne is symmetrical except for the label, the reactions of the $^-NH_2$ at the two carbons of the triple bond are equally likely. Consequently, the products are formed in identical amounts.

18.89 Deduce the structure of the "very interesting intermediate" by mentally imagining a "reverse Diels–Alder" reaction of triptycene that yields anthracene and the intermediate, which is benzyne:

triptycene   "reverse Diels-Alder"   benzyne   anthracene

The Grignard reagent has carbanion character, and this "carbanion" is a strong base. Elimination of the weaker base fluoride gives benzyne:

benzyne

Because alkynes require linear geometry, it is difficult to incorporate them into six-membered rings. Therefore, benzyne is highly strained and, although it is a neutral molecule, it is very unstable. (Benzyne is about 205 kJ mol^{-1} (49 kcal mol^{-1}) more unstable than an ordinary alkyne.) Indeed, benzyne has been too reactive to isolate except at temperatures near absolute zero. (See also the solution to Problem 18.88 in this manual.)

18.90 The six-proton singlet in the NMR spectrum indicates that the two methyl groups ortho to the oxygen have been retained; this resonance cannot be due to one ortho methyl and one para methyl, because these would not be chemically equivalent. The two-proton singlet at δ 5.49 indicates two alkene protons, and the two-proton singlet at δ 6.76 is accounted for by the two ring protons, which evidently remain. The total number of protons accounted for by the NMR spectrum are two fewer than are present in the starting material. The conditions are much like those for oxidation of a phenol to a quinone. The structure below for compound *A* fits the data; the singlets are all broad because of very slight splitting over more than three bonds.

δ 1.90 H$_3$C   CH$_3$ δ 1.90

δ 6.76 H   H δ 6.76

compound *A*

δ 5.49 H   C   H δ 5.49

18.91 The two reactions are intramolecular Friedel–Crafts acylations. The first equivalent of AlBr$_3$, a strong Lewis acid, forms a complex with the acid chloride group to generate the electrophile, an acylium ion, as discussed in Sec.

16.4F of the text, and is also consumed by formation of a complex with the product. As the lower reaction in the problem shows, the acylium ion reacts at a position para to a methoxy group, because a methoxy group is both ortho, para-directing and strongly activating. The question, then, is why reaction occurs at the ring that does not have methoxy substituents when more Lewis acid is present. The answer is that the second and third equivalents of Lewis acid form complexes with the methoxy groups:

A complexed methoxy group cannot donate a nonbonding electron pair by resonance, because its nonbonding pair is already "donated" to the aluminum. Consequently, the methoxy group, through this complexation, is transformed into an electronegative group that is incapable of a resonance effect. An electronegative group that cannot donate electrons by resonance is a deactivating group, and a ring containing such a group (or two such groups) undergoes substitution more slowly than a ring without such substituents. In other words, acylation of the ring that does not contain such deactivating substituents is the faster acylation.

18.92 (a) We consider only the cross-metathesis products. The pairwise mechanism predicts a 1:1 reaction between the two alkenes and thus produces only one product.

In contrast, the metathesis of cyclopentane with (E)-2-pentene produces three products in a ratio of 1:2:1. Cyclopentene can produce only one "fragment" (fragment A), a "double-headed" fragment, while (E)-2-pentene produces two (fragments B and C). Thus, fragment A can combine with two fragment Bs, two fragment Cs, or one of each (in two different combinations): $B + A + B$ (2,7-nonadiene), $C + A + C$ (3,8-undecadiene), and $B + A + C$ and $C + A + B$ (2,7-decadiene). Note that $B + A + C$ and $C + A + B$ produce the same product and thus twice as much of that product is formed.

(b) The pairwise mechanism can only give ethylene and ethylene-d_4, because at no point in the mechanism does one of the CH_2 (or CD_2) groups become detached from its parent molecule:

To see the results of the metallacycle mechanism, note that $M=CH_2$ and $M=CD_2$ (where M = the ruthenium and its ligands) are formed in the first catalytic cycle. (We leave it to you to fill in the details.)

Now, ethylene-d_2 can form in two ways: by the reaction of $M=CH_2$ with the d_4-diene starting material, or by the reaction of $M=CD_2$ with the undeuterated diene:

However, ethylene-d_4 can form in only one way—the reaction of $M=CD_2$ with the d_4-diene starting material; and ethylene-d_0 can form in only one way—the reaction of $M=CH_2$ with the undeuterated diene starting material. Consequently, the formation of ethylene-d_2 enjoys a two-fold statistical advantage, and the ratio of the three ethylenes is therefore $H_2C=CH_2 : H_2C=CD_2 : D_2C=CD_2 = 1 : 2 : 1$. The result of this very elegant experiment was that the three ethylenes were formed in this statistical ratio.

18.93 (a) The vinylic boronic acid reacts with bromine via a bromonium ion to give the dibromo product. Rotation around the C—C bond positions the boronic acid anti to the bromine on the adjacent carbon. Hydroxide ion (a Brønsted base) reacts with the boron (a Brønsted acid) to form borate, and excellent leaving group. Anti-elimination of the borate and bromide affords the vinylic bromide; the overall reaction thus occurs with *inversion* of configuration.

(b) Iodine adds to the vinylic boronic acid to form an iodonium ion, to which hydroxide ion adds to form an iodohydrin (see Sec. 5.2B of the text). Rotation around the C—C bond positions the boronic acid syn to the hydroxy group on the adjacent carbon. The hydroxy group (a Brønsted base) reacts with the boron (a

Brønsted acid) to form a cyclic borate that breaks down by a syn-elimination to afford the vinylic iodide with overall *retention* of configuration.

18.94 First, recognize that the bombykol carbon skeleton is comprised of only the 10-undecyne-1-ol and 1-pentyne carbon skeletons, thus a "coupling" of the two starting materials would give the desired product. Suzuki coupling involves a vinylic halide and a vinylic boronic acid. But which starting material should be converted to the boronic acid and which should be converted to the vinylic halide? Recall that reaction of a borohydride, such as catecholborane (see Eq. 18.57, text p. 986), with an alkyne produces a trans alkene boronic acid product via syn-addition of the borohydride. Inspecting the stereochemistry of bombykol, the hydroxynonyl group [—(CH₂)₉OH] is trans to the rest of the molecule while the propyl group is cis. Thus, reaction of the boronic acid formed by the addition of catecholborane to 1-pentyne followed by bromine addition in dichloromethane and the resulting product treated with aqueous sodium hydroxide [see Problem 18.93(a), text p. 1023] produces a *cis*-vinylic bromide. Coupling the *trans*-boronic acid with the *cis*-vinylic bromide with a palladium catalyst in the presence of sodium ethoxide will produce bombykol.

18.95 The hydride of potassium tri(isopropoxy)borohydride forms a Lewis acid–base association product with the boron. This positions the hydride for backside substitution at the vinylic carbon, which occurs with inversion of configuration.

CH₃CH₂CH₂CH₂ ... Br: ... C=C ... H ... B(O*i*Pr)₂ ... K⁺ (*i*PrO)₃B̄—H

⟶

CH₃CH₂CH₂CH₂ ... Br: ... C=C ... H ... B(O*i*Pr)₂ ... K⁺ + (*i*PrO)₃B

⟶

CH₃CH₂CH₂CH₂ ... B(O*i*Pr)₂ ... C=C ... H ... H ... + K⁺ :Br:⁻

This mechanism would appear to violate the principle developed on text pp. 962–963—that backside substitution does not occur at a vinylic carbon. However, such a mechanism is reasonable in this case, because, first, the hydride is small and suffers little steric repulsion with the other hydrogen on the same side of the double bond; and second, the reaction should be accelerated because it is intramolecular (Sec. 12.8). This acceleration counterbalances the normally slow substitution at a vinylic carbon.

18.96   Whether the nucleophile occurs in biological systems or in the laboratory, nucleophilic substitution does not occur on aryl derivatives (see Sec. 18.1 of the text) unless the ring is activated by electron-withdrawing groups (such as carbonyl groups or nitro groups) that stabilize the Meisenheimer complex in nucleophilic aromatic substitution. (See Sec. 18.4 of the text.) Furthermore, both fluoride and cyanide are fairly strong bases (see Table 3.2, text p. 142) and are therefore very poor leaving groups.

19

The Chemistry of Aldehydes and Ketones
Carbonyl-Addition Reactions

CHAPTER 19 SUMMARY

- The functional group in aldehydes and ketones is the carbonyl group.

- The carbon of the carbonyl groups is sp^2-hybridized, and the oxygen, as a terminal atom, is sp-hybridized. The C=O bond of the carbonyl consists of an sp^2–sp single bond and a carbon–oxygen π bond. The C=O bond, with a typical length of 120 pm, is considerably shorter than an alkene C=C bond (typically 133 pm).

- Aldehydes and ketones are polar molecules because of the substantial bond dipole of the C=O bond. Because of their polarity, the boiling points of simple aldehydes and ketones are higher than those of hydrocarbons but lower than those of alcohols. The aldehydes and ketones of low molecular mass are very soluble in water.

- The carbonyl stretching absorption near 1700 cm^{-1} is the most important infrared absorption of aldehydes and ketones. The proton NMR spectra of aldehydes have distinctive absorptions at δ 9–11 for the aldehydic protons, and the α-protons of both aldehydes and ketones absorb near δ 2.5. The most characteristic absorptions in the ^{13}C NMR spectra of aldehydes and ketones are the carbonyl carbon resonances at δ 190–220. Aldehydes and ketones have weak $n \longrightarrow \pi^*$ UV absorptions, and compounds that contain double bonds conjugated with the carbonyl group have $\pi \longrightarrow \pi^*$ absorptions. α-Cleavage, inductive cleavage, and the McLafferty rearrangement are the important fragmentation modes observed in the mass spectra of aldehydes and ketones.

- Aldehydes and ketones are weak bases and are protonated on their carbonyl oxygens to give α-hydroxy carbocations ($pK_a \approx -7$). The interaction of a proton or Lewis acid with a carbonyl group activates it toward addition reactions. Proton-catalyzed additions, Grignard additions, and hydride reductions provide examples of such activation.

- The most characteristic carbonyl-group reactions of aldehydes and ketones are carbonyl-addition reactions, which, depending on the reaction, occur under acidic or basic conditions or both. Cyanohydrin formation and hydration are examples of simple reversible carbonyl additions. Hydride reductions and Grignard reactions are examples of simple additions that are irreversible. Addition occurs by reaction of a nucleophile at the carbonyl carbon. The direction of approach of the nucleophile, from above or below the carbonyl plane, is governed by its interaction with the π^* MO of the carbonyl group.

- Hydride addition is an important process in biological reductions. NADH and its phosphorylated analog, NADPH, are the most common biological hydride reducing agents.

- Acetal formation is an example of addition to the carbonyl group followed by substitution. Acetals can be formed in acidic alcohol solvents and converted back into aldehydes or ketones in aqueous acid, but they are stable to base. Hemiacetals are intermediates in acetal formation. Acetals are often used as protecting groups for aldehydes and ketones under basic conditions.

- Imine (Schiff base) formation, enamine formation, and the Wittig alkene synthesis are examples of addition to the carbonyl group followed by elimination.

- Imines are used in biology to tether biologically important aldehydes, such as retinal and pyridoxal phosphate, to proteins; the nucleophilic amine that reacts with the aldehyde is typically the amino group of a lysine residue of the protein.

■ The carbonyl group of an aldehyde or ketone can be reduced to a methylene group by either the Wolff–Kishner reduction (a strongly basic procedure) or the Clemmensen reduction (an acidic procedure).

■ Aldehydes are readily oxidized to carboxylic acids; ketones cannot be oxidized without breaking carbon–carbon bonds.

■ Aldehydes and ketones can be converted into alcohols (Grignard reactions, hydrogenation, and hydride reductions), alkenes (Wittig synthesis), and alkanes (Wolff–Kishner and Clemmensen reductions).

STUDY GUIDE LINKS

19.1 Lewis Acid Catalysis

Equation 16.23c on text p. 885 focused on Lewis acid interactions with leaving groups. What you should notice about both Grignard reagents and lithium aluminum hydride is the important role of Lewis acid catalysis in promoting reactions of these reagents with carbonyl compounds (as well as with epoxides; Eq. 12.45a, text p. 642). Each reagent has its own "built-in" Lewis acid catalyst. As explained in the text, the Lewis acid catalyst in $LiAlH_4$ reductions is the lithium ion, Li^+; and the Lewis acid catalyst in Grignard reactions is the magnesium of the Grignard reagent.

Why not use proton (Brønsted) acids as catalysts in these reactions? Don't forget that both $LiAlH_4$ and Grignard reagents react instantaneously with protons of even weak acids such as water and alcohols. Consequently, the use of proton acids is not an option, because such acids destroy the reagents. Only *after* the addition of these reagents takes place can a proton source be added to the reaction mixture to replace the Lewis acid (Li^+ or ^+MgBr) with a hydrogen.

What you should notice here is that the Lewis acid serves the role of a "proton substitute"—or "oversized proton," as one of the author's colleagues likes to call it—and fulfills *exactly* the same catalytic role that a proton would fulfill if a proton *could* be used.

19.2 Hemiacetal Protonation

You might ask why, in Eq. 19.68b, text p. 1059, the hemiacetal is protonated on the —OH oxygen rather than on the —OCH₃ oxygen. This is a reasonable question because ethers and alcohols have similar basicities. The answer is that protonation on the —OCH₃ oxygen *does* occur, as does loss of methanol from the resulting protonated species. However, these steps are merely the reverse of hemiacetal formation and lead back to aldehyde or ketone and the starting alcohol. Under conditions of excess alcohol and removal of water, the equilibrium shifts to favor the alcohol.

When writing mechanisms, we usually do not write out reasonable steps that are not important in the formation of the product of interest.

19.3 Mechanism of Carbinolamine Formation

You are asked to write the mechanism for acid-catalyzed carbinolamine formation because it is another example of *carbonyl addition*. It is important to understand this reaction mechanistically because you'll see it repeatedly in organic chemistry and biochemistry. If you *really* want to make progress, take these suggestions seriously.

Are you having trouble getting started? In any carbonyl addition, *identify the nucleophile*. The nucleophilic atom is the nitrogen of the amine. *The nucleophile reacts at the carbonyl carbon* because electrons can flow onto the electronegative oxygen. (See Sec. 19.8A, "Carbonyl Resonance and Carbonyl Reactivity," p. 1045 in the text.)

(SG19.1)

Perhaps you added a proton to the carbonyl oxygen first, and *then* let the amine react as a nucleophile. *This is a reasonable step for a beginning student to take,* because it follows the mechanism for acid-catalyzed hydration.

(SG19.2)

Once the reaction of the carbonyl compound with the amine has occurred, a proton is transferred *from* H_3O^+ *to* the carbonyl oxygen and *from* the nitrogen to H_2O. Use *two separate steps* for these transfers. It doesn't matter which you write first. Notice that the oxygen, because of its negative charge, is an alkoxide, which is a rather strong base, and is efficiently protonated by rather low concentrations of H_3O^+.

(SG19.3)

The final step is deprotonation of the nitrogen. Because the nitrogen of the carbinolamine is a base, this final step is an equilibrium. If the acid concentration is high enough, a significant amount of the protonated carbinolamine could be present.

(SG19.4)

Notice that when H_3O^+ is the acid used in the mechanism, then its conjugate base H_2O must be used as the base—*don't make the mistake of using a strong base like HO^- in this step.*

As illustrated here, the mechanism, including the proton-transfer steps, is written *one step at a time.* Students are often tempted to try to show everything in one step. This is not correct because simultaneous collisions of more than two molecules are highly improbable. More important is that consolidating several steps into one can lead to confusion. Remember to write mechanisms *one step at a time.*

Also remember again that all heterolytic (non-radical) mechanisms involve elementary steps that can only be Lewis acid–base associations, Lewis acid-base dissociations, or electron–pair displacements, of which Brønsted acid–base reactions are a special case. (See Sec. 3.4A of the text.)

19.4   Dehydration of Carbinolamines

You should also try to write the mechanism for dehydration of a carbinolamine in acid, using the dehydration of alcohols as your guide. (See Sec. 11.2 of the text.) When you do this, notice that the carbocation intermediate is resonance-stabilized. This is one reason why the dehydration is so rapid.

Dehydration of carbinolamines, *unlike* dehydration of ordinary alcohols, can also occur in base. This could be a concerted β-elimination reaction—essentially an E2 reaction—wiyth hydroxide as the leaving group.

A concerted process avoids formation of a very basic anion on the nitrogen. Notice that hydroxide ion, ⁻OH, acts as a leaving group in the second step of the mechanism. Hydroxide cannot act as a leaving group in the dehydrations of ordinary alcohols because the carbon–oxygen bond is strong and because hydroxide ion is too basic. In this case, however, the formation of the strong C=N double bond (which is considerably stronger than a C=C double bond) provides an additional driving force for the reaction. In addition, the extra electron pair on the nitrogen anion is very basic (its conjugate-acid pK_a is about 25) and thus provides a strong electronic "push" to expel the leaving group.

19.5   Mechanism of the Wolff–Kishner Reaction

The text indicates that the mechanism of the Wolff–Kishner reaction is a series of acid–base reactions. Here they are. Can you supply the curved-arrow notation?

Research has suggested that the neutral diazene intermediate may not actually exist—the proton transfers that precede and follow it are concerted (occur together), possibly involving multiple solvent molecules. However, that's a detail.

Notice that the last intermediate is a carbanion. Carbanions, with conjugate-acid pK_a values in the 55–60 range, are generally too unstable to exist. If this one does really exist, it is probably because a very stable molecule (dinitrogen) is produced along with it. That is, the stability of dinitrogen offsets the instability of the carbanion.

FURTHER EXPLORATIONS

19.1 IR Absorptions of Cyclic Ketones

The increase of the infrared carbonyl absorption frequency with decreasing ring size is a very useful trend that can be used to diagnose the presence of small rings. The main reason for this trend is a coupling phenomenon. In the presentation of IR spectroscopy in Chapter 13, IR absorption is treated as a phenomenon associated with isolated bonds. Even though this is a highly useful approximation, the effects of nearby bonds in some cases cannot be ignored, and the carbonyl stretching frequencies of the cyclic ketones is one of those cases.

Imagine two extreme situations. In the first, the carbonyl group forms an angle of 180° with adjacent C—C bonds (as in ketene, H_2C=C=O). In such a situation, because the vibrations of both bonds occur along the same line, expansion of the carbon–oxygen bond that occurs during a vibration also compresses the carbon-carbon bond; and compression of the carbon-oxygen bond also expands the carbon–carbon bond. The two stretching vibrations are *coupled:*

$$H_2C = C = O \xleftarrow{\text{expansion of C=O bond;}}_{\text{compression of C=C bond}} \longrightarrow H_2C = C = O$$

compression of C=O bond;
expansion of C=C bond

$$H_2C = C = O$$

Both double bonds involve sp-hybridized carbon and oxygen atoms and are therefore relatively strong. The additional energy required to cause both strong bonds to vibrate at the same time is reflected in a higher vibrational frequency for both bonds.

Now consider a hypothetical situation in which the C=O bond is perpendicular to adjacent bonds. In such a situation, the C=O bond can stretch and compress without stretching the adjacent C—C bond:

$$\underset{H_2C \quad CH_2}{\overset{O}{\|}}{C} \xleftarrow{\text{C=O bond stretch}} H_2C - \underset{}{\overset{O}{\|}}{C} - CH_2$$

(Remember that bending of the C—C bond requires much less energy than stretching.) At this angle, the stretching vibrations of the two bonds are completely uncoupled. Hence, it requires much less energy to stretch the C=O bond in this situation than when the two bonds are co-linear.

An intermediate situation occurs at bond angles between 180° and 90°: the stretching vibrations are partially coupled. The more closely the C—C=O angle approaches 180°, the greater the coupling and the higher the vibration frequency. Notice that the bond angle in question is the one between the C=O bond and the ring C—C bond, not the internal angle within the ring. This angle varies from 120° in cyclohexanone to greater values for the ketones with smaller rings. As the data in Display 19.15 on text p. 1033 show, the carbonyl stretching frequencies increase toward smaller ring sizes.

Another factor that contributes to carbonyl stretching frequencies is the relative strengths of the C=O bonds. The C=O bond in ketene involves an sp-hybridized carbon atom, whereas the C=O bond in cyclohexanone involves an sp^2-hybridized carbon. The amount of s character in the C=O σ bond is therefore greater in ketene, and the amount of s character increases from cyclohexanone through the smaller rings. As with C—H bonds (Display 3.69 and discussion, Sec. 3.7C, text p. 153; Table 10.2, p. 520), the strengths of C=O bonds, and consequently their absorption frequencies, increase with increasing s character.

The coupling effect discussed above is believed to dominate the trend toward higher frequencies, with the bond–strength effect making a secondary contribution.

A similar trend can be seen with carbon–carbon double bond stretching frequencies:

C=C stretching frequencies: 1650 cm^{-1} 1672 cm^{-1} 1781 cm^{-1}

19.2   Use of Imines for Characterization of Aldehydes and Ketones

As noted in the text, another use for imines, which was more important before the advent of spectroscopy than it is now, is in the characterization of aldehydes and ketones. When a new compound was synthesized, it was typically characterized by conversion into two or more crystalline compounds called *derivatives*. These derivatives served as the basis for subsequent identification of the new compound when it was isolated from another source or from a different reaction. It was important to prepare derivatives because they eliminated the ambiguity that could arise if two compounds have very similar melting points or boiling points. It is relatively improbable that two compounds with the same melting or boiling points will give two crystalline derivatives with the same melting points.

Certain imines were frequently used as solid derivatives for characterization of aldehydes and ketones. These imines, and the amines from which they are derived, are listed in Table 19.3 on text p. 1065. For example, the 2,4-DNP derivative of acetone is prepared by formation of an imine with 2,4-dintrophenylhydrazine:

acetone   2,4-dinitrophenylhydrazine   a 2,4-dinitrophenylhydrazone
               (2,4-DNP)   (2,4-DNP derivative of acetone)

To illustrate how such derivatives might have been used in structure verification, let's suppose that a chemist had isolated a liquid that could be either 6-methyl-2-cyclohexenone or 2-methyl-2-cyclohexenone. The boiling points of these compounds are too similar for an unambiguous identification. Yet the melting point of either a 2,4-DNP derivative or a semicarbazone (see text Table 19.3) would quickly establish which compound had been isolated.

boiling point	68–70 °C (16 mm)	65–70 °C (16 mm)
semicarbazone, mp	177–178 °C	207–208 °C
2,4-DNP derivative, mp	162–164 °C	207–208 °C

Today, the identity of the compound could be readily established by modern spectroscopic techniques (explain how). But it is still important to be familiar with the imine derivatives in text Table 19.3 because references to the use of such derivatives are commonplace in the older chemical literature.

REACTION REVIEW

I. CARBONYL-ADDITION REACTIONS

A. HYDRATION OF ALDEHYDES AND KETONES

1. Aldehydes and ketones react reversibly with water; this reaction is called hydration.

$$
\underset{\text{C}}{\overset{\text{O}}{\|}} + H_2O \xrightarrow{H_3O^+ \text{ or } {}^-OH} \underset{\text{OH}}{\overset{\text{OH}}{\underset{|}{\overset{|}{-C-}}}}
$$

a carbonyl hydrate

Acid-catalyzed hydration of aldehydes and ketones is reminiscent of acid-catalyzed hydration of alkenes, although hydration of aldehydes and ketones is much faster.

a. The first step is protonation of the carbonyl oxygen; the protonated carbonyl compound is a much stronger Lewis acid (electron acceptor) than an unprotonated carbonyl compound.

b. The electron-deficient carbon reacts with H_2O, a nucleophile.

c. Loss of a proton to the solvent (water) completes the reaction.

a carbonyl hydrate

2. Hydration of aldehydes and ketones also occurs in neutral and basic solution.

B. CYANOHYDRIN FORMATION

1. Hydrogen cyanide, HCN, reacts reversibly with aldehydes and ketones by a nucleophilic addition mechanism to give cyanohydrins.

a. A cyanide ion, formed by ionization of the weak acid HCN, reacts with the carbonyl group at the carbonyl carbon; the carbonyl oxygen becomes negatively charged.

b. The negatively charged oxygen is a relatively strong base, and is protonated by either water or HCN to complete the addition.

a cyanohydrin

2. Cyanohydrin formation favors the cyanohydrin addition product in the case of aldehydes and methyl ketones, but not in the case of aryl ketones.

3. Cyanohydrin formation is another method of forming carbon–carbon bonds.

C. *REDUCTION OF ALDEHYDES AND KETONES TO ALCOHOLS*

1. Aldehydes and ketones are reduced to alcohols with either lithium aluminum hydride ($LiAlH_4$) followed by protonolysis, or sodium borohydride ($NaBH_4$) in a protic solvent. Both $LiAlH_4$ and $NaBH_4$ serve as sources of nucleophilic hydride ion (H:⁻).

    a. These reactions, which are not reversible, result in the net addition of the elements of H_2 across the $C{=}O$ bond.

    b. Reduction of an aldehyde gives a primary alcohol.

    c. Reduction of ketone gives a secondary alcohol.

2. $LiAlH_4$ and $NaBH_4$ reductions are generally referred to as hydride reductions and are examples of nucleophilic addition.

3. The reaction of $LiAlH_4$ with aldehydes and ketones involves the nucleophilic reaction of hydride (delivered from ⁻AlH_4) at the carbonyl carbon; all four hydride equivalents of $LiAlH_4$ are active in this reaction.

    a. A lithium ion coordinated to the carbonyl oxygen acts as a Lewis-acid catalyst.

    b. Hydride reacts with the carbonyl carbon to give an alkoxide addition compound.

    c. The alkoxide salt is converted by protonation in a separate step into the alcohol product; the proton source is water (or an aqueous solution of a weak acid such as ⁺NH_4⁻Cl).

    d. $LiAlH_4$ reacts violently with water and therefore must be used in dry solvents such as anhydrous ether or THF.

4. The reaction of $NaBH_4$ with aldehydes and ketones involves the nucleophilic reaction of hydride at the carbonyl carbon; all four hydride equivalents of $NaBH_4$ are active in this reaction.

    a. The sodium ion of $NaBH_4$ does not form as strong a bond to the carbonyl oxygen as the lithium ion.

    b. $NaBH_4$ reductions are carried out in protic solvents, such as alcohols; hydrogen bonding between the alcohol solvent and the carbonyl group serves as a weak acid catalysis that activates the carbonyl group.

    c. $NaBH_4$ reacts only slowly with alcohols, and it can be used in water if the solution is not acidic.

5. LiAlH$_4$ is a much more reactive reagent than NaBH$_4$.

   a. A number of functional groups react with LiAlH$_4$ but not NaBH$_4$. Among these are alkyl halides, alkyl tosylates and other sulfonate esters, esters, and nitro groups.

   b. The greater selectivity and safety of NaBH$_4$ make it the preferred reagent in many applications.

6. Aldehydes and ketones can also be reduced to alcohols by catalytic hydrogenation.

   a. This reaction is analogous to the catalytic hydrogenation of an alkene.

$$\underset{\substack{H_3C \qquad CH(CH_3)_2}}{\overset{O}{\underset{\|}{C}}} \xrightarrow[\text{pressure}]{\substack{H_2 \\ \text{Ni catalyst}}} \underset{\substack{\\ \quad}}{H_3C-\overset{OH}{\underset{|}{CH}}-CH(CH_3)_2}$$

   b. It is usually possible to use catalytic hydrogenation for the selective reduction of an alkene double bond in the presence of a carbonyl group; palladium catalysts are particularly effective for this purpose.

$$Ph-CH{=}CH-CH{=}O \xrightarrow[\text{5\% Pd/C}]{H_2} Ph-CH_2-CH_2-CH{=}O$$

D. REACTIONS OF ALDEHYDES AND KETONES WITH GRIGNARD AND RELATED REAGENTS

1 The reaction of Grignard reagents with carbonyl groups is the most important application of Grignard reagents in organic chemistry; addition to aldehydes and ketones in an ether solvent, followed by protonolysis, gives alcohols. (See Appendix V, Section F, text page A-10, for a review of other syntheses of alcohols.)

$$\underset{\substack{\text{an aldehyde} \\ \text{or ketone}}}{\overset{O}{\underset{\|}{C}}} + R{-}MgX \longrightarrow \underset{\substack{\\ \text{a halomagnesium} \\ \text{alkoxide}}}{-\overset{O^- \ \overset{+}{M}gX}{\underset{\underset{R}{|}}{C}}-} \xrightarrow{H_3O^+} \underset{\substack{\\ \text{an alcohol}}}{-\overset{OH}{\underset{\underset{R}{|}}{C}}-} + Mg^{2+} + X^-$$

2. The net effect of the Grignard reaction, followed by protonolysis, is addition of R—H (R = an alkyl or aryl group) across the C=O double bond; this addition is not reversible.

   a. Primary alcohols are synthesized by the addition of Grignard reagents to formaldehyde.

formaldehyde $\quad \underset{H \qquad H}{\overset{O}{\underset{\|}{C}}} \xrightarrow[\text{2) H}_3\text{O}^+]{\text{1) R–MgX}} \underset{\substack{\\ \quad H}}{H-\overset{OH}{\underset{\underset{H}{|}}{C}}-R} \quad$ a primary alcohol

   b. Secondary alcohols are synthesized by the addition of Grignard reagents to aldehydes other than formaldehyde.

an aldehyde $\quad \underset{H \qquad R}{\overset{O}{\underset{\|}{C}}} \xrightarrow[\text{2) H}_3\text{O}^+]{\text{1) R'–MgX}} \underset{\substack{\\ \quad R}}{H-\overset{OH}{\underset{\underset{R}{|}}{C}}-R'} \quad$ a secondary alcohol

   c. Tertiary alcohols are synthesized by the addition of Grignard reagents to ketones.

a ketone → a tertiary alcohol

1) R''–MgX
2) H_3O^+

3. The Grignard synthesis of a tertiary alcohol, or in some cases a secondary alcohol, can be extended to an alkene synthesis by dehydration of the alcohol with strong acid during the protonation step.

1) CH_3MgI/Et_2O
2) H_2SO_4/H_2O

4. The Grignard reaction is an excellent method of carbon–carbon bond formation. (See Appendix VI, text page A-13, for a review of other reactions used to form carbon–carbon bonds.)

5. In the mechanism of the Grignard addition—

   a. the magnesium of the Grignard reagent, a Lewis acid, bonds to the carbonyl oxygen; this bonding makes the carbonyl carbon more electrophilic.

   b. the carbon group of the Grignard reagent reacts as a nucleophile at the carbonyl carbon to form a halomagnesium alkoxide.

   c. addition of dilute acid in a separate protonolysis step to the reaction mixture gives an alcohol.

6. The reactions of organolithium and sodium acetylide reagents with aldehydes and ketones are fundamentally similar to the Grignard reaction.

E. PREPARATION AND HYDROLYSIS OF ACETALS

1. When an aldehyde or ketone reacts with a large excess of an alcohol in the presence of a trace of strong acid, an acetal is formed.

   a. An acetal is a di-ether in which both ether oxygens are bound to the same carbon.

   b. Acetals are ethers of hydrates or *gem*-diols.

$2\ R{-}OH\ +$ ... $\xrightarrow{H_3O^+}$ ... an acetal $+\ H_2O$

2. Two equivalents of alcohol are consumed in the formation of acetals, but one equivalent of a 1,2- or 1,3-diol can react to form a cyclic acetal, in which the acetal group is part of a five- or six-membered ring, respectively.

a cyclic acetal

3. Acetal formation is reversible and involves an acid-catalyzed carbonyl addition followed by a substitution that occurs by an S_N1 mechanism:

a. The first step in the mechanism of acetal formation is **acid-catalyzed addition of the alcohol to the** carbonyl group to give a hemiacetal (a compound with an —OR and —OH group on the same carbon).

b. The hemiacetal is converted into an acetal by acid-catalyzed **substitution of the —OH group by** another —OR group.

4. The reaction is driven to the right by applying Le Chatelier's **principle in one or both of the following** ways:

a. Use of excess alcohol as the solvent.

b. Removal of the water by-product.

5. Acetals in the presence of acid and excess water are transformed **rapidly back into the corresponding** carbonyl compounds and alcohols.

a. This process is called acetal hydrolysis.

b. By the principle of microscopic reversibility, the **mechanism of acetal hydrolysis is the reverse of** the mechanism of acetal formation.

c. Acetal hydrolysis, like hemiacetal formation, is acid-catalyzed.

6. The interconversion of hemiacetals and aldehydes is catalyzed not only by acids, **but by bases as well;** however, the interconversion of hemiacetals and acetals is **catalyzed only by acids. For this reason, the** hydrolysis of acetals is catalyzed only by acids; acetals are **stable in basic and neutral solutions.**

7. Hemiacetals in most cases cannot be isolated because they react **further to yield acetals or decompose to** aldehydes or ketones plus water.

a. Simple aldehydes form appreciable amounts of hemiacetals in alcohol solution, just as they form appreciable amounts of hydrates in water.

b. Five- and six-membered cyclic hemiacetals form spontaneously **from the corresponding hydroxy** aldehydes, and most are stable, isolable compounds.

II.   OXIDATION OF ALDEHYDES AND KETONES

A.   OXIDATION OF ALDEHYDES

1. Aldehydes can be oxidized to carboxylic acids.

2. Common oxidants, such as aqueous Cr(VI) reagents, nitric acid, or aqueous $KMnO_4$/NaOH, can be used in the oxidation of aldehydes to carboxylic acids; these oxidizing agents are the same ones used for oxidizing alcohols.

3. Some aldehyde oxidations begin as addition reactions.

   a. In the oxidation of aldehydes by Cr(VI) reagents, the hydrate, not the aldehyde, is actually the species oxidized.

4. In the laboratory, aldehydes can be conveniently oxidized to carboxylic acids with Ag(I) reagents.

   a. Aldehydes that contain double bonds or alcohol —OH groups (functional groups that react with other oxidizing reagents) are oxidized by Ag_2O without oxidizing these other functional groups.

   b. If the silver ion is solubilized as its ammonia complex, $^+Ag(NH_3)_2$, oxidation of the aldehyde is accompanied by the deposition of a metallic silver mirror on the walls of the reaction vessel. (This observation can be used as a convenient test for aldehydes, known as the Tollens test.)

ammonium salt of
a carboxylic acid

5. Many aldehydes are oxidized by the oxygen in air upon standing for long periods of time; this process is another example of autoxidation.

B. *OXIDATION OF KETONES*

   1. Ketones cannot be oxidized without breaking carbon–carbon bonds.
   2. Ketones are resistant to mild oxidation with Cr(VI) reagents.

III. REACTIONS OF ALDEHYDES AND KETONES WITH AMINES

A. *IMINE FORMATION WITH PRIMARY AMINES*

   1. Imines are prepared by the reaction of aldehydes or ketones with primary amines.

      a. Formation of imines is reversible and generally takes place with acid or base catalysis, or with heat.

an imine

   b. Imine formation is typically driven to completion by either precipitation of the imine product, removal of water, or both.

   2. Imine formation consists of a carbonyl-addition reaction followed by β-elimination.

      a. The first step of the mechanism is a nucleophilic addition to the carbonyl group in which the nucleophile (an amine) reacts with an aldehyde or ketone to give an unstable addition compound called a carbinolamine.

      b. The carbinolamine undergoes acid-catalyzed dehydration to form an imine.

      c. Dehydration of the carbinolamine is typically the rate-limiting step of imine formation.

3. Certain types of imine adducts sometimes find use as derivatives of aldehydes and ketones because they are in most cases solids with well-defined melting points. (See Table 19.3, Sec. 19.12A, text page 1065, for the list of derivatives, and Further Exploration 19.2 on p. 613 of this manual.)

B. *ENAMINE FORMATION WITH SECONDARY AMINES*

1. An enamine is formed when a secondary amine reacts with an aldehyde or ketone that has an α-hydrogen.

an aldehyde or ketone   a secondary   an enamine
with an α-hydrogen   amine

a. Just as most aldehydes and ketones are more stable than their corresponding enols, most imines are more stable than their corresponding enamines.

b. Because secondary amines cannot form imines, they form enamines instead.

2. Like imine formation, enamine formation is reversible and must be driven to completion by the removal of one of the reaction products (usually water).

a. The mechanism of enamine formation begins like the mechanism of imine formation, as a nucleophilic addition to give a carbinolamine intermediate.

b. Dehydration of a carbinolamine involves proton loss from an adjacent carbon.

a carbinolamine   an enamine

3. Enamines, like imines, revert to the corresponding carbonyl compounds and amines in aqueous acid.

C. *REACTIONS OF ALDEHYDES AND KETONES WITH TERTIARY AMINES*

Tertiary amines do not react with aldehydes and ketones to form stable derivatives.

IV.   CONVERSION OF KETONES AND ALDEHYDES INTO ALKANES AND ALKENES

A. *REDUCTION OF CARBONYL GROUPS TO METHYLENE GROUPS*

1. A carbonyl group of an aldehyde or ketone can be reduced completely to a methylene ($-CH_2-$) group.

2. One procedure for effecting this transformation involves heating the aldehyde or ketone with hydrazine (H_2N-NH_2) and strong base.

a. This reaction, called the Wolff–Kishner reduction, typically utilizes ethylene glycol or similar compounds as co-solvents.

b. The high boiling points of these solvents allow the reaction mixtures to reach the high temperatures required for the reduction to take place at a reasonable rate.

3. The Wolff–Kishner reduction is an extension of imine formation; an intermediate in the reduction is a hydrazone (an imine of hydrazine).

4. The Wolff–Kishner reduction takes place under strongly basic conditions.

5. The same overall transformation can be achieved under acidic conditions by a reaction called the Clemmensen reduction, which involves reduction of an aldehyde or ketone with zinc amalgam (a solution of zinc metal in mercury) in the presence of HCl.

6. The Wolff–Kishner and Clemmensen reductions are particularly useful for the introduction of alkyl substituents into benzene rings by reduction of the product of Friedel–Crafts acylation.

B.   THE WITTIG ALKENE SYNTHESIS

1. Another addition–elimination reaction, called the Wittig reaction, is an important method for preparing alkenes from aldehydes and ketones.

2. The nucleophile in the Wittig reaction is a type of ylid.

a. An ylid is any compound with opposite charges on adjacent covalently bound atoms, each of which has an electronic octet.

b. Although the structures of phosphorus ylids are sometimes written as hypervalent structures with phosphorus–carbon double bonds, the charged structures, in which each atom has an octet of electrons, are very important contributors.

$$\left[Ph_3\overset{+}{P}-\overset{..}{C}H_2 \longleftrightarrow Ph_3P=CH_2 \right]$$
an ylid

3. Preparation of phosphorus ylids:

   a. An alkyl halide reacts with triphenylphosphine (Ph₃P) in an S_N2 reaction to give a phosphonium salt. (Because the alkylation of triphenylphosphine is a typical S_N2 reaction, it is limited for the most part to methyl, benzylic, allylic, and primary alkyl halides.)

   b. The phosphonium salt is then allowed to react with a strong base such as an organolithium reagent (butyllithium, abbreviated BuLi in the example below) to give the ylid.

$$Ph_3\ddot{P} + \ ^{\backslash}CH_3-\overset{\frown}{I} \longrightarrow Ph_3\overset{+}{P}-CH_3 \ \ I^- \xrightarrow{\text{BuLi}} Ph_3\overset{+}{P}-\overset{..}{C}H_2 + Li^+\,I^- + \text{butane}$$

4. The mechanism of the Wittig reaction:

   a. A nucleophile (the anionic carbon of the ylid) reacts with the carbonyl carbon to give an oxaphosphetane intermediate.

   b. Under the usual reaction conditions, the oxaphosphetane spontaneously decomposes to the alkene and the by-product triphenylphosphine oxide.

an oxaphosphetane   triphenylphosphine oxide

5. The Wittig reaction is especially important because it gives alkenes in which the position of the alkene double bond is unambiguous.

6. To plan the synthesis of an alkene by the Wittig reaction, consider the origin of each part of the product, and then reason deductively.

   a. One carbon of the alkene double bond originates from the alkyl halide used to prepare the ylid; the other is the carbonyl carbon of the aldehyde or ketone.

   b. In principle, two Wittig syntheses are possible for any given alkene.

   c. Most Wittig syntheses are planned so that the most reactive alkyl halide (a methyl or primary alkyl halide) can be used as one of the starting materials. [See (3a) above.]

   c. The Wittig reaction in many cases gives mixtures of E and Z isomers, although certain modifications of the Wittig reaction that avoid this problem have been developed.

SOLUTIONS TO PROBLEMS

Solutions to Focused Problems

19.1

(a)

(CH₃)₂CHCH$=$O

(b)

(c)

ClCH₂CH₂CH₂CH with carbonyl O

(d)

CH₃CH₂CHC— with Cl substituent, attached to benzene ring with —CH aldehyde

(e)

(f)

19.2

(a)	2-,4-Dimethyl-3-pentanone	(b)	3-Methylcyclobutanecarbaldehyde
(c)	3-Allyl-2,4-pentanedione	(d)	(*E*)-3-Ethoxy-2-propenal
(e)	Benzene-1,4-dicarbaldehyde	(f)	4,4-Dimethyl-2,5-cyclohexadienone

19.3

(a)   The alkene double bond of the aldehyde has a C$=$C stretching absorption in the 1665 cm⁻¹ region, whereas the ketone has no alkene absorption; if the alkene has the *E* configuration, it should have a prominent C—H bending absorption in the 970 cm⁻¹ region. In addition, the aldehyde has a somewhat higher carbonyl stretching frequency, and it has a C—H stretching absorption near 2700 cm⁻¹ that is not present in the IR spectrum of the ketone.

cyclohexanone   CH₃CH$=$CH₂CH₂CH₂CH$=$O
                                    4-hexenal

(b)   2-Cyclohexenone has a lower carbonyl stretching frequency because its two double bonds are conjugated.

2-cyclohexenone   **3-cyclohexenone**
has conjugated double bonds

(c)   3-Buten-2-ol has both O—H and C$=$C stretching absorptions and 2-butanone does not; 2-butanone has a carbonyl stretching absorption, and 3-buten-2-ol does not.

3-buten-2-ol   **2-butanone**

19.4 (a) The IR spectrum suggests an aldehyde (carbonyl stretch at 1720 cm^{-1} and aldehyde C—H stretch at 2710 cm^{-1}, and the low-field absorption at δ 9.65 confirms the presence of the aldehyde hydrogen. The six-proton doublet at δ 1.2 and the septet at δ 2.4 indicate an isopropyl group $(CH_3)_2CH$—. Because the aldehyde carbon is the only remaining carbon, the structure (at right) is completely determined. (The CH of the isopropyl group and the aldehyde hydrogen split each other slightly.)

$(CH_3)_2CH-CH$ (with O double bonded to CH)

isobutyraldehyde (methylpropanal)

(b) The compound is 2-butanone:

$CH_3CCH_2CH_3$ (with O double bonded)

2-butanone

(c) The high frequency of the carbonyl absorption suggests a strained ring. (See Display 19.15, text Sec. 19.4A, p. 1033.) In fact, cyclobutanone matches the IR stretching frequency perfectly and the NMR fits as well:

δ 3.09 (triplet), δ 2.01 (quintet), δ 3.09 (triplet)

molecular mass = 70.1

cyclobutanone

(d) The molecular formula indicates an unsaturation number of 5. The IR spectrum suggests a conjugated carbonyl group and an aromatic ring, and these two features fully account for the unsaturation. The NMR indicates the partial structures at left; and these define the complete structure at right.

CH_3CH_2-O- $-CCH_3$

CH_3CH_2O-⟨ring⟩$-CCH_3$

p-ethoxyacetophenone

19.5 The low intensities of the carbonyl carbon absorption at δ 212.6 and the α-carbon absorption at δ 44.2 show that these two carbons have no attached hydrogens; therefore this compound must have the partial structure at right. Because all six carbons and their connectivities are accounted for, the only possible choice is to add the appropriate number of hydrogens. The compound is 3,3-dimethyl-2-butanone. shown at right.

3,3-dimethyl-2-butanone

19.6 The structure and carbon NMR assignments of 2-ethylbutanal are shown below. The two methyl groups are chemically equivalent, and the two methylene groups are chemically equivalent; all carbons with different carbon NMR chemical shifts are chemically nonequivalent.

δ 21.7 CH_3CH_2, δ 204.7, δ 11.5, δ 55.2 CH—CH=O, CH_3CH_2 δ 21.7

2-ethylbutanal

19.7 (a) The double bonds in 2-cyclohexenone are conjugated, but the double bonds in 3-cyclohexenone are not. Consequently, 2-cyclohexenone has the UV spectrum with the greater λ_{max}.

2-cyclohexenone
(conjugated; has UV spectrum
with the greater λ_{max})

3-cyclohexenone

(b) The first compound has more conjugated double bonds and therefore will have the UV spectrum with the greater λ_{max} and larger extinction coefficient ε.

(c) In 1-phenyl-2-propanone, the aromatic ring is not conjugated with the carbonyl group; in *p*-methylaceto-phenone the aromatic ring and the carbonyl group are conjugated. Because of its additional conjugation, *p*-methylacetophenone has the UV spectrum with the greater λ_{max} and larger extinction coefficient ε.

1-phenyl-2-propanone

p-methylacetophenone
(has the UV spectrum with
the greater ε and λ_{max})

(How could you distinguish between these two structures with IR spectroscopy?)

19.8 (a) In both ketones, inductive cleavage of the molecular ion to give an acyl radical and a carbocation accounts for the $m/z = 57$ peak:

molecular ion of 2-hexanone

butyl cation
($m/z = 57$)

molecular ion of
3,3-dimethyl-2-butanone

***tert*-butyl cation**
($m/z = 57$)

Because the carbocation formed from 3,3-dimethyl-2-butanone is more stable (why?), more of this fragmentation mode is observed for 3,3-dimethyl-2-butanone; that is, the $m/z = 57$ peak is more abundant.

(b) Its even mass suggests that the $m/z = 58$ fragment is an odd-electron ion. Such an ion can be readily produced by a McLafferty rearrangement from the molecular ion of 2-hexanone. However, in the mass spectrum of 3,3-dimethyl-2-butanone, because there is no hydrogen in a 1,6-relationship with the carbonyl oxygen, a McLafferty rearrangement fragment would not be observed.

molecular ion of 2-hexanone

($m/z = 58$)

19.9 The mass spectrum of 2-heptanone should have major peaks at $m/z = 43$ (from α-cleavage), 71 (from inductive cleavage), and 58 (from McLafferty rearrangement). The mass spectrum of 3-heptanone should have a major peak resulting from both inductive cleavage and α-cleavage at $m/z = 57$ and a major peak resulting from McLafferty rearrangement at $m/z = 72$. Notice that the position of the even-mass, odd-electron ion is a major distinguishing feature. We leave it to you to draw out these McLafferty-rearrangement fragmentations.

2-heptanone
McLafferty rearrangement at $m/z = 58$

3-heptanone
McLafferty rearrangement at $m/z = 72$

19.10 (a) The solvolysis mechanism of (chloromethoxy)methane in ethanol:

(b) (Chloromethoxy)methane undergoes more rapid solvolysis than 1-chlorobutane because the carbocation intermediate is resonance-stabilized [as shown in the mechanism of part (a)] and is therefore more stable relative to starting material than the ordinary primary carbocation intermediate derived from 1-chlorobutane. (In fact, 1-chlorobutane probably solvolyzes by an S_N2 mechanism; the S_N1 reaction would no doubt be even slower than the reaction indicated in the problem.)

19.11 The products follow from a consideration of the mechanism in Study Problem 19.2 on text pp. 1040–1041. The first steps of this mechanism consist of protonation of an —OH group and loss of water to form a carbocation intermediate.

(a) When the two —OH groups are chemically nonequivalent, the —OH group lost is the one that gives the more stable carbocation intermediate—in this case, the tertiary —OH.

(b) The reaction is exactly like the one shown in Study Problem 19.2 on pp. 1040–1041 with phenyl instead of methyl substituents. The product is the ketone at right.

(c) In this case, rearrangement involves a ring expansion. If reasoning through this type of rearrangement gives you difficulty, be sure to review Study Problem 11.2 (text Sec. 11.2, p. 572). The product is the ketone shown at right.

19.12 (a) Protonated p-methoxybenzaldehyde is essentially a benzylic carbocation; as we showed in Chapter 17, benzylic carbocations are stabilized by substituents that can donate electrons by resonance, and this cation is no exception; the resonance effect of the p-methoxy group stabilizes the cationic conjugate acid and increases the basicity of the aldehyde.

conjugate acid of *p*-methoxybenzaldehyde

However, the nitro group in *p*-nitrobenzaldehyde cannot donate electrons by resonance to stabilize the conjugate-base cation by resonance; moreover, one of the remaining resonance structures is less important because it contains a destabilizing interaction of two positive charges.

conjugate acid of *p*-nitrobenzaldehyde

Therefore, the conjugate acid of *p*-nitrobenzaldehyde is relatively less stable; as a result, *p*-nitrobenzaldehyde is the less basic of the two aldehydes.

(b)   The conjugate acid of 3-buten-2-one has more important resonance structures (3—see below) than the conjugate acid of 2-butanone (2—draw these) and is therefore more stable relative to unprotonated ketone than the conjugate acid of 2-butanone. Greater stability of the conjugate acid means that the ketone is more basic.

conjugate acid of 3-buten-2-one

19.13   (a)   The mechanism of the hydroxide-catalyzed hydration of acetaldehyde:

(b)   The curved-arrow mechanism for the decomposition of acetone cyanohydrin in aqueous hydroxide:

The ketone–cyanohydrin equilibrium favors the ketone at high pH because hydroxide ion is a much stronger base than cyanide ion. (The pK_a of water is 14.0; the pK_a of hydrogen cyanide is 9.40; see Table 3.2, text Sec. 3.6A, p. 142.) The higher the pH, the higher the ⁻OH concentration. Once the pH is a couple of units higher than 9.4, the cyanide ion is fully ionized; so higher pH, while increasing ⁻OH concentration, has no effect on cyanide concentration, and the equilibrium shifts to the right.

Remember that we typically invoke as a base the conjugate base of whatever acid is involved in the mechanism. Thus, if H_3O^+ is involved in a mechanism, its conjugate base (H_2O) acts as the base. We should not invoke both H_3O^+ and ^-OH in the same mechanism because a strong acid and a strong base cannot coexist in solution in significant concentrations.

19.14   (a)

(b)   The methoxide-catalyzed addition of methanol to benzaldehyde is very similar to the hydroxide-catalyzed hydration mechanism (shown in the solution to Problem 19.13a):

19.15   The data in Table 19.2, text p. 1046, show that hydration of benzaldehyde is less than 0.01 times as favorable as hydration of an unconjugated aliphatic aldehyde such as acetaldehyde. The assumption that the same principles apply to cyanohydrin formation leads to the prediction that propanal should have the greater proportion of cyanohydrin at equilibrium. The structure of this cyanohydrin is shown at right.

cyanohydrin derivative of propanal

19.16   We use the same principles to predict reactivity that we use to predict relative equilibrium constants. The more a carbonyl compound is prone to form an addition product, the more reactive it is.

(a)   Compound *B*, bromoacetone, is more reactive because the electronegative atom is closer to the carbonyl group.

(b)   Compound *A*, *p*-nitrobenzaldehyde, is more reactive, because the *p*-nitro group raises the energy of the molecule by an unfavorable interaction of the positive charge on the nitrogen with the positive charge on the carbonyl carbon.

destabilizing interaction of two positive charges

some resonance structures for *p*-nitrobenzaldehyde

In contrast, the *p*-methoxy group in the second compound, *p*-methoxybenzaldehyde, stabilizes the molecule by a resonance interaction of the nonbonding electrons on the oxygen of the methoxy substituent with the positive charge on the carbonyl carbon.

some resonance structures of *p*-methoxybenzaldehyde

(c)   Compound *A*, 2,3-butanedione, is more reactive, because the partial positive charge on one carbonyl destabilizes the molecule by its repulsive interaction with the partial positive charge on the other.

19.17   (a)

2-butanone

(b)

cyclopentanecarbaldehyde

(c)

19.18   The tertiary alcohol *C* could not be synthesized by a hydride reduction, because only primary and secondary alcohols can be prepared by this method.

19.19   (a)   The zinc ion acts as a Lewis acid catalyst by bonding to the carbonyl oxygen, making the carbonyl carbon more electrophilic, as shown in the answer to 19.19(b). The role of the zinc ion is analogous to the role of the lithium ion in reductions with lithium aluminum hydride (see Eq. 19.45a, text p. 1050).

(b)

19.20   All three compounds below will react with hydrogen and a catalyst to give the product shown in the problem. Recall from Section 7.8E (text p. 371) that catalytic hydrogenation is a syn addition. This fact is important when considering how the third compound in this group reacts.

The following two compounds may appear to work; however each would produce a mixture of the desired product and a stereoisomer.

(We leave it to you to show the two stereoisomers that would form from **each of the two starting materials above.**)

19.21 In all of the parts we need the Grignard reagent ethylmagnesium bromide, **which is prepared from bromoethane** (ethyl bromide):

$$CH_3CH_2-Br + Mg \xrightarrow{\text{ether}} CH_3CH_2-MgBr$$

bromoethane   **ethylmagnesium bromide**

(a)

benzaldehyde   **1-phenyl-1-propanol**

(b)

2-butanone   **3-methyl-3-pentanol**

(c)

ethylene oxide (oxirane)   **1-butanol**

(d)

(e)

19.22 Any of the three alkyl groups bound to the α-carbon of the alcohol can **in principle originate from the Grignard** reagent.

Synthesis #1:

3-hexanone   **3-methyl-3-hexanol**

Synthesis #2:

2-butanone   **3-methyl-3-hexanol**

Synthesis #3:

2-pentanone   **3-methyl-3-hexanol**

19.23   (a)   (b)

cyclopentanone diethyl acetal
(1,1-diethoxycyclopentane)

OCH(CH₃)₂
|
CH₃CH₂CH₂CH
|
OCH(CH₃)₂

butyraldehyde diisopropyl acetal
(1,1-diisopropoxybutane)

 Note that acetals are named as ethers in the IUPAC system. In the acetals derived from ketones in the common system were once called *ketals*. Thus, in older common nomenclature, the compound in the solution to part (a) would have been called *cyclopentanone diethyl ketal*.

19.24   (a)

(b)

19.25   (a)   The formula indicates addition of two carbon atoms, and it indicates one degree of unsaturation. If a diethyl acetal were formed and the ring were opened, the formula of the product would be $C_9H_{20}O_3$. The one degree of unsaturation suggests that the ring is intact. The product is a mixed acetal, shown at right.

(b)   When five- and six-membered rings can be formed, diols generally react with aldehydes and ketones to give cyclic acetals, and this case is no exception, as shown at right.

19.26   We protect the ketone as an acetal, and then prepare the Grignard reagent, which is allowed to react with acetaldehyde. The protonolysis step produces the alcohol and removes the acetal protecting group. (In the following synthesis, any common alcohol or ethylene glycol can be used instead of methanol.)

p-bromoacetophenone

19.27 (a)

(b)

CH₃CH₂CH₂CH₂CHCH=NCH₂CH₃
                    |
                   CH₃

19.28 First, the carbinolamine intermediate is formed. This intermediate then undergoes base-catalyzed dehydration to give the oxime.

Note that in writing curved-arrow mechanisms, you can write the loss of water and formation of the carbon–nitrogen double bond as one step to save time, even though the reaction is not actually concerted. In this case, we also save the effort of drawing resonance structures.

19.29 Imine hydrolysis is the reverse of imine formation. Therefore, retrace the steps of imine formation as illustrated in the solution to Problem 19.28, starting with the imine and working backwards to the aldehyde and the amine.

19.30 The imine linkage of pyridoxal phosphate with a protein:

19.31 (a)

(b)

19.32 Any compound with a carbonyl group on any of the prospective alkyl carbons could in principle serve as a starting material. (This answer is restricted to compounds containing only one carbonyl group.)

19.33 Carry out a Friedel–Crafts acylation, and then apply the Wolff–Kishner reaction (or the Clemmensen reduction). Note that Friedel–Crafts alkylation of the ring with $ClCH_2CH_2CH_3$ would give some rearrangement product in addition to the desired product; see the first reaction in Study Problem 19.5 in the text.

**1,4-dimethoxybenzene
(hydroquinone dimethyl ether)**

**1,4-dimethoxy-2-
propylbenzene**

19.34 (a)

2-methyl-2-butene

(b)

styrene

19.35   (a)   Either "half" of the alkene can in principle be derived from an aldehyde.

The first synthesis:

The second synthesis:

(b)

(c)

19.36   The starting material has five degrees of unsaturation, four of which are accounted for by a benzene ring. Consequently, the compound must be both an aldehyde and an alcohol (at right).

19.37   Silver(I) oxide oxidizes the aldehyde selectively to the carboxylic acid shown at right.

19.38 (a) This is a one-carbon chain extension via a Grignard reaction with formaldehyde.

(b) The product comes from oxidation of a secondary alcohol that, in turn, comes from the reaction of the Grignard reagent in part (a) with acetaldehyde.

from cyclopentanol
as shown in part (a)

Solutions to Skills Objectives Problems

19.39 (a) (*E*)-4-Cyclohexyl-1-methoxy-3-buten-2-one

(b) Once the carbonyl carbon is assigned the number 1 (because it is part of the principal group), the numbering of the ring carbons is determined by alphabetical citation. The name is (2*R*,6*S*)-2-ethyl-6-methylcyclohexanone.

(c) 5-Methyl-3-hexynal

(d) 4-Chloro-1-phenyl-1-butanone. (Many chemists would use the common name, γ-chlorobutyrophenone.)

19.40 (a) (b) (c)

valerophenone   **3-hydroxy-2-butanone**   **2-oxocyclopentanecarbaldehyde**

19.41 (a) In the IR spectrum we expect to confirm the presence of an aldehyde with the presence of a carbonyl stretching absorption at 1720–1725 cm^{-1} and the aldehyde C—H stretching absorption at about 2710 cm^{-1}.

(b) The only informative feature of the UV spectrum (other than the *absence* of strong absorptions) is the presence of a weak $n \longrightarrow \pi^*$ absorption in the 270–300 nm region.

(c) The proton NMR spectrum should show an aldehydic C—H at ~δ 10 integrating for 1*H*, which should be a triplet. Another absorption that should stand out is the α-hydrogen resonance, which should be a multiplet at about δ 2.3 integrating for 2*H* and consisting of a doublet of triplets or possibly a quartet (in the case of equal splittings). The remaining resonances in the alkyl group region could be a complex multiplet integrating for 9*H*. A spectrum taken at higher applied field might resolve the β-protons as a 2*H* quartet (or doublet of triplets, depending on relative coupling constants) at about δ 1.5.

(d) The carbon-13 NMR should show a typical carbonyl carbon resonance at δ 190–200. The entire spectrum should consist of five lines, which confirm the equivalence of the two methyl groups.

(e) The EI mass spectrum might show a peak at *m/z* = 99 (that is, M – 1) resulting from α-cleavage at the carbonyl-hydrogen bond and *m/z* = 71 resulting from inductive cleavage at the carbonyl– CH$_2$ bond. There should be a peak for an even-electron ion at *m/z* = 44 that results from McLafferty rearrangement. (The actual spectrum shows the latter two peaks.)

(f) While the CI mass spectrum could show some of the same peaks, it should display an M + 1 resonance at *m/z* = 101.

19.42 (a) The NMR spectrum suggests a great degree of symmetry: four aromatic protons, all equivalent, and two methyl groups adjacent to a carbonyl group. The IR spectrum suggests the presence of a carbonyl group conjugated with the aromatic ring. The chemical shift of the ring protons is greater than usual because of the electronegative carbonyl groups. Because all four ring protons are chemically equivalent, they appear in the NMR spectrum as a singlet. The structure is shown at right.

p-diacetylbenzene

(b) The NMR spectrum indicates the presence of a *tert*-butyl group and an aldehyde. The compound is

$(CH_3)_3C$—CH=O 2,2-dimethylpropanal (pivalaldehyde)

(c) The NMR indicates two vinylic hydrogens as well as an aldehyde adjacent to a CH. The IR absorption at 970 cm^{-1} shows that the vinylic hydrogens are trans. The double bond and the aldehyde account for all of the unsaturation. The UV absorption at 215 nm indicates that the alkene and aldehyde double bonds are conjugated. The allylic CH_2 absorption at δ 2.2 is essentially a quartet, which suggests three hydrogens on adjacent carbons, that is, a vinylic hydrogen (because the absorption is allylic) and two others. The triplet near δ 1 integrates for three protons, and thus represents a methyl group; its splitting suggests an adjacent CH_2 group. The δ 1.4 absorption implies a —CH_2— adjacent to five protons. The structure at right is defined by the data.

(*E*)-2-hexenal

19.43 The double ions of equal intensity separated by 2 mass units in the mass spectrum indicate the presence of one bromine. The IR absorptions indicate a ketone in which the carbonyl group is conjugated to a benzene ring (1678 cm^{-1}) as well as aromatic "double bonds" (1600 cm^{-1}). The NMR spectrum indicates a *para*-disubstituted benzene ring and a methyl group adjacent to a carbonyl group (that is, a methyl ketone). The compound has structure *A* below (*p*-bromoacetophenone). This compound has the molecular mass (198 for ^{79}Br) indicated by the mass spectrum.

p-bromoacetophenone
A

B

If you proposed structure *B*, it is not a bad answer, because you probably don't yet know about the IR spectra of acid halides. However, the IR carbonyl absorption of such a compound (an acid halide) occurs at much greater frequency. The mass spectrum, moreover, is more consistent with structure *A*. The *m/z* = 155 and 157 peaks imply a loss of a neutral acetyl radical (43 mass units) by inductive cleavage to give a cation *C*:

molecular ion of *p*-bromoacetophenone

inductive cleavage

cation *C*
m/z = 155, 157

acetyl radical

Loss of an acylium ion by α-cleavage is shown by the *m/z* = 43 peak.

molecular ion of *p*-bromoacetophenone

α-cleavage

an acylium ion
m/z = 43

19.44   (a)   (b)   (c)   (d)

, AlCl₃ catalyst

Me_2CH CH_3

Me_2CH CH_2 H

19.45   (a)   This is an addition reaction in which the sulfur of the bisulfite ion reacts as a nucleophile at the carbonyl carbon of the aldehyde. (The detailed Lewis structure of bisulfite ion is shown in the problem.) In the last step of the mechanism, the initially formed addition product *A* ionizes because it is a fairly strong acid.

$$RCH + HSO_3^-$$

A

(b)   Acid, that is, H_3O^+, destroys sodium bisulfite by reacting with it and converting it into H_2SO_3, which spontaneously decomposes into SO_2, a gas, and H_2O. Destruction of bisulfite in this manner eliminates a reactant in the equilibria shown above and pulls the reaction to the left, that is, toward free aldehyde. Hydroxide ion reacts with bisulfite and converts it into sulfite ion, $(OSO_2)^{2-}$; because bisulfite is removed as a reactant by this process as well, the equilibrium is also pulled to the left.

In the mechanisms presented to this point in this Study Guide and for the most part in the text, nonbonding electron pairs have been drawn in explicitly. Omitting these electron pairs can save some time and can eliminate a great deal of tedium in writing mechanisms; most chemists do not draw the nonbonding electron pairs explicitly. This shortcut is useful only if you realize that *the nonbonding electron pairs which are not shown are understood to be present*. In many of the mechanisms shown subsequently, nonbonding electron pairs are omitted in the most common situations. (In unusual or uncommon situations, the electron pairs are shown.) Here are some typical situations in which nonbonding electron pairs should not be necessary:

H_2O →   C═O   → $H_2\overset{+}{O}$—C—O⁻   means $H_2\overset{..}{O}$:→   C═$\overset{..}{O}$ → $H_2\overset{+}{O}$—C—$\overset{..}{\underset{..}{O}}$⁻

$(CH_3)_3C$—Br   → $(CH_3)_3C^+$ Br⁻   means $(CH_3)_3C$—$\overset{..}{\underset{..}{Br}}$:   → $(CH_3)_3C^+$:$\overset{..}{\underset{..}{Br}}$⁻

R_3N means R_3N: R_3C^- means $R_3\bar{C}$:

An oxygen with three bonds and a positive charge has one nonbonding electron pair; an oxygen with one bond and a negative charge has three nonbonding electron pairs; an uncharged nitrogen with three bonds or a negatively charged carbon with three bonds has one nonbonding electron pair; and a halide ion has four nonbonding electron pairs. (See Sec. 1.3C in the text, "Common Bonding Patterns.")

The reason we do not have to show nonbonding valence-electron pairs is that if the formal charge on an atom and the number of bonds to an atom are known, the number of nonbonding valence electrons are automatically known. It follows that the formal charge *must* be shown; otherwise, the valence electron count of an atom is undetermined.

Although most of the mechanisms shown subsequently will be written without showing the nonbonding electron pairs, if you ever become confused by the absence of nonbonding electron pairs, *do not hesitate to draw them in explicitly.*

(c)   The bisulfite addition product of 2-methylpentanal:

$$CH_3CH_2CH_2CHCH(OH) - S(=O)(=O) - O^- \ Na^+$$

(with a CH₃ substituent on the CH and S bearing OH... structure: CH₃CH₂CH₂CHCH—S—O⁻ Na⁺ with OH and O on the carbon/sulfur, and CH₃ below the chain, O below the S)

19.46   A curved-arrow mechanism for the formation of benzaldehyde dimethyl acetal ([dimethoxymethyl]benzene):

(We suggest that you draw in the missing nonbonding electron pairs for the first few steps.)

19.47   (a)   The more the carbonyl compound is stabilized by resonance (or other effects), the less reactive it is. For compound *A* we can draw four resonance structures for interaction of the aromatic π electrons with the ring. However, one structure is not very important because of the unfavorable interaction of two positive charges. Therefore, compound *A* has three important resonance structures.

charge repulsion destabilizes this structure

three important structures for compound *A*

Compound *B* has *five* important resonance structures because of the delocalization of the nonbonding electron pair and charge.

five important resonance structures for compound *B*

Compound *C* has four important resonance structures:

four important resonance structures for compound *C*

Because compound *B* has the greatest resonance stabilization, it is *least* reactive; because compound *A* has the poorest resonance stabilization it is most reactive. The reactivity order is *B* < *C* < *A*.

(b) Because aldehydes are more reactive than ketones, all other things being equal, compound *B* is most reactive. Of the ketones, compound *D* is least reactive because it lacks the fluorine substituent. The fluorine substituent in compounds *A* and *B* increases reactivity because the compounds are destabilized by the unfavorable interaction of the C=O and C—F bond dipoles. Of these two, the *tert*-butyl substituent decreases reactivity because of increased crowding in the tetrahedral intermediate and because of steric interference with reaction of a nucleophile. The reactivity order is *D* < *C* < *A* < *B*.

(c) The ideal C—C—C bond angle at a carbonyl carbon is 120°. Therefore, compound *B*, with a bond angle of about 60°, is highly strained. Reaction of a nucleophile changes the preferred bond angle to about 109.5°. Therefore, a large relief of strain occurs when compound *B* reacts with a nucleophile; *much* less strain relief occurs when compound *A* reacts. Therefore, the reactivity order is *A* << *B*.

19.48 The central carbonyl group is hydrated for the following reason. The carbon of any carbonyl group bears a partial positive charge. The central carbonyl group is therefore bonded to two partially positive carbonyl carbons, and the positive charge on each of these carbons has an unfavorable repulsive interaction with the partial positive charge on the central carbon. In addition, the other two carbonyl groups are conjugated with the benzene ring. As explained in the text, adjacent electronegative groups make carbonyl-addition reactions more favorable, and groups that are conjugated with the carbonyl group make carbonyl-addition reactions less favorable. It follows that hydration of the central carbonyl group is the more favorable process. The structure of the hydrate is shown above at right.

hydrate of ninhydrin

19.49

(g)

$$\underset{\underset{CH_3}{|}}{\overset{\overset{OH}{|}}{CH_3CCH_3}}$$

(h)

$$\underset{CH_3CHCH_3}{\overset{\overset{OH}{|}}{}}$$

(i)

$(CH_3)_2C\!=\!CH_2$
$+ \; Ph_3\overset{+}{P}\!-\!\overset{-}{O}$

(j)

$CH_3CH_2CH_3$

19.50 (a)

$$\underset{CH_3CH_2CH_2CHPh}{\overset{\overset{OH}{|}}{}}$$

(b)

$$\underset{CH_3CH_2CH_2CH_2}{\overset{\overset{OH}{|}}{}}$$

(c)

$$\underset{CH_3CH_2CH_2C}{\overset{\overset{O}{\|}}{}}\!-\!OH$$

(d)

$$\underset{CH_3CH_2CH_2C}{\overset{\overset{O}{\|}}{}}\!-\!OH$$

(e)

$$\underset{CH_3CH_2CH_2CH}{\overset{\overset{NOH}{\|}}{}}$$

(f)

$$\underset{CH_3CH_2CH_2C}{\overset{\overset{O}{\|}}{}}\!-\!OH$$

(g)

$CH_3CH_2CH_2CH_3$

(h)

$$\underset{H}{\overset{CH_3CH_2CH_2}{}}C\!=\!C\underset{CH_3}{\overset{H}{}} \quad + \quad \underset{CH_3CH_2CH_2}{\overset{H}{}}C\!=\!C\underset{CH_3}{\overset{H}{}}$$

$$+ \; Ph_3\overset{+}{P}\!-\!\overset{-}{O}$$

19.51 (a) Two stereoisomeric cyclic acetals are formed. They have different properties because they are diastereomers.

(b) Two diastereomeric alcohols are formed by reaction of hydride at either of the two possible faces of the carbonyl carbon. Diastereomers have different physical properties. (Notice that these are examples of diastereomers that are achiral.)

(c) The Wittig reaction gives a mixture of cis and trans alkenes, which, like all cis–trans isomers, are diastereomers and therefore have different properties.

$$\underset{H}{\overset{Ph}{}}C\!=\!C\underset{CH_3}{\overset{H}{}} \quad + \quad \underset{Ph}{\overset{H}{}}C\!=\!C\underset{CH_3}{\overset{H}{}}$$

19.52 (a) The reaction is formation of a hydrazone, a type of imine; see Table 19.3, text p. 1065.

(b) The reaction is a straightforward dimethyl acetal formation.

1,1-dimethoxy-4-methylcyclohexane

(c)   The formula shows that only one of the carbonyl groups is involved in acetal formation; the aldehyde carbonyl reacts selectively because aldehydes form addition compounds more rapidly than do ketones, and because addition reactions of aldehydes are thermodynamically more favorable than those of ketones.

1,1-dimethoxy-2-propanone

(d)   This reaction is internal acetal formation, in which the two alcohol groups form an acetal with the carbonyl group within the same molecule.

(e)   This is a Grignard addition to the ketone to give a tertiary alcohol that subsequently dehydrates under the acidic conditions to an alkene. Whether the dehydration occurs depends on the acid concentration and whether the conditions are designed to remove water. (See Eq. 19.62, Sec. 19.10, test p. 1057.)

1,1-diphenyl-1-propanol

(f)   This is a Wittig reaction.

(g)   From the formula, the overall sequence adds two carbons, most likely from the aldehyde used in the final step. The sequence introduces three additional degrees of unsaturation. If two of them come from formation of a diene in a "double" Wittig reaction, the third could be from formation of a ring. In fact, two Wittig reagents are formed within the same molecule, and each Wittig reagent reacts at a different carbonyl group of the same aldehyde molecule to form a cyclic diene.

19.53   (a)   This is a substituted phenylhydrazone (Table 19.3, text p. 1065) formed from the aldehyde pentanal and *p*-methoxyphenylhydrazine.

$CH_3CH_2CH_2CH_2CH{=}O$ and H_2NNH—⟨⟩—OCH_3

pentanal (valeraldehyde)   **p-methoxyphenylhydrazine**

(b)   Because a cyclic imine is formed, the reaction must be intramolecular; that is, the amine and the carbonyl group that react must be in the same molecule.

19.54   A protecting group is necessary to prevent the ketone from reacting with the Grignard reagent formed in a second molecule.

If you proposed protonolysis of the Grignard reagent in D_3O^+ you have the right idea. However, use of D_3O^+ would actually result in the following product:

The reason is that acidic D_2O brings about the exchange of protons that are α to an aldehyde or ketone carbonyl group. (We discuss this reaction in Chapter 22.) To avoid this exchange use D_2O for protonolysis of the Grignard reagent (as shown in the foregoing solution), which gives $Mg(OD)_2$ as a by-product. Because this by-product maintains basic conditions, the acetal is left intact. Hydrolysis of the acetal is then carried out with H_3O^+ in H_2O.

19.55   As always, bear in mind that there is sometimes more than one acceptable synthesis that fits the parameters given.

(a)

(b)

(c)

(d)

$$PhCH=O \xrightarrow[CH_3OH]{NaBH_4} PhCH_2OH \xrightarrow[pyridine]{SOCl_2}$$

$$\xrightarrow[-H_2]{NaH} PhCH_2O^- Na^+ \xrightarrow[pyridine]{PhCH_2Cl} PhCH_2OCH_2PH$$

dibenzyl ether

(e)

3-methyl-2-hexanone   **2,3-dimethyl-2-hexene**

(f)

3-methyl-2-hexanone   **2,3-dimethyl-1-hexene**

(g)

cyclohexene   **1,6-hexanediol**

(h)   First prepare the Grignard reagent benzylmagnesium bromide from benzaldehyde:

Then let this Grignard reagent react with **benzaldehyde** and oxidize the resulting alcohol:

(i)

anisole   **1-butyl-4-methoxybenzene**

(j)   The first hint indicates that a protecting group must be used. The second hint is meant to call attention to the hemiacetal linkage which, because it is a five-membered *cyclic* hemiacetal, will form spontaneously from the corresponding aldehyde.

19.56   (a)

(*Z*)-5,6-dimethyl-
5-decen-1,10-diol

(b)

(*Z*)-5,6-dimethyl-
5-decen-1,10-diol

(c)

(d)   Prepare the triflate of the phenol and subject it to a Stille reaction (Sec. 18.10B) with tetrabutylstannane.

hydroquinone monomethyl ether
(*p*-methoxyphenol)

19.57   (a)

(b)

(c) Hydrogen-bond donation from the tyrosine and serine OH groups delocalizes and stabilizes the developing negative charge on the carbonyl oxygen of the ketone and therefore stabilizes the transition state of the reaction.

transition state

Solutions to Integrated Problems

19.58 (a) Glycerol has three hydroxy groups. Two possible cyclic acetals can be formed; one (A) contains a six-membered ring, and the other (B) contains a five-membered ring.

(b) Only compound B is chiral. (It has one asymmetric carbon, identified by *.) Therefore, only compound B can be resolved into enantiomers, so compound B is the observed compound.

19.59 (a) A Lewis acid–base dissociation of the bromide—essentially an S_N1 reaction—gives an α-hydroxy carbocation, which is, by resonance, also the conjugate acid of acetophenone. Deprotonation by bromide ion gives the ketone and HBr.

(b) Part (a) of this problem shows that α-bromo alcohols break down spontaneously to aldehydes or ketones:

Osmium tetroxide reacts to form the glycol A, which is also an α-bromo alcohol. This breaks down to the aldehyde as shown by the foregoing reaction:

$$H_2C{=}CH{-}Br \xrightarrow{OsO_4,\ H_2O} H_2C{-}CH{-}Br \longrightarrow H_2C{-}CH + H{-}Br$$

with OH OH over the middle species labeled *A*, and OH O over the product.

19.60   The molecular formula of compound *A* shows that two equivalents of methanol are added to the alkyne. The fact that the **product hydrolyzes to acetophenone** indicates that the two methoxy groups of compound *A* are on the same carbon—that is, that compound *A* is an acetal.

[reaction scheme showing mechanism with steps (a) and (b), ending in]

compound *A*
**acetophenone
dimethyl acetal**

In the step labeled (*a*), the acid $CH_3\overset{+}{O}H_2$ is used to protonate the alkyne because it is the major acidic species present when H_2SO_4 is dissolved in methanol (just as H_3O^+ is the major acidic species present when H_2SO_4 is dissolved in water). Protonation occurs on the terminal carbon because it gives a carbocation that is benzylic and resonance-stabilized. In the step labeled (*b*), protonation again occurs on the terminal carbon because the resulting carbocation is resonance-stabilized by electron donation from both the benzene ring and the neighboring oxygen.

19.61   **(a)**   A THP ether is a cyclic acetal. In aqueous acid, the cyclic acetal undergoes hydrolysis to give, besides the ROH, 5-hydroxypentanal, which is in equilibrium with the cyclic hemiacetal 2-hydroxytetrahydropyran. The equilibrium between the two favors the cyclic hemiacetal. (See Eq. 19.70 on text p. 1060.)

[reaction scheme]   **2-hydroxytetrahydropyran**   **5-hydroxypentanal**

(b)   The synthesis of 2-methyl-2,6-hexanediol using the THP protecting group:

[reaction scheme]   **4-bromo-1-butanol**

(solution continues on the following page)

2-methyl-2,6-hexanediol

19.62   (a)   In this case, $LiAlD_4$ serves as a source of nucleophilic isotopic hydride (deuteride); deuteride opens the epoxide with inversion of configuration.

(racemate)

(b)   Hydride ion, delivered from $LiAlH_4$, opens the epoxide with inversion of configuration. Because the epoxide carbons are enantiotopic and therefore chemically equivalent, the nucleophilic reaction of hydride at each of these carbons occurs at the same rate. As a result, the racemic product is formed. An equivalent view is that the starting epoxide is a meso compound. Hence, the chiral product must be a racemate.

(formed as the racemate)

19.63   (a)   Thumbs wants a Grignard reagent to react selectively with a ketone in the presence of an aldehyde. Because aldehydes are more reactive than ketones, the aldehyde, not the ketone, will react most rapidly.

(b)   Thumbs is trying to form a Wittig reagent by first carrying out an S_N2 reaction of triphenylphosphine (Ph_3P) with neopentyl bromide. Because neopentyl bromide is virtually unreactive in S_N2 reactions, this reaction will not occur. (See Fig. 9.4b on text p. 465 and associated discussion.)

19.64   (a)   The synthesis of 4-methyl-3-heptanol:

$$CH_3CH_2CH_2CHBr \xrightarrow[\text{ether}]{Mg} CH_3CH_2CH_2CHMgBr \xrightarrow[\text{2) } H_3O^+]{\text{1) } O=CHCH_2CH_3} CH_3CH_2CH_2\overset{*}{C}HCHCH_2CH_3$$

with CH_3 substituents, yielding **4-methyl-3-heptanol**

(b)   Because 4-methyl-3-heptanol contains two asymmetric carbons [marked with asterisks (*) in the equation], it can exist as diastereomers. Even if the starting alkyl halide were enantiomerically pure, and even if the Grignard reagent could be prevented from racemizing, the product would likely be a mixture of diastereomers.

19.65   (a)   Subject the ketone benzyl phenyl ketone to an excess of isotopically enriched water in the presence of an acidic or basic catalyst. Several cycles of formation and decomposition of the hydrate will eventually "wash out" the isotope of lower atomic mass. (Note that $*O = {}^{18}O$)

$$\underset{\substack{\textbf{benzyl phenyl} \\ \textbf{ketone}}}{PhCCH_2Ph} + \underset{\substack{\text{(large} \\ \text{excess)}}}{H_2\overset{*}{O}} \rightleftharpoons \underset{OH}{\overset{OH}{PhCCH_2Ph}} \longrightarrow \overset{O^*}{PhCCH_2Ph} + H_2O$$

The last step is effectively irreversible only because of the large excess of isotopically enriched water; the unenriched water is present at very low concentration at all times. The ratio of enriched to unenriched ketone will be essentially the same as the ratio of enriched to unenriched water at equilibrium.

An alternative synthesis is a Hg^{2+}-catalyzed hydration of diphenylacetylene (PhC≡CPh) with H$_2$O*.

(b)   Reduce benzyl phenyl ketone [the starting material in part (a)] with LiAlD$_4$.

O
‖
PhCCH$_2$Ph $\xrightarrow[\text{2) H}_3\text{O}^+]{\text{1) LiAlD}_4}$ PhCDCH$_2$Ph   (OH above)

benzyl phenyl ketone

19.66   (a)   The cyclic trimer can exist as a mixture of diastereomers:

β-form   α-form
two diastereomers of the cyclic trimer of chloral

(You were not expected to specify initially which form is which.)

(b)   The higher-melting form has the more symmetrical structure, in this case, the β-form. See Sec. 8.5D.

(c)   The analysis of the NMR spectra assumes that the two chair conformations of each form are in rapid equilibrium. The α-form has two chemically nonequivalent sets of protons; therefore, its proton NMR spectrum has two resonances in a 2:1 ratio. Because all protons in the β-form are chemically equivalent, this form has only one resonance in its NMR spectrum. Related conclusions apply to the carbon NMR spectra: the α-form has four lines in its spectrum, whereas the β-form has only two.

19.67   (a)   In 1,2-cyclopentanedione the C=O bond dipoles are constrained by the geometry of the ring to be aligned at an angle of about 72°. This alignment provides a significant resultant dipole, as shown by diagram *A*:

1,2-cyclopentanedione (*A*)

biacetyl (*B*)

In biacetyl (2,3-butanedione), rotation about the central carbon–carbon bond is possible. Evidently, the two bond dipoles are oriented so that their resultant is less than the resultant in 1,2-cyclopentanedione. Notice that if their dihedral angle were 180°, their resultant would be zero. Hence, the bond dipoles are aligned at an angle somewhere between 72° and 180°, as shown in the Newman projection of diagram *B* above. (The angle turns out to be about 135°.) Aligned bond dipoles cause electrostatic repulsion. By undergoing internal rotation, biacetyl reduces this destabilizing effect.

 The alignment in biacetyl occurs because it is the alignment of lowest energy; there are two reasons for this reduced energy. First, the alignment reduces the electrostatic repulsion between the negative ends of the bond dipoles. Second, it prevents overlap between the π-electron systems of the two carbonyl groups that would occur at an angle of either 0° or 180°. This overlap is unfavorable because it creates a charge distribution in the molecule that involves an electron-deficient oxygen, as symbolized by the following resonance structures:

electron-deficient oxygen

The information given in the problem plus a simplifying assumption that the cyclopentane ring has the shape of a regular pentagon allows you to use the law of cosines to calculate the bond dipole of a C=O bond as well as the angle between the two carbonyl bond dipoles in biacetyl (given above as about 135°). Try it!

(b) The $n \rightarrow \pi^*$ absorption is characteristic of the carbonyl group. This absorption disappears because a reaction occurs in which the carbonyl group is converted into another group that does not have this absorption. This reaction is addition of ethanethiol to give the sulfur analog of a hemiacetal:

$$H_3C\!-\!CH\!=\!O \ + \ CH_3CH_2SH \ \rightleftharpoons \ H_3C\!-\!CH\!-\!OH$$

has a carbonyl group; has $n \rightarrow \pi^*$ absorption

SCH₂CH₃ → SCH_2CH_3

has no carbonyl group; has no $n \rightarrow \pi^*$ absorption

(c) Compound *A* exhibits a very weak carbonyl stretching absorption in its IR spectrum because very little of it is in the aldehyde form; rather, the molecule exists largely as a cyclic hemiacetal.

19.68 The molecular formula requires three degrees of unsaturation. The six-proton singlet suggests two chemically equivalent methyl groups. The IR indicates the presence of a carbonyl group. The two doublets at about δ 6 and δ 7.5 are vinylic hydrogens, one of which is significantly shifted by some electronegative group, no doubt the carbonyl group. If the vinylic hydrogens account for one double bond and a carbonyl group accounts for another, we are still missing one degree of unsaturation, which therefore must be a ring. The two remaining protons are in the chemical-shift region for protons α to a carbonyl. The data leave only one possibility, shown at right above.

4,4-dimethyl-2-cyclopentenone

If we return to Display 19.15 on text p. 1033 we see that cyclopentanones should have IR carbonyl absorptions at about 1745 cm⁻¹. However, conjugation reduces this frequency by 40–45 cm⁻¹, as shown by the trends in Display 19.13 on text p. 1032. In other words, the observed carbonyl frequency of 1703 cm⁻¹ is exactly what we would expect for this compound. Furthermore, the small vinylic coupling constant is expected for two protons in a cis relationship (see Table 14.3 on text p. 765).

19.69 The formula indicates one degree of unsaturation which is accounted for by a ketone (IR absorption at 1710 cm⁻¹, CMR absorption at δ 204.9 with zero attached hydrogens). The remaining oxygens cannot be alcohols (absence of O—H stretch in the IR) or other carbonyls. The CMR-DEPT data indicate the presence of at least one methyl ether (δ 53.5) as well as a carbon bearing one hydrogen (δ 101.7) that could also be attached to two oxygens; thus, an acetal group is a possibility. The

4,4-dimethoxy-2-butanone

proton NMR data indicate a methyl group attached to a carbonyl. The only carbon unaccounted for is a methylene group with a CMR absorption at δ 47.2, which places it adjacent to the carbonyl and near, but not adjacent to, the acetal. All the data conspire to define the structure on the right.

19.70   The oxygens of the OH and OCH_3 groups in the neutral compounds have very similar electronic character: two nonbonded electron pairs conjugated with the benzene ring and carbonyl group. However, in base, the OH group of *p*-hydroxyacetophenone ionizes but *p*-methoxyacetophenone is unaffected. The change in the UV-vis spectrum of *p*-hydroxyacetophenone reflects this conversion. The conjugate-base anion of *p*-hydroxyacetophenone has significant important resonance structures that provide significant delocalization of charge. This delocalization is reflected in a more intense UV-vis spectrum with the longer-wavelength λ_{max}.

19.71   (a)   In this reaction, the carbonyl group is reduced to an alcohol, which is subsequently ionized by NaH. The resulting intermediate is essentially a Wittig oxaphosphetane intermediate, as the hint suggests; this intermediate decomposes to form an alkene.

(b)   An initially formed imine reacts with sodium borohydride to form an amine. (Although this reaction is discussed in Sec. 23.7B, text p. 1294, the product follows directly from the hint.)

19.72 **(a)**

(b)

an oxaphosphetane

(c)

(d)

(e)

paraldehyde

protonated acetaldehyde; reacts to give more paraldehyde

19.73 The data indicate that compound *A* is a benzene derivative with an additional degree of unsaturation. Its reactivity in the Clemmensen reduction and the formation of a xylene suggests that it has two substituents on a benzene ring, one of which is a methyl group and one of which is an aldehyde. Only *p*-methylbenzaldehyde would give, after Clemmensen reduction, a compound (*p*-xylene) that in turn gives, because of its symmetry, one and only one monobromination product, 2-bromo-1,4-dimethylbenzene.

4--methylbenzaldehyde (compound *A*) $\xrightarrow{\text{Zn/Hg, HCl}}$ ***p*-xylene** $\xrightarrow{\text{Br}_2,\text{ Fe}}$ **2-bromo-1,4-dimethylbenzene**

19.74 Because compound *A* has one carbon more than the product *D*, a carbon is lost at some point; the most likely place is the benzylic oxidation that gives compound *D* itself. The structure of compound *D* shows that all compounds contain benzene rings with four substituent groups. That compound *A* can be regenerated by oxidation of optically active *B* suggests that *B* is an alcohol and *A* is a ketone. Compound *B* cannot be a primary alcohol because such an alcohol could not be optically active. (The chirality has to be associated with the alcohol functionality because optical activity is lost when the alcohol is oxidized.) It cannot be a tertiary alcohol because it could not be formed by LiAlH₄ reduction and it could not be oxidized. Compound *A* has one degree of unsaturation in addition to the four accounted for by its benzene ring and the one accounted for by its carbonyl group; the only way to accommodate all the data is for compound *A* to contain a ring. The correct structures are as follows:

A　　　　　*B*　　　　　*C*

19.75 Compound *A* is an aldehyde because it is oxidized with Ag(I). Because there is no additional unsaturation, the remaining oxygen is accounted for by either an alcohol or an ether functional group. Because Clemmensen reduction of the CrO_3 oxidation product gives a compound without oxygens, the CrO_3 oxidation product of *A* must be a keto aldehyde or a dialdehyde. Because compound *A* is oxidized to a dicarboxylic acid by H_2CrO_4, it must contain a primary alcohol; hence, its CrO_3 oxidation product must be a dialdehyde. Compound *A*, then, is a chiral hydroxy aldehyde that is oxidized by CrO_3 to an achiral dialdehyde. The Clemmensen-reduction product shows that all compounds have the carbon skeleton of 3-methylpentane. The compounds with this carbon skeleton that fit all the data are the following.

19.76 (a) The mechanism below begins with the protonated aldehyde chloral, which serves as a carbocation electrophile in the ring alkylation of chlorobenzene. The resulting product, an alcohol, dehydrates under the acidic conditions to give another carbocation that alkylates a second chlorobenzene molecule and thus forms the product.

DDT

(b)

a resonance-stabilized
carbocation

(resonance stablized)

+ H_2O

bisphenol A

19.77   The first step is the formation of an imine, by the standard mechanism (see the solution to Focused Problem 19.28 in this manual). The imine carbon-nitrogen double bond can be thought of as a less reactive version of a carbonyl carbon-oxygen double bond; draw a resonance structure of the imine to convince yourself of the partial positive charge on the carbon. An intramolecular Friedel-Crafts "alkylation" occurs, closing the ring. Regeneration of the aromatic ring by a base yields salsolinol.

dopamine

imine

salsolinol

The Chemistry of Carboxylic Acids

CHAPTER 20 SUMMARY

- The carboxy group is the characteristic functional group of carboxylic acids.

- In substitutive nomenclature of carboxylic acids the carboxy group takes precedence as a principal group over the carbonyl groups of aldehydes and ketones, the hydroxy group, and the mercapto group.

- Carboxylic acids are solids or high-boiling liquids, and carboxylic acids of relatively low molecular mass are very soluble in water.

- The carbonyl C=O stretching absorption and the broad, strong OH absorption are the most important infrared absorptions of carboxylic acids. In proton NMR spectra the α-hydrogens of carboxylic acids absorb in the δ 2.0–2.5 region, and the OH protons, which can be exchanged with D_2O, absorb in the δ 9–13 region. The carbonyl carbon resonances in the ^{13}C NMR spectra of carboxylic acids occur at δ 170–180, about 20 ppm less than the carbonyl-carbon resonances of aldehydes and ketones.

- Typical carboxylic acids have pK_a values between 4 and 5, although pK_a values are influenced significantly by polar effects. Sulfonic acids are even more acidic. The acidity of carboxylic acids is due to a combination of polar and resonance effects. The conjugate base of a carboxylic acid is a carboxylate ion.

- Carboxylic acids are weakly basic at the carbonyl oxygen with conjugate-acid pK_a values of about −6. Carbonyl protonation plays an important role in many acid-catalyzed reactions of carboxylic acids.

- Carboxylic acids with long unbranched carbon chains are called fatty acids. The alkali-metal salts of fatty acids are soaps. Soaps and other detergents are surfactants; they form micelles in aqueous solution.

- Because of their acidities, carboxylic acids dissolve not only in aqueous NaOH but also in aqueous solutions of weaker bases such as sodium bicarbonate.

- The reaction of Grignard reagents with CO_2 serves as both a synthesis of carboxylic acids and a method of carbon–carbon bond formation. (Other methods of carboxylic acid synthesis are summarized in Appendix VI, Section N.)

- The reactivity of the carbonyl carbon toward nucleophiles plays an important role in many reactions of carboxylic acids and their derivatives. In these reactions, a nucleophile approaches the carbonyl carbon from above or below the plane of the carbonyl group and reacts to form a tetrahedral addition intermediate, which then breaks down by loss of a leaving group. The result is a net substitution reaction at the carbonyl carbon. Acid-catalyzed esterification and lithium aluminum hydride reduction are two examples of nucleophilic carbonyl substitution. In lithium aluminum hydride reduction, the substitution product, an aldehyde, reacts further in an addition reaction to give, after protonolysis, an alcohol.

- The nucleophilic reactivity of the carboxylate oxygen is important in some reactions of carboxylic acids, such as ester formation by the alkylation of carboxylic acids with diazomethane or the alkylation of carboxylate salts with alkyl halides.

- Carboxylic acids are readily converted into acid chlorides with phosphorus pentachloride or thionyl chloride, and into anhydrides with P_2O_5. Cyclic anhydrides containing five- and six-membered rings are readily formed on heating the corresponding dicarboxylic acids or upon heating with acetic anhydride.

■ β-Keto acids.as well as derivatives of malonic acid and carbonic acid decarboxylate in acidic solution; most malonic acid derivatives require heating. The decarboxylation reactions of β-keto acids, and probably those of malonic acids as well, involve enol intermediates.

■ Decarboxylation is an important reaction in biology. Some decarboxylations involve thiamin pyrophosphate (TPP) as a coenzyme. The positively charged nitrogen of TPP serves as the ultimate destination ("electron sink") for the electrons provided by the decarboxylation reaction.

 STUDY GUIDE LINKS

 ### 20.1   Reactions of Bases with Carboxylic Acids

In Eq. 20.10 (text Sec. 20.4A, p. 1094), water is the base that reacts with a carboxylic acid, because water is the most abundant base present in the aqueous solution of an acid. However, you should not forget that carboxylic acids can in principle react with *any* base. (For example, write the reaction of a carboxylic acid with one equivalent of a Grignard reagent.)

Another point worth special note is that the reaction of strong bases not only with carboxylic acids, but also with alcohols and water, is very fast—so fast that a reaction occurs at virtually every encounter between an acid molecule and a base molecule. This becomes very important in situations in which proton transfer competes with other reactions for a limiting amount of a base. Suppose, for example, we want the aldehyde group in the following molecule to react with a Grignard reagent. The first equivalent of Grignard reagent reacts not with the aldehyde, but rather with the carboxylic acid, because proton transfer is such a fast reaction.

To achieve a reaction of the aldehyde, a second equivalent of the Grignard reagent must be added:

(The aldehyde carbonyl is much more reactive than the carboxylate carbonyl for reasons that are discussed in Chapter 21.)

In summary, proton transfers to strong bases from O—H groups in most cases are much faster than other reactions of those bases.

20.2 Resonance Effect on Carboxylic Acid Acidity

The resonance stabilization of carboxylate ions was once considered the *major* effect responsible for the enhanced acidity of carboxylic acids. It is now believed to be less important than originally thought (although still significant) because *carboxylic acids are also resonance-stabilized.*

Display 3.74 on text p. 155 shows that stabilizing a conjugate-base carboxylate decreases the pKa of the corresponding carboxylic acid. It follows that stabilizing a carboxylic acid, that is, lowering its energy relative to that of its conjugate-base carboxylate, *increases* its pK_a. Therefore, the resonance stabilization of an acid and the resonance stabilization of its carboxylate are opposing effects. However, *resonance is more important in the carboxylate ion* because the carboxylate resonance structures *disperse* or *delocalize* charge, whereas the carboxylic acid resonance structures *separate* charge. Structures that disperse a single type of charge are more important than those that separate opposite charges. (See Sec. 3.3C, Rule 3, text p. 128, and Display 3.29, p. 129.) For this reason, a carboxylate ion is stabilized *more* by resonance than its conjugate-acid carboxylic acid. If the two were equally stabilized by resonance, the resonance effect would cancel when the free energies of the carboxylic acid and the carboxylate are compared, and it would have no effect on the pK_a.

20.3 Mechanism of Acid Chloride Formation

The formation of acid chlorides follows a mechanism that is fundamentally the same as the mechanism for esterification. In esterification, the carbonyl group is activated by protonation. In acid chloride formation, it is activated by reaction with the Lewis acid $SOCl_2$. A chloride ion is displaced from $SOCl_2$ in the process:

As a result of this reaction the carbonyl oxygen is more electrophilic. In other words, the Lewis acid $SOCl_2$ activates the carbonyl carbon toward nucleophilic attack in much the same sense that a proton does. (You've seen that other Lewis acids, such as the lithium ion of $LiAlH_4$ or the magnesium of a Grignard reagent, can also activate nucleophilic reaction at the carbonyl carbon.) The nucleophile chloride ion produced in the step above reacts at the carbonyl carbon:

tetrahedral addition
intermediate

Notice the formation of a tetrahedral addition intermediate, as in esterification. *This is a common mechanistic thread that runs throughout carbonyl chemistry.* The —OSOCl group is an excellent leaving group and departs to give a protonated acid chloride. Loss of the —OSOCl group is irreversible, because it decomposes to SO₂ (a gas) and chloride ion:

Loss of a proton to chloride ion gives HCl (another gaseous by-product) and the acid chloride:

20.4   More on Synthetic Equivalents

Study Guide Link 17.1 (p. 511 of this manual) discussed the notion of a *synthetic equivalent.* Acid chlorides present another opportunity to think about this concept. Recall that acid chlorides are the acylating agents used in Friedel–Crafts acylation (Sec. 16.4F). Because acid chlorides are often prepared from carboxylic acids, it follows, then, that the acyl group of an aromatic ketone can originate from a carboxylic acid. Thus, a carboxylic acid or acid chloride can be viewed as the synthetic equivalent of the acyl group in an aromatic ketone.

20.5   Mechanism of Anhydride Formation

Working through the mechanism of anhydride formation will allow you to see again the patterns involved in carbonyl substitution reactions. Consider the mechanism for the formation of the anhydride of a general carboxylic acid, RCO₂H, by acetic anhydride. The acetic anhydride serves as a Lewis acid to activate the carbonyl group of an R—CO₂H molecule:

The activated carbonyl group reacts with another molecule of carboxylic acid and loses a proton to the acetate ion formed in the previous step to give a tetrahedral addition intermediate:

tetrahedral addition intermediate

The breakdown of the tetrahedral intermediate, possibly with a concerted proton transfer, gives the anhydride.

The mechanisms of anhydride formation with P_2O_5 and with $POCl_3$ are similar.

Anhydride formation is an equilibrium. When acetic anhydride is used as the reagent, this equilibrium is driven to the right by the use of a large excess of acetic anhydride (Le Chatelier's principle in operation). When cyclic anhydrides are being formed, the preference for intramolecular reactions that form five- and six-membered rings (Sec. 12.8) provides not only a kinetic advantage to the reaction—a greater reaction rate—but also an equilibrium advantage. Consequently, reactions that form five- and six-membered cyclic anhydrides in many cases require milder conditions.

FURTHER EXPLORATIONS

20.1   Chemical Shifts of Carbonyl Carbons

As pointed out in the text, the electronegativity of oxygen might lead one to expect the chemical shift for the carbonyl group of acetic acid to be *greater* than that of acetone. An additional puzzle is provided by the carbonyl chemical shift of the acetate ion, the conjugate base of acetic acid:

acetic acid   **acetate ion**

If chemical shifts in carbon NMR spectroscopy correlate with electronegativity, as they do in proton NMR, it seems that the acetate ion, because of its relatively electron-rich —O⁻, should have a *smaller* carbonyl chemical shift than acetic acid. The data above show that this is not the case.

A related phenomenon is the chemical shift of the O—H proton, which was discussed in Sec. 14.7E on text p. 770. In the gas phase (in which hydrogen bonding and other solvent effects are absent), this proton has a chemical shift of δ 0.8, a much smaller value than would be expected for a proton directly attached to an electronegative atom.

These cases illustrate the point that *chemical shifts are not determined solely by electronegativity of nearby atoms.* You learned, for example, that induced circulation of π electrons in the presence of the applied field accounts for the significant chemical shifts of both aromatic and alkene carbons as well as the shifts of aromatic and alkene protons—shifts far greater than would be predicted on the basis of the electronegativity of an sp^2 hybridized carbon (Fig. 14.15 on text p. 765 and Fig. 16.2 on text p. 873). In the cases discussed here, induced electron circulation also accounts for the results, except that the circulation involved is that of the nonbonding electrons of the H—O oxygen; as a result, a lower frequency is required for resonance of these atoms. The result is a smaller chemical shift. The α-carbon atoms of acetone have no such contribution, because they have no nonbonding electrons. Therefore, the carbonyl carbon chemical shifts of ketones in carbon NMR are greater than those of carboxylic acids. In acetate ion, the contribution of one nonbonding electron pair is partially offset by that of another, because the two orbitals are directed differently in space with respect to the applied field; therefore, acetate absorbs at a chemical shift that is closer to the "normal" carbonyl value.

20.2   More on Surfactants

The word "surfactant" means "surface-active compound." To say that something is *surface active* means that is lowers the *surface tension* of a liquid. Without getting into a technical discussion of surface tension, we can say that it is the resistance of a surface to penetration by another object. Water has a particularly high surface tension. (If you've ever seen someone float a coin on the surface of a glass of water, you've witnessed the effects of surface tension.) The high surface tension of water is due to the tight network of strong hydrogen bonds at the water surface. Because water cannot form hydrogen bonds to air, water molecules form particularly strong hydrogen bonds to each other at the air–water interface.

When a surfactant is dissolved in water, the surfactant molecules tend to collect at the surface of the water with the hydrocarbon tails pointed toward the air–water interface, as shown in **Fig. SG20.1** on the following page. This orientation of surfactant molecules at the surface allows the solvent water molecules to interact with the polar head groups and places the nonpolar tails in contact with air, which is composed mostly of other nonpolar molecules. In this situation the network of tight hydrogen bonds at the surface of pure water is no longer present. As a result, the surface tension of the solution is lowered. One perceptible manifestation of this phenomenon is aqueous solutions of surfactants (for example, soap solutions) are slippery to the touch.

20.3   Orthoesters

We can ask whether esters react further under the esterification conditions to form compounds analogous to acetals. (See Sec. 19.11A.)

$$R-\underset{\underset{OCH_3}{\|}}{\overset{\overset{O}{\|}}{C}} \quad + \quad 2\ CH_3OH \quad \overset{acid}{\underset{\times}{\longrightarrow}} \quad R-\underset{\underset{OCH_3}{|}}{\overset{\overset{OCH_3}{|}}{C}}-OCH_3 \quad + \quad H_2O$$

an orthoester
(not observed)

Figure SG20.1 A diagram showing schematically the orientation of surfactant molecules at the interface of an aqueous solution and air. The nonpolar tails are oriented towards the interface.

Such compounds, called orthoesters, are indeed known. However, they *cannot* be formed in this manner because the equilibrium constants for addition to the carbonyl groups of acid derivatives are only about 10^{-8} times those for formation of aldehyde addition compounds. Such equilibria are simply too unfavorable to be driven to completion in the usual ways.

An ester has more resonance structures and is more stable relative to its addition products; addition reactions are less favorable.

An aldehyde or ketone has fewer resonance structures and is less stable relative to its addition products; addition reactions are more favorable.

Therefore, the equilibrium constant for an addition to a ketone is more favorable than the equilibrium constant for the same addition to an ester. This point is illustrated in **Fig. SG20.2**.

Figure SG20.2 The greater resonance stabilization of esters causes addition reactions of esters to have much smaller equilibrium constants than the corresponding addition reactions of ketones.

20.4   Mechanism of the LiAlH₄ Reduction of Carboxylic Acids

The mechanism of the LiAlH₄ reduction of carboxylic acids is presented in the text only in outline. Further Exploration fills in the details.

Reduction of the carboxylate ion by AlH₃ (step *a* in the following equation) is analogous to the addition of BH₃ to the double bond of an alkene. (BH₃ also reduces carboxylic acids to primary alcohols.)

Notice that a tetrahedral addition intermediate is formed in step *a*. Breakdown of this intermediate (step *b*) expels $Li^+ H_2AlO^-$, which also can act as a reducing agent. Further reduction of the aldehyde to the alcohol (Sec. 19.9A) can occur either by this species or by excess LiAlH₄ present in the reaction mixture.

x

REACTION REVIEW

I.   SYNTHESIS AND REACTIVITY OF CARBOXYLIC ACIDS

A.   SYNTHESIS OF CARBOXYLIC ACIDS—REVIEW

1. Methods for preparing carboxylic acids introduced in previous chapters:

    a. Oxidation of primary alcohols (see Sec. 11.6B, text p. 592).

a primary alcohol   a carboxylic acid

a primary alcohol   a carboxylic acid

    b. Side-chain oxidation of alkylbenzenes (see Sec. 17.5C, text p. 937).

an aromatic hydrocarbon
with benzylic hydrogens

an aromatic
carboxylic acid

an aromatic hydrocarbon
with benzylic hydrogens

an aromatic
carboxylic acid

c. Oxidation of aldehydes (see Sec. 19.15, text p. 1074).

an aldehyde

a carboxylic acid

an aldehyde

an aldehyde hydrate

a carboxylic acid

d. Ozonolysis of alkenes followed by oxidative workup with H_2O_2 (Sec. 5.8, text page 259).

B.   SYNTHESIS OF CARBOXYLIC ACIDS WITH GRIGNARD REAGENTS

1. The reaction of a Grignard reagent with carbon dioxide, followed by protonolysis, gives a carboxylic acid.

   a. Addition of a Grignard reagent to carbon dioxide gives the halomagnesium salt of a carboxylic acid.

   b. When aqueous acid is added to the reaction mixture in a separate reaction step, the free carboxylic acid is formed.

2. The reaction of Grignard reagents with CO_2 is another method for the formation of carbon–carbon bonds. (See Appendix VI, text page A-13, for a review of other carbon–carbon bond-forming reactions.)

C.   INTRODUCTION TO CARBOXYLIC ACID REACTIONS

1. Reactions of carboxylic acids can be categorized into four types:

   a. Reactions at the carbonyl group.

      i. The most typical reaction at the carbonyl group is substitution at the carbonyl carbon. In such a reaction, the —OH of the carboxy group is typically substituted by a nucleophilic group.

  ii. Reaction of the carbonyl oxygen with an electrophile (Lewis acid or Brønsted acid), that is, the reaction of the carbonyl oxygen as a base.

 b. Reactions at the carboxylate oxygen, such as reactions of the carboxylate oxygen as a nucleophile.

 c. Loss of the carboxy group as CO_2 (decarboxylation).

 d. Reactions involving the α-carbon. (These reactions are considered in Chapter 22.)

2. Many substitution reactions at the carbonyl carbon are acid catalyzed; that is, the reactions of nucleophiles at the carbonyl carbon are catalyzed by the reactions of acids at the carbonyl oxygen.

II. REACTION OF CARBOXYLIC ACIDS AT THE CARBONYL GROUP

A. *ACID-CATALYZED ESTERIFICATION OF CARBOXYLIC ACIDS*

1. Esters are carboxylic acid derivatives in which the proton on the carboxylate oxygen in effect has been replaced by an alkyl or aryl group.

2. When a carboxylic acid is treated with a large excess of an alcohol in the presence of a strong acid catalyst, an ester is formed; this reaction is called acid-catalyzed esterification (or Fischer esterification).

an ester

 a. As shown by the asterisks in the foregoing equation, acid-catalyzed esterification is a substitution of —OH at the carbonyl group of the acid by the —OR group of the alcohol; the —OH group leaves as water.

 b. The equilibrium constants for esterification with most primary alcohols are near 1.

  i. The reaction is driven toward completion by applying Le Chatelier's principle.

  ii. Use of the reactant alcohol as the solvent, which ensures that the alcohol is present in large excess, drives the equilibrium toward the ester product.

 c. Acid-catalyzed esterification cannot be applied to the synthesis of esters from phenols or tertiary alcohols.

3. The mechanism of acid-catalyzed esterification serves as a model for the mechanisms of other acid-catalyzed reactions of carboxylic acids and their derivatives.

 a. The first step of the mechanism is protonation of the carboxyl oxygen.

  i. The catalyzing acid is the conjugate acid of the solvent, formed by reaction of the solvent with a strong acid such as H_2SO_4.

  ii. Protonation of a carbonyl oxygen makes the carbonyl carbon more electrophilic because the carbonyl oxygen becomes a better electron acceptor.

 b. Reaction of the alcohol with the carbonyl carbon of this carbocation, followed by loss of a proton, gives a tetrahedral addition intermediate.

tetrahedral
addition intermediate

c. The tetrahedral addition intermediate, after protonation, loses water to give the conjugate acid of the ester.

d. Loss of a proton gives the ester product and regenerates the acid catalyst.

B.   SYNTHESIS OF ACID CHLORIDES

1. Acid chlorides are carboxylic acid derivatives in which the —OH group has been replaced by —Cl.

    a. Acid chlorides are often prepared from the corresponding carboxylic acids.

    b. Two reagents used for this purpose are thionyl chloride, $SOCl_2$, and phosphorus pentachloride, PCl_5.

2. Acid chloride synthesis fits the general pattern of substitution reactions at a carbonyl group; in this case, —OH is substituted by —Cl. (For the mechanism, see Study Guide Link 20.3 on p. 658 of this manual.)

3. Sulfonyl chlorides are the acid chlorides of sulfonic acids.

    a. Sulfonyl chlorides are prepared by treatment of sulfonic acids or their sodium salts with PCl_5.

    b. Aromatic sulfonyl chlorides can be prepared directly by the reaction of aromatic compounds with chlorosulfonic acid, $ClSO_3H$.

        i. This reaction is a variation of aromatic sulfonation, an electrophilic aromatic substitution reaction.

        ii. Chlorosulfonic acid, the acid chloride of sulfuric acid, acts as an electrophile in this reaction.

        iii. The sulfonic acid produced in the reaction is converted into the sulfonyl chloride by reaction with another equivalent of chlorosulfonic acid.

C.   SYNTHESIS OF ANHYDRIDES

1. Carboxylic acid anhydrides are carboxylic acid derivatives in which the hydroxy group has been replaced by an acyloxy group.

    a. Anhydrides are prepared by treatment of carboxylic acids with strong dehydrating agents (usually P_2O_5).

an anhydride

b.   Most anhydrides can themselves be used as reagents for the preparation of other anhydrides.

2.   As shown in the foregoing example, some dicarboxylic acids react with acetic anhydride to form cyclic anhydrides—compounds in which the anhydride group is part of a ring.

a.   Cyclic anhydrides containing five- and six-membered rings are readily prepared from the corresponding dicarboxylic acids.

b.   Formation of cyclic anhydrides with five- or six-membered rings is so facile that in some cases it occurs on heating the dicarboxylic acid.

3.   Anhydrides, like acid chlorides, are used in the synthesis of other carboxylic acid derivatives.

D.   *REDUCTION OF CARBOXYLIC ACIDS TO ALCOHOLS*

1.   When a carboxylic acid is treated with $LiAlH_4$, then with dilute acid, a primary alcohol is formed.

a.   Before the reduction itself takes place, $LiAlH_4$ reacts with the acidic hydrogen of the carboxylic acid to give the lithium salt of the carboxylic acid and one equivalent of hydrogen gas. (The lithium salt of the carboxylic acid is the species that is actually reduced.)

2.   The reduction occurs in two stages:

a. AlH_3 (formed from the reaction of $LiAlH_4$ with the acidic proton of the carboxylic acid) reduces the carboxylate ion to an aldehyde.

b.   The aldehyde is rapidly reduced further to give, after protonolysis, the primary alcohol. (Because the aldehyde is more reactive than the carboxylate salt, it cannot be isolated.)

3.   The $LiAlH_4$ reduction of a carboxylic acid incorporates two different types of carbonyl reactions:

a.   A net substitution reaction at the carbonyl group to give an aldehyde intermediate.

b.   An addition to the aldehyde.

4.   Sodium borohydride, $NaBH_4$, does not reduce carboxylic acids, although it does react with the acidic hydrogens of carboxylic acids.

III. REACTION OF CARBOXYLIC ACIDS AT THE OXYGEN OF THE OH GROUP

A. ESTERIFICATION BY ALKYLATION

1. When carboxylic acids are treated with diazomethane in ether solution, they are rapidly converted into their methyl esters.

   a. Protonation of diazomethane by the carboxylic acid gives the methyldiazonium ion; this ion has one of the best leaving groups, dinitrogen.

   b. An S_N2 reaction of the methyldiazonium ion with the carboxylate oxygen results in the displacement of N_2 and formation of the ester.

2. The reaction of certain alkyl halides with carboxylate ions give esters.

   a. This is an S_N2 reaction in which the carboxylate ion, formed by a Brønsted acid–base reaction of the acid and a base (such as K_2CO_3), acts as the nucleophile that reacts with the alkyl halide.

   b. This reaction works best on alkyl halides that are especially reactive in the S_N2 reaction, such as methyl iodide and benzylic or allylic halides, because carboxylate ions are relatively weak bases and therefore relatively poor nucleophiles.

IV. DECARBOXYLATION OF CARBOXYLIC ACIDS

1. The loss of carbon dioxide from a carboxylic acid is called decarboxylation.

2. Certain types of carboxylic acids are readily decarboxylated:

   a. β-Keto acids (carboxylic acids with a keto group in the β-position)

   b. Malonic acid derivatives

   c. Derivatives of carbonic acid

3. β-Keto acids readily decarboxylate at room temperature in acidic solution.

   a. Decarboxylation of a β-keto acid involves an enol intermediate that is formed by an internal proton transfer from the carboxylic acid group to the carbonyl oxygen atom of the ketone.

b.   The enol is transformed spontaneously into the corresponding ketone.

enol intermediate

4   The acid form of the β-keto acid decarboxylates more readily than the conjugate-base carboxylate form because the latter has no acidic proton that can be donated to the β-carbonyl oxygen.

5.   Malonic acid and its derivatives readily decarboxylate upon heating in acid solution. This reaction, which does not occur in base, bears a close resemblance to the decarboxylation of β-keto acids, because both types of acid have a carbonyl group β to the carboxy group.

a malonic acid derivative

6.   Carbonic acid is unstable and decarboxylates spontaneously in acid solution to carbon dioxide and water.

a.   Carbonic acid derivatives in which only one carboxylate oxygen is involved in ester or amide formation also decarboxylate under acidic conditions.

b.   Under basic conditions, carbonic acid is converted into its salts bicarbonate and carbonate, which do not decarboxylate.

c.   Carbonic acid derivatives in which both carboxylic acid groups are involved in ester or amide formation are stable.

   i.   Carbonate esters are diesters of carbonic acid.

   ii.   Ureas are diamides of carbonic acid.

   iii.   Phosgene is the acid chloride of carbonic acid

a carbonate ester   a urea   **phosgene**

SOLUTIONS TO PROBLEMS

Solutions to Focused Problems

20.1 (a)

$$HOCH_2CH_2CH_2CO_2H$$

γ-hydroxybutyric acid

(b)

$$Cl_2CHCH_2CO_2H$$

β,β-dichloropropionic acid

(c)

$$CH_3CH_2 \quad CH_2CO_2H$$
$$C=C$$
$$H \quad\quad H$$

(Z)-3-hexenoic acid

(d)

$$CH_3CH_2CHCH_2CH_2CO_2H$$
$$|$$
$$CH_3$$

4-methylhexanoic acid

20.2 (a) 2,2-Dimethylbutanoic acid. *(substitutive)* α,α-Dimethylbutyric acid *(common)*
(b) 2,4-Dichlorobenzoic acid
(c) Isopropylpropanedioic acid *(substitutive)* Isopropylmalonic acid *(common)*
(d) Cyclopropanecarboxylic acid

20.3 The dimer owes its stability to intermolecular hydrogen bonds. The solvent carbon tetrachloride has no donor or acceptor atoms to compete for these hydrogen bonds. However, the solvent water can compete as both a donor and an acceptor for hydrogen bonds; so, the dimer has less of an energetic advantage in water and, as a result, there is a lower concentration of dimer (in fact, essentially none) in aqueous solution and a higher concentration in CCl_4.

20.4 (a) The IR spectrum indicates the presence of a carboxy group. A carboxy group accounts for 45 mass units; this leaves 43 mass units unaccounted for. The NMR spectrum indicates seven hydrogens in addition to the carboxy hydrogen at δ 10. A group with seven hydrogens and 43 mass units must be either a propyl or an isopropyl group; the splitting in the NMR spectrum indicates the latter. The compound is 2-methylpropanoic acid (structure at right).

$$(CH_3)_2CH-C\begin{smallmatrix}O\\ \\OH\end{smallmatrix}$$

isobutyric acid
(2-methylpropanoic acid)

20.5 The NMR data indicate a para-substituted benzoic acid derivative; given this deduction, the para substituent must be a chlorine, so the structure is the one shown at right.

$$Cl-\langle\text{benzene ring}\rangle-CO_2H$$

***p*-chlorobenzoic acid**
(4-chlorobenzoic acid)

20.6 First write the structures of the two compounds.

$$HO_2C-\overset{\overset{\displaystyle CH_3}{|}}{\underset{\underset{\displaystyle CH_3}{|}}{C}}-CH_2-CO_2H \qquad HO_2CCH_2CH_2CH_2CH_2CO_2H$$

α,α-dimethylsuccinic acid   **adipic acid**

(a) α,α-Dimethylsuccinic acid should have five resonances in its CMR spectrum because it has five chemically nonequivalent sets of carbons. Adipic acid, in contrast, should have three resonances because it has only three chemically nonequivalent sets of carbons.

(b) The proton NMR spectrum of only α,α-dimethylsuccinic acid consists of three singlets. (The protons of the OH groups, because of chemical exchange, will occur at the same chemical shift.) The proton NMR spectrum of adipic acid is more complex.

20.7 (a)   The first and second ionizations of succinic acid:

$$HO_2CCH_2CH_2CO_2H \underset{H_3O^+}{\overset{H_2O}{\rightleftarrows}} {}^-O_2CCH_2CH_2CO_2H \underset{H_3O^+}{\overset{H_2O}{\rightleftarrows}} {}^-O_2CCH_2CH_2CO_2^-$$

$$pK_a = 4.21 \qquad\qquad pK_a = 5.64$$

(b)   The first pK_a is lower because the electron-withdrawing polar effect of an un-ionized carboxy group stabilizes the anion formed by ionization of the other carboxy group. The second pK_a is higher because the negative charge of an ionized carboxy group destabilizes the anion formed by the second ionization. This destabilization is the result of the electrostatic repulsion of like charges.

20.8   Extract a methylene chloride solution of the two compounds with an aqueous solution of $NaHCO_3$, Na_2CO_3, or NaOH. The acid will ionize and its conjugate-base anion will dissolve in the aqueous layer as the sodium salt; *p*-bromotoluene will remain in the methylene chloride layer from which it can be isolated by evaporating the solvent. After isolating the aqueous layer, acidify it with concentrated HCl; neutral *p*-bromobenzoic acid will precipitate.

20.9 (a)

An alternate route to extend a carbon chain by one carbon is to let the Grignard reagent react with formaldehyde to give a primary alcohol, and then oxidize the primary alcohol to a carboxylic acid. The Grignard route shown above (and used for the remainder of this chapter) is shorter.

(b)

The conversion of a terminal alkene to a primary alkyl bromide (*A* to *B* in the foregoing synthesis) can also be carried out by the addition of HBr in the presence of peroxides.

20.10 (a)

CH₃CH₂CH₂CHCH₂COH (3-methylhexanoic acid), with CH₃ substituent → H₂SO₄ (cat.), CH₃CH₂OH (excess) → CH₃CH₂CH₂CHCH₂COCH₂CH₃ (ethyl 3-methylhexanoate), with CH₃ substituent

3-methylhexanoic acid → **ethyl 3-methylhexanoate**

(b)

HOC(CH₂)₄COH (adipic acid) → H₂SO₄ (cat.), CH₃CH₂CH₂OH → CH₃CH₂CH₂OC(CH₂)₄COCH₂CH₂CH₃ (dipropyl adipate)

adipic acid → **dipropyl adipate**

20.11 (a) Follow the reverse of the steps shown in Eqs. 20.34a–d, text pp. 1103–1104, with R— = Ph—.

(The reaction mechanism showing the acid-catalyzed hydrolysis of PhC(=O)OCH₃ through protonation, addition of H₂O, proton transfers, and loss of HOCH₃ to give PhC(=O)OH and H—⁺OH₂.)

(b) To favor ester hydrolysis rather than ester formation, use a large excess of water as solvent rather than an alcohol. By Le Chatelier's principle, this drives the carboxylic acid–ester equilibrium toward the carboxylic acid.

20.12 (a) The mechanism of hydroxide-promoted ester hydrolysis:

(The reaction mechanism showing hydroxide attack on PhC(=O)OCH₃, forming the tetrahedral intermediate PhC(O⁻)(OH)OCH₃, then collapse and proton transfer to give PhC(=O)O⁻ + H—OCH₃.)

Notice that under the basic conditions the carboxylic acid product is ionized in the last step. This drives base-promoted ester hydrolysis to the right.

The base-promoted hydrolysis of esters is called *saponification*, because soaps are made this way from the base-promoted hydrolysis of fats. (Fats are esters of the fatty acids.) Saponification is discussed in Sec. 21.7A, text p. 1147. However, if you were able to construct this mechanism, you're ahead of the game—you understand the reaction already!

(b)   As shown in the last step of the mechanism in the solution to part (a), methoxide reacts with the *acidic proton* of benzoic acid to form the benzoate ion much more rapidly than it reacts at the carbonyl carbon. (See Study Guide Link 20.1 on p. 657 of this manual for a discussion of this point.)

20.13   (a)   (b)   (c)   (d)

| methyl isobutyrate | dimethyl succinate | benzyl isobutyrate | allyl benzoate |

20.14   In this reaction the carboxylic acid is alkylated by the *tert*-butyl cation, which is formed by protonation of methyl-propene.

A variation on this mechanism proposed by many students is alkylation of the carboxylate oxygen rather than the carbonyl oxygen by the *tert*-butyl cation:

Although this appears to be a more direct route to the ester, the mechanism given above is more likely to be correct for the following reason. Just as a carboxylic acid is *protonated* on the carbonyl oxygen rather than the carboxylate oxygen (why? see Sec. 20.4B, p. 1097 of the text), it should be *alkylated* on the carbonyl oxygen for the same reason.

20.15   (a)   (b)   (c)

| 2-methylbutanoyl chloride | *p*-methoxybenzoyl chloride (anisoyl chloride) | 1-propanesulfonyl chloride |

20.16   (a)

ethanesulfonyl chloride

(b)

benzoyl chloride

(c)

p-toluenesulfonyl chloride
(tosyl chloride)

20.17   Study Guide Link 20.4 in this manual, points out that a carboxylic acid is the synthetic equivalent of an acyl group in an aromatic ketone. This concept is realized by applying the Friedel–Crafts acylation reaction to the corresponding acid chloride. This synthesis takes advantage of the fact that the methoxy group is an ortho, para-directing substituent in electrophilic aromatic substitution.

benzoic acid $\xrightarrow{SOCl_2}$ benzoyl chloride $\xrightarrow[\text{2) H}_3\text{O}^+]{\text{1) CH}_3\text{O}-\bigcirc, \text{AlCl}_3}$ *p*-methoxybenzophenone

20.18   The corresponding anhydride is formed in each case.

(a)

chloroacetic anhydride

(b)

p-chlorobenzoic anhydride

20.19   (a)   Maleic acid forms a cyclic anhydride because its carboxy groups are cis. A cyclic anhydride of fumaric acid would require that a five-membered ring contain a trans double bond, which would introduce a large amount of ring strain. (See Sec. 7.6C, text p. 352.)

maleic anhydride   fumaric acid

This distance is too
great to be bridged by
a five-membered ring.

(b)   On heating, 2,3-dimethylbutanedioic acid forms a cyclic anhydride containing a five-membered ring. Because a cyclic anhydride of methylmalonic acid would contain a very strained four-membered ring, it is not formed on heating.

2,3-dimethylbutanedioic acid → (heat, $-H_2O$) → **2,3-dimethylbutanedioic anhydride**

methylmalonic acid → (heat, $-H_2O$ crossed out) → **methylmalonic anhydride**

20.20 (a) The compound must contain three oxygens and two degrees of unsaturation in addition to the four degrees of unsaturation of the benzene ring. *p*-Formylbenzoic acid meets the indicated criteria.

$$HC(=O)-C_6H_4-COOH \xrightarrow[\text{2) } H_3O^+]{\text{1) LiAlH}_4} HOCH_2-C_6H_4-CH_2OH$$

p-formylbenzoic acid

(b) 1,4-Benzenedicarboxylic acid (terephthalic acid) is a compound with the formula $C_8H_6O_4$ that gives the indicated diol on treatment with $LiAlH_4$ followed by protonolysis.

$$HOC(=O)-C_6H_4-COOH \xrightarrow[\text{2) } H_3O^+]{\text{1) LiAlH}_4} HOCH_2-C_6H_4-CH_2OH$$

1,4-benzenedicarboxylic acid (terephthalic acid)

20.21 (a) This synthesis requires the addition of one carbon. Follow the general scheme in Study Problem 20.2, text pp. 1112–1113.

$$PhCO_2H \xrightarrow[\text{2) } H_3O^+]{\text{1) LiAlH}_4} PhCH_2OH \xrightarrow{\text{conc.HBr}} PhCH_2Br \xrightarrow[\text{ether}]{\text{Mg}} PhCH_2MgBr \xrightarrow[\text{2) } H_3O^+]{\text{1) CO}_2} PhCH_2CO_2H$$

benzoic acid   **phenylacetic acid**

(b) This synthesis requires the addition of two carbons, which can be accomplished by the reaction of a Grignard reagent with oxirane (Sec. 12.5C, text p. 641).

$$PhCH_2MgBr \xrightarrow[\text{2) } H_3O^+]{\text{1) oxirane}} PhCH_2CH_2CH_2OH \xrightarrow[\text{2) } H_3O^+]{\text{1) KMnO}_4, \text{ } ^-OH} PhCH_2CH_2CO_2H$$

[from part (a)]   **3-phenylpropanoic acid**

(Burning away the side chain with $KMnO_4$ to give benzoic acid requires extreme conditions; see Eq. 17.46, Sec. 17.5C, text p. 938.)

20.22 (a)

$$Ph-C(=O)-CH(CH_3)_2$$

(b)

$$\text{cyclopentyl}-CO_2H$$

(c)

$$CH_3CH_2NH_2 \underset{}{\overset{H_3O^+}{\rightleftharpoons}} CH_3CH_2\overset{+}{N}H_3$$

20.23 The following β-keto acids (and their enantiomers) will decarboxylate to give 2-methylcyclohexanone:

**1-methyl-2-oxocyclo-
hexanecarboxylic acid**   *cis*-**3-methyl-2-oxocyclo-
hexanecarboxylic acid**   *trans*-**3-methyl-2-oxocyclo-
hexanecarboxylic acid**

20.24 Were this carboxylic acid to decarboxylate by the mechanism shown in Eq. 20.58b on p. 1114 of the text, the enol shown at right would be formed as an intermediate. This enol is very unstable because it violates Bredt's rule (Sec. 7.6C, text p. 352); that is, the double bond is twisted to such an extent that the 2p orbitals on each carbon cannot overlap. The resistance of the β-keto acid shown in the problem to decarboxylation, in fact, is considered to be evidence that the enol is an obligatory intermediate in the decarboxylation reaction.

20.25 (a) The second resonance structure contains a carbene carbon—a very unstable electronic configuration for carbon that, while neutral, has fewer than eight electrons. Even though the first structure has opposite charges on adjacent carbons, it is important because every atom has an electronic octet.

   (b) The nitrogen in the second structure violates the octet rule. As if that isn't bad enough, the second structure contains two sp-hybridized atoms (the carbon and nitrogen of the triple bond) within a five-membered ring, which requires a bond angle of 180° at each atom, an impossibility in a five-membered ring.

20.26 In the first step, the hydroxy group of malic acid is oxidized by NAD^+. The mechanism of an analogous enzymatic reaction mediated by NAD^+, the oxidation of ethanol, is discussed in Sec. 11.8 in the text (p. 598).

malic acid   **oxalacetate**

Oxalacetate is a β-keto carboxylic acid, which spontaneously decarboxylates (loses carbon dioxide) to an enol, which then isomerizes into a ketone, in this case pyruvate:

a β-keto acid

oxalacetate   **an enol**   **pyruvate**

Finally, the NADH that was generated in the first step serves as a reducing agent to reduce the pyruvate to lactic acid (see Sec. 19.9B in the text). In the process, NADH is oxidized back to NAD^+, which can reenter the enzymatic cycle in step one, above.

pyruvate   **lactic acid**   regenerated

The oxidation step is required to provide the "electron sink" for the decarboxylation reaction.

Solutions to Skills Objectives Problems

20.27 (a)

1, 4-cyclohexanedicarboxylic acid

(b)

$HO_2CCCH_2CH_2CH_2CO_2H$

α,α-dichloroadipic acid

(c)

$HO_2C \!-\! CO_2H$

oxalic acid

20.28 (a) (*E*)-9-Methyl-4-decenoic acid
(b) 3-Hydroxy-4-oxocyclohexanecarboxylic acid
(c) (2*R*,3*R*)-2-Ethoxy-3-formylpentanedioic acid

20.29 Oxygen *a*: sp²; carbon *b*: sp²; oxygen *c*: sp (a terminal atom); carbon *d*: sp²; carbon *e*: sp³. The bond angles *abc* and *abd* are both 120°.

20.30 Increasing bond lengths: *g* < *c* < *f* < *b* < *d* < *a*

20.31 (a) *C* < *A* < *B*. Intermolecular hydrogen bonds cause the higher boiling points of *A* and *B*. The strongly hydrogen-bonded dimer of the carboxylic acid causes its boiling point to be greater than that of the alcohol.
(b) *A* < *B* < *C*. Intermolecular hydrogen bonding cause a higher boiling point for *C*. The greater molecular mass of *B* and its greater polarity, cause its higher boiling point than that of *A*.

20.32 *B* < *A* << *C*. (Though it is less symmetrical than *A*, *C* is an ionic compound.)

20.33 The IR spectrum of compound *B* would show the broad O—H absorption and carbonyl C═O stretching absorption characteristic of a carboxylic acid. The IR spectrum of *A* would show a carbonyl C═O stretching absorption and an O—H stretching absorption which is sharper and much less intense than the corresponding absorption in *B*. The proton NMR spectrum of *A* will show a 3*H* singlet at roughly δ 2.5 that is not present in the spectrum of *B*. The proton NMR spectrum of *A* will show a 1*H* triplet in the δ 4.5–5 region corresponding to the proton that is α to both the benzene ring and the OH group. There will be no resonances in this region of the spectrum in compound *B*.

20.34 (a) The IR spectrum of *A* indicates the presence of a carboxylic acid, and the NMR indicates the presence of an ethoxy group (—OCH₂CH₃) and a para-disubstituted benzene ring. The compound is 4-ethoxybenzoic acid (structure at right).

4-ethoxybenzoic acid (compound *A*)

(b) The presence of two exchangeable hydrogens suggests that compound *B* is a dicarboxylic acid. The carbon NMR chemical shifts suggest the presence of a methyl group, a carbonyl group, and a carbon bound to a carbonyl group, and, with six carbons in the molecular formula, two chemically equivalent groups of each type. Because compound *B* is a dicarboxylic acid, each carbon NMR resonance must correspond to two carbons. Compound *B* is 2,3-dimethylbutanedioic acid; the fact that it can be resolved into enantiomers shows that it is the racemate rather than the *meso* diastereomer.

(±)-2,3-dimethylbutanedioic acid (compound *B*)

20.35 (a) *A*, *B*, *C*, and *D*   (b) *A* and *D* (Soaps and detergents do not occur naturally.)

(c)   *B* and *C*   (d)   *C*   (e)   *C* (specifically, a cationic detergent)

(f)   *A*   (g)   *A*

20.36   The typical range is (b), 4–5, although some carboxylic acids with electron-withdrawing substituents have pK_a values in the 1–3 range.

20.37   The pK_a order is $A > D > B > C$. (The acidity order is the reverse.) Compound *A* contains a negative charge, which lowers the stability of the di-anion formed when the other carboxy group ionizes. Compound *B* is a typical carboxylic acid. The electronegative methoxy group increases the stability of the conjugate-base anion. Finally, the cationic substituent in *C* stabilizes the conjugate-base anion.

20.38   Only one equivalent of NaOH is present, so there is only 0.01 mol of the conjugate-base anion(s) formed. The two chlorines exert an electron-withdrawing polar effect that stabilizes both possible conjugate-base anions; but the effect decreases with increasing distance, so ionization of the nearer carboxylic acid group is favored. The predominant species in solution is the mono-anion shown at right.

20.39   As the chain length becomes larger the distance between the two carboxy groups becomes greater. When the carboxy groups are close together, the pK_a of the first group is lowered by the electron-withdrawing polar effect of the other, and the pK_a of the second carboxy group is raised by charge repulsion between the conjugate-base anions of the two carboxylic acid groups. Because polar effects and charge repulsions decrease with distance, these effects are negligible when the chain length is great, and the two pK_a values are nearly equal. (However, they are not *exactly* equal—they differ by log(4) = 0.6. For an explanation, see the icon comment following the solution to Problem 3.65(e) in this manual.)

20.40   (a)   There are two possible sites of protonation of a carboxylic acid—the two oxygens. Only protonation of the carbonyl oxygen gives a resonance-stabilized cation *AH*. The other possible cation is not formed.

cation *AH* (resonance-stabilized)

not resonance-stabilized;
not formed

(b)   By the same reasoning, protonation of the carbonyl oxygen gives a resonance-stabilized cation *BH*. Protonation of the nitrogen doesn't occur because the resulting cation is not resonance stabilized. (It is estimated that the pK_a difference between the two cations is about 7–8.)

cation *BH* (resonance-stabilized)

not resonance-stabilized;
not formed

20.41   The relative rates are $D \ll B < A < C$. The enamine from decarboxylation of compound *D* has a bridgehead double bond and violates Bredt's rule; this compound in fact does not decarboxylate under ordinary conditions. Malonic acid derivatives decarboxylate more slowly than β-keto acids, so *B* decarboxylates relative slowly. The relative rates of decarboxylation of compound *A* and *C* depend on the stability of the enols.

Enols formed by decarboxylation:

from compound *D*
(does not form)

from *B*

from *A*

from *C*

The enol from decarboxylation of compound *A* has a double bond in a small ring, and is therefore somewhat strained. The enol from decarboxylation of compound *C* is the enol of an ordinary ketone, and decarboxylates most rapidly.

20.42 (a) Decarboxylation of the imino acid zwitterion gives an enamine in addition to carbon dioxide. Notice that the protonated nitrogen serves as an electron sink for the decarboxylation.

an enamine

In the presence of aqueous acid or base, enamines hydrolyze to aldehydes or ketones (the ketone acetone in this case) and the corresponding amine (aniline in this case). We show the acid-catalyzed mechanism.

(b) When the nitrogen is protonated it can best serve as an electron sink for the decarboxylation, and when the carboxylic acid is deprotonated—that is, in its conjugate-base anion form—the oxygen serves as a better source of electrons for the decarboxylation. The pK_a values of both the carboxylic acid and the iminium ion are about 3–4. Therefore, a pH of about 3–4 is the best compromise for the maximum amount of the form of the molecule that is most active in decarboxylation.

lower pH

raise pH

most reactive form
in the decarboxylation reaction;
present in greatest amount at pH ≈ 3–4

(c) A primary amine catalyzes the decarboxylation of a β-keto acid by forming the imine, which is in equilibrium with the protonated imine. This species (as the previous two parts of this solution show) is very reactive in decarboxylation at moderate pH values. Once the decarboxylation occurs, the amine is regenerated and can react with another molecule of β-keto acid. Therefore, the amine is a true catalyst of the decarboxylation, with the caveat that the pH must be in the proper range.

20.43 In each case, the base reacts with acetic acid in a Brønsted acid–base reaction to form the conjugate base of acetic acid (acetate ion) and a by-product.

(a)

$$CH_3C(=O)-O-H \quad {}^-OCH_3 \ Na^+ \longrightarrow CH_3C(=O)-O^- \ Na^+ \ + \ H-OCH_3$$

(b)

$$CH_3C(=O)-O-H \quad {}^-OH \quad Cs^+ \longrightarrow CH_3C(=O)-O^- \ Cs^+ \ + \ H_2O$$

(c)

$$CH_3C(=O)-O-H \quad CH_3CH_2-MgBr \longrightarrow CH_3C(=O)-O^- \ {}^+MgBr \ + \ CH_3CH_3$$

(d)

$$CH_3C(=O)-O-H \quad CH_3-Li \longrightarrow CH_3C(=O)-O^- \ Li^+ \ + \ CH_4$$

(e)

$$CH_3C(=O)-O-H \quad \ddot{N}H_3 \longrightarrow CH_3C(=O)-O^- \ {}^+NH_4$$

(f)

$$CH_3C(=O)-O-H \quad H-Na \longrightarrow CH_3C(=O)-O^- \ Na^+ \ + \ H_2$$

(g)

$$CH_3C(=O)-O-H \quad {}^-O-C(=O)-OH \ (Na^+) \ \rightleftharpoons \ CH_3C(=O)-O^- \ Na^+ \ + \ H-O-C(=O)-OH \longrightarrow H_2O \ + \ CO_2$$

bicarbonate ion **carbonic acid**

20.44

(a)

$$CH_3CH_2CH_2COCH_2CH_3$$

ethyl butyrate

(b)

$$CH_3CH_2CH_2CO^- \ Na^+$$

sodium butyrate

(c)

$$CH_3CH_2CH_2CH_2OH$$

1-butanol

(d)

$$CH_3CH_2CH_2CCl \ (=O)$$

butyryl chloride

(e)

$$CH_3CH_2CH_2COCH_3$$

methyl butyrate

(f)

$$(CH_3CH_2CH_2CH_2O)_2CHCH_3$$

acetaldehyde dibutyl acetal

(g)

$$CH_3CH_2CH_2CPh \ (=O)$$

butyrophenone

(h)

$$CH_3CH_2CH_2CH_2Ph$$

butylbenzene

20.45 (a)

allyl benzoate

(b)

benzoyl chloride

(c)

m-nitrobenzoic acid

(d)

benzoic anhydride

20.46 Compound A, a malonic acid derivative, will decarboxylate on heating and therefore will give off CO_2 on heating, forming compound A'. Compound B can form a cyclic anhydride (B') containing a five-membered ring on heating; therefore, it gives off water (steam) on heating. While you might think that compound C could also form a cyclic anhydride, the stereochemistry of the two carboxylic acid groups is such that the six-membered ring in the cyclic anhydride C' is constrained by the ring to be very strained. (Build a model if this isn't obvious.) Therefore, compound C loses neither H_2O nor CO_2 on heating. As a malonic acid derivative, compound D will decarboxylate and give off CO_2 on heating, forming ester D' as a product.

20.47 (a)

(b) The reaction can be driven to the right by applying Le Chatelier's principle. One way is to distill out the methanol by-product, which has a much lower boiling point than 1-butanol. The other way is to use an excess of 1-butanol, but this would require separating the 1-butanol from the ester product.

20.48 (a)

isobutyric acid
(2-methylpropanoic acid)

propyl isobutyrate

(b)

$(CH_3)_2CHCOH$ $\xrightarrow{\text{LiAlH}_4, \text{ then } H_3O^+}$ $(CH_3)_2CHCH_2OH$ $\xrightarrow{\text{HBr}}$ $(CH_3)_2CHCH_2Br$ $\xrightarrow{\text{Mg, ether}}$

$(CH_3)_2CHCH_2MgBr$ $\xrightarrow{\text{CO}_2, \text{ then } H_3O^+}$ $(CH_3)_2CHCH_2COH$ $\xrightarrow[\text{H}_2\text{SO}_4 \text{ catalyst}]{\text{CH}_3\text{OH (solvent),}}$ $(CH_3)_2CHCH_2COCH_3$

methyl isovalerate

(c)

$(CH_3)_2CHCH_2MgBr$ $\xrightarrow[\text{2) } H_3O^+]{\text{1) } \triangle \text{O}}$ $(CH_3)_2CHCH_2CH_2CH_2OH$ $\xrightarrow[\text{2) } H_3O^+]{\text{1) KMnO}_4, {}^-\text{OH}}$ $(CH_3)_2CHCH_2CH_2COH$

from part (b)   **4-methylpentanoic acid**

(d)

$(CH_3)_2CHCOH$ $\xrightarrow{\text{SOCl}_2}$ $(CH_3)_2CHCCl$ $\xrightarrow[\text{2) } H_3O^+]{\text{1) benzene, AlCl}_3}$ $(CH_3)_2CHC$—Ph

isobutyrophenone
(2-methyl-1-phenyl-1-propanone)

(e)

$(CH_3)_2CHCH_2OH$ $\xrightarrow{\text{PCC}}$ $(CH_3)_2CHCH{=}O$ $\xrightarrow[\text{2) } H_3O^+]{\text{1) CH}_3\text{MgI}}$ $(CH_3)_2CHCHCH_3$ (OH) $\xrightarrow{\text{PCC}}$ $(CH_3)_2CHCCH_3$

from part (b)   **3-methyl-2-butanone**

(f)

$(CH_3)_2CHCH_2OH$ $\xrightarrow{\text{CrO}_3(\text{pyridine})_2}$ $(CH_3)_2CHCH{=}O$ $\xrightarrow[\text{(Wittig reaction)}]{{}^-CH_2{-}{}^+PPh_3}$ $(CH_3)_2CHCH{=}CH_2$

from part (b)   **3-methyl-1-butene**

20.49 Use a method that does not involve loss of oxygen from benzoic acid. Thus, acid-catalyzed esterification should not be used, because, as Eq. 20.34c on text p. 1104 shows, this method results in cleavage of the bond between the carbonyl carbon and the carboxylate oxygen. Esterification with either diazomethane or methyl iodide and K_2CO_3 would be the preferred method because, as the discussion in Sec. 20.8B, text pp. 1105–1106, shows, these methods do not involve loss of oxygen. ($O^* = {}^{18}O$.)

preferred method: Ph—C(=O*)—OH* $\xrightarrow[\text{CH}_3\text{I/acetone/K}_2\text{CO}_3]{\text{CH}_2\text{N}_2, \text{ether or}}$ Ph—C(=O*)—OCH₃ (all isotopic oxygen is retained)

inferior method: Ph—C(=O*)—OH* $\xrightarrow[\text{H}_2\text{SO}_4 \text{ (catalyst)}]{\text{CH}_3\text{OH (solvent)}}$ Ph—C(=O*)—OCH₃ + H₂O* (some isotopic oxygen is lost)

20.50 (a) The substitutive name of valproic acid is 2-propylpentanoic acid.
      (b) The common name of valproic acid is α-propylvaleric acid.

(c)   A number of syntheses are possible. Here are two:

Synthesis #1:

Synthesis #2

Solutions to Integrated Problems

20.51 (a)   First calculate the number of moles of 0.1 M NaOH required for the neutralization:

$$\text{mol NaOH} = (0.100 \text{ mol NaOH L}^{-1})(8.61 \times 10^{-3} \text{ L}) = 8.61 \times 10^{-4}$$

The same molar amount of the carboxylic acid must be present because one mole of NaOH is required to neutralize one mole of a monocarboxylic acid. The molecular mass is

$$\text{molecular mass of the acid} = (100 \times 10^{-3} \text{ g}) \div (8.61 \times 10^{-4} \text{ mol}) = 116 \text{ g mol}^{-1}$$

(b)   First, calculate the number of millimoles of succinic acid (see Table 20.1, text p. 1006) in 100 mg of succinic acid:

$$\text{mmol succinic acid} = \frac{100 \text{ mg succinic acid}}{118.10 \text{ mg mmol}^{-1} \text{ succinic acid}} = 0.847 \text{ mmol}$$

Using the facts that sodium hydroxide reacts with the carboxylic acid groups and that succinic acid has two dicarboxylic acid groups per molecule, calculate the millimoles of carboxylic acid groups present in the sample:

$$\text{mmol carboxylic acid groups} = \frac{2.00 \text{ mmol carboxylic acid groups}}{\text{mmol succinic acid}} \times 0.847 \text{ mmol succinic acid} = 1.69 \text{ mmol}$$

Finally, calculate the volume 0.1 M NaOH required for the neutralization:

mL NaOH solution =

$$\frac{1 \text{ mL of NaOH solution}}{0.1 \text{ mmol of NaOH solution}} \times \frac{0.1 \text{ mmol of NaOH}}{0.1 \text{ mmol of carboxylic acid}} \times 1.69 \text{ mmol of carboxylic acid} = 16.9 \text{ mL}$$

20.52 Carry out a calculation of pH using the dissociation-constant expression for each acid. (See Eq. 3.51, text p. 140.) Let the concentration of the undissociated acid be $0.1 - x$; the concentrations of both H_3O^+ and $CH_3CO_2^-$ are then x.

$$K_a = 10^{-4.76} = 1.74 \times 10^{-5} = \frac{[H_3O^+][CH_3CO_2^-]}{[CH_3CO_2H]} = \frac{x^2}{0.1 - x}$$

Because the dissociation constant of acetic acid is small, use the approximation that x is small compared with the amount of acetic acid present.

$$(x^2)/0.1 = 1.74 \times 10^{-5}$$
$$x = [H_3O^+] = 1.32 \times 10^{-3}$$
$$-\log x = \text{pH} = 2.88$$

Therefore, the pH is less than 3, so acetic acid therefore turns litmus paper red.

An analogous calculation reveals that the pH of a 0.1 M solution of phenol ($pK_a = 9.95$) is 5.48. Because this pH is greater than 3, a 0.1 M solution of phenol ("carbolic acid") does not turn litmus paper red.

20.53 (a) Consider the following equilibrium, in which AH is an acid, and S is the solvent:

$$\text{AH} + \text{S} \rightleftharpoons \overset{+}{\text{S}}\text{H} + \text{A}^-$$

Suppose that AH is more acidic than $^+$SH. Then this equilibrium lies to the right, and $^+$SH, the conjugate acid of the solvent, is the major species present. In other words, the strongest acid that can exist is the conjugate acid of the solvent.

(b) Use the result in part (a). When acetic acid is the solvent, the most acidic species that can exist is the *conjugate acid* of acetic acid (that is, protonated acetic acid; see structure in Eq. 20.18, text p. 1097, with R—— = H$_3$C——), which has a pK_a of about –6 (Eq. 20.18). Table 3.2 on text p. 142 gives the pK_a of HBr as –8 to –9.5. Consequently, the HBr–acetic acid system has an effective pK_a of about –6. In aqueous solution, HBr is dissociated to give H_3O^+, the conjugate acid of water, which has a pK_a of 0; therefore the effective pK_a of aqueous HBr is 0. Therefore, HBr in acetic acid is far more acidic than HBr in water because of the greater acidity of the conjugate acid of acetic acid, the solvent.

20.54 The essence of this solution is to determine the ionization state of penicillin-G at the different pH values. The principles involved are discussed in Sec. 3.6E and described by Eq. 3.67a–b on text p. 148. Because penicillin G is a carboxylic acid, its pK_a should be in the 3–5 range. Because the pH of blood, 7.4, is considerably higher than the pK_a of the drug, penicillin-G is ionized in blood. Because the pH of stomach acid is lower than the pK_a of penicillin-G, the penicillin is largely un-ionized in stomach acid. Because carboxylate ions are generally more soluble in aqueous solution than un-ionized carboxylic acids, penicillin-G is more soluble in blood than it is in stomach acid.

20.55 Dissolve the mixture in a suitable solvent; because the company specializes in chlorinated organic compounds, methylene chloride might be readily available for the purpose. Extract with 5% NaHCO$_3$ solution. *p*-Chloro-benzoic acid is extracted as its conjugate-base anion into the aqueous layer. Acidification of the aqueous layer will give *p*-chlorobenzoic acid itself as a precipitate, which can be filtered. Then extract the methylene chloride solution with 5% aqueous NaOH solution. *p*-Chlorophenol will be extracted into the aqueous layer as its conjugate-base, *p*-chlorophenolate anion; acidification will afford *p*-chlorophenol. The basis for this separation is Eq. 3.67a–b on text p. 148. When the pK_a of an acid is much lower than the pH of the solution, the acid is converted into its conjugate base; that is, the ratio of [RCO$_2^-$] to [RCO$_2$H] is large. The pK_a of *p*-chlorobenzoic acid, which is about 4 (see Table 20.2 on text p. 1095 of the text), is much lower than the pH of aqueous sodium bicarbonate, which is about 8.5. Therefore, *p*-chlorobenzoic acid is more than 99.99% ionized by the sodium bicarbonate solution. The pK_a of *p*-chlorophenol, 9.4, is higher than the pH of aqueous sodium bicarbonate, and therefore this phenol is not appreciably ionized. However, the pK_a of the phenol is much lower than the pH of 0.1 *M* aqueous NaOH solution, which is about 13; hence, the phenol is ionized by, and thus extracted into, this solution.

Once the acid and the phenol are removed, evaporation of the methylene chloride gives a mixture of 4-chlorocyclohexanol and chlorocyclohexane. The boiling point of the alcohol is much higher because of hydrogen bonding; consequently, these two remaining substances can be separated by fractional distillation.

20.56 (a) The first ionization of squaric acid has the lower pK_a, just as the first ionization of a dicarboxylic acid has a lower pK_a than the second. (For the reason, see the solution to Problem 20.39 in this manual.)

(b) One reason for the greater acidity of squaric acid is the electron-withdrawing polar effect of the two carbonyl groups, which stabilizes the anions. Another reason for the greater acidity of squaric acid is the resonance stabilization of the conjugate-base anions that results from interaction of the nonbonding electron pairs on each of the oxygens with a carbonyl group.

In the conjugate base of an enol, in contrast, the stabilizing polar effect of a carbonyl oxygen is not present, and the only resonance stabilization results from interaction of the electron pair with the double bond.

conjugate base of an enol

(c) The reactions of squaric acid with a number of reagents are analogous to the corresponding reactions of carboxylic acids. Thus, reaction with SOCl$_2$ results in the formation of the "di-acid chloride" *A*; and reaction with ethanol and an acid catalyst results in formation of the "di-ester" *B*.

20.57 (a) For this problem, let the *p*-dimethylamino group be abbreviated as shown at right. The color of crystal violet is due to the overlap of the empty 2p orbital of the positively charged carbon with the π orbitals of the benzene rings. The bleaching is due to the reaction of the carbocation with hydroxide to form the alcohol. In the alcohol, there is no longer an empty 2p orbital; therefore, the conjugation decreases and the color disappears.

$$Ar_3C^+ \quad \overset{\frown}{^-}OH \longrightarrow Ar_3C{-}OH$$

crystal violet   (colorless)
(colored)

(b)   SDS, a typical detergent, forms micelles. Despite its charge, crystal violet is incorporated into the interior of the SDS micelles because the "greasy" benzene rings interact favorably with the "greasy" hydrocarbon tails of SDS. (Think of this as a solubility phenomenon; benzene is more soluble in hydrocarbons than it is in water.) In the interior of a micelle, crystal violet is no longer accessible to hydroxide ion. Because hydroxide is strongly solvated by water, hydroxide ion cannot easily enter the interior of the micelles, and therefore it cannot react with the entrapped crystal violet. The very slow reaction that does occur is probably due to hydroxide reacting with the very small amount of crystal violet in solution. As this crystal violet reacts, more is pulled out of the micelle (by Le Chatelier's principle); however, the loss of crystal violet from the micelle interior is slow.

20.58   (a)   (b)   (c)

meso-α,β-dimethylsuccinic anhydride

20.59

HO₂C(CH₂)₄CO₂H

**hexanedioic acid
(adipic acid)
(1)**

HO₂CCH₂CH₂CHCH₃
                 |
                 CO₂H

**2-methylpentanedioic acid
(α-methylglutaric acid)
(2)**

HO₂CCH₂CHCH₂CH₃
              |
              CO₂H

**2-ethylbutanedioic acid
(α-ethylsuccinic acid)
(3)**

HO₂CCHCH₂CH₂CH₃
        |
        CO₂H

**2-propylpropanedioic acid
(α-propylmalonic acid,
propylmalonic acid)
(4)**

          CH₃
          |
HO₂CCCO₂H
          |
          CH₂CH₃

**2-ethyl-2-methylpropanedioic acid
(α-ethyl-α-methylmalonic acid,
ethylmethylmalonic acid)
(5)**

HO₂CCHCH(CH₃)₂
        |
        CO₂H

**2-isopropylpropanedioic acid
(α-isopropylmalonic acid,
isopropylmalonic acid)
(6)**

HO₂CCH₂CHCH₂CO₂H
              |
              CH₃

**3-methylpentanedioic acid
(β-methylglutaric acid)
(7)**

HO₂CCHCHCO₂H
       | |
      H₃C CH₃

**2,3-dimethylbutanedioic acid
(α,β-dimethylsuccinic acid)
(±) and *meso*
(8)   (9)**

          CH₃
          |
HO₂CCCH₂CO₂H
          |
          CH₃

**2,2-dimethylbutanedioic acid
(α,α-dimethylsuccinic acid)
(10)**

Compounds (2), (3), and (8) are chiral. Compounds (2), (3), (7), (8), (9), and (10) would readily form cyclic anhydrides on heating, because those anhydrides contain five- or six-membered rings. Compounds (4), (5), and (6) would readily decarboxylate on heating because they are malonic acid derivatives.

20.60   (a)   The two different compounds are the diastereomers (at right) that result respectively from loss of each of the two carboxy groups from compound *A*.

(b)   The question is whether loss of the different carboxy groups gives rise to the same compound or to different compounds. Draw the two possible structures and determine whether they are different or identical. In fact, the two are identical; therefore, only one product is formed when compound *B* decarboxylates.

20.61   (a)   The reaction of PLP with *meso*-1,6-diaminopimelate:

In the *meso*-diamine, the asymmetric carbons have opposite configurations. The PLP must react specifically at the nitrogen on the carbon with the *R* configuration (as shown) to give lysine ultimately that has the *S* stereochemistry shown in the problem.

(b)   The positively charged nitrogen within the ring acts as the electron sink necessary for decarboxylation.

(c)   The driving force for this reaction is the regeneration of aromaticity. You can check your answer by showing the product of the reaction of lysine with PLP, which also produces imine *Y*, as stated in the problem.

20.62 (a) The formation of phosphate suggests that it serves as a leaving group and that the reaction proceeds via solvolysis, as described in Sec. 9.6, text p. 487.

isopentenyl pyrophosphate

Because this is an E1-like mechanism (with loss of CO₂ instead of loss of a β-proton), the stability of the carbocation intermediate has a direct effect on the rate of the reaction. The methyl group stabilizes the carbocation by hyperconjugation. The fluoromethyl group withdraws electron density because of the very electronegative fluorine atom, which destabilizes the electron-deficient carbon. Therefore, the carbocation that results is much higher in energy and the reaction rate is slower.

(b) The positively charged carbon of the carbocation is electron-deficient and acts as the electron sink necessary for decarboxylation.

20.63 Always remember that correct syntheses other than the one given may be possible. Be sure to ask your professor or teaching assistant if your synthesis differs from the one given and you think it could be correct.

(a)

(b)

$H_2C{=}CHCH_2OH$ $\xrightarrow{\text{H}_2\text{, cat.}}$ $CH_3CH_2CH_2OH$ $\xrightarrow[\text{2) H}_3\text{O}^+]{\text{1) KMnO}_4,\ ^-\text{OH}}$ $CH_3CH_2CO_2H$

allyl alcohol

$\xrightarrow{\text{Ph}_3\overset{+}{\text{P}}\text{Br Br}^-}$ $H_2C{=}CHCH_2Br$ $\longrightarrow$ $\xleftarrow{\text{K}_2\text{CO}_3\text{, acetone}}$ $CH_3CH_3\overset{\displaystyle O}{\overset{\|}{C}}OCH_2CH{=}CH_2$

allyl bromide **allyl propanoate**

We use triphenphosphine dibromide (Sec. 11.4D, text p. 582) to prepare allyl bromide from allyl alcohol because concentrated HBr might also add to the double bond.

(c)

$CH_3(CH_2)_3CO_2H$ $\xrightarrow[\text{2) H}_3\text{O}^+]{\text{1) LiAlH}_4}$ $CH_3(CH_2)_3CH_2OH$ $\xrightarrow[\text{H}_2\text{SO}_4]{\text{conc. HBr}}$ $CH_3(CH_2)_3CH_2Br$ $\xrightarrow[\text{ether}]{\text{Mg}}$

pentanoic acid

$CH_3(CH_2)_3CH_2MgBr$ $\xrightarrow[\text{2) H}_3\text{O}^+]{\text{1) (CH}_3\text{)}_2\text{C}{=}\text{O}}$ $CH_3(CH_2)_3CH_2\overset{\displaystyle OH}{\underset{\displaystyle }{C}}(CH_3)_2$ $\xrightarrow{\text{HBr}}$

A

$CH_3(CH_2)_3CH_2\overset{\displaystyle Br}{\underset{\displaystyle |}{C}}(CH_3)_2$ $\xrightarrow[\text{ether}]{\text{Mg}}$ $CH_3(CH_2)_3CH_2\overset{\displaystyle MgBr}{\underset{\displaystyle |}{C}}(CH_3)_2$ $\xrightarrow{\text{H}_3\text{O}^+}$ $CH_3(CH_2)_3CH_2CH(CH_3)_2$

2-methylheptane

An alternative is to convert alcohol A into an alkene (or mixture of alkenes) by acid-catalyzed dehydration, and then catalytically hydrogenate the alkenes to the target alkane.

(d)

CH₃ [benzene ring with CH₃ substituent] $\xrightarrow[\text{2) H}_3\text{O}^+]{\text{1) KMnO}_4,\ ^-\text{OH, heat}}$ CO₂H [benzene ring with CO₂H] $\xrightarrow{\text{HNO}_3\text{, H}_2\text{SO}_4}$ CO₂H [benzene ring with CO₂H and NO₂] NO₂

toluene ***m*-nitrobenzoic acid**

(e)

$PhCO_2H$ $\xrightarrow[\text{2) H}_3\text{O}^+]{\text{1) LiAlH}_4}$ $PhCH_2OH$ $\xrightarrow{\text{conc. HBr}}$ $PhCH_2Br$ $\xrightarrow[\text{ether}]{\text{Mg}}$ $PhCH_2MgBr$ $\xrightarrow[\text{2) H}_2\text{SO}_4,\ -\text{H}_2\text{O}]{\text{1) O}{=}\text{CHCH}_2\text{Ph}}$

$PhCH{=}CHCHPh$ $\xrightarrow{\text{H}_2\text{, cat.}}$ $PhCH_2CH_2CH_2Ph$

(3-phenylpropyl)benzene

 Although the rigorously correct substitutive name for this compound is (3-phenylpropyl)benzene, most chemists would call the product in part (e) 1,3-diphenylpropane.

(f) First, reduce the ketone to an alcohol to form a γ-hydroxy acid. γ-Hydroxy acids spontaneously cyclize to form cyclic esters, or lactones. See Sec. 20.8A in the text for a discussion of the reaction mechanism of esterification from alcohols and carboxylic acids. Although the general reaction is catalyzed with an acid catalyst, the cyclization in the reaction below, which results in the formation of a five-membered ring, requires no catalyst because it proceeds at an appreciable rate on its own. [This is a manifestation of the *proximity effect* (Sec. 12.8)].

4-oxopentanoic acid → **γ-valerolactone** + H_2O

(g)

1) O_3
2) H_2O_2, H_2O

***cis*-1,3-cyclopentanedicarboxylic acid**

(h)

$CH_3C(CH_2)_3Br$
5-bromo-2-pentanone

$\xrightarrow{\begin{array}{c}HOCH_2CH_2OH\\ HCl \text{ (catalyst)}\end{array}}$

$\xrightarrow[\text{ether}]{Mg}$

$\xrightarrow{\begin{array}{c}1) CO_2\\ 2) H_3O^+\end{array}}$

$CH_3C(CH_2)_3CO_2H$
5-oxohexanoic acid

(i)

$Ph-C\equiv C-H \xrightarrow{C_2H_5MgBr} Ph-C\equiv CMgBr + CH_3CH_3$

$\xrightarrow{\begin{array}{c}1) CO_2\\ 2) H_3O^+\end{array}}$

$Ph-C\equiv C-CO_2H \xrightarrow[\text{THF}]{\begin{array}{c}H_2\\ \text{Lindlar catalyst}\end{array}}$

***cis*-cinnamic acid**

20.64 Convert each enantiomer of 4-methylhexanoic acid into 3-methylhexane by any process that does not involve breaking a bond to the asymmetric carbon. Because each enantiomer of 4-methylhexanoic acid has known absolute configuration, the configuration of the corresponding enantiomer of 3-methylhexane will be determined. Suppose, for example, that (*S*)-4-methylhexanoic acid gives (–)-3-methylhexane. Then the two molecules should have corresponding configurations, and the configuration of (–)-3-methylhexane would then be *S*, as shown at right. The following sequence of reactions would work for such a conversion:

(S)-4-methylhexanoic acid → **(S)-3-methylhexane**

$CH_3CH_2CHCH_2CH_2CO_2H \xrightarrow{\begin{array}{c}1) LiAlH_4\\ 2) H_3O^+\end{array}} CH_3CH_2CHCH_2CH_2CH_2OH \xrightarrow{\begin{array}{c}\text{conc. HBr}\\ H_2SO_4\end{array}}$
with CH_3 substituent

$CH_3CH_2CHCH_2CH_2CH_2Br \xrightarrow[\text{ether}]{Mg} CH_3CH_2CHCH_2CH_2CH_2MgBr \xrightarrow{H_3O^+} CH_3CH_2CHCH_2CH_2CH_3$
with CH_3 substituent

20.65   To confirm the labeling pattern, oxidize phenylacetic acid vigorously with $KMnO_4$ or other oxidizing agent that will convert it into benzoic acid. ($^*C = {}^{14}C$.)

If all the radiolabel of phenylacetic acid is at the carbonyl carbon, the product benzoic acid should be completely devoid of radioactivity. However, if there is labeling elsewhere, the fraction of radioactivity located in carbons other than the carbonyl carbon is equal to the amount of radioactivity in the product benzoic acid divided by the total radioactivity in the starting phenylacetic acid.

20.66   (a)   The reagent involved is essentially a diazomethane molecule in which the hydrogens of diazomethane have been replaced by phenyl groups. Just as diazomethane forms methyl esters, this reagent forms a diphenylmethyl ester.

diphenylmethyl 4-methylbenzoate

(b)   The KOH converts benzoic acid into its conjugate-base benzoate anion, which is alkylated by benzyl chloride to give benzyl benzoate.

potassium benzoate   **benzyl benzoate**

(c)   Ethylene glycol is a diol and the acid, terephthalic acid (1,4-benzenedicarboxylic acid), is a dicarboxylic acid. Each end of both the diacid and the diol can be esterified; the result is a polymeric ester, poly(ethylene terephthalate). You may be wearing some of this compound, which is known in the industrial world as *polyester*.

poly(ethylene terephthalate) (polyester)

(d)   This is an oxymercuration-reduction reaction in which acetic acid rather than water serves as the nucleophile that opens the mercurinium ion. (See the solution to Problem 5.38(c) in this manual.)

1-methylcyclohexyl acetate

(e)   The conjugate-base carboxylate of propionic acid acts as the nucleophile in a ring-opening reaction of the epoxide. Because the conditions are basic, reaction occurs at the less branched carbon of the epoxide. Note that KOH reacts much more rapidly with the acidic hydrogen of the acid than it does with the epoxide; because only one equivalent of KOH is present, reaction of hydroxide ion with the epoxide is not a competing reaction.

2-hydroxy-1-propyl propanoate
(α-hydroxypropyl propionate)

(f)   This is an intramolecular variation of the esterification shown in Eq. 20.39, text p. 1106. In this case, potassium carbonate converts the carboxylic acid into its conjugate-base potassium carboxylate, which is then intramolecularly alkylated to form the cyclic ester (lactone).

(g)   Potassium carbonate converts both the carboxy group and the phenol —OH groups into their respective conjugate-base anions; these are all alkylated by dimethyl sulfate.

methyl 3,5-dimethoxy-4-methylbenzoate

(h)   Chlorosulfonation of chlorobenzene gives electrophilic aromatic substitution at the para position. (See Eq. 20.46a–b on text pp. 1108.)

p-chlorobenzenesulfonyl chloride

20.67   The proton and the π bond move back and forth between the two oxygens so rapidly that the two oxygens retain their identities for only a tiny fraction of a second. The mechanism varies with the situation. In a solvent (for example, water) in which the acid readily dissociates, even to a small extent, the two oxygens in the conjugate-base carboxylate ion are equivalent and are protonated with essentially equal frequency; these reactions are very fast. (In the equations that follow, one of the oxygens is marked with an asterisk (*) so that its fate can be traced throughout the reaction.)

In solvents in which the carboxylic acid molecule cannot readily dissociate, protons can jump back and forth within a carboxylic acid hydrogen-bonded dimer (Display 20.8, text p. 1092):

20.68 (a)   The hint refers to the fact that organolithium reagents, like Grignard reagents, are strong bases and react instantaneously with carboxylic acids to give their conjugate-base lithium carboxylate salts:

A second equivalent of organolithium reagent undergoes a carbonyl addition with the carboxylate salt to give compound *A*, which is the conjugate-base di-anion of a ketone hydrate. Since compound *A* does not contain a good leaving group, it is stable until acid is added. Protonation forms the hydrate, which then decomposes to the ketone by the reverse of the mechanism shown for hydrate formation in Eqs. 19.36a–b on text p. 1044.

(b)   By analogy to the reaction discussed in part (a), the organic products are the ethyl ketone and ethane.

20.69 (a)   Orthoesters have the same relationship to esters that acetals have to ketones, and the hydrolysis mechanisms of both orthoesters and acetals are virtually identical.

(continued next page)

The mechanism schemes and reaction structures are shown as chemical diagrams.

(b) The mechanism involves neighboring-group participation of the carboxylic acid group, which reacts with the protonated ketone. (Recall that many intramolecular reactions that form small rings are faster than related intermolecular reactions; see Sec. 12.8A–B, text p. 602.)

(c) Sulfuric acid protonates the alkene to give a carbocation, which reacts with carbon monoxide to give an acylium ion (Eq 16.31, Sec. 16.4F, text p. 887). This ion, in turn, reacts with water to give, after appropriate proton transfers, the carboxylic acid.

an acylium ion

(d)   The pattern is much like that in the solution to part (c), except that acetic acid rather than carbon monoxide reacts with the carbocation intermediate.

(e)   Acid-promoted opening of the epoxide gives a tertiary and benzylic carbocation, which decarboxylates to the enol. The enol then forms the aldehyde by a mechanism like that shown in Eqs. 5.28a–b, Sec. 5.5, on text pp. 249–250.

The other C—O bond of the epoxide does not rupture instead because (1) the carbocation that would result is secondary, and (2) carbocations α to carbonyl groups are very unstable. (The latter point is considered in Sec. 22.3D.)

(f)   The carbon of diazomethane reacts as a nucleophile at the carbonyl carbon to form tetrahedral adduct *T*. This intermediate has two fates. First, the loss of dinitrogen and a rearrangement can occur in a concerted manner to give the ketone. The rearrangement involves migration of a ring bond, so a seven-membered ring results.

T

Alternatively, an internal substitution reaction occurs in which the anionic oxygen of *T* reacts as an intramolecular nucleophile with dinitrogen as a leaving group to give the minor product.

T

(g)   The alkene reacts with iodine to form an iodonium ion, formed in the same way as the bromonium ion formation discussed in Sec. 5.2A, text p. 240. The OH group of the carboxylic acid opens the iodonium ion in an intramolecular nucleophilic substitution, analogous to halohydrin formation discussed in Sec. 5.2B, Eq. 5.12a–b, text p. 242. The result is an iodine-containing lactone (cyclic ester). (This process is nicknamed *iodolactonization*.)

20.70   (a)   The even mass of the $m/z = 74$ ion shows that the ion is an odd-electron ion. McLafferty rearrangement is a common mechanism for formation of such ions in carbonyl compounds. (See Sec. 19.4E, Eq. 19.22, text p. 1037.)

molecular ion $m/z = 74$

(b)   The molecular mass of benzoic acid is 122; therefore, the $m/z = 105$ peak in its mass spectrum represents a mass loss of 17 units, which corresponds to loss of an OH group. This loss can occur by an α-cleavage mechanism.

molecular ion

The $m/z = 77$ peak corresponds to a phenyl cation, which can be lost by inductive cleavage at the other side of the carboxy group.

molecular ion

You may recall that aryl cations are very unstable. (See Fig. 18.2, text p. 965.) The large electron energies involved in EI mass spectroscopy enable the formation of such intermediates that would ordinarily not form in solution.

20.71 (a) The 2300–3200 cm^{-1} and 1710 cm^{-1} absorptions clearly indicate a carboxylic acid, and this is confirmed by the δ 11.7 absorption in the NMR. The NMR shows five aromatic protons; so, there is a monosubstituted benzene ring. The two mutually split triplets indicate a partial structure —CH$_2$CH$_2$—. We've accounted for all of the carbons and all of the unsaturation. However, what about the third oxygen? The chemical shift of the triplet at δ 4.1 shows that these protons must be adjacent to an oxygen. Therefore, our part structure should be modified to —O—CH$_2$CH$_2$—. All that it left is to put the benzene ring at one end of the chain and the carboxy group at the other. This can be done in two ways:

Only structure *A* is a stable compound. (Why can't *B* exist? See Eq. 20.63a on text p. 1115.) Structure *A* is the correct answer.

(b) The IR spectrum indicates a carboxylic acid; the carbonyl absorption indicates that the carbonyl could be conjugated, and this is consistent with the UV absorption. The IR also shows *trans*-alkene absorption at 981 cm^{-1}. We use the mass spectrum to see how much mass remains after subtracting the masses of the structural features we just deduced: 114 – 45 (carboxy group) – 13 (alkene CH) – 13 (alkene CH) = 43. A mass of 43 units is characteristic of a propyl group or an isopropyl group. A propyl group (CH$_3$CH$_2$CH$_2$—) would require a three-proton methyl triplet; this is observed at about δ 1. An isopropyl group would require a six-proton methyl doublet; this is not observed. The resonances at δ 1.5 and δ 2.1 could be the remaining protons of a propyl group. All the data are consistent with the structure at right. (Try to account for the remaining splitting that was not discussed above.)

(E)-2-hexenoic acid

20.72 A hydrocarbon cannot have an odd mass; hence, the peak in the mass spectrum at $m/z = 67$ cannot be the molecular ion. Given that compound *B* is a dicarboxylic acid, compound *A* must contain a double bond within a ring. Calculate the number of moles of carboxy groups titrated as follows:

$$\text{moles of carboxy groups} = (13.7 \text{ mL})(0.100 \text{ mol L}^{-1})(1.00 \times 10^{-3} \text{ L}^{-1}) = 1.37 \times 10^{-3}$$

Noting that there are two carboxy groups per mole of *B*, we calculate the molecular mass of compound *B* as follows:

$$\text{molecular mass of } B = \frac{(100 \times 10^{-3} \text{ g})(2.00 \text{ moles of carboxy groups per mole of } B)}{1.37 \times 10^{-3} \text{ moles of carboxy groups}} = 146 \text{ g mol}^{-1}$$

From the following transformations, the atomic masses of the four oxygens introduced, and the molecular mass of compound *B*, the molecular mass of compound *A* can be deduced as follows:

A
molecular mass = *x*

1) OsO$_4$
2) H$_5$IO$_6$
3) Ag$_2$O

B
molecular mass = *x* + 64 = 146

$x = 146 - 64 = 82$

If the molecular mass of compound *A* is 82, then the $m/z = 67$ peak in its mass spectrum corresponds to loss of 15 mass units, which corresponds to the loss of a methyl group. Evidently, compounds *A* and *B* contain a ring with a methyl branch. This branch is positioned such that there are three allylic hydrogens (from the NMR integration data) and it is also positioned such that dicarboxylic acid *B* is chiral. Compound *A* is 3-methylcyclopentene.

A
3-methylcyclopentene

B
2-methylpentanedioic acid

20.73 The titration data suggest that compound *A* is both a carboxylic acid and a phenol, and the IR spectrum shows both —OH and carbonyl absorptions. The change in the UV spectrum at high pH is also consistent with the presence of a group that ionizes at high pH and is conjugated with a benzene ring, namely, the phenolic —OH group. The resonances near δ 7 indicate a *para*-disubstituted benzene ring. The two triplets at δ 2.4 and δ 2.7 integrates for four hydrogens; and they indicate a partial structure —CH₂CH₂— . Add the masses of the groups known to be present: the *para*-disubstituted benzene ring (76 units), the phenol —OH group (17 units), the carboxy group (45 units), and the —CH₂CH₂— group (28 units), and we have accounted for the *m/z* = 166 peak in the mass spectrum as the molecular ion. Compound *A* is 3-(4-hydroxyphenyl)propanoic acid.

3-(4-hydroxyphenyl)propanoic acid
(compound *A*)

The Chemistry of Carboxylic Acid Derivatives

CHAPTER 21 SUMMARY

■ The common carboxylic acid derivatives are esters, thioesters, amides, anhydrides, acid chlorides and nitriles. The nomenclature of carboxylic acid derivatives is discussed in Sec. 21.1.

■ Cyclic esters are called lactones; cyclic amides are called lactams.

■ Carboxylic acid derivatives are polar molecules; except for amides, they have low water solubilities and low-to-moderate boiling points. Because of their capacity for hydrogen bonding, amides, in contrast, have high boiling points (many are solids) and moderate solubilities in water.

■ The proton NMR spectra of carboxylic acid derivatives show typical δ 2–3 absorptions for the α-protons, as well as other characteristic absorptions: δ 3.5–4.5 for the O-alkyl groups of esters, δ 2.6–3 for the N-alkyl groups of amides, and δ 7.5–8.5 for the NH protons of amides. Because internal rotation about the C—N bond of amides is relatively slow, E and Z isomers of amides can sometimes be observed separately in NMR.

■ The carbonyl C=O stretching absorptions (and the C≡N stretching absorption of nitriles) are the most important absorptions in the infrared spectra of carboxylic acid derivatives (Table 21.3).

■ In ^{13}C NMR spectra, the chemical shifts of carbonyl carbons are in the δ 170 region, and those of nitrile carbons are in the δ 115–120 range.

■ Carboxylic acid derivatives, like carboxylic acids themselves, are weak bases that can be protonated by strong acids on the carbonyl oxygen or the nitrile nitrogen.

■ The most characteristic reaction of carboxylic acid derivatives is nucleophilic acyl substitution. Hydro-lysis reactions and the reactions of acid chlorides, anhydrides, and esters with nucleophiles are examples of this type of reactivity. Nucleophilic acyl substitution typically involves addition of a nucleophile at the carbonyl carbon to form a tetrahedral addition intermediate, which then loses a leaving group to form the substitution product.

■ The relative reactivities of carbonyl compounds in nucleophilic addition are determined by the relative free energies of the carbonyl compounds and the tetrahedral addition intermediates, which are used as models for the transition states for addition or substitution. For carboxylic acid derivatives, resonance stabilization of the carbonyl compound and polar stabilization of the transition state are important factors in determining relative reactivity.

■ The relative reactivities of carbonyl compounds and nitriles with nucleophiles increase in the order nitriles < amides < esters << ketones < aldehydes < acid chlorides. The reactivity of anhydrides is considerably greater than that of esters but less than that of acid chlorides.

■ A number of important reactions of carboxylic acid derivatives are discussed in the chapter. (See the Reaction Summary for details.)

• The conversion of acid chlorides into esters, amides, and anhydrides, and the conversion of esters into amides and into other esters, by nucleophilic carbonyl substitution reactions.

• The conversion of acid chlorides into aldehydes by hydride reduction, and into ketones with lithium dialkylcuprate reagents.

- The conversion of esters into alcohols with LiAlH$_4$ and with Grignard or organolithium reagents,.

- The reduction of amides and nitriles to amines.

■ Nitriles react by addition, much like aldehydes and ketones. In some reactions, the product of addition undergoes a second addition (as in reduction to primary amines); and in other cases it undergoes a substitution reaction (as in hydrolysis).

■ The preparation of carboxylic acid derivatives is summarized in Sec. 21.11. In most cases, carboxylic

acid derivatives are prepared from carboxylic acids. An exception is the preparation of nitriles by the S$_N$2 rection of cyanide ion with alkyl halides.

■ Polyesters, such as poly(ethylene terephthalate), and polyamides, such as nylon, are important commercial examples of condensative-addition polymers that are also carboxylic acid derivatives. Proteins are naturally occurring polyamides. In nature, waxes, fats, and phospholipids are important natural examples of esters.

STUDY GUIDE LINKS

21.1 Solving Structure Problems Involving Nitrogen-Containing Compounds

Amides and nitriles are the first nitrogen-containing functional groups that are considered in the text. When you solve structure problems you should remember certain things about compounds that contain nitrogen. First, a compound that contains an odd number of nitrogen atoms has an *odd molecular mass*. If you studied mass spectrometry (Chapter 13), you will appreciate that this means that the parent ion in the mass spectrum of such a compound will occur at an odd *m/z* value. Moreover, even-electron ions have an even mass and odd-electron fragment ions have an odd mass. (This is the so-called *nitrogen rule*.) Second, the special formula given below (see also Eq. 4.18 in Sec. 4.3, on text p. 193) is used to calculate the unsaturation number *U* of a compound containing nitrogen. If *C* is the number of carbons, *H* is the number of hydrogens, *X* is the number of halogens, and *N* is the number of nitrogens, then *U* is given by

$$U = \frac{(2C+2)+N-(H+X)}{2}$$

The implication of this formula is that every nitrogen increases the number of hydrogens in a fully saturated molecule by one relative to the number of hydrogens in the corresponding hydrocarbon.

21.2 Basicity of Nitriles

Let's place the basicity of nitriles within the context of other basicities that you've studied. Consider the following series:

increasing s character in lone-pair orbital ⟶

conjugate bases:

Notice that, as the amount of s character in the lone-pair orbital of the conjugate base increases, the pK_a of the conjugate acid decreases. Although the pK_a values are much different, this is exactly the same trend observed in the acidities of hydrocarbons (Sec. 3.7C, Display 3.69, text p. 153):

increasing *s* character in lone-pair orbital ⟶

In other words, just as an acetylenic anion, with its lone pair in an sp orbital, is the least basic hydrocarbon anion, a nitrile, also with a lone pair in an sp orbital, is the least basic among the analogous series of nitrogen compounds.

21.3   Mechanism of Ester Hydrolysis

Some students try to combine the features of the acid-catalyzed hydrolysis and saponification reactions by using strong acid to protonate the carbonyl oxygen and hydroxide to act as the nucleophile.

Implicit in such a scheme is the incorrect notion that when the base is added to the acidic solution, the carbonyl oxygen of the ester would remain protonated and that all hydronium ions would not dissociate. However, both protonated esters and hydronium ions are strong acids. Remember: *the deprotonation of strong acids by strong bases is instantaneous.* (See Study Guide Link 20.1 on p. 657 of this manual.) The very first event that would occur on addition of base to the solution would be deprotonation of the protonated carbonyl oxygen and rapid neutralization of the acid (H_3O^+) in solution. The nucleophilic reaction of ⁻OH on the ester carbonyl carbon is a much slower reaction.

All of the hydronium ions in solution would react with the hydroxide ion before hydroxide could react to a significant extent at the carbonyl carbon.

Remember: *in mechanisms, an acid and its conjugate base act in tandem.* If H_3O^+ is the acid, H_2O should be used as the base (and the nucleophile). If ^-OH is the base (or nucleophile), H_2O should be used as the acid.

21.4 Another Look at the Friedel–Crafts Reaction

The Friedel–Crafts acylation reaction is first classified in the text as an *electrophilic aromatic substitution* reaction (Sec. 16.4F on text p. 887) because the reaction is first considered as a reaction of benzene derivatives. However, in terms of what happens to the acid chloride, the reaction can also be viewed as a nucleophilic acyl substitution reaction.

If the Friedel–Crafts acylation is considered to be a nucleophilic acyl substitution reaction, the nucleophilic electrons are the π electrons of the aromatic ring. Because benzene and its derivatives are very weak nucleophiles, a strong Lewis acid such as $AlCl_3$ is required to activate the acid chloride. Review again the mechanism of Friedel–Crafts acylation (Eq. 16.31 and 16.32 on text pp. 887-888), paying particular attention to what happens to the acid chloride.

21.5 Esters and Nucleophiles

Notice that most esters of *carboxylic acids* react with nucleophiles at the carbonyl carbon—that is, they undergo nucleophilic acyl substitution reactions. The result is that the bond between the carbonyl group and the ester oxygen—the *acyl–oxygen bond*—is cleaved. (Further Exploration 21.2 discusses one exception to this generalization.)

Esters of *sulfonic acids* react in a fundamentally different manner. These reactions are presented in Sec. 11.4A (text pp. 578–580), where the focus is on the sulfonate ester groups acting essentially like halide leaving groups. As the discussion in the text shows, most sulfonate esters react with nucleophiles not by the reaction of the nucleophile at the sulfur, but by S_N2 reaction of the nucleophile on the *alkyl group* (if the alkyl group is primary or secondary); in such a reaction, the bond between the alkyl group and the ester oxygen is cleaved. (See Eq. 11.34 on text p. 578.)

You may have noticed that we are returning with increasing frequency to earlier parts of the text and looking at reactions you've studied earlier in somewhat different contexts. If you willingly participate in this type of exercise, you'll gain a lot of insight about organic chemistry. In particular, you'll see how the parts begin to "fit together," that is, you'll begin to see the "forest" and not just the individual "trees."

FURTHER EXPLORATIONS

21.1   NMR Evidence for Internal Rotation in Amides

Internal rotation about the carbonyl-nitrogen bond of amides can be conveniently studied by NMR using the principles discussed in Sec. 14.8 on p. 772 of the text. Consider, for example, internal rotation in *N,N*-dimethylacetamide:

these methyl groups are diastereotopic and therefore chemically nonequivalent

The *N*-methyl groups in this compound are diastereotopic; one is cis to the carbonyl oxygen, and the other is trans. At low temperature, rotation about the C—N bond is slow enough that these methyl groups are observed as separate singlets in the NMR spectrum. However, when the temperature is raised, these signals broaden and, at about 60 °C, they coalesce into one singlet at a chemical shift that is the average of the two individual methyl chemical shifts. This is illustrated in **Fig. SG21.1** on the following page.

21.2   Cleavage of Tertiary Esters and Carbonless Carbon Paper

The hydrolysis of tertiary esters has some noteworthy differences from the hydrolysis of other esters discussed in Sec. 21.7A. First, the saponification of tertiary esters is considerably slower than that of primary and secondary esters. For example, the saponification of *tert*-butyl acetate occurs at about 0.01 times the rate of the saponification of methyl acetate. The reason is that in the transition state for saponification, the methyl branches in the *tert*-butyl group are involved in van der Waals repulsions with the hydroxide nucleophile as it approaches the carbonyl carbon. A practical consequence of this rate difference is that a methyl or ethyl ester can generally be saponified without affecting a *tert*-butyl ester in the same molecule.

tertiary ester is not hydrolyzed

methyl ester is hydrolyzed

A second difference is that the acid-catalyzed hydrolysis of tertiary esters occurs by a mechanism that is completely different from that of primary and secondary esters. The first step in the hydrolysis mechanism is the same—protonation of the carbonyl oxygen. But there the similarity ends. The protonated ester dissociates by an S_N1 mechanism into a tertiary carbocation and the carboxylic acid.

(continued on p. 705)

Figure SG21.1 The 60 MHz NMR spectrum of *N,N*-dimethylacetamide changes with increasing temperature as the rate of rotation about the carbonyl–nitrogen bond increases. The full spectrum of this compound is shown in the lower part of the figure, and the pair of singlets near δ 2.9 (173.3 Hz) is expanded in the upper part of the figure and shown as a function of temperature. The two *N*-methyl groups are individually observable at temperatures at or below 52° (lower two spectra) because internal rotation about the carbonyl–nitrogen bond is relatively slow at these temperatures. As the temperature is raised, the two resonances broaden and coalesce into one because the internal rotation becomes too rapid for the NMR experiment to resolve the individual lines. (See Fig. 14.21 in Sec. 14.8 on text p. 773 for a related phenomenon.) The *k* values are the first-order rate constants for internal rotation; a *k* of 1.45 s⁻¹ corresponds to an approximate lifetime of 0.5 s for the individual conformations.

a protonated ester

The Lewis acid–base association of water (or any other nucleophile that might be present) occurs at the tertiary carbocation.

Notice that hydrolysis of tertiary esters by this mechanism involves breaking the *alkyl–oxygen bond.* In contrast, the mechanism of hydrolysis of primary and secondary esters involves breaking the *carbonyl–oxygen bond.* Consequently, the hydrolysis of tertiary esters is *not* a nucleophilic acyl substitution reaction. Rather, it is more like the S_N1 mechanism for substitution of a tertiary alcohol (see Eqs. 11.25a–c, Sec. 11.3, on text p. 574). Esters of primary and secondary alcohols in most cases do not cleave by an S_N1 mechanism because the carbocation intermediates that would be involved are much less stable.

"Carbonless carbon paper" (sometimes called NCR paper, for "No Carbon Required") is a very ingenious commercial application of tertiary ester cleavage as applied to lactones. (This is often used in multi-copy forms and in personal checkbooks that provide the user with carbonless check copies.) The active component in "carbonless carbon paper" is the colorless lactone *A*, which is encapsulated within microscopic particles of a phenol–formaldehyde resin (Sec. 19.17 on text p. 1076) and deposited on a paper base that also contains an acidic layer. When the particles are broken by the pressure of a pen or (in earlier times) a typewriter, the lactone is brought into contact with the acidic layer. This causes an S_N1-like opening of the lactone ring to give a relatively stable carbocation *B*. This carbocation has a blue-violet color.

A (colorless)

B (blue-violet)

It takes some time for color to develop because the reaction is not instantaneous. As you may know from experience, the color in carbonless carbon paper intensifies with time.

21.3 Reaction of Tertiary Amines with Acid Chloride

Tertiary amines react with acid chlorides to form *acylammonium salts.*

pyridine
(a tertiary amine)

an acylammonium salt

Acylammonium salts, because of their positive charge adjacent to the carbonyl carbon, are very reactive and are converted into amides by reaction with primary or secondary amines.

pyridinium chloride
(the HCl salt of pyridine)

If the acid chloride has acidic α-hydrogens, another possible reaction is the formation of a *ketene* by a β-elimination reaction:

a ketene

Ketenes also react rapidly with primary and secondary amines to yield amides.

Despite these side reactions, then, tertiary amines can be used as catalysts for amide formation with primary and secondary amines because the products of these side reactions—acylammonium salts and ketenes—themselves react with amines to form amides.

REACTION REVIEW

I.   HYDROLYSIS OF CARBOXYLIC ACID DERIVATIVES

A.   HYDROLYSIS OF ACID CHLORIDES AND ANHYDRIDES

1.   Acid chlorides and anhydrides react rapidly with water, even in the absence of acids or bases.

2.   The hydrolysis reactions of acid chlorides and anhydrides are almost never used for the preparation of carboxylic acids because these derivatives are themselves usually prepared from acids.

B.   BASIC HYDROLYSIS (SAPONIFICATION) OF ESTERS AND LACTONES

1.   Esters hydrolyze in aqueous hydroxide.

2.   This process is called *saponification*; the term saponification is sometimes used to refer to hydrolysis in base of any acid derivative.

3.   The mechanism of ester saponification involves the following steps:

   a.   A nucleophilic reaction of hydroxide anion at the carbonyl carbon to give a tetrahedral addition intermediate from which an alkoxide ion is expelled.

   b.   The alkoxide ion thus formed reacts with the carboxylic acid to give the carboxylate salt and the alcohol.

4.   The equilibrium in this reaction lies far to the right because the carboxylic acid is a much stronger acid than the liberated alcohol; saponification is effectively irreversible.

5.   Many esters can be saponified with just one equivalent of ⁻OH, although an excess of ⁻OH is often used as a matter of convenience.

6.   Saponification converts a lactone completely into the carboxylate salt of the corresponding hydroxy acid; upon acidification, the hydroxy acid forms.

a. If a hydroxy acid is allowed to stand in acidic solution, it comes to equilibrium with the corresponding lactone. This reaction is an acid-catalyzed, intramolecular esterification.

b. Lactones containing five- and six-membered rings are favored at equilibrium over their corresponding hydroxy acids; those with ring sizes smaller than five or larger than six are less stable than their corresponding hydroxy acids.

7. Ester and lactone saponifications are examples of acyl substitution.

a. The mechanisms of these reactions are classified as nucleophilic acyl substitution mechanisms.

b. In a nucleophilic acyl substitution reaction, the substituting group reacts at the carbonyl carbon as a nucleophile, and a leaving group (another base) departs, usually in a separate step.

C. *ACID-CATALYZED HYDROLYSIS OF ESTERS*

1. Esters can be hydrolyzed to carboxylic acids in aqueous solutions of strong acids; in most cases this reaction is slower than base-promoted hydrolysis and must be carried out with an excess of water (in which esters are insoluble).

2. By the principle of microscopic reversibility, the mechanism of acid-catalyzed hydrolysis is the exact reverse of acid-catalyzed esterification.

a. The ester is first protonated by the acid catalyst; protonation makes the carbonyl carbon more electrophilic by making the carbonyl oxygen a better acceptor of electrons.

b. Water, as the nucleophile, reacts as a nucleophile at the carbonyl carbon and then loses a proton to give the tetrahedral intermediate.

c. Protonation of the leaving oxygen converts it into a better leaving group.

d. Loss of the protonated leaving group gives a protonated carboxylic acid, from which a proton is removed to give the carboxylic acid itself.

3. Ester hydrolysis is another example of nucleophilic acyl substitution.

4. Saponification, followed by acidification, is a much more convenient method for hydrolysis of most esters than acid-catalyzed hydrolysis for the following reasons:

a. It is faster.

b. It is irreversible.

c. It can be carried out not only in water but also in a variety of solvents, even in alcohols.

D. *HYDROLYSIS OF AMIDES*

1. Amides can be hydrolyzed to carboxylic acids and ammonia or amines by heating them in acidic or basic solution.

2. In acid, protonation of the ammonia or amine by-product drives the hydrolysis equilibrium to completion; the amine can be isolated, if desired, by addition of base to the reaction mixture following hydrolysis.

tetrahedral
addition intermediate

3. Hydrolysis of amides in base is analogous to saponification of esters; the reaction is driven to completion by formation of the carboxylic acid salt.

4. The conditions of both acid- and base-promoted amide hydrolysis are considerably more severe than the corresponding reactions of esters.

5. The mechanisms of amide hydrolysis are typical nucleophilic acyl substitution mechanisms.

E. *HYDROLYSIS OF NITRILES*

1. Nitriles are converted by hydrolysis into carboxylic acids and ammonia by heating them in acidic or basic solution.

2. The conditions of nitrile hydrolysis are considerably more severe than the corresponding reactions of esters and amides.

3. The mechanism of nitrile hydrolysis in acidic solution involves the following steps:

    a. Protonation of the nitrogen, which makes the nitrile carbon more electrophilic.

    b. Reaction of the nucleophile water at the nitrile carbon and loss of a proton, which give an intermediate called an imidic acid, the nitrogen analog of an enol.

    c. Conversion of the unstable imidic acid under the reaction conditions into an amide.

    d. Hydrolysis of the amide to a carboxylic acid and ammonium ion. (See Eqs. 21.43a–d on text pp. 1151–1152 for the detailed mechanism.)

4. The mechanism of nitrile hydrolysis in base consists of the following steps:

    a. The nitrile group reacts with aqueous hydroxide at the nitrile carbon and, as a result, the electronegative nitrogen assumes a negative charge.

    b. Proton transfer gives the conjugate base of an imidic acid.

    c. The imidic acid reacts further to give the corresponding amide.

    d. The amide hydrolyzes under the basic reaction conditions to the carboxylate salt of the corresponding carboxylic acid. (See Eqs. 21.44a–b on text pp. 1152–1153 for the detailed mechanism.)

5. The hydrolysis of nitriles is a useful way to prepare carboxylic acids because nitriles, unlike many other carboxylic acid derivatives, are generally synthesized from compounds other than the acids themselves, such as alkyl halides and sulfonates. The formation of nitriles from alkyl halides also constructs a new carbon–carbon bond.

II.   REACTIONS OF CARBOXYLIC ACID DERIVATIVES WITH ALCOHOLS

A.   REACTION OF ACID CHLORIDES AND ANHYDRIDES WITH ALCOHOLS AND PHENOLS

1. Esters are formed rapidly when acid chlorides react with alcohols or phenols, usually in the presence of a tertiary amine such as pyridine, or a related base.

tetrahedral addition intermediate

pyridinium chloride

2. Esters of tertiary alcohols and phenols, which cannot be prepared by acid-catalyzed esterification, can be prepared by this method.

3. Anhydrides react with alcohols and phenols in much the same way as acid chlorides. Cyclic anhydrides react with alcohols and phenols to give half-esters.

a cyclic anhydride

a half-ester

4. Sulfonate esters are prepared by the analogous reactions of sulfonyl chlorides with alcohols.

B. *REACTION OF ESTERS WITH ALCOHOLS*

1. When an ester reacts with an alcohol under acidic conditions, or with an alkoxide under basic conditions, a new ester is formed.

(removed by distillation)

2. This type of reaction, called transesterification, typically has an equilibrium constant near unity.

3. The reaction is driven to completion by the use of an excess of the displacing alcohol or by removal of a relatively volatile alcohol by-product as it is formed.

III. REACTIONS OF CARBOXYLIC AID DERIVATIVES WITH AMINES

A. *REACTION OF ACID CHLORIDES AND ANHYDRIDES WITH AMMONIA AND AMINES*

1. Acid chlorides react rapidly and irreversibly with ammonia or amines by a nucleophilic acyl substitution reaction mechanism to give amides. (Amide classification is discussed in Sec. 21.1E, text p. 1133.)

   a. Reaction with ammonia yields a primary amide.

   b. Reaction with a primary amine yields a secondary amide.

   c. Reaction with a secondary amine yields a tertiary amide.

2. The amine reacts as a nucleophile at the carbonyl group to form a tetrahedral intermediate, which expels chloride ion; a proton-transfer step yields the amide.

    a. An important aspect of amide formation is the proton transfer in the last step of the mechanism.

    b. Unless another base is added to the reaction mixture, the starting amine acts as the base in this step; if the only base present is the amine nucleophile, then at least two equivalents of the amine must be used— one equivalent as the nucleophile, and one equivalent as the base in the final proton-transfer step.

3. In the Schotten–Baumann technique for amide formation, the reaction is run with a water-insoluble acid chloride and an amine together in a separate organic layer over an aqueous solution of NaOH. The NaOH neutralizes HCl produced as a by-product of the reaction.

4. A tertiary amine such as pyridine can be used as the second amine equivalent. The presence of a tertiary amine does not interfere with amide formation because a tertiary amine itself cannot form an amide.

5. Anhydrides react with amines in much the same way as acid chlorides.

    a. Half-amides of dicarboxylic acids are produced in analogous reactions of amines and cyclic anhydrides.

    b. These compounds can be cyclized to imides by treatment with dehydrating agents, or in some cases, simply by heating.

a cyclic anhydride   a half-amide   a cyclic imide

B. *REACTION OF ESTERS WITH AMINES*

1. The reaction of an ester with ammonia or amines yields an amide.

tetrahedral
addition
intermediate

2. The reaction of esters with hydroxylamine (NH_2OH) gives *N*-hydroxyamides; these compounds are known as hydroxamic acids.

   a. This chemistry forms the basis for the hydroxamate test, used mostly for esters.

   b. The hydroxamic acid products are easily recognized because they form highly colored complexes with ferric ion.

a hydroxamic acid

IV. REDUCTION OF CARBOXYLIC ACID DERIVATIVES

A. *REDUCTION OF ACID CHLORIDES TO ALDEHYDES AND ALCOHOLS*

1. Acid chlorides can be reduced to aldehydes by either of two procedures:

   a. Hydrogenation over a catalyst that has been deactivated, or poisoned, with an amine, such as quinoline, that has been heated with sulfur.

   i. This reaction is called the *Rosenmund reduction*.

   ii. The poisoning of the catalyst prevents further reduction of the aldehyde product.

   b. Reaction at low temperature with lithium tri(*tert*-butoxy)aluminum hydride.

tetrahedral
addition
intermediate

   i. The hydride reagent used in this reduction is derived by the replacement of three hydrogens of $LiAlH_4$ by *tert*-butoxy groups; this reagent is much less reactive, and is more selective, than $LiAlH_4$.

   ii. Because acid chlorides are more reactive than aldehydes toward nucleophiles, the reagent reacts preferentially with the acid chloride reactant rather than with the product aldehyde.

2. Acid chlorides (and anhydrides) react with LiAlH$_4$ to give primary alcohols. but this is not a widely used reaction.

B. *REDUCTION OF ESTERS TO PRIMARY ALCOHOLS*

1. Esters can be reduced to alcohols with LiAlH$_4$. This reaction involves a nucleophilic acyl substitution reaction followed by a carbonyl-addition reaction.

a. The active nucleophile in LiAlH$_4$ reductions is the hydride ion, H:⁻, which replaces the alkoxide at the carbonyl group of the ester to give an aldehyde.

b. The aldehyde cannot be isolated because it reacts rapidly with LiAlH$_4$ to give, after protonolysis, the alcohol.

c. Two alcohols are formed: one is derived from the acyl group of the ester, and the other is derived from the alkoxy group (usually methanol or ethanol, which is discarded).

2. Sodium borohydride reacts very sluggishly or not at all with most esters.

C. *REDUCTION OF AMIDES TO AMINES*

1. Amides are reduced to amines with LiAlH$_4$.

a. Primary amines are obtained from primary amides.

b. Secondary amines are obtained from secondary amides.

c. Tertiary amines are obtained from tertiary amides.

2. The mechanism of the reaction of a secondary amide with lithium aluminum hydride involves the following steps:

a. Formation of the lithium salt of the amide by reaction of the amide N—H proton with an equivalent of hydride, a strong base.

b. Reaction of the lithium salt of the amide, a Lewis base, with AlH$_3$, a Lewis acid, to give a tetrahedral addition intermediate.

c. Loss of ⁻OAlH$_2$, a good leaving group (better than the nitrogen group), to form an imine.

d. Reduction of the imine and protonolysis in a separate step yield the product amine.

(d)

$$\underset{\text{an imine}}{\overset{\displaystyle H}{\underset{\displaystyle R}{\big\backslash}}C=NR'} \quad \xrightarrow[\substack{\text{1) LiAlH}_4 \\ \text{2) H}_3\text{O}^+/\text{H}_2\text{O} \\ \text{3)} \ ^-\text{OH}}]{} \quad \underset{\text{an amine}}{R-CH_2-NHR'}$$

    i.   An excess of acid will convert the amine, which is a base, into its conjugate-acid ammonium ion.

    ii.   Hydroxide is required to neutralize this ammonium salt and thus give the neutral amine.

3.   The reductions of primary and tertiary amides involve somewhat different mechanisms, but they too involve loss of oxygen rather than nitrogen as a leaving group.

D.   *REDUCTION OF NITRILES TO AMINES*

1.   Nitriles are reduced to primary amines by reaction with LiAlH$_4$, followed by protonolysis in a separate step.

$$R-C\equiv N \quad \xrightarrow[\substack{\text{1) LiAlH}_4 \\ \text{2) H}_3\text{O}^+/\text{H}_2\text{O}}]{} \quad R-CH_2-\overset{+}{N}H_3 \quad \xrightarrow{\ ^-\text{OH}\ } \quad R-CH_2-NH_2$$

    a.   An excess of acid will convert the amine, which is a base, into its conjugate-acid ammonium ion.

    b.   Hydroxide is required to neutralize this ammonium salt to give the neutral amine.

2.   Nitriles are also reduced to primary amines by catalytic hydrogenation. An intermediate in the reaction is the imine, which is not isolated but is hydrogenated to the amine product.

$$R-C\equiv N \quad \xrightarrow{\text{H}_2, \text{ (cat.)}} \quad R-CH=NH \quad \xrightarrow[\substack{\text{H}_2 \\ \text{Raney Ni catalyst} \\ \text{high pressure}}]{} \quad R-CH_2-NH_2$$

V.   REACTIONS OF CARBOXYLIC ACID DERIVATIVES WITH ORGANOMETALLIC REAGENTS

A.   *REACTION OF ACID CHLORIDES WITH LITHIUM DIALKYLCUPRATES*

1.   The reaction of lithium dialkylcuprates with acid chlorides gives ketones in excellent yield; lithium dialkylcuprates typically react with acid chlorides and aldehydes, very slowly with ketones, and not at all with esters.

$$\underset{R}{\overset{\displaystyle O}{\underset{\displaystyle }{\big\Vert}}}\!\!\!C\!\!-\!\!Cl \;+\; Li^+ \ ^-CuR'_2 \quad \longrightarrow \quad \underset{R}{\overset{\displaystyle O}{\underset{\displaystyle }{\big\Vert}}}\!\!\!C\!\!-\!\!R' \;+\; LiCl \;+\; CuR'$$

2.   Lithium dialkylcuprates react much like Grignard or lithium reagents, but are less reactive; a lithium dialkylcuprate can be considered conceptually as an alkyl anion complexed with copper.

    a.   Lithium dialkylcuprate reagents are prepared by the reaction of two equivalents of an organolithium reagent with one equivalent of a cuprous halide such as CuBr.

$$RLi \;+\; CuI \quad \longrightarrow \quad LiI \;+\; CuR \quad \xrightarrow{RLi} \quad LiCuR_2 \quad \text{(Think of LiCuR}_2 \text{ as if it were } Li^+ \ CuR \ \text{:}\bar{R}\text{)}$$

    b.   The first equivalent forms an alkylcopper compound; the driving force for this reaction is the preference of lithium, the more electropositive metal, to exist as an ion (Li$^+$).

    c.   The copper of an alkylcopper reagent is a Lewis acid and reacts accordingly with "alkyl anion" from a second equivalent of the organolithium reagent; the product of this reaction is a lithium dialkylcuprate.

B.   *REACTION OF ESTERS WITH GRIGNARD REAGENTS*

1.   In the reaction of esters with Grignard reagents, tertiary alcohols are formed after protonolysis. (Secondary alcohols are formed from esters of formic acid after protonolysis.)

   a.   Two equivalents of organometallic reagent react per mole of ester.

   b.   A second alcohol (derived from the alkoxy group, usually methanol or ethanol) is produced in the reaction (R″—OH in the following equation) and is typically discarded.

2.   This reaction is a nucleophilic acyl substitution followed by an addition.

   a.   A ketone is formed in the substitution step along with an alkoxide leaving group.

   b.   The ketone intermediate is not isolated because ketones are more reactive than esters toward nucleophilic reagents.

   c.   The ketone reacts with a second equivalent of the Grignard reagent to form, after protonolysis, the alcohol.

3.   This reaction is a very important method for the synthesis of alcohols in which at least two of the groups on the α-carbon of the alcohol product are identical.

VI.   REACTIONS OF CARBOXYLIC ACID DERIVATIVES WITH OTHER NUCLEOPHILES

A.   *REACTION OF ACID CHLORIDES WITH CARBOXYLATE SALTS*

1.   Even though carboxylate salts are weak nucleophiles, acid chlorides are reactive enough to react with carboxylate salts to give anhydrides.

2.   The reaction of acid chlorides with carboxylate salts can be used to prepare mixed anhydrides.

VII.   SYNTHESIS OF CARBOXYLIC ACID DERIVATIVES—REVIEW

A.   *SYNTHESIS OF ESTERS*

1.   Acid-catalyzed esterification of carboxylic acids: see the Reaction Review in Chapter 20 of this manual, Sec. II.A.

2.   Reaction of carboxylic acids with diazomethane, and reaction of carboxylates with reactive alkyl halides: see Chapter 20 of this manual, Sec. III.A.

3.   Reaction of acid chlorides and anhydrides with alcohols and phenols: see Sec. II.A of this Reaction Review.

4.   Transesterification of esters: see Sec. II.B of this Reaction Review.

B.   *SYNTHESIS OF ACID CHLORIDES AND ANHYDRIDES*

1.   Synthesis of acid chlorides and anhydrides from carboxylic acids: see the Reaction Review in Chapter 20 of this manual, Secs. II.B and II.C.

2. Synthesis of anhydrides from carboxylates and acid chlorides: see Sec. VI.A of this Reaction Review.

C.   *SYNTHESIS OF AMIDES*

1. Reaction of acid chlorides and anhydrides with amines: see Sec. III.A in this Reaction Review.
2. Reaction of esters with amines: see Sec. III.B of this Reaction Review.

D.   *SYNTHESIS OF NITRILES*

1. S_N2 reaction of cyanide ion with alkyl halides or sulfonate esters. In this reaction, primary or unbranched secondary alkyl halides or sulfonate esters are required.

$$R-CH_2-X + {}^-C{\equiv}N \longrightarrow R-CH_2-C{\equiv}N + X^- \quad (X^- = \text{halide ion, } {}^-OTs)$$

2. Cyanohydrin formation (Sec. 19.8, text p. 1042).

SOLUTIONS TO PROBLEMS

Solutions to Focused Problems

21.1   (a)
$$N{\equiv}CCH_2CH_2CH_2CH_2CO_2H$$
5-cyanopentanoic acid

(b)
$$CH_3CH_2CH_2CH_2\overset{\overset{\displaystyle O}{\|}}{C}OCH(CH_3)_2$$
isopropyl valerate

(c)
ethyl methyl malonate

(d)
$$CH_3\overset{\overset{\displaystyle O}{\|}}{C}O{-}\text{(cyclohexyl)}$$
cyclohexyl acetate

(e)
$$H\overset{\overset{\displaystyle O}{\|}}{C}N(CH_3)_2$$
N, N-**dimethylformamide**

(f)
γ-valerolactone

(g)
glutarimide

(h)
$$(CH_3)_2\overset{\overset{\displaystyle O}{\|}}{C}CCl$$
$$\underset{Cl}{|}$$
α-chloroisobutyryl chloride

(i)
$$HO_2CCH_2CHCH_2CH_2CO_2H$$
3-ethoxycarbonylhexanedioic acid

21.2   (a)   Butanenitrile (*common:* butyronitrile)

(b) Cyclopropanecarbonyl chloride
(c) Isopentyl 3-methylbutanoate (*common*: isoamyl isovalerate)

 The *isoamyl* group is the same as an isopentyl or 3-methylbutyl group:

(d) *N*,*N*-Dimethylbenzamide
(e) (*Z*)-2-Butenedioic anhydride (*common name more often used*: maleic anhydride)
(f) Ethyl 3-oxopentanoate (*common*: ethyl β-ketovalerate)
(g) 1-Methyl-3-butenyl propanoate. (Notice the numbering from the point of attachment to the carboxylate oxygen.)

21.3 (a) It is the *E* conformation. (b)

Z conformation

21.4 Any lactam containing a relatively small ring (fewer than eight members) must exist in an *E* conformation. An example is δ-valerolactam. A small-ring lactam cannot exist in a *Z* conformation, because such a conformation would result in a transoid bond within the ring, a very strained and therefore unstable arrangement. (This is the same reason that small-ring (*E*)-cycloalkenes are unstable.) If you try to construct a model of a small-ring lactam with its amide bond in a *Z* conformation, the strain of such a conformation should be clear.

δ-valerolactam

amide bond in the *Z* conformation

This distance requires as least four more carbons.

21.5 As shown by the data above the problem, a carboxylic acid has a higher boiling point than an ester of the same molecular mass because it can both donate and accept hydrogen bonds within its liquid state; hydrogen bonding does not occur in the ester. Consequently, pentanoic acid (valeric acid) has a higher boiling point than methyl butanoate. The actual data are shown at right.

$CH_3CH_2CH_2CH_2CO_2H$ $CH_3CH_2CH_2CO_2CH_3$
pentanoic acid   **methyl butanoate**
(valeric acid)   **(methyl butyarate)**
bp 186 °C   bp 98 °C

21.6 (a) 2-Butanone should have the higher dipole moment because, other things being equal, molecules with higher dipole moments have higher boiling points.
(b) The dipole-moment data require that esters, like amides, exist predominantly in the *Z* conformation. In this conformation of an ester, the resultant of the bond dipoles of the two carbon–oxygen single bonds opposes the bond dipole of the C=O bond; the overall dipole moment of the ester is thus reduced. In the ketone, the C=O bond dipole is not opposed by any other significant bond dipole.

If the ester were in an *E* conformation, the resultant bond dipoles of the two carbon–oxygen single bonds would *augment* the bond dipole of the C=O bond. In this case, the dipole moment of the ester would be larger than that of the ketone.

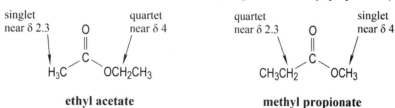

The experimental facts bear out these predictions. γ-Butyrolactone, an "ester" that is locked into an *E* conformation, has a greater dipole moment than 2-butanone, which has a greater dipole moment than methyl acetate, an ester which exists mostly in the *Z* conformation. The boiling points increase with increasing dipole moments.

	methyl acetate	2-butanone	γ-butyrolactone
	CH_3COCH_3	$CH_3CCH_2CH_3$	
dipole moments:	1.5 D	2.8 D	4.1 D
boiling points:	57 °C	80 °C	204 °C

21.7 (a) The carbonyl absorption of the ester occurs at higher wavenumber, and only the carboxylic acid has the characteristic strong, broad O—H stretching absorption in 2400–3600 cm^{-1} region.

(b) In the proton NMR spectra, the amide has a single methyl resonance that is a doublet, whereas the nitrile has two methyl resonances that are singlets. The amide spectrum loses a resonance with the D_2O shake. In addition, the integral of the aromatic proton absorptions in the nitrile is smaller in proportion to the integral of the other absorptions because there are fewer aromatic protons.

N-methylbenzamide

2,4-dimethylbenzonitrile

(c) In methyl propionate, the singlet for the methyl of the CH_3O— group has a greater chemical shift (near δ 4) than the methyl singlet in ethyl acetate (near δ 2.3). In ethyl acetate, the quartet for the —CH_2— group has a greater chemical shift (near δ 4) than the —CH_2— quartet in methyl propionate (near δ 2.3).

ethyl acetate

methyl propionate

(d) In *N*-methylpropanamide, the *N*-methyl group is a doublet at about δ 3. *N*-Ethylacetamide has no doublet resonances. In *N*-methylpropanamide, the α-protons are a quartet near δ 2.5. In *N*-ethylacetamide, the α-protons are a singlet at δ 2. The NMR spectrum of *N*-methylpropanamide has no singlets. (The proton NMR spectrum of *N*-ethylacetamide is shown in Focused Problem 21.8 and Fig. 21.3, p. 1143 of the text.)

N-methylpropanamide   **N-ethylacetamide**

(e) Ethyl isobutyrate, $(CH_3)_2CHCO_2CH_2CH_3$, has five CMR resonances, whereas ethyl butyrate, $CH_3CH_2CH_2CO_2CH_2CH_3$, has six.

21.8 That the compound is an amide is confirmed by the N—H absorption at 3300 cm^{-1} and by the carbonyl absorption at 1650 cm^{-1}. The compound is *N*-ethylacetamide (structure at right); the —CH$_2$— resonance of the ethyl group is split by both the adjacent methyl protons and the N—H proton.

N-ethylacetamide

21.9 (a) The first ester is more basic because its conjugate acid is stabilized not only by resonance interaction with the ester oxygen, but also by resonance interaction with the double bond; that is, the conjugate acid of the first ester has one more important resonance structure than the conjugate acid of the second.

resonance interaction of the protonated carbonyl
group with the double bond

(b) The first ester is more basic because its conjugate acid is stabilized by resonance interaction of the *p*-methoxy oxygen.

The meta position of the methoxy group in the second ester does not allow a similar resonance interaction; consequently, only its electron-withdrawing polar effect operates. Moreover, this polar effect destabilizes the conjugate acid of the second ester.

(c) The second ester is more basic because its conjugate acid is stabilized through the polar effect (Sec 3.7E, text p. 157) of the attached electron-donating methyl group. The first ester is less basic because its conjugate acid is destabilized by the polar effect of the attached electron-withdrawing trifluoromethyl group.

21.10 (a) The hydrolysis mechanism in acidic solution:

(b)   The hydrolysis mechanism in basic solution:

21.11   The hydrolysis products consist of the conjugate base of the carboxylic acid and an amine. In (b) the amine and carboxylic acid are part of the same molecule, and they are cis substituents on the ring.

(a)

isobutyrate ion   pyrrolidine

(b)

21.12   *Resonance effects*: The application of the resonance effect for predicting the relative rates of acid-catalyzed ester and amide hydrolyses is exactly the same as it is for the corresponding base-promoted hydrolyses. (See Display 21.49, text p. 1154 and the discussion in Sec. 21.7E of the text.)

   Leaving-group basicities: In the acid-catalyzed hydrolysis of an ester, the leaving group is an alcohol; in the acid-catalyzed hydrolysis of an amide, the leaving group is ammonia or an amine. Alcohols are much weaker bases (conjugate-acid pK_a = –2 to –3) than amines or ammonia (conjugate-acid pK_a = 9–10).Because weaker bases are better leaving groups, esters should hydrolyze more rapidly, and they do.

21.13   First, consider the stability of the starting materials. Acetyl fluoride is more stabilized by resonance than acetyl chloride, because overlap of the 2p orbitals of fluorine and the carbonyl carbon is more effective than the overlap of chlorine 3p orbitals with the carbonyl carbon.

unimportant
structure

Next, consider the stabilities of the intermediates. The tetrahedral intermediates derived from acid fluorides are stabilized more than the intermediates derived from acid chlorides because of the greater effective positive charge on the carbonyl carbon attached to the fluorine caused by the larger carbon–fluorine bond dipole.

If the stability of the starting material is the dominant factor in determining the rate of the reaction, the acid chloride should react faster because it is less stable. If the stability of the intermediate is the dominant factor, then the acid fluoride should react faster, because its intermediate (and, by Hammond's Postulate, its transition state leading to the intermediate) would be lower in energy. Because we know from the question that the acid chloride reacts faster, it must be the stabilities of the starting materials that determine the rate of the reaction.

21.14 (a) Because esters hydrolyze much more rapidly than nitriles, one equivalent of base brings about the selective hydrolysis of the ester.

$$N\equiv C-CH_2-C\overset{O}{\underset{O^-}{}} \ + \ HOCH_3$$

(b) This is an ordinary ester saponification followed by acidification.

$$F-\bigcirc-C\overset{O}{\underset{OH}{}} \ + \ HOCH_3$$

(c) Hydrolysis of one amide gives carbamic acid, which spontaneously decarboxylates; see Sec. 20.11, Eq. 20.63b on text p. 1115.

$$H_2N\overset{O}{\underset{}{C}}NH_2 \ + \ H_2O \xrightarrow{H_3O^+,\ heat} \overset{+}{N}H_4 \ + \ H_2N\overset{O}{\underset{}{C}}OH \xrightarrow{H_3O^+} \overset{+}{N}H_4 \ + \ CO_2$$

carbamic acid

21.15 (a)

$$CH_3CH_2\overset{O}{\underset{}{C}}N(CH_3)_2$$
$$+ \ H_2\overset{+}{N}(CH_3)_2$$

(b)

$$\triangle\overset{O}{\underset{}{C}}O\triangleright$$

(c)

$$PhCH_2\overset{O}{\underset{}{C}}SEt$$

(d)

$$CH_3CH_2CH_2\overset{O}{\underset{}{C}}O\overset{O}{\underset{}{C}}CH_3$$

(e)

$$CH_3O\overset{O}{\underset{}{C}}Cl$$

(f)

$$CH_3O\overset{O}{\underset{}{C}}OCH_3$$

(g)

$$O\overset{O}{\underset{}{C}}O$$

(h)

$$\bigcirc\overset{O}{\underset{}{C}}OCH_3 \ \overset{O}{\underset{}{C}}OH$$

21.16   (a)

acetyl chloride   *p*-nitrophenol   *p*-nitrophenyl acetate

(b)

l-phenylethanol

p-toluenesulfonyl chloride   l-phenylethyl tosylate

(c)

catechol   phosgene

(d)

tert-butyl alcohol

malonyl dichloride   di-*tert*-butyl malonate

21.17   There are two cyclic amide (or lactam) functional groups present in the compound. The lactam contained within the four-membered ring, because of its ring strain, is considerably more reactive than the other lactam. So, one equivalent of sodium methoxide will react preferentially at the more reactive lactam, to give the product shown in (a). In strong acid, both lactams would be hydrolyzed to amines and carboxylic acids as shown in (b), and the amines would be protonated to give their conjugate acids.

(a)   (b)

21.18   (a)   Carry out one of the reductions in Sec. 21.9D:

benzoyl chloride   benzaldehyde

(b)   Reduce the tertiary amide with LiAlH$_4$.

benzoyl chloride

pyrrolidine
(excess)

N-benzylpyrrolidine

21.19   (a)   PhCH$_2$CH$_2$NH$_2$

(b)   Although the ester reacts most rapidly, the nitrile also reacts because an excess of LiAlH$_4$ is present. The product is the amino alcohol HOCH$_2$CH$_2$CH$_2$NH$_2$ plus the by-product ethanol, C$_2$H$_5$OH.

(c)   Both esters are reduced.

ethanol

1-phenyl-1,2-ethanediol

(from reduction
of both the acetoxy
group and the ethyl ester)

21.20   Two compounds that would give (CH$_3$)$_2$CHCH$_2$CH$_2$CH$_2$NH$_2$ as the result of LiAlH$_4$ reduction are the following:

(CH$_3$)$_2$CHCH$_2$CH$_2$C≡N   (CH$_3$)$_2$CHCH$_2$CH$_2$C—NH$_2$

4-methylpentanenitrile   **4-methylpentanamide**

21.21   (a)   As shown in Eq. 21.85b, text p. 1170, the reduction involves an imine intermediate. This intermediate *A* can react with a molecule of amine product to form the nitrogen analog of an acetal, which then can lose ammonia to form a different imine *B*, which in turn is reduced to the by-product. (You should be able to provide the curved-arrow notation for the reactions involved in the *A* ⇌ *B* equilibrium in the overall process outlined below.)

(b)   By Le Chatelier's principle, ammonia drives the equilibria in the above equation away from imine *B* and back toward imine *A*, which is eventually reduced.

21.22   (a)

PhCO$_2$H

benzoic acid

PhCOCH$_2$CH$_3$

Ph$_3$C—OH

triphenylmethanol

(b)

$$CH_3CH_2CH_2COH \xrightarrow{PCl_5} CH_3CH_2CH_2CCl \xrightarrow[\text{2) } H_3O^+]{\text{1) Li}^+ \ ^-Cu(CH_3)_2} CH_3CH_2CH_2CCH_3 \xrightarrow[\text{2) } H_3O^+]{\text{1) } CH_3CH_2Mgl} CH_3CH_2CH_2CCH_2CH_3$$

butyric acid

3-methyl-3-hexanol

(c)   First, the nitrile is hydrolyzed to isobutyric acid, which is converted into its acid chloride:

$$(CH_3)_2CHC{\equiv}N \xrightarrow{H_2O, \ H_3O^+, \ heat} (CH_3)_2CHC-OH \xrightarrow{SOCl_2} (CH_3)_2CHC-Cl$$

isobutyronitrile   **isobutyric acid**   **isobutyryl chloride**

One synthesis involves reaction of the acid chloride with lithium dimethylcuprate:

$$(CH_3)_2CHC-Cl \xrightarrow[\text{2) } H_3O^+]{\text{1) Li}^+ \ ^-Cu(CH_3)_2} (CH_3)_2CHC-CH_3 \xrightarrow[\text{2) } H_3O^+]{\text{1) } CH_3Mgl} (CH_3)_2CHC-CH_3$$

2,3 dimethyl-2-butanol

The second synthesis involves the Grignard reaction of an ester:

$$(CH_3)_2CHC-Cl \xrightarrow[\text{(solvent)}]{C_2H_5OH} (CH_3)_2CHC-OC_2H_5 \xrightarrow[\text{2) } H_3O^+]{\text{1) } CH_3Mgl \text{ (excess)}} (CH_3)_2CHC-CH_3$$

2,3-dimethyl-2-butanol

 Although you have not studied this reaction explicitly, a more direct route to the alcohol is the **reaction of the acid chloride** with excess CH_3Mgl followed by protonolysis.

$$(CH_3)_2CHC-Cl \xrightarrow[\text{2) } H_3O+]{\text{1) } CH_3Mgl \text{ (excess)}} (CH_3)_2CHC-CH_3$$

Can you see why acid chlorides should react the same way as esters in this reaction?

(d)

$$CH_3CH_2COH \xrightarrow{PCl_5} CH_3CH_2CCl \xrightarrow[\text{2) } H_3O^+]{\text{1) Li}^+ \ ^-Cu(C_2H_5)_2} CH_3CH_2CCH_2CH_3$$

propionic acid

3-pentanone

21.23 (a)   The reaction of a Grignard reagent with ethyl formate gives a secondary alcohol in **which the two alkyl** groups at the α-carbon are identical.

$$CH_3CH_2OCH + 2 \ RMgBr \longrightarrow \xrightarrow{H_3O^+} R-CH-R + CH_3CH_2OH$$

ethyl formate

(b)

$$CH_3CH_2O\overset{\overset{\displaystyle O}{\|}}{C}H + 2\ CH_3CH_2MgBr \longrightarrow \xrightarrow{H_3O^+} CH_3CH_2\overset{\overset{\displaystyle OH}{|}}{C}HCH_2CH_3 + CH_3CH_2OH$$

ethyl formate   **3-pentanol**

21.24 In both parts, only the acid chloride reacts with the organometallic reagent.

(a)   (b)

$$N\equiv C(CH_2)_{10}\overset{\overset{\displaystyle O}{\|}}{C}CH_3 \qquad\qquad CH_3(CH_2)_3O\overset{\overset{\displaystyle O}{\|}}{C}(CH_2)_4\overset{\overset{\displaystyle O}{\|}}{C}CH_3$$

21.25 The first method involves carbonation of the corresponding Grignard reagent; the second involves hydrolysis of the nitrile.

$$(CH_3)_2CH(CH_2)_3Br \begin{cases} \xrightarrow[\text{ether}]{Mg} \xrightarrow[\text{2) }H_3O^+]{\text{1) }CO_2} (CH_3)_2CH(CH_2)_3CO_2H \\[2ex] \xrightarrow[\text{DMSO}]{Na^+\ {}^-CN} (CH_3)_2CH(CH_2)_3C\equiv N \xrightarrow{H_2O,\ H_3O^+,\ \text{heat}} \end{cases}$$

1-bromo-4-methylpentane   **5-methylhexanoic acid**

21.26 The required nitrile is 2-hydroxypropanenitrile. This is a cyanohydrin, and this cyanohydrin can be prepared by addition of cyanide to acetaldehyde.

$$CH_3\overset{\overset{\displaystyle O}{\|}}{C}H \xrightarrow{NaCN,\ H_2O} CH_3\overset{\overset{\displaystyle OH}{|}}{\underset{\underset{\displaystyle C\equiv N}{|}}{C}}H \xrightarrow{H_2O,\ H_3O^+,\ \text{heat}} CH_3\overset{\overset{\displaystyle OH}{|}}{\underset{\underset{\displaystyle CO_2H}{|}}{C}}H$$

acetaldehyde   **2-hydroxypropanenitrile**   **2-hydroxypropanoic acid**

 Notice that the alcohol does *not* dehydrate under the rather severe conditions of nitrile hydrolysis. The reason is that the carbocation intermediate that would be involved is destabilized by both the cyano group and the carboxy group of the product. Can you think of a reason why?

21.27 Amides are more resistant to base-promoted hydrolysis than esters; the same is true of *polymeric* amides and esters. Nylon is more resistant to hydrolysis than polyester.

21.28 Process (*a*) is catalytic hydrogenation. Because furan is aromatic, high pressure and/or heat might be required. Process (*b*) is ether cleavage with HCl and ZnCl$_2$ (or other acidic catalyst) and heat. Process (*c*) consists of the S$_N$2 reactions of cyanide ion with the dichloride. Process (*d*) is catalytic hydrogenation or LiAlH$_4$ reduction followed by protonolysis. Process (*e*) is nitrile hydrolysis with aqueous acid and heat. Finally, process (*f*) is to mix the amine and the carboxylic acid and heat (Eq. 21.104, text p. 1179), or to form the di-acid chloride and allow it to react with the amine.

21.29 First, water opens the lactam to an amino acid *A*. The amino group of this compound serves as a nucleophile to open another lactam molecule; and the resulting amino group thus liberated repeats the process, thus growing the polymer chain:

Solutions to Skills Objectives Problems

21.30 (a)

ethyl *p*-acetylbenzoate

(b)

$(CH_3)_2CH—O—\overset{\overset{\textstyle O}{\|}}{C}—OCH_3$

isopropyl methylcarbonate

(c)

β-valerolactone

(d)

3-bromocyclopentanecarbonitrile

(e)

N-ethyl-N-methylcyclohexanecarboxamide

21.31 (a) Ethyl 3-formylbenzoate, or ethyl *m*-formylbenzoate (*common*)
(b) Pentanediamide, or glutaramide (*common*)
(c) *Tert*-butyl 2-Bromo-7-cyanoheptanoate
(d) 4-Chloroformyl-3,3-dimethylbutanoic acid

21.32 (a) The two conformations of *N*-methylbenzamide and methyl benzoate:

Z conformation   *A*   *E* conformation   *Z* conformation   *B*   *E* conformation

(b) The second resonance structure of benzamide shows that there is significant double-bond character in the carbon–nitrogen double bond. Because internal rotation about a double bond is slow, to the extent that this resonance structure is important, internal rotation is slow.

(c) The second resonance structure in an ester is less important than the second structure in an amide because it places positive charge on a more electronegative atom. Because this structure is less important in the ester, internal rotation in the ester is faster.

21.33 The order of increasing boiling point is $D < B < C < A$. (The actual boiling points are 116, 177, 199, and 275 °C, respectively.) Hydrogen bonding between molecules in the carboxylic acid accounts for its higher boiling point. The polarity of the ketone and the ester cause their boiling points to be higher than the hydrocarbon. Because the ketone has a greater dipole moment than the ester, its boiling point is greater. (See the solution to Focused Problem 21.6 for an explanation of the relative dipole moments.)

21.34 (a) Although the C≡C and C≡N triple bonds do have IR absorptions at different frequencies, a more telling difference is that the absorption of the internal alkyne will not be observed because the triple bond is symmetrically placed within the molecule. (See Sec. 13.3B, text p. 696.)

(b) The lactones B and C have carbonyl absorptions at a greater wavenumber than the ketone A. Because the carbonyl absorption frequencies of lactones increase with a decrease in ring size, compound C will have the carbonyl absorption at greatest wavenumber.

(c) Carbonyl absorption frequency increases in the order amide < ketone < ester (or lactone), all other things being equal. This means that the compounds can be differentiated on the basis of carbonyl absorption frequency: $C < A < B$.

21.35 (a) The odd molecular mass indicates the presence of an odd number of nitrogens, and the hydroxamate test and the IR carbonyl absorption at 1733 cm^{-1} indicate the presence of an ester. The IR absorption at 2237 cm^{-1} indicates a nitrile. The triplet–quartet pattern in the NMR spectrum clearly indicates that compound B is an ethyl ester. Subtracting the masses of all the atoms accounted for leaves 14 mass units, the mass of a CH$_2$ group. Compound A is ethyl cyanoacetate.

ethyl cyanoacetate (compound A)

(b) From its IR absorption, compound B is an ester. The formula indicates an unsaturation number of 1, which is completely accounted for by the ester carbonyl group. The chemical shift and splitting of the δ 4.0 resonance indicates the part-structure —CH$_2$CH$_2$O—. The hint in the figure caption indicates that the δ 1.6 resonance is the other CH$_2$ of this part-structure; because this consists of six lines, then a more complete part-structure for this group is CH$_3$CH$_2$CH$_2$O—. The triplet at δ 0.95 is consistent with the methyl group of this part structure. This leaves an ethyl group unaccounted for; from its chemical shift of δ 2.3, the CH$_2$ of the ethyl group is α to a carbonyl group. Compound B is propyl propionate.

propyl propionate (compound B)

(c)   The odd molecular mass indicates the presence of nitrogen. The IR spectrum indicates the presence of both a nitrile and an alcohol, and there are no carbonyl absorptions. The broad, D_2O-exchangeable resonance at δ 3.4 in the NMR spectrum confirms the presence of an O—H group, and the pair of triplets indicates a —CH_2CH_2— group. Compound C is 3-hydroxypropanenitrile (β-hydroxypropionitrile).

$$HO—CH_2CH_2C≡N$$

**3-hydroxypropanenitrile
(β-hydroxypropionitrile)**
(compound C)

(d)   The IR absorption indicates the presence of an ester. The doubled peaks in the mass spectrum indicate the presence of a single bromine. Consequently, compound D contains the elements CO_2 (from the ester) and Br. The remaining mass could correspond to C_4H_9. Therefore, adopt $C_5H_9O_2Br$ as a provisional formula. In the NMR spectrum, the triplet and the doublet at low chemical shift cannot be splitting each other, because their integrations are not consistent with such a situation. That is, a mutually split triplet (which indicates two adjacent hydrogens) and doublet (which indicates one adjacent hydrogen) must have an integral ratio of 1:2, not 1:1 as observed. Evidently, the two absorptions at δ 4.23 and δ 4.37 are the resonances that are split by the two triplets at lower chemical shift. From their chemical shifts, some of these protons are probably α to the Br and some are probably α to the oxygen. If the ester is an ethyl ester, then the high-field triplet is the resonance of the methyl group, and, from its integration, the doublet is also due to a methyl group on a carbon adjacent to a CH. These deductions conspire to identify compound D as ethyl 2-bromopropanoate.

**ethyl 2-bromopropanoate
(ethyl α-bromopropionate)**
(compound D)

(e)   The mass spectrum shows an odd mass, which indicates the presence of nitrogen. The IR spectrum could indicate the presence of a nitrile, although, if so, the C≡N absorption is about 50 cm^{-1} lower in frequency than the C≡N absorption of an ordinary nitrile. The IR spectrum also indicates a trans alkene. The UV spectrum indicates extensive conjugation. This shift of the nitrile IR absorption is similar to that observed with conjugated carbonyl groups, and it suggests that the nitrile group is conjugated with the double bond. The NMR shows resonances for 7 protons. The large splitting in the resonances at δ 5.85 and δ 7.35 confirms the presence of the trans alkene. Evidently, the apparent singlet at δ 7.4 corresponds to five protons, that is, to a monosubstituted benzene ring. Compound E is (E)-3-phenylpropenenitrile.

(E)-3-phenylpropenenitrile
(compound E)

(f)   The unsaturation number is 5, and the NMR resonances in the δ 6.8–7.4 region indicate a para-substituted benzene ring. The broad resonance at δ 7.6 could be an amide N—H, and the IR absorption at 1659 cm^{-1} is confirmatory. The triplet–quartet pattern at δ 1.4 and δ 4.0 could indicate an ethyl ester, but we reject this idea because there is no ester carbonyl absorption. Therefore, this quartet is evidently due to a *para*-CH_3CH_2O— substituent on the ring. The singlet at δ 2.1 is at a chemical shift consistent possibly with an *N*-methyl group or with protons α to a carbonyl group. Here are two possible structures:

4-ethoxy-*N*-methylbenzamide

**N-4-ethoxyphenylacetamide
(*p*-ethoxyacetanilide)**
(compound F)

The absence of splitting in the δ 2.1 resonance rules out the first structure, for which we would expect a doublet for the *N*-methyl group. Therefore, the second structure is the correct one.

(g)   The IR absorptions indicate the presence of a primary amide; the odd mass is consistent with the presence of nitrogen. The weak CMR resonances are probably due to carbons that do not bear hydrogens. The δ 180.5 resonance is the amide carbonyl group; the resonance at δ 38.0 is a quaternary carbon, which, from its chemical shift, could be α to the carbonyl carbon. Subtracting the mass of the carboxamide group from the mass of the molecule gives the remaining mass as 57 units, which is the mass of a C_4H_9 group. Only a *tert*-

**2, 2-dimethylpropanamide
(pivalamide)**
(compound G)

butyl group has the symmetry required to fit the CMR data. Compound *G* is 2,2-dimethylpropanamide (pivalamide).

21.36 The order of increasing acidity (decreasing pK_a, or increasingly negative pK_a) is *B* < *E* < *D* < *A* < *C*, with uncertainty about the relative order of *A* and *C*. The order of *B*, *E*, and *D* follows from the resonance stabilization of the conjugate acids shown; the more important the resonance stabilization, the more stable is the structure, and the less acidic it is. Resonance results in a partial positive charge on the N, O, or S. Resonance is more important for *B* than for *E* because the nitrogen is less electronegative than oxygen. Although the sulfur in *D* is less electronegative than nitrogen, resonance stabilization is less important because sulfur must use 3p orbitals to overlap with the 2p orbitals of the carbonyl group. and this overlap (and thus resonance stabilization) is poor. The protonated nitrile has a pK_a of about –10, which makes it a stronger acid than any of the protonated carbonyl compounds, whose pK_a values range from –1 to –6. The pK_a of *A* is not known, but this ion is destabilized by the powerful polar effect of the carbonyl group; notice that it is *not* stabilized by resonance. (The pK_a has been estimated to be about –8; see Eq. 21.28b, text p. 1144.)

21.37 In the mechanisms that follow, simple proton transfers are not shown explicitly with the curved-arrow notation.

(a) The aminolysis of a thioester under basic conditions:

(b) The hydrolysis of a thioester under acidic conditions:

21.38 (a)
$$CH_3CH_2CO_2H$$

(b)

(c)

(d)

(e)

(f)

$$CH_3CH_2C\overset{O}{\underset{\parallel}{}}\text{—}\bigcirc\text{—}CH_3$$

(g)

$$CH_3CH_2\overset{O}{\underset{\parallel}{C}}NHCH(CH_3)_2$$

$$+ \ Cl^- \ H_3\overset{+}{N}CH(CH_3)_2$$

(h)

$$CH_3CH_2\overset{O}{\underset{\parallel}{C}}\overset{O}{\underset{\parallel}{O}}CPh$$

(i)

$$CH_3CH_2\overset{O}{\underset{\parallel}{C}}O\text{—}\bigcirc\text{—}CH_3$$

$$+ \ \bigcirc\overset{+}{N}H \ Cl^-$$

21.39 (a)

PhCO$_2$H
+ CH$_3$CH$_2$OH

(b)

PhCO$_2^-$ Na$^+$
+ CH$_3$CH$_2$OH

(c)

$$Ph\overset{O}{\underset{\parallel}{C}}\text{—}NH_2$$

$$+ \ CH_3CH_2OH$$

(d)

PhCH$_2$OH
+ CH$_3$CH$_2$OH

(e)

$$\underset{|}{\overset{OH}{|}}$$
$$Ph\overset{|}{C}CH_2CH_2CH_3$$
$$\underset{|}{CH_2CH_2CH_3}$$
$$+ \ CH_3CH_2OH$$

(f)

$$\underset{|}{\overset{O}{\overset{\parallel}{O}CCH_3}}$$
$$Ph\overset{|}{C}CH_2CH_2CH_3$$
$$\underset{|}{CH_2CH_2CH_3}$$
$$+ \ \bigcirc\overset{+}{N}H \ Cl^-$$

(g)

$$\overset{O}{\underset{O}{\overset{\diagdown}{S}}}\overset{O}{\diagup}Ph$$
$$Ph\overset{|}{C}CH_2CH_2CH_3$$
$$\underset{|}{CH_2CH_2CH_3}$$

(h)   Although ethoxide reacts at the carbonyl group as a nucleophile, there is no *net* reaction because ethoxide is also displaced as a leaving group.

21.40 (a)

CH$_3$CH$_2$C≡N

propionitrile

(b)

$$CH_3CH_2\overset{O}{\underset{\parallel}{C}}NH_2$$

propionamide

(c)

$$CH_3\overset{O}{\underset{\parallel}{C}}OCH_2CH_2CH_2CH_3$$

butyl acetate

(d)

$$CH_3(CH_2)_4\overset{O}{\underset{\parallel}{C}}O\underset{|}{\overset{}{C}}H(CH_2)_3CH_3$$
$$\underset{|}{CH_3}$$

1-methylpentyl hexanoate

(e)

N≡CCH$_2$C≡N

malononitrile

In (e), malononitrile hydrolyzes to give malonic acid and two equivalents of ammonium ion; the malonic acid then decarboxylates to give CO$_2$ and acetic acid.

21.41 (a)   In aqueous base the ester groups are saponified to give glycerol (1,2,3-propanetriol) and three equivalents of sodium oleate, the sodium salt of oleic acid.

sodium oleate   **glycerol**

(b)   The product of hydrogenation is glyceryl tristearate. Because unsaturated fats are oils and saturated fats are solids, glyceryl trioleate is a liquid, whereas the product of hydrogenation is a solid. In fact, glyceryl trioleate is a major component of olive oil, whereas glyceryl tristearate is a major component of lard.

$$CH_3(CH_2)_{16}\overset{O}{\underset{\parallel}{C}}O\text{—}CH_2$$
$$CH_3(CH_2)_{16}\overset{O}{\underset{\parallel}{C}}O\text{—}CH$$
$$CH_3(CH_2)_{16}\overset{O}{\underset{\parallel}{C}}O\text{—}CH_2$$

glyceryl tristearate

21.42 The order of increasing hydrolysis rate is $C < A < B$. Acid chlorides such as B react much faster than esters. Between the two esters, the methyl branches of the *tert*-butyl group in C provide steric interference to the approach of a nucleophile to the carbonyl carbon.

21.43 (a) Resonance stabilization of the carbonyl compound depends on the importance of the resonance contributor Y in each case. The resonance structure Y is much less important for thioesters (X = S) because the resonance interaction depends on the overlap of sulfur 3p orbitals with the 2p orbitals of the carbonyl group. Because this overlap is poor, resonance contributor Y is much less important for thioesters than it is for oxygen esters. This means that resonance stabilizes thioesters less than it does oxygen esters, and, on this basis, thioesters should be more reactive.

(b) Because of its greater electronegativity, oxygen in the OEt group exerts a greater stabilizing polar effect on the tetrahedral intermediate than the less electronegative sulfur in the SEt group.

> larger bond dipole;
> greater + charge on carbon;
> greater stabilization

> smaller bond dipole;
> smaller + charge on carbon;
> less stabilization

tetrahedral addition intermediates
(models for the addition transition state)

Therefore, on the basis of the polar effect, the oxygen ester should be more reactive.

The effects in (a) and (b) have opposite effects on rate, so the effects tend to cancel. In fact, the two types of ester undergo saponification at very similar rates.

21.44 (a)

CH₃CH₂CH₂COH → (SOCl₂) → CH₃CH₂CH₂CCl → 1) Li⁺ (CH₃)₂Cu⁻ 2) H₃O⁺ →

butyric acid

CH₃CH₂CH₂CCH₃ → 1) CH₃CH₂CH₂MgBr 2) H₃O⁺ → CH₃CH₂CH₂C(OH)(CH₃)CH₂CH₂CH₃

4-methyl-4-heptanol

(b)

CH₃CH₂CH₂COH → (C₂H₅OH, H₂SO₄) → CH₃CH₂CH₂CO₂C₂H₅ → 1) CH₃MgI 2) H₃O⁺ → CH₃CH₂CH₂C(OH)(CH₃)CH₃

2-methyl-2-pentanol

(c)

CH₃CH₂CH₂COH → (SOCl₂) → CH₃CH₂CH₂CCl → (H₂, quinoline, S, Pd/C) → CH₃CH₂CH₂CH → 1) CH₃CH₂CH₂MgBr 2) H₃O⁺ →

OH
|
CH₃CH₂CH₂CHCH₂CH₂CH₃
$$\text{OH} \atop \text{CH}_3\text{CH}_2\text{CH}_2\text{CHCH}_2\text{CH}_2\text{CH}_3$$

4-heptanol

(d)

$$\text{CH}_3\text{CH}_2\text{CH}_2\overset{\text{O}}{\overset{\|}{\text{C}}}\text{OH} \xrightarrow[\text{2) H}_3\text{O}^+]{\text{1) LiAlH}_4} \text{CH}_3\text{CH}_2\text{CH}_2\text{CH}_2\text{OH} \xrightarrow[\text{H}_2\text{SO}_4]{\text{conc. HBr}} \text{CH}_3\text{CH}_2\text{CH}_2\text{CH}_2\text{Br} \xrightarrow[\text{ether}]{\text{Mg}}$$

$$\text{CH}_3\text{CH}_2\text{CH}_2\text{CH}_2\text{MgBr} \xrightarrow[\text{2) H}_3\text{O}^+]{\text{1) }\triangle\text{O}} \text{CH}_3\text{CH}_2\text{CH}_2\text{CH}_2\text{CH}_2\text{CH}_2\text{OH} \xrightarrow[\text{2) H}_3\text{O}^+]{\text{1) KMnO}_4,\ ^-\text{OH}}$$

$$\text{CH}_3\text{CH}_2\text{CH}_2\text{CH}_2\text{CH}_2\text{CO}_2\text{H} \xrightarrow{\text{SOCl}_2} \text{CH}_3\text{CH}_2\text{CH}_2\text{CH}_2\text{CH}_2\overset{\text{O}}{\overset{\|}{\text{C}}}\text{Cl} \xrightarrow{\text{NH}_3\text{ (excess)}}$$

$$\text{CH}_3\text{CH}_2\text{CH}_2\text{CH}_2\text{CH}_2\overset{\text{O}}{\overset{\|}{\text{C}}}\text{NH}_2 \xrightarrow[\text{2) H}_2\text{O}^+]{\text{1) LiAlH}_4} \text{CH}_3\text{CH}_2\text{CH}_2\text{CH}_2\text{CH}_2\text{CH}_2\text{NH}_2$$

1-hexanamine (hexylamine)

(e)

$$\text{CH}_3\text{CH}_2\text{CH}_2\overset{\text{O}}{\overset{\|}{\text{C}}}\text{OH} \xrightarrow[\text{2) H}_3\text{O}^+]{\text{1) LiAlH}_4} \text{CH}_3\text{CH}_2\text{CH}_2\text{CH}_2\text{OH} \xrightarrow[\text{H}_2\text{SO}_4]{\text{conc. HBr}} \underset{A}{\text{CH}_3\text{CH}_2\text{CH}_2\text{CH}_2\text{Br}} \xrightarrow[\text{ether}]{\text{Mg}}$$

$$\text{CH}_3\text{CH}_2\text{CH}_2\text{CH}_2\text{MgBr} \xrightarrow[\text{2) H}_3\text{O}^+]{\text{1) CO}_2} \text{CH}_3\text{CH}_2\text{CH}_2\text{CH}_2\text{CO}_2\text{H} \xrightarrow{\text{SOCl}_2} \text{CH}_3\text{CH}_2\text{CH}_2\text{CH}_2\overset{\text{O}}{\overset{\|}{\text{C}}}\text{Cl} \xrightarrow{\text{NH}_3\text{ (excess)}}$$

$$\text{CH}_3\text{CH}_2\text{CH}_2\text{CH}_2\overset{\text{O}}{\overset{\|}{\text{C}}}\text{NH}_2 \xrightarrow[\text{2) H}_2\text{O}]{\text{1) LiAlH}_4} \text{CH}_3\text{CH}_2\text{CH}_2\text{CH}_2\text{CH}_2\text{NH}_2$$

1-pentanamine (pentylamine)

An alternative is to convert the bromoalkane *A* into the nitrile with cyanide ion, and then produce the target molecule by catalytic hydrogenation of this nitrile with Raney nickel at high pressure.

(f)

$$\text{CH}_3\text{CH}_2\text{CH}_2\overset{\text{O}}{\overset{\|}{\text{C}}}\text{OH} \xrightarrow{\text{SOCl}_2} \text{CH}_3\text{CH}_2\text{CH}_2\overset{\text{O}}{\overset{\|}{\text{C}}}\text{Cl} \xrightarrow[\text{(excess)}]{\text{NH}_3} \text{CH}_3\text{CH}_2\text{CH}_2\overset{\text{O}}{\overset{\|}{\text{C}}}\text{NH}_2 \xrightarrow[\text{2) H}_2\text{O}]{\text{1) LiAlH}_4} \text{CH}_3\text{CH}_2\text{CH}_2\text{CH}_2\text{NH}_2$$

1-butanamine (butylamine)

21.45 (a)

$$\text{CH}_3(\text{CH}_2)_4\overset{\text{O}}{\overset{\|}{\text{C}}}-\text{OH} \xrightarrow{\text{SOCl}_2} \text{CH}_3(\text{CH}_2)_4\overset{\text{O}}{\overset{\|}{\text{C}}}-\text{Cl} \xrightarrow[\text{(excess)}]{\text{H}_2\text{NCH}_3} \text{CH}_3(\text{CH}_2)_4\overset{\text{O}}{\overset{\|}{\text{C}}}-\text{NHCH}_3$$

hexanoic acid   *N*-**methylhexanamide**

(b)

o-bromotoluene

o-methylbutyrophenone

(c)

bromocyclohexane

1-cyclohexyl-2-methyl-2-propanol

(d)

ethyl 7-oxooctanoate

8-hydroxy-2-octanone

(e)

isovaleric acid

(f)

cyclohexanone

(g)

toluene

$\xrightarrow{\text{Br}_2,\ \text{FeBr}_3}$

$\xrightarrow[\text{ether}]{\text{Mg}}$

MgBr

1) 3-pentanone

2) H_3O^+

$\xrightarrow{\text{acid},\ -H_2O}$

(h)

benzoic acid

$\xrightarrow[\text{H}_2\text{SO}_4]{\text{HNO}_3,}$

CO_2H / NO_2

$\xrightarrow{\text{PCl}_5}$

NO_2

$\downarrow$ NaOH

$CO_2^-\ Na^+$

21.46   (a)   Synthesis of nylon-4,6 requires mixing 1,4-butanediamine and adipoyl dichloride.

$$H_2N(CH_2)_4NH_2\ +\ ClC(CH_2)_4CCl \longrightarrow \text{nylon-4,6}$$

1,4-butanediamine   adipoyl dichloride

(Alternatively, 1,4-butanediamine can be heated with adipic acid; see Eq. 21.104 on text p. 1179.) The preparations of these materials are as follows:

$$\text{HOCCH}_2\text{CH}_2\text{COH} \xrightarrow{\text{SOCl}_2} \text{ClCCH}_2\text{CH}_2\text{CCl} \xrightarrow{\text{NH}_3\ (\text{excess})} \text{H}_2\text{NCCH}_2\text{CH}_2\text{CNH}_2 \xrightarrow[\substack{2)\ \text{H}_2\text{O}}]{1)\ \text{LiAlH}_4} \text{H}_2\text{N(CH}_2)_4\text{NH}_2$$

succinic acid   **1,4-butanediamine**

$$\text{HOC(CH}_2)_4\text{COH} \xrightarrow{\text{SOCl}_2} \text{ClC(CH}_2)_4\text{CCl}$$

adipic acid   **adipoyl dichloride**

 Why do you think heating the dicarboxylic acid with the diamine is preferred industrially to reaction of the diamine with the dicarboxylic acid dichloride? (There are at least two reasons.)

(b)   The preparation of this polyester requires heating 1,4-butanediol with glutaric acid. (Alternatively, the di-acid chloride of glutaric acid could be mixed with 1,4-butanediol.)

21.47 (a)

(b)

21.48 (a) Compound *A* is a fat. Compound *B* is a phospholipid. Compound *C* is a soap, and therefore none of the classifications given. Compound *D* is a wax.

(b) Compounds *B* and *C* have both polar and nonpolar ends and are therefore the amphipathic molecules.

Solutions to Integrated Problems

21.49 (a) First, two sets of constitutional isomers can be obtained from the reaction of 1-phenylethanol at either of the nonequivalent carbonyl groups of the anhydride:

Each of these products has two asymmetric carbon stereocenters and can thus exist as four stereoisomers. All stereoisomers of *A* can in principle be separated from all stereoisomers of *B* because constitutional isomers have different properties. The four stereoisomers of *A* can be separated into two pairs of enantiomers; the same is true of *B*. Consequently, four compounds can in principle be obtained from this reaction mixture without enantiomeric resolution.

(b)   The same two constitutional isomers would be obtained, and each would be obtained as a pair of diastereomers. However, only one enantiomer of each diastereomer (the one with the *S* configuration at the carbon derived from the alcohol) would be formed.

If the diastereomeric esters from part (b) were separated from each other and saponified, the α-phenylglutaric acid formed would be enantiomerically pure. The process described here would therefore be useful as an enantiomeric resolution of α-phenylglutaric acid (or any other chiral dicarboxylic acid from which a cyclic anhydride can be readily formed).

21.50   Because the two carbonyl groups of the anhydride are not equivalent, two different products are possible; each corresponds to reaction with the ethanol at a different carbonyl group. Because ethyl propionate has a higher molecular mass than ethyl acetate, ethyl propionate should have the higher boiling point. Since it is given that the ester with the lower boiling point is formed in greater amount, then ethyl acetate is the major product.

$$
\underset{\textbf{acetic propionic anhydride}}{CH_3\overset{\displaystyle O}{\overset{\|}{C}}-O-\overset{\displaystyle O}{\overset{\|}{C}}CH_2CH_3} \;+\; CH_3CH_2OH \longrightarrow
$$

$$
\underset{\textbf{ethyl acetate}}{CH_3\overset{\displaystyle O}{\overset{\|}{C}}-OCH_2CH_3} \;+\; HO-\overset{\displaystyle O}{\overset{\|}{C}}CH_2CH_3 \;+\; CH_3CH_2O-\overset{\displaystyle O}{\overset{\|}{C}}CH_2CH_3 \;+\; \underset{\textbf{ethyl propionate}}{CH_3\overset{\displaystyle O}{\overset{\|}{C}}-OH}
$$

major products

21.51   The easiest way to work this problem is to start with the structure and work backwards. Because carbonation of a Grignard reagent gives a carboxylic acid, compound *D* must be the corresponding alkyl bromide. Compound *E* must be the hydrocarbon that gives compound *D* on benzylic bromination. The Wolff–Kishner reaction produces compound *E* from compound *B*, which therefore must be a ketone. Reduction of ketone *B* with sodium borohydride gives alcohol *C*, which is converted into alkyl bromide *D*. Finally, the structure of compound *B* shows that reaction *A* must be a Friedel–Crafts acylation. In summary:

$$
\underset{A}{Cl\overset{\displaystyle O}{\overset{\|}{C}}CH_2CH_3,\ AlCl_3,\ then\ H_3O^+} \qquad \underset{B}{Ph\overset{\displaystyle O}{\overset{\|}{C}}CH_2CH_3} \qquad \underset{C}{Ph\overset{OH}{\overset{|}{C}H}CH_2CH_3} \qquad \underset{D}{Ph\overset{Br}{\overset{|}{C}H}CH_2CH_3} \qquad \underset{E}{PhCH_2CH_2CH_3}
$$

21.52   The product is (*R*)-(–)-1-phenyl-1,2-ethanediol. Notice that the asymmetric carbon is unaffected by these transformations.

$$
\underset{\textbf{(R) - (–)-mandelic acid}}{H\overset{\displaystyle OH}{\underset{\displaystyle Ph}{\overset{|}{\underset{|}{C}}}}CO_2H} \xrightarrow[H_2SO_4]{CH_3OH,} \underset{}{H\overset{\displaystyle OH}{\underset{\displaystyle Ph}{\overset{|}{\underset{|}{C}}}}CO_2CH_3} \xrightarrow[2)\ H_3O^+]{1)\ LiAlH_4} \underset{\substack{\textbf{(R) - (–)-1-phenyl-1,2-ethanediol} \\ \text{(a 1,2-glycol; reacts with periodic acid)}}}{H\overset{\displaystyle OH}{\underset{\displaystyle Ph}{\overset{|}{\underset{|}{C}}}}CH_2OH}
$$

21.53   Lithium aluminum hydride reduces both the carboxylic acid and the ketone [as shown in [a] below), whereas sodium borohydride reduces the ketone [as shown in (b) below] but reacts only with the acidic hydrogen of the carboxylic acid to give the sodium carboxylate, which re-forms the carboxylic acid when acid is added.

1,4-pentanediol

levulinic acid

(a)
1) LiAlH₄ (excess)
2) H₃O⁺

(b)
1) NaBH₄ (excess)
2) H₃O⁺

4-hydroxypentanoic acid

21.54 (a) Because phenols are very reactive in electrophilic substitution reactions, bromine in CCl₄ will effect bromination of the phenol ring as well as addition to the alkene double bond. (See Eq. 18.99a–c in Sec. 18.9, on text p. 1005.)

bromine addition occurs with anti stereochemistry to give the enantiomeric (S,R) and (R,S) stereoisomers in equal amounts.

(b) Dilute aqueous NaOH will bring about ionization of the phenol to its conjugate-base phenolate ion. The amide is unaffected, because amide hydrolysis requires heat and strong base.

(c) No significant reaction occurs with dilute aqueous HCl. (The carbonyl oxygen of the amide is protonated to a very small extent.)
(d) Catalytic hydrogenation brings about addition of hydrogen to the alkene double bond

Depending on the conditions, hydrogenation of the *single bond* between the benzylic CH₂ and the nitrogen can also occur; such a reaction is called *hydrogenolysis*. Because you have not studied this reaction, it is ignored.

(e)   Heating the product of (d) in 6 *M* HCl brings about amide hydrolysis. (It is also conceivable that some cleavage of the methyl ether will occur to give methyl chloride and the diphenol.)

(f)   The phenolate ion formed in part (b) is alkylated to give a second methyl ether group.

(g)   On heating, the same amide hydrolysis observed in part (e) will take place, and the methyl ether will cleave.

21.55   The IR spectrum of compound *A* indicates that it is an anhydride, and probably a cyclic anhydride containing a five-membered ring. (Compare the carbonyl absorptions of compound *A* with those in Table 21.3 in Sec. 21.4A on text p. 1140; the 1050 cm^{-1} absorption is a C—O stretching absorption). Addition of methanol to a cyclic anhydride should give a methyl half-ester, compound *B*. Indeed, the IR spectrum of compound *B* indicates the presence of both a carboxylic acid and an ester group. Subtracting the elements of methanol (CH_4O) from the formula of compound *B* gives the formula of compound *A*, $C_4H_4O_3$. Compound *A* is succinic anhydride.

Two points about the NMR spectra are worth noting. First, the methylene protons of compound *B* give a complex resonance at δ 2.7 because they are chemically nonequivalent but have very similar chemical shifts. This is the condition under which complex, non-first-order splitting can occur. (See text Sec. 14.5B, pp. 759–762.) That these protons should have similar chemical shifts is a reasonable observation because both sets of protons are in very similar electronic environments: both are on carbons that are α to carbonyl groups. The second point is that the proton of the carboxy group of compound *B* is not observed because the spectrum is taken in D_2O. Recall (Sec. 14.7E, text pp. 770–771) that O—H protons are rapidly exchanged for deuterium when D_2O is present, and that deuterium nuclei are "silent" in proton NMR.

succinic anhydride
(compound *A*)

methyl hydrogen succinate
(compound *B*)

21.56 The high-frequency IR absorption and the reaction with water to give a dicarboxylic acid suggest that this compound is the anhydride of dimethylmalonic acid. The formation of a half ester in the reaction with methanol also supports the hypothesis that this compound is an anhydride.

dimethylmalonic anhydride

21.57 Use the titration data to calculate the molecular mass of the ester. The amount of 1 M NaOH solution consumed in the saponification reaction is (15.00 mL – 5.30 mL) = 9.70 mL. Make the provisional assumption that there is one ester group per molecule of compound A. Then one mole of NaOH is consumed per mole of ester saponified, and the molecular mass of compound A is then

$$\text{molecular mass of } A = \frac{2.00 \text{ g}}{(9.70 \times 10^{-3} \text{ L})(1.00 \text{ mol L}^{-1})} = 206 \text{ g mol}^{-1}$$

Next, try to deduce a structure from the data in the problem. The structure of compound C, the alcohol that results from the saponification of compound A, can be determined; it is the alcohol that can be oxidized to acetophenone. Therefore, compound C is 1-phenylethanol.

1-phenylethanol (compound C) $\xrightarrow{\text{K}_2\text{Cr}_2\text{O}_7}$ **acetophenone**

The molecular mass of compound C is 122 g mol^{-1}. Now, from the overall transformation of ester hydrolysis, the molecular mass of ester A must equal the sum of the molecular masses of compounds B and C minus the molecular mass of H_2O, that is,

$$\text{molecular mass of } A = 206 \text{ g mol}^{-1} = \text{molecular mass of } B + 122 \text{ g mol}^{-1} - 18 \text{ g mol}^{-1}$$

or

$$\text{molecular mass of } B = 206 \text{ g mol}^{-1} + 18 \text{ g mol}^{-1} - 122 \text{ g mol}^{-1} = 102 \text{ g mol}^{-1}$$

Compound B is a carboxylic acid; its carboxy group accounts for 45 molecular mass units. Therefore, the remainder of the carboxylic acid has a mass of 57 units, which is the mass of a butyl group. Because acid B is optically active, it is chiral; therefore the butyl group must contain an asymmetric carbon stereocenter. Only a *sec*-butyl group meets this criterion. The structures of compounds B and A are therefore

2-methylbutanoic acid (compound B)

1-phenylethyl 2-methylbutanoate (compound A)

As noted in the problem, the absolute configurations of the asymmetric carbons cannot be determined from the data. Hence, compounds B and C are each one of two possible enantiomers, and compound A is the particular stereoisomer with the corresponding absolute configurations of its asymmetric carbons. Compound D is the racemate of compound C, and compound E is the diastereomer of compound A with the following configurations:

same configuration as compound A

opposite configuration to compound A

compound E

21.58   Determine the structure of compound B and reason backwards. Compound B is a diol because it gives a diacetate derivative C. Because compound B is the diol that can be oxidized to β-methylglutaric acid, compound B is 3-methyl-1,5-pentanediol.

Compound A can be neither a hydroxy acid, a dicarboxylic acid, an oxo acid, nor an ester of these, because it contains only two oxygens. If it has only one carboxy group (either as a carboxylic acid or an ester) then it must contain either a ring or a carbon–carbon double bond. The structure of the LiAlH$_4$ reduction product B rules out a carbon–carbon double bond; therefore, compound A contains a ring. Therefore, compound A is a lactone, and X is the sodium salt of the corresponding hydroxy acid, which is soluble in aqueous base. Compound A appears to dissolve when it is converted into the water-soluble compound X.

The absolute configuration of compound A is not defined by the problem.

21.59   (a)   The very weak acid HCN is not significantly dissociated. Consequently, the solution is insufficiently acidic to effect protonation of the —OH group of the alcohol. This protonation is necessary to convert this group into a good leaving group. Furthermore, there is virtually no cyanide ion ($^-$C≡N) present, so there is virtually no nucleophile to displace the —OH group.

(b)   First of all, an excess of an alcohol is generally required to drive acid-catalyzed esterification to completion. However, even if some of the adipic acid is converted into its ester, there is no reason why this monoester would not be essentially as reactive as adipic acid itself, and the reaction mixture would ultimately contain a mixture of adipic acid, its monomethyl ester, and its dimethyl ester. The yield of the desired monoester would be poor, and it would have to be separated from both the di-ester and the unreacted adipic acid.

(c)   There is no reason why acetic acid should form an anhydride only with benzoic acid, and vice-versa; substantial amounts of acetic anhydride and benzoic anhydride should also be obtained. If all the reactions were totally random, the desired unsymmetrical anhydride would be formed in a maximum of 50% yield.

(d)   The hydroxide ion reacts *much* more rapidly with the O—H proton of the phenol than it does with the ester. (See Study Guide Link 20.1 on p. 657 of this manual for a discussion of this point.) Consequently, the one equivalent of hydroxide is consumed by this reaction, and no base is left to saponify the ester.

(e)   Several functional groups in the β-lactam molecule are more reactive toward acid hydrolysis than the amide indicated. Because esters are more reactive than amides, the acetate ester will undoubtedly also hydrolyze under the reaction conditions. The β-lactam ring itself is an amide, and because hydrolysis of this amide relieves substantial ring strain, this amide should also hydrolyze more rapidly. Once the β-lactam hydrolyzes, the resulting enamine also should hydrolyze readily in acid, and the N—C—S linkage is also unstable toward hydrolysis in the same sense that an acetal is unstable. With so many faster competing processes, Klutz has no hope for a selective reaction.

21.60 (a) The "alcohol" formed as one of the saponification products is an enol, which spontaneously reverts to the corresponding ketone. (See Eqs. 5.27b–c in Sec. 5.5 on text p. 249.)

(b) Transesterification of the formate ester occurs.

(c) The formula shows that the product includes the carbonyl group of phosgene (the acid chloride) minus two HCl molecules. Because the product has an unsaturation number of 7, a ring must be formed. Because both amines and carboxylic acids can react with acid chlorides, the product, called *isatoic anhydride*, is simultaneously a cyclic anhydride and a cyclic amide (lactam).

isatoic anhydride

(d) Both amino groups are acylated by the excess of the acid chloride.

(e) Both the amide and the nitrile are **hydrolyzed.**

$$PhCO_2H \ + \ H_2\overset{+}{N}(CH_2)_5CO_2H \ + \ \overset{+}{N}H_4$$

(f) The cyclic amide is reduced to a cyclic amine.

(g) The Grignard reactions of lactones are much like those of esters, except that the alcohol displaced by the nucleophilic acyl substitution reaction remains as part of the same molecule.

(h) Diethyl carbonate reacts first to give an ester, then a ketone, and then a tertiary alcohol (after protonolysis) in which all three groups attached to the α-carbon are the same.

From this pattern, the product of the reaction given is 3-ethyl-3-pentanol (R = ethyl in the preceding tertiary alcohol structure).

$$CH_3CH_2\!-\!\underset{\underset{CH_2CH_3}{|}}{\overset{\overset{OH}{|}}{C}}\!-\!CH_2CH_3 \quad \textbf{3-ethyl-3-pentanol}$$

(i) Lithium aluminum hydride reacts with acid chlorides in the same way that it reacts with esters, only faster. The product is $(CH_3)_3CCH_2OH$ (2,2-dimethyl-1-propanol).

(j) Because aldehydes are much more reactive than esters, only the aldehyde reacts when only one equivalent of the Grignard reagent is used.

$$CH_3O\overset{\overset{O}{\|}}{C}(CH_2)_3\overset{\overset{OH}{|}}{C}HPh$$

methyl 5-hydroxy-5-phenylpentanoate

(k) Transesterification occurs to give three equivalents of methyl stearate and one of glycerol.

$$3\ CH_3(CH_2)_{16}CO_2CH_3\ +\ \underset{}{CH_2\!-\!CH\!-\!CH_2}$$

methyl stearate   **glycerol**

(l) The ester group undergoes intramolecular aminolysis to form a γ-lactam. Such intramolecular nucleophilic substitution reactions are very common when five- and six-membered rings can be formed. This is another manifestation of the greater rate of intramolecular reactions (Sec. 12.8 of the text).

$+\ CH_3OH$

21.61 (a) Compound *A* has the formula $C_{11}H_{13}NO$. The formula of compound *B* indicates that it differs from *A* by the addition of the elements of water. The fact that compound *C* is a lactone with an oxygen at carbon-4 and an additional hydrogen at carbon-3 suggests that addition of water to the double bond of compound *A*, that is, alkene hydration, occurs metabolically to give an alcohol *B*. The lactone *C* is formed by cyclization of compound *B*, that is, an intramolecular esterification. The structure of compound *B* is given in the mechanism of part (b).

(b) The curved-arrow mechanism for conversion of compound *B* into compound *C*:

4-hydroxy-3-methyl-4-phenylbutanamide
(compound B)

+ OH$_2$

+ NH$_3$

(protonated under the acidic conditions)

C

21.62 (a) The reversible hydration of carbon dioxide to carbonic acid is an acid-catalyzed addition of water that is mechanistically identical to the hydration of aldehydes and ketones shown in Eq. 19.36a–b, Sec. 19.8A, text p. 1044.

(b) The first step forms a Grignard reagent. The carbon-magnesium bond is highly polarized towards carbon, and reacts to form an alkene and displace the carboxylate leaving group.

(c) Protonation of the carbonyl oxygen produces a good leaving group which is displaced by bromide ion. The carboxy group is then esterified by the usual mechanism (Eqs. 20.34a–d, Sec. 20.6A, text pp. 1103–1104).

$$Br \diagup\diagdown\diagup\overset{O}{\underset{}{C}}-OH \xrightarrow[\text{(esterification)}]{C_2H_5OH, HBr} Br \diagup\diagdown\diagup\overset{O}{\underset{}{C}}-OC_2H_5$$

An alternative reasonable mechanism involves transesterification of the lactone by ethanol to give a hydroxy ester, followed by conversion of the alcohol to a bromide by an acid-catalyzed S_N2 reaction. (See Eqs. 11.25a–c, Sec. 11.3, text p. 574.) However, it is likely that the mechanism shown above is the correct one, because the equilibrium between lactone and hydroxy ester is likely to strongly favor the lactone, and because a protonated carboxy group is an excellent leaving group.

(d) A mercurinium ion forms by the standard mechanism for alkene oxymercuration (Eq. 5.20a, Sec. 5.4A, p. 246), which generates an equivalent of trifluoroacetate. The nonbonding electrons on the carbonyl oxygen react with the alkene, which. in turn, opens the mercurinium ion in a series of steps that are likely concerted, giving compound A. Sodium borohydride then exchanges the mercury for hydrogen. Notice the anti addition to the double bond.

(e) Protonation of the ring double bond gives a tertiary carbocation that reacts with the carbonyl oxygen of the ester to give the lactone.

As an alternative to the step marked (*), the water, instead of reacting at the methyl group, could react as a nucleophile at the carbon marked (†) to form a tetrahedral addition intermediate with eventual loss of methanol and formation of the carbonyl group in a manner similar to ester hydrolysis. (We leave it to you to show this alternative.)

Yet another alternative is hydration of the double bond to form a tertiary alcohol, which then transesterifies the methyl ester to form a lactone.

We chose the first alternative shown because the reaction of the carbocation with the ester carbonyl oxygen is an intramolecular reaction, whereas the reaction with water is an intermolecular reaction; recall (Sec. 12.8) that intramolecular reactions typically have a significant kinetic advantage.

21.63 Addition of the Grignard reagent to the carbon–nitrogen triple bond is expected by analogy with carbonyl-group reactions. The product A is an imine.

The $C=\ddot{N}$: cannot add another equivalent of the Grignard reagent; why?

This imine hydrolyzes in acidic solution to benzophenone:

21.64 Compound A is an O-substituted oxime, which is formed by the mechanism summarized in Eqs. 19.76a–b in Sec. 19.12A on text p. 1064. Compound A undergoes a base-promoted β-elimination to give the nitrile. (In the following mechanism, the 2,4-dinitrophenyl group is abbreviated Ar.)

$$Ph-C \equiv N + H-OH + \bar{}OAr$$

A

21.65   (a)   Structure of the Weinreb amide:

Weinreb amide

(b)   The first reason that this intermediate doesn't break down is because the leaving group is a very strong base (conjugate acid $pK_a \approx 25$–30); strong bases are very poor leaving groups.

strong amide base, therefore poor leaving group

Additionally, the lithium ion (or magnesium, if this were a Grignard reaction) forms a complex with the intermediate, thus stabilizing it and helping it resist decomposition.

(c)   The mechanism for decomposition of the tetrahedral intermediate in acidic solution:

(d)   The product of this Weinreb reaction is 1-cyclohexyl-1-butanone:

A   *B*   **1-cyclohexyl-1-butanone**

(e)   The point of the Weinreb process is that, as shown in part (b), the tetrahedral intermediate does not break down until water is added. In the reaction of a Grignard or organolithium reagent with an alkyl halide, the ketone is formed in the reaction mixture, and it can react further to give an alcohol.

21.66   Compound *A* is an oxime (see Table 19.3 in Sec. 19.12A on text p. 1065). It is formed by the mechanism for imine formation summarized in Eqs. 19.76a–b in Sec. 19.12A on text p. 1064. Protonation of the —OH group and loss of water *(a)* gives an electron-deficient nitrogen, a very unstable species. Subsequent rearrangement *(b)* gives a carbocation that is somewhat resonance stabilized. (We say "somewhat" because incorporation of a triple bond within a seven-membered ring introduces considerable strain.) This cation reacts with water, and, after several proton-transfer steps, ε-caprolactam is formed.

cyclohexanone
oxime
(compound *A*)

an imidic acid

ε-caprolactam

21.67   (a)

3-methylbenzaldehyde

N,N-diethyl-3-methylbenzamide
compound *A*

(b)

2-bromobenzoic acid

compound *B*

(c)

phthalic acid

(d)

(e)

$$+ \ HOC_2H_5$$

(f)

p-methylanisole
(*p*-methoxytoluene,
1-methoxy-4-methylbenzene)

(g)

phosgene
(excess)

methyl chlorocarbonate

The remaining phosgene (bp = 8.3 °C), a gas at room temperature, is easily removed from methyl chlorocarbonate (bp 73 °C), the product of the first reaction. (The toxic phosgene can be passed into aqueous NaOH in which it decomposes to sodium carbonate and sodium chloride.)

(h)

p-bromotoluene

21.68 We follow the discussion in Study Guide Link 21.5, or the part of Sec. 11.4 beginning on text page 568. Alkyl esters of sulfonic acids react with nucleophiles at *the carbon of the alkyl group,* whereas carboxylic acid esters react at *the carbonyl carbon.* The benzyl mesylate ester *A* reacts at the carbon of the benzyl group to give ^{18}O-labeled benzyl alcohol and unlabeled methane sulfonate anion. Benzyl acetate, in contrast, gives ^{18}O-labeled acetate ion by ester saponification and unlabeled benzyl alcohol. (*O = ^{18}O)

21.69 (a) Because the carbonyl group has trigonal-planar geometry, it must undergo internal rotation about the bond to the ring so that its plane is perpendicular to the plane of the benzene ring in order to avoid significant van der Waals repulsions with the ortho methyl groups. This occurs even though conjugation between the carbonyl π electrons and the ring π electrons is lost.

conformation of
lower energy

(b) Remember that nucleophiles approach a carbonyl carbon from above or below the plane of the carbonyl group. (See Fig. 21.4 in Sec. 21.7A on text p. 1149.) In the "perpendicular" conformation shown on the right, the path of a nucleophile to the carbonyl carbon is blocked on both sides by the ortho methyl groups. Because ester hydrolysis occurs by the nucleophilic reaction of a water molecule at the carbonyl carbon, this mechanism, and thus the hydrolysis reaction, cannot occur.

(c) For the same reason that ester *hydrolysis* cannot occur, ester *formation* also cannot occur by any mechanism that involves the reaction of a nucleophile at the carbonyl carbon. Because acid-catalyzed esterification involves the nucleophilic reaction of an alcohol molecule at the carbon of a protonated carbonyl group, this reaction does not take place. However, esterification with diazomethane can occur because the carboxylate oxygen reacts as a nucleophile with protonated diazomethane (see Eq. 20.38b in Sec. 20.8B on text p. 1106). The reaction is less susceptible to van der Waals repulsions because it takes place at an atom that is more remote from the ortho methyl groups.

21.70 (a) Phthalic anhydride can be formed by the nucleophilic reaction of a carboxy group at the carbonyl carbon of the protonated amide. (Note that the amide carbonyl oxygen is about 10^5 times as basic as the carbonyl oxygen of the carboxylic acid; see Eqs. 21.28a–b on text p. 1144.)

phthalic anhydride

(b)   The phthalic anhydride is labeled with ^{13}C on only one of its two carbonyl carbons. However, the two carbonyl groups, except for isotopic differences, are chemically indistinguishable and therefore react at essentially equal rates with isotopic water:

from the reaction of   from the reaction of
isotopic water at   isotopic water at
carbonyl (*a*)   carbonyl (*b*)

(c)   Anchimeric assistance (neighboring-group participation) by the carboxylic acid group provides a rate acceleration. (See Sec. 12.8A, text p. 653.)

22 | The Chemistry of Enolate Ions, Enols, and α,β-Unsaturated Carbonyl Compounds

CHAPTER 22 SUMMARY

■ Hydrogens on carbon atoms α to carbonyl groups and cyano groups are acidic. Ionization of these hydrogens gives enolate ions.

■ Most carbonyl compounds with α-hydrogens are in equilibrium with small amounts of enol (vinylic alcohol) isomers. Generally, carbonyl–enol equilibria favor the carbonyl compounds, but there are important exceptions, such as phenols and β-diketones, in which enol isomers are the major forms.

■ The equilibria between carbonyl compounds and their enol isomers are catalyzed by acids and bases.

■ Enolate ions can act as nucleophiles in the following types of reactions:

- α-Halogenation (haloform reactions)

- Addition to carbonyl groups (aldol addition and condensation, Reformatsky reaction, and Knoevenagel condensation)

- Nucleophilic carbonyl-substitution reactions (Claisen and Dieckmann condensations)

- Nucleophilic substitution reactions with alkyl halides (enolate alkylation, as in the malonic ester synthesis and the acetoacetic ester synthesis)

- Addition to α,β-unsaturated carbonyl compounds (Michael addition)

■ Enols of aldehydes and ketones undergo α-halogenation and aldol condensation reactions in acidic solution.

■ The aldol addition and the Claisen condensation are reversible reactions. The aldol addition is generally favorable for aldehydes, but unfavorable for most ketones. It can be driven to completion by dehydration of the addition product, a β-hydroxy carbonyl compound (aldol condensation). The Claisen condensation is driven to completion by ionization of the product; for that reason, one full equivalent of base must be used in the reaction.

■ Aldol additions to aldehydes and ketones can be carried out with lithium and zinc enolates (Reformatsky reagent) of esters and with lithium enolates of ketones (followed by protonation) because these reactions are irreversible.

■ Two types of addition to α,β-unsaturated carbonyl compounds are possible: addition to the carbonyl group and addition to the double bond (conjugate addition, or 1,4-addition). When carbonyl addition is reversible, conjugate addition is observed because it gives the more stable product. When carbonyl addition is irreversible, it is usually observed instead because it is faster.

■ Reactions closely resembling aldol additions and Claisen condensations are observed in living systems. Many such reactions employ acetyl-CoA as a starting material.

STUDY GUIDE LINKS

22.1   Ionization versus Nucleophilic Reaction at the Carbonyl Carbon

One of the things you learned about Brønsted bases is that they can usually act as nucleophiles, and that nucleophiles react at the carbon atoms of carbonyl groups. Yet in Eqs. 22.3 and 22.4 of the text (p. 1194), bases are shown removing protons. You might be wondering, "What is the major reaction: a nucleophilic reaction at the carbonyl carbon, or the removal of an α-hydrogen to give an enolate ion?" The answer is that both can occur. This section of the text focuses on the proton removal, but later sections of this chapter show the interplay between both types of reactions. Nevertheless, it might help you to see just one example here.

Suppose an ester is treated with aqueous NaOH. Recall (Sec. 21.7A) that NaOH saponifies esters. Saponification results from a nucleophilic reaction of hydroxide ion at the carbonyl carbon of an ester. However, while saponification is taking place, hydroxide ion removes the α-protons of a few ester molecules to form some enolate ions. This reaction has no consequences for saponification because saponification occurs much more rapidly than the reactions of enolate ions (many of which are discussed in this chapter). For this reason, it was not important to consider enolate-ion formation in the discussion of saponification.

The point of this chapter is to focus on situations in which enolate ion formation is significant, and ultimately on what happens when *both* proton removal and nucleophilic reaction at a carbonyl carbon can occur.

22.2   Kinetic versus Thermodynamic Stability of Enols

When we think of unstable organic compounds we often think of molecules containing strained rings or molecules containing van der Waals repulsions. The structures of enols have no such destabilizing features. Yet enols in most cases cannot exist for significant lengths of time, while other less stable molecules can be put in bottles and stored for years. Why this seeming contradiction?

The essence of this problem is the meaning of the word *stability* as applied to chemical substances. *Thermodynamic stability* is a term used to describe the relative energy contents of molecules. In general, thermodynamic stability has very little to do with whether a compound can be isolated. *Kinetic stability* is a term used to describe how rapidly molecules revert to more stable molecules. In order for a molecule to be *kinetically unstable*, a mechanism must exist for its conversion into a more stable molecule. For a compound to be kinetically unstable, it must be thermodynamically unstable, but the converse need not be true. Some *thermodynamically unstable compounds* are easy to isolate because they are *kinetically stable*. For example, almost all hydrocarbons in the presence of oxygen are thermodynamically unstable relative to CO_2 and H_2O by hundreds of kJ mol^{-1}; yet hydrocarbons do not rapidly revert to CO_2 and H_2O because there is no readily available mechanism by which this conversion can be accomplished. If we supply a flame, however, such a conversion mechanism becomes available—we call it *combustion*—and we see very graphically how unstable hydrocarbons really are!

Examples of kinetic stability and instability in carbonyl chemistry are provided by acetals (Sec. 19.11, text p. 1058) and carbonyl hydrates (Sec. 19.8B, text p. 1046). The hydrate of cyclohexanone is both thermodynamically and kinetically unstable with respect to the ketone and water:

cyclohexanone hydrate   **cyclohexanone**

Yet replacing the O—H hydrogens with alkyl groups gives compounds called *acetals* that are thermodynamically unstable, but *kinetically stable*, in neutral and basic solution.

cyclohexanone dimethyl acetal   **cyclohexanone**
(1,1-dimethoxycyclohexane)

The decomposition of hydrates in base requires the presence of an O—H hydrogen, which acetals lack. Hence, the base-catalyzed pathway of hydrate decomposition cannot operate with acetals. Therefore, acetals, although thermodynamically unstable relative to their corresponding carbonyl compounds, are kinetically stable in basic solution. This is why acetals are good protecting groups for carbonyl groups under basic or neutral conditions. Under acidic conditions, however, acetals are kinetically unstable because an acid-catalyzed mechanism exists for their conversion into ketones.

The thermodynamic stabilities of most enols are not much less than the stabilities of their corresponding carbonyl isomers—typically, about 30–40 kJ mol^{-1} (7–10 kcal mol^{-1}). What makes enols difficult to isolate is their *kinetic* instabilities. Simple paths, or mechanisms, exist for their rapid transformation into carbonyl compounds in the presence of acids or bases; these mechanisms are shown in Eqs. 22.21a–b in the text (p. 1202). Enols are kinetically unstable molecules because the conditions under which they are converted into carbonyl compounds—the presence of acids and/or bases—are so prevalent. It's hard to generate an enol without an acid or base around. In fact, enols themselves are weak acids, and can catalyze their own slow conversion into ketones. Enols simply don't stand much of a chance for survival.

In the mid-1970s, chemists figured out ways to generate simple enols in solution in the absence of other acids and bases and, indeed, found that they could be isolated and observed for significant periods of time.

22.3   Dehydration of β-Hydroxy Carbonyl Compounds

At first sight, the dehydration of β-hydroxy carbonyl compounds would appear to be an ordinary acid-catalyzed dehydration as discussed in Sec. 11.2 of the text. Recall that, in the mechanism of such a dehydration, protonation of the hydroxy group is followed by a Lewis acid–base dissociation to form a carbocation and water:

Loss of a C—H proton from carbocation A gives the α,β-unsaturated ketone. However, a different mechanism—and, as it turns out, the correct one—is possible. This mechanism involves dehydration not of the ketone, but instead of the *enol*.

The dehydration involves protonation of the —OH group, formation of carbocation B, and loss of a proton from this carbocation:

The reason the enol mechanism is favored is that *carbocation B is much more stable than carbocation A*. Can you see why? Draw resonance structure(s) for carbocation B. Is carbocation A resonance-stabilized? What is the *polar effect* of the carbonyl group on the stability of carbocation A?

22.4   Understanding Condensation Reactions

Condensations such as the aldol condensation and its variants, as well as the other reactions you'll encounter in Chapter 22, present a particular challenge to the beginning student, because the products are somewhat more complex than the products of many typical organic reactions, and because more than one functional-group transformation is involved (an addition followed by a dehydration). The way to develop an understanding of these reactions is to write the mechanism of each new reaction you see. First, identify the sources of the atoms in the product. Use the base-catalyzed aldol reaction of acetaldehyde (Eqs. 22.45a–b and 2.46, text pp. 1210–1211, and the acid-catalyzed aldol reaction of acetone (Eqs. 22.52a–c, text pp. 1213–1214) as models. If you still have difficulty, *seek assistance!* You should have taken this approach with the crossed-aldol products in Study Problem 22.1 on text p. 1215. Be sure you understand the reactions shown in Eqs. 22.53 and 22.54 on text pp. 1215–1216. If you work through these examples, the reaction in Eq. 22.61 on text p. 1219 should be clear. The only thing different in this case is that the condensation occurs internally.

Just as in a sport, or in music, or in any other worthwhile endeavor, improving your skill requires practice. If you are concerned or discouraged that these transformations are not immediately obvious, don't be! They are more likely to become easier to understand, however, if you'll take the approach described here.

22.5   Variants of the Aldol and Claisen Condensations

Many variations of the aldol and Claisen condensations exist. The text has considered a few: the Claisen–Schmidt, Reformatsky, and Knoevenagel variations of the aldol condensation, and the Dieckmann variation of the Claisen condensation. Possibly you'll hear of the Perkin condensation, the Stobbe condensation, or other variations that have no name. To cover all these in a text of reasonable (or even unreasonable) size is neither possible nor even necessary. For if you master the principles of reactivity contained in Chapters 21 and 22, understanding variations of these condensation reactions will only be a matter of re-applying the principles. In fact, mastering these principles might even allow you to invent your own condensation reaction!

22.6 More Target Analysis for the Claisen Condensation

If you imagine the Claisen condensation on an ester of general structure *A*, you'll see that saponification and decarboxylation *must* yield a *symmetrical ketone* of general structure *C*. (A symmetrical ketone is a ketone in which the two groups attached to the carbonyl carbon are the same.)

$$2\ RCH_2COEt \xrightarrow{\text{Claisen}} \underset{\underset{B}{\underset{|}{R}}}{RCH_2CCHCO_2Et} \xrightarrow{\text{saponification}} \xrightarrow{\text{decarboxylation}} \underset{C}{RCH_2CCH_2R}$$

$$A \qquad\qquad\qquad\qquad\qquad\qquad\qquad\qquad\qquad\qquad \text{(a symmetrical ketone)}$$

(Be sure to go through this case in detail to verify the foregoing statement.) Alkylation of the Claisen product *B* can lead to a variety of *substituted derivatives* of *C*. For example, treatment of *B* with NaOEt, then a general alkyl halide R′—I, would give, after saponification, acidification, and decarboxylation, a general derivative *D*.

$$\underset{\underset{R'}{\underset{|}{}}}{RCH_2 - \overset{\overset{O}{\|}}{C} - CHR} \longleftarrow \text{symmetrical ketone structure}$$

$$D$$

Notice the symmetrical ketone structure "lurking" as the circled part of *D*.

When you are asked to prepare a ketone by the Claisen condensation, look for such symmetrical-ketone "substructures" within the target molecule. Focused Problems 22.36 and 22.37 (text p. 1243) give you practice in applying this strategy.

22.7 Synthetic Equivalents in Conjugate Addition

Study Guide Links 17.1 (p. 511 of this manual) and 20.4 (p. 659 of this manual) discussed the notion of *synthetic equivalents* (and the corresponding property, *synthetic equivalence*), that is, thinking of one group in terms of other groups from which it might be formed. This notion is particularly useful in planning conjugate additions. In Study Problem 22.6 on text p. 1257, for example, the di(ethoxycarbonyl)methyl anion ($^-$:CH(CO$_2$Et)$_2$, the conjugate base of diethyl malonate,) is used as the *synthetic equivalent* of a carboxymethyl anion ($^-$:CH$_2$CO$_2$H), which itself cannot exist. (Why can't this anion exist?).

What are some other synthetic equivalents that are useful in conjugate-addition reactions?

$$^-\text{:CN is equivalent to } {}^-\overset{\overset{O}{\|}}{\text{:C}}-\text{OH} \qquad\qquad \text{(see Eq. 22.110 on text p. 1250)}$$

$$^-\text{:CN is also equivalent to } {}^-\text{:CH}_2-\text{NH}_2 \qquad\qquad \text{(see Sec. 21.9C on text p. 1169)}$$

$$\underset{\underset{CO_2Et}{\underset{|}{}}}{{}^-\text{:CH} - \overset{\overset{O}{\|}}{C} - R}\ \text{is equivalent to } {}^-\text{:CH}_2 - \overset{\overset{O}{\|}}{C} - R \qquad \text{(see Display 22.96 on text p. 1241)}$$

(You should show how each group can be converted into the other, for example, how a —C≡N group can be converted into a —CH$_2$NH$_2$ group.)

When you can start thinking in terms of synthetic equivalents you really have started to master the art of organic synthesis.

FURTHER EXPLORATIONS

22.1   Malonic Ester Alkylations

In Eq. 22.88b of the text (p. 1238), a *secondary* alkyl halide is used to alkylate the enolate ion of malonic ester. When you studied the S_N2 reaction, you learned that a side reaction is E2 elimination, and that this side reaction is particularly prevalent with secondary and tertiary alkyl halides. Because the product of Eq. 22.88b is formed in high yield, elimination is evidently not a problem in this case, and one might ask why.

Recall (Sec. 9.5G) that, although most strong bases are good nucleophiles, a number of relatively weak bases are also good nucleophiles, and that branched or secondary alkyl halides undergo S_N2 reactions with such nucleophiles with little interference from E2 reactions. (Note particularly the example in Eq. 9.55 on text p. 485.) The enolate ion of diethyl malonate is in this category. Chemists have found that delocalized (that is, resonance-stabilized) carbanions such as enolate ions are particularly good nucleophiles (for reasons that are not completely understood). Consequently, the S_N2 reaction is particularly rapid for these anions, and the E2 reaction does not interfere unless highly branched secondary or tertiary alkyl halides are used.

22.2   Alkylation of Enolate Ions Derived from Ketones

The acetoacetic ester synthesis requires the presence of the ester group of a β-keto ester; this group is then "thrown away" by hydrolysis and decarboxylation at the end of the synthesis and replaced by a hydrogen. For the same reasons that carboxylic acids with quaternary α-carbons cannot be prepared by the malonic ester synthesis, ketones with quaternary α-carbons cannot be prepared by the acetoacetic ester synthesis. For example, suppose we want to prepare 2,2-dimethylcyclo-hexanone:

quaternary carbon

2,2-dimethylcyclohexanone

Although 2-methylcyclohexanone is easily prepared by a Dieckmann condensation–alkylation sequence, introduction of the second methyl group is not possible in such a reaction scheme.

Why not, then, resort to the strategy used in Sec. 22.8B of the text? 2-Methylcyclohexanone could be alkylated *directly* by forming its enolate ion with a strong base such as LDA or LCHIA and treatment of that enolate with methyl iodide:

Can you see the problem with this approach? With an ester, only one enolate ion is possible. But with a ketone, in general *two* enolate ions are possible:

The problem, then, is how to form one enolate and not the other, and to keep the two enolate ions from equilibrating once the desired one is formed.

But the problems don't stop here. Suppose a way could be found to form the desired enolate ion. An additional problem is that enolate ions can react in more than one way. The enolate ions of ketones are not only nucleophilic at carbon; they are also nucleophilic at *oxygen*:

a vinyl ether

Alkylation at oxygen gives a vinyl ether—sometimes called an *enol ether,* because conceptually it is derived by replacement of the —OH hydrogen of an enol with an alkyl group. In general, both products are formed. (Alkylation of oxygen is not usually observed with esters because enols of esters—and enol ethers of esters—are *much* less stable relative to their parent esters than enols of ketones are; see Eq. 22.18 in Sec. 22.2 on text p. 1200.)

Direct alkylation of ketones, then, is a much more complex problem than alkylation of esters. Nevertheless, chemists have discovered ways to control, in many cases, which enolate ion is formed, and to control which alkylation product is produced. (The selective formation of enolates was discussed in Section 22.4C, text p. 1216–1219.) These solutions, although beyond the scope of an introductory text, are conceptually not difficult to understand if you understand the principles of enolate-ion chemistry.

22.3   Conjugate Addition of Organocuprate Reagents

Although the mechanism presented in the text is conceptually useful, it does not explain the effect of copper in promoting conjugate addition. Because carbonyl addition rather than conjugate addition is observed in the *absence* of copper, it is evident that copper has a special effect. A substantial body of evidence has shown that conjugate-addition reactions of lithium organocuprate reagents involve free-radical intermediates, and that conjugate addition is a consequence of a free-radical mechanism.

REACTION REVIEW

I. α-HALOGENATION OF CARBONYL COMPOUNDS

A. *ACID-CATALYZED α-HALOGENATION OF ALDEHYDES AND KETONES*

1. Halogenation of an aldehyde or ketone in acidic solution usually results in the replacement of one α-hydrogen by halogen.

   a   Enols are reactive intermediates in these reactions.

   b. Enols add only one halogen atom.

$$\underset{\text{rate-limiting step}}{\overset{\displaystyle -\overset{\displaystyle \overset{O}{\|}}{C}-\underset{|}{CH}- \quad \xrightarrow{H_3O^+}}{}} \quad -\overset{\displaystyle \overset{OH}{|}}{C}=\underset{\text{enol}}{C}- \quad \xrightarrow{Br_2} \quad -\overset{\displaystyle \overset{O}{\|}}{C}-\overset{\displaystyle \overset{Br}{|}}{\underset{|}{C}}- \ + \ HBr$$

2. The rate law implies that even though the reaction is a halogenation, the rate is independent of the halogen concentration.

$$\text{rate} = k[\text{ketone}][H_3O^+]$$

   a. Halogens are not involved in the transition state for the rate-limiting step of the reaction.

   b. Enol formation is the rate-limiting process in acid-catalyzed halogenation of aldehydes and ketones.

3. Introduction of a second halogen is much slower than introduction of the first halogen.

B. *HALOGENATION OF ALDEHYDES AND KETONES IN BASE; THE HALOFORM REACTION*

1. Halogenation of aldehydes and ketones with α-hydrogens also occurs in base; all α-hydrogens are substituted by halogen.

$$-CH_2-\overset{\displaystyle \overset{O}{\|}}{C}-\underset{|}{CH}- \quad \xrightarrow[^-OH]{Br_2} \quad -\overset{\displaystyle \overset{Br}{|}}{\underset{|}{C}}-\overset{\displaystyle \overset{O}{\|}}{C}-\overset{\displaystyle \overset{Br}{|}}{\underset{|}{C}}-$$

2. When the aldehyde or ketone starting material is either acetaldehyde or a methyl ketone, the trihalomethyl carbonyl compound is not stable under the reaction conditions, and it reacts further to give, after acidification of the reaction mixture, a carboxylic acid and a haloform.

$$-\overset{\displaystyle \overset{O}{\|}}{C}\diagdown_{CH_3} \quad \xrightarrow{Br_2,\ NaOH} \quad -\overset{\displaystyle \overset{O}{\|}}{C}\diagdown_{CBr_3} \quad \xrightarrow{^-OH} \quad -\overset{\displaystyle \overset{O}{\|}}{C}\diagdown_{O^-} \ + \ \underset{\substack{\textbf{bromoform}\\ \text{(a haloform)}}}{HCBr_3}$$
$$+\ 3\ NaBr$$
$$+\ 3\ H_2O$$

   a. This reaction is called the *haloform reaction*.

   b. A carbon–carbon bond is broken.

3. The mechanism of the haloform reaction involves the formation of an enolate ion as a reactive intermediate.

   a. The enolate ion reacts as a nucleophile with halogen to give an α-halo carbonyl compound.

   b. A dihalo and trihalo carbonyl compound are formed more rapidly in successive halogenations.

c. A carbon–carbon bond is broken when the trihalo carbonyl compound undergoes a nucleophilic acyl substitution reaction.

  i. The leaving group in this reaction is a trihalomethyl anion, which is much less basic than ordinary carbanions.

  ii. The trihalomethyl anion reacts irreversibly with the carboxylic acid by-product to drive the overall haloform reaction to completion.

  iii. The carboxylic acid itself can be isolated by acidifying the reaction mixture.

4. The haloform reaction can be used to prepare carboxylic acids from readily available methyl ketones.

5. The haloform reaction is used as a qualitative test for methyl ketones, called the iodoform test.

  a. A compound of unknown structure is mixed with alkaline I_2.

  b. A yellow precipitate of iodoform (HCI_3) is taken as evidence for a methyl ketone (or acetaldehyde).

  c. Alcohols of the structure $R-\overset{\underset{|}{OH}}{C}H-CH_3$ also give a positive iodoform test because they are oxidized to methyl ketones (or to acetaldehyde, in the case of ethanol) by the basic iodine solution.

C. *α-BROMINATION OF CARBOXYLIC ACIDS: THE HELL–VOLHARD–ZELINSKY REACTION*

1. A bromine is substituted for an α-hydrogen when a carboxylic acid is treated with Br_2 and a catalytic amount of red phosphorus or PBr_3; this reaction is called the Hell–Volhard–Zelinsky (HVZ) reaction.

2. The first stage in the mechanism of this reaction is the conversion of the carboxylic acid into a small amount of acid bromide by the catalyst PBr_3.

  a. The acid bromide enolizes in the presence of acid; this reaction is similar to the acid-catalyzed bromination of ketones.

  b. The enol of the acid bromide is the species that actually brominates.

  i. When a small amount of PBr_3 catalyst is used, the α-bromo acid bromide reacts with the carboxylic acid to form more acid bromide, which is then brominated.

  ii. If one full equivalent of PBr_3 catalyst is used, the α-bromo acid bromide is the reaction product.

  iii. The reaction mixture can be treated with an alcohol or a phenol to give an α-bromo ester.

D.   REACTIONS OF α-HALO CARBONYL COMPOUNDS

1.   Most α-halo carbonyl compounds are very reactive in S_N2 reactions, and can be used to prepare other α-substituted carbonyl compounds.

2.   In the case of α-halo ketones, nucleophiles used in these reactions must not be too basic.

   a.   Stronger bases promote enolate-ion formation; and the enolate ions of α-halo ketones undergo other reactions.

   b.   More basic nucleophiles can be used with α-halo acids because, under basic conditions, α-halo acids are ionized to form their carboxylate conjugate-base anions, and these do not undergo the side-reactions that α-halo ketones undergo.

3.   α-Halo carbonyl compounds react so slowly by the S_N1 mechanism that this reaction is not useful; reactions that require the formation of carbocations alpha to carbonyl groups generally do not occur.

II.   CONDENSATIONS INVOLVING ENOLATE IONS AND ENOLS OF ALDEHYDES AND KETONES

A.   BASE-CATALYZED ALDOL ADDITION AND CONDENSATION REACTIONS

1.   In aqueous base, aldehydes and ketones with α-hydrogens undergo a reaction called the aldol addition.

an aldehyde or ketone   a β-hydroxy aldehyde or ketone

   a.   The term aldol is both a trivial name for 3-hydroxybutanal and a generic name for β-hydroxy aldehydes.

   b.   The aldol addition is a very important and general reaction of aldehydes and ketones that have α-hydrogens.

   c.   The aldol addition is another nucleophilic addition to a carbonyl group.

2.   The base-catalyzed aldol addition involves an enolate ion as an intermediate; the enolate ion adds to a second molecule of aldehyde or ketone.

an enolate ion

aldol

3.   The aldol addition is reversible.

   a.   The equilibrium for the aldol addition is much more favorable for aldehydes than for ketones.

   b.   For example, the aldol-addition product of acetone is obtained only if an apparatus is used that allows the product to be removed from the base catalyst as it is formed.

4.   Under severe conditions (higher base concentration and/or heat), the product of aldol addition undergoes a dehydration reaction. (The sequence of reactions consisting of the aldol addition followed by

dehydration is called the aldol condensation; a condensation is a reaction in which two molecules combine to form a larger molecule with the elimination of a small molecule, in many cases water.)

$$2\,R-CH_2-\overset{\displaystyle O}{\overset{\|}{C}}-H \underset{\substack{\text{addition}\\ \textit{via the enolate}}}{\overset{^-OH}{\rightleftarrows}} R-CH_2-\overset{OH}{\underset{}{CH}}-\underset{R}{CH}-\overset{\displaystyle O}{\overset{\|}{C}}-H \xrightarrow[\text{dehydration}]{^-OH} R-CH_2-CH=\underset{R}{\overset{\overset{\displaystyle O}{\overset{\|}{C}}-H}{C}}$$

a. The dehydration part of the aldol condensation is catalyzed by base, and occurs in two distinct steps through a carbanion intermediate. (This mechanism is nicknamed the E1cB mechanism.)

b. Base-catalyzed dehydration reactions of alcohols are unusual; β-hydroxy aldehydes and β-hydroxy ketones undergo base-catalyzed dehydration for two reasons:

    i. The α-hydrogen is relatively acidic.

    ii. The product is conjugated and therefore particularly stable.

5. The product of an aldol addition–dehydration sequence is an α,β-unsaturated carbonyl compound.

6. Whether the aldol addition product or the condensation product is formed depends on the reaction conditions, which must be determined on a case-by-case basis. (See Section 22.4A, text p. 1210.)

B. ACID-CATALYZED ALDOL CONDENSATION

1. Aldol condensations are also catalyzed by acid.

    a. Acid-catalyzed aldol condensations generally give α,β-unsaturated carbonyl compounds as products.

    b. The addition products usually cannot be isolated.

2. Both the conjugate acid and the enol isomer of the aldehyde or ketone are key reactive intermediates. (See Eqs. 22.52a–b on text page 1213–1214 for a detailed mechanism.)

3. The α-hydroxy aldehyde or ketone formed in the reaction spontaneously undergoes acid-catalyzed dehydration to give an α,β-unsaturated carbonyl compound; this dehydration drives the aldol condensation to completion.

$$2\,RCH_2-\overset{\displaystyle O}{\overset{\|}{C}}-CH_2R \underset{\substack{\text{addition}\\ \textit{via enol}}}{\overset{H_3O^+}{\rightleftarrows}} RCH_2-\underset{RCH_2}{\overset{OH}{\underset{}{C}}}-\underset{R}{CH}-\overset{\displaystyle O}{\overset{\|}{C}}-CH_2R \xrightarrow[\text{dehydration}]{H_3O^+} RCH_2-\underset{RCH_2}{C}=\underset{R}{C}-\overset{\displaystyle O}{\overset{\|}{C}}-CH_2R$$

C. SPECIAL TYPES OF ALDOL CONDENSATION

1. When two different carbonyl compounds are used, a crossed aldol condensation occurs; in many cases, the result is a difficult-to-separate mixture.

2. Under the usual conditions (aqueous or alcoholic acid or base), useful crossed aldol condensations as a practical matter are limited to situations in which a ketone with α-hydrogens is condensed with an aldehyde without α-hydrogens.

3. An important example of this type of crossed aldol condensation is the Claisen–Schmidt condensation.

    a. A ketone with α-hydrogens is condensed with an aromatic aldehyde that has no α-hydrogens.

$$Ph-\overset{\displaystyle O}{\overset{\|}{C}}-H \;+\; H_3C-\overset{\displaystyle O}{\overset{\|}{C}}-CH_3 \underset{\substack{\text{condensation}\\ \textit{via the enol}}}{\overset{H_3O^+}{\rightleftarrows}} Ph-\underset{}{\overset{OH}{\underset{}{CH}}}-CH_2-\overset{\displaystyle O}{\overset{\|}{C}}-CH_3 \xrightarrow[\text{dehydration}]{H_3O^+} Ph-CH=CH-\overset{\displaystyle O}{\overset{\|}{C}}-CH_3$$

not isolated

    b. The Claisen–Schmidt condensation occurs instead of other competing processes for three reasons:

    i. Because the aldehyde has no α-hydrogens, it cannot act as the enolate or enol component of the aldol condensation.

    ii. Addition of the ketone enolate or enol to a ketone occurs more slowly than addition to an aldehyde; aldehydes are more reactive than ketones.

    iii The aldol addition reaction of two ketone molecules is reversible; thus, to the extent that this side reaction occurs, it is reversed under the reaction conditions.

  c. The Claisen–Schmidt condensation, like other aldol condensations, can be catalyzed by either acid or base.

4. When a molecule contains more than one aldehyde or ketone group, an intramolecular reaction (a reaction within the same molecule) is possible; the resulting formation of a ring is particularly favorable when five- and six-membered rings can be formed.

5. Directed aldol reactions involve pre-forming the enolate of a ketone at low temperatures, then adding an aldehyde. Only the desired reaction is promoted under these conditions.

  a. A strong, non-nucleophilic amide base such as LDA is used to generate the first enolate ion selectively (see IV.C.1. in this section).

  b. A different aldehyde (or ketone) is added to the solution containing the enolate ion, and only one aldol reaction occurs.

III. CONDENSATION REACTIONS INVOLVING ESTER ENOLATE IONS

A. CLAISEN CONDENSATION

1. The Claisen condensation is used for the formation of β-dicarbonyl compounds (compounds with two carbonyl groups in a β-relationship), such as the following:

  a. β-Keto esters (compounds in which a ketone carbonyl group is β to an ester carbonyl group)

  b. β-Diketones (See Display 22.79. Sec. 22.6D, text p. 1232.)

2. The mechanism of the Claisen condensation:

  a. An enolate ion is formed by the reaction of an ester with ethoxide as the base. (Although the ester enolate ion is formed in low concentration, it is a strong base and good nucleophile.)

b. The ester enolate undergoes a nucleophilic acyl substitution reaction with a second molecule of ester.

c. The overall equilibrium favors the reactants; that is, all β-keto esters are less stable than the esters from which they are derived.

d. The Claisen condensation has to be driven to completion by applying Le Chatelier's principle.

    i.   The most common technique is to use one equivalent of alkoxide base.

    ii. The hydrogens adjacent to both carbonyl groups in the β-keto ester product are especially acidic ($pK_a \approx 10$–11); the alkoxide (conjugate acid $pK_a \approx 16$) removes one of these hydrogens to form quantitatively the conjugate base of the product.

e. The un-ionized β-keto ester product is formed when acid is added subsequently to the reaction mixture.

3. Attempts to condense an ester that has only one α-hydrogen result in little or no product, because the desired condensation product, formed reversibly, has no α-hydrogens acidic enough to react completely with the alkoxide.

B. *DIECKMANN CONDENSATION*

1. Intramolecular Claisen condensations take place readily when five- or six-membered rings can be formed.

2.   The intramolecular Claisen condensation reaction is called the Dieckmann condensation.

3.   As with the Claisen condensation, one equivalent of base is necessary; the α-proton must be removed from the initially formed product in order for the reaction to be driven to completion.

C.   CROSSED CLAISEN CONDENSATION

1.   The Claisen condensation of two different esters is called a crossed Claisen condensation; the crossed Claisen condensation of two esters that both have α-hydrogens gives a mixture of compounds that are typically difficult to separate.

2.   Crossed Claisen condensations are useful if one ester is especially reactive or has no α-hydrogens. Formate esters and diethyl carbonate (a less reactive ester that is used in excess) are frequently used in this type of reaction.

3.   Another type of crossed Claisen condensation is the reaction of ketones with esters.

a.   In this type of reaction the enolate ion of a ketone reacts as a nucleophile at the carbonyl group of an ester.

   i.   The equilibrium for the aldol addition of two ketones favors the reactants, whereas the Claisen condensation is irreversible because one equivalent of base is used to form the enolate ion of the product.

   ii.   Ketones are far more acidic than esters; the enolate ion of the ketone is formed in much greater concentration than the enolate ion of the ester.

b.   The product is a β-dicarbonyl compound, which is especially acidic and is ionized completely by the one equivalent of base; ionization makes this reaction irreversible.

IV.   ALKYLATION OF ESTER ENOLATE IONS

A.   MALONIC ESTER SYNTHESIS

1.   Diethyl malonate (malonic ester), like many other β-dicarbonyl compounds, has unusually acidic α-hydrogens ($pK_a = 12.9$).

a.   Its conjugate-base enolate ion can be formed completely with alkoxide bases such as sodium ethoxide, whose conjugate acid, ethanol, has a pK_a of about 16.

b.   The conjugate-base anion of diethyl malonate is nucleophilic, and it reacts with alkyl halides and sulfonate esters (RX, R′X in the following equation) in typical S_N2 reactions.

diethyl malonate
$pK_a = 12.9$

$+ EtOH + X^-$

$+ EtOH + X^-$

    c.   This reaction can be used to introduce alkyl groups at the α-position of malonic ester.

2.   The malonic ester synthesis can be extended to the preparation of carboxylic acids. (See Eqs. 22.88a–c, Sec. 22.8A, pp. 1238–1239.)

    a.   Saponification of the diester and acidification of the resulting solution give a substituted malonic acid derivative.

    b.   Heating any malonic acid derivative causes it to decarboxylate.

saponification

$+ 2 EtOH$

decarboxylation

$+ CO_2$

    c.   The result of alkylation, saponification, and decarboxylation is a carboxylic acid that conceptually is a substituted acetic acid—an acetic acid molecule with one or two alkyl substituents on its α-carbon.

3.   The overall sequence of ionization, alkylation, saponification, and decarboxylation starting from diethyl malonate is called the *malonic ester synthesis*.

    a.   The alkylation step of the malonic ester synthesis results in the formation of one or two new carbon–carbon bonds.

    b.   The anion of malonic acid can be alkylated twice in two successive reactions with different alkyl halides (if desired) to give, after hydrolysis and decarboxylation, a disubstituted acetic acid.

B.   *ACETOACETIC ESTER SYNTHESIS*

1.   β-Keto esters ($pK_a = 10$–11), which are substantially more acidic than ordinary esters ($pK_a = 25$), are completely ionized by alkoxide bases, which have conjugate-acid pK_a values near 16.

    a.   The resulting enolate ions can be alkylated by alkyl halides or sulfonate esters.

an alkyl or aryl group

a β-keto ester
$pK_a = 10$–11

$+ EtOH + X^-$

$+ EtOH + X^-$

    b.   Dialkylation of β-keto esters is also possible.

    c.   The alkylated derivatives of ethyl acetoacetate can be hydrolyzed and decarboxylated to give ketones.

    d.   The decarboxylation of β-keto acids occurs at room temperature.

2. The alkylation of ethyl acetoacetate followed by saponification, protonation, and decarboxylation to give a ketone is called the acetoacetic ester synthesis and involves the construction of one or two new carbon–carbon bonds.

3. Whether a target ketone can be prepared by the acetoacetic ester synthesis can be determined by mentally reversing the synthesis (see Display 22.96, Sec. 22.8B, text p. 1241).

   a. Replace an α-hydrogen on the target ketone with a —CO$_2$Et group.

   b. Determine whether the resulting β-keto ester can be prepared by a Claisen condensation or from other β-keto esters by alkylation or di-alkylation with appropriate alkyl halides.

C. *DIRECT ALKYLATION OF ENOLATE IONS DERIVED FROM MONOESTERS*

1. A family of very strong, highly branched nitrogen bases can be used to form stable enolate ions rapidly at –78 °C from esters and ketones. (See also II.C.5.a.)

   a. The nitrogen bases themselves are generated from the corresponding amines and butyllithium at –78 °C in tetrahydrofuran (THF).

$$CH_3CH_2CH_2CH_2\text{—}Li + i\text{-}Pr_2NH \xrightarrow[-78\,°C]{} i\text{-}Pr_2N^- Li^+ + CH_3CH_2CH_2CH_3$$

   pK$_a$ = 32–35   pK$_a$ = 55–60

   b. These amide bases are strong enough to convert esters rapidly and irreversibly into their conjugate-base enolate ions.

   i. The ester is added to the base; the reaction of esters with strong amide bases is much faster, even at –78 °C, than the Claisen condensation; the enolate is formed rapidly.

   ii. A nucleophilic reaction of the amide base on the carbonyl carbon of the ester, a reaction that competes with proton removal, is retarded by van der Waals repulsions between groups on the carbonyl compound and the large branched groups on the amide base.

   iii. This type of van der Waals repulsions has been termed F-strain, or "front-strain."

   c. The ester enolate anions formed with these bases can be alkylated with primary alkyl halides.

   d. Esters with quaternary α-carbon atoms can be prepared with this method.

2. This method of ester alkylation is considerably more expensive than the malonic ester synthesis and requires special inert-atmosphere techniques.

D. *ALDOL REACTIONS OF ESTER ENOLATES*

1. Ester enolates can serve as nucleophiles in aldol reactions.

2. The Reformatsky reaction utilizes an organozinc compound as the nucleophile, which behaves analogously to a Grignard reagent. It is, however, less reactive that a Grignard reagent.

    a. As shown below, the organozinc compound (Reformatsky reagent) does not necessarily have to be pre-formed in a second step. It can be formed in the presence of the aldehyde or ketone.

    b. Organozinc compounds are less reactive than Grignard reagents; therefore. they can be formed from esters.

    c. Although the actual structure of the Reformatsky reagent is more complex than is shown, it behaves as if it were the simple reagent depicted.

3. The Knoevenagel reaction is an aldol reaction that utilizes the enolate of a malonic ester or acetoacetic ester as the nucleophile. The product of this reactions is an α,β-unsaturated carbonyl compound.

    a. First, the base piperidine deprotonates the malonic ester.

    b. Then, a typical aldol reaction proceeds normally, and the addition product is protonated by the piperidinium cation.

    c. Finally, the addition product, which is not isolated, is deprotonated by the piperidine, which drives the reaction to completion through a dehydration reaction by the E1cB mechanism. The hydroxide by-product then reacts with the piperidinium ion to regenerate the catalyst.

V. CONJUGATE-ADDITION REACTIONS

A. CONJUGATE ADDITION TO α,β-UNSATURATED CARBONYL COMPOUNDS

1. Addition of a nucleophile (Nuc⁻ in the following equation) to the double bond of an α,β-unsaturated carbonyl compound is an example of conjugate addition (1,4-addition); nucleophilic conjugate addition has no parallel in the reactions of simple conjugated dienes.

2. Nucleophilic addition to the carbon–carbon double bonds of α,β-unsaturated aldehydes, ketones, esters, and nitriles is a rather general reaction that can be observed with a variety of nucleophiles.

   a. The addition of a nucleophile to acrylonitrile (H_2C=CH—C≡N) is a useful reaction called *cyanoethylation.*

   b. Quinones are α,β-unsaturated carbonyl compounds, and they also undergo similar addition reactions.

   c. When cyanide is the nucleophile, a new carbon–carbon bond is formed; the nitrile group can then be converted into a carboxylic acid group by acid-catalyzed hydrolysis.

3. Nucleophilic addition to the double bond in an α,β-unsaturated carbonyl compound occurs because it gives a resonance-stabilized enolate ion intermediate.

4. Acid-catalyzed conjugate additions to α,β-unsaturated carbonyl compounds are also known.

B. CONJUGATE ADDITION OF ENOLATE IONS: MICHAEL ADDITION

1. Enolate ions derived from malonic ester derivatives, β-keto esters, and the like undergo conjugate-addition reactions with α,β-unsaturated carbonyl compounds.

a. Conjugate additions of carbanions to α,β-unsaturated carbonyl compounds are called *Michael additions*.

b. Many Michael additions can originate from either of two pairs of reactants.

c. To maximize conjugate addition, choose the pair of reactants with the less basic enolate ion.

2. In one useful variation of the Michael addition, called the *Robinson annulation*, the immediate product of the reaction can be subjected to an aldol condensation that closes a ring. (An annulation is a ring-forming reaction.)

VI. OTHER REACTIONS OF α,β-UNSATURATED CARBONYL COMPOUNDS

A. REDUCTION OF α,β-UNSATURATED CARBONYL COMPOUNDS

1. The carbonyl group of an α,β-unsaturated aldehyde or ketone is reduced to an alcohol with lithium aluminum hydride.

a. This transformation involves the nucleophilic reaction of hydride at the carbonyl carbon and is therefore a carbonyl addition.

b. Carbonyl addition is not only faster than conjugate addition but, in this case, also irreversible; reduction of the carbonyl group with $LiAlH_4$ is a kinetically controlled reaction.

2. Many α,β-unsaturated carbonyl compounds are reduced by $NaBH_4$ to give mixtures of both carbonyl-addition products and conjugate-addition products.

3. The carbon–carbon double bond of an α,β-unsaturated carbonyl compound can in most cases be reduced selectively by catalytic hydrogenation.

B. ADDITION OF ORGANOLITHIUM REAGENTS TO THE CARBONYL GROUP

1. Organolithium reagents react with α,β-unsaturated carbonyl compounds to yield products of carbonyl addition.

2. The product is the result of kinetic control, because carbonyl addition is more rapid than conjugate addition and it is also irreversible.

C. *CONJUGATE ADDITION OF LITHIUM DIALKYLCUPRATE REAGENTS*

1. Lithium dialkylcuprate reagents give exclusively products of conjugate addition when they react with α,β-unsaturated esters and ketones.

2. Although conjugate addition of lithium dialkylcuprates actually involves a radical mechanism (see Further Exploration 22.3 on p. 753 of this manual), the reaction can be conceptualized mechanistically to be similar to other conjugate additions:

   a. A nucleophilic reaction of an "alkyl anion" of the dialkylcuprate at the β-carbon of the double bond gives a resonance-stabilized enolate ion.

   b. Protonation of the enolate ion by the addition of water gives the conjugate-addition product.

3. Conjugate addition also occurs with Grignard reagents in the presence of added CuCl.

SOLUTIONS TO PROBLEMS

Solutions to Focused Problems

22.1 In both parts, the acidic hydrogens are the α-hydrogens on the carbons between the two carbonyl groups. (—OEt = ethoxy group = —OCH$_2$CH$_3$.)

acidic hydrogens

O   O
‖   ‖
EtOC — CH$_2$ — COEt

diethyl malonate

acidic hydrogens

O   O
‖   ‖
CH$_3$C — CH$_2$ — COEt

ethyl acetoacetate

The reason that these hydrogens are particularly acidic is that their conjugate-base enolate ions are stabilized by the polar effects and resonance effects of two carbonyl groups, whereas the conjugate-base enolate ion of an ordinary ester is stabilized by the corresponding effects of only one carbonyl group. For example, the resonance structures of the conjugate-base enolate ion of diethyl malonate are as follows.

resonance structures for the conjugate-base enolate ion of diethyl malonate

(We leave it to you to draw analogous structures for the enolate conjugate base of ethyl acetoacetate.)

22.2   The imide of succinic acid (succinimide) is more acidic, because its conjugate-base anion is stabilized by the polar effects and resonance effects of two adjacent carbonyl groups. The conjugate-base anion of succinic acid diamide is stabilized by resonance interaction with only one carbonyl group.

conjugate-base anion of succinimide

conjugate-base anion of succinic acid diamide

22.3   A mechanism for replacement of one hydrogen in the reaction of Eq. 22.11, text p. 1198, is shown in the following equation. (The mechanisms for replacement of the others are identical.) Only the α-hydrogens are replaced because the enolate ion is the only carbanion stable enough to be formed. The carbanion intermediates required in a similar mechanism for the replacement of the hydrogens other than the α-hydrogens are not resonance-stabilized.

a resonance-stabilized
enolate ion

22.4   The proton NMR spectrum would change as follows:

δ 2.0 (s)   δ 1.0 (t)   δ 1.0 (s)

H_3C—C—CH_2—CH_3 $\xrightarrow{\text{base, }D_2O}$ D_3C—C—CD_2—CH_3

δ 2.4 (q)

no proton NMR
resonances

That is, in basic D_2O, the resonances for the α-protons disappear and the triplet at δ 1.0 becomes a singlet.

22.5   Exchange does not occur because the orbital containing the nonbonding electron pair on the α-carbon of the enolate ion is perpendicular to the π-electron system of the carbonyl group; therefore, the enolate ion is not stabilized by resonance. Another way to view this situation is suggested by the hint: the resonance structure of the enolate ion violates Bredt's rule (Sec. 7.6C, text p. 353).

violates Bredt's rule

22.6 (a) All α-hydrogens are exchanged for deuterium. (The methyl hydrogens of the *tert*-butyl group are not α-hydrogens.) The structure after deuterium exchange is shown at right.

(b) All α-hydrogens as well as the O—H hydrogen are exchanged for deuterium:

(c) The only possible enolate ion is not resonance-stabilized for the same reason that the enolate ion in Focused Problem 22.5 is not resonance-stabilized; consequently, it is too unstable to form. Because deuterium exchange of the α-hydrogens would involve an enolate ion intermediate, deuterium exchange cannot occur.

22.7 (a) The enol isomers of 2-methylcyclohexanone:

2-methylcyclohexanone   enol isomers of 2-methylcyclohexanone

(b) The enol isomer of 2-methylpentanoic acid:

2-methylpentanoic acid   enol isomer of 2-methylpentanoic acid

(c) Benzaldehyde does not have an enol isomer because it has no α-hydrogen.

no α-hydrogen

benzaldehyde

(d) The enol isomer of *N,N*-dimethylacetamide:

N,N-**dimethylacetamide**   enol isomer of *N,N*-dimethylacetamide

(e) *N,N*-Dimethylformamide has no enol isomer because it has no α-hydrogen.

$$H_3C-\overset{\overset{\displaystyle O}{\|}}{\underset{\underset{\displaystyle CH_3}{|}}{N}}-\overset{\displaystyle C}{}-H$$

N,N-dimethylformamide

22.8   The enol forms of 2-butanone.

$$H_3CCCH_2CH_3 \longleftrightarrow H_2C=\overset{\overset{\displaystyle OH}{|}}{C}CH_2CH_3 + \underset{H_3C}{\overset{HO}{}}C=C\underset{H}{\overset{CH_3}{}} + \underset{H_3C}{\overset{HO}{}}C=C\underset{CH_3}{\overset{H}{}}$$

| **2-butanone** | *least stable* | (Z) | (E) |

most stable

enol forms of 2-butanone

Double bonds are stabilized by alkyl substitution (see Sec. 4.5B, text pp. 196–198). Therefore, the enol on the left in the foregoing equation is the least stable. The relative stability of the *E* and *Z* stereoisomers assumes that a methyl group is larger than an OH group.

22.9   (a)   The "enol" form of nitromethane:

$$H_2C=\overset{+}{N}\overset{:\ddot{O}H}{\underset{:\underset{\cdot\cdot}{\ddot{O}}:}{}}$$

"enol" form of nitromethane
(an *aci*-nitro compound)

(b)   The "enol" form of an amide that has a carbon–nitrogen double bond is called an *imidic* acid; see Eq. 21.43b, Sec. 21.7c, text p. 1151. Notice that this particular amide cannot enolize toward the α-carbon because it has no *α*-hydrogens on that carbon.

$$Ph-\overset{\overset{\displaystyle OH}{|}}{C}\!\!\diagup\!\!\!\diagdown^{NH}$$

"enol" form of benzamide
(an imidic acid)

22.10   (a)   One source of stabilization of the enols of β-diketones is their intramolecular hydrogen bonds; see Display 22.20c on text p. 1201. In aqueous solution, the solvent can serve as a hydrogen-bond donor to the carbonyl oxygen; so, the energetic advantage of the enol that results from intramolecular hydrogen bonding is reduced in aqueous solution. In a solvent that cannot donate hydrogen bonds, such as hexane, there are no external hydrogen bonds, and the internal hydrogen bond thus provides an energetic advantage to the enol.

(b)   As the structure in Display 22.20c shows, the enol form of 2,4-pentanedione has a conjugated π-electron system involving a carbonyl double bond and a carbon–carbon double bond. This conjugated system in the enol is the source of the UV spectrum. The absorption is stronger in hexane because there is a greater proportion of the enol form in hexane for the reasons discussed in the solution to part (a) of this problem.

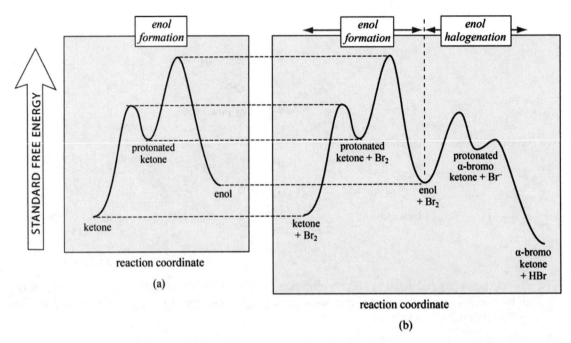

Figure SG22.1   Reaction–free energy diagrams to accompany the solution to Focused Problem 22.12. (a) Reaction–free energy diagram for acid-catalyzed enolization of a ketone. (b) Reaction–free energy diagram for α-bromination of a ketone incorporating the diagram for enolization from part (a) as the rate-limiting step. (The heights of the last two barriers are speculative because these steps do not affect the overall rate; however, they must be well below the barrier for the rate-limiting step.)

22.11   The acid-catalyzed formation of an enol occurs by the mechanism shown in Eq. 22.21b on text p. 1202. The enol is achiral, and it can be protonated at either face of the double bond at the same rate to give the racemate of the protonated ketone.

22.12   (a)   The reaction–free energy diagram for acid-catalyzed enolization of a ketone is shown in **Fig, 22.1a.**
        (b)   The reaction–free energy diagram for acid-catalyzed bromination of a ketone incorporating the diagram for enolization as the rate-limiting step is shown in **Fig. 22.1b.**

22.13   (a)   Because enolization is the process responsible for acid-catalyzed racemization, and because the rate-limiting process in acid-catalyzed ketone halogenation is enolization, the rates of the two processes are the same—the rate of enolization.
        (b)   Because the rate of ketone halogenation is independent of the halogen concentration, the rates of halogenation of the same ketone with two different halogens are also independent of halogen concentration and therefore independent of the identity of the halogen itself. In fact, the rate in both cases is the rate of enolization, as in part (a). The two processes are compared at the same acid concentration because the enolization process is acid-catalyzed (Eq. 22.21b, text p. 1202).

22.14 (a)

+ HCBr₃

(c)

+ HCBr₃

In part (b), benzophenone does not halogenate because it cannot form an enolate ion (it has no α-hydrogens), and the aromatic rings are deactivated by the carbonyl group toward electrophilic halogenation.

In part (c), the alcohol is first oxidized to the corresponding ketone, acetophenone, which then undergoes the haloform reaction.

22.15 2,5-Hexanedione gives succinic acid and iodoform when subjected to the iodoform reaction.

22.16 (a) The α-bromo acid bromide reacts with ethanol to give the α-bromo ester.

In part (a), neutral ethanol is not basic enough to act as a nucleophile and displace the α-bromine.

(b) The α-bromo acid bromide is formed first. Then ammonia reacts as a nucleophile with it in both an acyl substitution reaction to give the amide, and in an S_N2 reaction at the α-carbon to give the α-amino amide of the α-amino acid alanine.

22.17 In (a), the base pyridine displaces the α-bromine in an S_N2 reaction. In (b), The base acetate ion displaces the α-bromine to give a compound that is an example of a *phenacyl ester*.

(a)

(b)

benzoylmethyl acetate
(a phenacyl ester)

22.18 One equivalent of NaOH is consumed by its reaction with the carboxylic acid. The second equivalent forms the 2,4-dichlorophenolate ion, which is the nucleophile in the ensuing S_N2 reaction.

The final step of Eq. 22.38 in the text (protonation of the carboxylate to give the carboxylic acid) does not take place until acid is added to the reaction mixture; under the basic conditions of the substitution reaction the product is ionized.

22.19 (a) The aldol addition reaction of phenylacetaldehyde:

3-hydroxy-2,4-diphenylbutanal

(b) The aldol addition reaction of propionaldehyde:

3-hydroxy-2-methylpentanal

22.20 (a)

2,2-dimethylcyclohexanone   simplified enolate structure

aldol addition product   aldol dehydration product

(b) Both the addition product and the dehydration product can exist as diastereomers that are formed in different amounts in each case.

aldol addition product

two asymmetric carbons (*);
two diastereomers possible;
two enantiomers of each

aldol dehydration product

Z isomer   E isomer

22.21 (a) The problem states that equal molar amounts of each component are used. (Otherwise two moles of the aldehyde might react—one mole at each set of ketone α-hydrogens.) The enolate ion of the ketone reacts with the aldehyde because aldehyde carbonyl groups are more reactive in additions than ketone carbonyl groups. The product is shown at right.

(b) As in part (a), the enolate ion of the ketone adds to the aldehyde carbonyl group to give the product below because aldehyde carbonyl groups are more reactive than ketone carbonyl groups.

(c) This is a directed aldol condensation. The ketone enolate is generated at the methyl carbon at low temperatures and the aldehyde is added after. The addition product can be isolated, but heating affords the α,β-unsaturated ketone.

addition product   dehydration product

(d) This is an intramolecular aldol condensation.

22.22 (a) In this case, the enolate ion of a symmetrical ketone, 3-pentanone, would have to react with *p*-methoxybenzaldehyde. There is only one possible enolate ion, and the aldehyde carbonyl group is much more reactive than the ketone carbonyl group. Therefore, the starting materials shown at right are reasonable.

p-methoxybenzaldehyde

3-pentanone

(b)   The product shown would require the starting materials shown at right. However, because the enolate ion required for the synthesis is only one of two that could form, at least two isomeric products are possible of which the desired product is only one. The desired product would therefore be formed as one component of a mixture of isomers.

an enolate ion could form at either carbon

(c)   This product would require the intramolecular aldol condensation of the following keto aldehyde. In this case, the enolate ion formed at the α-carbon of the aldehyde would have to react with the ketone carbonyl group. Because the aldehyde carbonyl group is more reactive, and because its intramolecular reaction with an enolate ion of the ketone would also give a six-membered ring, synthesis of the compound shown in the problem is not reasonable. The reaction shown at right would likely occur instead.

This compound is required for the product given

This compound would be formed instead

(d)   This product is the result of an intramolecular reaction of a ketone enolate and an aldehyde carbonyl group; therefore, its synthesis from the starting material at right is reasonable.

(e)   A double Claisen–Schmidt condensation of the starting materials shown at right would give the desired product. The reaction is particularly favorable because the product is highly conjugated.

(f)   This is a reaction of two equivalents of benzaldehyde with one of acetone (shown. at right). Because an aldehyde without α-hydrogens is involved, the reaction is a reasonable example of the Claisen–Schmidt condensation.

$2\,PhCH{=}O + CH_3CCH_3$

(g)   This product could be formed by a directed aldol reaction. The enolate is generated first, at low temperatures, and the aldehyde is added afterwards.

(h)   Because the diketone starting material is symmetrical, the product shown is the only possible one, and thus the synthesis is reasonable from the following starting material.

22.23   First, break the carbon–carbon double bond of the product as shown in Eq. 22.61 on text p. 1219 to reveal the possible starting material (at right). The product shown in the problem results from the reaction of an enol formed at carbon *a* with carbonyl group *B*. Reaction of the enol formed at carbon *b* with carbonyl *B* would give a strained four-membered ring (product *bB* below) and is thus not likely. Reaction of an enol formed at carbon *c* with carbonyl *A* also would require the formation of a strained four-membered ring (product *cA* below). Reaction of an enol formed at carbon *d* with carbonyl *A* would give bicyclic compound *dA*:

product *bB*   product *cA*   product *dA*

Compound *dA* is a modest violation of Bredt's rule (Sec. 7.6C, text p. 352; build a model to verify that it is strained). The aldol *addition* product could form, thus avoiding dehydration to the strained alkene, but aldol additions of ketones are highly reversible. The product shown in the problem is thus the only one that can readily be driven to completion by dehydration.

22.24   (a)   Pyruvate forms an enamine with the lysine residue of the enzyme in an analogous process to the one shown in Eqs. 22.65a–b on text p. 1223. When (*R*)-glyceraldehyde-3-phosphate (G3P) enters the enzyme's active site, the α-carbon of the "enolate" reacts at the carbonyl carbon of G3P in an aldol-like addition reaction. Acids (HA) and their conjugate bases (A⁻) within the active site protonate and deprotonate the intermediates as needed (see Eqs. 22.65c–d on text pp. 1223–1224).

enamine form of the
enzyme-pyruvate complex

Once the carbon-carbon bond is formed as a result of the aldol addition reaction, the imine is cleaved by the usual mechanism (see Eq. 22.65e for an analogous reaction, and Study Guide Links 19.3 and 19.4 on pp. 607–609 of this manual for details of the reaction mechanism.)

KDPG

free enzyme

(b)   The two diastereomeric cyclic acetals (and the mechanism of their formation):

22.25   (a)   First, form an enamine (compound *X*) by the standard mechanism (see Sec. 19.12C, text p. 1066) at the carbonyl carbon marked with the asterisks.

an enamine

(b)   Next, use one of the resonance structures of the enamine to do an intramolecular aldol addition reaction (Sec. 22.4C, text pp. 1219–1220). After proton transfers, another enamine is regenerated.

(c)   Treatment with acid hydrolyzes the enamine back to the ketone (Sec. 19.12C, text p. 1066–1068). Continued treatment with acid will dehydrate the β-hydroxyketone (Sec. 22.4B) yielding an α,β-unsaturated ketone, or the aldol condensation product.

22.26   (a)

ethyl 2,4-diphenyl-3-oxobutyrate

(b)

ethyl 2-ethyl-3-oxohexanoate

22.27 Hydroxide ion would *not* be a good base for the Claisen condensation because it would saponify ethyl acetate and convert it into acetate ion. The role of the ethoxy group in ethyl acetate is to serve as a leaving group; acetate ion lacks this leaving group and therefore will not react in the Claisen condensation.

22.28 (a) An ordinary Claisen condensation is driven to completion by ionization of an α-hydrogen more acidic than ethanol, the conjugate acid of the base ethoxide. Compound *A* lacks such an acidic α-hydrogen. Consequently, ethoxide reacts with the ketone carbonyl group and converts it into *B*. This reaction is driven to completion by ionization of the acidic α-hydrogen of compound *B*. Compound *B* itself is formed on acidification of the reaction mixture.

(b) The mechanism of the conversion of *A* to *B*:

(c) Use reasoning similar to that used in solving part (a). Two possible enolate ions, *X* and *Y*, can form; the product derived from ion *X* is ultimately observed because only this product is formed irreversibly by ionization.

EtO$_2$CC̄H(CH$_2$)$_2$CHCO$_2$Et ⇌ EtO$_2$CCH$_2$(CH$_2$)$_2$C̄CO$_2$Et
          | |
         CH$_3$ CH$_3$

              X   *Y*

(observed product; acidification
gives the keto ester)

22.29 (a)

O   O
‖   ‖
EtOCCH$_2$CCMe$_3$

(b)

O   O
‖   ‖
PhCCH$_2$CPh

(c)

In all three parts, ethanol is a by-product. In part (c), the enolate ion formed at carbon *a* in the structure at right leads to the product. Hydrogens *b* are more acidic than hydrogens *a*, but a carbanion formed at carbon *b* cannot lead to a product containing a ring of reasonable size. Therefore, the two anions are in an equilibrium that is driven by ionization of the product (Le Chatelier's principle).

$$\underset{a}{\overset{O}{\|}}\ \underset{}{\overset{Me}{\underset{|}{\ }}}\ \overset{b}{\ }$$

CH₃CCH₂CCH(CO₂Et)₂
|
Me

22.30 **(a)** Apply the analysis shown in Eq. 22.79 on text p. 1232. The two possible sets of starting materials are as follows:

$$\underset{A}{\underset{}{CH_3\overset{O}{\overset{\|}{C}}OEt\ +\ CH_3CH_2\overset{O}{\overset{\|}{C}}CH_2CH_3}}\quad and\quad \underset{B}{\underset{}{CH_3\overset{O}{\overset{\|}{C}}CH_2CH_3\ +\ EtO\overset{O}{\overset{\|}{C}}CH_2CH_3}}$$

Set *A* could work provided that an excess of the ester is used. The analysis is similar to the analysis of Eq. 22.77 on text p. 1230. Because the ketone is symmetrical, only one enolate ion is possible. Set *B* would give a mixture of products because the ketone could form two different conjugate-base enolate ions, and there is no reason for one to be strongly preferred over the other.

(b) The two possible sets of starting materials are as follows:

$$\underset{A}{\underset{}{CH_3CH_2\overset{O}{\overset{\|}{C}}CH_2CH_3\ +\ EtO\overset{O}{\overset{\|}{C}}OEt}}\quad and\quad \underset{B}{\underset{}{2\ CH_3CH_2\overset{O}{\overset{\|}{C}}OEt}}$$

Either set would work. In set *A*, a large excess of diethyl carbonate must be used. (See Eq. 22.75, text p. 1230, and the discussion that follows.) However, set *B* would be more convenient because only one ester is required as a starting material and because we would not have to separate the product from an excess of starting material, as in set *A*.

(c) Apply the analysis shown in Eq. 22.78 on text p. 1231. The two possible sets of starting materials are as follows:

Neither set of starting materials is a good choice. Because the desired product has no α-hydrogen more acidic than ethanol, it cannot be formed in a Claisen condensation because the equilibrium would be unfavorable. In fact, set *A could* react in a Claisen condensation, but a different product would be formed. (Can you give its structure?) Reactants *B* are also not a good choice for the same reason: the desired product is not acidic enough to be ionized by ethoxide. Furthermore, in this case a different product would also be formed.

(d) The two possible sets of starting materials are as follows:

Set *A* consists of two esters with α-hydrogens; four different products are possible. Furthermore, the α-hydrogens of ethyl phenylacetate are considerably more acidic than those of ethyl valerate (why?);

therefore, the major product is likely to be derived from the enolate ion formed at this position, but this is not the desired product. In set *B*, there are two sets of α-hydrogens, but the hydrogens that are α to the phenyl ring are considerably more acidic (why?). Unfortunately, the desired product is *not* derived from the enolate ion formed at this position. Therefore, both sets of starting materials are unsatisfactory. In summary, the desired product cannot be made by a Claisen condensation.

22.31   (a)   The two possible sets of starting materials are as follows:

Either answer is satisfactory, but perhaps compound *B* would be a somewhat better choice because, in compound *A*, reaction of the ketone enolate as a nucleophile at the ester carbonyl carbon might be somewhat retarded by van der Waals repulsions with the α-methyl group.

(b)   Only one starting material would provide the desired product in a Dieckmann condensation.

ethyl 2-acetylbenzoate

22.32   Acetyl CoA is the ultimate source of carbon atoms in fatty acid biosynthesis, as noted in the text (Eqs. 22.81 – 22.86). If the carbonyl carbon of acetyl CoA is labeled, then the carbonyl carbon and every other carbon in a long chain fatty acid such as palmitic acid will carry the label.

palmitate
(conjugate-base anion of **palmitic acid**)

22.33   (a)   Species *X* is the "enzyme thioester":

species *X*

(b)   Let H—A and A⁻ be acids and bases available inside the active site of the enzyme that assist in generating *X* and the acetylated CoA.

Compound *X* then reacts with CoA–SH to form the acetylated product and to regenerate the free enzyme.

(c)   The first reaction is a reverse Claisen condensation.

22.34   (a)   3-Phenylpropanoic acid can be prepared by a malonic ester synthesis:

Recall that benzyl chloride, PhCH$_2$Cl, is exceptionally reactive in S$_N$2 reactions. (See Eq. 17.33a–b, Sec. 17.4, text p. 931.)

(b)   2-Ethylbutanoic acid can be prepared by a malonic ester synthesis:

(c)   3,3-Dimethylbutanoic acid cannot be prepared by a malonic ester synthesis because the requisite alkyl halide, *tert*-butyl bromide, is virtually unreactive in S$_N$2 reactions. (See Sec. 9.4D, p. 463, in the text.)

3,3-dimethylbutanoic   ***tert*-butyl bromide**
acid

22.35   (a)   On the assumption that hydrolysis of the ester groups and decarboxylation of one carboxy group have occurred, the unsaturation number (2) of the product calculated from the molecular formula indicates that a ring has been formed. The formation of cyclobutanecarboxylic acid by the following reaction sequence accounts for the results:

The ring closure occurs by an internal "S$_N$2" reaction in which the enolate ion is the nucleophile:

$$CICH_2CH_2CH_2CH(CO_2Et)_2 \xrightarrow{NaOEt} CI\text{—}CH_2CH_2CH_2\ddot{C}\begin{smallmatrix}CO_2Et\\CO_2Et\end{smallmatrix} \longrightarrow \square\begin{smallmatrix}CO_2Et\\CO_2Et\end{smallmatrix} + CI^-$$

(b)   The two alkylations must occur as two separate reactions. If the dibromide had been used, two equivalents of diethyl malonate could have reacted with some of the 1,3-dibromopropane molecules. Because alkyl bromides are considerably more reactive than alkyl chlorides, the reaction conditions can be controlled so that only one diethyl malonate molecule reacts with the dihalide at the "alkyl bromide end" of the molecule. Once the enolate is formed in the second reaction step, the proximity effect (Sec. 12.8) accelerates the reaction rate of the alkyl chloride so that the ring closure occurs at a satisfactory rate.

22.36   (a)

$$CH_3CCH_2CO_2Et \xrightarrow[\substack{isobutyl\ bromide}]{\substack{1)\ NaOEt,\ EtOH \\ 2)\ BrCH_2CH(CH_3)_2}} CH_3CCHCO_2Et \xrightarrow[\substack{2)\ H_3O^+,\ -CO_2}]{1)\ NaOH} CH_3CCH_2CH_2CH(CH_3)_2$$

ethyl acetoacetate   (with $CH_2CH(CH_3)_2$ substituent)   **5-methyl-2-hexanone**

(b)

$$CH_3CCH_2CO_2Et \xrightarrow[\substack{benzyl\ bromide}]{\substack{1)\ NaOEt,\ EtOH \\ 2)\ BrCH_2Ph}} CH_3CCHCO_2Et \xrightarrow[\substack{2)\ H_3O^+,\ -CO_2}]{1)\ NaOH} CH_3CCH_2CH_2Ph$$

ethyl acetoacetate   (with CH_2Ph substituent)   **4-phenyl-2-butanone**

22.37   (a)   Follow the hint in Study Guide Link 22.6 on p. 755 of this manual.

$$EtO_2C\text{—}CH(CH_3)\text{—}C(=O)\text{—}CH_2CH_3 \xrightarrow[2)\ PhCH_2Br]{1)\ NaOEt,\ EtOH} EtO_2C\text{—}C(PhCH_2)(CH_3)\text{—}C(=O)\text{—}CH_2CH_3 \xrightarrow[\substack{2)\ H_3O^+,\ -CO_2}]{1)\ NaOH} CH_3CHCCH_2CH_3$$
(product with $PhCH_2$ substituent)

The starting β-keto ester is prepared by a Claisen condensation of ethyl propionate:

$$2\ EtOCCH_2CH_3 \xrightarrow[2)\ H_3O^+]{1)\ NaOEt,\ EtOH} EtO_2C\text{—}CH(CH_3)\text{—}C(=O)\text{—}CH_2CH_3$$

ethyl propionate

(b)

$$EtO_2C\text{—}CH(Ph)\text{—}C(=O)\text{—}CH_2Ph \xrightarrow[2)\ CH_3I]{1)\ NaOEt,\ EtOH} EtO_2C\text{—}C(Ph)(CH_3)\text{—}C(=O)\text{—}CH_2Ph \xrightarrow[\substack{2)\ H_3O^+,\ -CO_2}]{1)\ NaOH} PhCHCCH_2Ph$$
(product with CH_3 substituent)

The starting β-keto ester is prepared by a Claisen condensation of ethyl phenylacetate:

$$2\ EtOCCH_2Ph \xrightarrow[2)\ H_3O^+]{1)\ NaOEt,\ EtOH} EtO_2C\text{—}CH(Ph)\text{—}C(=O)\text{—}CH_2Ph$$

ethyl phenylacetate

22.38 Species *A* is the conjugate-base enolate ion of diethyl malonate. Reaction of this enolate ion as a nucleophile at the methylene carbon of the epoxide gives the alkoxide *B*, which, in turn, undergoes an intramolecular transesterification to give the final product, a lactone.

22.39 Because part (c) involves formation of a carboxylic acid with a quaternary α-carbon, and because such compounds cannot be prepared by a malonic ester synthesis, we'll use our one allowed reaction of amide bases to form this compound. All of the other compounds can be prepared by the malonic ester synthesis.

(a)

(b)

(c)

22.40 Once the enolate has been generated, it can react with either enantiomer of 2-phenylpropanal (reaction with only the *R* enantiomer is shown). The products are diastereomers, formed in unequal amounts. The enantiomer of each product shown is also formed.

22.41   (a)   Treatment of the α-bromo ester with Zn yields a Reformatsky reagent (text Sec. 22.8D), an ester enolate. An aldol reaction with the ketone gives the zinc alkoxide, *A*. Treatment of *A* with acid would yield an aldol addition product.

(b)   Compound *B* is an aldol condensation product, formed by a Knoevenagel reaction (text Sec. 22.8D). The mechanism is analogous to that shown in Eqs. 22.104a–c, text p. 1246–1247. Compound *C* is a carboxylic acid formed by hydrolysis and decarboxylation of the diester.

22.42   A transthioesterification transfers the acetyl group from acetyl-CoA onto the thiol of the enzyme. (The two curved arrows on the carbonyl group in the first structure are a shorthand for formation and decomposition of the tetrahedral addition intermediate.)

An aldol addition reaction links the acetoacetyl-CoA to the enzyme. Finally, hydrolysis of the thioester releases the HMG-CoA from the enzyme.

22.43 (a)

H$_2$C=C—CO$_2$CH$_3$ $\xrightarrow[\text{MeOH}]{^-\text{CN/HCN}}$ N≡CCH$_2$—CH—CO$_2$CH$_3$
          | |
          CH$_3$ CH$_3$

methyl methacrylate

(b)

C$_2$H$_5$SCH$_2$—CH—C(=O)—OCH$_3$
               |
               CH$_3$

(c)

BrCH$_2$—CH—CO$_2$CH$_3$
        |
        CH$_3$

(d)

H$_2$C=C—C(=O)—O$^-$ + CH$_3$OH
      |
      CH$_3$

In part (d), saponification occurs instead of conjugate addition because saponification is an *irreversible* nucleophilic acyl substitution reaction.

When considering acid-catalyzed additions to α,β-unsaturated carbonyl compounds, pay careful attention to the mechanism in Eq. 22.117 on text p. 1253. In part (c), it is tempting to add HBr to the double bond to give the tertiary carbocation A; however, such a mechanism would lead to a product B that is different from the product observed:

H$_2$C=C—C(=O)—OCH$_3$ ✗ H$_3$C—C$^+$—C(=O)—OCH$_3$ ⟶ H$_3$C—C—C(=O)—OCH$_3$
      | | |
      CH$_3$ CH$_3$ Br CH$_3$
                                      A   *B*

Remember that carbocations α to carbonyl groups are quite unstable and are almost never formed; see Eq. 22.38 on text p. 1119. In contrast, the mechanism in Eq. 22.117 on text p. 1253 involves a ketone that is protonated on the carbonyl oxygen; this is a resonance-stabilized species in which positive charge is shared at the terminal carbon. Reaction of bromide with this carbon leads to the observed product.

22.44 (a) This is a conjugate addition of the benzenethiolate anion to the carbon–carbon double bond. The benzenethiolate nucleophile, PhS$^-$, is formed by the reaction of benzenethiol, PhSH, with $^-$OEt. A mixture of diastereomers is formed because the final protonation step can occur from the same face as the methyl group or from the face opposite the methyl group.

(mixture of cis and trans)

(b) The product results from two successive conjugate additions. (Proton transfers are shown for simplicity as intramolecular processes, but they could be mediated by other acids and bases in solution.)

(c) The conjugate-addition product has no carbon–carbon double bond. Internal rotation within this adduct, followed by reversal of the addition, gives the isomerized product.

22.45 (a) The nucleophile must be a species that can be converted into a carboxymethyl (HO_2CCH_2—) group. Because the nucleophile ends up as a "substituted acetic acid," the conjugate-base enolate ion of diethyl malonate can serve as species X. Addition of X to the ester gives compound A. All ester groups of A are hydrolyzed and one of the resulting carboxy groups is decarboxylated when A is heated in aqueous acid.

(b) The key is to recognize that the desired product results from the addition of *two equivalents* of acrylonitrile, which end up as the two —$CH_2CH_2CO_2H$ groups of the product. The conjugate-base anion of ethyl acetoacetate would be a suitable nucleophile Y. Reaction of Y with one equivalent of acrylonitrile would give compound B. Ionization of B gives anion C, which reacts with a second equivalent of acrylonitrile to give compound D. The nitrile groups are hydrolyzed to carboxy groups and the ester group is hydrolyzed and decarboxylated to give the final product shown in the problem when compound D is heated in acid.

$CH_3\overset{O}{\overset{\|}{C}}\overset{-}{C}HCO_2Et$ $\xrightarrow[\substack{\text{EtOH}\\ \text{(Michael}\\ \text{addition)}}]{\substack{H_2C=CHCN\\ \textbf{acrylonitrile}}}$ $CH_3\overset{O}{\overset{\|}{C}}-\overset{CH_2CH_2CN}{\underset{|}{C}H}-CO_2Et$ $\longrightarrow$ $CH_3\overset{O}{\overset{\|}{C}}-\overset{CH_2CH_2CN}{\underset{|}{\overset{-}{C}}}-CO_2Et$ $\xrightarrow[\text{EtOH}]{\substack{H_2C=CHCN\\ \textbf{acrylonitrile}}}$ $CH_3\overset{O}{\overset{\|}{C}}-\overset{CH_2CH_2CN}{\underset{|}{\underset{CH_2CH_2CN}{C}}}-CO_2Et$

Y
enolate ion from
ethyl acetoacetate

B

C

D

22.46 (a)   The first part of the mechanism is a Michael addition followed by loss of formic acid to give, following protonation, intermediate *A*.

Compound *A* then undergoes an intramolecular aldol condensation reaction.

(b)   A Michael addition of the diethyl malonate conjugate-base anion gives compound *B*, which undergoes an intramolecular Claisen (Dieckmann) condensation. This Dieckmann condensation, like all Claisen-type condensations, is driven to completion by ionization of the product; the un-ionized product is formed when acid is added. These final steps are not shown in the mechanism below. Note that the tertiary hydrogen of compound *B* is the most acidic hydrogen; although the anion resulting from removal of this proton is formed in the reaction mixture, it is depleted by the reaction that gives the product.

B (redrawn)

(ionizes under the
basic conditions, and
is regenerated when acid
is added.)

 In the foregoing mechanisms many of the steps are reversible and should rigorously be shown with equilibrium arrows. However, the focus of the problem is on the mechanism of the forward reaction and not on the reversibility of the reaction. In such cases, forward arrows only are shown.

22.47   (a)   The conjugate addition of the hydride from NADPH is promoted by protonation of the carbonyl oxygen. The initial product is the enol which isomerizes to the product thioester. We use H_3O^+ and H_2O as acid and base, respectively, but acidic and basic groups in the active site of the enzyme are undoubtedly involved in these transformations. (See the mechanism in **Fig. SG22.2** on the following page.)

(b)   Because the hydride (H^-) was added to the β-carbon and a proton to the carbonyl oxygen, the reaction is a conjugate addition, as shown by the mechanism. Overall the reaction *looks* like a 1,2-addition, but that is because the initially formed 1,4-ddition product—the enol—subsequently isomerizes to the thioester.

(c)   To solve this problem, first draw the structure of (2*S*,3*R*)-butyryl-ACP-2,3-*d₂*.

Next, internally rotate along the C2–C3 bond into the eclipsed conformation. Then, rotate the structure so that the deuteriums are in the plane of the page. "Remove" the two deuteriums and analyze the resulting alkene. Since the alkene is trans, it confirms that the deuteriums were added syn.

trans-crotonyl-ACP

 Had the alkene that resulted from this analysis been cis, then we would have deduced that the addition of deuteriums was anti. This technique will always work for these types of problems

Figure SG22.2   The mechanism to accompany the solution to Focused Problem 22.47(a).

22.48   (a)

$$CH_3CH=CHCOEt \xrightarrow[\text{catalyst}]{H_2} CH_3CH_2CH_2COEt$$

ethyl 2-butenoate   **ethyl butanoate**

(b)

$$CH_3CH=CHCOEt \xrightarrow[\text{2) } H_3O^+]{\text{1) LiAlH}_4} CH_3CH=CHCH_2OH$$

ethyl 2-butenoate   **2-buten-1-ol**

$$+ \text{ EtOH}$$

22.49   (a)

$$(CH_3)_2C=CHCCH_3 \xrightarrow[\text{2) } H_3O^+]{\text{1) Li}^+ (CH_3)_2Cu^-} (CH_3)_3CCH_2CCH_3$$

mesityl oxide   **4,4-dimethyl-2-pentanone**
(4-methyl-3-penten-2-one)

(b)

$$(CH_3)_2C=CHCCH_3 \xrightarrow[\text{2) dil. } H_3O^+]{\text{1) CH}_3\text{Li}} (CH_3)_2C=CHCCH_3$$
$$\underset{CH_3}{\overset{OH}{|}}$$

mesityl oxide
(4-methyl-3-penten-2-one)

2,4-dimethyl-2-penten-1-ol

(c)

$$(CH_3)_2C=CHCCH_3 \xrightarrow[\text{2) } H_3O^+]{\text{1) Li}^+ (C_2H_5)_2Cu^-} \underset{CH_3}{\overset{CH_3}{C_2H_5CCH_2CCH_3}} \xrightarrow[\text{2) } H_2SO_4]{\text{1) CH}_3\text{MgI}} \underset{CH_3}{\overset{CH_3}{C_2H_5CCH=C(CH_3)_2}}$$

2,4,4-trimethyl-2-hexene

22.50   (a)   The question in this problem is what type of conjugate addition occurs (1,4 or 1,6). The 1,6-addition occurs because it gives the enolate ion intermediate with the greater number of resonance structures and therefore, the more stable enolate ion.

1,6-addition product
(observed)

enolate ion intermediate (more resonance structures)

1,4-addition product

enolate ion intermediate (fewer resonance structures)

(b) Conjugate addition also occurs with α,β-unsaturated compounds in which a carbon–carbon triple bond is part of the conjugated π-electron system. The product is $(CH_3)_2C=CHCO_2Me$ (methyl 3-methyl-2-butenoate).

22.51 (a) An analysis similar to that used in Study Problem 22.7 on text p. 1264 reveals that either a methyl group or an ethyl group can be added in the conjugate addition. (Addition of an ethyl group is illustrated here.)

3-methyl-3-penten-2-one

3,4-dimethyl-2-hexanone

(b) We use the same method as in part (a), except that we introduce the methyl group with an organocuprate reagent.

3-methyl-3-hexen-2-one

3,4-dimethyl-2-hexanone

(c) An analysis along the lines of Study Problem 22.7, text p. 1264, reveals the following two possibilities:

A practical synthetic equivalent to the anion in *(a)* is the conjugate base of ethyl acetoacetate:

conjugate-base
enolate ion of
ethyl acetoacetate

ethyl acrylate
(ethyl 2-propenoate)

5-oxohexanoic acid

The practical synthetic equivalent to the anion in *(b)* is the conjugate base of diethyl malonate:

(d)   An analysis similar to that in Study Problem 22.7 suggests the following possibilities:

A practical equivalent for the anion in (*a*) might be an acetylide anion, $^-C{\equiv}CH$. The resulting acetylene could then be hydrated to the desired ketone. Unfortunately, acetylenic cuprate reagents do not work in conjugate additions because the acetylenic ligands do not transfer from the copper. Nevertheless, if you came up with this possibility, you are analyzing the problem with considerable sophistication.

Anion (*b*) does have a simple practical equivalent: the cyanide ion:

Solutions to Skills Objectives Problems

22.52 (a)   The two α-hydrogens are acidic enough to be replaced. These hydrogens are shown as deuteriums in the following structure.

(b)   The α-hydrogens and the N—H hydrogen are acidic enough to be replaced. These hydrogens are shown as deuteriums in the structure at right. The stereochemistry of the deuterium at the ring junction could be a mixture of "up" and "down" (that is, trans and cis), because the planar carbanion intermediate could be protonated from either face. However, because the *trans*-decalin derivatives are more stable than *cis*-decalin derivatives, and because the reaction conditions provide a path for equilibration of the two isomers, the stereoisomer shown should be the predominant one.

22.53   Racemization occurs because the enolate-ion intermediate can be protonated at either face of the ring.

22.54 (a) The isomerization favors compound *B* because trans-decalin derivatives are more stable than *cis*-decalin derivatives. (See Sec. 7.6B, text p. 350.) The isomerization mechanism involves formation of an enolate-ion intermediate, which can be protonated on either face to give *A* or *B*, respectively.

A enolate ion intermediate *B*

(b) The analogous isomerization of compound *C* cannot occur because there is no α-hydrogen at the ring junction. Only the removal of an acidic hydrogen at a ring junction could lead to isomerization, and the hydrogen at the other carbon of the ring junction is not acidic enough to ionize under the conditions given.

22.55 2-Nitropropane is acidic because the conjugate-base anion is stabilized both by resonance and by the electron-withdrawing polar effect of the nitro group.

the conjugate-base anion of 2-nitropropane

22.56 (a) The phenyl group stabilizes anions at the carbon α to the ring (benzylic anions; see Display 17.23, Sec. 17.3, text p. 927). Compound *B* is most acidic because the conjugate-base anion (at right) is both benzylic and is α to two carbonyl groups.

conjugate-base anion of *B*

(b) Compound *A* is most acidic because the conjugate-base anion is stabilized by resonance interaction with, and the electron-withdrawing polar effects of, both the carbonyl group and the carbon–carbon double bond. Compound *B* lacks the carbon–carbon double bond.

(c) Compound *A* is most acidic because ketones are more acidic than esters for the reasons given in Sec. 22.1A. (See Fig. 22.2 on text p. 1197.)

22.57 (a) This compound is an "ynol"—the acetylenic analog of an enol—and it converts spontaneously into methylpropenone ("methylketene"), the corresponding ketone (shown at right).

$H_3C-HC=C=O$ **methylpropenone (methylketene)**

(b) This compound, 1,3,5-cyclohexanetrione, exists as its "triene-triol" isomer phloroglucinol because the latter is aromatic.

1,3,5-benzenetriol (phloroglucinol)

22.58 (a) The conjugate bases of the two compounds, because they are resonance structures, are identical.

(b)   Because compound B (phenol) is aromatic, it is more stable, and thus more energy is required to convert it into the conjugate base shown in part (a). From Eq. 3.61b–c, Sec. 3.6D on text p. 145, $\Delta G_a^{\circ} = 2.3RT(pK_a)$ Therefore, compound B has the greater pK_a (it is less acidic). This conclusion is demonstrated in **Fig. SG22.3**.

22.59   (a)   Removal of the O—H proton from A and a C—H proton from B gives conjugate-base anions that are resonance structures and therefore identical. (B:⁻ = a general base.)

Figure SG22.3   Figure to accompany the solution to Problem 22.58(b). Because phenol is aromatic, it is more stable than the ketone, and thus more energy is required to convert it into the conjugate base, which is common to both compounds.

(b)   Because the ketone is more stable than its enol, more energy must be expended for it to ionize. Consequently, the ketone is less acidic. The logic is similar to that used in the solution to the previous problem. (We leave it to you to construct the corresponding free-energy diagram.)

22.60   The enol forms of phenylacetone:

Enols *A* and *B* are more stable than enol *C* because there are more substituents on the double bond, and because of the conjugation of the carbon–carbon double bond with the phenyl group. If we assume that a methyl group is somewhat larger than an OH group, enol *A* is most stable because of the stabilizing conjugation of the carbon–carbon double bond with the phenyl group. Because of the *E* stereochemistry of enol *B*, it is somewhat less stable than its *Z* isomer, enol *A*. (The stability difference between *A* and *B*, if any, is probably not great.)

22.61   (a)   The mechanism of the acid-catalyzed enolization of phenylacetone (A—H = a general acid):

(b)   The mechanism of the base-catalyzed enolization of phenylacetone.

22.62   (a)   The likely reaction:

(Addition of acid in a separate step would be required to give the neutral β-keto ester.) Because the product is the relatively stable anion, a second condensation (at the other α-position of the 5-membered ring) is unlikely because it would require formation of a much less stable enolate of the ketone (unless a second equivalent of base were used). An aldol condensation of the ketone with itself is reversible and unfavorable, so this product will not be observed. A Claisen condensation of the ester with itself is not likely because the ketone is much more acidic than the ester; because there is an excess of the ester, the ketone enolate will react with the ester.

(b)   The likely reaction:

(Whether the last dehydration occurs depends on the severity of the reaction conditions.)

The only possible competing reaction is the aldol reaction of acetaldehyde with itself. This is not likely because, when the aldehyde is added to the reaction mixture it can react with the enolate conjugate base of diethyl malonate in two ways. The aldehyde can lose an α-proton to the diethyl malonate enolate, which is a very unfavorable reaction ($K_{eq} \sim 10^{-7}$), or it can undergo an aldol addition with the enolate. The aldol addition will likely occur because the reaction is driven to completion by the formation of the relatively stable anion X.

22.63 (a) (b) (c) (d) (e)

22.64 (a) $BrCH_2CH_2CCH_3$ (b) $CH_3CH_2CCH_3$ (c) $H_2C=CHCHCH_3$ (d) $N\equiv CCH_2CH_2CCH_3$ (e) $CH_3CH_2CH_2CH_2CCH_3$ (f) $(EtO_2C)_2CHCH_2CH_2CCH_3$ (g) (h)

22.65

by a Reformatsky reaction
(Sec. 22.8D)

by a Knoevenagel reaction
(Sec. 22.8D)

22.66

(a) HO₂CCHCH₂CO₂H with CH₃

(b) (CH₃)₂NCHCH₂CO₂Et with CH₃

(e) CH₃CH₂CH₂CO₂Et

(both compounds are racemates)

22.67 (a) The dibromo derivative is a mixture of diastereomers: the (±)-diastereomer (that is, the racemate) and the meso diastereomer.

(±)-1,3-dibromo-1,3-diphenyl-2-propanone

meso-1,3-dibromo-1,3-diphenyl-2-propanone

(b) The is an HVZ reaction that gives the α-bromo acid bromide, which reacts with methanol to give the methyl ester.

(c) This is an aldol addition; dehydration cannot occur because the resulting alkene would violate Bredt's rule. The mechanism below begins with the enolate ion of the methyl ketone, which is formed when a proton is removed by potassium carbonate.

(d)   The lactone, a cyclic ester, can form an enolate ion, and this ion is alkylated by methyl iodide.

(e)   Lithium aluminum hydride reduces the carbonyl group to an alcohol.

(f)   This is a Michael addition in which the conjugate-base enolate ion of diethyl methylmalonate is the nucleophile.

(g)   This is a Michael addition in which the conjugate-base enolate ion of acetoacetic ester is the nucleophile. If the acid hydrolysis step is accompanied by heat, ester hydrolysis and decarboxylation of the resulting β-keto acid can also occur. Either answer is correct.

(h)   The conjugate-base enolate ion is alkylated by the alkyl halide. Then the conjugate-base enolate ion of the alkylation product is formed and reacts intramolecularly with the second alkyl halide group to form a ring. Hydrolysis and decarboxylation of the ester group give the final product.

This would be the major product if the first enolate alkylation is faster than the S_N2 reaction of the base ethoxide with the alkyl halide, 1,4-dibromobutane. (The second alkylation should be very fast because of the proximity effect.) However, it is very likely that some alkylation of ethoxide by the dibromide will occur. In that case, the following side products would also be expected:

$$BrCH_2CH_2CH_2CH_2OEt \qquad EtOCH_2CH_2CH_2CH_2OEt \qquad CH_3\overset{\displaystyle O}{\overset{\|}{C}}CH_2CH_2CH_2CH_2OEt$$

(i) This is an aldol condensation of the Knoevenagel variety.

(j) Conjugate addition of cyanide, then hydrolysis of both the nitrile and the ethyl esters to a substituted malonic acid and decarboxylation gives a dicarboxylic acid with the correct formula.

(k) This is a conjugate addition of the magnesium organocuprate reagent. Assume that the isopropenyl group (the group delivered from the organocuprate reagent) enters trans to the angular methyl group to avoid van der Waals repulsions.

22.68 (a) The gas is H_2 and the species A is the sodium salt of the conjugate-base enolate ion of 2,4-pentanedione. As the resonance structures indicate, the negative charge, and therefore the nucleophilic character, of this anion is shared by both the anionic carbon and the oxygens.

(b) The three species that are formed all result from reaction with a nucleophilic atom with methyl iodide in an S_N2 reaction. The products are the two stereoisomeric ethers B and C and the alkylated β-diketone D.

22.69 (a) The mechanism shown below begins with an α-bromo enolate ion, formed by the removal of an α-proton by NaH. This ion displaces a bromine in an internal nucleophilic substitution reaction to form the ring. NaH forms another enolate ion that undergoes elimination to give the product. (Recall that β-elimination reactions are particularly rapid when the proton undergoing elimination is particularly acidic; see Sec. 17.3B, text pp. 929–930.)

(b) This is a double Claisen condensation followed by hydrolysis and decarboxylation of the ester groups. The mechanism below begins with the conjugate-base enolate ion of the diester.

Compound *A* forms a conjugate-base enolate ion under the basic conditions but is regenerated when acid is added. Hydrolysis of the two ester groups and decarboxylation of the resulting β-keto acid groups give the product. In the step labeled *(a)*, the hydrogen that is removed is not the most acidic hydrogen in the molecule. (Which hydrogen is most acidic?) However, removal of the most acidic hydrogen does not lead to the observed product. Removal of the hydrogen shown, although occurring less frequently, leads rapidly to product.

Note also that the second condensation reaction could involve the reaction of a different enolate ion with a different ester group so that the β-keto ester shown on the right would be formed:

alternate enolate for
second condensation

Ester hydrolysis and decarboxylation of this compound would also give the observed product.

(c) This reaction is a crossed aldol (Claisen–Schmidt) condensation followed by an intramolecular crossed aldol condensation followed by the same sequence with another molecule of dialdehyde. The following mechanism begins with the conjugate-base enolate ion of the diketone. Although the dehydration step is shown for brevity as a concerted process, it probably involves an enolate-ion intermediate in an E1cB mechanism. (See Eq. 22.49, Sec. 22.4A, text p. 1212.)

(d) An intramolecular ring closure within the tetrahedral addition intermediate of the initial enolate addition reaction provides the cyclic ether.

This unusual product probably forms because a significant steric interaction between the two methyl substituents and the cyclohexane ring forces the —O⁻ group and the carbonyl carbon together.

 There are many other examples in which steric interactions between backbone substituents accelerate an intramolecular reaction. This effect is known in organic chemistry as the *"gem-dimethyl effect."*

(e) This reaction is a double crossed-Claisen condensation followed by an intramolecular transesterification reaction in which the anionic oxygen of an enolate ion serves as the nucleophile. (The oxygen, rather than the anionic carbon, acts as the nucleophile because a six-membered ring is formed; reaction with the anionic carbon would give a strained four-membered ring.) The mechanism that follows begins with the conjugate-base enolate ion of acetone. Acid-catalyzed hydrolysis of the diester by the usual mechanism gives chelidonic acid.

22.70   (a)   A conjugate addition of hydroxide to (R)-pulegone gives a β-hydroxy ketone, which then undergoes the reverse of an aldol addition to give acetone and (R)-3-methylcyclohexanone.

(R)-3-methylcyclohexanone

(b)   The starting material is a β-hydroxy ketone. Ionization of the hydroxy group followed by the reverse of an aldol addition gives the product. The reaction is driven by relief of strain in the four-membered ring.

22.71 In the first step, the conjugate-base anion of diethyl malonate reacts with mesityl oxide, an α,β-unsaturated ketone, in a Michael addition.

A crossed-Claisen condensation between the enolate ion of the methyl ketone and one of the ester groups then ensues.

Dimedone results from the saponification and decarboxylation of the remaining ester group (a β-keto ester) in compound *B*.

22.72   (a)

2-cyclopentenone → 1) Li⁺ (C₂H₅)₂Cu⁻ 2) H₃O⁺ → (3-ethyl ketone) → NaBH₄ methanol → 3-ethylcyclopentanol (four stereoisomers are possible)

(b)

3-methyl-2-cyclohexenone → 1) Li⁺ (CH₃)₂Cu⁻ 2) H₃O⁺ → 1) CH₃MgI 2) H₃O⁺ → **1, 3, 3-trimethylcyclohexanol**

(c)

EtO₂C(CH₂)₅CO₂Et **diethyl pimelate** → NaOEt EtOH → (cyclohexanone, CO₂Et) → ClCH₂Ph → (CO₂Et, CH₂Ph) → 1) NaOH 2) H₃O⁺, −CO₂ → **2-benzylcyclohexanone** (CH₂Ph)

(d)

EtO₂CCH₂CO₂Et **diethyl malonate** → 1) NaOEt, EtOH 2) CH₃I → EtO₂CCHCO₂Et (CH₃) → 1) NaOEt, EtOH 2) CH₃I → EtO₂CCCO₂Et (CH₃, CH₃) → 1) LiAlH₄ 2) H₃O⁺ → HOCH₂CCH₂OH (CH₃, CH₃) **2,2-dimethyl-1,3-propanediol**

(e)

H₂C=CHCO₂Et **ethyl acrylate** → HCN → N≡CCH₂CH₂CO₂Et → LiAlH₄ → H₃O⁺ → ⁻OH → H₂NCH₂CH₂CH₂CH₂OH **4-amino-1-butanol**

(f)

H₂C=CH—CN **acrylonitrile** → (Et)₂NH → (Et)₂NCH₂CH₂CN → LiAlH₄ → H₃O⁺ → ⁻OH → (Et)₂NCH₂CH₂CH₂NH₂

(g)

PhCH₂COH (O) **phenylacetic acid** → H₂SO₄ (catalyst) EtOH (solvent) → PhCH₂COEt (O) → EtOCOEt (excess) NaOEt → PhCHCOEt (O) (CO₂Et) → 1) NaOEt, EtOH 2) CH₃CH₂Br →

PhC—COEt (CH₃CH₂, O) (CO₂Et) → H₃O⁺, heat −EtOH −CO₂ → CH₃CH₂CHCO₂H (Ph) **2-phenylbutanoic acid**

(h)

(i)

(j)

(k)

(l)

(m)

22.73 (a) Use a conjugate addition of cyanide to an α,β-unsaturated ester, and then hydrolyze. (Either stereoisomer of the starting ester can be used.)

(b) Take the same approach as in part (a), but with a different ester.

22.74 We begin with hexanoyl-ACP, which is transferred by a transthioesterification to the "Synth" protein. The ACP protein is then charged with a malonyl group from malonyl-CoA, and the series of reactions ensues that produces hexanoylacetyl-ACP. This β-keto ester is reduced in three steps to octanoyl-ACP. This entire process is shown in **Fig. SG22.4** on the following page.

22.75 This mechanism is shown in **Fig. SG22.5** on p. 813. This is an enzyme-catalyzed aldol addition of the enolate of acetyl-CoA to the keto group of oxaloacetate followed by thioester hydrolysis. H—A^1, H—A^2, and H—A^3 are hypothetical general acid groups, and $^-$A^1, $^-$A^2, and $^-$A^3 their conjugate bases, in the enzyme active site. (You could also use H$_3$O$^+$ and water molecules as acids and bases for purposes of this mechanism.) Hydrogen bonding to the carbonyl oxygen of the enolate secures it within the active site. Once the aldol addition is complete, the concerted action of a water nucleophile, acids and bases facilitates the hydrolysis of the thio ester.

Solutions to Integrated Problems

22.76 (a)   (b)

(either enantiomer)   (two of several possibilities)

(Solution continues on the following page)

Figure SG22.4   The conversion of hexanoyl-ACP to octanoyl-ACP during fatty-acid biosynthesis.

(Solution to Problem 22.76, continued)

(c)

and many others

(Each could be any one of several stereoisomers—
six possible stereoisomers for the middle structure
and four possible stereoisomers for the others two.)

(d)

$$CH_3CH_2CHCH_2CH=O$$
$$\quad\quad\; | \quad\quad$$
$$\quad\quad CH_3$$

Figure SG22.5 A mechanism for the conversion of oxaloacetate to citrate for the solution to Problem 22.75. H—A¹, H—A², and H—A³ are hypothetical general acid groups, and ⁻A¹, ⁻A², and ⁻A³ their conjugate bases, in the enzyme active site.

22.77 (a)

(b)

22.78 (a)

(b) The hydrolysis reaction of *A* is much more favorable than that of *B* because the alcohol hydrolysis product of *A* is an enol. Ketones are much more stable than their isomeric enols, so the ketone isomerization adds an extra free-energy increment in favor of the hydrolysis product. Although the product of reaction *B* is also an "enol," it is not transformed further because it is aromatic..

22.79 The order of increasing acidity (decreasing pK_a) is as follows. [Approximate pK_a values are in brackets.]

toluene [42] < ethyl acetate ≈ phenylacetylene [25] < Isobutyramide [15] < phenol [10] < octanoic acid [4.5]

Toluene is much less acidic than phenol, and amides are much less acidic than carboxylic acids (element effect). The pK_a values of 1-alkynes and esters are very similar (about 25). The remaining rankings require knowledge of approximate pK_a values. The point of this problem is that it is very important to know the typical pK_a values of organic compounds because so much of organic reactivity is based on acid–base principles.

22.80   (a)   Bromination can occur at either of the two α-carbons; compounds *B* and *C* are diastereomers. (Each of the three compounds can also exist as enantiomers.)

A   *B*   *C*

(b)   The enol intermediate in the formation of *A* is the more stable one because it has the greater number of alkyl substituents on the double bond. Therefore, compound *A* is the α-bromo derivative that is formed.

enol intermediate involved in the formation of *A* (more stable enol)

enol intermediate involved in the formation of compounds *B* and *C*

22.81   The discussion surrounding Eq. 22.27, text pp. 1204, explains that enols are intermediates in the acid-catalyzed bromination of aldehydes and ketones, and that enol formation is the rate-limiting process. As explained in Sec. 20.11A, an enol intermediate is also involved in the decarboxylation of a β-keto acid. Therefore, the bromination reaction described in the problem evidently involves an enol intermediate, and formation of the enol is the rate-limiting process. Hence, the rate of the bromination reaction is zero-order in bromine.

22.82   A reasonable hypothesis is that the carbon atom of the iodoform must originate from the central carbon of the β-diketone:

Base-promoted iodination occurs at the central carbon by the usual mechanism involving enolate ions to give the diiodo ketone *A*. Hydroxide then displaces an α-diiodo enolate anion *B* in a nucleophilic acyl substitution reaction; this resonance-stabilized anion is a good leaving group because the iodine substituents reduce its basicity. Anion *B* is then iodinated, and the resulting triiodo ketone *C* undergoes the final steps of the haloform reaction illustrated in Eq. 22.31e in Sec. 22.3B, text p. 1206, to give iodoform and benzoate anion. The two equivalents of benzoate anion produced in this sequence give benzoic acid when the reaction mixture is acidified.

22.83 (a) The equilibrium favors the α,β-unsaturated isomer because it is conjugated. Conjugation is a stabilizing effect because of the additional π-electron overlap that is possible in conjugated compounds. (See Sec. 15.1A, text pp. 801–804.)

(b) The mechanism in aqueous base involves a resonance-stabilized enolate ion intermediate, which can be protonated on either the α-carbon or the γ-carbon to give the respective unsaturated ketones.

(c) The mechanism below commences with the dienol 1,3-cyclohexadienol, which is formed by the usual acid-catalyzed enolization mechanism (Eq. 22.21b on text p. 1202). Protonation of this dienol gives a resonance-stabilized carbocation, which is identical to the conjugate acid of the α,β-unsaturated ketone. Loss of the O—H proton gives the α,β-unsaturated ketone itself.

(d) The equilibrium constant for conversion of 4-methyl-3-cyclohexenone into 4-methyl-2-cyclohexenone should be smaller, because the additional alkyl substituent on the β,γ-double bond tends to offset the stabilizing effect of conjugation in the α,β-unsaturated derivative.

22.84 (a) The exchange of H^a occurs by the usual enolate-ion mechanism.

Removal of either hydrogen H^c or H^d gives an anion that has resonance structures which show that it is an enolate ion. Therefore, these hydrogens are acidic enough to undergo exchange. The exchange of H^c is shown explicitly below; you should show the exchange of H^d, which occurs by essentially the same mechanism.

(b)   Although H^b is an α-proton, it is not acidic because the anion that would result from its removal is not resonance-stabilized. The reason is that the orbital containing the nonbonding electron pair in this anion is perpendicular to the π-electron system of the carbonyl group and cannot overlap with it. An equivalent resonance argument is that if the overlap which lies at the basis of resonance were to occur, the nuclei would have to move, and a cumulated double bond and the attendant large amount of strain would be introduced into the six-membered ring.

overlap introduces
a strained cumulated
double bond

(c)   Hydrogen H^b is not acidic in the α,β-unsaturated ketone. However, as shown in the foregoing Problem 22.83, the α,β-unsaturated ketone readily isomerizes to a β,γ-unsaturated ketone. Ionization of H^b in this compound gives a resonance-stabilized anion, as shown in the solution to Problem 22.83b in this manual. Therefore, H^b exchanges, and the β,γ-unsaturated ketone isomerizes back to the α,β-unsaturated ketone.

22.85   The protons on carbons indicated by arrows and the one attached to the oxygen would be exchanged for deuterium under these conditions.

Protons *a* are ordinary α-protons. Protons *c,* when removed, give a resonance stabilized ion; one resonance form is an enolate. The α,β-double bond in base isomerizes to a β,γ-double bond as described in the solution in Problem 22.84(c). When this happens, protons at *b* become α-protons and are exchanged in base. Finally, protons *d* are ordinary OH protons which rapidly exchange in D_2O. (This is the basis of the "D_2O shake" in proton NMR spectroscopy; see Sec. 14.6 in the text.)

22.86   (a)   Although ethers are usually more basic than carbonyl oxygens, in compound *A* the carbonyl oxygen is protonated because the resulting carbocation has $4n + 2$ π electrons; that is, it is aromatic. Remember that increasing the stability of the conjugate acid in an acid–base equilibrium decreases its acidity (or increases the basicity of the conjugate base). The aromaticity of its conjugate acid therefore explains why compound *A* is more basic than compound *B*. Compound *B* is protonated on the ether oxygen and has the basicity typical of an ether.

conjugate acid of compound *A*
(an aromatic species)

conjugate acid
of compound *B*

> The relative basicities of these two compounds are deceptive because they involve protonation on different oxygens. The conjugate-acid pK_a of a ketone is typically –6. Therefore, aromaticity actually raises the pK_a of the conjugate acid of *A* by about 6 units.

(b)   Tropone is unusually basic because its conjugate acid is an aromatic carbocation.

conjugate acid of tropone
(an aromatic α-hydroxy carbocation)

The consequences of aromaticity on the pK_a of a conjugate acid were explored in part (a), and these apply equally here.

22.87 (a) As shown in Fig. SG22.3 on page 798 of this manual, the more stable a species is relative to its conjugate base the less acidic it is. Consequently, resonance stabilization of an α,β-unsaturated carboxylic acid decreases its acidity, assuming that the resonance structures for the conjugate base are less important.

(b) The pK_a of the first compound, butanoic acid, is the reference value—the pK_a to be expected for a four-carbon carboxylic acid containing no carbon–carbon double bonds. In the second compound, 3-butenoic acid, the carbon-carbon double bond is not conjugated with the carbonyl group. Therefore, the difference between its pK_a value and that of butanoic acid is due to the electron-withdrawing polar effect of the carbon–carbon double bond. If the polar effect were the only effect operating in the last compound, *trans*-2-butenoic acid (crotonic acid), it should be the most acidic compound in the series, because the polar effect of any group on acidity increases when the group, in this case a carbon–carbon double bond, is closer to the site of ionization. Because *trans*-2-butenoic acid is in fact *less* acidic than 3-butenoic acid, some acid-weakening effect must also be operating, and this is the resonance effect discussed in part (a).

22.88 (a) The conjugate base of 2-nitropropane was discussed in the solution to Problem 22.55. Protonation of this conjugate base can occur either on the α-carbon, to regenerate the nitro compound, or on an oxygen of the nitro group, to give an isomer, an *aci*-nitro compound, which is the "enol" analog of a nitro compound.

(b) Because 2-nitropropane (pK_a = 10) is much more acidic than ethanol, it is completely converted into its anion by sodium ethoxide. This anion, like many other "enolate ions," undergoes a Michael addition to ethyl acrylate to give ethyl 4-methyl-4-nitropentanoate.

22.89 Intermediate *A* is the anionic product of a crossed Claisen condensation. This anion undergoes a Michael addition to the α,β-unsaturated ketone, and the product of that reaction undergoes an intramolecular aldol condensation to give compound *B*. Hydroxide ion effects a reverse Claisen condensation that removes the formyl group as formate ion. The carbanion formed in this step is resonance-stabilized, and one of the resonance forms is an enolate ion. The steps are outlined below; you should provide the mechanistic details using the curved-arrow notation.

22.90 (a) One full equivalent of base is required in the Claisen or Dieckmann condensation to ionize the product. Without this ionization, the condensation equilibrium is unfavorable. (See Sec. 22.6A, text pp. 1226–1227.)

(b) Sodium phenoxide, the conjugate base of phenol, is about 10^{-5} times as basic as sodium ethoxide. Therefore, sodium phenoxide is not basic enough to ionize completely the Claisen condensation product. Without such an ionization, the condensation equilibrium is unfavorable.

(c) The equilibrium for the Claisen condensation is rather unfavorable and requires the final ionization of the product to proceed to completion. Under acidic conditions, no base is available to cause this ionization. In addition, an acid-catalyzed Claisen condensation would require the enol of an ester, because enols are the α-carbon nucleophiles under acidic conditions. As shown in Eqs. 22.17 and 22.18, Sec. 22.2, text p. 1200, the formation of enols from esters is *much* less favorable than the formation of enols from aldehydes or ketones. The much lower concentration of reactive intermediate would translate into a much-reduced rate. In summary, both the rate of the reaction is very low, and its equilibrium constant is highly unfavorable; so, no reaction is observed.

22.91 (a) An alkylation of this type requires a base that will rapidly and completely convert the ester into its conjugate-base enolate anion so that the enolate ion is not present simultaneously with the ester. In the presence of sodium ethoxide, however, small amounts of the enolate ion are present together with large amounts of the un-ionized ester; consequently, the Claisen condensation will occur as the major reaction. In all likelihood methyl iodide will alkylate the anionic product of the Claisen condensation to give the product shown at right.:

(b) The conjugate-base anion of diethyl malonate would be formed with sodium ethoxide, but it would not react with bromobenzene because the latter compound, an aryl halide, does not undergo S_N2 reactions. (See text Sec. 18.1 on text p. 962.)

(c) Alcohol dehydration involves a carbocation intermediate, and carbocations α to carbonyl groups are particularly unstable. The desired dehydration will not occur. The more likely reaction would be polymerization of the ester by reaction of the hydroxy group of one molecule with the carbonyl of the other, hydrolysis of the ester group to a carboxylic acid, or some of both.

(d) The first step will effect α-halogenation of the carboxylic acid. However, an attempt to form the Grignard reagent will fail, because the Grignard reagent will be destroyed instantly by a protonolysis reaction with the α-bromo acid. Because the Grignard reagent does not survive, its reaction with acetaldehyde cannot occur.

(e) This is a crossed aldol condensation in which the conjugate-base enolate ion of the ketone reacts with the aldehyde. A mixture of isomeric products will result because there are two possible conjugate-base enolate ions of the ketone that can react. Moreover, a very significant, if not predominant, competing reaction will be the aldol condensation of acetaldehyde with itself.

(f) In addition to bromination of the benzene ring, the acidic conditions will also promote α-bromination at the methyl group. Lewis acids can catalyze this reaction just as Brønsted acids can. In addition, HBr is a by-product of ring halogenation, and this can also catalyze α-halogenation.

22.92 This reaction is essentially a type of crossed-Claisen condensation in which the "enolate ion" is the conjugate-base anion of urea. A second reaction of the same type closes the ring. The following mechanism starts with the reaction of the ionized urea with one of the ester groups.

As in other Claisen-type condensation reactions, the product Veronal is ionized by the ethoxide base, and Veronal is regenerated when acid is added. (These final steps are not shown in the foregoing mechanism.)

22.93 The synthesis of pentothal involves, first, preparation of the substituted malonic ester derivative, and then condensing it with thiourea, the sulfur analog of urea.

22.94 Compound B is an aldehyde (it gives a positive Tollens test); therefore, compound A is a primary alcohol. Compound B is conjugated (it has a strong UV spectrum), and it has the same carbon skeleton as octanoic acid (an unbranched chain of eight carbons). Therefore, compound B is 2-octenal, and compound A is 2-octenol. (The data do not differentiate between the E and Z stereoisomers.) In order to form 2-octenol, lithium dibutylcuprate must undergo a *conjugate addition* to the epoxide:

2-octenol (compound A)   **2-octenal (compound B)**

22.95   (a)   A crossed Claisen condensation is followed by a reverse Claisen condensation, driven by expulsion of the more volatile ester, ethyl acetate, from the reaction mixture. The resulting anion *A*, which is both the enolate conjugate base of a β-keto ester and a benzylic anion, is a much weaker base than ethoxide, so it doesn't protonate under the basic conditions. Its conjugate acid, ethyl 3-oxo-3-phenylpropanoate, is formed only when acid is added at the end of the reaction.

ethyl 3-oxo-3-phenylpropanoate

(b)   The conjugate-base anion of diethyl malonate serves as the nucleophile in an epoxide-opening reaction. Normally, epoxide opening under basic conditions involves reaction with the nucleophile at the less branched carbon. In this case, however, reaction also occurs at the more branched carbon because it is benzylic and therefore activated toward nucleophilic substitution. (Only reaction at the benzylic carbon is shown below; reaction at the other carbon occurs by a similar mechanism.) The resulting alkoxide undergoes intramolecular transesterification to give a lactone; hydrolysis and decarboxylation of the remaining ester give the product. The lactone does not hydrolyze because the equilibrium favors lactone formation when five- and six-membered rings are involved; see Eq. 21.36a on text p. 1149.

(c)   This is an acid-catalyzed conjugate addition of water followed by elimination of phenol. The resulting enol is then spontaneously converted into the corresponding aldehyde.

(d)   This mechanism is much like that in part (c), except that basic conditions rather than acidic conditions are involved in all steps but the last.

(e)   Enolization of the α,β-unsaturated ketone gives a dienol. A carbon–carbon double bond of the dienol on one ring reacts as a nucleophile with the protonated carbonyl group on the other. The mechanism below begins with a protonated α,β-unsaturated ketone.

(f)   Ionization of the benzylic hydrogen gives a carbanion that not only is allylic and benzylic but also is stabilized by the polar effect of the α-oxygen. The O—H hydrogen is also acidic, and removal of it occurs reversibly, but the resulting anion does not lead to a reaction. The equilibrium in the reaction is pulled to the right by formation of the ketone from the enol, and the overall reaction is favorable because aldehydes and ketones have a lower energy than their isomeric allylic alcohols. (See the heats of formation in Eq. 22.121 on text p. 1255 in Sec. 22.9B,.) In addition, the ketone product in this case is conjugated with the benzene ring, whereas the double bond in the starting material is not. This provides an additional driving force for the reaction.

(g)   Conjugate addition with the nitrogen of hydroxylamine as a nucleophile gives compound *A*, which undergoes intramolecular transesterification.

(h)   α-Cyano esters have acidities comparable to those of β-keto esters. (Why?) The conjugate-base enolate ion of the α-cyano ester adds to the aldehyde carbonyl group. Elimination of water by an E1cB mechanism gives an α,β-unsaturated nitrile, which undergoes conjugate addition with cyanide ion. Hydrolysis of the ester and decarboxylation of the resulting acid give the product.

(i)   This reaction is a "conjugate addition" of cyanide ion to a cyclopropane. Notice that the reaction occurs at the least branched carbon of the cyclopropane and that the immediate product of ring opening is a very stable anion. Heating in acid protonates this anion, hydrolyzes the ester and nitrile groups, and decarboxylates the resulting β-keto acid.

(j)   Conjugate addition of triphenylphosphine to the carbon–carbon double bond of maleic anhydride followed by proton transfer to the initially formed anion gives an ylid *A*, which undergoes a Wittig reaction with the aldehyde butanal. Hydrolysis of the resulting anhydride gives the dicarboxylic acid.

22.96 (a)   The conjugate-base enolate ion of the α-bromo ester, formed by reaction of the α-bromo ester with potassium *tert*-butoxide, reacts at the carbonyl carbon of the aldehyde to give β-bromo alkoxide *A*, the conjugate base of a bromohydrin. As shown in Eqs. 12.25 and 12.26, Sec. 12.3B, text p. 633, an alkoxide of this type readily cyclizes to an epoxide by an internal nucleophilic substitution mechanism.

(b)   One of the two possible conjugate-base enolate ions undergoes an intramolecular nucleophilic substitution reaction to give the cyclopropane. Although the other enolate (from loss of a proton of the methyl group) is undoubtedly also formed, its cyclization would lead to a different product. In addition, the enolate that leads to the observed product is the more stable enolate. (Why?)

22.97   The deuterium incorporation results show that the carbon–carbon double bond is protonated. This is exactly the way that enols are protonated (see the reverse of Eq. 22.21b in Sec. 22.2 on text p. 1202). As a result, a relatively stable α-alkoxy carbocation is formed, and this reacts with water to give a hemiacetal. The hemiacetal breaks down by the usual mechanism (see text Sec. 19.11A) to give acetaldehyde and the alcohol. Hydrolysis of vinyl ethers is much faster than hydrolysis of ordinary ethers because vinyl ether hydrolysis involves a relatively stable carbocation intermediate, an α-alkoxy carbocation.

    In the following mechanism, the asterisk indicates the position of the deuterium that is incorporated into the product if the reaction were run in D_2O. (All the other "hydrogens" of the OH group and H_2O and H_3O^+ would be deuteriums as well, but these are not shown in the mechanism.)

22.98   The Wittig reaction gives a vinylic ether, which, as shown in the previous problem, hydrolyzes rapidly to the corresponding aldehyde.

22.99   (a)   This is a nucleophilic aromatic substitution reaction (text Sec. 18.4) in which the nucleophile is an enolate ion. The ester groups are hydrolyzed, and one of the resulting carboxylic acids is decarboxylated, by the hot acidic aqueous conditions.

initial substitution
product

(b)   The diethylmalonate enolate ion, like other strong bases, displaces the chloride of the acid chloride. After hydrolysis of the esters, the resulting β-keto diacid loses two equivalents of carboxylic acid by decarboxylation to yield a ketone as a final product.

(c)   The magnesium organocuprate reagent undergoes conjugate addition; the acetal group hydrolyzes; and an acid-catalyzed intramolecular aldol condensation ensues.

product of
conjugate addition

(d)   The first reaction is a Knoevenagel reaction of acetone (compound *A*) with diethyl malonate. The product of this reaction undergoes a conjugate addition with the methyl organomagnesium cuprate reagent to give diethyl *tert*-butylmalonate. . Ester hydrolysis and decarboxylation gives the final product *B*.

diethyl *tert*-butylmalonate

$(CH_3)_3C-CH_2-CO_2H$ + EtOH + CO_2

**3,3-dimethylbutanoic acid
("*tert*-butyl acetic acid")**

B

(e)   The benzene π electrons serve as the nucleophile in an acid-catalyzed conjugate addition, which can also be viewed as a variation of Friedel–Crafts alkylation.

(f)   Under ordinary conditions, aryl halides do not react with nucleophiles such as an enolate; however, in the presence of a transition-metal catalyst, such as Pd[P(*t*-Bu)₃]₄, the reaction proceeds in excellent yield. Treatment with acid hydrolyzes the three ester groups to produce a malonic acid derivative, which under heat, decarboxylates to the product shown at right.

4-carboxymethylbenzoic acid

22.100   Conjugate addition of lithium dimethylcuprate to the α,β-unsaturated ketone can in principle give two diastereomers. Only product *A* and its Clemmensen reduction product *C* could be optically active. Compound *B* and its reduction product are meso compounds and are therefore achiral. Therefore, compound *A*, which has the (3*S*,5*S*) configuration, is the product of conjugate addition, and product *C*, which has the (1*R*,3*R*) configuration, is the Clemmensen reduction product.

22.101   (a)   This reaction is essentially an E2 reaction.

(b)   As shown in part (a), the proton on the carbon adjacent to the carbon bearing the phosphate leaving group is removed by a base. E2 reactions are accelerated when the β-proton is unusually acidic. In order to enhance the acidity of this proton, step 1 converts the other adjacent carbon, bearing an OH group, into a carbonyl group. As you know, protons on carbons α to a carbonyl are particularly acidic. After the proton has been removed in step 2, the carbonyl is converted back into an OH group in step 3.

(c)   Reaction 4 also bears resemblance to an E2 reaction, or to the collapse of a tetrahedral intermediate, expelling a leaving group, as in nucleophilic acyl substitution reactions.

Reaction 5 is an aldol addition reaction.

3-dehydroquinate

22.102 The following mechanism accounts for the formation of *A*.

A

The following mechanism shows the formation of *B* and the structure of by-product *C*.

A

B

C

The Chemistry of Amines

CHAPTER 23 SUMMARY

- Amines are classified as primary, secondary, or tertiary. Quaternary ammonium salts are compounds in which all four hydrogens of the ammonium ion are replaced by alkyl or aryl groups.

- Most amines undergo rapid inversion at nitrogen. This inversion interconverts a chiral amine to its enantiomer (or, if there are other stereocenters) to a diastereomer.

- Simple amines are liquids with unpleasant odors. Amines of low molecular mass are miscible with water.

- The N—H stretching absorptions are the most important infrared absorptions of primary and secondary amines. In the NMR spectra of amines, the α-protons have chemical shifts in the δ 2.5–3.0 range. The ^{13}C NMR chemical shifts of amine α-carbons are in the δ 30–50 range.

- Basicity is one of the most important chemical properties of amines. The basicity of an amine is expressed by the pK_a of its conjugate-acid ammonium ion. Ammonium-ion pK_a values are affected by alkyl substitution on the nitrogen, the polar effects of nearby substituent groups, and resonance interactions of the amine's nonbonding electron pair with an adjacent aromatic ring.

- Because amines are basic, they are also nucleophilic. The nucleophilicity of amines is important in many of their reactions, such as alkylation, imine formation, and acylation.

- Amines can serve as leaving groups. Quaternary ammonium hydroxides, when heated, undergo the Hofmann elimination, in which an amine is lost from the α-carbon and a proton is lost from a β-

carbon atom. However, because amines are basic, they are relatively poor leaving groups.

- Amines are weak acids with pK_a values in the 32–40 range. The conjugate bases of amines, called amide bases, are used to form other strong bases, such as enolate ions.

- The amino group is ortho, para directing in electrophilic aromatic substitution reactions. However, in many reactions of this type, such as nitration and sulfonation, protection of the amine nitrogen as an amide is required to prevent its protonation under the acidic reaction conditions.

- Treatment of primary amines with nitrous acid gives diazonium ions. Aliphatic diazonium ions decompose under the diazotization conditions to give complex mixtures of products. Aryldiazonium ions, however, can be used in a number of substitution reactions: the Sandmeyer reaction (substitution of N_2 by halide or cyanide groups) or hydrolysis (substitution with —OH). Because aryldiazonium ions are electrophiles, they react with activated aromatic compounds, such as aromatic amines and phenols, to give substituted azobenzenes, some of which are dyes. Secondary amines react with nitrous acid to give nitrosamines. Under acidic conditions, tertiary amines do not react, although aromatic tertiary amines give ring-nitrosated products.

- Amines can be synthesized from amides, nitriles, nitro compounds, and other amines, as summarized in Sec. 23.11E. The Curtius and Hofmann rearrangements yield amines with one less carbon atom. The Curtius rearrangement can also be used to prepare isocyanates, which react with nucleophiles by addition across the C=N bond to give carbamic acid derivatives.

STUDY GUIDE LINKS

23.1 Nitration of Aniline

Remember that rates of reactions are affected by both intrinsic reactivity (reflected in the magnitude of the rate constant) and concentrations of the species involved (Sec. 9.3A, text p. 456). Under the acidic conditions of nitration, the highly reactive free aniline is present in very small concentration; the much less reactive protonated aniline is present in much greater concentration.

$$\text{rate of formation of } para\text{-nitroaniline} = k_{\text{para}} \, [\text{PhNH}_2]$$

very large ⟵ ⟶ very small

$$\text{rate of formation of } meta\text{-nitroaniline} = k_{\text{meta}} \, [\text{Ph}\overset{+}{\text{N}}\text{H}_3]$$

very small ⟵ ⟶ very large

These two competing rates must be about the same, because comparable amounts of *meta-* and *para-* substituted products are formed in text Eq. 23.49 (text p. 1299).

23.2 Mechanism of the Curtius Rearrangement

Although the Curtius rearrangement is concerted, as shown in the text, it is helpful to *think* of the reaction in a stepwise fashion to see why the rearrangement occurs. There is a strong driving force for loss of dinitrogen, a very stable molecule:

Loss of dinitrogen leaves behind an electron-deficient nitrogen; it is electron-deficient because it is short one electron pair of an octet.

Just as rearrangements of alkyl groups occur to the electron-deficient carbons of carbocations, they also occur to electron-deficient nitrogens.

In other words, the Curtius rearrangement is driven by the very strong tendency of nitrogen to have a complete octet.

23.3 Formation and Decarboxylation of Carbamic Acids

The formation of carbamic acids by hydration of isocyanates is an acid-catalyzed addition to the C=N double bond:

Hydration can also occur in base by the nucleophilic reaction of ⁻OH at the carbonyl carbon. (Write the mechanism of this reaction.) The hydration in basic solution gives the carbamate ion, the conjugate base of a carbamic acid. The hydration of isocyanates under basic conditions is important in the Hofmann rearrangement, which is also discussed in Sec. 23.11D of the text.

Decarboxylation is a concerted reaction in which the same water molecule acts as both an acid and base catalyst at the same time.

The decarboxylation reaction is reversible; that is, amines react readily with CO_2 to give carbamic acids. However, the decarboxylation reaction in acidic solution is driven to completion by two factors: (1) protonation of the product amine, and (2) loss of CO_2 from solution as a gas.

Decarboxylation is fastest in acidic solution, but also occurs in base:

This reaction is driven to completion by reaction of the hydroxide ion with CO_2 to form bicarbonate ion:

bicarbonate ion

If the solution is strongly basic, bicarbonate ion itself ionizes further to carbonate ion. Decarboxylation of carbamates in basic solution is important in the Hofmann rearrangement.

FURTHER EXPLORATIONS

23.1 Alkyl Group Polarization in Ionization Reactions

The text points out that alkyl groups can stabilize either positive charge or negative charge, as evidenced by studies of gas-phase acidity of alcohols (Sec. 11.1D) and gas-phase basicity of amines (this chapter). The purpose of this section is to give you a more concrete feeling for the nature of these alkyl-group effects.

Consider first the stabilization of negative charge by alkyl groups. Let's consider the ionization reaction of *tert*-butyl alcohol:

$$(CH_3)_3C-OH \rightleftharpoons (CH_3)_3C-O^- + H^+$$

tert-butyl alcohol conjugate base of *tert*-butyl alcohol

Using a quantum-mechanics program, we can examine the charge on each atom of the molecule before and after the reaction. These charges aren't *formal charges*, which, as you recall, are calculated by a bookkeeping procedure, but are the *actual charges* in the molecule that are present as a result of bond dipoles.

tert-butyl alcohol   conjugate base of *tert*-butyl alcohol

(You can add the charges on the left to verify that they add to zero, which is the overall charge on the molecule; and you can add the charges on the right to verify that they add to –1. [A slight difference from –1.0 is due to rounding error.] Notice that two of the three methyl groups are not shown). In response to the greater negative charge on oxygen following ionization, the α-carbon (bold numbers) takes on greater positive charge, and the positive charge on the hydrogens is reduced. This is another way of saying that *electrons flow away from the oxygen onto the hydrogens of the methyl groups*.

Now let's look at the protonation of trimethylamine in a similar way:

$$(CH_3)_3N: + H^+ \rightleftharpoons (CH_3)_3\overset{+}{N}-H$$

trimethylamine   conjugate acid of trimethylamine

Here is the result of the charge calculation:

Notice in this case that introduction of positive charge on the electronegative nitrogen causes the α-carbons (bold numbers) to become more *negative*; the additional negative charge is clearly coming from the hydrogens, which have become more positive.

These examples show graphically how the electron distribution in methyl groups can be polarized by nearby charges. Polarization can occur either way, and can therefore stabilize both positive and negative charges.

23.2 Structures of Amide Bases

Although the structures of lithium amide bases are written as ionic compounds for conceptual simplicity, research has shown that the structures of these species are considerably more complex. For example, lithium diisopropylamide in tetrahydrofuran (THF) has a symmetrical dimer structure (that is, it is a symmetrical aggregate of two units):

Each lithium is partially bonded to two nitrogens with "half bonds" (dashed lines in the structure above). The nitrogen–lithium bonds are covalent, although they have significant ionic character. Each lithium is strongly solvated by two solvent molecules (THF in this case). The degree to which these amide bases are aggregated into larger structures varies with their concentration and with the nature of the solvent.

23.3 Mechanism of Diazotization

Diazotization at first sight may seem to be a very strange reaction, but once you understand the reactivity of nitrous acid, diazotization should make more sense.

Nitrous acid in solution is in equilibrium with its anhydride, dinitrogen trioxide.

$$2\,HNO_2 \rightleftharpoons O{=}N{-}O{-}N{=}O + H_2O$$

dinitrogen trioxide

Dinitrogen trioxide is the actual diazotizing agent under conditions of dilute acid; under conditions of very strong acid, the diazotizing agent is the nitrosyl cation, ^+NO, formed by dehydration of nitrous acid. The dinitrogen trioxide mechanism is explored here.

The N=O double bonds of dinitrogen trioxide have a reactivity somewhat analogous to that of a carbonyl group. They react at nitrogen with nucleophiles such as amines with loss of nitrite.

$$RNH_2 + N-O-N=O \longrightarrow RNH_2-N-\ddot{O}-N=O \longrightarrow$$

$$RNH-N + \ddot{O}-N=O \longrightarrow RNH-N + H-\ddot{O}-N=O$$

nitrite anion a nitrosamine **nitrous acid**

(The pK_a of nitrous acid is 3.2; therefore, the nitrite anion is a weak base, and a good leaving group.) The product of this sequence is a *nitrosamine*. Because the amine nitrogen loses only one hydrogen in forming a nitrosamine, secondary amines, as well as primary amines, form nitrosamines when they react with nitrous acid (Sec. 23.10C).

In the case of primary amines, nitrosamines react further under the acidic conditions. First, they undergo isomerization to give another intermediate called a diazoic acid. (You should draw the curved-arrow notation for this step, which is analogous to enol formation from a ketone.)

$$RNH-N=O \underset{}{\overset{H_2O, H_3O^+}{\rightleftharpoons}} R-N=N-OH$$

a nitrosamine   a diazoic acid

Under the acidic conditions of diazotization, diazoic acids dehydrate to give diazonium ions:

$$R-\ddot{N}=\ddot{N}-\ddot{O}H \rightleftharpoons \ddot{O}H_2 + R-\ddot{N}=\ddot{N}-\ddot{O}H \longrightarrow R-\overset{+}{N}\equiv N: + \ddot{O}H_2$$

a diazonium ion

REACTION REVIEW

I.   AMINES AS BRØNSTED BASES AND ACIDS

A.   BASICITY OF AMINES

1. Alkylamines react with Brønsted acids to give fully ionic compounds called alkylammonium salts.

    a. The conjugate-acid pK_a of alkylamines is in the 9–11 range.

$$R_3N: + H-Cl \longrightarrow R_3\overset{+}{N}-H \ Cl^-$$

a trialkylammonium salt
p$K_a \approx$ 9–11

    b. The conjugate-acid pK_a of arylamines is in the 4–5 range.

$$\langle\!\!\!\bigcirc\!\!\!\rangle\!\!-\!\ddot{N}H_2 \; + \; H\!-\!OAc \; \rightleftharpoons \; \langle\!\!\!\bigcirc\!\!\!\rangle\!\!-\!\overset{+}{N}H_3 \; {}^-OAc$$

anilinium ion
$pK_a = 4.62$

2. Amines can be extracted from organic solutions with aqueous acid.

B. *ACIDITY OF AMINES*

1. Alkylamines have pK_a values in the 32–35 range.

2. Alkylamines react with organometallic reagents such as butyllithium to give amide bases. (The term "amide" should not be confused with the carboxylic acid derivative of the same name.)

diisopropylamine
$pK_a \approx 35$

lithium diisopropylamide (LDA)

3. Lithium amides find frequent use as strong bases for forming the lithium enolates of carbonyl compounds. (See Secs. 22.4C, text pp. 1216–1218, and 22.8C, text pp. 1243–1244,)

II. SYNTHESIS OF AMINES; ALKYLATION REACTIONS

A. *GENERAL*

1. Amines are good nucleophiles (Lewis bases).
2. Three reactions of nucleophiles have been studied in previous sections:

   a. S_N2 reaction with alkyl halides, sulfonate esters, or epoxides (Sec. 9.1A, text p. 451; Sec. 9.4, text p. 459; Secs. 11.4A–C, text pp. 576–581; and Sec. 12.5A, text p. 636).

   b. Addition to aldehydes, ketones, and α,β-unsaturated carbonyl compounds (Sec. 19.8, text p. 1042; Sec. 19.12, text p. 1063; and Sec. 22.9A, text p. 1249).

   c. Nucleophilic acyl substitution at the carbonyl groups of carboxylic acid derivatives (Sec. 21.8, text p. 1157).

B. *DIRECT ALKYLATION OF AMINES*

1. Treatment of ammonia or an amine with an alkyl halide or other alkylating agent (RX in the following equation) results in alkylation of the nitrogen. This is an example of an S_N2 reaction in which the amine acts as the nucleophile.

$$NH_3 \; + \; RX \longrightarrow RNH_2 \; + \; R_2NH \; + \; R_3N \; + \; R_4N^+ \, X^-$$

   a. The immediate product of the reaction is an alkylammonium ion.

b. If this ammonium ion has N—H bonds, further alkylations can take place to give a complex product mixture.

c. A mixture of products is formed because the alkylammonium ion produced initially can be deprotonated by the amine starting material.

d. The monoalkylation product can be formed in good yield if a large excess of ammonia is used.

2. Epoxides, as well as α,β-unsaturated carbonyl compounds and α,β-unsaturated nitriles, also react with amines or ammonia; multiple alkylations can occur with these alkylating agents as well.

$$NH_3 \xrightarrow{\triangle} HOCH_2CH_2NH_2 \xrightarrow{\triangle} (HOCH_2CH_2)_2NH \xrightarrow{\triangle} (HOCH_2CH_2)_3N$$

$$RNH_2 \xrightarrow{H_2C=CH-CN} RNH-CH_2CH_2CN \xrightarrow[\text{excess}]{H_2C=CH-CN} RN(CH_2CH_2CN)_2$$

3. In an alkylation reaction, the exact amount of each product obtained depends on the precise reaction conditions and on the relative amounts of starting amine and alkyl halide; the utility of alkylation as a preparative method for amines is limited.

C. QUATERNIZATION OF AMINES

1. Amines can be converted into quaternary ammonium salts with excess alkyl halide or sulfonate ester (R—X) under forcing conditions.

a. This process, called quaternization, is one of the most important synthetic applications of amine alkylation.

b. The reaction is particularly useful when especially reactive alkyl halides are used.

2. Conversion of an amine into a quaternary ammonium salt with excess methyl iodide is called exhaustive methylation.

D. REDUCTIVE AMINATION

1. Primary and secondary amines form imines and enamines, respectively, when they react with aldehydes and ketones (Sec. 19.12 in the text). In the presence of a reducing agent, imines and enamines are reduced to amines.

a. Reduction of the C=N double bond is analogous to reduction of the C=O double bond.

b. The imine or enamine does not have to be isolated; because imines and enamines are reduced much more rapidly than carbonyl compounds, reduction of the carbonyl compound is not a competing reaction.

2. The process in which an amine reacts with a carbonyl compound and the resulting imine is reduced to an amine is called reductive amination. (It has also been referred to as reductive alkylation.)

    a. When the reduction step is carried out by hydride reducing agents such as $NaBH_4$, $NaBH_3CN$, or $NaBH(OAc)_3$, a proton source such as a protic solvent or dilute acid is required.

    b. An iminium ion is formed in solution by protonation and dehydration of the carbinolamine intermediate in the reductive amination reaction of a secondary amine and formaldehyde; this ion is rapidly and irreversibly reduced.

$$H_2C = O \; \underset{}{\overset{R_2NH}{\rightleftharpoons}} \; R_2N - CH_2 - OH \; \underset{}{\overset{H_3O^+}{\rightleftharpoons}}$$

a carbinolamine

$$R_2\overset{+}{N} = CH_2 \;+\; H_2O \; \xrightarrow[\text{reduction}]{NaBH_3CN} \; R_2N - CH_3$$

an iminium ion

    c. The reaction of an amine with an excess of formaldehyde is a useful way to introduce methyl groups to the level of a tertiary amine; quaternization does not occur in this reaction.

3. When the reducing agent is sodium cyanoborohydride ($NaBH_3CN$), the reaction is known as the Borch reaction.

$$\underset{}{\overset{}{C}} = O \; \xrightarrow{RNH_2} \; \underset{}{\overset{R}{C}} = N \; \xrightarrow[\text{MeOH, HCl}]{NaBH_3CN} \; \underset{}{\overset{H}{-C}} - NH - R$$

4. To determine the required starting materials for reductive amination:

    a. Starting with the desired compound, mentally reverse the reductive amination process by breaking one of the C—N bonds:

        i.   On the nitrogen side, replace it with an N—H bond.

        ii. On the carbon side, remove a hydrogen from the carbon and add a carbonyl oxygen.

    b. Repeat this process for each C—N bond and evaluate the best pathway.

III. OTHER SYNTHESES OF AMINES

A. GABRIEL SYNTHESIS OF PRIMARY AMINES

1. The alkylation of phthalimide anion followed by hydrolysis of the alkylated derivative to the primary amine is called the Gabriel synthesis.

phthalimide

a. The nitrogen in phthalimide has only one acidic hydrogen, so it can be alkylated only once.

   i. The conjugate base of phthalimide ($pK_a \approx 9$) is easily formed with KOH or NaOH.

   ii. The phthalimide anion can be alkylated by alkyl halides or sulfonate esters (primary or unbranched secondary, shown as RX in the preceding equation) in S_N2 reactions.

   iii. The *N*-alkylated phthalimide is converted into the free amine by amide hydrolysis in either strong acid or base.

b. Although *N*-alkyl phthalimides also have a pair of nonbonding electrons on nitrogen, they do not alkylate further, because neutral imides are much less basic, and therefore less nucleophilic, than the phthalimide anion.

2. Multiple alkylation, which occurs in the direct alkylation of ammonia, does not occur in the Gabriel synthesis.

B. *STAUDINGER REACTION FOR SYNTHESIS OF PRIMARY AMINES*

1. Primary alkyl azides form phosphazides upon reaction with triphenylphosphine. Phosphazides spontaneously rearrange to iminophosphoranes, which can be hydrolyzed with water to yield primary amines.

a. The alkyl azide starting materials are typically synthesized from alkyl halides or sulfonate esters by an S_N2 reaction, so relatively unhindered amines are produced by this method.

b. The rearrangement of the phosphazide to an iminophosphorane is shown in Eq. 23.69b on p. 1308 of the text, and is analogous to the rearrangement of oxaphosphetanes in the Wittig alkene synthesis (Eq. 19.94b, text p.1072).

2. This method allows for the preparation of primary amines without multiple alkylations of the nitrogen.

C. *REDUCTION OF NITRO COMPOUNDS*

1. Nitro compounds can be reduced to amines under a variety of conditions.

a. The nitro group is usually reduced very easily by catalytic hydrogenation.

b. The reduction of aromatic nitro compounds to primary amines can also be accomplished with finely divided metal powders and HCl.

   i. Iron or tin powder is frequently used.

   ii. In this reaction the nitro compound is reduced at nitrogen, and the metal, which is oxidized to a metal ion, is the reducing agent.

2. Aromatic nitro compounds react with $LiAlH_4$, but the reduction products are azobenzenes, not amines.

3. Nitro groups do not react at all with $NaBH_4$ under the usual conditions.

D. *AMINATION OF ARYL HALIDES AND ARYL TRIFLATES*

1. Arylamines can be prepared by the direct amination of aryl chlorides and aryl bromides in the presence of a base and a Pd(0) catalyst.

    a. Direct amination of aryl halides is sometimes called Buchwald–Hartwig amination.

    b. The catalysts are typically of the form PdL_2, where L is a sterically demanding ligand, formed by mixing $Pd(OAc)_2$ and two equivalents of the ligand.

2. These amination reactions operate by more than one mechanism, but they are conceptually similar to the Heck (Sec. 18.6A), Suzuki (Sec. 18.6B), and Stille (18.10B) reactions.

    a. The amine starting material must lose a hydrogen.

        i. Secondary amines produce tertiary amine products.

        ii. Primary amines produce secondary amines that may react further unless the primary amine is itself an aryl amine or contains a large, highly branched alkyl group.

        iii. Nitro groups may be present.

E. *HOFMANN REARRANGEMENT*

1. Treatment of a primary amide with bromine in base gives rise to a rearrangement called the Hofmann rearrangement or Hofmann hypobromite reaction.

2. The first step in the mechanism is ionization of the amide N—H; the resulting anion is then brominated.

    a. The *N*-bromoamide product is more acidic than the amide starting material and it ionizes.

        i. The *N*-bromo anion rearranges rapidly to an isocyanate. (See Eq. 23.92a–d, text p. 1314-1315, for details.)

        ii. Because the Hofmann rearrangement is carried out in aqueous base, the isocyanate cannot be isolated.

    b. The isocyanate formed spontaneously hydrates to a carbamate ion, which then decarboxylates to the amine product under the strongly basic reaction conditions.

an amide   an *N*-bromoamide   an isocyanate

a carbamate   an amine

3. The Hofmann rearrangement takes place with retention of stereochemical configuration in the migrating alkyl group. (See the foregoing equation for an example.)

F. CURTIUS REARRANGEMENT

1. When an acyl azide is heated in an inert solvent such as benzene or toluene, it is transformed with loss of nitrogen gas into an isocyanate; this concerted reaction is called the Curtius rearrangement.

2. The overall transformation that occurs as a result of the Curtius rearrangement followed by hydration is the removal of the carbonyl carbon of the acyl azide as CO_2.

   a. The isocyanate product of the Curtius rearrangement can be transformed into a carbamic acid by hydration in either acid or base.

   b. Spontaneous decarboxylation of the carbamic acid gives the amine, which is protonated under the acidic conditions of the reaction.

   c. The free amine is obtained by neutralization.

3. The Curtius rearrangement takes place with retention of stereochemical configuration in the migrating alkyl group. (See the preceding equation.)

4. The Curtius reaction can be run under mild, neutral conditions.

   a. The isocyanate can be isolated if desired.

   b. Some acyl azides in the pure state can detonate without warning, and extreme caution is required in handling them.

5. An important use of the Curtius rearrangement is for the preparation of carbamic acid derivatives.

   a. Reaction of isocyanates with alcohols or phenols yields carbamate esters.

   b. Reaction of isocyanates with amines yields ureas.

6. Acyl azides can be prepared in two ways:

   a. The most straightforward method is the reaction of an acid chloride with sodium azide.

   b. An ester can be converted into an acyl derivative of hydrazine by aminolysis; the resulting amide, an acyl hydrazide, is then diazotized with nitrous acid to give the acyl azide.

G. *AROMATIC SUBSTITUTION REACTIONS OF ANILINE DERIVATIVES*

   1. Aromatic amines can undergo electrophilic aromatic substitution reactions on the aromatic ring.

   a. The amino group is one of the most powerful ortho, para-directing groups in electrophilic aromatic substitution provided that it is not protonated under the reaction conditions.

   b. A protonated amino group does not have the nonbonding electron pair on nitrogen that gives rise through resonance to the activating, ortho, para-directing effect of a free amino group.

   c. Ammonium groups are meta-directing groups.

   2. Aniline can be nitrated regioselectively at the para position if the amino group is first protected from protonation by conversion into an acetamide derivative by acetylation.

H. SYNTHESIS OF AMINES: SUMMARY

1. Methods of amine synthesis:
   a. Reduction of amides with LiAlH₄ (Sec. 21.9B, text p. 1167).
   b. Reduction of nitriles with LiAlH₄ (Sec. 21.9C, text p. 1169).
   c. Direct alkylation of amines (Sec. 23.7A, text p. 1292). This reaction is of limited utility, but it is useful for preparing quaternary ammonium salts.
   d. Reductive amination (Sec. 23.7B, text p. 1294).
   e. Aromatic substitution reactions of anilines (Sec. 23.9, text p. 1299).
   f. Gabriel synthesis of primary amines (Sec. 23.11A, text p. 1306).
   g. Staudinger reaction for synthesis of primary amines (Sec. 23.11A, text p. 1306).
   h. Reduction of nitro compounds (Sec. 23.11B, text p. 1308).
   i. Amination of aryl chlorides or aryl triflates in the presence of a Pd(0) catalyst (Sec. 23.11C, text p. 1309).
   j. Hofmann and Curtius rearrangements (Sec. 23.11D, text p. 1312).

2. Methods c, d, e, and i represent methods of preparing amines from other amines. Because amides can be prepared from amines, method a can be placed in this category as well.
3. Methods b, f, g, h, and j are limited to the preparation of primary amines.
4. Methods a, b, d, h, and j can be used to obtain amines from other functional groups.

IV. SYNTHESES OF ORGANIC COMPOUNDS WITH AMINES

A. ACYLATION OF AMINES

1. Primary and secondary amines can be converted into amides by acylation with acid chlorides, anhydrides or esters.

2. The reaction of an amine with an acid chloride or anhydride requires either two equivalents of the amine or one equivalent of the amine and an additional equivalent of another base such as a tertiary amine or hydroxide ion to react with the equivalent of protons produced in the reaction.

B. HOFMANN ELIMINATION OF QUATERNARY AMMONIUM HYDROXIDES

1. Alkenes can be formed from amines by a three-step process:
   a. Exhaustive methylation
   b. Conversion of the ammonium salt to the hydroxide
   c. Hofmann elimination

$$\underset{\overset{|}{H}}{\overset{\overset{NH_2}{|}}{-C}}-\overset{|}{\underset{|}{C}}- \quad \xrightarrow{MeI} \quad \underset{\overset{|}{H}}{\overset{\overset{\overset{+}{N}Me_3 \ I^-}{|}}{-C}}-\overset{|}{\underset{|}{C}}- \quad \xrightarrow{AgOH} \quad \underset{\overset{|}{H}}{\overset{\overset{\overset{+}{N}Me_3 \ HO^-}{|}}{-C}}-\overset{|}{\underset{|}{C}}- \quad \xrightarrow[-NMe_3]{heat} \quad \overset{\diagdown}{\diagup}C=C\overset{\diagup}{\diagdown} \ + \ H_2O$$

2. The Hofmann elimination involves a quaternary ammonium hydroxide as the starting material; an amine acts as the leaving group.

   a. When a quaternary ammonium hydroxide is heated, a β-elimination reaction takes place to give an alkene, which distills from the reaction mixture.

   b. A quaternary ammonium hydroxide is formed by treating a quaternary ammonium salt with silver hydroxide (AgOH, formed from water and silver oxide, Ag_2O).

   c. The Hofmann elimination is conceptually analogous to the E2 reaction of alkyl halides, in which a proton and a halide ion are eliminated; in the Hofmann elimination, a proton and a tertiary amine are eliminated.

   d. Hofmann elimination generally occurs as an anti-elimination.

   e. The conditions of the Hofmann elimination are typically harsh.

C. *DIAZOTIZATION; REACTIONS OF DIAZONIUM IONS*

   1. Oxidation of amines generally occurs at the amino nitrogen; an important oxidation reaction of amines is called diazotization.

$$\underset{\substack{\text{a primary} \\ \text{amine}}}{R-NH_2} + \underset{\substack{\textbf{nitrous} \\ \textbf{acid}}}{HNO_2} + HCl \longrightarrow \underset{\text{a diazonium salt}}{R-\overset{+}{N}\equiv N \ \ Cl^-} + 2 \ H_2O$$

   a. Diazotization is the reaction of a primary amine with nitrous acid (HNO_2) to form a diazonium salt.

   b. A diazonium salt is a compound of the form $R-\overset{+}{N}\equiv N: \ X^-$, in which X^- is a typical anion.

   c. Both aliphatic and aromatic primary amines are readily diazotized.

   2. Diazonium ions incorporate one of the best leaving groups, molecular nitrogen.

   3. Aliphatic diazonium ions react immediately as they are formed by S_N1, E1, and/or S_N2 mechanisms to give substitution and elimination products along with nitrogen gas. (The rapid liberation of nitrogen gas on treatment with nitrous acid is a qualitative test for primary alkylamines.)

   4. Aromatic diazonium salts may be isolated and used in a variety of reactions; however, they are usually prepared in solution at 0–5 °C and used without isolation, because they lose nitrogen on heating and they are explosive in the dry state.

D. *AROMATIC SUBSTITUTION WITH DIAZONIUM IONS*

   1. Aryldiazonium ions react with aromatic compounds containing strongly activating substituent groups, such as amines and phenols, to give substituted azobenzenes.

   2. This is an electrophilic aromatic substitution reaction in which the terminal nitrogen of the diazonium ion is the electrophile. (See Eqs. 23.61a–b, text p. 1304, for a detailed mechanism.)

   3. The azobenzene derivatives formed in these reactions have extensive conjugated π-electron systems, and most of them are colored.

   4. Some of these compounds are used as dyes and indicators; as a class they are known as *azo dyes*. (An azo dye is a colored derivative of azobenzene.)

E. SUBSTITUTION REACTIONS OF ARYLDIAZONIUM SALTS

1. Among the most important reactions of aryldiazonium salts are substitution reactions with cuprous halides; in these reactions the diazonium group is replaced by a halogen.

   a. An analogous reaction occurs with CuCN.
      i. This reaction is another way of forming a carbon–carbon bond, in this case to an aromatic ring.
      ii. The resulting nitrile can be converted by hydrolysis into a carboxylic acid.
   b. The reaction of an aryldiazonium ion with a cuprous salt is called the Sandmeyer reaction.
   c. This reaction is an important method for the synthesis of aryl halides and nitriles.

2. Aryl iodides can also be made by the reaction of diazonium salts with potassium iodide; a cuprous salt is not required.

3. Diazonium salts can be hydrolyzed to phenols by heating them with water or by treating them with cuprous oxide (Cu_2O) and an excess of aqueous cupric nitrate [$Cu(NO_3)_2$] at room temperature.

4. The diazonium group is replaced by hydrogen when the diazonium salt is treated with hypophosphorous acid, H_3PO_2.

5. All of the diazonium salt reactions listed above are substitution reactions but none are S_N2 or S_N1 reactions because aromatic rings do not undergo substitution by these mechanisms.
   a. The Sandmeyer and related reactions occur by radical-like mechanisms mediated by the copper.
   b. The reaction of diazonium salts with KI probably occurs by a similar mechanism.
   c. The reaction of diazonium salts with H_3PO_2 has been shown definitely to be a free-radical chain reaction.

6. The substitution reactions of diazonium salts achieve ring-substitution patterns that cannot be obtained in other ways.

F. *REACTIONS OF SECONDARY AND TERTIARY AMINES WITH NITROUS ACID*

1. Secondary amines react with nitrous acid to yield *N*-nitrosoamines, usually called simply nitrosamines. Nitrosamines are known to be potent carcinogens.

$$R_2NH + HNO_2 \longrightarrow R_2N-N=O \quad \text{a nitrosamine}$$

2. The nitrogen of tertiary amines does not react under the strongly acidic conditions used in diazotization reactions.

3. *N,N*-Disubstituted aromatic amines undergo electrophilic aromatic substitution on the benzene ring; the electrophile is the nitrosyl cation, which is generated from nitrous acid under acidic conditions.

SOLUTIONS TO PROBLEMS

Solutions to Focused Problems

23.1 (a) *N*-isopropylaniline

(b) 3-methoxypiperidine

(c) ethyl 2-(diethylamino)pentanoate

(d) *N,N*-diethyl-3-heptanamine

23.2 (a) *N*-ethyl-*N*-methylisobutylamine (common), or *N*-ethyl-*N*,2-dimethyl-1-propanamine
(b) *N,N*-dimethyl-*p*-nitroaniline (common), or 1-dimethylamino-4-nitrobenzene

23.3 (a) The principle to apply is that the more double-bond character the carbon–nitrogen bond has, the shorter it is. The order of decreasing double-bond character, and therefore increasing carbon–nitrogen bond length, is

p-nitroaniline < aniline < cyclohexylamine

Because of resonance interaction of their nitrogen nonbonding electron pairs with the benzene ring, aniline and *p*-nitroaniline have the most double-bond character; cyclohexylamine lacks this interaction and therefore has the least carbon–nitrogen double-bond character. In *p*-nitroaniline, the nonbonding electron pair on the amine nitrogen can be delocalized into the *p*-nitro group:

This interaction is in addition to the resonance interaction of the nonbonding pair with the ring of the type shown in Eq. 23.8 on p. 1281 of the text. This additional delocalization increases the carbon–nitrogen double-bond character and decreases the carbon–nitrogen bond length.

(b)   Apply the principle used in part (a). The order of increasing carbon–nitrogen bond length is

$$A < C < B$$

Compound B has a true carbon–nitrogen single bond, so it has the longest carbon–nitrogen bond. Compound A has a true carbon–nitrogen double bond, so it has the shortest carbon–nitrogen bond. The carbon–nitrogen bond of C has some double-bond character because of resonance interaction of the nitrogen nonbonding pair with the double bond. (Draw the appropriate resonance structure.) Another way to compare the C—N bond lengths of compounds B and C is to note that the C—N bond in B is an sp^3–sp^3 single bond, whereas the C—N bond in C is an sp^2–sp^3 single bond, and to remember that σ bonds with greater s character are shorter. (See discussion in Secs. 4.1B and 15.1B of the text.)

23.4   The five-proton resonance at δ 7.18 in the NMR spectrum indicates the presence of a monosubstituted benzene ring (which accounts for 77 mass units), and the mutually split triplet–quartet pattern at δ 1.07 and δ 2.60 indicates an ethyl group (which accounts for 29 mass units). The IR spectrum indicates the presence of an amine; if it is a secondary amine, as suggested by the integral of the δ 0.91 resonance in the NMR spectrum, the NH accounts for 15 mass units, leaving 14 mass units unaccounted for. A CH_2 group could account for this remaining mass, and the two-proton singlet at δ 3.70 could correspond to such a resonance. On the assumption that the N—H proton is undergoing rapid exchange and does not show splitting with neighboring protons, *N*-ethylbenzylamine fits the data.

$-CH_2-NH-CH_2CH_3$ *N*-ethylbenzylamine

The predicted chemical shift (Fig. 14.4, text p. 738) for the benzylic CH_2 group is greater than that of ordinary benzylic protons because these protons are also α to a nitrogen; the chemical shift should also be greater than protons that are only α to a nitrogen. The chemical shift of δ 3.6 is quite reasonable. The $m/z = 120$ peak in the mass spectrum corresponds to loss of a methyl radical by α-cleavage; and the $m/z = 91$ peak corresponds to a benzyl cation, which is formed by inductive cleavage. (See Sec. 12.6D of the text for a discussion of these cleavage mechanisms.)

23.5   The NMR spectrum indicates the presence of only two phenyl hydrogens, and the singlet absorptions at δ 2.07 and δ 2.16 indicate the presence of chemically nonequivalent methyl groups in the ratio 2:1. Only compound (2), 2,4,6-trimethylaniline, fits the data. The two resonances at δ 2.07 and δ 2.16 correspond to the ortho and para methyl groups, respectively, and the δ 3.19 resonance to the —NH_2 protons of the amine.

23.6   (a)   2,2-Dimethyl-1-propanamine (neopentylamine, $(CH_3)_3CCH_2NH_2$) has a maximum of three resonances in its ^{13}C NMR spectrum. 2-Methyl-2-butanamine, $CH_3CH_2C(CH_3)_2NH_2$, should have four resonances in its ^{13}C NMR spectrum.

(b)   *Trans*-1,4-cyclohexanediamine has only two nonequivalent sets of carbons and thus should have only two resonances in its ^{13}C NMR spectrum. *Trans*-1,2-cyclohexanediamine has three nonequivalent sets of carbons and therefore should have three resonances in its ^{13}C NMR spectrum. [Chemically equivalent (and in these cases, homotopic) carbons in each compound below are indicated by the same symbol.]

NH$_2$ · · · · (structure) · · · ·
NH$_2$

trans-1,4-cyclohexanediamine
two nonequivalent sets of carbons

NH$_2$ · · · NH$_2$ · · · (structure) · · · § §

trans-1,2-cyclohexanediamine
three nonequivalent sets of carbons

23.7   (a)   The order of increasing basicity (and increasing conjugate-acid pK_a) is NH$_3$ (ammonia) < PrNH$_2$ (propylamine) < Pr$_2$NH$_2$ (dipropylamine). As the text discussion in Sec. 23.5B, text pp. 1284-1289, shows, within a series of amines of increasing alkyl substitution, basicity increases from ammonia to the secondary amine.

(b)   The order of increasing basicity is

$$H_3\overset{+}{N}CH_2CH_2NH_2 \;<\; H_2NCH_2CH_2CO_2Me \;<\; CH_3CHCH_2CH_3$$

methyl
3-aminopropanoate

NH$_2$

sec-butylamine

(Note that _basicity_ refers to the basicity of the uncharged amino group in the first compound, not the basicity of the _conjugate base_ of the ammonium ion.) The electron-withdrawing polar effect of the partial positive charge on the carbonyl carbon of the ester reduces the basicity of its β-amino group, and the polar effect of the full positive charge in the ammonium ion reduces the basicity of its β-amino group even more. The polar effect of a substituent containing a fully charged atom is generally greater than the polar effect of a substituent containing a dipolar group.

(c)   The order of increasing basicity is

MeO$_2$C—⟨ring⟩—NH$_2$ < ⟨ring, MeO$_2$C⟩—NH$_2$ < ⟨ring⟩—NH$_2$

methyl _p_-aminobenzoate   **methyl _m_-aminobenzoate**   **aniline**

The methoxycarbonyl group (that is, the methyl ester group) decreases basicity in the first two derivatives by a polar effect. Furthermore, methyl _p_-aminobenzoate is stabilized by resonance interaction of the nitrogen nonbonding pair with the carbonyl group:

$$\left[\; \begin{array}{c} O \\ \| \\ C \\ | \\ CH_3O \end{array} \text{—⟨ring⟩—}\ddot{N}H_2 \;\longleftrightarrow\; \begin{array}{c} {}^-O \\ | \\ C \\ | \\ CH_3O \end{array} \text{=⟨ring⟩=}\overset{+}{N}H_2 \;\right]$$

Because there is no nonbonding pair on nitrogen in the conjugate-acid ammonium ion, such resonance stabilization of the conjugate acid is not possible. As Fig. 23.2 on text p. 1286 shows, anything that stabilizes an amine relative to its conjugate-acid ammonium ion lowers its basicity. Note that resonance interaction between the carbonyl group and a _meta_-amino group is not possible.

(d)   The order of increasing basicity is

⟨ring⟩—NH$_2$ < O$_2$N—⟨ring⟩—CH$_2$NH$_2$ < ⟨ring⟩—CH$_2$NH$_2$ < ⟨ring⟩—NH$_2$

aniline   **_p_-nitrobenzylamine**   **benzylamine**   **cyclohexylamine**

The resonance and electron-withdrawing polar effect of the benzene ring severely reduces the nitrogen basicity of aniline relative to that of an alkylamine. The electron-withdrawing polar effect of the nitro group of *p*-nitrobenzylamine reduces its basicity relative to that of benzylamine. Finally, the electron-withdrawing polar effect of the benzene rings reduces the basicity of both benzylamines relative to cyclohexylamine, an ordinary alkylamine.

23.8 From Table 23.1, text p. 1285, we see that the basicity order is as follows:

$$A < B < C$$

The basicities of all these amines are reduced by resonance interaction of the nitrogen nonbonding electron pair with the ring. The electron-withdrawing polar effect of the nitro group on *A* and *B* lowers basicity further, and resonance interaction of the nitrogen nonbonding pair with the *p*-nitro group stabilizes *p*-nitroaniline (*A*) and lowers its basicity even more. (Resonance structures depicting this interaction are shown in the solution to Problem 23.3(a) above.)

23.9 Dissolve the mixture in a low-boiling, water-insoluble organic solvent such as methylene chloride. Extract this solution with 5% sodium bicarbonate or 5% sodium hydroxide solution; *p*-chlorobenzoic acid will dissolve in the aqueous layer as its sodium salt. It can be recovered as the free acid by acidifying the aqueous extracts.

The methylene chloride solution now contains only *p*-chloroaniline and *p*-chlorotoluene. Extract the methylene chloride solution with 5% HCl solution. *p*-Chloroaniline will dissolve in the aqueous layer as its hydrochloride salt. It can be recovered as the free amine by neutralizing the aqueous layer with dilute NaOH.

p-Chlorotoluene remains in the methylene chloride layer; it can be isolated by drying the solution and evaporating the methylene chloride. (Note that the order of the acid and base extractions could be interchanged.)

23.10 Mix the racemic acid with one equivalent of the enantiomerically pure amine. (The *S* enantiomer of the amine is utilized in the equation below, but either *pure* enantiomer of the amine could be used.) The carboxylic acids will react with the amine to give a mixture of *diastereomeric* salts.

Because these salts have different properties, they can be separated by differential solubility, for example, by fractional crystallization from some solvent, probably an alcohol. After separating the salts, aqueous acid is added to each. The amine resolving agent dissolves in the aqueous acid solution (from which it can be recovered), leaving the enantiomerically pure free carboxylic acid as a solid that can be isolated by filtration and recrystallized.

Notice that not only is amine basicity important in forming the initial salt, but also it is important in separating the amine resolving agent from the resolved carboxylic acid.

23.11 (a)

$$(CH_3CH_2)_4\overset{+}{N} \ F^-$$

tetraethylammonium fluoride

(b)

$$(PhCH_2)_2\overset{+}{N}(CH_3)_2 \ Br^-$$

dibenzyldimethylammonium bromide

23.12 Compound *A* has an asymmetric nitrogen stereocenter; therefore, it is chiral and can exist as two enantiomers. Compound *B* also contains an asymmetric nitrogen; however, compound *B* undergoes a Brønsted acid–base equilibrium, by which it is in equilibrium with a very small amount of the free amine:

compound *B*

This equilibrium takes place so rapidly, and the minuscule amount of free amine undergoes nitrogen inversion so quickly, that the ammonium salt is rapidly racemized. Because compound *A* does not have a hydrogen that can be involved in an equilibrium of this type, it is not racemized.

23.13 (a) As detailed in text Sec. 12.5A, nucleophiles can react with epoxides to form alcohols. Water serves to both deprotonate the protonated nitrogen after the initial reaction, and as an acid to protonate the resulting alkoxide oxygen.

Because the resulting product is also nucleophilic at nitrogen, a second addition can occur.

(b) As shown by Eq. 22.112 on p. 1250 of the text, amines can act as nucleophiles in conjugate addition reactions. Ammonia is the nucleophile in this reaction:

Because the resulting product is also nucleophilic at nitrogen, a second addition can occur, leading to the dialkylated product formed in 57% yield as shown in Eq. 23.29. Write the mechanism, using the one shown in part (a) as a guide.

(c)   A third alkylation doesn't occur in high yields most likely due to sterics, or the concentrations of the reagents. Increasing the concentration of acrylonitrile (2-propenenitrile) relative to ammonia would favor the formation of the trisubstituted compound.

23.14   The reaction of a primary amine and a ketone yields an imine, by the standard mechanism. This mechanism is shown in Study Guide Links 19.3 and 19.4 in this manual.

23.15   Under the acidic conditions shown in Eqs. 23.34 (acetic acid) and 23.35 (HCl), the initial product is protonated on the basic amine nitrogen. To isolate the neutral amine, base (NaOH) is added to remove the acidic proton.

23.16   The products of Eqs. 23.37 and 23.38 are both tertiary amines. Tertiary amines form unstable products with ketones and aldehydes, that quickly break back down into starting materials (see Eq. 19.54, Sec. 19.12C, text p. 1068).

Because no imine or enamine can form, there is no π-bond to reduce. Consequently, quaternization cannot occur.

23.17   Either the ethyl group can originate from acetaldehyde, or the cyclohexyl group can originate from cyclohexanone.

23.18   As mentioned in Study Problem 23.3, the ethyl group could originate from acetaldehyde:

23.19   (a)   Prepare the tertiary amine by reductive amination and alkylate it to give the ammonium salt.

An alternative synthesis would be to reductively aminate the same amine with formaldehyde to give trimethylamine and then alkylate this tertiary amine with benzyl bromide (PhCH$_2$Br).

(b)   Exhaustively methylate benzylamine with a large excess of methyl bromide.

benzylamine

23.20 (a) Caleb has forgotten, if he ever knew, that aryl halides such as bromobenzene do not undergo S_N2 reactions (Sec. 18.1).

(b) A Pd(0) catalyst of the type PdL_2. This reaction is discussed in Sec. 23.11C, where L is:

or

23.21 The dimethylammonium ion can lose a proton to give dimethylamine, which can then be alkylated by methyl iodide.

trimethylammonium iodide

23.22 (a)

benzoyl chloride   aniline

N-phenylbenzamide

(b)

propanoyl chloride   benzylethylamine   *N*-benzyl-*N*-ethylpropanamide

23.23 (a) Draw the molecule in a conformation in which the trimethylammonium group is anti to the β-hydrogen, and examine the relative positions of the phenyl groups. In this part, the alkene product must have the *Z* configuration.

note that phenyl groups are gauche ⇨ phenyl groups are cis; therefore, alkene has *Z* configuration

(b) Draw the molecule in a conformation in which the trimethylammonium group is anti to the β-hydrogen, and examine the relative positions of the phenyl groups. In this part, the alkene product must have the *E* configuration.

note that phenyl groups are anti ⇨ phenyl groups are trans; therefore, alkene has *E* configuration

23.24 (a) The product of the exhaustive methylation followed by heating with silver oxide is 3-methylcyclohexene, not 1-methylcyclohexene. Because the amino and methyl groups are in a trans orientation on the cyclohexane ring and Hofmann elimination generally occurs as an anti-elimination, the only β-hydrogen in an anti relationship to the amine group is the hydrogen on C6 of the ring.

β-hydrogen anti to the leaving group   β-hydrogen *not* anti to the leaving group

(b) After methylation, four hydrogens are β to the nitrogen leaving group. The hydrogen α to the carbonyl group is removed because it is considerably more acidic than the other β-hydrogens.

this β-hydrogen is especially acidic because it is α to a carbonyl

23.25 Coniine has one degree of unsaturation. Because the exhaustive methylation–Hofmann elimination sequence introduces a double bond, and because each of the products *A–C* contains only one double bond, it follows that coniine is a cyclic amine. Product *C* shows one point of attachment of the nitrogen in coniine, and shows that the coniine ring has a propyl side chain. Products *A* and *B* show the other point of attachment of the nitrogen: the alkene carbon that is common to both structures. Because five carbons intervene between points of attachment of the nitrogen, coniine must be a piperidine derivative (see text Eq. 23.4 on p. 1279):

2-propylpiperidine (coniine)

Exhaustive methylation, Ag₂O treatment, and Hofmann elimination give three alkenes that result from the three possible points of β-elimination:

elimination of this
β-hydrogen gives *B*

elimination of this ⟶ H
β-hydrogen gives *C*

elimination of this
β-hydrogen gives *A*

product resulting from exhaustive methylation
and Ag₂O treatment of coniine

 In many problems, the text uses Ag₂O [silver(I) oxide] for the conversion of quaternary ammonium halides into quaternary ammonium hydroxides. Attempts to balance these reactions will show that a hydrogen is unaccounted for. The resolution of this problem is that silver(I) oxide is typically used in a hydrated form; hydrated Ag₂O is equivalent to AgOH. Consequently, you should assume in all problems involving Hofmann elimination that Ag₂O is operationally the same thing as AgOH.

23.26

repeat two more
times at the positions
ortho to the amine group

23.27 Use an acetyl protecting group as outlined in Study Problem 23.4, text p. 1300.

**p-acetamidobenzenesulfonyl
chloride**

sulfanilamide

If the acetyl protecting group were not used, the amine group on one molecule would react with a chlorosulfonyl group on the other and a polymer would be formed.

23.28 (a) Begin with *p*-nitroacetanilide, prepared as shown in Study Problem 23.4, text pp. 1300. Then convert it into 2,4 dinitroaniline. The acetyl protecting group is used to avoid protonation of the nitrogen, which would result in a significant amount of unwanted meta substitution. (See Eq. 23.49 on text p. 1299.)

p-nitroacetanilide   2,4-dinitroaniline

(b) Begin with *p*-acetamidobenzenesulfonyl chloride, prepared as shown in the solution to Problem 23.27 above.

p-acetamidobenzenesulfonyl
chloride

sulfathiazole

23.29 (a)

o-toluidine   2-bromobenzoic acid

(b)

aniline   2,4,6-tribromobenzoic
acid

23.30 (a) Because diazotization does not break the carbon–nitrogen bond, the diazonium ion intermediate, like the starting amine, has the *R* configuration. Therefore, inversion of stereochemical configuration occurs in the reaction with water.

(b)   Because inversion of stereochemical configuration is observed, an S_N2 reaction in which water is the nucleophile and N_2 is the leaving group is likely to be involved. An S_N1 mechanism involving a carbocation intermediate would be expected to result in some racemization. (Why?)

23.31   (a)   (b)

23.32   *p*-Aminobenzenesulfonic acid is prepared and diazotized to give the diazonium ion *A*. This synthesis begins with *p*-acetamidobenzenesulfonyl chloride, prepared as shown in the solution to Problem 23.27 above. Formaldehyde is reductively aminated with aniline to give *N,N*-dimethylaniline, *B*. Compounds *A* and *B* are coupled to give methyl orange.

p-acetamidobenzenesulfonyl chloride (see solution to Problem 23.27)

methyl orange

23.33   The following compounds would react to give FD & C Yellow No. 6:

FD & C yellow #6

The position next to the ring junction (called the α-position) in a naphthalene ring is more reactive in electrophilic substitution than the ring position one carbon removed. Therefore, of the two positions "ortho" to the phenolic —OH group, the α-position is the more reactive and gives the observed product.

23.34 (a) Protonation of the —OH oxygen and loss of water give the nitrosyl cation. (The chloride counter-ion is not shown.)

O=N nitrosyl cation

(b) The loss of H_2O from protonated nitrous acid provides the nitrosyl cation, as shown in the solution to part (a) of this problem. This cation then serves as the electrophile in an electrophilic aromatic substitution reaction:

resonance-stabilized
carbocation intermediate

23.35 (a) In both the Gabriel synthesis and Staudinger reaction, only primary or unbranched secondary alkyl halides or sulfonates will react since each process involves an S_N2 mechanism. [See the discussion of this point in part (d).]

(b) A possible alkyl halide starting material for this amine is 1-bromo-3-methylpentane.

1-bromo-3-methylpentane

3-methyl-1-pentanamine

(c)

1-bromo-3-methylpentane

3-methyl-1-pentanamine

(d)   The alkyl halide or sulfonate required to produce 2,2-dimethyl-1-propanamine would be a neopentyl halide or sulfonate ester; these are virtually inert in S_N2 reactions. An aryl halide would be required to produce *N*-butylaniline; these do not undergo S_N2 reactions. Therefore, 3-methyl-1-pentanamine is the only amine of the three that can be prepared by either method.

23.36 (a)   The nitration of phenol is discussed in text Sec. 18.9.

phenol

p-iodoanisole

Note that formation of the ether is carried out before reduction of the nitro group because, otherwise, the amino group could also be alkylated with methyl iodide.

(b)   The tin metal reduction of the nitro group is shown in text Eq. 23.72.

nitrobenzene

m-bromoiodobenzene

23.37 (a)

chlorobenzene

N-(*sec*-butyl)-
N-ethylamine

N-(*sec*-butyl)-*N*-ethylaniline

(b)   (TfOTf = triflic anhydride; see Sec. 18.10B in the text for the structure.)

(c)

[reaction scheme: Ph—C(=O)—Cl + chlorobenzene, with AlCl₃/CH₂Cl₂, then H₃O⁺, giving 4-chlorobenzophenone; then PhNH₂, Na⁺ ⁻Ot-Bu, Pd(0) catalyst, toluene, giving the diarylamine product]

23.38 (a)   *Tert*-butylamine cannot be prepared by the Gabriel synthesis because it would require that the nitrogen of phthalimide be alkylated with *tert*-butyl bromide in an S$_N$2 reaction. Tertiary alkyl halides generally do not undergo S$_N$2 reactions but, under basic conditions, undergo E2 reactions instead.

(b)   The Hofmann or Curtius rearrangement would be an ideal way to prepare *tert*-butylamine. (The Hofmann rearrangement is illustrated below.)

$$(CH_3)_3C-C(=O)-NH_2 \xrightarrow{Br_2,\ NaOH} (CH_3)_3C-NH_2$$

2,2-dimethylpropanamide (pivalamide)   ***tert*-butylamine**

23.39 (a)   The curved-arrow mechanism for the conversion of ethyl isocyanate into ethyl *N*-ethylcarbamate:

[curved-arrow mechanism: HO—Et attacking Et—N=C=O (ethyl isocyanate), through bracketed resonance intermediates, proton transfer, giving Et—NH—C(=O)—OEt, ethyl *N*-ethylcarbamate]

ethyl isocyanate   **ethyl *N*-ethylcarbamate**

Note that the proton-transfer step, shown in the foregoing reaction for brevity as an intramolecular process, probably is intermolecular. Intramolecular proton transfers are possible, but they tend to take place fastest through 6-, 7-, and 8-membered rings. Nevertheless, we can abbreviate the more complex intermolecular processes as intramolecular processes, and we'll sometimes follow this practice for convenience. Sometimes the curved arrows will not be shown at all, but the proton transfer process will simply be indicated under the reaction arrow.

(b)   The curved-arrow mechanism for the reaction of ethylamine with ethyl isocyanate to give *N,N′*-diethylurea:

[curved-arrow mechanism: Et—N=C=O reacting with H₂N—Et, through bracketed resonance intermediates, proton transfer, giving Et—NH—C(=O)—HNEt, *N,N*-diethylurea]

***N,N*-diethylurea**

23.40   (a)   In ethanol solvent, ethanol (or its conjugate base ethoxide) is the most abundant nucleophile present, and thus an ethyl carbamate is obtained as the product.

2-methylpropanamide   **ethyl *N*-isopropylcarbamate**

(b)   In aqueous NaOH, a carbamate anion is formed, which decarboxylates to form the primary amine.

2-methylpropanamide   **isopropylamine**

23.41   Pentyl isocyanate is an intermediate in the reaction. Because it is largely insoluble in water, it forms a separate phase. A small amount of pentyl isocyanate reacts with the aqueous NaOH to form pentylamine, which is then extracted into the isocyanate, with which it reacts to give the urea.

hexanamide   *A*

N,*N*′-dipentylurea
B

23.42   (a)

2-cyclopentyl-*N*,*N*-dimethylethanamine

(b)

(c)

(d)   One synthesis involves a reductive amination of the aldehyde:

A second synthesis involves oxidizing the aldehyde to the carboxylic acid starting material of part (a) with aqueous H_2CrO_4, $KMnO_4$, or other common oxidant, and then proceeding to the product as in part (a).

23.43   In morphine, the nitrogen of the tertiary amine is the most basic atom. The most basic atom in mescaline is the amine nitrogen.

(a)

conjugate acid
of morphine

(b)

conjugate acid
of mescaline

Solutions to Skills Objectives Problems

23.44   (a)

$(CH_3)_3C-NH_2$

***tert*-butylamine**

(b)

2,2-dimethyl-3-hexanamine

(c)

***N*,*N*-dimethyl-2-hexanamine**

(d)

***N*-ethyl-*N*-methylisobutylamine**

23.45   (a)   Dicyclohexylamine (common), or *N*-cyclohexylcyclohexanamine
(b)   3-Methylamino-1-pentanol. Note that the —OH group is the principal group.
(c)   1-(2-Chloroethyl)-3-propylpyrrolidine

23.46   (a)   Dimethylamine has a higher boiling point than propane because the amine molecules can form hydrogen bonds and dipole-dipole interactions with each other. These interactions must be overcome for the amine molecules to enter the gas phase.
(b)   1-Propanol has a higher boiling point than 1-propanamine. Although both can form hydrogen bonds and both have dipole-dipole interactions, the hydrogen bonds and bond dipoles involving oxygen have greater partial charges (because of the greater electronegativity of oxygen). The greater partial charges result in

greater electrostatic attractions, and therefore greater intermolecular forces for the alcohol. As a result, the alcohol has the higher boiling point.

(c)   *N*-Methylethanamine has a higher boiling point than trimethylamine because it can both donate and accept hydrogen bonds. The tertiary amine cannot form hydrogen bonds and will have a lower boiling point.

23.47   Assume that the —NH$_2$ or —NH protons of the amine rapidly exchange, so that they are not coupled to adjacent protons on carbon.

Compound *C* should have the resonance at greatest chemical shift, because the proton α to the nitrogen is also benzylic and is a methine proton. This proton should be a triplet. The methyl resonance should also be a triplet.

Compound *B* should have the resonance at the next greatest chemical shift, because the protons α to the nitrogen are also benzylic, but are methylenes, which have smaller chemical shifts than methine protons by about 0.5–0.7 ppm. These α-protons should be a singlet. The relative integrations of the α-protons of *B* and *C* are different, as well. Compound *B* is the only one with a six-proton singlet in its proton NMR spectrum.

Between compounds *A* and *D*, compound *A* has the resonance at smallest chemical shift—the methyl group, which should be a doublet. These methyl protons are neither benzylic nor α to a nitrogen, so they should have a resonance not far from δ 1.

All of the resonances of compound *D* are in the δ 2–3.5 range. This spectrum should consist of two triplets and a singlet (plus the NH proton).

23.48   The unsaturation number of compound *A* is 0. The instability of *A* in aqueous acid and the presence of two oxygens is consistent with the presence of an acetal. Further evidence of an acetal is presented in the hint: after exposure to acid, an aldehyde carbonyl stretch is visible in the IR spectrum. The six-proton singlet at δ 3.27 could be due to the methyl protons of a dimethyl acetal; the δ 53.2 resonance in the ^{13}C NMR spectrum could be due to the carbons of these methyl groups, and the δ 102.4 resonance could be due to the carbon of the acetal group, that is, —O—CH—O— . If so, the resonance in the proton NMR at δ 4.50 is due to the proton of this CH group, and its splitting indicates two adjacent protons. The resonance for these two protons is found at δ 2.45. This leaves a six-proton singlet at δ 2.30 unaccounted for; this could be due to the six methyl protons of a dimethylamino group. Compound *A* is

$$(CH_3)_2N — CH_2 — \overset{\overset{\displaystyle OCH_3}{|}}{\underset{\underset{\displaystyle OCH_3}{|}}{CH}}$$

dimethylaminoacetaldehyde dimethyl acetal
(2,2-dimethoxy-*N,N*-dimethylethanamine)

23.49   (a)   In aqueous solution, adding alkyl groups to an amine increases its basicity up to dialkylamines. Therefore the order of increasing basicity is ammonia < ethylamine < ethylmethylamine.

(b)   The diammonium ion has no nonbonding electron pair, so it cannot act as a base. Positive charges will destabilize the conjugate acids of amines, so the most basic amine is the neutral ethane-1,2-diamine. The full basicity trend is as follows:

$$H_3\overset{+}{N}CH_2CH_2\overset{+}{N}H_3 \quad < \quad H_3\overset{+}{N}CH_2CH_2NH_2 \quad < \quad H_2NCH_2CH_2NH_2$$

(c)   The aromatic rings, and the electron-withdrawing groups on them, will decrease the basicity of the aniline derivatives. The decrease in basicity will be even more significant if the nitrogen nonbonding electron pair can be delocalized onto the nitro groups via resonance, as it can with the nitro group is in the 2-, 4-, or 6-position. The complete trend is as follows:

23.50   The aromatic ring lowers the basicity of aniline and its chloro derivatives. The effect of the chloro substituents results from their acid-strengthening polar effect on the stability of the conjugate acid. (There is no additional resonance effect due to either chlorine.) This effect decreases with increasing distance of the chlorine substituent from the —$\overset{+}{N}H_3$ group. As can be seen from the data in Table 23.1 in the text, *o*-chloroaniline (*B*) is the weakest base (and its conjugate acid is the strongest acid) because the *o*-chloro substituent is closest to the nonbonding pair on nitrogen. *p*-Chloroaniline (*A*) has a greater distance between the electron-withdrawing chlorine and the nitrogen and is therefore a slightly stronger base than *B*. Aniline (*C*) is the most basic of the three because it has no additional electron-withdrawing group on the aromatic ring.

23.51   The basicity of trifluralin is much less than that of *N,N*-diethylaniline. The electron-withdrawing polar effects of both the nitro substituents and the trifluoromethyl substituent destabilize the conjugate acid of trifluralin, and resonance interaction of the nitrogen nonbonding electron pair with the two nitro groups stabilizes trifluralin itself; this interaction is absent in the ammonium ion. (In fact, trifluralin is not basic enough to dissolve in dilute HCl.)

some resonance structures of trifluralin

23.52   (a)   The amine nitrogen (but *not* the amide nitrogen) is the most basic atom in the molecule. Therefore, labetalol hydrochloride is the ammonium ion that results from protonation of this nitrogen.

   (b)   When labetalol hydrochloride is treated with one molar equivalent of NaOH, the proton on the positively charged nitrogen is removed and neutral labetalol is formed.

   (c)   Hot NaOH ionizes the phenol group and also brings about hydrolysis of the amide group to a carboxylate.

(d)   Hot 6 *M* aqueous HCl solution results in protonation of the amine nitrogen, as in part (a), and hydrolysis of the amide.

Depending on the severity of the conditions, the alcohol could dehydrate to an enamine, which is protonated on nitrogen to form its conjugate acid *A*. This, like all enamines, would hydrolyze to a dialkylammonium ion and an aldehyde under the aqueous conditions.

23.53   The quaternary ammonium cation has an asymmetric nitrogen stereocenter; therefore, it is chiral and can exist as two enantiomers. The neutral amine also contains an asymmetric nitrogen, but the nonbonding electron pair on the nitrogen undergoes inversion so quickly that the amine is rapidly racemized and is optically inactive.

23.54   The structure of *p*-chloroaniline is

p-chloroaniline

(a)   (b)   (c)   (d)

(e)   (f)   (g)   (h)   (i)

Sulfonamides are acidic enough to be ionized by NaOH. How would you know this? An amide has a pK_a of about 16. We know that sulfonic acids are stronger than carboxylic acids by about 6 pK_a units; assuming the same difference holds for sulfonamides versus carboxamides, then a sulfonamide should have an N—H pK_a of about 16 − 6 = 10. This is about as acidic as a phenol, and acidic enough to be fully ionized by NaOH (which has a conjugate-acid pK_a = 14). Furthermore, the conjugate-base anion of this particular sulfonamide is resonance-stabilized by the phenyl group. We know that phenols are more acidic than alcohols by about 5 pK_a units; if the phenyl group affects the acidity of the sulfonamide in the same way, the pK_a of this compound could be as low as 5.

23.55 The structure of *N*-methylaniline is

23.56 The structure of isopropylamine is $(CH_3)_2CH—NH_2$

23.57   The Gabriel synthesis of benzylamine requires an S_N2 alkylation of phthalimide anion with benzyl bromide, followed by hydrolysis of the alkylated derivative to afford the primary amine.

phthalimide   **benzylamine**

The Staudinger synthesis of benzylamine begins with the conversion of benzyl bromide into benzyl azide, followed by a reaction with triphenylphosphine to form a phosphazide. The phosphazide spontaneously rearranges to an iminophosphorane, which is hydrolyzed with water to yield benzylamine and triphenylphosphine oxide.

benzyl bromide

benzylamine

23.58   Benzylamine can be synthesized from either benzamide or benzonitrile by reduction with lithium aluminum hydride.

benzamide   **benzylamine**   **benzonitrile**

23.59   (a)   This reaction is related to the Hofmann rearrangement. Reaction with hydroxide on the carbonyl carbon of *N*-bromosuccinimide gives an *N*-bromo anion, which undergoes a Hofmann rearrangement to the corresponding isocyanate. The isocyanate hydrolyzes and decarboxylates to the amine product.

(b)   This reaction involves a Curtius rearrangement (text Sec. 23.11D), followed by intramolecular formation of a cyclic urea (text Eq. 23.87).

Solutions to Integrated Problems

23.60 (a)

2-cyclobutenamine   or   2-methyl-2-cyclo-propenamine

(b) CH₃CHCH₂CH₃ with NH₂

sec-butylamine
(2-butanamine)

(c) NHCH(CH₃)₂   and   NHCH₂CH₂CH₃

N-isopropylaniline   *N*-propylaniline

(d)   The required compound could be either of the two imines (CH₃)₂C=NCH₃ or (CH₃)₂CHN=CH₂

23.61 (a)

CH₃CH₂CH₂CH₂CH₂CNHCH₃   CH₃CH₂CH₂CH₂CH₂CH₂CH₂CH₂NH₂   CH₃CH₂CH₂CH₂CH₂CH₂N(CH₃)₂

N-methylhexanamide   **1-octanamine**   **N,N-dimethyl-1-hexanamine**

N-Methylhexanamide is an amide, and it is therefore not significantly soluble in dilute HCl; the two amines are soluble as their conjugate-acid ammonium salts in dilute HCl solution. 1-Octanamine, like all primary alkylamines, gives off a gas (N_2) when diazotized with NaNO₂ in aqueous acid. *N,N*-Dimethyl-1-hexanamine does not give off a gas when treated with nitrous acid.

(b)

p-methylaniline
(p-toluidine)   **benzylamine**   **p-cresol**   **anisole**

Only the two amines, *p*-methylaniline and benzylamine, are soluble (as their conjugate-acid ammonium ions) in dilute HCl. Only the phenol *p*-cresol is soluble (as its conjugate-base phenolate) in dilute NaOH solution. Anisole is soluble in neither dilute acid nor dilute base. Only benzylamine gives off a gas (N_2) when diazotized at 0° with HNO₂. *p*-Methylaniline gives off N_2 when the corresponding diazonium salt is heated.

23.62 (a)   In 1 *M* HCl solution, cocaine is protonated on its amino nitrogen.

and/or

(b)   Hot aqueous NaOH would bring about saponification of the ester groups:

(c)   Concentrated, hot aqueous HCl would protonate the amine nitrogen, as in part (a), and would bring about hydrolysis of the two ester groups.

23.63   Dissolve the mixture in a suitable low-boiling solvent such as ether or dichloromethane. Extract with 5% aqueous sodium bicarbonate solution. The carboxylic acid *p*-nitrobenzoic acid will dissolve in the aqueous solution as its sodium salt. Isolate this aqueous solution and acidify it with HCl to precipitate the carboxylic acid. Then extract the remaining organic solution with 5% aqueous NaOH solution. *p*-Chlorophenol will dissolve in the aqueous solution as its conjugate-base phenolate ion. Isolate this aqueous layer and acidify it to obtain the neutral phenol. Then extract the organic solution with 5% aqueous HCl solution. The amine aniline will dissolve in the aqueous layer as its conjugate-acid ammonium ion. Separate the aqueous layer and neutralize it with NaOH. Aniline will form a separate layer that can be separated, dried, and distilled. Finally, dry the remaining organic layer and remove the solvent to obtain nitrobenzene. (Alternatively, the acidic extraction could be carried out as the first step. The order of the other steps would remain the same.)

23.64   (a)   Nitrous acid forms a diazonium ion from anthranilic acid. This diazonium ion undergoes coupling with *N,N*-dimethylaniline to give methyl red.

methyl red

(b)   In acidic solution, methyl red is protonated to give its conjugate acid. Protonation occurs on one of the diazo nitrogens because the resulting conjugate acid is resonance-stabilized.

You may have noticed that the pK_a of the carboxy group is unusually low for the pK_a of a carboxylic acid. Can you explain why?

23.65 (a) First prepare *p*-nitroaniline from aniline as shown in Study Problem 23.4 on text p. 1300. Then diazotize this amine to give the diazonium salt. Couple the diazonium salt to salicylic acid to obtain alizarin yellow R.

p-nitroaniline

alizarin yellow R

(b) From the solution to the Problem 23.64 above, or from the information in the problem, the pK_a of a protonated diazo group should be around 5. Consequently, at a pH value well above 5, the diazo nitrogens are unprotonated. The pK_a of the carboxy group should be around 4. Therefore, at pH = 9, the carboxy group should be ionized. The group that ionizes between 10 and 12 should affect the conjugated π-electron system because it affects the color. Because this group is not the diazo group, it must be the phenol. If the phenol has a pK_a of about 11, then at pH = 9 it is un-ionized.

structure of alizarin yellow R
at pH = 9

(c) At pH > 12, both the carboxy group and the phenol are ionized. (Show the resonance structures for the delocalization of the phenolate ion throughout the π-electron system of the rings, the diazo group, and the *p*-nitro group. This interaction is the basis for the color change.)

structure of alizarin yellow R
at pH > 12

23.66 (a) Amanda is evidently trying to run a Friedel–Crafts acylation on aniline. This cannot work as shown for two reasons. First, the aluminum trichloride, a Lewis acid, will react with the nonbonding electron pair on the amino group, thus deactivating the ring. (This is much like its effect on phenol acylation; see Eq. 18.103, Sec. 18.9, text p. 1007.) Second, a competing reaction is acylation of the amino group to form acetanilide, a reaction that will undoubtedly be catalyzed by the aluminum trichloride.

(b) Excess methyl iodide will form the quaternary salt $(CH_3)_3\overset{+}{C}N(CH_3)_3$ I^-; the reaction will not stop at the secondary amine shown in the problem.

(c) Amanda is attempting to nitrate the ring and is relying on the activating, ortho, para-directing effect of the dimethylamino group to direct the substitution to the para position. The fallacy here is that under the very acidic conditions the dimethylamino group is protonated, and the protonated dimethylamino group is expected to be a meta director. (See text Eq. 23.49 and the related discussion.)

(d) Although the aldehyde will indeed be reduced, so will the nitro group. (See text Eq. 23.73.)

(e) Amanda is attempting to displace water from the protonated alcohol with dimethylamine, which is a good nucleophile. Unfortunately, the acidic conditions required to convert the —OH group of the alcohol into a good leaving group also protonate the amine, which is much more basic than the alcohol. The protonated amine no longer has the nonbonding electron pair that is responsible for its nucleophilicity.

(f) Amanda is attempting to carry out an S_N2 reaction on a tertiary alkyl halide. Unfortunately, tertiary alkyl halides do not undergo S_N2 reactions. If a reaction occurs at all, it will be the competing E2 process (a β-elimination) to form an alkene.

23.67   (a)

PhNH$_2$ $\xrightarrow[\text{2) CuCN}]{\text{1) NaNO}_2\text{/HCl}}$ PhC≡N $\xrightarrow{\text{H}_2\text{, cat.}}$ PhCH$_2$NH$_2$

benzylamine

(b)

PhNH$_2$ $\xrightarrow[\text{2) CuCN}]{\text{1) NaNO}_2\text{/HCl}}$ PhC≡N $\xrightarrow{\text{H}_3\text{O}^+\text{, H}_2\text{O, heat}}$ PhCOH $\xrightarrow[\text{2) H}_3\text{O}^+]{\text{1) LiAlH}_4}$ PhCH$_2$OH

benzyl alcohol

(c)

PhC≡N $\xrightarrow{\text{H}_3\text{O}^+\text{, H}_2\text{O, heat}}$ PhCOH $\xrightarrow[\text{2) H}_3\text{O}^+]{\text{1) LiAlH}_4}$ PhCH$_2$OH $\xrightarrow{\text{conc. HBr}}$ PhCH$_2$Br $\xrightarrow[\text{ether}]{\text{Mg}}$ $\xrightarrow[\text{2) H}_3\text{O}^+]{\text{1) CO}_2}$

prepared in
part (a)

PhCH$_2$COH $\xrightarrow{\text{SOCl}_2}$ PhCH$_2$CCl $\xrightarrow{\text{NH}_3\text{ (excess)}}$ PhCH$_2$CNH$_2$ $\xrightarrow[\text{3) neutralize with }^-\text{OH}]{\substack{\text{1) LiAlH}_4 \\ \text{2) H}_3\text{O}^+}}$ PhCH$_2$CH$_2$NH$_2$

2-phenylethanamine

(d)

PhNH$_2$ + CH$_3$CCH$_2$CH$_3$ $\xrightarrow[\text{CH}_3\text{OH}]{\text{NaCNBH}_3}$ PhNHCHCH$_2$CH$_3$
                                                                                          |
                                                                                         CH$_3$

***N*-phenyl-2-butanamine**

(e)

—NH$_2$ $\xrightarrow{\text{Ac}_2\text{O}}$ —NHAc $\xrightarrow{\text{Cl}_2\text{, FeCl}_3}$ Cl——NHAc $\xrightarrow[\text{2) NaOH (dilute)}]{\text{1) H}_3\text{O}^+\text{, H}_2\text{O, heat}}$

Cl——NH$_2$ $\xrightarrow[\text{2) CuCN}]{\text{1) NaNO}_2\text{/HCl}}$ Cl——C≡N $\xrightarrow[\text{H}_2\text{O}]{\text{H}_3\text{O}^+\text{, heat}}$ Cl——CO$_2$H

***p*-chlorobenzoic acid**

(f)

PhNH$_2$ $\xrightarrow[\text{2) CuCl}]{\text{1) NaNO}_2\text{/HCl}}$ PhCl $\xrightarrow[\text{toluene}]{\substack{\text{PhNH}_2 \\ \text{Na}^+ \ ^-\text{O}t\text{-Bu} \\ \text{Pd(0) catalyst}}}$ (diphenylamine structure)
                                                                                                                                  NH

diphenylamine

23.68   The acetic anhydride treatment acetylates either the amino group or the hydroxy group. Because the initially formed product *A* can form an ethyl ether, its —OH group is not affected in the first reaction. Consequently, compound *A* is *p*-acetamidophenol and compound *B* is its ethyl ether. This is reasonable because, so long as the hydroxy group is not ionized, the amino group is the most basic group in the molecule, and is thus the more nucleophilic group. Compound *A* dissolves in base because the hydroxy proton can ionize.

The reaction sequence at the top of the page:

H_2N—⟨benzene ring⟩—OH $\xrightarrow{Ac_2O}$ AcHN—⟨benzene ring⟩—OH $\xrightleftharpoons{NaOH}$ AcHN—⟨benzene ring⟩—O^- $\xrightarrow{Et-I}$

p-aminophenol

p-acetamidophenol
(4-acetylaminophenol,
or **acetaminophen)**
(compound *A*)

AcHN—⟨benzene ring⟩—OEt

N-(4-ethoxyphenyl)acetamide
(compound *B*)

In *p*-acetamidophenol the acidity of the amide N—H is likely to be comparable to that of the phenol (why?). Therefore, the conjugate-base anion of the amide is in equilibrium with the phenoxide ion. If both ions are indeed present, evidently the conjugate base phenolate is alkylated more rapidly. Alkylation of the oxygen then pulls the equilibrium between the anions toward the phenoxide:

$CH_3C(=O)$—NH—⟨benzene ring⟩—OH $\xrightleftharpoons{NaOH}$ $CH_3C(=O)$—NH—⟨benzene ring⟩—O^- $\xrightarrow{Et-I}$ $CH_3C(=O)$—NH—⟨benzene ring⟩—OEt

p-acetamidophenol

N-(4-ethoxyphenyl)acetamide

$CH_3C(=O)$—$\bar{N}$—⟨benzene ring⟩—OH

23.69 First identify compound *B* by completing the reaction sequence:

H_2C=$CHCH_2CH_2COH$ (with C=O) $\xrightarrow{SOCl_2}$ H_2C=$CHCH_2CH_2CCl$ (with C=O) $\xrightarrow[\text{piperidine}]{HN⟨piperidine⟩}$

4-pentenoic acid

H_2C=$CHCH_2CH_2C$—N⟨ring⟩ (with C=O) $\xrightarrow[\text{3) dilute NaOH}]{\text{1) LiAlH}_4 \text{, 2) H}_3O^+}$ H_2C=$CHCH_2CH_2CH_2N$⟨ring⟩

compound *B*

Compound *A*, because of its reaction with Ag^+ to give AgBr, appears to be an ionic compound. The formula of compound *B* shows that it must be derived from two molecules of 1,5-dibromopentane. Because the alkene *B* is obtained by an elimination reaction, it is reasonable to suppose that compound *A* is a quaternary ammonium salt. A reasonable structure of such a salt that could result from the reaction of two equivalents of 1,5-dibromopentane and one of ammonia is as follows. (This equation also shows the reaction of compound *A* with hydroxide ion to give compound *B*.)

⟨spiro ammonium structure with $^-$OH, N^+⟩ Br^- ⟶ ⟨bicyclic N structure⟩ + H—OH + Br^-

compound *A*

compound *B*

The formation of compound *A* from 1,5-dibromopentane and ammonia thus must involve the following steps:

$$Br(CH_2)_5Br + NH_3 \longrightarrow Br(CH_2)_5\overset{+}{N}H_3\ Br^- \underset{}{\overset{NH_3}{\rightleftharpoons}} Br(CH_2)_5NH_2 \xrightarrow[\text{alkylation}]{\text{intramolecular}} \overset{+}{N}H_2\ Br^- \overset{NH_3}{\rightleftharpoons}$$

$$+ \overset{+}{N}H_4\ Br^-$$

[reaction scheme: piperidine (NH) + $Br(CH_2)_5Br$ → ... $\xrightarrow[\text{alkylation}]{\text{intramolecular}}$ compound *A*]

$+ \overset{+}{N}H_4\ Br^-$ $+ \overset{+}{N}H_4\ Br^-$

23.70 (a) The barrier to internal rotation about the *N*-phenyl bond in *N*-methyl-*p*-nitroaniline is greater because there is more double-bond character in this bond than there is in *N*-methylaniline. There is more double-bond character because of the resonance interaction between the nonbonding electron pair on the amine nitrogen and the *p*-nitro group.

double-bond character

[resonance structures of *N*-methyl-*p*-nitroaniline]

(b) *Cis*-and *trans*-1,3-dimethylpyrrolidine rapidly interconvert because all that is required for this process is amine inversion, which is very rapid.

[structures of trans and cis 1,3-dimethylpyrrolidine]

trans   cis

(c) This compound is the nitrogen analog of an acetal and, by a mechanism similar to that for the hydrolysis of an acetal, decomposes to formaldehyde and two equivalents of methylamine in aqueous solution.

$$CH_3NH-CH_2-NHCH_3 \rightleftharpoons CH_3\overset{}{N}H-CH_2-\overset{+}{N}H_2CH_3\ {}^-OH \rightleftharpoons H_2NCH_3 + CH_3\overset{+}{N}H=CH_2 \rightleftharpoons$$

$${}^-OH$$

$$CH_3NH-CH_2-OH \rightleftharpoons CH_3\overset{+}{N}H_2-CH_2-\overset{-}{O} \rightleftharpoons CH_3NH_2 + H_2C=O + H-OH$$

(d) This compound exists as an enamine for the same reason that β-diketones exist as enols: internal hydrogen bonding and conjugation stabilize the enamine structure.

internal hydrogen bond

[structure of enamine showing conjugated double bonds]

conjugated double bonds

(e) The diazonium salt does not decompose to a carbocation because the carbocation has a $4n$ π-electron system and is therefore antiaromatic and very unstable.

2,4-cyclopentadien-1-amine

antiaromatic; unstable

23.71 First, resolve 2-phenylbutanoic acid into enantiomers using an enantiomerically pure amine as the resolving agent. Then allow the appropriate enantiomer of the carboxylic acid to undergo the reactions shown in each part below. The key element in all these syntheses is that the Hofmann and Curtius rearrangements take place with retention of stereochemical configuration.

(a)

(b) In this case, esterify the S enantiomer of the carboxylic acid that was obtained from the enantiomeric resolution.

(c) In this part, the isocyanate product from the Curtius rearrangement is divided in half, and part of it is converted into an amine. This amine is then allowed to react with the remaining isocyanate.

(d) Use the same reactions in part (c), except prepare the (S)-amine from the (S)-carboxylic acid. Then let the (R)-isocyanate react with the (S)-amine to give the meso product. (The same product would be obtained from the reaction of the (S)-isocyanate with the (R)-amine.)

(e) Use the same reactions in part (c) to prepare (R)-1-phenylpropanamine. Then couple the (R)-amine with ethyl p-chlorobenzoate in the presence of a Pd(0) catalyst and Na$^+$ $^-$Ot-Bu (a Buchwald–Hartwig amination) followed by hydrolysis of the ester.

23.72   (a)   Sodium benzenethiolate, Na⁺ PhS⁻, is a salt and thus soluble in water but not an organic solvent; 1-bromooctane is relatively nonpolar and thus insoluble in water. No reaction occurs initially because the two reactants cannot interact; however, with the addition of the phase-transfer catalyst tetrabutylammonium bromide, $(CH_3CH_2CH_2CH_2)_4N^+$ Br^-, the two reactants can interact to form octyl phenyl sulfide. (See text Sec. 23.6.)

$$Ph-S-CH_2(CH_2)_6CH_3$$

octyl phenyl sulfide

(b)   Phenylacetonitrile and 1,4-dibromobutane are not soluble in water, especially with an ionic compound like sodium hydroxide in the aqueous layer. Phenylacetonitrile and 1,4-dibromobutane are soluble in dichloromethane; the tetrabutylammonium bromide is a phase-transfer catalyst and the rapid stirring brings the hydroxide ion into the organic phase. The hydrogens of phenylacetonitrile that are α to both the phenyl ring and the nitrile group are acidic and can be removed by hydroxide ion. The resulting conjugate base reacts with the 1,4-dibromobutane in an S_N2 reaction:

This product reacts with another equivalent of hydroxide to give the conjugate base, which undergoes an intramolecular S_N2 reaction to form the $C_{12}H_{13}N$ product in high yield.

(c)   The reaction between morpholine and bromobenzene in the presence of aqueous NaOH and the catalyst $Pd[P(t-Bu)_3]_2$ is a Buchwald–Hartwig amination reaction. The cetyltrimethylammonium bromide is a phase-transfer catalyst that brings the water-insoluble bromobenzene into contact with the other reactants.

23.73   Compound *A* appears to be an *N*-benzoyl amide of compound *B* because it is hydrolyzed to benzoic acid, and because it is re-formed when *B* is treated with benzoyl chloride. Compound *A* cannot be an ester because it contains only one oxygen. Compound *B* is therefore an achiral primary or secondary amine. It cannot be a primary amine because it does not liberate nitrogen gas after treatment with nitrous acid; thus, it must be secondary because a tertiary amine cannot form an amide. The formula of B is obtained by subtracting the formula of a benzoyl group (Ph—CO—, C_7H_5O) from the formula of A and adding a hydrogen. Thus, the formula of B is $C_{15}H_{23}N$. Compound B therefore contains five degrees of unsaturation. One degree of unsaturation is accounted for by a cyclohexane ring, reflected in the degradation product D, and the remaining four by a phenyl ring, reflected in the degradation product styrene, which is formed along with product C. There are only two structures for compound B, shown below as $B1$ and $B2$, that would give the indicated degradation products:

B1   *B2*

Although both are reasonable answers, structure $B1$ is better because it would lead to *one* alkene D, whereas $B2$ would lead to *two* different alkenes of which D is the major one. Taking $B1$ as the correct structure for B, the identities of the unknown compounds are as follows:

A   *B*

C   *D*

23.74 The oxidation data reveal that compound A contains a monosubstituted benzene ring, compounds B and C contain *para*-disubstituted benzene rings, and compound C additionally has an amino group directly attached to the ring. The pK_a of compound C additionally confirms that it is an aniline derivative, and the fact that compound C, following diazotization, requires warming to evolve a gas (N_2) also indicates that it is an aniline derivative. Compound A is chiral, and furthermore, it must be the racemate, because it evidently forms diastereomeric salts with (+)-tartaric acid. Because six carbons are involved in a benzene ring, the problem is to determine how the remaining two carbons are arranged so that the resulting structures are consistent with the data. The following structures are the only reasonable possibilities:

A   *B*   *C*

23.75 (a) Since diazotization of R—NH_2 gives N_2 and products derived from R^+, then diazotization of H—NH_2 should give H^+, that is, H_3O^+, and N_2.

    (b) Exhaustive methylation is followed by Hofmann elimination in which a proton is removed from a β-carbon.

A   *B*   *C*

Although it is not discussed in the text, the regioselectivity of the Hofmann elimination tends to favor removal of a proton at the β-carbon with fewest alkyl substituents. Therefore, *A* is the predominant product.

(c)   The triamide resulting from treatment of the acid chloride with dimethyl amine is reduced with LiAlH₄ to give triamine *A*. (See structures below.)

(d)   The triamine *A* from part (c) is exhaustively methylated and then subjected to Hofmann elimination to give the product *B*, which is an isomer of benzene. Three equivalents of trimethylamine will also be formed.

A   *B*

(e)   The nitrosyl cation acts as the electrophile in an electrophilic aromatic substitution reaction of phenol.

(f)   In this variation of the Sandmeyer reaction, copper(I) nitrite introduces a nitro group into the ring.

p-Dinitrobenzene would be very difficult to prepare by electrophilic aromatic substitution; can you see why?

(g)   The amino group is alkylated twice by ethylene oxide to give $CH_3CH_2CH_2CH_2N(CH_2CH_2OH)_2$.

(h)   The amine reacts with the epoxide at the carbon with fewer alkyl substituents.

$$(CH_3)_2\overset{\overset{\displaystyle OH}{|}}{C}-CH_2NEt_2$$

(i)   The conjugate-base anion of phthalimide reacts at the carbon of the epoxide with no alkyl substituents to give compound *A*; hydrolysis of the amide groups gives the amine product.

A

(j)   The amine reacts with the epoxide with inversion of configuration. Because both starting materials are achiral (the epoxide is the meso stereoisomer), the product is the racemate; one enantiomer results from reaction at one carbon of the epoxide, and the other enantiomer results from reaction at the other carbon.

(k)   Vinylic halides are inert in S_N2 reactions, and allylic halides are very reactive. Because an excess of the amine is used, only one halide molecule reacts with each amine, and the allylic bromine is displaced.

(l)   Catalytic hydrogenation reduces nitrobenzene to aniline, which then reductively aminates the aldehyde; because excess aldehyde is present, the resulting secondary amine then reductively aminates a second molecule of aldehyde. The final product is the tertiary amine N,N-dibutylaniline, $PhN(CH_2CH_2CH_2CH_3)_2$.

23.76   (a)

(b)

(c)

(d)

(e)

(f)

O=CHCH₂CH₂CH₃ $\xrightarrow[\text{CH}_3\text{OH}]{\begin{array}{c}\text{(CH}_3)_2\text{NH}\\\text{NaCNBH}_3\end{array}}$ (CH₃)₂NCH₂CH₂CH₂CH₃ $\xrightarrow{\text{PhCH}_2\text{Br}}$ PhCH₂—$\overset{\overset{\displaystyle CH_3}{|}}{\underset{\underset{\displaystyle CH_3}{|}}{\overset{+}{N}}}$—CH₂CH₂CH₂CH₃ Br⁻

butyraldehyde

(g)

PhCH₃ $\xrightarrow[\text{peroxides}]{\text{NBS, CCl}_4}$ PhCH₂Br $\xrightarrow[\text{ether}]{\text{Mg}}$ $\xrightarrow[\text{2) H}_3\text{O}^+]{1) \triangle\!O}$ PhCH₂CH₂CH₂OH $\xrightarrow[\text{2) H}_3\text{O}^+]{1) \text{KMnO}_4, ^-\text{OH}}$

toluene

PhCH₂CH₂COH $\xrightarrow{\text{SOCl}_2}$ PhCH₂CH₂CCl $\xrightarrow[\text{(excess)}]{\text{H}_2\text{NEt}}$ PhCH₂CH₂CNHEt $\xrightarrow[\begin{array}{c}\text{2) H}_3\text{O}^+\\\text{3) NaOH}\end{array}]{1) \text{LiAlH}_4}$ PhCH₂CH₂CH₂NHEt

N-ethyl-3-phenyl-1-propanamine

(h)

CH₂(CO₂Et)₂ $\xrightarrow[\text{2) CH}_3\text{CH}_2\text{CH}_2\text{Br}]{1) \text{NaOEt, EtOH}}$ CH₃CH₂CH₂CH(CO₂Et)₂ $\xrightarrow[\text{2) CH}_3\text{I}]{1) \text{NaOEt, EtOH}}$

diethyl malonate

CH₃CH₂CH₂$\underset{\underset{\displaystyle CH_3}{|}}{C}$(CO₂Et)₂ $\xrightarrow{\text{H}_3\text{O}^+, \text{H}_2\text{O, heat}}$ CH₃CH₂CH₂$\underset{\underset{\displaystyle CH_3}{|}}{CH}$COOH $\xrightarrow[\text{2) NH}_3]{1) \text{SOCl}_2}$

CH₃CH₂CH₂$\underset{\underset{\displaystyle CH_3}{|}}{CH}$CNH₂ $\xrightarrow[\text{H}_2\text{O}]{\text{Br}_2, \text{NaOH}}$ CH₃CH₂CH₂$\underset{\underset{\displaystyle CH_3}{|}}{CH}$NH₂

2-pentanamine

(i)

(CH₃)₂C=O $\xrightarrow[\text{2) H}_3\text{O}^+]{1) \text{LiAlH}_4}$ (CH₃)₂CHOH $\xrightarrow{\text{conc. HBr}}$ (CH₃)₂CHBr $\xrightarrow[\text{ether}]{\text{Mg}}$ $\xrightarrow[\text{2) H}_3\text{O}^+]{1) \triangle\!O}$

acetone

(CH₃)₂CHCH₂CH₂OH $\xrightarrow[\text{2) H}_3\text{O}^+]{1) \text{KMnO}_4, ^-\text{OH}}$ (CH₃)₂CHCH₂COH $\xrightarrow[\text{2) NH}_3]{1) \text{SOCl}_2}$

(CH₃)₂CHCH₂CNH₂ $\xrightarrow[\text{H}_2\text{O}]{\text{Br}_2, \text{NaOH}}$ (CH₃)₂CHCH₂NH₂

isobutylamine

(j)

$$(CH_3)_2C{=}O \xrightarrow[\text{2) H}_3\text{O}^+]{\text{1) LiAlH}_4} (CH_3)_2CHOH \xrightarrow{\text{conc. HBr}} (CH_3)_2CHBr \xrightarrow[\text{ether}]{\text{Mg}} \xrightarrow[\text{2) H}_3\text{O}^+]{\text{1)} \triangle\text{O}}$$

acetone

$$(CH_3)_2CHCH_2CH_2OH \xrightarrow[\text{2) H}_3\text{O}^+]{\text{1) KMnO}_4,\ {}^-\text{OH}} (CH_3)_2CHCH_2\overset{\text{O}}{\overset{\|}{C}}OH \xrightarrow[\text{2) NH}_3]{\text{1) SOCl}_2}$$

$$(CH_3)_2CHCH_2\overset{\text{O}}{\overset{\|}{C}}NH_2 \xrightarrow[\text{3) NaOH}]{\substack{\text{1) LiAlH}_4 \\ \text{2) H}_3\text{O}^+}} (CH_3)_2CHCH_2CH_2NH_2$$

isopentylamine

(k)

nitrobenzene $\xrightarrow[\text{heat}]{\text{Br}_2,\ \text{FeBr}_3}$ (NO_2, Br) $\xrightarrow[\text{2) NaOH}]{\text{1) Sn/HCl}}$ (NH_2, Br) $\xrightarrow[\text{2) CuCl}]{\text{1) NaNO}_2/\text{HCl}}$ **m-bromochlorobenzene** (Cl, Br)

(l)

nitrobenzene $\xrightarrow[\text{2) NaOH}]{\text{1) Sn/HCl}}$ (NH_2) $\xrightarrow[\text{2) CuCl}]{\text{1) NaNO}_2/\text{HCl}}$ (Cl) $\xrightarrow{\text{Br}_2,\ \text{FeBr}_3}$ (Cl, Br) **p-bromochlorobenzene**

(m) For the first reaction below, see Eq. 18.101 on text p. 1006.

phenol $\xrightarrow[\text{CHCl}_3]{\text{HNO}_3}$ (OH, NO_2) $\xrightarrow[\text{2) CH}_3\text{I}]{\text{1) NaOH}}$ (OCH_3, NO_2) $\xrightarrow{\text{H}_2,\ \text{Pd/C}}$ (OCH_3, NH_2) $\xrightarrow[\text{2) CuCN}]{\text{1) NaNO}_2/\text{HCl}}$ (OCH_3, CN) **p-methoxybenzonitrile**

(n)

$$\underset{\text{CH}_3}{\overset{\text{D}}{\underset{|}{\overset{|}{C}}}}\text{(H, OH)} \xrightarrow[\text{2) NaCN}]{\text{1) TsCl, pyridine}} N{\equiv}C{-}\underset{\text{CH}_3}{\overset{\text{D}}{\underset{|}{\overset{|}{C}}}}\text{H} \xrightarrow[\text{2) H}_2\text{O}]{\text{1) LiAlH}_4} H_2NCH_2{-}\underset{\text{CH}_3}{\overset{\text{D}}{\underset{|}{\overset{|}{C}}}}\text{H}$$

Note that the S_N2 reaction of cyanide ion with the tosylate occurs with inversion of configuration.

(o)

$$CH_3-C(=O)-CH_2-CH_2-C(=O)-OH \xrightarrow{CH_2N_2} CH_3-C(=O)-CH_2-CH_2-C(=O)-OCH_3 \xrightarrow[CH_3OH]{\begin{array}{c}CH_3NH_2\\ NaCNBH_3\end{array}}$$

$$CH_3-CH(NHCH_3)-CH_2-CH_2-C(=O)-OCH_3 \xrightarrow[-CH_3OH]{heat} H_3C\text{-(N-methyl pyrrolidinone)}$$

(p)   Carbaryl can be prepared by the reaction of 1-naphthol with methyl isocyanate.

1-naphthol (OH) + $H_3C-N=C=O$ (methyl isocyanate) $\longrightarrow$ carbaryl (O-C(=O)-NHCH_3)

23.77 (a) Following the loss of an electron, the molecular ion of 2-methyl-2-heptanamine undergoes an α-cleavage to give $(CH_3)_2C=\overset{+}{N}H_2$, which has $m/z = 58$. Following the loss of an electron, the molecular ion N-ethyl-4-methylpentanamine undergoes an α-cleavage to give $CH_3CH=\overset{+}{N}HEt$, which has $m/z = 72$.

$$CH_3CH_2CH_2CH_2CH_2-\underset{CH_3}{\underset{|}{C}}(NH_2)-CH_3 \xrightarrow{-e^-} \cdots \longrightarrow CH_3CH_2CH_2\dot{C}H_2 + H_3C-\overset{NH_2}{\overset{||}{C}}-CH_3 \quad m/z = 58$$

2-methyl-2-heptanamine

$$(CH_3)_2CH_2CH_2-\underset{}{CH}(NHCH_2CH_3)-CH_3 \xrightarrow{-e^-} \cdots \longrightarrow (CH_3)_2CH_2\dot{C}H_2 + \overset{\overset{+}{N}HCH_2CH_3}{\overset{||}{CH}}-CH_3 \quad m/z = 72$$

N-ethyl-4-methyl-2-pentanamine

(b)   The α-cleavage of tributylamine produces a major peak at $m/z = 142$:

$$CH_3CH_2CH_2-CH_2-\overset{+}{N}Bu_2 \longrightarrow CH_3CH_2\dot{C}H_2 + CH_2=\overset{+}{N}Bu_2 \quad m/z = 142$$

23.78 At pH 1, methylamine exists almost completely as the methylammonium ion, $CH_3-\overset{+}{N}H_3$. The methyl quartet is expected for the presence of three neighboring nitrogen protons provided that chemical exchange is slow. As the pH is raised, more free methylamine is present and chemical exchange of the protons on the nitrogen can occur rapidly by the following mechanism. (The asterisk is used to differentiate the two nitrogens.)

$$\overset{+}{CH_3\overset{*}{N}H_3} + CH_3NH_2 \rightleftharpoons CH_3\overset{*}{N}H_2 + CH_3\overset{+}{N}H_3$$

Evidently, this exchange is so fast on the NMR time scale that splitting is obliterated. (See Secs. 14.7E and 14.8 in the text.)

23.79   The λ_{max} of aniline occurs at longer wavelength than that of benzene because of the conjugation of the aniline nonbonding electron pair with the π-electron system of the benzene ring. Recall that the more atoms are involved in a conjugated π-electron system, the greater is the intensity of the UV spectrum and the greater is the λ_{max} (Sec. 15.2C in the text). When aniline is protonated, the nitrogen no longer has a nonbonding pair, and the conjugated π-electron system is restricted to the phenyl ring, in which case the UV spectrum looks much as it does in benzene. In other words, protonated aniline, as far as the π-electron system and the UV spectrum are concerned, is essentially the same as benzene.

23.80   (a)   With an unsaturation number of 4 and clear evidence for a *para*-substituted benzene ring (the four-proton pair of doublets centered at about δ 7), no other unsaturations are possible. The NMR indicates partial structures —OCH₃ and —CH₂CH₂—. With no other unsaturation, the nitrogen has to be an amine, and the broad, two-proton resonance near δ 1 suggests an —NH₂ group. The only structure that fits all the data is shown at right.

2-(4-methoxyphenyl)ethylamine
(compound *A*)

(b)   This compound has only one degree of unsaturation, and both the NMR (δ 4.9) and the IR (898 cm⁻¹) suggest a vinylic methylene group (H₂C=) in which the protons are coupled to no other protons. The chemical shift of the three-proton singlet (δ 1.8) suggests an allylic methyl group. We have a two-proton singlet at δ 3.2 that is clearly α to the nitrogen, and an ethyl absorption in which the —CH₂— group is also α to the nitrogen. These data define the structure at right. The absorptions of the two diastereotopic vinylic protons consist of two singlets near δ 4.8; geminal splitting is typically very small (Table 14.3, text p. 765) and evidently is not observed within the resolution of this spectrum.

N-ethyl-2-methyl-2-propenamine
(compound *B*)

(c)   The compound has no unsaturation and therefore must be an amine. The mutually split triplet–quartet pattern suggests an ethyl group, and the integration suggests two identical ethyl groups. If so, the δ 1.1 resonance has buried under it (as suggested by the hint) a broad resonance for two additional protons. Indeed, the chemical shift is reasonable for an —NH₂ group. Because there are two amine nitrogens, these have to be two chemically identical secondary amines. Two resonances have four identical protons; this suggests two chemical identical —CH₂— groups for each resonance. One of these resonances is a quartet and is part of the ethyl group discussed above; this confirms that there are two chemically equivalent ethyl groups. The other —CH₂— resonance is a singlet; therefore, this corresponds to two —CH₂— groups in identical environments that do not give rise to splitting.

N,N'-diethyl-1,2-ethanediamine
(compound *C*)

This molecule must have some internal symmetry. The structure is shown at right. Although the two —CH₂— groups in the center of the molecule are adjacent, they do not show any splitting because they are chemically equivalent.

23.81   (a)   The chemical data indicate the presence of a primary amine. The compound has one degree of unsaturation, and the chemical shifts in the ^{13}C NMR are too small for any carbon to be involved in a double bond. Hence, compound *A* contains a ring, and only CH and CH$_2$ groups are present. The presence of only three resonances means that two carbons are equivalent. Compound *A* is cyclobutylamine (at right). Cyclopropylmethanamine is ruled out by the fact that the carbon bound to one hydrogen has the largest chemical shift and therefore must be bound to the nitrogen.

cyclobutylamine
(compound *A*)

cyclopropylmethanamine
(ruled out)

   (b)   The IR data for this compound indicate a —CH=CH$_2$ group. The molecule is neither a primary amine (chemical data) nor a tertiary amine (it has an N—H). Hence it is a secondary amine. Two structures worth considering are *A* and *B* are shown at right. Because the ^{13}C NMR spectrum contains no resonances near δ 15, which would be expected for the methyl carbon of compound *A*, the structure must be *B*, *N*-methyl-2-propenamine. The chemical shifts of the carbon resonances are consistent with this structure.

CH$_3$CH$_2$NHCH=CH$_2$ CH$_3$NHCH$_2$CH=CH$_2$

A   *B*

21.82   Evidently, an anion is generated at the α-carbon of the ester, and this then adds to the aldehyde. Although one might be tempted to use the amine as a base to remove an α-hydrogen, vinylic hydrogens that are α to carbonyl groups are not particularly acidic, because the nonbonding electron pair of the enolate ion that would form and the double bond cannot simultaneously be conjugated with the carbonyl. Rather, the amine acts as a nucleophile in a conjugate addition, which generates the required anion. Once the addition occurs, the amine is expelled again to regenerate the α,β-unsaturated product.

23.83   An intermediate in the formation of 1-pentanamine is the imine *A*. The other products arise from the fact that 1-pentanamine, *B*, can react with either the starting aldehyde, pentanal, or with imine *A* to give a new imine *C*. (Give the mechanism for this reaction.) Imine *C* can also undergo hydrogenation to the second product, dipentylamine, *D*. Dipentylamine can also react with other imine intermediates to form iminium ion *E*, which can undergo hydrogenation to form tripentylamine *F*, the tertiary amine by-product.

pentanal   *A*   *B*

The yield of pentylamine (compound *B*) can be increased, and the amounts of the other by-products reduced, by including an excess of ammonia in the reaction mixture. Can you see why? (*Hint*: Apply Le Chatelier's principle.)

23.84 Both compound *A* and *C* undergo Hofmann elimination after exhaustive methylation followed by heating with Ag$_2$O. Compound *A* is converted to cyclooctatetraene (COT), compound *B*, and compound *C* is converted to "1,3,5-cyclohexatriene," which is actually benzene. COT contains $4n$ π electrons and is not aromatic; benzene contains $4n + 2$ π electrons and is aromatic (see discussion in Sec. 15.7A in the text). Under mild conditions, benzene does not react with bromine, while COT behaves as a typical cycloalkene and does react with bromine (see Eq. 15.56, text p. 840).

23.85 (a) Amides, like amines, can be diazotized. The result is an acyldiazonium ion *A*, which rapidly hydrolyzes; remember that N$_2$ is a superb leaving group. This produces carbamic acid, which decomposes to ammonia and CO$_2$. Then ammonia is itself diazotized to give N$_2$ as shown in the solution to Problem 23.75(a) above.

(b) The amine reacts intramolecularly with the carbonyl group to give a carbinolamine, which then reacts to form a cyclic iminium ion *A*, which is reduced by the sodium borohydride.

A

(c)   Cyanide ion acts as a base to promote a β-elimination of trimethylamine and form an α,β-unsaturated ketone *A*. This undergoes conjugate addition of cyanide ion to give the product (Sec. 22.9A in the text).

(d)   The acyl azide *A* undergoes Curtius rearrangement to an enamine *B*, which hydrolyzes under the aqueous reaction conditions to the product ketone.

The mechanism of the Curtius rearrangement is shown in text Sec. 23.11D, and the hydrolysis of the enamine occurs as follows:

(a resonance-stabilized cation)

(a resonance-stabilized cation)

(e) The diazonium ion *A* formed from the starting material loses nitrogen to give a carbocation *B*. This carbocation undergoes a rearrangement to the product. (See Study Problem 19.2 on text pp. 1040-1041 for a discussion of a closely related reaction.)

(f) This reaction, called the *Lossen rearrangement*, is related to the Hofmann rearrangement, on p. 1314 of the text.

23.86 The conjugate-acid pK_a of amine *C* is a normal conjugate-acid pK_a for a tertiary amine. (See, for example, the pK_a of the conjugate acid of triethylamine in Table 23.1 on text p. 1285.) This observation shows that incorporation of an amine into a bicyclic structure has no significant effect on its conjugate-acid pK_a. The conjugate-acid pK_a of amine *A* is not unusual for an *N,N*-dialkylaniline. (See, for example, the pK_a of PhN(CH$_3$)$_2$ in Table 23.1.) Recall (text p. 1287) that the conjugate-acid pK_a values of aniline derivatives are considerably lower than those of tertiary alkylamines because of resonance interaction of the nonbonding electron pair on the nitrogen with the benzene ring. The conjugate-acid pK_a of compound *B*, then, has the unusual value that requires explanation. In this compound, the nonbonding pair on nitrogen cannot overlap with the π-electron system of the benzene ring because the π orbitals of the benzene ring and the orbital on the nitrogen are forced by the bicyclic structure out of coplanarity.

perpendicular to π-electron system of the benzene ring

Consequently, the basicity-lowering resonance effect is absent. For this reason, the conjugate-acid pK_a of compound *B* is higher than that of compound *A*. However, the basicity-lowering polar effect of the benzene ring remains; for this reason, the conjugate-acid pK_a of compound *B* is lower than that of compound *C*.

 The pK_a data in the problem provide good estimates of the polar and resonance effects of a benzene ring on amine basicity. These data show that polar and resonance effects are about equally important.

23.87   The data show that the conjugate-acid pK_a values of a bicyclic tertiary amine C (which is the same as amine C in the previous problem) and a monocyclic tertiary amine B are not very different. Therefore, the additional ring has no unusual effect on pK_a. The conjugate-acid pK_a of amide A is also fairly normal for an amide that is protonated on the carbonyl oxygen. (See text Sec. 21.5.) The pK_a of amide D, then, is unusual. Normally, amides protonate on the carbonyl oxygen because the resulting carbocation is resonance-stabilized. In this case, however, the bicyclic structure of the ring prevents the overlap of the nitrogen nonbonding electron pair with the π-electron system of the carbonyl group. (See structure at right.) Therefore, the electronic character of this nitrogen is more like that of an amine with a neighboring electron-withdrawing substituent (the carbonyl group) than it is like that of an amide. Consequently, this amide behaves like an amine: it protonates on nitrogen. What about the hydrolysis rate? Recall that ordinary amides hydrolyze slowly because they are resonance-stabilized. (See Sec. 21.7E in the text.) When resonance stabilization is not present, as in the case of compound D, an amide hydrolyzes rapidly. In acidic solution, when the nitrogen is protonated, its hydrolysis rate is more like that of an acid chloride! (See also the sidebar on text p. 1145 for a discussion of this case.)

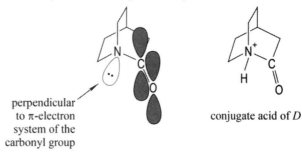

perpendicular to π-electron system of the carbonyl group

conjugate acid of D

23.88   (a)   The phosphine-containing probe compound reacts with the azide-containing protein as in a standard Staudinger reaction, as shown in Eqs. 23.69a–b on text p. 1308, releasing nitrogen as a by-product.

(b)   The negatively-charged nitrogen reacts with the ester in a nucleophilic acylation reaction to form an amide (see Sec. 21.8C for related reactions). The iminophosphorane is then hydrolyzed by water to give the probe-labeled protein.

(c)   Had the ester group not been present in proximity to the iminophosphorane nitrogen, the reaction in part (b) would not occur, and instead, water would simply hydrolyze the iminophosphorane into the triphenylphosphine-derivative and the amine linked to the protein, as in a standard Staudinger reaction (see Eq. 23.70 on text p. 1308).

24 | Carbohydrates

CHAPTER 24 SUMMARY

- Carbohydrates are aldehydes and ketones that contain a number of hydroxy groups on an unbranched carbon chain, as well as their chemical derivatives.

- The structures of chiral compounds can be drawn in planar representations called Fischer projections, which are especially useful for depicting molecules that contain contiguous asymmetric carbons in an unbranched chain. A Fischer projection is derived from an eclipsed conformation in which all asymmetric carbons are aligned vertically. Asymmetric carbons are represented as the intersection points of vertical and horizontal lines. All vertical bonds at each asymmetric carbon are assumed to be oriented away from the observer and all horizontal bonds toward the observer. Several valid Fischer projections can be drawn for any chiral molecule. These are derived by the rules given in Sec. 24.2.

- A Fischer projection shows the configuration of each asymmetric carbon in a chiral molecule but implies nothing about its conformation.

- The D,L system is an older but widely used method for specifying carbohydrate enantiomers. The D enantiomer is the one in which the asymmetric carbon of highest number has the same configuration as (R)-glyceraldehyde (D-glyceraldehyde).

- Monosaccharides exist in cyclic furanose or pyranose forms in which a hydroxy group and the carbonyl group of the aldehyde or ketone have reacted to form a cyclic hemiacetal.

- The cyclic forms of monosaccharides are in equilibrium with small amounts of their respective aldehydes or ketones and can therefore undergo a number of aldehyde and ketone reactions. These include oxidation (bromine water or dilute nitric acid); reduction with sodium borohydride; cyanohydrin formation (the first step in the Kiliani–Fischer synthesis); and base-catalyzed enolization and enolate-ion formation (the Lobry de Bruyn–Alberda van Eckenstein reaction).

- The —OH groups of carbohydrates undergo many typical reactions of alcohols and glycols, such as ether formation, ester formation, and glycol cleavage with periodate.

- Because the hemiacetal carbons of monosaccharides are asymmetric, the cyclic forms of monosaccharides exist as diastereomers called anomers. The equilibration of anomers is the reason that carbohydrates undergo mutarotation.

- In a glycoside, the —OH group at the anomeric carbon of a carbohydrate is substituted with an ether (—OR) group. In disaccharides or polysaccharides, the —OR group is derived from another saccharide residue. The —OR group of glycosides can be replaced with an —OH group by hydrolysis. Thus, higher saccharides can be hydrolyzed to their component monosaccharides in aqueous acid.

- Disaccharides, trisaccharides, and so on, can be classified as reducing or nonreducing sugars. Reducing sugars have at least one free hemiacetal group. In nonreducing sugars, all anomeric carbons are involved in glycosidic linkages.

STUDY GUIDE LINKS

24.1 Acid Catalysis of Carbohydrate Reactions

Acid catalysis of glycoside formation involves protonation of the hydroxy group at the anomeric carbon (Eq. 24.39a in text Sec. 24.6); acid catalysis of mutarotation involves protonation of the ring oxygen (Focused Problem 24.9 in text Sec. 24.4). You may be asking why these specific oxygens are protonated—why not others? The answer is that *any one* of the oxygens can indeed be protonated to small extents in acidic solution. However, *only* protonation at the ring oxygen leads to mutarotation, and *only* protonation of the —OH group at the anomeric carbon leads to glycoside formation. In other words, the various protonated forms are in rapid equilibrium and are present whether they are shown explicitly or not. Only certain of the protonated forms can react further, and these are the ones shown in the equations.

24.2 Configurations of Aldaric Acids

Consideration of the stereochemistry of aldaric acids reveals a subtle yet important aspect of the D,L configurational system. Certain aldaric acids can be derived from either a D or an L carbohydrate.

$$\text{D-glucaric acid (from D-glucose) or L-gularic acid (from L-gulose)}$$
(the same compound)

In this case, whether the aldaric acid is classified as D or L depends on which carbon is specified as carbon-1. Because the two ends of the molecule are *constitutionally equivalent* (see text Sec. 11.9A), the choice is completely arbitrary! The names D-glucaric acid and L-gularic acid are both correct names for this aldaric acid. The two names reflect the fact that this aldaric acid can be formed by nitric acid oxidation of either D-glucose or L-gulose.

This situation arises because the —OH groups on the endmost asymmetric carbons are on the same side of the Fischer projection. However, when these —OH groups are on opposite sides, the configuration is unambiguous. For example, D-tartaric acid also has constitutionally equivalent ends, but it has the D configuration no matter how it is turned.

$$\text{D-tartaric acid}$$

As you might expect, a similar situation arises in other derivatives with constitutionally equivalent ends, such as alditols. This issue is explored in Focused Problem 24.21 in text Sec. 24.8B.

 # FURTHER EXPLORATIONS

 ## 24.1   Nomenclature of Anomers

Long before the actual structures of the anomeric forms of many carbohydrates were known, the α-anomer of a D-carbohydrate was simply defined as "the more dextrorotatory of the two anomers." This convention, proposed by C. S. Hudson (1881–1952), an American carbohydrate chemist, is cumbersome because, in order to apply it, one has to know the optical rotations of the two anomers. When the structures of several anomers became known, it was found that in most cases the more dextrorotatory anomer of each D-carbohydrate has the configuration shown in the text for the α-anomer. Hudson's definition in terms of a physical property—optical rotation—was subsequently replaced by the structural definition used in the text.

 ## 24.2   More on the Fischer Proof

The Fischer proof of glucose stereochemistry has a number of interesting facets that are not discussed in the text. One of the most important experimental aspects of carbohydrate chemistry is that many carbohydrates are notoriously difficult to crystallize; they are frequently isolated as "syrups" and "gums." Because chemists in Fischer's day did not have available the sophisticated purification methods available today, they had to rely almost exclusively on crystallization as a purification method. Because carbohydrates were difficult to crystallize, they were very difficult to purify. Few chemists were interested in working on problems fraught with so many experimental difficulties; consequently, little progress had been made in the field of carbohydrate chemistry when Fischer began his work. In fact, Fischer himself, in some of his correspondence, deplored the dreadfully slow pace of carbohydrate research in his own laboratory resulting from the experimental difficulties of handling carbohydrates.

Two solutions to the problem of carbohydrate crystallization emerged from Fischer's long-standing interest in phenylhydrazine, $PhNHNH_2$. He had developed an indole synthesis (today called the *Fischer indole synthesis*) using phenylhydrazine as a starting material. When he allowed phenylhydrazine to react with aldohexoses, he found that very interesting compounds called *osazones* were obtained.

$$
\begin{array}{c}
CH{=}O \\
H{-}\!\!-OH \\
HO{-}\!\!-H \\
H{-}\!\!-OH \\
H{-}\!\!-OH \\
CH_2OH
\end{array}
\;+\; 3\ PhNHNH_2 \;\longrightarrow\;
\begin{array}{c}
CH{=}N{-}NHPh \\
C{=}N{-}NHPh \\
HO{-}\!\!-H \\
H{-}\!\!-OH \\
H{-}\!\!-OH \\
CH_2OH
\end{array}
\;+\; PhNH_2 \;+\; NH_3 \;+\; H_2O
$$

(+)-glucose   glucosazone

This reaction is very reminiscent of phenylhydrazone formation (Table 19.3 in text Sec. 19.12A). However, an unusual aspect of this reaction is that one equivalent of phenylhydrazine serves as an oxidizing agent, with the net result that a second substituted imine group is introduced. (For the mechanism of this reaction, see Loudon, *Organic Chemistry,* First edition, Addison–Wesley Publishing Co., 1984, pp. 1417–1418.) Osazones proved to be much more easily crystallized, and thus more easily characterized, than aldoses themselves. Furthermore, they served to confirm some of Fischer's stereochemical deductions. Notice that *neither carbon-1 nor carbon-2 are asymmetric* in the osazone. Consequently, isomeric aldoses that are epimeric only at carbon-2 give the same osazone. Fischer found that (+)-glucose and (+)-mannose give *different* phenylhydrazones (formed under milder conditions with phenylhydrazine), but give the same osazone. This fact, along with their simultaneous preparation in the Kiliani–Fischer synthesis from (–)-arabinose, showed that they are epimers at carbon-2.

Phenylhydrazine figured prominently in Fischer's research in a second way when Fischer found that the aminolysis reactions of aldonic acids (undoubtedly reacting as their lactones) with phenylhydrazine give phenylhydrazides, which, unlike the parent acids, are beautifully crystalline compounds.

D-gluconic acid   D-γ-gluconolactone   D-gluconic acid
                                                       phenylhydrazide

One other interesting aspect of the Fischer proof has to do with the last step, which was made possible by a remarkable bit of serendipity. It is noted in the text that Fischer had prepared L-(+)-gulose in the course of his research; recall that it was the oxidation of D-glucose and L-gulose to the same aldaric acid that completed the Fischer proof. Equation 24.51 in text Sec. 24.8C shows that D-glucaric acid can exist as a mixture of two different lactones (Step 4, text Sec. 24.10A). Fortuitously, Fischer's procedure for isolating the lactones gave mainly the 1,4-lactone. His synthesis of L-gulose began with this lactone.

D-glucaric acid 1,4-lactone   D-glucuronic acid

$$\text{L-gulonic acid} \xrightarrow[\text{- H}_2\text{O}]{} \text{L-gulonolactone} \xrightarrow{\text{Na/Hg}} \text{L-gulose}$$

"Na/Hg" is sodium amalgam, a solution of sodium in mercury, which was widely used as a reducing agent until lithium aluminum hydride and sodium borohydride were developed. The sodium–amalgam reductions shown above can be carried out in stages by controlling the conditions of each step. Recall that D-glucose is readily oxidized to D-glucaric acid-1,4 lactone with dilute nitric acid. Consequently, beginning with D-glucose, the overall sequence of glucose oxidation followed by the sodium amalgam reductions effects the *net interchange* of the CH=O and CH$_2$OH groups, and thus gives L-gulose from D-glucose. The serendipitous element was this: had Fischer isolated the 3,6-lactone of D-glucaric acid, it would have given back D-glucose when taken through this sequence of reactions, and the last step of the proof would not have been possible.

An intriguing account of Fischer's life and work, including details of the Fischer proof, was written by F. W. Lichtenthaler and published in *Angewandte Chemie International Edition in English,* vol. 31, No. 12, pp. 1541–1596 (1992). This journal is available in most chemistry libraries.

REACTION REVIEW

I.   REACTIONS OF CARBOHYDRATES

A.   BASE-CATALYZED ISOMERIZATION OF CARBOHYDRATES

1. In base, aldoses and ketoses rapidly equilibrate to mixtures of other aldose and ketoses; this transformation is an example of the Lobry de Bruyn–Alberda van Ekenstein reaction.

    a. An aldose can ionize to give a small amount of its enolate ion in base.

    b. Protonation of this enolate ion at one face gives back the aldose; protonation at the other face gives an epimer.

$$\text{D-(+)-glucose} \xrightleftharpoons[\text{H}_2\text{O}]{\text{HO}^-} \text{enolate ion} \xrightleftharpoons[\text{HO}^-]{\text{H}_2\text{O}} \text{D-(+)-mannose}$$

c. The enolate ion can also protonate on the oxygen to give a new enol, called an enediol, which contains a hydroxy group at both ends of a double bond.

d. The enediol is simultaneously the enol of two epimeric aldoses and an isomeric ketose, and the enediol is transformed into a mixture of these.

enolate ion of
D-(+)-glucose or D-(+)-mannose
an enediol
D-(+)-fructose

2. Several transformations of this type are important in metabolism.

B. *Ester and Ether Derivatives of Carbohydrates*

1. In the presence of concentrated base, carbohydrates are converted into ethers by reactive alkylating agents such as dimethyl sulfate, methyl iodide, or benzyl chloride (all examples of the Williamson ether synthesis).

methyl 2,3,4,6-tetra-*O*-methyl-D-glucopyranoside

2. Methyl ethers of carbohydrates can also be formed using various other reagents, including CH_3I and Ag_2O, or $NaNH_2$ in NH_3 followed by CH_3I.

3. The alkoxy group at the anomeric carbon differs from other alkoxy groups in the alkylated carbohydrate.

a. The alkoxy group at the anomeric carbon is part of a glycosidic linkage.

b. Because it is an acetal, it can be hydrolyzed in aqueous acid under mild conditions; the other alkoxy groups are ordinary ethers and do not hydrolyze under these conditions.

2,3,4,6-tetra-*O*-methyl-D-glucopyranose

4. The hydroxy groups of carbohydrates can be esterified; the resulting esters can be saponified in base or can be removed by transesterification with an alkoxide.

4. Ethers and esters are used as protecting groups in reactions involving carbohydrates; furthermore, they have broader solubility characteristics and greater volatility than the carbohydrates themselves.

C.   OXIDATION AND REDUCTION REACTIONS OF CARBOHYDRATES

1. Treatment of an aldose with bromine water oxidizes aldoses to aldonic acids; this reaction is a useful test for aldoses. (Carbohydrates that can be oxidized by bromine water are called reducing sugars.)

    a.   Aldonic acids exist in acidic solution as five-membered lactones called aldonolactones.

an aldonic acid   an aldolactone

    b.   Glycosides are not oxidized by bromine water because the aldehyde carbonyl group is protected as an acetal. (Carbohydrates that cannot be oxidized by bromine water are called nonreducing sugars.)

2. Aldoses can also be oxidized with the Tollens reagent; the alkaline conditions of Tollens test, however, also promote the equilibration of ketoses and aldoses. For this reason, ketoses also give a positive Tollens test.

3. Both ends of an aldose are oxidized to carboxylic acid groups by dilute HNO_3, but the secondary alcohol groups are not affected; the oxidation product is an aldaric acid.

    a.   Aldaric acids in acidic solution form lactones.

    b.   Two different five-membered lactones are possible from a hexose or pentose, depending on which carboxylic acid group undergoes lactonization.

    c.   Under certain conditions, some aldaric acids derived from aldohexoses can be isolated as dilactones, in which both carboxylic acid groups are lactonized. (See Eq. 24.51 in text Sec. 24.8B, p. 1357.)

4. Many carbohydrates contain vicinal glycol units and are oxidized by periodic acid.

a. α-Hydroxy aldehydes are oxidized to formic acid and another aldehyde with one fewer carbon. (See Eq. 24.52 in text Sec. 24.8C.)

b. Hydroxymethyl ketones are oxidized to formaldehyde and a carboxylic acid.

$$HOCH_2-\overset{O}{\overset{\|}{C}}-R \xrightarrow{H_5IO_6} \overset{O}{\overset{\|}{CH_2}} + HO-\overset{O}{\overset{\|}{C}}-R$$

a hydroxymethyl ketone

c. Because it is possible to determine accurately both the amount of periodic acid consumed and the amount of formic acid produced, periodic acid oxidation can be used to differentiate between pyranose and furanose structures for saccharide derivatives.

5. Aldoses and ketoses undergo many of the usual carbonyl reductions, such as catalytic hydrogenation and reaction with $NaBH_4$; an aldose is reduced to a primary alcohol called an alditol.

II. SYNTHESIS OF SUGARS FROM OTHER SUGARS

A. KILIANI–FISCHER SYNTHESIS; INCREASING THE LENGTH OF THE ALDOSE CHAIN

1. Addition of hydrogen cyanide to aldoses gives cyanohydrins; because the cyanohydrin product of such a reaction has an additional asymmetric carbon, it is formed as a mixture of two epimers.

   a. These epimers are diastereomers and are typically formed in different amounts.

   b. The mixture of cyanohydrins can be converted by catalytic hydrogenation into a mixture of imines, which are not isolated.

   c. The imines would normally hydrogenate to amines, but the acidic aqueous reaction conditions intercept them and convert them into aldehydes, which are the aldose products.

2. The cyanohydrin formation–reduction sequence converts an aldose into two epimeric aldoses with one additional carbon; this process is known as the Kiliani–Fischer synthesis.

B. RUFF DEGRADATION; DECREASING THE LENGTH OF THE ALDOSE CHAIN

1. In the Ruff degradation, the calcium salt of an aldonic acid is oxidized with hydrogen peroxide in the presence of Fe^{3+}.

   a. An aldose is degraded to another aldose with one less carbon atom.

b. The stereochemistry of the remaining groups remains the same.

CHO
HO——H
H——OH
H——OH
CH₂OH

1) Br₂/H₂O
2) Ca(OH)₂
3) Fe(OAc)₃/30% H₂O₂
→

CHO
H——OH
H——OH
CH₂OH

SOLUTIONS TO PROBLEMS

Solutions to Focused Problems

24.1 (a) Many Fischer projections are possible for a compound with three asymmetric carbons. The two shown here are related by a 180° rotation in the plane of the page. If yours don't look like these, try to use the allowed manipulations of Fischer projections to see whether yours are equivalent. (There are eighteen possible correct Fischer projections!)

H
HO——CO₂H
HO——H
HOCH₂——OH
H

rotate 180° →

H
HO——CH₂OH
H——OH
HO₂C——OH
H

(b) Shown here are two of the nine possible Fischer projections of (S)-2-butanol. These two are related by a cyclic permutation of the ethyl, —OH, and the methyl group.

CH₃CH₂——CH₃
H

CH₃CH₂
H₃C——OH
H

24.2 (a) The way to solve a problem of this sort is to manipulate one of the Fischer projections so that two groups in one projection are in the same positions as the corresponding groups in the other; then the relationship between the two molecules should be clear. Thus, manipulate the projection on the right so that the OH groups are in the vertical positions. It then becomes apparent that this structure and the one on the left in the problem are enantiomers.

CH₃
HO₂C——OH
HO——CO₂H
CH₃

two cyclic permulations →

OH
H₃C——CO₂H
HO₂C——CH₃
OH

OH
HO₂C——CH₃
H₃C——CO₂H
OH

← left structure in problem

enantiomers

(b) Turning either projection 180° in the plane of the page establishes that it is identical to the other projection.

24.3 It is relatively simple to determine if a compound is meso using Fisher projections. First, perform a cyclic permutation at either end of the Fisher projection (see Rule 4, Text Sec. 24.2, Displays 24.9-10) to give similar groups at each end, if possible, as in *C* and *D* below. Then draw an imaginary horizontal line through the center of the Fisher projection. If the top half is a mirror image of the bottom half, the compound is meso; if the two halves are not mirror images, then the compound is not meso. Compound *A* has an aldehyde group on top and a primary alcohol group on the bottom, so it cannot be a meso compound. Compound *B* is a meso compound. This structure (as well as structure *D*) illustrates that the "mirror plane" can go through certain atoms. After a cyclic permutation at the top carbon of compound *C*, the OH group is on the right while the lower carbon has an OH group on the left; compound *C* is not a meso compound. After cyclic permutations at the top and bottom carbons of compound *D*, all three OH groups are on the same side; compound *D* is a meso compound.

CH=O	CH₂OH	CH₂OH	CH₂OH
H—OH	HO—H	H—OH	H—OH
H—OH	H—OH	HO—H	H—OH
H—OH	HO—H	HO—H	H—OH
CH₂OH	CH₂OH	CH₂OH	CH₂OH
A	*B*	*C*	*D*

24.4 **(a)** Use a cyclic permutation to transform the given Fischer projection into a standard form in which all of the backbone carbons are in a vertical line. This will show that the —OH group on carbon-5 is on the right. Thus, this aldose has the D-configuration.

H₃C—H   ⇨   H—OH

carbon-5 OH   CH₃

OH on the right; therefore D-configuration

(b) An analysis of the carbon-3 configuration of D-glucose shows that carbon-3 has the *S* configuration. Consequently, the L-enantiomer of glucose has the *R* configuration at carbon-3.

⚠ Remember that the configuration of an asymmetric carbon in a Fischer projection can be analyzed directly. To review, see Text Sec. 24.2, Displays 24.12-13.

24.5 Use cyclic permutations to put all of the aldoses in the standard Fischer projection to see the relative locations of corresponding groups. Compounds *B* and *C* are enantiomers. Compound *A* is an epimer of *B* (at carbon-5) and a diastereomer of both *B* and *C*. (Diastereomers that differ in configuration at only one of several asymmetric carbons are called *epimers*.)

CH=O	CH=O	CH=O
H—OH	H—OH	HO—H
HO—H	HO—H	H—OH
H—OH	H—OH	HO—H
HO—H	H—OH	HO—H
CH₂OH	CH₂OH	CH₂OH
A	*B*	*C*

24.6 **(a)**

H₃C—CH—CH₂CH₂CH₂—CH=O ⇌ ⇌
|
HO

(b)

24.7   (a)   Recognize that α-D-glucopyranose is an anomer of β-D-glucopyranose, and convert the β-D-glucopyranose structures in Study Problem 24.2 in text Sec. 24.3B to α-D-glucopyranose structures by inverting the configuration of carbon-1.

α-D-glucopyranose
Fischer projection

α-D-glucopyranose
Haworth projection

α-D-glucopyranose
the two chair conformations

(b)   Recognize that D-mannose is epimeric to D-glucose at carbon-2, and convert the β-D-glucopyranose structures in Display 24.22 in text Sec. 24.3B to β-D-mannopyranose structures by inverting the configuration of carbon-2.

β-D-mannopyranose
Fischer projection

β-D-mannopyranose
Haworth projection

β-D-mannopyranose
the two chair conformations

(c)   Start with the structure of D-xylose in Figure 24.3 (text Sec. 24.3A) and perform the usual manipulations. An easier solution is to recognize that D-xylose and D-ribose are epimeric at carbon-3. Therefore, the structures for β-D-ribofuranose (text Sec. 24.3B, Display 14.23b) can be used if the configuration of carbon-3 is inverted.

β-D-xylofuranose
Fischer projection

β-D-xylofuranose
Haworth projection

(d) Start with the Fischer projection for D-fructose from text Display 24.17. Form a hemiacetal with the carbonyl group at C–2, placing the OH group in the α-position.

D-fructose
Fischer projection

α-D-fructopyranose
Fischer projection

α-D-fructopyranose
Haworth projection

α-D-fructopyranose
chair conformations

(e) Draw the enantiomers of the corresponding structures for α-D-glucopyranose, which are shown in the answers to part (a) of this problem.

OH is on the same side as the oxygen at the configurational carbon; therefore α

cyclic permutation

configurational carbon

α-L-glucopyranose
Fischer projection

α-L-glucopyranose
Haworth projections

α-L-glucopyranose
chair conformations

(f)   Start with the structures of α-D-glucopyranose in the solution for part (e) and make the bond to the anomeric carbon a "squiggly bond." (For the Fischer projection, do not show the configuration. "Squiggly bonds" are not used in Fischer projections.)

α-L-glucopyranose
and β-L-glucopyranose
Fischer projection
indicating a mixture of anomers

L-glucopyranose
mixture of anomers
Haworth projections

L-glucopyranose
mixture of anomers
chair conformations

24.8   (a)   Perform the reverse of the analysis in Study Problem 24.2 (text Sec. 24.3B) to reveal that this is α-D-ribose. But you can also look at Display 24.23 in the text and see that the compound given is the anomer of β-D-ribose—that is, α-D-ribose.

(b)   This is the β-anomer of a D-hexopyranose that differs from D-glucose in its configuration at carbons 3 and 4; therefore, it is β-D-gulopyranose.

(c)   This is β-D-arabinose.

24.9   The following mechanism begins with the α-anomer protonated on the ring oxygen and ends with the β-anomer, also protonated on the ring oxygen.

(loss of a proton to
water gives the β-anomer)

24.10 The mechanism for base-catalyzed mutarotation of glucose:

24.11 The β-pyranose forms of D-glucose and D-talose are as follows:

β-D-glucopyranose   **β-D-talopyranose**

Both chair forms of the β-D-talopyranose molecule have a significantly greater number of 1,3-diaxial interactions than β-D-glucopyranose. These unfavorable interactions in β-D-talopyranose raise its energy and thus lower the relative amounts of these forms at equilibrium. Therefore, talose contains a lower percentage of the β-pyranose form (and a lower percentage of the α-pyranose form for the same reason) and a higher percentage of the furanose and aldehyde forms at equilibrium.

24.12 (a) The easiest way to do this is to consider the relationship of D-allose to D-glucose and modify the conformational representation of β-D-glucopyranose (in which all of the ring substituents are equatorial) accordingly. Thus, Fig. 24.3 in text Sec. 24.3A shows that D-allose is epimeric to D-glucose at carbon-3. Consequently, configurational inversion of the β-D-glucopyranose structure at carbon-3 gives the desired conformational representation of β-D-allopyranose.

CH₂OH ... HO ... HO ... OH ... OH

β-D-glucopyranose

invert
C-3

CH₂OH ... HO ... OH ... OH ... HO

β-D-allopyranose

(b)   Follow the procedure in Study Problem 24.2 (text Sec. 24.3B). Note that in the second and third Fischer projections the configuration of carbon-5 cannot be represented without violating the Fischer-projection conventions. We must remember that this carbon has the R configuration, and this configuration must be shown explicitly in the conformational representation.

CH=O / HO—H / H—OH / HO—H / H—OH / CH₂OH

⇨

CH=O / HO—H / H—OH / H—CH(OH)—CH₂OH / OH | R

⇨

H—OH / HO—H / H—OH / H—CH(OH)—CH₂OH | R

⇨

α-D-idofuranose

24.13   Each form of the sugar contributes its own optical rotation in proportion to the amount that is present. Let N_i = the fraction of form i. (The dimensions of the specific rotations are abbreviated simply as degrees.)

$$N_{\text{total}} = 52.7° = N_\alpha(112.0°) + N_\beta(18.7°)$$

Because $N_\alpha + N_\beta = 1$, substitute $N_\beta = (1 - N_\alpha)$ and obtain

$$52.7° = N_\alpha(112.0°) + (1 - N_\alpha)(18.7°)$$

$$N_\alpha(112.0° - 18.7°) = 52.7° - 18.7° = 34.0°$$

$$N_\alpha = \frac{34.0°}{93.3°} = 0.36$$

$$N_\beta = 1 - N_\alpha = 0.64$$

Table 24.1 (text Sec. 24.4) confirms that D-glucopyranose at equilibrium contains 36% of the α-form and 64% of the β-form.

24.14   (a)   D-Galactose would be transformed into the aldohexose that is epimeric at carbon-2, namely, D-talose. (See the structures in Fig. 24.3, text Sec. 24.3A.) Also formed would be the ketose D-tagatose, which has the shown below.

(b)   D-Allose would be transformed by base into a mixture of D-altrose (structure in Fig. 24.3, text Sec. 24.3A) and the ketose D-psicose.

D-tagatose

D-psicose

24.15 (a) The compound is the β-*p*-nitrophenyl glycoside of D-galactopyranose; its name is *p*-nitrophenyl β-D-galactopyranoside.

(b) This compound is hydrolyzed in aqueous acid to *p*-nitrophenol and a (mutarotated) mixture of α- and β-D-galactopyranose.

α- and β-D-galactopyranose

p-nitrophenol

24.16 Because naturally occurring glycosides generally have bonds between an alcohol or phenol oxygen and the anomeric carbon of a carbohydrate, it is reasonable to propose that the β-D-glycoside of vanillin and glucose has the following structure:

24.17 (a) Start with the structure of β-D-fructofuranose (Display 24.30, text Sec. 24.4) and use it to draw the β-glycoside of methanol (structure below).

(b) Start with the structure of α-D-galactopyranose, which is the same as α-D-glucopyranose except that carbon-4 is inverted. Then substitute an isopropoxy group for the hydroxy group at the anomeric carbon.

methyl β-D-fructofuranoside

isopropyl α-D-galactopyranoside

24.18 The protonation of an oxygen of the acetal group (the glycosidic oxygen in the example below) and loss of an alcohol (methanol in the example below) gives a resonance-stabilized carbocation intermediate. A similar mechanism applied to ordinary ethers gives carbocations that are not resonance-stabilized. Hammond's postulate implies that the hydrolysis reaction that involves the more stable carbocation intermediate should occur more rapidly.

24.19   (a)   Prepare the 1-*O*-ethyl derivative and use the Williamson synthesis to introduce the methyl groups.

ethyl 2,3,4,6-tetra-*O*-methyl-
D-galactopyranoside

   (b)   Alkylate all of the hydroxy groups with benzyl chloride (PhCH₂Cl), a very reactive alkyl halide; then hydrolyze the benzyl glycoside. (Why is benzyl chloride very reactive? See text Sec. 17.4.)

2,3,4,6-tetra-*O*-benzyl-D-glucopyranose

24.20   (a)   (b)   Notice that ribitol is a meso compound.

D-galacturonic acid

ribitol

24.21 The structures of D-glucaric and L-gularic acids are as follows. Turning either structure 180° in the plane of the page shows that the two structures are identical. (This identity was a key element in the Fischer proof of glucose stereochemistry, which is discussed in text Sec. 24.10A and Further Exploration 24.2.)

24.22 (a) The structure of the aldaric acid derived from oxidation of D-galactose, and its 1,4-lactone; the Haworth projections were not requested, but they are shown also:

(b) The structure of the aldaric acid derived from oxidation of D-mannose, and the structure of its 1,4-lactone, are derived by recognizing that these compounds are epimeric at carbons 2 and 4 to the ones in part (a):

24.23   As is the case with carbohydrates, the primary alcohol groups are selectively oxidized by HNO_3.

(a)

trans-3-hydroxycyclopentane-1-carboxylic acid

(b)

$$HO_2C\text{---}CH_2CH_2\text{---}CO_2H$$

succinic acid

24.24   As Eq. 24.53a (text Sec. 24.8C) shows, periodate oxidation of a methyl pyranoside in the D-series gives a product containing carbons 1, 2, 4, 5, and 6 of the pyranoside. The secondary alcohols at carbons 2 and 4 are converted into aldehyde groups in all cases. In the oxidation products, only the carbons corresponding to carbons 1 and 5 in the pyranoside starting material are asymmetric. Because the configuration of carbon-5 determines whether the pyranoside has the D or the L configuration, and the configuration of carbon-1 determines whether the pyranoside is α or β, it follows that all methyl α-D-pyranosides give the same oxidation product with periodate.

fragment obtained from
periodate oxidation of any
methyl *α*-D-pyranoside

fragment obtained from
periodate oxidation of any
methyl *β*-D-pyranoside

(See the solution to Problem 24.66 in this manual for a more detailed consideration of the structures of the periodate oxidation products of carbohydrates.)

24.25   The discussion in the solution to Focused Problem 24.24 shows that the fragment obtained from oxidation of a β-D-pyranoside is the *diastereomer* of the fragment obtained from oxidation of an α-D-pyranoside. Diastereomers have different properties. Therefore, these fragments, which can in principle be identified by their properties, including their optical rotations, can be used to determine whether the original pyranoside was α or β. This method works not only for the methyl galactopyranosides, but for *all* pyranosides. (Periodate oxidation was used to determine the anomeric configurations of many pyranosides before modern spectroscopic tools were available, and these results have withstood the scrutiny of modern analytical methods.)

24.26   The aldopentose *A* is D-ribose, and the aldohexose *B* is D-allose. The other hexose formed in the Kiliani–Fischer synthesis is D-altrose. (See Fig. 24.3, text Sec. 24.3A, for the structures of these aldoses. Note also in this figure the "Kiliani–Fischer relationships" among the aldoses.) Although D-xylose would also be oxidized to an optically inactive aldaric acid, its Kiliani–Fischer products, D-gulose and D-idose, would *both* be oxidized to optically active aldaric acids.

24.27   The aldopentose *A* is D-lyxose, and the aldohexose *B* is D-galactose. The other hexose *C* formed in the Kiliani–Fischer synthesis is D-talose. (See Fig. 24.3, text Sec. 24.3A, for the structures of these aldoses.) Although D-arabinose would also be oxidized to an optically active aldaric acid, its Kiliani–Fischer products, D-glucose and D-mannose, would both be oxidized to optically active aldaric acids.

24.28   Compound *A* is D-gulose, because it gives the same aldaric acid as L-glucose. The fact that some aldaric acids can be derived from two different aldoses is also addressed in the solution to Focused Problem 24.21 in this manual. The same point is discussed in Study Guide Link 24.2. In contrast, D-idose is the only aldohexose that can be oxidized to its aldaric acid.

24.29 Oxidize both (with dilute nitric acid) to their respective aldaric acids (which in this case are stereoisomeric tartaric acids). The compound that gives the optically active tartaric acid is D-threose; the compound that gives *meso*-tartaric acid (which is achiral and therefore optically inactive) is D-erythrose.

24.30 In such a case, the structure of D-(+)-glucose, indeed, the structures of *all* D-carbohydrates, would be the enantiomers of the structures shown in Fig. 24.3, text Sec. 24.3A.

24.31 (a) Aqueous HCl (1 *M*) would bring about the hydrolysis of lactobionic acid into one equivalent each of D-galactose and D-gluconic acid, which, under the acidic conditions, exists primarily as D-γ-gluconolactone (structure in Eq. 24.49, text Sec. 24.8A).

(b) Dimethyl sulfate in the presence of NaOH methylates all of the hydroxy groups.

(c) Hydrolysis of all acetal and hemiacetal groups (but not the other ethers) takes place.

24.32 Because cellobiose is a dimer of D-glucose, it is hydrolyzed by aqueous acid into two molar equivalents of D-glucose.

24.33 The products would be a small amount of 2,3,4,6-tetra-*O*-methyl-D-glucopyranose (*A*, from the residue at the nonreducing end of the polymer) and mostly 2,3,6-tri-*O*-methyl-D-glucopyranose (*B*):

Solutions to Skills Objectives Problems

24.34 (a)   (b)   (c)   (d)

24.35 (a) (b) (c) (d)

24.36 (a) (b) (c) (d)

24.37 (a) (b) (c) (d)

D-(+)-glyceraldehyde

D-(−)-threose

D-(−)-arabinose

D-(+)-mannose

24.38 "Turning over" a Fischer projection effectively switches two groups on each asymmetric carbon. You can see this by converting each of the structures given in Eq. 24.8 into line-and-wedge structures, and comparing the first structure to the "turned over" structure. The second structure is a noncongruent mirror image of the first; therefore, turning a Fisher projection over converts it into something else, specifically, its enantiomer.

24.39 In each, the anomeric carbon is carbon-1, indicated with an arrow. The carbons that are epimeric relative to glucose in each structure are indicated with a (*).

β-D-glucopyranose

β-D-galactopyranose

β-D-allopyranose

β-D-altropyranose

β-D-mannopyranose

β-D-gulopyranose   β-D-idopyranose   β-D-talopyranose

24.40

β-D-glucopyranose   β-D-galactopyranose

β-D-allopyranose   β-D-altropyranose   β-D-mannopyranose

β-D-gulopyranose   β-D-idopyranose   β-D-talopyranose

24.41

β-D-ribofuranose   β-D-arabinofuranose   β-D-xylofuranose   β-D-lyxofuranose

24.42

β-D-ribofuranose   β-D-arabinofuranose   β-D-xylofuranose   β-D-lyxofuranose

24.43

(protonated form of
β-D-galactose)

+ H—O⁺H₂

180° internal rotation

+ OH₂

(loss of a proton to
water gives the α-anomer)

24.44

(α-anomer)

+ H—OH

180° internal rotation

+ ⁻OH

(β-anomer)

24.45 Start with one of the anomers of glucose. We've started at the protonated form of the β-anomer. Opening the hemiacetal at the anomeric carbon reveals the hydroxy group at C5. Do an internal rotation around the C3–C4 to put the C4 hydroxy group in position to close a new hemiacetal. Deprotonation gives β-D-glucofuranose.

120° internal rotation

+ H₃O⁺

β-D-glucofuranose

The analogous mechanism, starting instead with the α-anomer, yields α-D-glucofuranose:

α-D-glucofuranose

24.46 (a) D-Gulose would be transformed into the aldohexose that is epimeric at carbon-2, namely, D-idose. Also formed would be the ketose D-sorbose:

D-idose

D-sorbose

(b) D-Ribose would be transformed by base into a mixture of D-arabinose and the ketose D-ribulose:

D-arabinose

D-ribulose

24.47 Note that carboxylic acids derived from carbohydrates usually exist as lactones, and aldoses exist primarily as cyclic hemiacetals. For simplicity they are represented below in their noncyclic forms.

(a)

(b)

(c)

(d)

(e)   (This is a Ruff degradation)   (f)

24.48   Note that carboxylic acids derived from carbohydrates usually exist as lactones, and aldoses exist primarily as cyclic hemiacetals. For simplicity they are represented below in their noncyclic forms.

(a)   (b)   (c)

(d)

(e)

24.49   The structure of 3-*O*-β-D-glucopyranosyl-α-D-arabinofuranose:

3-*O*-β-D-glucopyranosyl-α-D-arabinofuranose

Solutions to Integrated Problems

24.50 (a)

α-ᴅ-talopyranose

(b)

propyl β-ʟ-arabinopyranoside

Remember that there are several valid ways to draw the chair conformations of pyranoses. Because the text probably has made you accustomed to seeing β-anomers with equatorial groups at carbon-1, you may be thinking that the chair conformation in the solution to part (b) shows the α-anomer. However, this *is* a correct representation of the β-anomer. Whether the β-anomer has an equatorial or axial group at the anomeric carbon depends on which chair conformation you choose to draw. The following equally valid representations of a different chair conformation of the same compound do have the group at the anomeric carbon in an equatorial position.

the other conformation of **propyl β-ʟ-arabinopyranoside**, from different perspectives

24.51 (a) Turning this structure over 180° shows that it is the noncongruent mirror image of β-ᴅ-galactopyranose; therefore it is β-ʟ-galactopyranose.

β-ʟ-galactopyranose   β-ᴅ-galactopyranose

enantiomers

(b) This pyranose is epimeric to β-ᴅ-glucopyranose at carbon-3. Consult Fig. 24.3, text Sec. 24.3A, to discover that the aldohexose epimeric to glucose at carbon-3 is allose. Therefore, this compound is β-ᴅ-allo-pyranose.

(c) First, notice that carbon-5 of this furanose has the *R* configuration. Because the ᴅ-hexoses have the *R* configuration at carbon-5 (the configurational carbon), and because the hemiacetal carbon bears a hydrogen, this must be a ᴅ-aldohexose. This can be converted into a Fischer projection as follows:

This shows that this compound is a form of D-glucose; it is β-D-glucofuranose.

24.52 (a) The D-series of the stereoisomeric 2-ketohexoses are shown below. (There is, of course, an enantiomeric L-series.) The problem didn't ask for the names, but they are given as part of our policy of extra value at no added cost.

D-psicose   **D-fructose**   **D-sorbose**   **D-tagatose**

(b) An achiral ketopentose must be a meso compound:

(c) The structure of α-D-galactofuranose is derived from the Fischer projection as follows:

(d) Derivation of the structure of β-D-idofuranose from the Fischer projection:

β-D-idofuranose
(Fischer projection)

β-D-idofuranose
(Haworth projection)

This is identical to the structure in the solution to Focused Problem 24.12b except for the configuration at the anomeric carbon.

24.53 (a) Diastereomers, epimers, and anomers. Epimers and anomers are special types of diastereomers.
   (b) Diastereomers and epimers
   (c) Enantiomers
   (d) Constitutional isomers
   (e) Constitutional isomers
   (f) The two names describe the same structure.

methyl α-D-fructofuranoside
(2-*O*-methyl-α-D-fructofuranose)

24.54 (a) L-Sorbose is a hexose.
   (b) L-Sorbose is a ketohexose, or hexulose.
   (c) L-Sorbose is not a glycoside because it is not involved in acetal formation with an alcohol or a phenol.
   (d) L-Sorbose is not an aldohexose; it is a ketohexose, as noted in the solution to part (b).
   (e) This is another valid Fischer projection of L-sorbose obtained by a cyclic permutation at carbon-5.
   (f) This compound is epimeric to L-sorbose at carbon-5, and thus it is not a form of L-sorbose. In fact, it is D-fructose.
   (g) This is the α-furanose form of L-sorbose (α-L-sorbofuranose). The simplest way to see this is to work from the Haworth projection given in the problem to the Fischer projection:

establishes that
this is an α-furanose

establishes that
this is the L configuration

L-sorbose

   (h) This structure is the enantiomer of the structure in part (g), and thus it is not a form of L-sorbose. (It is the α-furanose form of D-sorbose.)

(i)   The relationship of this furanose to the one shown in part (g) is best discerned by turning either of the structures over 180°. This operation reveals that this structure differs in configuration from the one in part (g) at carbons 3 and 4. Therefore, this is not a form of L-sorbose.

(j)   This is an aldopyranose and therefore cannot be a form of L-sorbose. (It is β-D-glucopyranose.)

(k)   This is the α-pyranose form of L-sorbose, α-L-sorbopyranose. As with part (g), the easiest way to see this is by working from the chair to the Haworth to the Fischer.

establishes that
this is an α-pyranose

L-sorbose

24.55 (a)   Raffinose is a nonreducing sugar because it has no hemiacetal groups; the saccharide residues at both ends are involved in acetal linkages.

(b)   The glycoside linkages in raffinose:

α(1,6) glycoside linkage

α1 to β2 glycoside linkage

The type of linkage in the furanose residue can be determined systematically by using the Haworth projection to form the standard Fischer projection. (The glycosidic linkage is indicated as RO—.)

(c)   Raffinose is hydrolyzed into equal amounts of D-galactose, D-glucose, and D-fructose, and each is a mixture of anomers.

(d)   The products of methylation followed by hydrolysis are the following; each is a mixture of anomers.

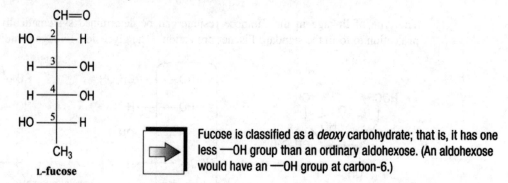

2,3,4,6-tetra-*O*-methyl-
D-galactopyranose

2,3,4-tri-*O*-methyl-
D-glucopyranose

1,3,4,6-tetra-*O*-methyl-
D-fructofuranose

24.56   (a)   Create the Haworth projection, and then rotate it 180° about an axis perpendicular to the page to get another Haworth projection in a more familiar perspective. Then convert this to a Fischer projection as shown. These manipulations reveal that, when we draw the Fischer projection in standard form (carbon-6 vertical), the oxygen at carbon-5 is on the left. Therefore, this enantiomer of fucose has the L configuration.

oxygen on left at C-5;
therefore L configuration

(b)   Because the —OH at carbon-1 in this standard representation is on the same side as the oxygen at carbon-5, this is by definition the α-anomer.

(c)   Fucose is an aldose, because the hemiacetal carbon bears a hydrogen; thus, in the carbonyl form, there is also a hydrogen at carbon-1. Therefore, the carbonyl form must be an aldehyde.

(d)   A Fischer projection of the aldehyde form of L-fucose:

L-fucose

Fucose is classified as a *deoxy* carbohydrate; that is, it has one less —OH group than an ordinary aldohexose. (An aldohexose would have an —OH group at carbon-6.)

24.57   (a)   Because the osazones from the two carbohydrates are the same, and given that both are aldohexoses, the only stereochemical difference between them—a difference that is obliterated by osazone formation—is at carbon-2. Consequently, the two carbohydrates must be epimeric at carbon-2.

(b) D-Ribose gives the same osazone as D-arabinose because D-ribose is epimeric to D-arabinose at carbon-2.

24.58 (a) Just as the acetal units in carbohydrates undergo *hydrolysis* in water, they also undergo *methanolysis* in methanol. The result is a mixture of the methyl glycosides along with the by-product phenol.

phenyl β-D-glucopyranoside   methyl D-glucopyranoside
                                              (mixture α and β)

(b) Any 3- or 4-hydroxy aldehyde, like a carbohydrate, will form a cyclic acetal in acidic methanol.

(c) Osmium tetroxide forms *cis*-1,2-cyclohexanediol; periodic acid cleaves this diol to form the dialdehyde 1,6-hexanedial, $O{=}CHCH_2CH_2CH_2CH_2CH{=}O$; and this dialdehyde is reduced by sodium borohydride to 1,6-hexanediol, $HOCH_2CH_2CH_2CH_2CH_2CH_2OH$.

(d) Cyclohexene reacts with OsO_4 to give *cis*-1,2-cyclohexanediol, which forms a cyclic acetal with acetone.

(e) All of the free hydroxy groups in sucrose are methylated. (For the structure of (+)-sucrose see Display 24.72 in text Sec. 24.11A.)

(f) Ethanolysis of the acetal linkage occurs in ethanol just as hydrolysis occurs in water. An ethyl pyranoside is also formed from the glucose residue at the reducing end of lactose.

(g) This reaction is not unlike the one in part (a), in which one acetal is converted into another in methanol. Acid-catalyzed loss of methanol is followed by cyclization to give a mixture of the two methyl pyranosides.

(Note that $CH_3\overset{+}{O}H_2$ Cl^-, the conjugate acid that results from protonation of the solvent by HCl, is the effective acid catalyst.)

24.59 (a) This is an isotopically labeled analog of glucose, and is prepared by the Kiliani–Fischer synthesis shown in Eqs. 24.56 and 24.57 in text Sec. 24.9, except that radioactive sodium cyanide ($Na^{14}CN$) is used instead of ordinary NaCN. In this synthesis, ^{14}C-D-mannose will be a by-product.

(b) This compound is prepared in the same way as the compound in (a), except that ordinary (unlabeled) NaCN is used, and the reduction is carried out with tritium-enriched hydrogen ($^3H_2$). In this synthesis, ^3H-D-mannose will be isolated as a by-product.

(c) The synthesis of compound (c) requires removal of an *unlabeled* carbon from D-arabinose, replacing it with a labeled carbon, and then completion of an ordinary Kiliani–Fischer synthesis to give labeled mannose (* = ^{14}C).

The first step is a Ruff degradation (see Display 24.65, text Sec. 24.10B). Each of the last two reactions also gives epimeric sugars—D-ribose and D-glucose, respectively—as by-products.

24.60 The results of the oxidation of *A* indicate that compound *A* is not a methyl pyranoside or furanoside, that is, the methyl ether is not at carbon-1. The Ruff degradation yields another reducing sugar. This means that the methyl ether cannot be at carbon-2. Only if the ether is at carbon-3 can one of the Kiliani–Fischer products be oxidized to an optically inactive (that is, a meso) compound:

B (left structure, from Br₂/H₂O):

CO₂H
HO — OH
CH₃O — H
H — OH
H — OH
CH₂OH

B

Br₂
H₂O (arrow pointing left)

A (center structure):

CH=O
H — OH
CH₃O — H
H — OH
H — OH
CH₂OH

A

Kiliani–Fischer synthesis (arrow pointing right)

(two product structures):

CH=O
H — OH
H — OH
CH₃O — H
H — OH
H — OH
CH₂OH

+

CH=O
HO — H
H — OH
CH₃O — H
H — OH
H — OH
CH₂OH

HNO₃ oxidation
gives a meso compound

24.61 (a)

CH=O
H — OH
HO — H
HO — H
H — OH
CH₂OH

D-(+)-galactose

+ (Ph₃P)₃RhCl ⟶

CH₂OH
HO — H
HO — H
H — OH
CH₂OH

D-lyxitol

+ (Ph₃P)₂Rh(CO) + Ph₃P

(b) D-Galactose, an aldohexose, exists predominantly in the pyranose form. The decarbonylation reaction only occurs when D-galactose is in the aldohexose form; the increased temperature (130 °C) is required to increase the rate at which the reactions in the pyranose ⇌ aldohexose equilibrium occur. This strategy ensures that, once the small amount of the aldehyde form reacts with the organometallic reagent, the aldehyde form in the equilibrium is rapidly replenished.

CH=O
H — OH
HO — H
HO — H
H — OH
CH₂OH

D-galactose
(reacts with the
organometallic reagent)

⇌

β-D-galactopyranose
(does not react)

(c) In the following mechanism, L = Ph₃P (the triphenylphosphino ligand).

rhodium (I)
a 16e⁻ complex

ligand
dissociation
⟶

rhodium (I)
a 14e⁻ complex

+ L

rhodium (III)
a 16e⁻ complex

rhodium (I)
a 16e⁻ complex

By using special ligands and by altering the reaction conditions, investigators have been able to turn this into a catalytic process rather than one that is stoichiometric in rhodium.

24.62 Because the product of the decarbonylation reaction is optically inactive, compound B must be *meso*-(2R,3S)-butane-1,2,3,4-tetraol (erythritol) in which all of the —OH groups are on the right side in the Fisher projection. Thus, the aldopentose A is either D-ribose or D-arabinose (see Fig. 24.3, text Sec. 24.3A, p. 1336).

The products from the Kiliani–Fisher synthesis would then be either D-allose and D-altrose or D-glucose and D-mannose (see Fig. 24.3, text Sec. 24.3A). Because both of the products from the nitric acid oxidation are optically active, that means that D-allose could not be a product from the Kiliani–Fisher synthesis because it would produce an optically inactive aldaric acid. Thus, A is D-arabinose, C and D are D-glucose and D-mannose, and E and F are D-glucaric acid and D-mannaric acid. (See Eqs. 24.56 and 24.57 in text Sec. 24.9, for the Kiliani–Fisher synthesis using D-arabinose as the starting aldose.)

B
erythritol
(a meso compound)
optically inactive

A
D-arabinose

B
erythritol

C
D-glucose

D
D-mannose

E
D-glucaric acid

F
D-mannaric acid

24.63 The extract of mouse spleen evidently contains enzymes that catalyze the removal of a phosphate as well as the conversion of ribose into an optically inactive isomer. Ribitol and xylitol, formed by $NaBH_4$ reduction of X, are the alditols derived from ribose and xylose, respectively, and therefore are epimeric at carbon-3. Compound X has one degree of unsaturation, and, because reduction gives alcohols that are epimeric at carbon-3, compound X is a 3-ketopentose. The structure for X shown in the following transformation fits the data:

H_2O + ribose-5-phosphate —extract of mouse spleen→ compound X + $HOPO_3^{2-}$ —periodic acid→ (products) + 2 $H_2C=O$

ribose-5-phosphate   **compound X**

24.64 Compound A is the corresponding oxime (see Table 19.3 in text Sec. 19.12A), and compound B results from acetylation of the hydroxy groups of the oxime. Compound C results from a sodium acetate-promoted β-elimination reaction. The conversion of compound C to D-arabinose is the result of a transesterification reaction, which removes all acetyl groups as methyl acetate, followed by the reverse of cyanohydrin formation.

Identity of compounds A and B:

A B

The mechanism for the conversion of compound B into compound C:

B → AcO—H + C

The mechanism for the conversion of compound C into D-arabinose:

C → $CH_3O-C-CH_3$ (**methyl acetate**) + → $H-C=O$ + $^-C\equiv N$

D-arabinose

Note that the other acetate groups are removed in transesterification reactions, the mechanisms of which are identical to the first two steps of the foregoing mechanism.

24.65   Compound *A* is the lactone of the corresponding aldonic acid, and compound *B* is the amide that results from ester (lactone) ammonolysis. Chlorine and NaOH bring about a Hofmann rearrangement of amide *B* to give *C*, a carbinolamine, which, under the aqueous reaction conditions, breaks down spontaneously to ammonia and an aldehyde. This aldehyde is the aldose that has one fewer carbon atom than the starting aldose. (This is illustrated below with D-glucose, but any aldose whose aldonic acid can form a lactone would undergo the same chemistry.)

*(Reaction scheme: D-glucose Fischer projection → 1) Br_2/H_2O 2) H_3O^+ → lactone **A** → NH_3 → amide **B** → Cl_2, NaOH →)*

A

B

*(Carbinolamine **C** Fischer projection → aldose Fischer projection + NH_3)*

C
(a carbinolamine)

24.66   The product *X* might be reasonably expected to result from treatment of methyl α-L-rhamnopyranoside with periodic acid. (See the solution to Focused Problem 24.24 in this chapter.)

*(Reaction scheme: methyl α-L-rhamnopyranoside → H_5IO_6 → **X** + HCO_2H **formic acid**)*

methyl α-L-rhamnopyranoside   *X*

The formula of *X*, $C_6H_{10}O_4$, is short of the formula of actual product *A* by the elements of H_2O. Nevertheless, reduction of this product by $NaBH_4$ would give compound *C*. Evidently, compound *A* has two —OH groups, because methylation introduces two carbon atoms. Compound *A* results, first, from hydration of one of the carbonyl groups of *X*. Recall that periodic acid (Sec.12.6B) contains water (it can be written as H_5IO_6 or $HIO_4 \cdot 2H_2O$), so water is available to generate the hydrate of the dialdehyde *X*. An —OH group of the hydrate then reacts with the other carbonyl carbon to give a "double hemiacetal" *A*. The structure of *A* and its methylation product *B* are as follows:

H_2O + X ⇌

a hydrate of X

(Compounds A and B are each undoubtedly a mixture of stereoisomers.) Because compound A is a hemiacetal, it shows no carbonyl absorption. However, it is in equilibrium with a small amount of the dialdehyde X, which is reduced by $NaBH_4$ to give compound C. The reduction pulls the equilibrium between A and X toward X until all of this compound has been reduced.

24.67 The residue at the nonreducing ("left") end undergoes the following transformation:

glycerol   hydroxy-
             acetaldehyde

The internal residues undergo the following transformation:

erythritol   hydroxy-
               acetaldehyde

The residue at the reducing ("right") end gives erythritol, R′OH, and two equivalents of formic acid (HCO_2H). (Demonstrate this for yourself; see Eq. 24.52, text Sec. 24.8C.) Erythritol therefore results from the cleavage reactions of all residues except for the one at the nonreducing end; and glycerol results from cleavage of that residue. Thus, glycerol is produced from one residue, and erythritol is produced from $n - 2 + 1 = n - 1$ residues. Because $n = 12$, eleven moles of erythritol per mole of glycerol are formed.

24.68 The hydrolysis of maltose by α-amylase shows that maltose contains two glucose residues connected by an α-glycosidic bond. Because maltose is a reducing sugar, a free hemiacetal group must be at carbon-1 of one of the glucose units. The 2,3,4,6-tetra-O-methyl-D-glucose must arise from the glucose residue at the nonreducing end. The question is which oxygen in the glucose residue at the reducing end is involved in the glycosidic linkage; the other product of methylation-hydrolysis provides evidence on this point. Because the oxygens at carbons 4 and 5 are not methylated in this product, one of these oxygens is involved in pyranoside or furanoside ring formation, and the other is involved in the glycosidic linkage. Two structures satisfy these requirements:

A and *B*

Methylation of maltobionic acid occurs under basic conditions; under these conditions any lactone present would be saponified. Therefore, the additional oxygen that is methylated under these conditions must be the oxygen that was *within* the glycoside ring at the reducing end of maltose. In other words, oxidation and saponification expose this oxygen as an —OH group and thus make it available for methylation. Because the oxygen at carbon-5 is methylated, it must have been the oxygen within the saccharide ring. Since the oxygen at carbon-4 is *not* methylated, it must have been the oxygen involved in the glycosidic bond. Therefore, structure *A* is the correct one for maltose. To summarize the chemistry using the correct structure:

maltobionic acid (lactone) $(CH_3)_2SO_4$, NaOH H_3O^+

24.69   The hydrolysis of planteose by almond emulsin establishes that an intact sucrose unit is connected to a galactose residue by an α-galactosyl glycosidic bond. The question, then, is which oxygen of the sucrose residue is involved in the glycosidic linkage to the galactose. Methylation followed by hydrolysis provides the answer. (It will be helpful to refer to the structure of sucrose; see Display 24.72 in text Sec. 24.11A.) If the fructose residue of sucrose were *not* connected to the galactose, it would have four oxygens available for methylation: the oxygens at carbons 1, 3, 4, and 6. Because the oxygen at carbon-6 is not methylated, it must be the point of attachment of the galactose residue. Therefore, the structure of planteose is as follows:

planteose

24.70 (a) The hydroxy groups of cellulose react with the anhydride to introduce succinate half-esters into the cellulose structure, as follows. (In these equations. HO—cellulose means any of the hydroxy groups of cellulose.)

$pK_a \approx 4\text{--}5$

(b) Sizing introduces a large number of alkyl groups into the cellulose structure. These resist solvation by water for the same reasons that hydrocarbons are insoluble in water; they are "hydrophobic" groups. Thus, these groups tend to resist the incorporation of solvent water into the cellulose structure. Furthermore, many of the —OH groups in the cellulose, which are hydrogen-bond donors before the reaction, become ester oxygens as a result of the reaction and therefore can no longer donate hydrogen bonds.

(c) As the reaction in the solution to part (a) shows, the carboxylic acid groups introduced by the reaction are ionized by the pH 7 treatment because their pK_a values are well below the pH of the solution. When the paper is dried, these groups are left in their ionized state (with an appropriate counter-ion). The carboxylate groups are the slightly basic groups.

   For example, when they come into contact with water, they react with water to produce a small amount of hydroxide ion:

You can use the pK_a value for the carboxylic acid and the ion-product constant of water to calculate that the pH of an aqueous solution of a carboxylic acid salt should have a pH of around 9. Therefore, when the "sized" paper comes into contact with acid, the acid is neutralized by the small amount of hydroxide ion present. So many carboxylate groups are present that the hydroxide ion is immediately replenished, by Le Chatelier's principle.

24.71 The reaction of an amine with an aldose gives a carbinolamine, which dehydrates to an iminium ion A. This ion is trapped by the hydroxy group at carbon-5 to give an aminoglycoside, which is shown as its conjugate acid B. (The p-methylphenyl group of the amine is abbreviated R.)

This reaction is called the *Maillard reaction*. From here on, only the part of D-glucose that is involved in the reaction is shown explicitly. The iminium ion *A* (which, by resonance, is also a carbocation) isomerizes to an enamine *C*, which is also an enol. Acid-catalyzed formation of the isomeric ketone completes the reaction. This isomerization is called the *Amadori rearrangement*.

This reaction is similar to the Lobry de Bruyn–Alberda van Ekenstein rearrangement (text pp. 1251–2), except that imine and enamine intermediates are involved rather than aldehyde and enol intermediates. The driving force for the rearrangement is that the C=N double bond is lost and a stronger C=O double bond is formed.

24.72 This transformation is essentially a Lobry de Bruyn–Alberda van Ekenstein reaction, the detailed mechanism of which is shown in Eqs. 24.33–24.34b, text Sec. 24.5. The aldehyde form of D-glucosamine is in equilibrium with a small amount of its enol *A*. This enol is also an enamine, which is in equilibrium with an imine *B*. Because imine formation is reversible in aqueous acid and base, the imine hydrolyzes to give the corresponding ketose, D-fructose, along with the by-product ammonia. Under the basic conditions, D-fructose is also in equilibrium with D-glucose and D-mannose, as shown by Eq. 24.32, text Sec. 24.5.

24.73 (a) The acidic hydrogen is the one that ionizes to give a conjugate-base anion in which the negative charge is delocalized into the carbonyl group. This anion is stabilized by the electron-withdrawing polar effects of nearby oxygens and by the same type of resonance interactions that are present in a carboxylate ion. Consequently, ascorbic acid is about as acidic as a carboxylic acid.

(b) The reaction sequence with the missing structures added begins with compound *A*. The fact that carbon-2 of L-sorbitol (or carbon-5 of D-glucitol) becomes the acetal carbon as a result of the reaction labeled *C* in the problem shows that the —OH group on this carbon is the one that is oxidized; this carbon is therefore at the ketone level of oxidation. $KMnO_4$ oxidizes the —CH_2OH group to a carboxylate; because the other —OH groups are protected as acetals, they are stable under the basic conditions of permanganate oxidation. Hydrolysis of the oxidation product in acid removes the acetal protecting groups, protonates the carboxylate group, and promotes lactone formation.

Can you think of one or more reasons why ascorbic acid should exist as an enediol rather than as an α-keto lactone? (See Sec. 22.2, and the discussion of carbonyl dipoles at the end of text Sec. 18.8A.)

24.74 (a) The anhydro form of D-idose has the following chair conformation:

1,6-anhydro-D-idopyranose

(b) The issue is the relative stabilities of the anhydro and ordinary pyranose forms of D-idose and D-glucose. In the chair conformation of 1,6-anhydro-D-glucopyranose, all of the hydroxy groups are axial. The many resulting 1,3-diaxial interactions destabilize this form of glucose. In the more stable chair form of D-gluco-pyranose, in contrast, all of the hydroxy groups are equatorial. Thus, the pyranose form of D-glucose is much more stable than the 1,6-anhydro form, and D-glucose exists primarily as an ordinary pyranose. In 1,6-anhydro-D-idopyranose, all of the hydroxy groups are equatorial. In the more stable of the chair conformations of the ordinary pyranose form of D-idose, the —CH$_2$OH group is forced into an axial position. Formation of the ether bond between carbons 1 and 6 relieves one set of 1,3-diaxial interactions in this form of D-idose. Thus, the 1,6-anhydro form of D-idose is more stable than the pyranose form, and therefore D-idose exists mostly in the 1,6-anhydro form.

β-D-glucopyranose ⟷ (structure) $\xrightarrow{-H_2O}$ 1,6-anhydro-D-glucopyranose

1,3-diaxial interaction

β-D-idopyranose $\xrightarrow{-H_2O}$ 1,6-anhydro-D-idopyranose

24.75 (a) The resonance at δ 5.24 is from the α-anomer, and the resonance at δ 4.66 is from the β-anomer. We know this from the coupling constants. The Karplus curve (Fig. 14.9, Sec. 14.5A, text p. 754) indicates that the protons that have a dihedral angle of 180° (the C1 and C2 protons of the β-anomer) should have the larger coupling constant.

(b) The integral of the β-anomer is 66.3% of the total, and the integral of the α-anomer is 34.7% of the total. These are the percentages of the two anomers. (See Table 24.1, text Sec. 24.4; compare this with the percentages of the two pyranose anomers of glucose.)

The Chemistry of Thioesters, Phosphate Esters, and Phosphate Anhydrides

CHAPTER 25 SUMMARY

- Thioesters are conceptually similar to esters. However, thioesters have much less resonance stabilization than esters because of the relatively poor orbital overlap between sulfur 3p orbitals and the 2p orbitals of the carbonyl group. The weaker resonance stabilization of thioesters accounts for their status as "high-energy" compounds.

- In spite of their weaker resonance stabilization, thioesters have about the same reactivity as esters in hydrolysis reactions because the sulfur, being much less electronegative than oxygen, does not stabilize the transition state for hydrolysis as well as oxygen.

- Acetyl-CoA is the most important and widely occurring biological thioester. Thioesters formed between acyl groups and the thiol groups of cysteine residues of proteins are also important intermediates in biology.

- Phosphoric acid derivatives have tetrahedral geometry. One, two, or all three of the OH groups can be derivatized, so phosphoric acid can have monoester, diester, and triester derivatives. Monoester and diester derivatives have acidic protons.

- The NMR spectra of phosphoric acid derivatives are similar to those of carboxylic acid derivatives except that ^{31}P (nuclear spin = $\pm\frac{1}{2}$) can cause splitting of nearby ^{13}C and ^{1}H nuclei. Phosphorus NMR can be used to detect ^{31}P.

- Pyrophosphate monoesters in biology typically react by S_N1 or S_N2 substitution at the carbon of the ester, and the resulting pyrophosphate leaving group is hydrolyzed to two molecules of phosphate in a reaction catalyzed by the enzyme pyrophosphatase. The biosynthesis reactions of di- and polysaccharides involve conceptually similar nucleophilic substitutions by sugar nucleophiles at an anomeric carbon in which a nucleotide diphosphate such as UDP acts as a leaving group.

- Phosphate triesters have roughly the same reactivity in hydrolysis as carboxylic acid esters. Phosphate diesters are much less reactive in base-promoted hydrolysis because of repulsion between negative charges in the transition state. Phosphate monoester di-anions are even less reactive; they react very slowly by an acid-catalyzed dissociative mechanism to give a very unstable metaphosphate intermediate, which rapidly reacts with nucleophiles.

- The catalysis of hydrolysis reactions of phosphate diesters and monoesters by enzymes typically involves the following features:

  - Reaction of the nucleophile at phosphorus, and cleavage of the bond between phosphorus and the leaving group with inversion of configuration at phosphorus (called "in-line" substitution).

  - Catalysis by proximity: all of the reactive groups are bound in an ideal or nearly ideal arrangement for reaction in the enzyme active site.

  - Base catalysis: a proton is removed from a nucleophilic water or alcohol molecule by a basic group of the enzyme.

  - Acid catalysis: the leaving-group basicity is reduced by partial protonation by an acidic group of the enzyme or by interaction with a metal ion (typically Mg^{2+}).

  - Electrophilic catalysis: the negative charge in the transition state for hydrolysis is neutralized by divalent metal ions, by hydrogen-bond donation, and/or by positively charged groups in the enzyme active site.

- Phosphate monoester hydrolysis is catalyzed by phosphatases. Some of these reactions occur with inversion of stereochemical configuration at phosphorus, but others involve net retention of configuration at phosphorus. This stereochemical outcome is invariably the result of two successive opposite-side displacements involving a phospho-enzyme intermediate.

- Phosphate anhydrides, although much more reactive than phosphate esters, are stable under physiological conditions. Pyrophosphate hydrolysis is catalyzed by inorganic pyrophosphatase. This is an important reaction that is used to drive many biological processes to completion.

- ATP can react with nucleophiles at any of its three phosphorus atoms; the particular reaction observed depends on the enzyme catalyst. Reaction at the γ-phosphorus, catalyzed by kinases, results in phosphorylation of the nucleophile and loss of ADP as a leaving group. Reaction at the β-phosphorus results in loss of AMP and pyrophosphorylation of the nucleophile. Reaction at the α-phosphorus results in adenylation of the nucleophile and loss of pyrophosphate as a leaving group. The mechanistic and stereochemical aspects of these reactions are much like those observed in the enzyme-catalyzed hydrolysis of phosphate esters.

- The $\Delta G°$ most commonly used in biology is $\Delta G°′$, which is the $\Delta G°$ value adjusted to pH = 7 standard state.

- "High-energy" compounds release a substantial amount of free energy (as measured by $\Delta G°′$) when they are hydrolyzed. Examples in laboratory chemistry are carboxylic anhydrides and carboxylic acid chlorides. Examples in biology are ATP and other nucleoside triphosphates, thioesters, and acyl phosphates. The reactions of these compounds with ordinary nucleophiles such as alcohols are typically irreversible.

- Most biochemical reactions are not at equilibrium in living systems. The driving force for biochemical reactions at pH = 7 is $\Delta G′$, the actual free-energy change of the reaction under nonequilibrium conditions (Eq. 25.57, text p. 1418).

 # FURTHER EXPLORATIONS

 ## 25.1 Pentacovalent Intermediates in Phosphate Ester Hydrolysis

The evidence for pentacovalent intermediates in phosphate ester hydrolysis hinges on three key points:

1. Compounds of pentacovalent phosphorus contain two types of bonds, called **apical** (or **axial**) bonds, and **equatorial** bonds. The apical bonds are somewhat longer; they have a bond angle of 180° with each other, and a bond angle of 90° with the equatorial bonds. The equatorial bonds have bond angles of 120° with each other and define a plane that contains the phosphorus at the center.

(The grey triangle indicates the plane of the equatorial bonds.)

2. In pentacovalent compounds derived from phosphate esters in which the phosphorus is part of a five-membered ring, one O—P ring bond must be apical and the other must be axial. The reason is that the preferred O—P—O bond angle is close to 90°.

preferred geometry
O—P—O angle = 90°
(the ring bridges apical and equatorial positions)

strained
O—P—O angle = 120°
(the ring bridges two equatorial positions)

too strained to exist
O—P—O angle = 180°
(the ring bridges two axial positions)

If the ring bridges two equatorial positions, the O—P—O angle is 120° and the ring is strained. A molecule in which the ring bridges two apical positions has impossibly long bond lengths and is too strained to exist.

3. Leaving groups depart from apical positions. Because, in the reverse reaction, the roles of nucleophiles and leaving groups are reversed, nucleophiles must also enter from apical positions.

Thus, when a nucleophile reacts at the P=O bond in a tetrahedral phosphate ester, it reacts with phosphorus along a path that is approximately perpendicular to the P=O bond, and the new bond to the nucleophile is apical in the pentacovalent intermediate. The bond to the leaving group is also apical so that the leaving group departs from the pentacovalent intermediate along a path that is also perpendicular to the P=O bond that is formed. This situation is much the same as in carbonyl chemistry (text Fig. 21.4, p. 1149), and it is the result of the positioning of the antibonding molecular orbitals that serve as the destination for the nucleophile electrons or departure point for the leaving-group electrons.

Now we're ready to examine the evidence for pentacovalent intermediates. When ethylene methyl phosphate reacts with water at pH 5.5, two products are formed corresponding to loss of the two possible leaving groups:

ethylene methyl phosphate

A
(70% of product)

B
(30% of product)

Let's consider the possibility of a pentacovalent intermediate. The first pentacovalent intermediate has to be the following, from points 2 and 3:

P1   *P2*   *P3*

where the three forms of the intermediate *P1*, *P2*, and *P3* differ only in the position of a proton. Because one bond to a ring oxygen is apical, that oxygen can become a leaving group. This accounts for the "70% product."

P3

A
(70% of product)

Although this result is consistent with a pentacovalent intermediate, it could also occur in a concerted manner, with proton transfers occurring before, during, or after the substitution.

The "30% product" *B* is the key to the argument for pentacovalent intermediates. For the OCH_3 group to be the leaving group, it must occupy an apical position in a pentacovalent intermediate. However, one apical position is occupied by the nucleophile and the other by a P—O ring bond. For the OCH_3 group to become apical, the molecule undergoes a very subtle internal reorganization called a **pseudorotation**. In a pseudorotation, two of the equatorial bonds lengthen, their bond angle to phosphorus opens to 180°, and they become apical: these would have to be the bond to the OCH_3 group (because it is to become the leaving group) and the formerly equatorial P—O ring bond; and two of the apical bonds shorten, their bond angles close to 120°, and they become equatorial: these would have to be formerly apical P—O ring bond, and the bond to the nucleophilic oxygen. The P—O⁻ bond (which was a P=O bond in the starting ester) remains equatorial; this bond is termed the *pivot bond*. Pseudorotation on *P2* gives *P4*, which, after proton transfers to form *P4′*, loses apical methanol to give product *B*.

Here is the logic: *there can't be a pseudorotation if there isn't an actual pentacovalent intermediate that lasts long enough for pseudorotation to take place.* Therefore, it seems that hydrolysis of this cyclic ester must involve a pentacovalent intermediate. Presumably the products *A* and *B* arise from two different pentacovalent intermediates (*P2* and *P2′*) that are related by a pseudorotation.

It is interesting that saponification of ethylene methyl phosphate in base ($^-$OH) gives only product *A*. Therefore, there is no direct evidence for pentacovalent intermediates under these conditions.

Pseudorotation itself is a well-documented phenomenon in phosphorus chemistry. For example, at low temperature the apical and equatorial fluorines of PF_5 are separately observable at different chemical shifts in ^{19}F NMR. When the temperature is raised, the resonances for the different fluorines coalesce into a single resonance. This shows that the apical and equatorial fluorines are equivalent over time. Pseudorotation is the mechanism by which this equilibration takes place. In other words, there is no question that pseudorotation can occur. The issue in this section is whether it occurs during substitution reactions of phosphate esters.

REACTION REVIEW

I.   REACTIONS OF THIOESTERS WITH NUCLEOPHILES

   A.   HYDROLYSIS OF THIOESTERS

   1. Thioesters undergo saponification like esters.

   B.   REACTIONS OF THIOESTERS WITH OTHER NUCLEOPHILES

   1. Thioesters, like esters, can react with other nucleophiles in acyl substitution reactions. (Nuc:$^-$ = nucleophile)

2. Thioesters can also be reduced to alcohols and thiols with standard laboratory reducing agents ($NaBH_4$ and $LiAlH_4$) and biological reducing agents like NADPH.

II.   REACTIONS OF PHOSPHATE ESTERS

A.   HYDROLYSIS OF PHOSPHATE ESTERS

1. Phosphate triesters can react with a nucleophile (such as hydroxide) at carbon, cleaving the C—O bond, in a reaction analogous to the S_N2 reaction of sulfonate esters (see Sec. 11.4A).

C—O cleavage occurs when the alkyl group is very reactive in S_N2 reactions: methyl, benzyl, allyl, and so on.

2. Most ordinary phosphate triesters also react with nucleophiles at phosphorous.

a. Two mechanisms are possible, and either is acceptable.

   i. An addition-elimination reaction mechanism, analogous to carboxylate ester saponification, is one mode of hydrolysis. This mechanism proceeds through a pentacovalent *intermediate*.

pentacovalent
intermediate

(ionizes to give
the product)

ii. A concerted reaction mechanism, analogous to an S_N2 reaction, is another mode of hydrolysis. This mechanism proceeds through a pentacovalent *transition state*.

transition state

(ionizes to give
the product)

This mechanism is sometimes called "in-line displacement." (This is analogous to "opposite-side substitution" in the S_N2 mechanism of nucleophilic substitution.)

3. Phosphate diesters and monoesters can also be hydrolyzed, but at a much lower rate than triesters because of the effect of repulsive negative charges (see Eq. 25.33, text p. 1397).

4. Enzymes can accelerate monoester and diester hydrolysis, so enzyme-catalyzed hydrolysis of these esters is common in biology.

III. REACTIONS OF PHOSPHATE ANHYDRIDES

A. HYDROLYSIS AND REACTION AT PHOSPHORUS WITH NUCLEOPHILES

1. Adenosine triphosphate (ATP) is an important phosphate anhydride that can be hydrolyzed or can react with nucleophiles at either the α-, β-, or γ-phosphorus in enzyme-catalyzed reactions (see Sec. 25.7A).

ATP

B. CARBONYL SUBSTITUTION REACTIONS OF ACYL PHOSPHATES WITH NUCLEOPHILES

1. Acyl phosphates, like other carboxylic acid derivatives, react with nucleophiles in carbonyl-substitution reactions.

phosphate

C. REACTION OF ALKYL PYROPHOSPHATES AT CARBON

1. Alkyl pyrophosphates can also react with nucleophiles at carbon. This reaction generates a pyrophosphate leaving group.

The reaction at the top of the page:

$$H_2O + R-CH_2-O-\overset{\displaystyle O}{\underset{\displaystyle O^-}{P}}-O-\overset{\displaystyle O}{\underset{\displaystyle O^-}{P}}-O^- \longrightarrow R-CH_2-Nuc + {}^-O-\overset{\displaystyle O}{\underset{\displaystyle O^-}{P}}-O-\overset{\displaystyle O}{\underset{\displaystyle O^-}{P}}-OH + {}^-OH$$

with Nuc:⁻

2. Depending on the alkyl-group structure, these reactions can proceed by either an S_N2 or S_N1 mechanism. In biology the alkyl groups are typically allylic pyrophosphates, which are particularly reactive in S_N1 and S_N2 reactions.

3. The pyrophosphate leaving group is invariably chelated to divalent metal ions such as Mg^{2+} or Mn^{2+}, which make the pyrophosphate ion a better leaving group by neutralizing much of its charge and thereby reducing its basicity.

4. Pyrophosphate generated in biology is in many cases hydrolyzed further to phosphate under the influence of pyrophosphatases. This hydrolysis reaction is employed to pull biological processes to completion by Le Chatelier's principle.

5. Other reactions in this category include the reactions of nucleoside diphosphate sugars such as UDP-galactose with nucleophiles such as other sugars.

SOLUTIONS TO PROBLEMS

Solutions to Focused Problems

25.1 In part (a), the *S*- prefix is understood if no prefix is given, as discussed in the text.

(a)

cyclohexyl thiobenzoate
(*S*-cyclohexyl thiobenzoate)

(b)

S-isopropyl butanethioate

(c)

S-phenyl cyclohexanecarbothioate

25.2 Allyl butanethioate (common) and *S*-2-propenyl butanethioate (IUPAC).

25.3 To solve this problem, review the discussion on text p. 149, and utilize Eqs. 3.67a–b on text p. 148. Two protons can dissociate from methyl phosphate, so there are three forms: the un-ionized form (charge = 0), the mono-anion form (charge = –1) and the di-anion form (charge = –2). Consider each proton's pK_a, one at a time. The most acidic proton has a pK_a of 2.3. Statement 3 on p. 149 says that when the pK_a is much lower than the pH, than we can treat that proton as effectively being 100% dissociated from the acid. Therefore, there is *essentially none* of the un-ionized, neutral form of methyl phosphate at pH = 7.4; it's either all mono-anion, all di-anion, or a mixture.

Next, consider the second proton, with a $pK_a = 6.7$. Statement 3 tells us that since this pK_a is less than the pH, it's going to be partly dissociated. If the pK_a were 1 unit less than the pH, it would be 90% dissociated, and if the pK_a were equal to the pH, it would be 50% dissociated. The difference, Δ, is equal to 0.7, so the ratio of the –2 form to the –1 form must be somewhere between 50%–90%, but probably closer to 90%. To calculate the amount of the di-anion (the –2 form) exactly, use Eq. 3.67a. Solving for f_A (in this case, f_A = the fraction of di-anion relative to mono-ion) gives 0.83, or 83%. The rest (17%) must be the mono-anion (you can check your work by using Eq. 3.67b and solving for f_{AH}).

So, at this pH, there is approximately 83% of the di-anionic, 17% of the mono-anionic, and none of the neutral form of methyl phosphate.

25.4 The structure of serine phosphate:

25.5 The phosphorus-containing functional groups in acetyl-CoA

25.6 The 120° P—O—C bond angle suggests that the oxygen is sp^2-hybridized. This is the predicted hybridization. (See Display 1.80, Sec. 1.7E, text p. 49.)

25.7 The two enantiomers of this diester are in such rapid equilibrium that the molecule cannot sustain optical activity. The exact mechanism of the equilibration depends on conditions, but in aqueous solution (or a solution in any protic solvent), the acid can ionize; in the ionization product the two oxygens become enantiotopic. Although the two oxygens are still distinguishable by a *chiral* object, such as an enzyme, they are chemically equivalent in ordinary aqueous solution, so the proton can return to either oxygen with equal probability.

25.8 The CH_2 proton resonance would be split into five lines, or a quintet, by the $n + 1$ rule: three hydrogens on the adjacent carbon and one phosphorous on the adjacent oxygen.

25.9 (a) The thioester undergoes aminolysis to give the *N*-ethyl amide:

(b)   Under acidic conditions, the amine is protonated to give the *N*-ethylammonium ion, $H_3\overset{+}{N}Et$. Because this ion has no nonbonding electron pair, it is no longer nucleophilic. Therefore, the reaction in part (a) cannot take place.

25.10   Hydrogen bonding from a protonated amino group (or other proton donor) in the active site of the enzyme activates the aldehyde carbonyl group toward addition. (This is much like the carbonyl activation caused by hydrogen-bond donation to the carbonyl oxygen from protic solvents in sodium borohydride reductions; see Eq. 19.46, Sec. 19.9A, text p. 1051.) Nucleophilic hydride is delivered from NADPH to the carbonyl carbon.

25.11   (a)   Because there are only three isotopes of oxygen, ^{16}O, ^{17}O, and ^{18}O, studying the stereochemistry of the reaction is impossible when the nucleophile is water (in this case, water labeled with ^{17}O). The product would contain two ^{17}O atoms, and with the proton free to associate with any of the oxygens, regardless of the isotope present, the phosphorus would only contain three different attached groups, and would thus be achiral. (See the solution to Focused Problem 25.7.)

Sulfur effectively provides a "fourth isotope" of oxygen if we assume that it doesn't change the overall mechanism of the reaction. With four "oxygens" (S, ^{16}O, ^{17}O, and ^{18}O), and thus four different groups attached to the tetrahedral phosphorous in the product, we can probe the stereochemistry of the reaction mechanism by comparing the *R/S* configuration of the starting material and of the product. This experiment is explored in part (b).

(b)   The mechanism of F16BP is described in Fig 25.5, text p. 1401. Assuming inversion of configuration, the product of the reaction and its stereochemistry is shown below.

25.12 This mechanism involves a double inversion.

25.13 (a) The ionized carboxylate of the glutamic acid residue provides the base (*1*) to deprotonate a water molecule, thus making the water more nucleophilic. The various magnesium ions (*2*) stabilize the negative charges that develop as a result of the reaction, as discussed in Point 2(b) on text p. 1402. The arginine and tyrosine residues provide hydrogen bonds (*3*) to the oxygens of the leaving group, thus making the ion less basic and a better leaving group. This hydrogen bonding also helps to hold the pyrophosphate ion within the enzyme–substrate complex. In total, all features in the active site of the enzyme serve to accelerate the reaction by both minimizing or enhancing charges where needed, and aligning reactants.

(b) From the reaction mechanism shown, inversion at the phosphorous would be observed.

25.14 **(a)**   First, the fatty acid is adenylated by ATP, as shown in Eq. 25.45c on text p. 1407.

ATP

pyrophosphate

Next, the fatty acyl adenylate reacts with the thiol group of coenzyme A, followed by a proton transfer step, to produce a fatty acyl-coenzyme A.

a fatty acyl-CoA

AMP

(The double curved arrows on the carbonyl group are a shorthand for formation and breakdown of a tetrahedral addition intermediate.)

(b)   As described by Eq. 25.44 and the accompanying text on p. 1405, after pyrophosphate is generated in a reaction, inorganic phosphatase catalyzes its hydrolysis to yield two equivalents of phosphate. This reaction is necessary to prevent the reverse of the reaction shown in the second step described in part (a), and to drive the reaction forward.

25.15 **(a)**   The reaction is an enzyme-catalyzed S_N2 reaction in which the sulfur of the amino acid methionine reacts at the primary carbon of the ribose moiety of ATP, displacing the "double anhydride" triphosphate as a leaving group.

ATP

methionine

SAM synthase (enzyme)

S-adenosylmethionine (SAM)

triphosphate

(b)   Since the triphosphate leaving group has many negative charges, the magnesium ions likely neutralize these charges as the S_N2 reaction is taking place, thereby making the triphosphate less basic and a better leaving group..

25.16   (a)   The structure of UDP-glucose:

UDP-glucose

(b)   Examine the structure of sucrose, and imagine breaking the glycosidic bond, forming the individual monosaccharide pieces.

From this analysis, it appears that the fructose moiety of sucrose must approach from the axial orientation relative to the glucose—the same direction just occupied by the leaving UDP. The reaction therefore proceeds with either retention of configuration at the anomeric carbon or by a double displacement. Assuming substitution by an S_N1 mechanism, the carbocation is stabilized by resonance with the adjacent oxygen.

Fructose can then act as a nucleophile as shown below, forming an α-glycosidic bond with glucose. Deprotonation yields sucrose and UDP.

(c)   These positively-charged residues likely stabilize the negatively-charged UDP leaving group. Not only do they hold the UDP within the active site, but they also reduce its charge, thereby making it less basic and making it a better leaving group.

25.17   (a)   The free-energy diagram is shown in **Fig. SG25.1**.

Figure SG25.1   A free-energy diagram that relates the standard free energy $\Delta G^{\circ\prime}$ for acetyl-CoA hydrolysis (on left) and acetylcholine hydrolysis (on right) to the $\Delta G^{\circ\prime}$ for acetylcholine biosynthesis (on right).

(b) The problem gives values for the $\Delta G^{o\prime}$ of hydrolysis of acetylcholine and acetyl-CoA. We want to find $\Delta G^{o\prime}$ for the conversion of choline and acetyl-CoA into acetylcholine and HS–CoA. For the two given reactions to sum to the reaction under inspection, reverse the reaction for hydrolysis of acetylcholine, and change the sign of $\Delta G^{o\prime}$ for that reaction. Canceling the reactants that appear on both sides of the equilibrium gives the reaction for the synthesis of acetylcholine. Summing the $\Delta G^{o\prime}$ values gives –6.3 kJ mol^{-1} for the overall reaction.

$$\text{acetyl CoA} + H_2O \rightleftharpoons CH_3CO_2^- + \text{HSCoA} + H^+ \qquad \Delta G^{o\prime} = -31.4 \text{ kJ mol}^{-1}$$

$$CH_3CO_2^- + \text{choline} + H^+ \rightleftharpoons \text{acetylcholine} + H_2O \qquad \Delta G^{o\prime} = +25.1 \text{ kJ mol}^{-1}$$

$$\text{acetyl CoA} + \text{choline} \rightleftharpoons \text{acetylcholine} + \text{HSCoA} \qquad \Delta G^{o\prime} = -6.3 \text{ kJ mol}^{-1}$$

(c) With a temperature of 37 °C (310 K), use $\Delta G^{o\prime} = -2.3RT \log K_{eq}$, and obtain $K_{eq} = 11.5$.

25.18 (a) We use the following equations. (The $\Delta G^{o\prime}$ for hydrolysis of ATP to ADP and phosphate comes from Eq. 25.56, text p. 1415.)

	$\Delta G^{o\prime}$ kJ mol^{-1}	kcal mol^{-1}
phosphocreatine + ADP ⟶ creatine + ATP	–12.5	–2.99
ATP + H$_2$O ⟶ ADP + phosphate	–30.5	–7.29
phosphocreatine + H$_2$O ⟶ creatine + phosphate	–43.0	–10.3

This large value for hydrolysis of phosphocreatine places it in the "high-energy compound" category.

(b) If we use the important octet structure for phosphorus, we see that phosphocreatine has significant charge repulsion in one of its resonance structures:

two adjacent
positive charges

Hydrolysis to creatine and phosphate gives a resonance-stabilized creatine molecule in which this charge repulsion is absent.

creatine

The loss of this charge repulsion is responsible for the release of energy in this hydrolysis reaction.

25.19 Use Eq. 25.57 on text p. 1418 with ATP, ADP, and phosphate = 10^{-3} M; we use the unit-activity standard state for water and physiological temperature 310 K. This equation becomes

$$\Delta G' = \Delta G^{o\prime} + 2.3RT \log \frac{[\text{ADP}][\text{phosphate}]}{[\text{ATP}]}$$

$$= -30.5 + (5.93) \log (10^{-3})$$

$$= -30.5 + (5.93)(-3.00) = -48.3 \text{ kJ mol}^{-1} \ (-11.5 \text{ kcal mol}^{-1})$$

25.20   There are *two* high-energy phosphate bonds in triphosphate; it is a "double anhydride." The hydrolysis of triphosphate to phosphate and pyrophosphate captures the energy of one high-energy bond, and the conversion of pyrophosphate to phosphate by inorganic pyrophosphatase captures the energy of the second high-energy bond, both being used to drive the synthesis of SAM to completion. If methionine were present in very small concentration, the coupling of two high-energy phosphates makes many reactions very favorable by coupling them to this irreversible reaction. If ADP were used as an adenosyl donor, then we would only capture the energy of one high-energy bond in pyrophosphate. Essentially, the beneficial effects of hydrolyzing one of the high-energy phosphate bonds is lost if the starting material were ADP instead of ATP. We now can see why even small concentrations of methionine can be completely converted into SAM–the high-energy releasing processes of conversion of ATP to SAM and triphosphate, followed by triphosphate to pyrophosphate and phosphate, followed finally by pyrophosphate to phosphate are very favorable processes.

> The cellular concentrations of methionine and SAM are 0.1 mM and 0.08 mM, and the $\Delta G^{\circ\prime}$ for the hypothetical conversion of methionine into SAM with non-phosphorylated adenosine is 34 kJ mol^{-1} (8.2 kcal mol^{-1})—that is, this is a very unfavorable conversion. Therefore, the hydrolysis of the two high-energy bonds is crucial to the biosynthesis of SAM even at the relatively high concentration of methionine present in the cell, let alone at even lower concentrations.

Solutions to Skills Objectives Problems

25.21   (a)   (b)   (c)

isobutyl thiobenzoate

S-ethyl-3-oxohexanethioate

ethyl 3-(ethylthio)hexanoate

25.22   (a)   *S*-Phenyl 4-methoxybenzenecarbothioate, *S*-phenyl 4-methoxythiobenzoate, phenyl *p*-methoxythiobenzoate

(b)   Methyl 4-(methylthio)benzoate. (Notice the use of parentheses to show that "thio" is part of the substituent rather than a thiobenzoate ester.)

25.23   (a)   There is no large difference in C=O bond lengths.

(b)   A C—S bond is considerably longer than a C—O bond.

(c)   The C—S—C bond angle is close to 90°, whereas the C—O—C bond angle is about 120°, as expected for an sp^2-hybridized oxygen.

25.24   For identification of the phosphorus-containing functional groups see **Fig. SG25.2** on the following page.

25.25   (a)   Phosphate triesters hydrolyze at rates similar to oxygen esters. Phosphate diesters hydrolyze more slowly, and phosphate monoesters even more slowly. The order of hydrolysis rates is $B < C < A$.

(b)   We would expect the relative rates to parallel the relative hydrolysis rates of analogous carboxylic acid derivatives—that is, amides (C) < esters (B) < anhydrides (A) < acid chlorides (D).

(c)   Because of the bulky alkoxy groups that restrict the access of nucleophiles to the phosphorus, we would expect A to react very slowly in P—O cleavage. However, we know that the rates of C—O and P—O cleavage are similar for *ordinary* phosphate triesters, and this ester has an allyl group that should be especially reactive in C—O cleavage. Esters C and B will hydrolyze more slowly because of the additional negative charge, but C should hydrolyze faster than B because the C—O cleavage pathway is activated by the allyl ester. The order is $B < C < A$.

Figure SG25.2   The phosphorus-containing functional groups in the solution to Problem 25.24.

25.26   Triesters hydrolyze rapidly enough that they would hydrolyze over time in biological media. In addition, they would be largely insoluble because they are uncharged. On the other hand, phosphate diesters and monoesters have negligible spontaneous hydrolysis rates unless their hydrolysis reactions are catalyzed by specific enzymes. Moreover, they are charged and therefore have water solubility.

25.27   The anomeric proton is split simultaneously by the proton on the adjacent carbon and by the phosphorus on the adjacent oxygen. The coupling constant of 3.5 Hz comes from the hydrogen, since hydrogens with a gauche relationship are typically split by this amount (see Fig. 14.9, Sec. 14.5A, text p. 754). The other 7.3 Hz coupling constant must then come from the phosphorous. Because the coupling constants are nonequivalent, a doublet of doublets is observed (Sec. 14.5A in the text).

25.28   (a)   The protons are all equivalent and therefore do not split each other. The protons are split into a doublet by the phosphorus.

   (b)   According to the $n + 1$ rule, the phosphorus is split into a 10-line pattern by the nine methyl protons.

25.29   (a)   The phosphorus resonance at smaller chemical shift originates from the phosphorus bonded to three carbons. [See (a) below.] The phosphorus resonance at larger chemical shift originates from the phosphorus bonded to three oxygens. Each phosphorus resonance is split into a septet because there are six equivalent hydrogens. The size of the coupling constant relates inversely to the distance of these hydrogens from the phosphorus.

   (b)   In the proton NMR spectrum, the hydrogens are split by the two phosphorus nuclei into a doublet of doublets with coupling constants of 2.6 and 8.9 Hz.

25.30 A thioester is replaced by an oxygen ester. Resonance stabilization in the thioester is largely nonexistent because sulfur 3p orbitals don't overlap well with carbon 2p orbitals. Resonance stabilization in the oxygen ester is important. The reaction replaces a less stable ester with a more stable one.

resonance is not important   resonance is important

25.31 Racemization of the product suggests that a single inversion by an in-line, or S_N2, mechanism is not operating. Instead, the formation of a planar and achiral metaphosphate ion (Eq. 25.37a, text p. 1400) could be trapped by the *t*-butyl alcohol to give the racemic product.

racemic product

Another possibility, however, is that the solvent acetonitrile could act as a nucleophile in a direct displacement of the leaving group with inversion, and other solvent molecules could carry out successive direct displacements that would give racemic intermediates that could react with *t*-butyl alcohol to give racemic products.

racemic mixture

25.32 We would expect to find the following features: the nucleophile and phosphate ester held in proximity in an arrangement that favors an in-line displacement, or S_N2, reaction; a base in the active site that removes a proton from the nucleophile; one or more acidic groups that deliver a proton to the phosphate leaving group as it separates as a leaving group; hydrogen-bond donors in the enzyme structure or metal ions chelated to both the phosphate and to carboxylate groups on the enzyme that hold the phosphate in place and increase its leaving-group potential.

25.33 We should expect to find in the active site a Mg^{2+} or Mn^{2+} ion associated with the diphosphate group in the UDP-diphosphate; a basic group such as a carboxylate group poised to remove a proton from the oxygen of the phenol nucleophile; binding of the phenol and the UDP-glucuronic acid in such a way that the phenol reacts at the anomeric carbon of the sugar with inversion of configuration; many hydrogen-bond donors and acceptors that can interact with the acceptor and donor sites of the sugars and holding these groups in the proper orientation for reaction; a protonated amino group or guanidino group that interacts electrostatically with the carboxylate group of the glucuronic acid.

25.34 Nucleophiles can react at any one of the three phosphates. Reaction at the γ-phosphate results in phosphorylation of the nucleophile; reaction at the β-phosphate results in pyrophosphorylation of the nucleophile; and reaction at the α-phosphate results in adenylation of the nucleophile. These reactions are described in Eqs. 25.45a–c in Sec. 25.7A on pp. 1407–1408 of the text.

25.35 (a) The hydroxy group of ribose-5-phosphate reacts as a nucleophile at the β-phosphate group. The by-product of the reaction is adenylic acid (that is, AMP):

AMP

(b) The reaction of ammonia with PRPP is an S_N2 reaction that occurs with inversion of configuration

5-phosphoribosyl-1-pyrophosphate
(PRPP)

pyrophosphate

(c) The pyrophosphate by-product can be converted into two equivalents of phosphate by the pyrophosphatase-catalyzed hydrolysis reaction. Removing pyrophosphate from the reaction in part (b) ensures that the reaction goes to completion, by Le Chatelier's principle.

25.36 (a) Aspartic acid is adenylated. This produces an acyl phosphate, which is very reactive toward nucleophiles, along with pyrophosphate, which can be further hydrolyzed to phosphate by pyrophosphatase.

aspartic acid

ATP

pyrophosphate

From here, there are two possible pathways. In one, the acyl phosphate reacts with the amide nitrogen of glutamine to give an imide, which is hydrolyzed to asparagine and glutamate. Amide nitrogens are weak nucleophiles, but this step is plausible because acyl phosphates are so reactive. (We leave it to you to provide the curved-arrow notation.)

glutamine

AMP

asparagine   glutamic acid

An issue with this mechanism is that the hydrolysis of the imide must be completely selective at two carbonyls that differ little in reactivity.

In fact, the actual mechanism is somewhat different and doesn't require this exquisite selectivity. The enzyme has two active sites, one at each end of the molecule. At one end, the enzyme hydrolyzes glutamine to give ammonia, which is not free in solution, but rather diffuses through a channel in the enzyme where it reacts with the adenylated aspartic acid.

glutamine   glutamate   routed as ammonia through a channel in the enzyme

(b) The equilibrium constant for equilibration of glutamine and asparagine in the absence of ATP should be about 1 because the two amides and acids have very similar structures and pK_a values. The coupling of ATP hydrolysis with this equilibration, as shown in part (a), drives the reaction in the direction of asparagine formation. When the pyrophosphate by-product is hydrolyzed by pyrophosphatase, the equilibrium in favor of asparagine is even more favorable.

(c) The $\Delta G°'$ for the overall reaction is determined by summing the individual reactions and cancelling like terms on opposite sides:

			$\Delta G°'$
Asp + NH₃	⇌	Asn + H₂O	+15.1
Gln + H₂O	⇌	Glu + NH₃	−14.2
ATP + H₂O	⇌	AMP + pyrophosphate	−45.6
Asp + Gln + ATP + H₂O	⇌	Asn + Glu + AMP + pyrophosphate	−44.7

(d)   As the previous part shows, the $\Delta G°'$ for ATP hydrolysis is a component of the overall $\Delta G°'$ for the overall reaction, even though the mechanism does not involve ATP hydrolysis explicitly. Remember that a fundamental idea of calculating enthalpies, entropies, and free energies is that the path taken for the calculation doesn't matter as long as the initial and final states are the same (Hess's law).

(e)   Hydrolysis of pyrophosphate catalyzed by pyrophosphatase, by removing a product of the synthesis reaction, further drives the reaction in favor of asparagine synthesis.

25.37   A double inversion mechanism must be operating if the product shows retention of stereochemistry. However, the problem states that no phosphoenzyme intermediate is formed. Therefore, we could postulate that another nucleophile does the initial displacement, inverting the stereochemistry at phosphorus, and the labeled water inverts the phosphorus back to its original configuration to yield the product. In fact, there is a nucleophile available in the same molecule nearby.

25.38   (a)   The mechanism involving a phosphohistidine intermediate:

(b)   Because inversion occurs twice at phosphorus, the product would contain phosphorous with retention of configuration.

25.39   The two species involves in the reaction whose concentrations are potentially affected by a difference in pH are phenol and acetic acid. Calculate the r values (Eq. 25.54b, text p. 1413) for these two species. For acetic acid, r = 0.0057 (Eq. 25.54c and discussion). For phenol, with pK_a = 9.95, the fraction unionized at pH = 7 is 0.001, so that $r \approx 1$. Therefore, Eq. 25.54b (text p. 1413) applies to this case, with $\Delta G° = -7.39$ kJ mol^{-1}. (T is taken as 25 °C or 298 K.)

$$\Delta G°' = \Delta G° + 2.3RT \log r_{CH_3CO_2H}$$
$$= -7.39 + 5.71 \log (0.0057)$$
$$= -7.39 - 12.81 = -20.20 \text{ kJ mol}^{-1} \ (-4.83 \text{ kcal mol}^{-1})$$

25.40   In this problem we have to correct both starting materials and products for their actual concentrations at pH = 7. We have to calculate the fraction of the ester conjugate acid (as shown in the problem) at pH 7 and the fraction at pH 0; r is then the ratio of these two. To calculate fractions of the starting ester conjugate acid at pH 0 and 7, we use Eq. 3.67b on text p. 148 for the fraction conjugate acid f at the two pH values. At pH 0, $f(\text{pH} = 0) = 1$; at pH 7, $f(\text{pH } 7) = 0.871$. The value r for the ester conjugate acid, $r_{\text{ester conjugate acid}}$, is then $0.871/1 = 0.871$.

The two pK_a values for the glycine conjugate acid are the actual measured pK_a values, which means that the amine conjugate-acid pK_a is measured when the carboxylic acid group is ionized, and the carboxylic acid pK_a is measured when the amine is fully protonated. Because there are *two* groups whose dissociation state is affected by pH, we calculate r for each group and then multiply the two together. For the amino group the fraction conjugate acid, $f(\text{pH } 0) = 1$, and $f(\text{pH } 7) = 0.998$, so that the value of r for the amino group conjugate acid is $f(\text{pH } 7)/f(\text{pH } 0) = 0.998$, which is essentially 1. For the carboxylic acid group, $f(\text{pH } 0) = 0.995$ and $f(\text{pH } 7) = 2.19 \times 10^{-5}$, so that the r value for this group is $f(\text{pH } 7)/f(\text{pH } 0) = 2.20 \times 10^{-5}$. The r for glycine conjugate acid, $r_{\text{glycine conjugate acid}}$, is the product of the two r values: $(2.20 \times 10^{-5})(0.998) = 2.20 \times 10^{-5}$

Now we calculate the $\Delta G°'$ value for the hydrolysis of glycine ethyl ester conjugate acid using the same equation as in the solution to the previous problem but with the appropriate r values. (Again we use $T = 298$ K.

$$\Delta G°' = \Delta G° + 2.3RT \log \left(\frac{r_{\text{glycine conjugate acid}}}{r_{\text{glycine ester conjugate acid}}} \right)$$

$$= -8.34 + 5.71 \log \left(\frac{220 \times 10^{-5}}{0.871} \right) = -8.34 + 5.71 \log (2.53 \times 10^{-5})$$

$$= -8.34 - 26.2 = -34.6 \text{ kJ mol}^{-1} \ (-8.27 \text{ kcal mol}^{-1})$$

From this calculation, we can see that the ionization of the carboxylic acid group has a marked effect on the calculation, as we should expect, because this ionization pulls the equilibrium toward the hydrolysis products.

25.41   (a)   Use a strategy similar to the one shown in the solution to Problem 25.18(a) in this manual.

a fatty acyl-CoA + ~~H₂O~~ ⇌ ~~RCO₂⁻~~ + HSCoA + ~~H⁺~~ $\Delta G°' = -31.4$ kJ mol^{-1}

~~RCO₂⁻~~ + carnitine + ~~H⁺~~ ⇌ a fatty acyl carnitine + ~~H₂O~~ $\Delta G°' = +17.4$ kJ mol^{-1}

a fatty acyl-CoA + carnitine ⇌ a fatty acyl carnitine + HSCoA $\Delta G°' = -14.0$ kJ mol^{-1}

(b)   With $\Delta G°' = -2.3RT \log K_{eq}$, and assuming a temperature of 37 °C (310 K), $K_{eq} = 229$. (At $T = 298$ K, K_{eq} = 283.)

25.42   Use a strategy similar to the ones shown in the solutions to problem 25.18(a) and 25.41(a) in this manual. The free-energy diagram is shown in **Fig. SG25.3** on the next page.

acetyl phosphate + ~~H₂O~~ ⇌ ~~CH₃CO₂⁻~~ + phosphate + ~~H⁺~~ $\Delta G°' = -43.1$ kJ mol^{-1}

~~RCO₂⁻~~ + HSCoA + ~~H⁺~~ ⇌ acetyl CoA + ~~H₂O~~ $\Delta G°' = +31.4$ kJ mol^{-1}

acetyl phosphate + HSCoA ⇌ acetyl CoA + phosphate $\Delta G°' = -11.7$ kJ mol^{-1}

Figure SG25.3   A free-energy diagram to accompany the solution to Problem 25.42.

Solutions to Integrated Problems

25.43   Part (a) is the reaction of an acid chloride with an excess of amine, which proceeds to a tri-amide and three equivalents of the ammonium salt by analogy with the similar reaction of carboxylic acid chlorides. Part (b) is the formation of a tosylate ester, then the chloride by S_N2 displacement of the tosylate in a polar aprotic solvent, and then displacement of the chloride with a pyrophosphate ion in a polar aprotic solvent. This is an important method of preparing alkyl pyrophosphates. In part (c), the ^{18}O-hydroxide nucleophile reacts at the carbonyl carbon to give ^{18}O-labeled acetic acid and phosphate ion. The acetic acid ionizes under the basic conditions so that the ^{18}O label is distributed evenly between the two oxygens of the resulting acetate ion.

25.44   When base (triethylamine) is present, the product is an amide C because the amine of 2-aminoethanol is much more basic, and therefore more nucleophilic, than the neutral alcohol.

When acid (*p*-toluenesulfonic acid) is present, the amine is protonated, and, because it no longer has a nonbonding electron pair, it is not nucleophilic. In that case, the alcohol end of the 2-aminoethanol reacts with the carbonyl of the thioester to yield an ester, *D*.

If the ester *D* is neutralized with base to give the free amino ester, it rapidly isomerizes to amide *C* by an intramolecular ester aminolysis.

25.45   The first hydride equivalent reacts with the thioester to convert it to an aldehyde by displacing thiolate as a leaving group. A second equivalent of hydride reduces the aldehyde to an alcohol. The overall process is analogous to the reduction of esters (Sec. 21.9A of the text).

When boron trifluoride is added, a boron-oxygen bond is formed that stabilizes the tetrahedral intermediate. Instead of breaking down by re-formation of the carbonyl and loss of the leaving group, as above, the tetrahedral intermediate loses the oxygen instead, because the resulting $^-OBF_2$ group is actually a good leaving group. The electrons on sulfur help to stabilize the resulting carbocation in a manner similar to the way that nitrogen operates in the reduction of amides, as shown in Eq. 21.82d, Sec. 21.9B on text p. 1169. The boron-containing leaving group likely breaks down into the aluminum hydride addition product with BOF_2 and LiF.

The sulfur-stabilized carbocation is then reduced with a second equivalent of hydride, again analogously to the reduction of the imine intermediate in the case with amides (see Eq. 21.82e, text p. 1169).

cyclohexyl ethyl sulfide

25.46 (a)   The *S* enantiomer of sarin: (b)   The reaction product of sarin with a serine residue:

The reaction with serine occurs with inversion of configuration; but, because the relative group priorities change, the inverted product (shown above) also has the *S* configuration.

25.47 (a)   The hydrolysis of phosphoenolpyruvate gives, first, an enol, which, like many other enols, is rapidly transformed into a ketone:

phosphoenolpyruvate (PEP)   **an enol**   **pyruvate**

(b)   This reaction is driven to completion by the fact that the initially formed "alcohol" hydrolysis product is actually an enol, which equilibrates into a much more stable ketone. Since these products are so much lower in energy than the products of the hydrolysis of an ordinary phosphate ester, this conversion contributes a significant negative $\Delta G°$ component to the standard free-energy change for the reaction.

25.48 (a)   The labeled methanol reacts at the sulfur of a sulfonyl chloride, but hydroxide (like most other nucleophiles) reacts at the alkyl group of a sulfonate ester. (Recall that sulfonate esters react like alkyl halides in S_N2 reaction; Sec. 11.4A of the text.)

A   *B*

(b)   Both the labeled methanol and hydroxide react as nucleophiles at the carbonyl group:

(c)   The methanol reacts at the phosphorus of the acid chloride to give a mixed phosphate triester. The sodium hydroxide could react as a nucleophile either at phosphorus or at one of the alkyl groups. Because the isopropyl groups hinder reaction at the phosphorus, an S_N2 reaction occurs at the methyl group.

25.49   Protonation of the phosphoryl oxygen (analogous to carbonyl protonation) converts the phosphate group into an excellent leaving group. An S_N2 reaction of the labeled water at one of the methyl groups provides labeled methanol and unlabeled dimethyl phosphate.

25.50   From the standpoint of the octet rule, there is no reason that sodium hydroxide couldn't add to the N=O bond of the nitrate ion, although the large accumulation of negative charge around a second-row atom could be a source of instability.

sodium orthonitrate

In fact, the trisodium salt, sodium orthonitrate, was first prepared in 1977 by heating sodium nitrate with sodium oxide (Na_2O, essentially a "dehydrated sodium hydroxide") with sodium nitrate at 300 °C for several days. The corresponding triacid, orthonitric acid (which is the nitrogen analog of phosphoric acid), is unknown.

25.51   If $\Delta G°'$ were redefined to be the standard $\Delta G°$ of the reaction at pH 11, it would be more negative than $\Delta G°'$ at pH 7. However—and this is the point of the question— $\Delta G°'$ is *defined* to be the $\Delta G°$ at the standard state of pH 7, and therefore it is *not* dependent on the pH at which the reaction is actually carried out. What *would* change, however, is the concentration ratio of reactants and products at equilibrium, because the hydrogen-ion concentration is 10^{-4} that of the standard state hydrogen-ion concentration. Because the SH group of CoASH would be fully ionized at pH 11, the reaction would be pulled toward hydrolysis products by this ionization.

The Chemistry of the Aromatic Heterocycles and Nucleic Acids

CHAPTER 26 SUMMARY

- Heterocyclic organic compounds contain rings consisting of carbon atoms and atoms of other elements.

- In aromatic heterocycles, a vinylic nitrogen contributes one electron to the aromatic π-electron system, and its nonbonding electron pair is in the plane of the ring. A nitrogen of this type (for example, in pyridine) is basic with $pK_a \sim 5$. An allylic nitrogen contributes its nonbonding electron pair to the aromatic π-electron system. A nitrogen of this type (for example, in pyrrole), is not basic, because protonation would disrupt the $4n + 2$ aromatic π-electron system. Furan has two nonbonding electron pairs on oxygen, one of each type.

- Pyrrole, furan, and thiophene are all more reactive than benzene in electrophilic aromatic substitution and undergo substitution predominantly at the 2-position. The reactivity order is pyrrole > furan > thiophene.

- Because furan has a relatively small empirical resonance energy, it undergoes some diene conjugate-addition reactions, such as the Diels–Alder reaction.

- Pyridine reacts very slowly in electrophilic aromatic substitution. Pyridine and its derivatives undergo electrophilic substitution at the 3-position. Electrophilic substitution reactions of pyridine-N-oxides, however, occur at the 4-position.

- Many side-chain reactions of heterocyclic compounds proceed normally without disrupting the heterocyclic rings.

- Pyridine derivatives undergo nucleophilic aromatic substitution reactions at the 2- and 4-positions. For example, pyridines react in the Chichibabin and related reactions; 2- and 4-halopyridines undergo nucleophilic aromatic substitution reactions. Pyridinium salts are even more reactive than pyridines in these reactions. The chemistry of the pyridine C=N linkage has some similarity to the chemistry of the carbonyl group.

- The "benzylic" hydrogens of 2-alkyl- and 4-alkylpyridines and especially pyridinium salts are acidic enough to be removed by bases. In biology, the reactions of NAD+ and pyridoxal phosphate are due to the electron-accepting ability of the pyridinium ions in these molecules.

- Ribonucleotides and deoxyribonucleotides, which are phosphorylated derivatives of ribonucleosides and deoxyribonucleosides, are the building blocks of RNA and DNA, respectively. These compounds are phosphorylated derivatives of either ribose or 2'-deoxyribose, respectively, and a purine or pyrimidine base, in which the base is connected to C-1 of the ribose with β stereochemistry. Adenine, guanine, and cytosine are bases in both DNA and RNA, whereas thymine is unique to DNA, and uracil is unique to RNA.

- An important conformation of DNA is the double helix, in which two right-handed helical strands of DNA running in opposite directions wrap around a common axis. The sugars and phosphate groups lie on the outside of the helix, and the bases are stacked in parallel planes on the inside. The two strands of the double helix are held together by Watson–Crick base pairs—that is, by hydrogen bonds between a purine base (A or G) and its complementary pyrimidine (T or C). A number of known carcinogens apparently modify DNA in such a way that this complementary hydrogen bonding is disrupted.

FURTHER EXPLORATIONS

 ### 26.1   Relative Acidities of 1,3-Cyclopentadiene and Pyrrole

It is interesting that pyrrole, a nitrogen acid, is somewhat *less* acidic than 1,3-cyclopentadiene, a carbon acid.

1,3-cyclopentadiene
$pK_a = 15$ pyrrole
$pK_a = 17$

This result stands in contrast to the general expectation from the *element effect* (Sec. 3.7B of the text) that hydrogens on more electronegative atoms are more acidic than those on less electronegative atoms. (For example, the pK_a of NH_3 is about 35; the pK_a of CH_4 is about 55; the N—H pK_a of amides is about 15; the α-hydrogen pK_a of a ketone is about 19.)

The greater acidity of 1,3-cyclopentadiene reflects the importance of aromaticity. 1,3-Cyclopentadiene itself is not aromatic, but its conjugate-base anion is. Consequently, 1,3-cyclopentadiene "gains aromaticity" when it ionizes. This effect lowers the pK_a of 1,3-cyclopentadiene significantly from what it otherwise would be. (The effect of aromaticity might be worth as much as 20 pK_a units.) In contrast, *both* pyrrole and its conjugate-base anion are aromatic; ionization of pyrrole does not gain any aromatic stability for the molecule. In summary, then, the increase in stability resulting from the formation of an aromatic species causes the pK_a of the carbon acid to be lower than that of the nitrogen acid.

 # REACTION REVIEW

I.   ELECTROPHILIC AROMATIC SUBSTITUTION REACTIONS OF AROMATIC HETEROCYCLIC COMPOUNDS

A.   *FURAN, PYRROLE, AND THIOPHENE*

1.   Furan, thiophene, and pyrrole undergo electrophilic substitution predominantly at the 2-position of the ring.

   a.   The carbocation resulting from substitution at carbon-2 has more important resonance structures, and is therefore more stable, than the carbocation resulting from reaction at carbon-3.

b. Some carbon-3 substitution product accompanies the major carbon-2 substitution product in many cases.

2. Furan, pyrrole, and thiophene are all much more reactive than benzene in electrophilic aromatic substitution reactions.

 a. The reactivity order of the heterocycles is a consequence of the relative abilities of the heteroatoms to stabilize positive charge in the intermediate carbocations.

pyrrole > furan > thiophene >> benzene

 i. Both pyrrole and furan have heteroatoms from the second period of the periodic table.

 ii. Because nitrogen is better than oxygen at delocalizing positive charge (it is less electronegative), pyrrole is more reactive than furan.

 iii. The sulfur of thiophene is a third-period element and, although it is less electronegative than oxygen, its 3p orbitals overlap less efficiently with the 2p orbitals of the aromatic π-electron system.

 b. The reactivity order of the heterocycles in aromatic substitution parallels the reactivity order of the corresponding substituted benzene derivatives:

$(CH_3)_2N$—Ph > CH_3O—Ph > CH_3S—Ph

3. Milder reaction conditions must be used with more reactive compounds.

4. The usual activating and directing effects of substituents in aromatic substitution apply; superimposed on these effects is the normal effect of the heterocyclic atom in directing substitution to the 2-position.

 a. Count around the carbon framework of the heterocycle, not through the heteroatom, when using the ortho, meta, para analogy.

 b. When the directing effects of substituents and the ring compete, it is not unusual to observe mixtures of products.

 c. If both 2-positions are occupied, 3-substitution takes place.

B. PYRIDINE

1. In general, pyridine has very low reactivity in electrophilic aromatic substitutions; it is much less reactive than benzene.

   a. An important reason for this low reactivity is that pyridine is protonated under the very acidic conditions of most electrophilic aromatic substitution reactions.

   b. The resulting positive charge on nitrogen makes it difficult to form a carbocation intermediate, which would place a second positive charge within the same ring.

2. Pyridine rings substituted with activating groups such as methyl groups do undergo electrophilic aromatic substitution reactions.

   a. When substitution in pyridine does occur, it generally takes place at the 3-position.

   b. The preference for 3-substitution in pyridines can be understood by considering the resonance structures for the possible carbocation intermediates.

      i. Substitution in the 3-position gives a carbocation with three resonance structures.

      ii. Substitution at the 4-position also involves an intermediate with three resonance structures, but the last one on the right is particularly unfavorable because the nitrogen, an electronegative atom, is also electron deficient.

no octet of electrons

3. Pyridine can be oxidized to give pyridine-N-oxide.

pyridine-N-oxide

4. Pyridine-N-oxide is more reactive than pyridine and undergoes nitration in the 4-position.

a.   The *N*-oxide function can be removed by catalytic hydrogenation, which will also reduce any nitro groups present.

b.   Reaction with trivalent phosphorous compounds, such as PCl_3, removes the *N*-oxide function without reducing nitro groups.

II.   NUCLEOPHILIC AROMATIC SUBSTITUTION REACTIONS OF AROMATIC HETEROCYCLIC COMPOUNDS

A.   PYRIDINE

1.   Treatment of a pyridine derivative with the strong base sodium amide (Na^+ $^-NH_2$) brings about the direct substitution of an amino group for a ring hydrogen; this reaction is called the Chichibabin reaction.

2.   A reaction similar to the Chichibabin reaction occurs with organolithium reagents.

3.   When pyridine is substituted with a better leaving group than hydride at the 2-position, it reacts more rapidly with nucleophiles. For example, the 2-halopyridines readily undergo substitution of the halogen by other nucleophiles under conditions that are much milder than those used in the Chichibabin reaction.

4.   Nucleophilic substitution reactions on pyridine rings can be classified as nucleophilic aromatic substitution reactions.

a.   The "electron-withdrawing group" in the reaction of pyridines is the pyridine nitrogen itself; consider the ring nitrogen of pyridine as if it were a nitro group attached to a benzene ring.

b   The tetrahedral addition intermediate is analogous to the Meisenheimer complex of nucleophilic aromatic substitution. (See Sec. 18.4 of the text.)

tetrahedral
addition intermediate

5. 3-Substituted pyridines are not reactive in nucleophilic substitution because the negative charge in the addition intermediate cannot be delocalized onto the electronegative nitrogen.

6. 2-Pyridone contains a significant amount of both carbonyl and hydroxy isomers, and it often undergoes some reactions reminiscent of carboxylic acids.

2-pyridone

B.   PYRIDINIUM SALTS AND THEIR REACTIONS

1. Pyridine is a nucleophile and reacts in S_N2 reactions with alkyl halides or sulfonate esters to form quaternary ammonium salts, called pyridinium salts.

2. Pyridinium salts are activated toward nucleophilic addition at the 2- and 4-positions of the ring much more than are pyridines themselves, because the positively charged nitrogen is much more electronegative than the neutral nitrogen of a pyridine.

a. When the nucleophiles in such addition reactions are anions, charge is neutralized.

b. Pyridine-N-oxides are in one sense pyridinium ions, and they react with nucleophiles in much the same way as quaternary pyridinium salts.

III.   OTHER REACTIONS OF HETEROCYCLIC AROMATIC COMPOUNDS

A.   ADDITION REACTIONS TO FURAN

1. Furan, pyrrole, or thiophene can be viewed as a 1,3-butadiene with its terminal carbons "tied down" by a heteroatom bridge.

←————1,3-butadiene unit

2. Of the three heterocyclic compounds furan, pyrrole, and thiophene, furan has the least resonance energy.

    a. Furan has the greatest tendency to behave like a conjugated diene.

    b. Furan undergoes some conjugate-addition reactions.

    c. Furan undergoes Diels–Alder reactions with reactive dienophiles.

B. SIDE-CHAIN REACTIONS OF FURAN, PYRROLE, AND THIOPHENE

1. Many reactions occur at the side chains of heterocyclic compounds without affecting the rings.

C. SIDE-CHAIN REACTIONS OF PYRIDINE DERIVATIVES

1. The "benzylic" hydrogens of an alkyl group at the 2- or 4-position of a pyridine ring are more acidic than ordinary benzylic hydrogens because the nonbonding electron pair (and charge) in the conjugate-base anion is delocalized onto the electronegative pyridine nitrogen.

    a. Strongly basic reagents such as organolithium reagents or $NaNH_2$ abstract the benzylic proton from 2- or 4-alkylpyridines.

    b. The anion formed in this way has a reactivity much like that of other organolithium reagents.

    c. These anions are somewhat analogous to enolate ions, so they undergo some of the reactions typical of enolate anions, such as aldol condensations.

2. The "benzylic" hydrogens of 2- or 4-alkylpyridinium salts are much more acidic than those of the analogous pyridines.

    a. One resonance form of the conjugate-base "anion" is an uncharged species.

    b. The conjugate base "anion" can be formed in useful concentrations by aqueous NaOH or amines.

3. Many side-chain reactions of pyridines, for example, side-chain oxidation, are analogous to those of the corresponding benzene derivatives.

D.   PYRIDINIUM IONS IN BIOLOGY: PYRIDOXAL PHOSPHATE

1. Pyridoxal phosphate is one of the several forms of vitamin B_6; the others are pyridoxal, pyridoxamine, and pyridoxol (pyridoxine). (See Fig. 26.3, text p. 1448, for structures.)

2. Pyridoxal phosphate is an essential reactant in several important biochemical transformations.

    a. Interconversion of α-amino acids and α-keto acids:

an α-amino acid   an α-keto acid   an α-keto acid   an α-amino acid

    b. Decarboxylation of α-amino acids:

an α-amino acid

    c. Loss of formaldehyde from serine:

serine
(an α-amino acid)   **formaldehyde**   **glycine**
(an α-amino acid)

3. Pyridoxal phosphate is an example of a *coenzyme*—a compound that, with an enzyme, is an essential co-catalyst in a biochemical reaction.

SOLUTIONS TO PROBLEMS

Solutions to Focused Problems

26.1 (a)

(b)

4-(dimethylamino)pyridine   **4-ethyl-2-nitroimidazole**

26.2 (a)   5-Bromo-2-methylthiazole   (b)   2-Bromo-4-nitropyrrole
(c)   Pyridine-4-carbaldehyde   (d)   8-Methoxyquinoline

26.3 The resonance structures of pyrrole:

26.4 (a) Because nitrogen is an electronegative atom, the C—N bond dipoles in pyrrolidine are directed towards the nitrogen, and their resultant is also directed towards the nitrogen:

It is given that the dipole moment of pyrrole is in the opposite direction, that is, directed away from the nitrogen. Although there is a dipole component directed towards the nitrogen as shown for pyrrolidine, the major component is due to the charge separation shown in the resonance structures given in the solution to Focused Problem 26.3. In all resonance structures but the first, positive charge resides on the nitrogen and negative charge is on one of the carbons. Evidently, these contributions significantly outweigh those from the carbon–nitrogen bond dipoles.

(b) Oxygen is more electronegative than nitrogen. Consequently, the contribution of the carbon–oxygen bond dipoles in furan is *greater* than the contribution of the carbon–nitrogen bond dipoles in pyrrole, and the contribution of the charge-separated resonance structures for furan is *smaller* than the contribution of the analogous structures for pyrrole. Notice that the resonance energy of furan is less than that of pyrrole (Table 26.1 on text p. 1429); this finding is consistent with the idea that the resonance structures of furan are less important than those of pyrrole.

In summary: In furan, the sum of the C—O bond dipoles outweighs the dipole–moment contribution of π-electron delocalization (see Display 26.4 on text p. 1428). In pyrrole, the dipole–moment contribution of π-electron delocalization outweighs the sum of its C—N bond dipoles.

(c) Because the dipole vector of pyrrole is directed away from the nitrogen, and because the resultant of the two carbon-chlorine bond dipoles is also directed away from the nitrogen, the dipole moment of 3,4-dichloropyrrole should be greater than that of pyrrole:

3,4-dichloropyrrole

Although the dipole moment of 3,4-dichloropyrrole has not been measured experimentally, high-level calculations that predict correctly the known dipole moments of other pyrroles give the dipole moment of 3,4-dichloropyrrole as 4.50 D; the dipole moment of pyrrole itself is 1.80 D. This verifies the prediction made by vector analysis above.

26.5   Both pyrrole and pyridine are aromatic, and therefore both should have ring current-induced chemical shifts. Consequently, the two larger shifts are associated with the two aromatic compounds, and the δ 2.82 resonance is that of pyrrolidine. This chemical shift is not far from the value predicted for the α-proton resonance of an alkylamine. (See Fig. 14.4 on text p. 738.) The resonance structures for pyrrole (see the solution to Problem 26.3, above) as well as its dipole moment (see the solution to Focused Problem 26.4(a), above) show that pyrrole has partial negative charge on carbon-2; in contrast, the resonance structures of pyridine (Display. 26.5, text p. 1428) show that carbon-2 has partial positive charge. Because electron deficiency is associated with large chemical shifts, it follows that the δ 8.51 resonance is that of pyridine and the δ 6.41 resonance is that of pyrrole.

26.6   (a)   The pyridine nitrogen is basic. Since bases can accept hydrogen bonds, pyridine can accept hydrogen bonds from water. This hydrogen-bonding capability increases the solubility of pyridine in water. Pyrrole, in contrast, is not basic, and the N—H hydrogen is not particularly acidic. Hence, pyrrole does not form strong hydrogen bonds with water, and it is therefore not very soluble.

    (b)   Imidazole has one nitrogen that has the electronic character of the nitrogen in pyrrole, and one that is like the nitrogen in pyridine, except that it is somewhat more basic than the nitrogen of pyridine (Study Problem 26.1 on text p. 1431). Because of its basicity, imidazole can accept hydrogen bonds from water, and, like pyridine, it is very soluble in water.

26.7   (a)   Protonation on the pyridine nitrogen occurs because the conjugate acid is resonance-stabilized.

conjugate acid of
4-(dimethylamino)pyridine

Protonation on the dimethylamino nitrogen, in contrast, gives a conjugate acid that is not resonance-stabilized. (Be sure to demonstrate this for yourself.) 4-(Dimethylamino)pyridine is more basic than pyridine because of the electron-donating resonance interaction of the dimethylamino substituent in the conjugate-acid cation, shown above. The conjugate-acid cation of pyridine itself lacks this stabilizing interaction.

    (b)   Because the pyridine nitrogen is the more basic nitrogen, it is also the more nucleophilic nitrogen. Consequently, alkylation occurs on this nitrogen.

26.8   The nonbonding electron pair on the nitrogen of aniline is conjugated with the π-electron system of the ring. In other words, the 2p orbital containing the nonbonding electron pair is part of the π molecular orbital system of the compound. With more 2p orbitals contributing to the system of π MOs, the absorption wavelength is increased.

   When aniline is protonated, this nonbonding electron pair is no longer available for conjugation with the ring. That is, the system of π orbitals contains one less 2p orbital. This reduction in conjugation causes a reduction in the λ_{max} in the UV spectrum.

   The pyridine nonbonding electron pair, in contrast, is not conjugated with the π-electron system of the ring because its orbital is in the plane of the ring. Consequently, protonation does not affect the conjugated system and therefore it does not affect the UV spectrum.

26.9   (See structures following the discussion.) In (a), the bromine directs the incoming nitro group to the "ortho" position, and the thiophene ring is more activated at the same position. Part (b) is a Friedel–Crafts acylation; both the methyl group and the ring direct substitution to the open "para" position (5-position). (The curved-arrow mechanism for formation of this product is shown below the structures.) The reaction part (c) is an aldol condensation. Part (d) is essentially a "benzylic" bromination: a free-radical bromination at a carbon adjacent to an aromatic ring (Sec. 17.2).

(a)

3-bromo-2-nitrothiophene

(b)

(c)

(d)

3-(bromometh-yl)thiophene

The mechanism for the formation of the product (b):

26.10   This is an electrophilic aromatic substitution reaction in which the electrophile is the protonated aldehyde. (The carbonyl-protonated aldehyde shown in the mechanism below is undoubtedly in equilibrium with its nitrogen-protonated isomer, but the latter is not shown since it is not involved directly in the mechanism.)

protonated aldehyde

26.11 The structures of the picolines (methylpyridines) are given in Sec. 26.4A in the text. α-Picoline is more reactive because both the methyl substituent group and the ring direct substitution to the 3- and 5-positions. The carbocation intermediate for substitution at the 3-position has a resonance structure in which the positive charge is on the carbon bearing the methyl group:

tertiary carbocation resonance structure

product of nitration at carbon-3

(The same is true for substitution at the 5-position.) In contrast, nitration of β-picoline gives a carbocation intermediate for which all the resonance structures are secondary carbocations.

(Nitration at the 2- or 4-position gives a carbocation that has a tertiary resonance structure; however, one of the resonance structures of this carbocation is unimportant because it has an electron-deficient nitrogen.)

26.12 Substitution at the 4-position gives a carbocation that is stabilized by resonance interaction with a nonbonding electron pair on the oxygen. Substitution at the 3-position gives a carbocation that is not stabilized by such an interaction. (E = electrophile)

two resonance structures of the carbocation intermediate that results from substitution at the 4-position

26.13 (a) The curved-arrow mechanism:

(b) Follow the steps described in Sec. 11.6A of the text. Calculate the oxidation level of each atom involved in the reaction, in this case, nitrogen and phosphorus. Nitrogen is reduced (it gains three electrons) and phosphorus is oxidized (it loses three electrons). This is a three-electron redox reaction. (Remember that a positive charge contributes +1 to the oxidation level, and a nonbonding electron pair contributes –1.)

26.14 The curved-arrow mechanism for the nucleophilic aromatic substitution reaction of 4-chloropyridine with methoxide ion is as follows. Notice that negative charge in the anionic intermediate is delocalized onto the nitrogen, an electronegative atom.

26.15 4-Bromopyridine undergoes nucleophilic aromatic substitution by phenolate ion. The anionic intermediate is stabilized because negative charge is delocalized onto the nitrogen, as shown in the solution to Focused Problem 26.14 as well as in Eq. 26.47b on text p. 1445 (with Y = Br, ⁻:Nuc = ⁻OPh). In the analogous substitution reaction of 3-bromopyridine, negative charge cannot be delocalized onto the nitrogen in the anionic intermediate; consequently, the intermediate is less stable, and the reaction doesn't occur.

4-bromopyridine   **4-phenoxypyridine**

26.16 (a)

quinoline-*N*-oxide

(b)

2-aminoquinoline

(c)

4-nitroquinoline-*N*-oxide

26.17 (a) Because nitration of a pyridine ring generally occurs in a 3- or 5-position, the pyridine must first be oxidized to a pyridine-*N*-oxide so that nitration is directed to the 4-position. Once nitrated, the *N*-oxide is reduced back to a substituted pyridine. (See Eq. 26.36 on text p. 1441.)

3-picoline
(β-picoline)

30% H_2O_2 →

HNO_3
H_2SO_4 →

PCl_3 →

3-methyl-4-nitropyridine

(b) In this case, the pyridine ring is nitrated on the 3-position by direct nitration.

4-picoline
(γ-picoline)

HNO_3
H_2SO_4 →

4-methyl-
3-nitropyridine

(c) The "anion" formed by abstraction of a methyl proton by butyllithium, any other organolithium reagent or Grignard reagent, can be carbonated with CO_2.

2-picoline
(α-picoline)

$LiCH_2CH_2CH_2CH_3$ →

+ $CH_3CH_2CH_2CH_3$

1) CO_2
2) H_3O^+ →

(2-pyridyl)acetic acid

(d) A Hofmann or Curtius rearrangement is used in the last step of this synthesis to introduce an amino group with loss of a carbon atom.

3-picoline
(β-picoline)

HNO_3, heat →

1) $SOCl_2$
2) NH_3 →

Br_2, NaOH →

3-aminopyridine

26.18 (a) The formula of the product shows that one methyl has been introduced. The conjugate-base anion of 3,4-dimethylpyridine is formed by proton removal from the 4-methyl group, and this anion is alkylated by methyl iodide. A 4-methyl hydrogen is considerably more acidic than a 3-methyl hydrogen because the anion *A* that results from removal of a 4-methyl hydrogen is more stable than the one formed by removal of a hydrogen from the 3-methyl. It is more stable because the negative charge can be delocalized to the pyridine nitrogen. (Draw the appropriate resonance structures.) The negative charge in the anion derived from ionization of the 3-methyl group cannot be delocalized to the nitrogen.

3,4-dimethylpyridine   *A*   **4-ethyl-3-methylpyridine**

(b)   The bromine at the 4-position is selectively displaced by ammonia for the reasons discussed in the solution to Focused Problem 26.14.

3,4-dibromopyridine   **4-amino-3-bromopyridine**

26.19   (a)   The pyridinium ion can facilitate the reaction by stabilizing a negative charge formed by loss of the α-proton (see Eq. 26.62b, text p. 1451).

pyridoxamine phosphate (protonated form)

an α-keto acid

(b)   Begin with the imine product of Eq. 26.60, text p. 1450, formed from pyridoxal phosphate and the amino group of serine. (We abbreviate the structure of pyridoxal phosphate as in the text.)

The first step is a β-elimination that produces formaldehyde and a resonance-stabilized "carbanion" in which the charge is essentially neutralized by delocalization of electrons onto the nitrogen of the pyridinium ring. (Be sure to show this!) This "carbanion" is then protonated to give an imine, which, in turn, hydrolyzes to glycine and pyridoxal phosphate.

formaldehyde

resonance-stabilized "anion"
(Draw the resonance structures!)

imine hydrolysis

glycine

pyridoxal phosphate

26.20   Isoniazid is a type of amide called a *hydrazide*. Semicarbazide (Table 19.3, text p. 1065) is the same type of compound. As Table 19.3 shows, such compounds form imine derivatives called *hydrazones*. In this case the structure of the hydrazone is as follows:

It is reasonable to ask why isoniazid does not harm the host organism (that is, humans) by destroying pyridoxal phosphate in the host. The reason is that the bacterium has a very low level of the enzyme that converts pyridoxal into pyridoxal phosphate, whereas the host has a much higher level. Therefore, replenishing the pool of pyridoxal phosphate takes a relatively long time in *Mycobacterium*. Isoniazid is typically co-dosed with large amounts of Vitamin B_6 (that is, unphosphorylated pyridoxal and its congeners), which the human body recruits to form pyridoxal phosphate much faster than *Mycobacterium* does. Therefore, *Mycobacterium*'s reserve of available pyridoxal phosphate is destroyed by the drug; unlike the human host, it cannot replenish pyridoxal phosphate fast enough to survive.

26.21 (a)

deoxythymidine monophosphate (*d*TMP)

(b)

guanosine diphosphate (GDP)

26.22 The syn and anti conformations of guanosine:

guanosine
(anti conformation
about C1′–N9)

180° internal
rotation

guanosine
(syn conformation
about C1′–N9)

26.23 The structure of a single strand of RNA with the 5′ ⟶ 3′ sequence AUCG is shown in **Fig. SG26.1** on the following page.

26.24 Chargaff's first parity rule does not apply to individual DNA strands because the rule is a consequence of the base–pair complementarity that occurs *between* strands.

26.25 (a) The structure of an *O*-4-methylated thymine residue in DNA is as follows. (For the numbering of purine and pyrimidine rings see Display 26.66 on text p. 1454.)

(b) Refer to Fig. 26.6(b), text p. 1459, and to Eq. 26.71 on text p. 1462. The answer to part (a) shows that *O*-methylation eliminates the hydrogen on nitrogen-3 of thymine that is involved in hydrogen bonding with nitrogen-1 of adenine. Because this hydrogen bonding is an essential part of base pairing, disruption of this hydrogen bond also disrupts base pairing.

Figure SG26.1   The structure of the DNA strand requested in Problem 26.23.

Solutions to Skills Objectives Problems

26.26 (a)   4-Methylfuran-2-carboxylic acid   (b)   Thiophene-3-carboxamide
          (c)   1-Methylpyrrole   (d)   4-Ethyl-2-isopropyl-5-methylpyridine

26.27 (a)   (b)   (c)

3,4-dibromothiophene   **2,5-furandicarboxylic acid**   **1-(2-furyl)-1-propanone**

26.28 (a)   2-Aminothiazole   (b)   5-Quinolinecarbaldehyde   (c)   1-Methylimidazole

26.29 (a)

4-oxazolecarbaldehyde

5-methoxyindole

3-quinolinecarboxylic acid

26.30 (a) Because alkylamines are more basic than pyridines (or quinolines), the conjugate acid of quinine is protonated on the nitrogen of the bridged bicyclic ring:

conjugate acid of quinine

(b) Alkylamines are more basic than quinolines or arylamines, so hydroxychloroquine is protonated on the nitrogen in the long chain.

conjugate acid of hydroxychloroquine

(c) Because alkylamines are more basic than pyridines, nicotine is protonated on the nitrogen of the 5-membered ring:

conjugate acid of nicotine

(d) Although it might seem that the nitrogen of an amino group might be more basic than the nitrogen of a pyridine ring, the pyridine nitrogen protonates because the conjugate acid is resonance-stabilized.

conjugate acid of 3,4-diaminopyridine

(e) One nitrogen resembles the nitrogen of indole, which is not very basic; the other resembles the nitrogen of pyridine, which is more basic. Furthermore, protonation of the pyridine nitrogen gives a conjugate-acid cation that is stabilized by resonance interaction with the nonbonding electron pair on the "indole" nitrogen.

(Draw the appropriate resonance structures.) In fact, the conjugate-acid pK_a of 1,4-diazaindene is 6.92, whereas the conjugate-acid pK_a of pyridine is 5.2. The greater basicity of 1,4-diazaindene reflects the resonance stabilization of its conjugate acid.

conjugate acid of 1,4-diazaindene

(f)   Think of a 1,2,3-triazole as an imidazole with an extra nitrogen. Protonation occurs on the terminal nitrogen of the double bond as it does in imidazole because a resonance-stabilized cation is formed:

conjugate acid of
1-methyl-1,2,3-benzotriazole

 This resonance stabilization notwithstanding, the conjugate-acid pK_a of benzotriazole is 1.6; the 1-methyl group will raise that somewhat, but not a great deal. Compare this with the conjugate-acid pK_a of benzimidazole, which is 5.52. Why do you think 1-methyl-1,2,3-benzotriazole is so much less basic?

(g)   The tertiary alkyl nitrogens are more basic than both the aromatic nitrogens and amide nitrogen. The two alkyl nitrogens should have very similar pK_a values, but the electron-withdrawing polar effect of the aromatic ring probably reduces the conjugate-acid pK_a of the tertiary amine closest to it. However, the compound may be a mixture of the two forms protonated on the each of the tertiary-amine nitrogens with the form shown below predominating.

conjugate acid of imatinib

 On p. 2 of the text we stated that you would learn to understand the structure and chemical properties of molecules like imatinib. You're there!

26.31   (a)   The order of increasing basicity (increasing conjugate-acid pK_a) is as follows:

5-methoxyindole $\ll$ 3-methoxypyridine $<$ pyridine $<$ 4-methoxypyridine

All pyridines are much more basic than the indole, which is not basic at nitrogen. 4-Methoxypyridine is more basic than either of the other pyridines, because a nonbonding electron pair of the methoxy substituent is available to stabilize the conjugate acid by resonance:

conjugate acid of
4-methoxypyridine

Such resonance is not possible when the methoxy group is in the 3-position; so, the effect of a 3-methoxy substituent is limited to its acid-strengthening (base weakening) polar effect. Pyridine itself has neither the base-strengthening resonance effect nor the base-weakening polar effect of a methoxy substituent. Therefore, its conjugate-acid pK_a is between those of the methoxypyridines.

(b)   The order of increasing basicity is

3-nitropyridine < 3-chloropyridine < pyridine

The order is a consequence of the relative polar effects of the nitro and chloro substituents, both of which are electron-withdrawing and base-weakening. As shown by the pK_a values in Table 23.1, text p. 1285, the electron-withdrawing polar effect of the nitro group on the pK_a of an ammonium ion is greater than that of the chloro group.

(c)   The first "anion" in the problem is actually a neutral compound.

A   *B*

That is, it is a carbonyl compound. It has much of the characteristics of an amide, because, as structures *A* and *B* above show, a nitrogen nonbonding electron pair is delocalized into the carbonyl group, as in an amide. This analogy suggests that the pK_a of its conjugate acid is near −1. (In fact, its conjugate-acid pK_a is 0.79.) The other anion in the problem is a phenolate ion; its conjugate acid is a phenol, with a pK_a of about 10. Consequently, the phenolate ion is considerably more basic.

(d)   The resonance structures of protonated imidazole are shown in Eq. 26.11 in Study Problem 26.1 on text p. 1431. Compare the structure of protonated imidazole with the structure of protonated oxazole:

conjugate acid of oxazole

In protonated oxazole, the positive charge is shared between an oxygen and a nitrogen; in protonated imidazole, the positive charge is shared between two nitrogens. Because oxygen is more electronegative than nitrogen, it supports positive charge less effectively; consequently, oxazole is less basic, and imidazole is more basic. (The conjugate-acid pK_a of imidazole is 6.95; that of oxazole is 0.38.)

(e)   See Study Problem 26.1 on text p. 1431 for the structure of protonated imidazole. In the case of thiazole, an electron pair of a sulfur is involved in the resonance stabilization of the conjugate acid:

conjugate acid of thiazole

The orbitals of sulfur overlap more poorly than those of nitrogen with carbon 2p orbitals because they are derived from a different quantum level; consequently, the resonance structure on the right is less important than the corresponding structure of protonated imidazole. The sulfur in thiazole is actually less electro-negative than the nitrogen in imidazole, but, as in electrophilic aromatic substitution, the electron-donating resonance effect of nitrogen far outweighs its electron-withdrawing polar effect. Consequently, imidazole is considerably more basic than thiazole. (The actual conjugate-acid pK_a of imidazole is 6.95, and that of thiazole is 2.44.)

26.32 (a) Electrophilic aromatic substitution on furan, pyrrole, and thiophene involves a carbocation intermediate that can be stabilized by resonance structures involving the heteroatom. (E = electrophile, X = O, S, or NH).

This equation shows substitution at carbon-2. Substitution at carbon-3 is also activated, although, because the carbocation intermediate has only two resonance structures, substitution at carbon-3 is not activated as much.

Because its nonbonding electron pair is not part of the aromatic π-electron system—it is an sp^2 orbital that is oriented in the plane of the ring—the nitrogen of pyridine cannot activate electrophilic substitution by resonance, and its electronegativity destabilizes the carbocation intermediate by its polar effect. Electrophilic substitution on the pyridine ring is therefore deactivated.

(b) In nucleophilic aromatic substitution, the nucleophile introduces additional electrons (and a negative charge) into the ring. When the nucleophile reacts at the 2- or 4-position, the resulting intermediate is stabilized because the additional electrons (and negative charge) are delocalized onto the nitrogen, an electronegative atom. (X = halogen, ⁻:Nuc = nucleophile)

In a halofuran, halopyrrole, or halothiophene, however, the negative charge in the intermediate cannot be delocalized onto the electronegative heteroatom whether the halogen is in the 2- or 3-position, Consequently, these compounds do not readily undergo nucleophilic aromatic substitution.

> If a halofuran, halopyrrole, or halothiophene is substituted with one or more nitro groups in the "ortho" or "para" position relative to the halogen, then a nucleophilic aromatic substitution reaction can occur for the same reason that it would occur in 2,4-dintrochlorobenzene—it is activated by the nitro substituents.

26.33 Because methyl groups activate electrophilic aromatic substitution, then *A* is less reactive than *D*. Because thiophenes are less reactive than furans, the methyl-substituted thiophene *D* is less reactive than the methyl-substituted furan *E*. Because furans and thiophenes are more reactive than benzene, and benzene is more reactive than any pyridine, the reactivity order is solved: $C < B < A < D < E$.

26.34 (a) The furan ring is more activated at the 5-position, and the alkyl substituents at carbons 1 and 2 (that is, the carbons of the six-membered ring) activate substitution. There are more resonance structures (3) for the intermediate *X* formed by substitution at the 5-position than there are for substitution at the 4-position.

(We leave it to you to draw the two resonance structures for substitution at the 3-position.)

(b) Under the highly acidic condition the pyridine nitrogen is undoubtedly completely protonated. Substitution at the 5-position provides a carbocation intermediate that can be resonance-stabilized by the nitrogen of the ethoxycarbonylamino substituent.

Nitration at carbon-3 or carbon-5 places the positive charge so that it cannot be delocalized by the substituent nitrogen. Furthermore, the positive charge would be delocalized onto the already positive pyridine nitrogen, giving it a +2 charge and electron deficiency. (We leave it to you to draw the relevant resonance structures.)

You may be surprised that we can put a second positive charge in the pyridine ring in the mechanism shown. This is, in fact, a very high-energy situation. However, notice that the reaction is run under forcing conditions, which ensures a reasonable rate even for such an unfavorable reaction.

(c) The oxide can stabilize the carbocation intermediate when substitution occurs at the 4-position

(d)   An alkyl-substituted furan is more reactive than an alkyl-substituted thiophene, and substitution of furan at the 5-position is favored because the carbocation intermediate has three resonance structures, whereas substitution at the 3- or 4-position gives a carbocation with only two resonance structures. (Only the furan "side" of the molecule is shown below.)

(e)   This is a Friedel-Crafts acylation of thiophene at the 2-position, which gives a carbocation intermediate with three resonance structures..(Substitution at the 3-position gives a carbocation with only two resonance structures. We leave it to you to draw these structures.)

(f)   Nitration of 2,6-dimethylpyridine occurs at the 3-position. (For resonance structures resulting from substitutions at the 2- and 3-positions of pyridine, see Eqs. 26.33a–b, text p. 1440.) The methyl groups also direct substitution to the positions that are both ortho and para to them, as they do in benzene derivatives.

26.35   The order of increasing reactivity is $B \ll A < D < C$. Nucleophilic substitution in a pyridine requires that the negative charge in the intermediate Meisenheimer complex be delocalized either into a nitro substituent that is ortho or para to the leaving group, onto the pyridine nitrogen, or both. In A the negative charge can be delocalized onto the pyridine nitrogen, and in D the negative charge can be delocalized into the nitro group. In C the negative charge can be delocalized both ways. (The relative order of A and D could be interchanged.)

26.36   This is a nucleophilic aromatic substitution. Only the chlorine at the 4-position is displaced because only in that case can the negative charge in the reactive intermediate be delocalized to the pyridine nitrogen.

26.37   (a)   (b)   (c)   (d)   no reaction

(e)   no reaction   (f)   (g)

26.38   (a)   (b)   no reaction   (c)   (d)

(e)   (f)   (g)   (h)

26.39   (a)   Although one might be tempted to propose a Chichibabin reaction, 4-alkylpyridines with α-hydrogens are acidic toward strong bases such as sodium amide because a resonance-stabilized ion is formed..

(b)   The text has stressed that many reactions of pyridines can be rationalized by thinking of the N≡C bond of the pyridine as if it were a carbonyl group. Therefore, the hydrogens of the methyl group, like the α-hydrogens of a methyl ketone, are acidic enough to exchange for deuterium in basic D_2O. A resonance-stabilized anionic intermediate is involved.

(c)   The indole N—H is acidic enough to be removed by strong bases such as phenyllithium:

(d)   The product results from hydrogenation of the double bond that is *not* part of the aromatic π-electron system:

(e)   This is an S_N2-type alkylation of pyridine with the very reactive benzyl bromide.

(f)   This is a Wolff-Kishner reduction of the ketone (Sec. 19.13A), an effective way of introducing alkyl substituents into the pyridine ring.

(g)   This is a Reformatsky reaction of the aldehyde with a zinc enolate formed from the α-bromo ester.

a Reformatsky reagent

If the conditions (acid, heat) are severe enough, the alcohol will undergo dehydration to the **trans-alkene** by the usual carbocation mechanism. Notice that the carbocation is resonance-stabilized by delocalization into the furan ring.

(h)   This is a Diels–Alder reaction in which furan acts as the diene (Sec. 26.3B). Two stereoisomers of the product are possible.

26.40 Form the imine intermediate as shown in Eq. 26.61 on p. 1451 of the text. A carbanion intermediate can be formed from this imine in two ways. Decarboxylation is one way, shown in Eq. 26.62a. However, that doesn't occur in this case. A different resonance-stabilized carbanion intermediate is formed by proton removal from the α-carbon of the amino acid:

This carbanion can be protonated from either face to give enantiomers. The enantiomeric amino acids are freed by imine hydrolysis.

26.41 (a) The bases of DNA and their classification:

pyrimidines — purines

cytosine (C)   thymine (T)   guanine (G)   adenine (A)

(b)   The bases of RNA and their classification:

26.42   (a)   (b)   (c)

uridine monophosphate
(UMP, uridylic acid)

2'-deoxycytidine

adenosine diphosphate (ADP)

26.43   (a)   Guanosine   (b) 3'-deoxyuridine triphosphate
         (c)   Adenosine monophosphate, adenylic acid, AMP

26.44   The structures are shown in **Fig. SG26.2** On the following page.

Solutions to Integrated Problems

26.45   The conjugate acid is formed by protonation on the nitrogen of the double bond. Protonation on either of the other nitrogens does not yield a resonance-stabilized cation.

25.46   Protonation occurs on the amino nitrogen. The *unprotonated* amine (that is, the conjugate base) with the most resonance structures involving the nitrogen nonbonding electron pair is the weaker base. Remember that stabilization of a conjugate base makes it less basic. (See Sec. 3.7D, Study Problem 3.10.) 4-Aminobenzofuran has five resonance structures, and one of them delocalizes charge into the furan ring; that is, the π-electrons of the furan ring help to stabilize the amine. 5-Aminobenzofuran, in contrast, has four resonance structures. We predict that 5-aminobenzofuran is the stronger base. Validation of our prediction has to await the actual measurement of the conjugate-acid pK_a values of these compounds.

(Solution continues on p. 985)

Figure SG26.2 The structures requested in Problem 26.44. Notice that the sequence complementarity in (b) runs from the 5′ end to the 3′end—that is, A is complementary to T in (a), C to G, and G to C.

(Solution continues from p. 983)

resonance structures for 4-aminobenzofuran

resonance structures for 5-aminobenzofuran

26.47   To understand this issue, remember that the oxygen has two nonequivalent nonbonding electron pairs. One is in a 2p orbital and is involved in the aromatic π-electron system. If protonation occurs on this electron pair, the aromaticity of furan is disrupted. This would have a very high energy cost. The other nonbonding electron pair is in an sp² orbital that is not part of the aromatic π-electron system. Protonation of this electron pair also disrupts furan's aromaticity, but in a different way. The resonance structures of furan (Display 26.2, Sec. 26.1B, text p. 1428) show that the oxygen of furan has a significant positive charge. If the sp² electron pair is protonated, then oxygen takes on a +2 charge in all but one of its resonance structures.

This means that *protonated* furan has extremely unfavorable resonance structures. Yet the resonance stabilization of furan is due to its aromaticity. Therefore, protonation of the oxygen on the other electron pair also disrupts furan's aromaticity by forcing the nonbonding electrons in the 2p orbital to localize on oxygen so that the +2 charge is avoided. In other words, the 2p electrons can't delocalize, and, if they can't delocalize, protonated furan is not aromatic.

   As to protonation on carbon, this also disrupts furan's aromaticity; however, the resulting carbocation is at least resonance-stabilized. This is essentially an electrophilic aromatic substitution on furan by a proton:

Consequently, this is the preferred mode of protonation, but this protonated furan, because aromaticity is disrupted, is a very strong acid.

26.48   (a)   The order of increasing S_N1 reactivity is $B < A < D < C$. These reactivities parallel the corresponding reactivities of the respective ring systems themselves in electrophilic aromatic substitution reactions. For example, furan is more reactive than benzene in electrophilic aromatic substitution, and likewise compound C is more reactive than compound A in S_N1 reactions because of the involvement of an oxygen nonbonding electron pair in the resonance stabilization of the carbocation intermediate:

C

Atoms that donate electrons by resonance within the ring also accelerate solvolysis reactions. In contrast, a pyridine ring reduces the rates of solvolysis reactions because the nitrogen nonbonding electron pair *cannot* be used to stabilize the carbocation by resonance; this electron pair is in the plane of the ring and cannot overlap with the rest of the π-electron system. Furthermore, one resonance structure of the carbocation intermediate places positive charge and electron deficiency on the nitrogen, an electronegative atom. Such a structure is not important. Thus, the nitrogen exerts only its rate-retarding polar effect. The reactivity of the pyridine derivative is further reduced by the formation of HCl as a by-product of the solvolysis reaction. The HCl protonates the pyridine nitrogen and further enhances its electronegativity, thus reducing the solvolysis rate even more.

The carbocation intermediate in the solvolysis of the thiophene derivative has resonance structures analogous to those shown above for the furan derivative. However, because sulfur uses orbitals from period 3, its overlap is less effective in stabilizing the carbocation, just as in electrophilic substitution.

(b)   Compound *A* is the same as compound *C* in part (a). As shown in the solution to part (a), the cation derived from compound *A* has four resonance structures. The carbocation from compound *B* has only three resonance structures. Therefore, compound *A* undergoes solvolysis faster than *B*, because the carbocation intermediate is more stable.

B

26.49   The hydroxy isomer is aromatic, and the carbonyl isomer is not. The principle to apply is that the compound with the greater resonance energy (Table 26.1 on text p. 1429) has the greater amount of hydroxy isomer.

(a)   Because pyrrole has more resonance energy (that is, more stabilization due to aromaticity) than furan, 2-hydroxypyrrole contains a greater percentage of hydroxy isomer than 2-hydroxyfuran.

2-hydroxyfuran

2-hydroxypyrrole

(b)   Because benzene has greater aromatic stabilization (greater resonance energy) than pyridine, phenol contains the greater percentage of hydroxy isomer and 4-hydroxypyridine contains more carbonyl isomer.

phenol

4-hydroxypyridine

A

B

Carbonyl isomer *A* is the preferred one because the nitrogen can participate in amide-like resonance—that is, the nonbonding electron pair on nitrogen can be delocalized into the carbonyl group, and the resulting resonance structure is aromatic. (See the discussion on text p. 1444 for the analogous situation for 2-pyridone and 2-hydroxypyridine.)

26.50 (a) The chemists at MakeAmines are attempting to apply a reaction of pyridines and quinolines to indole. Because the nucleophilic reaction by the amide ion on a double bond of indole cannot result in charge delocalization to the nitrogen, the reaction does not take place. The reaction that occurs instead is removal of the N—H proton of indole.

(b) The MakeAmines chemists are attempting the Chichibabin reaction on a pyridine derivative, but this derivative contains a chlorine in the 2-position, which, as chloride ion, is a vastly superior leaving group to hydride. The chlorine is lost rather than hydride to give 2-aminopyridine.

2-aminopyridine

26.51 The conjugate-base anion of 2-aminopyridine, formed as shown in Eq. 26.38d, text p. 1442, reacts with another pyridine molecule in a Chichibabin-like reaction:

26.52 (a) Protonation of one pyrrole molecule at carbon-2 gives a carbocation that reacts with a second pyrrole. The resulting product *X* undergoes an acid-catalyzed double-bond shift to give *A*. This rearrangement is favorable because, in compound *A*, the double bond is conjugated with a nitrogen nonbonding electron pair whereas, in compound *X*, such conjugation is absent.

X

A

(b) Compound *A* from part (a) is protonated to give a resonance-stabilized carbocation *Y*, which serves as the electrophile in an electrophilic aromatic substitution reaction at carbon-2 of pyrrole.

26.53 The *p*-nitrobenzenediazonium ion, (which can be formed in the diazotization reaction of *p*-nitroaniline), acts as the electrophile in an electrophilic aromatic substitution at carbon-3 of indole:

26.54 (a) Prepare pyridine-2-carbonitrile and reduce it to the amine.

(b) Use the *N*-oxide to direct nitration to the 4-position. Then convert the nitrated *N*-oxide into 4-chloropyridine, which is then displaced with ethanethiolate anion.

(c) Prepare 1-(2-furyl)-1-ethanone ("2-acetylfuran") and carry out an aldol condensation with furfural.

(d)   Oxidize furfural to the carboxylic acid and convert the acid into the ester. Notice that the acid chloride method of esterification is preferable to acid-catalyzed esterification with propanol because furans are somewhat sensitive to strongly acidic conditions. (See, for example, Problem 26.55, text p. 1470.)

(e)   Oxidize the methyl group to a carboxylic acid, convert it into an acyl azide, and carry out a Curtius rearrangement to the isocyanate. Divide the isocyanate in half; convert one part into the amine, and then let the amine react with the remaining isocyanate to give the urea.

(f)   Use the enhanced "benzylic" acidity of *N*-methylpyridinium salts to form a nucleophilic anion at the benzylic carbon of the ethyl group, which is then cyanoethylated twice. (Compare to the reactions in Eqs. 26.54 and 26.55, text p. 1447. Cyanoethylation is discussed in Sec. 22.9A, text p. 1251).

(g)   Alkylate the conjugate-base anion of 2-methylpyridine with 1-iodopropane; then convert the resulting 2-butylpyridine into its conjugate-base anion, which is carboxylated with CO_2 like any other organolithium or Grignard reagent.

26.55 A compound with two degrees of unsaturation that has an NMR spectrum consisting entirely of singlets which is completely obliterated in base, and that gives hexane as a Clemmensen reduction product, must be a six-carbon unbranched diketone in which all hydrogens are α-hydrogens, that is, 2,5-hexanedione. A mechanism for the formation of this compound from 2,5-dimethylfuran is as follows.

a resonance-stabilized carbocation

enol of *A*

2,4-hexanedione
(compound *A*)

26.56 The unsaturation number of compound *A* is 4; its oxidation to nicotinic acid shows that it contains a pyridine ring with a 3-substituent. The pyridine ring accounts for all four degrees of unsaturation. Because pyridine contains five carbons, the side-chain has three carbons. Compound *A* cannot be a primary alcohol, because the oxidation product of such an alcohol would be an aldehyde, which could not have five exchangeable α-hydrogens. (An aldehyde of the form RCH$_2$CH=O has only two α-hydrogens.) Therefore, compound *A* is a secondary alcohol, and ketone *B* is its oxidation product.

A   *B*

26.57 (a) By the principle of microscopic reversibility (see text p. 228), the mechanism for the desulfonation of an arylsulfonic acid will be the reverse of the mechanism for sulfonation (see Sec. 16.4D, Eq. 16.21, text p. 883). For sulfonation to take place, a high concentration of the electrophilic agent (SO_3) is used in the presence of H_2SO_4, a powerful acid and dehydrating agent to provide a low concentration of water. For desulfonation to take place, a high concentration of water is required to reverse the equilibrium of the reaction. (SO_3 reacts with water to give H_2SO_4.)

(b) During the nitration, the sulfonyl chloride is converted into the sulfonic acid by the aqueous reaction conditions.

**2-thiophene-
sulfonyl chloride**
A   **4-nitro-2-thiophene-
sulfonic acid**
B   **3-nitrothiophene**
C

(c) 3-Nitrothiophene cannot be made directly from thiophene because thiophene undergoes nitration (and electrophilic aromatic substitution in general) mostly in the 2-position to give 2-nitrothiophene.

26.58 (a) Diethylamine and formaldehyde react to form an iminium ion by a mechanism completely analogous to the one shown in Eq. 23.39 on text p. 1296. This ion then serves as the electrophile in an electrophilic aromatic substitution reaction at the 3-position of indole.

Indole, benzofuran, and benzothiophene tend to give a much greater percentage of 3-substitution than pyrrole, furan, and thiophene

(b) Exchange is initiated with protonation by the deuterated acid at carbon-3 of the indole. In any exchange reaction of this sort, the experiment is set up so that there is a large excess of deuterium in the solvent so that the hydrogen that washes out into the solvent is swamped by the large excess of deuterium. In the last step, hydrogen is lost much more rapidly than deuterium because of the primary isotope effect (Sec. 9.5D). However, the relatively few molecules from which deuterium is lost are continually recycled through the mechanism until hydrogen is eventually washed out into the solvent.

The hydrogen at position-2 would probably undergo exchange as well, albeit more slowly.

(c)   This reaction is a carbocation rearrangement known as the *indolenine rearrangement*. The driving force for this reaction is the aromaticity of the product.

(d)   Alkylation of the nitrogen to give compound *A* (a quaternary ammonium salt) activates the carbon–nitrogen double bond toward a nucleophilic reaction with hydroxide ion, which, in turn, leads to ring opening.

compound *A*

(e)   Sulfonylation of the *N*-oxide by the sulfonyl chloride activates the carbon–nitrogen double bond toward a nucleophilic reaction with chloride ion, which is produced in the sulfonation reaction. Hydroxide-promoted elimination forms a double bond and liberates chloride ion, which then reacts as a nucleophile at the terminal carbon of the double bond and displaces tosylate ion in an allylic variation of an S_N2 reaction (sometimes called an S_N2' reaction). This last step is undoubtedly driven by the aromatic stability of the product. The mechanism below begins with the sulfonate ester and chloride ion, produced in the formation of the sulfonate ester. (For the mechanism of sulfonate ester formation, see Further Exploration 11.2 on p. 290 of this manual.) The toluenesulfonate group is abbreviated —OTs.

(f)   Because of the three chlorines on its α-carbon, trichloroacetyl chloride is unusually reactive in nucleophilic acyl substitution reactions. (Recall that reactivity toward nucleophiles of a carbonyl compound is enhanced by electron-withdrawing substituents on the α-carbon; see Sec. 19.8B.) The π electrons of pyrrole are nucleophilic enough to be acylated by this acid chloride to give compound *B*. The remaining mechanistic steps are the same as the final steps of a haloform reaction (see Eq. 22.30b on text p. 1205), except that the base is ethoxide rather than hydroxide. Ethoxide displaces the trichloromethyl anion, which reacts with ethanol to give chloroform and ethoxide.

(g)   This reaction is a nucleophilic aromatic substitution reaction by the benzenethiolate anion (PhS⁻), which is formed by the reaction of benzenethiol with triethylamine. This reaction is like a nucleophilic aromatic substitution reaction on pyridine in the sense that a pyridine nitrogen stabilizes the anionic intermediate.

(h)   Acetic anhydride dehydrates the nitric acid to nitronium acetate, which is the source of nitronium ion:

Furan then undergoes a 1,4-addition with nitronium acetate to give intermediate *A*. (Nitronium ion is the electrophile, acetate ion is the nucleophile.)

Finally, pyridine, acting as a base, removes a proton to initiate a 1,4-elimination of acetic acid to form 2-nitrofuran.

(i)   A Diels-Alder reaction of the "furan" with maleic anhydride yields a bicyclic anhydride. (Only the endo isomer is shown, but the stereochemistry is immaterial to the ultimate outcome of the sequence.)

The anhydride is hydrolyzed to the dicarboxylic acid (mechanism not shown), and the dicarboxylic acid undergoes a successive acid-catalyzed ether cleavage and alcohol dehydration to give the product. The aromaticity of the final product is the driving force for the reaction.

26.59 The reactivity order is $C < A < B$. In compounds A and B, the leaving groups have resonance structures that place positive charge on the nitrogen that is bonded to the carbonyl.

resonance structures of A:
less important

resonance structures of B:
more important

Recall (Sec. 21.7E) that compounds with electronegative leaving groups are particularly reactive in nucleophilic substitution at a carbonyl carbon. Both of these groups, with a positively charged nitrogen bonded to the carbonyl, are considerably more reactive than C, which is an ordinary amide. Because the resonance structure of compound B *delocalizes* charge, its resonance structure is more important than the resonance structure of compound A, which *separates* charge. Therefore, compound B has a greater positive charge on the nitrogen bound to the carbonyl than compound A, so it is therefore more reactive.

26.60 (a)   For simplicity, represent the histidine side-chain as R; that is,

In this mechanism, the pyruvyl group is acting as a "surrogate" of pyridoxal phosphate. The first step is formation of the imine A of the pyruvyl group. This forms at the ketone carbonyl (why?). The mechanism of imine formation is outlined in Eq. 19.76a–b on text p. 1064. This imine then loses CO_2 to give carbanion B, which is stabilized because its electron pair is delocalized into the other carbonyl group. (**E** = the rest of the enzyme.)

Protonation of this carbanion yields a new imine, which hydrolyses by a mechanism that is the reverse of imine formation to give histamine:

(c)   The carbonyl group of carbanion *B* shown in part (a) serves as the electron sink for decarboxylation, as shown in the above mechanism.

26.61   (a)   First, serine forms an imine with pyridoxal phosphate as shown in Eq. 26.61 on text p. 1451. This imine undergoes a elimination of the elements of water to give the imine of dehydroserine (compound *A* in the problem). The role of imine formation is to make the eliminated proton quite acidic; recall (Sec. 17.3B) that E2 reactions involving acidic β-protons are particularly rapid. (Why is this β proton acidic?) A base on the enzyme B: abstracts the proton, and an acidic group on the enzyme $\overset{+}{B}$—H protonates the —OH group as it leaves to form water.

imine derivative of serine and
pyridoxal phosphate

compound *A*

$+ \ H_2O \ + \ \overset{+}{B}—H \ + \ :B$

(b)   Delocalization of electrons from the double bond on the colored carbon to the pyridinium nitrogen reveals that this carbon has carbocation character, and should thus be electrophilic:

(c)   An electron pair of the indole ring π electrons react as a nucleophile at the electrophilic carbon of the imine of β-alanine:

The resulting carbocation intermediate is resonance-stabilized by the electrons of the indole nitrogen. A basic group removes an O—H proton to start an electronic cascade that generates glyceraldehyde-3-phosphate and an imine *C*:

Finally, compound *C* is protonated on carbon to form an imine *D* between pyridoxal phosphate and tryptophan. Hydrolysis of this imine gives pyridoxal phosphate and the amino acid.

imine of tryptophan and pyridoxal phosphate

tryptophan

pyridoxal phosphate

26.62 (a) Exchange occurs when the conjugate-base anion is formed by abstraction of one of the colored protons, and this anion removes a deuteron (a deuterium nucleus) from the solvent. The anionic intermediate involved in this exchange is resonance-stabilized by delocalizing the negative charge onto two electronegative atoms—one of the nitrogens and the carbonyl oxygen. In contrast, removal of a proton from the other methyl group gives an anion in which the negative charge is not delocalized onto an electronegative atom; consequently, this anion is less stable. Therefore, exchange of a colored hydrogen occurs because this exchange involves the more stable anionic intermediate.

anionic intermediate involved in the exchange of the colored protons

The imide hydrogen is also exchanged by a base-catalyzed mechanism that involves the conjugate-base anion of the imide. (Imides, like other β-dicarbonyl compounds, are acidic; see Problem 22.2 on text p. 1197.)

(b) Removal of a hydrogen from the black methyl group gives an anion that is stabilized by the polar effect of the attached positively charged nitrogen and by resonance interaction with the adjacent double bond. However, the resonance structure of this anion (draw it) delocalizes charge to a carbon atom. In contrast, removal of a hydrogen from the colored methyl group gives an "anion" which is actually a neutral molecule; that is, the negative charge is delocalized to the positively charged nitrogen. Such charge neutralization is a significant stabilizing effect, and is the reason for the greater acidity of the colored hydrogen.

resonance structures of the "anion" formed by removal of a hydrogen from the colored methyl group

(c)   The positive charge on the nitrogen activates the carbon–nitrogen double bond towards a nucleophilic reaction with hydroxide ion. This forms a tetrahedral addition intermediate that breaks down to a substituted formamide *A*, which then undergoes amide hydrolysis to the products shown in the problem.

A

(ionizes under the basic conditions)

(d)   2-Pyridone does not hydrolyze because it undergoes a different reaction: ionization to its conjugate-base anion. The ion doesn't hydrolyze because the ionized amide is resonance-stabilized *and aromatic*. Remember that the more stable a carbonyl compound is, the slower it hydrolyzes (Sec. 21.7E).

Furthermore, hydrolysis would require that a second negative charge be introduced into the molecule when ⁻OH reacts at the carbonyl group. γ-Butyrolactam is much less acidic; consequently, ionization does not compete with the hydrolysis reaction.

conjugate-base anion of 2-pyridone, an aromatic species

(e)   The reaction of 4-chloropyridine with ammonia is a nucleophilic aromatic substitution reaction that occurs readily because the negative charge in the addition intermediate is delocalized to the pyridine nitrogen, an electronegative atom, as shown in Eq. 26.47b, text p. 1445 (with :Nuc = :NH₃ and Y = Cl). In contrast, the negative charge in the corresponding intermediate for nucleophilic aromatic substitution of 3-

chloropyridine can only be delocalized to carbon atoms. (See Eq. 26.47c, text page 1445.) As a result, 3-chloropyridine is much less reactive.

26.63 Step (1) is a nitration reaction. Because 2-pyridone derivatives such as this starting material contain significant amounts of their aromatic hydroxypyridine isomers (Eq. 26.42 on text p. 1444), the nitro group can be introduced by aromatic nitration. The conditions of step (1) are HNO_3 and H_2SO_4.

The reagent for step (2) is PCl_5 and heat. (See Eq. 26.44 on text p. 1444.)

Step (3) is a reduction of both the nitro group and the nitrile to amine groups; catalytic hydrogenation is the method of choice. Evidently, the chlorine is also removed under the conditions of this hydrogenation. To see why this is reasonable, recall that 2-chloropyridines have some of the reactivity characteristics of acid chlorides, and that acid chlorides are reduced to aldehydes (that is, the chlorine is replaced by a hydrogen) in a type of catalytic hydrogenation (Rosenmund reduction; see Sec. 21.9D of the text).

Step (4) is an ether cleavage, which can be effected with aqueous H_2SO_4; water is the nucleophile in this reaction. This ether is activated toward cleavage by either an S_N1 or an S_N2 mechanism because it is benzylic. Neutralization with aqueous hydroxide gives the free amine shown.

Finally, step (5) can be realized by diazotization with $NaNO_2$ in aqueous H_2SO_4, which converts the benzylic amine into a benzylic diazonium ion, and the arylamine into an aryldiazonium ion; nitrogen is rapidly displaced from the benzylic diazonium ion by water. Elimination, usually a competing reaction in the decomposition of alkyldiazonium ions, is not observed here because there are no β-hydrogens. The aryldiazonium ion is also converted into the phenolic hydroxy group by heating the aqueous solution. (See the discussion above Eq. 26.42 on text p. 1444.)

26.64 (a) Bromine adds to the double bond, and KOH saponifies the resulting dibromo lactone. An internal nucleophilic substitution reaction then closes the five-membered ring with loss of one bromine as bromide ion. (Recall that α-halo carbonyl compounds are particularly reactive in nucleophilic substitution reactions; see Sec. 22.3D.) An E2 reaction forms the double bond with loss of the second bromine as bromide ion. (This is a very favorable E2 reaction; why? See Sec. 17.3B.) The mechanism of bromine addition, not shown below, is exactly analogous to the mechanism shown in Eqs. 5.10a-b on text pp. 240–241; and the mechanism of saponification, also not shown below, is exactly analogous to the mechanism shown in Eqs. 21.33a–b on text p. 1147.)

(b) This reaction begins as an aldol condensation between the conjugate-base enolate ion A of the diester and the diketone. Normally esters undergo Claisen condensations, but, because 1,2-diketones are particularly reactive (why?), the aldol condensation is observed instead. Once the two molecules are joined by the first aldol condensation, a second aldol condensation by an analogous mechanism closes the ring.

(c)   The mechanism shown below starts with the enamine formed between the amino group and the carbonyl group of 2-butanone. (The mechanism of enamine formation is discussed in Sec. 19.12C of the text.) A key step is the nucleophilic reaction of this enamine at the carbon of the protonated carbonyl group. As this reaction demonstrates, enamines, like enolate ions, are nucleophilic at the α-carbon. (See part (e) for a more extensive discussion of this point.) Note that, although primary amines form imines with ketones, imines are in equilibrium with enamines just as ketones are in equilibrium with enols. (See Eq. 19.83a–b on text p. 1067–1068.) Water is produced as a by-product of enamine formation. The conjugate acid of water, H_3O^+, is shown as the catalyzing acid, but any of the other acids present could also catalyze the reaction.

(d)   Imine formation between the amine and one of the ketone groups is followed by an intramolecular electrophilic aromatic substitution reaction in which the protonated ketone serves as the electrophile. The mechanism below begins with the imine; the mechanism of imine formation is discussed in Sec. 19.12A in the text.

(e) Diethylamine serves as a base catalyst; its first function is to form the conjugate-base enolate ion *A* of ethyl acetoacetate:

The fact that a secondary amine rather than a tertiary amine is used suggests that the role of diethylamine is also to form the enamine, which, like an enol, can serve as a nucleophile:

enamine is nucleophilic
at the α-carbon

Diethylamine can also form an iminium ion by reaction with formaldehyde (Eq. 23.39, text p. 1296):

$$Et_2NH + H_2C{=}O \rightleftharpoons Et_2N{-}CH_2{-}OH \xrightarrow{Et_2\overset{+}{N}H_2} Et_2\overset{+}{N}{=}CH_2 + H_2O$$

iminium ion

The first stage of the mechanism is a crossed-aldol condensation in which enolate *A* (or the enamine shown above) condenses with formaldehyde (or with the iminium ion shown above). The mechanism of the aldol condensation is given in Sec. 22.4A, text pp. 1210–1211. The product of this aldol condensation then undergoes a conjugate-addition reaction (that is, a Michael addition) with a second enolate ion *A* (or a second enamine). Ammonia reacts with either one of the ketone functional groups to form an enamine, and the ring is closed when the amino group of that enamine forms an intramolecular enamine with the second ketone group. (The mechanism of enamine formation is discussed in Sec. 19.12C of the text; see the note about the equilibration of imines and enamines in the solution to part (c) of this problem.)

crossed aldol condensation product of ethyl acetoacetate and formaldehyde

Michael addition

A

+ HNEt₂

NH₃ enamine formation

intramolecular enamine formation

We leave it to you to complete the similar mechanism with an enamine as the nucleophile.

(f)   The *o*-nitro group is an essential element in the success of this reaction because its presence makes the benzylic methyl hydrogens acidic enough to be remove by ethoxide; the resulting anion is resonance-stabilized:

o-nitrotoluene

+ EtOH

In a variation of the Claisen condensation (see Sec. 22.6A), this nucleophilic anion reacts with a carbonyl group of diethyl oxalate, displacing ethoxide. Like the Claisen condensation, this reaction is driven to completion by ionization of the product. For this reason, at least one equivalent of the base must be used:

diethyl oxalate

+ EtO⁻ ⇌ + EtOH

The anion is neutralized by protonation in acetic acid, and the nitro group is converted into an amino group in a separate reduction step. [Catalytic hydrogenation (see Sec. 23.11B) is the reduction method used here.] The amino group thus formed reacts with the neighboring ketone to yield, after acid–base equilibria, an "enamine," that is, the aromatic indole.

B

C
ethyl 2-indolecarboxylate

(g)   The reaction proceeds by enamine formation (see Sec. 19.12C) followed by an intramolecular Heck reaction (see Sec. 18.6A) in the presence of a tertiary-amine base (DABCO, 1,4-diazabicyclo[2,2,2]octane, structure in the problem). The base is abbreviated here as RNH_2.

$$+ \ R\overset{+}{N}H_3 + \ ^-OH$$

$$\longrightarrow R\ddot{N}H_2 + H_2O$$

The catalytic intermediates in the Heck reaction are (*1*) the first-formed oxidative-addition product; (*2*) the product of ligand exchange with the alkene; (*3*) the 1,2-ligand insertion product. (L = a triarylphosphine ligand.)

(*1*) Oxidative-addition product   (*2*) Ligand-exchange product   (*3*) syn-1,2-Ligand-insertion product

Because subsequent β-elimination of hydride and the Pd must be syn, the initially formed product is probably not the product shown in the problem, but rather the isomer shown below. This then undergoes a isomerization catalyzed by the amine base to give the product. The anionic intermediate is both allylic and benzylic, and one of its resonance forms is aromatic. The isomerization equilibrium is driven by the aromaticity of the product.

26.65 (a) The pairing of an imine isomer of C with A: (b) The pairing of an enol isomer of T with G:

imine isomer
of cytosine

adenine

H₃C — enol isomer
of thymine — guanine

26.66 (a) Periodic acid cleaves the *cis*-2′,3′-diol group at the 3′-end of RNA to form a dialdehyde. (Note that only the 3′ residue of RNA contains a 2′,3′-diol.) The presence of an aldehyde carbonyl group makes the α-hydrogen at carbon-4′ more acidic. Removal of this hydrogen sparks an E2 reaction in which the internucleotide bond is cleaved.

(b) DNA cannot undergo this reaction because a cis 2′,3′-diol group is needed for the periodate cleavage reaction. DNA, of course, has no 2′-hydroxy group at the 3′ end, and thus, no 2′,3′-diol.

26.67 The reaction of RNA involves ionization of the 2′-hydroxy group within each residue and reaction of the resulting alkoxide ion as a nucleophile at the neighboring phosphorus. This reaction is particularly rapid because it is intramolecular; it is an example of neighboring-group participation. (See Sec. 12.8.) Loss of the leaving group splits the internucleotide bond. DNA cannot undergo this reaction because the reaction depends on the presence of the 2′-hydroxy group to act as an intramolecular nucleophile; DNA lacks this hydroxy group. (That is why it is called deoxyribonucleic acid.)

27 | Amino Acids, Peptides, and Proteins

 CHAPTER 27 SUMMARY

- Amino acids are compounds that contain both an amino group and a carboxylic acid group. Peptides and proteins are polymers of the α-amino acids.

- The common naturally occurring chiral amino acids have the L configuration, which is the same as the *S* configuration for all amino acids except cysteine.

- Amino acids, as well as peptides that contain both acidic and basic side-chain groups, exist at neutral pH as zwitterions.

- The charge on a peptide can be deduced from a knowledge of its isoelectric point relative to the pH of the solution. For example, a peptide with an isoelectric point >7 is positively charged at pH 7.

- Common methods for the synthesis of α-amino acids include the alkylation of ammonia with *a*-halo acids, the alkylation of acetamidomalonate esters followed by hydrolysis and decarboxylation, and the Strecker synthesis.

- Amino acids react both as amines and carboxylic acids. For example, the amino group can be acylated, and the carboxylic acid group can be esterified.

- The amide bonds of proteins and peptides can be hydrolyzed by heating them for several hours in 6 *N* HCl or 6 *N* NaOH solution. The constituent α-amino acids are formed. In amino acid analysis, the amino acids (or their derivatives) are separated by chromatography and quantified. In this way, the relative amount of each amino acid in the peptide or protein is determined.

- Peptide synthesis strategically involves the attachment of α-amino acids one at a time to a peptide chain beginning at the carboxy terminus. Amino-protecting groups such as the Fmoc group must be

used to prevent competing reactions. In solid-phase peptide synthesis, an amino-protected amino acid is covalently attached to an insoluble resin; the protecting group is removed; and an Fmoc-protected active ester (such as an HOBt ester) of the next α-amino acid is attached by an ester aminolysis reaction with the deprotected amino group. A coupling agent such as PyBOP (text p. 1499) is used to prepare the active esters. The new peptide is deprotected, and the cycle is continued. At the conclusion of the synthesis, the peptide is removed from the resin with trifluoroacetic acid.

- Proteins are synthetized biologically from the amino terminus. The genetic code from DNA is translated into messenger RNA (mRNA). Each three-base triplet in mRNA codes for one amino acid of a protein. Amino acids are introduced as aminoacyl-transfer RNA (tRNA) derivatives, which are carboxylate esters of the 3′-hydroxy group of a tRNA. Adjacent mRNA triplets dictate the binding of aminoacyl-tRNAs with complementary anticodons to adjacent sites on the ribosome, and the peptide bond is formed by an ester aminolysis reaction.

- Specific hydrolysis reactions catalyzed by enzymes are important in determining the primary structures of proteins. Peptides and proteins undergo trypsin-catalyzed hydrolysis at the carbonyl group of arginine and lysine residues. Chymotrypsin catalyzes the hydrolysis of peptides at hydrophobic residues such as Phe, Tyr, Trp, and occasionally Leu and Ile.

- The primary structure of a peptide or protein includes its amino-acid sequence and the location of its disulfide bonds. Peptides and proteins can be sequenced by tandem mass spectrometry (MS–MS) and by the Edman degradation.

- In MS–MS sequencing, the M + 1 ions of a peptide undergo fragmentation at the peptide bonds. In many cases, the sequence can be read directly from the mass differences of the fragment peaks.

- In the Edman degradation, a peptide is treated with phenyl isothiocyanate followed by acid. Each cycle of this degradation yields a phenylthiohydantoin (PTH) derivative of the amino-terminal amino acid plus a peptide that is one residue shorter, which is then subjected to the same procedure.

■ Posttranslational modifications of proteins are chemical alterations that occur following its biosynthesis. Two important examples are protein phosphorylation, which occurs mostly at Ser, Tyr, or Cys residues; and protein glycosylation, which occurs at the amide nitrogen of Asn or the oxygen of Ser or Thr residues.

■ The higher-order structure of proteins includes secondary, tertiary, and quaternary structure. The secondary structure of a protein is a description of the relative orientations of the planes of its peptide bonds. Common types of secondary structure are the α-helix and the β-sheet. Proteins are aggregates of localized regions of these secondary structures connected by regions of random structure. The secondary structure of a protein can be conveyed with a ribbon structure.

■ The tertiary structure of a protein is a complete description of its three-dimensional structure. The quaternary structure of a protein is the manner in which individual subunits of a protein aggregate to form larger structures. In addition to the peptide and the disulfide bonds, noncovalent forces determine the three-dimensional structures of proteins. The noncovalent attractions within the protein are balanced against the noncovalent attractions of the protein with the solvent or surrounding milieu.

■ Protein secondary, tertiary, and quaternary structures are disrupted when the protein is denatured by heat or by treatment with thiols, such as DTT, and 8 M aqueous urea.

■ Enzymes are highly specific and efficient biological catalysts. Most enzymes are proteins. A substrate is bound noncovalently at the enzyme active site before it is converted into products.

■ A compound that binds tightly to an enzyme active site without undergoing a reaction is called a competitive inhibitor. Competitive inhibitors block the access of substrates to the enzyme active site. The development of competitive inhibitors is an important basis of drug design. The anti-HIV drugs constitute one example of drugs based on the concept of competitive inhibition.

FURTHER EXPLORATIONS

27.1 Solid-Phase Peptide Synthesis

The reactions of solid-phase peptide synthesis must be carried out in solvents that *solvate* the Merrifield resin—that is, solvents that interact well with the groups of the resin. If these groups are not solvated, reagents in solution cannot penetrate the resin and no reaction takes place. Because the polystyrene-based Merrifield resin is essentially a polymeric ethylbenzene, it is mostly a hydrocarbon. Yet many of the reactions of solid-phase peptide synthesis involve ionic compounds and intermediates. Water and alcohols do not work as solvents in solid-phase peptide synthesis because they do not solvate the resin. Dimethylformamide, *N*-methylpyrrolidone, and methylene chloride (dichloromethane) have been found empirically to solvate the Merrifield resin adequately and to dissolve the reagents used in solid-phase peptide synthesis. In fact, when such solvents are added to a Merrifield resin, a visually perceptible swelling of the resin bed occurs; this is caused by incorporation of the solvent into the resin.

The use of a cesium salt in Eq. 27.38 on text p. 1497 is related to the requirement for aprotic solvents. In DMF, ionic compounds tend to form ion pairs and other aggregates, and carboxylate salts have metal–oxygen bonds with a significant amount of covalent character. Yet the alkylation shown in this equation is fastest when the carboxylate oxygen has as much *ionic character* as possible, because a carboxylate oxygen with a full negative charge is more nucleophilic and therefore reacts much more rapidly in S_N2 reactions than an ion-paired (or covalently bonded) carboxylate oxygen. Because cesium is the most electropositive of the readily available alkali metals, its carboxylate salts have more ionic character than lithium, sodium, or potassium salts and therefore are most reactive in S_N2 reactions.

REACTION REVIEW

I.   SYNTHESIS AND ENANTIOMERIC RESOLUTION OF α-AMINO ACIDS

    A.   ALKYLATION OF AMMONIA

        1.   Some α-amino acids can be prepared by alkylation of ammonia with α-bromo carboxylic acids.

$$\underset{\overset{|}{C}}{\overset{\overset{Br}{|}}{}}\!\!-CO_2H \xrightarrow{\ NH_3\text{ (large excess)}\ } \underset{\overset{|}{C}}{\overset{\overset{\overset{+}{N}H_3}{|}}{}}\!\!-CO_2^- + \overset{+}{N}H_4 + Br^-$$

            a.   This is an S_N2 reaction in which ammonia acts as the nucleophile.

            b.   The use of a large excess of ammonia in the synthesis favors monoalkylation.

        2.   Multiple alkylation is not a problem for two reasons:

            a.   Amino acids are less reactive toward alkylating agents than simple alkylamines because the amino groups of amino acids are less basic (nucleophilic) than ammonia and simple alkylamines.

            b.   Branching in amino acids provides a steric effect that retards further alkylation.

    B.   ALKYLATION OF AMINOMALONATE DERIVATIVES

        1.   Another method for preparing α-amino acids is a variation of the malonic ester synthesis.

$$\underset{\textbf{diethyl α-acetamidomalonate}}{H_3C\overset{\overset{O}{\|}}{-C}\diagdown NH\!\!-\underset{\overset{|}{CO_2Et}}{\overset{\overset{CO_2Et}{|}}{C}}\!\!-H} + X\!\!-R \longrightarrow \longrightarrow H_3\overset{+}{N}\!\!-\underset{\underset{\overset{|}{CO_2Et}}{X^-}}{CH}\!\!-R + CO_2 + 2\,EtOH + CH_3CO_2H$$

        2.   The malonic ester derivative used is one in which a protected amino group is already in place: diethyl α-acetamidomalonate.

            a.   Treatment of diethyl α-acetamidomalonate with sodium ethoxide in ethanol forms the conjugate base enolate ion, which is then alkylated with an alkyl halide.

            b.   The resulting compound is then treated with hot aqueous HCl or HBr which accomplishes three things:

                i.   The ester groups are hydrolyzed to give a substituted malonic acid.

                ii.   The malonic acid derivative is decarboxylated under the reaction conditions.

                iii.   The acetamido group, an amide, is also hydrolyzed.

c. Neutralization affords the α-amino acid.

$$\text{AcNH}-\underset{\underset{H}{|}}{\overset{\overset{CO_2Et}{|}}{C}}-CO_2Et \xrightarrow[\text{2) R}-X]{\text{1) NaOEt/HOEt}} \text{AcNH}-\underset{\underset{R}{|}}{\overset{\overset{CO_2Et}{|}}{C}}-CO_2Et \xrightarrow[\text{heat}]{H_3O^+} H_3\overset{+}{N}-\underset{\underset{R}{|}}{CH}-CO_2H + 2\,EtOH + CO_2\uparrow + HOAc$$

C. STRECKER SYNTHESIS

1. The preparation and hydrolysis of α-amino nitriles to α-amino acids is called the Strecker synthesis.

$$H_2N-\underset{\underset{R}{|}}{CH}-C\equiv N \xrightarrow[\text{heat}]{H_3O^+} H_3\overset{+}{N}-\underset{\underset{R}{|}}{CH}-CO_2^- + \overset{+}{N}H_4$$

2. α-Amino nitriles are prepared by treatment of aldehydes with ammonia in the presence of cyanide ion, a reaction that probably involves an imine intermediate.

   a. The conjugate acid of the imine reacts with cyanide under the conditions of the reaction to give the α-amino nitrile.

   b. The addition of cyanide to an imine is analogous to the formation of a cyanohydrin from an aldehyde or ketone.

$$O=\underset{\underset{R}{|}}{CH} \xrightarrow[\text{NaCN}]{\text{NH}_4\text{Cl}} HN=\underset{\underset{R}{|}}{CH} \longrightarrow H_2N-\underset{\underset{R}{|}}{CH}-C\equiv N$$

D. ENANTIOMERIC RESOLUTION OF α-AMINO ACIDS

1. α-Amino acids synthesized by common laboratory methods are racemic; since many applications require the pure enantiomers, the racemic compounds must be resolved.

2. An alternative approach to the preparation of enantiomerically pure amino acids is the synthesis of amino acids by microbiological fermentation.

3. Certain enzymes can be used to resolve racemic amino acid derivatives by selectively catalyzing a reaction of one enantiomer.

II. REACTIONS OF AMINO ACIDS AND SYNTHESIS OF PEPTIDES

A. ACYLATION AND ESTERIFICATION

1. Amino acids undergo many of the characteristic reactions of both amines and carboxylic acids.

   a. They can be acylated by on the amino group by acid chlorides or anhydrides.

   b. Their carboxylic acid groups are easily esterified by heating with an alcohol and a strong acid catalyst.

B. SOLID-PHASE PEPTIDE SYNTHESIS—GENERAL

1. In a method called solid-phase peptide synthesis, often abbreviated SPPS, the carboxy-terminal amino acid is covalently anchored to an insoluble polymer, and the peptide is "grown" by adding one residue at a time to this polymer.

   a. Solutions containing the appropriate reagents are allowed to contact the resin with shaking.

   b. At the conclusion of each step, the resin containing the peptide is simply filtered away from the solution, which contains soluble by-products and impurities.

   c. The completed peptide is removed from the resin by a reaction that cleaves the peptide–resin bond.

2. The advantage of the solid-phase method is the ease with which dissolved impurities and by-products are removed from the resin-bound peptide by simple filtration.

a. The same reagents used in solid-phase peptide synthesis can also be used for peptide synthesis in solution, but removal of by-products from the product peptide is sometimes difficult.

b. In order to avoid resin-bound impurities, each step in the solid-phase synthesis must occur with virtually 100% yield, an ideal that is often approached in practice.

C. *SOLID-PHASE PEPTIDE SYNTHESIS—PROTECTING GROUPS AND SPECIFIC REACTIONS*

1. In a solid-phase peptide synthesis, the amino group of the amino acids is protected with a special acyl group, the (9-fluorenylmethyloxy)carbonyl (Fmoc) group. (Fmoc is pronounced "Eff-mock.")

a. Fmoc-amino acids are prepared by allowing an *N*-hydroxysuccinimide ester derivative, Fmoc-NHS, to react with the amino group of the amino acid.

b. An S_N2 reaction between the cesium salt of an Fmoc-amino acid and the chloromethyl group of the resin results in the formation of an ester linkage to the resin.

2. Once the Fmoc-amino acid is anchored to the resin, the Fmoc-protecting group is removed by treatment with an amine base, piperidine, in an E2 reaction.

$$=CH_2 + CO_2 + H_2N-CH-C \quad (R, \quad O-CH_2-\bigcirc-resin)$$

resin-bound amino acid

a. The product of the β-elimination is a carbamate anion.

b. The carbamate anion decarboxylates under the reaction conditions.

3. Coupling of the next Fmoc-amino acid to the free amino group of the resin-bound amino acid is effected by the reagent PyBOP, which forms an ester of the amino acid with 1-hydroxy-1,2,3-benzotriazole (HOBt).

the Fmoc-amino acid
(conjugate base)

PyBOP

PyBOP ester of the
Fmoc amino acid

HOBt ester of the
Fmoc-amino acid

a. The HOBt ester of the Fmoc-α-amino acid formed in this way reacts with the amino group of the resin-bound amino acid.

b. The resulting ester aminolysis reaction results in a new peptide bond.

4. The process of deprotection (items 2 and 3 previous) is repeated many times to give an amino-protected peptide attached to the resin.

5. The ester linkage used to bind the peptide to the resin is broken by a carbocation mechanism using 50–60% trifluoroacetic acid (TFA) in dichloromethane.

resin-bound peptide

peptide

III. ANALYTICALLY IMPORTANT REACTIONS OF PEPTIDES

A. *HYDROLYSIS OF PEPTIDES; AMINO ACID ANALYSIS*

1. An important reaction used to determine the structures of unknown peptides is hydrolysis in strong acid of the amide bonds of the peptide to give its constituent amino acids.

   a. The amide bonds of glutamine (Gln) and asparagine (Asn) are also hydrolyzed.

   b. Aspartic acid determined by amino acid analysis is the sum of Asp and Asn residues; glutamic acid is the sum of Glu and Gln.

2. When a peptide or protein is hydrolyzed, the product amino acids can be separated, identified, and quantitated by a technique called amino acid analysis.

3. In the method most commonly used today, the amino acids in the hydrolyzed mixture are allowed to react with AQC-NHS to form AQC-amino acids.

1-[[(6-quinolylamino)carbonyl]-
oxy]-2,5-pyrrolidinedione
(AQC-NHS)

α-amino acid

AQC-amino acid

N-hydroxysuccinimide

   a. The mixture of AQC-amino acids are separated by passing them through a C18 HPLC column under very carefully defined conditions; the time at which each derivatized amino acid emerges from the column is accurately known.

   b. The AQC-amino acids with the greatest hydrocarbon character are eluted last.

   c. The AQC group is fluorescent; the absorption of light at 254 nm (UV detector) or emission of light at 395 nm (fluorescence detector) is used to determine the relative amounts of each amino acid present.

4. Amino acid analysis can determine the identity and relative amounts of amino acid residues, but not their relative order, within a peptide.

B. *SEQUENCE ANALYSIS OF PEPTIDES*

1. The actual arrangement, or sequential order, of amino acid residues in a peptide is called the amino acid sequence, or primary sequence, of the peptide.

2. Tandem mass spectrometry (MS-MS) can be used to determine the sequence of short peptides since peptides undergo residue-by-residue fragmentation, in many cases from both ends simultaneously, in a mass spectrometer.

   a. The most common sequencing method used today involves proteolytic cleavage of larger peptides into smaller peptides (see C. and D. below) followed by sequencing of the smaller peptides with MS-MS.

3. It is possible to remove one residue at a time chemically from the amino end of the peptide, identify it, and then repeat the process sequentially on the remaining peptide; the standard method for implementing this strategy is called the Edman degradation.

a. The peptide is treated with phenyl isothiocyanate (Edman reagent), with which it reacts at its amino groups to give a thiourea derivative. (Only the reaction at the terminal amino group is relevant to the degradation.)

b. Any remaining phenyl isothiocyanate is removed, and the modified peptide is then treated with anhydrous trifluoroacetic acid.

c. The sulfur of the thiourea, which is nucleophilic, displaces the amino group of the adjacent residue to yield a five-membered heterocycle called a thiazolinone; the other product of the reaction is a peptide that is one residue shorter.

d. When treated subsequently with aqueous acid, the thiazolinone derivative forms an isomer called a phenylthiohydantoin (PTH).

e. Because the PTH derivative carries the characteristic side chain of the amino-terminal residue, the structure of the PTH derivative identifies the amino acid residue that was removed.

4. In practice, this type of analysis is limited to 20–60 consecutive residues because the yields at each step are not perfectly quantitative; therefore, an increasingly complex mixture of peptides is formed with each successive step in the cleavage, and after a number of such steps the results become ambiguous.

C. *SPECIFIC CLEAVAGE OF PEPTIDES: INTRODUCTION*

1. The amino acid sequence of most large proteins is determined by breaking the protein into smaller peptides and sequencing these peptides individually. (The sequence of the parent protein is reconstructed from the sequences of the peptides.)

2. When breaking a larger protein into smaller peptides, it is desirable to use reactions that cleave the protein in high yield at well-defined points so that a relatively small number of peptides are obtained.

3. Although there are a few chemical methods to specifically cleave peptides, the method discussed in the text involves the use of proteolytic enzymes to catalyze peptide-bond hydrolysis.

D. *PEPTIDE CLEAVAGE WITH PROTEOLYTIC ENZYMES*

1. A number of enzymes (called proteases, peptidases, or proteolytic enzymes) catalyze the hydrolysis of peptide bonds at specific points in an amino acid sequence.

a. One of the most widely used proteases is the enzyme trypsin, which catalyzes the hydrolysis of peptides or proteins at the carbonyl group of arginine or lysine residues with the following caveats:

i. These residues are not at the amino end of the protein.

ii. These residues are not followed by a proline residue.

b. Chymotrypsin, a protein related to trypsin, is used to cleave peptides at the carbonyl side of amino acid residues with aromatic side chains and, to a lesser extent, residues with large hydrocarbon side chains. Chymotrypsin cleaves peptides at Phe, Trp, Tyr, and in some cases Leu and Ile residues.

c. Staphylococcus protease catalyzes peptide hydrolysis at the carbonyl side of glutamic acid residues in peptides.

2. Because these proteases catalyze the hydrolysis of peptides at internal rather than terminal residues, they are called *endopeptidases*. (Enzymes that cleave peptides only at terminal residues are termed *exopeptidases*.)

SOLUTIONS TO PROBLEMS

Solutions to Focused Problems

27.1 The peptides are drawn in the form that exists at neutral pH. Although the stereochemistry is not specified in the problem, the L stereochemistry of each chiral amino acid is assumed.

(a)

tryptophylglycylisoleucylaspartic acid
(Trp-Gly-Ile-Asp, or W-G-I-D)

(b)

Glu-Gln-Phe-Arg
(glutamylglutaminylphenylalanylarginine, or E-Q-F-R)

27.2 This peptide is His-Ile-Tyr-Met-Ser.

27.3 (a)

(b)

E—CH₂OH   (E = enzyme)

27.4 Three more of the many possible configurations of α-amino acids in addition to the ones shown at the beginning) of Sec. 27.2 on text p. 1482.

27.5 The configuration of L-isoleucine given in (a) on right. As noted on text p. 1483, the "allo" version of α-amino acids retains the L configuration at the α-carbon (which means the S configuration for all amino acids except cysteine). So, it must be the side chain of isoleucine that has the opposite configuration from the side chain of L-isoleucine. L-Alloisoleucine has the $2S$ configuration, that is, the same configuration at the α-carbon as L-isoleucine, but has a different configuration at the β-carbon, as shown in (b) at right.

(a) CH₃CH₂ ... CO₂⁻
L-isoleucine
(2*S*, 3*S*)

(b) CH₃CH₂ ... CO₂⁻
L-alloisoleucine
(2*S*, 3*R*)

27.6 The α-carbon of L-cysteine has the R configuration, as shown in (a) at right. In terms of the sequence rules for determining configuration, the two carbons attached to the α-carbons of cysteine or serine can be represented as C(O,O,O) and C(X,H,H), where X = S or O, respectively. When X = S (cysteine), the latter carbon receives priority. When X = O (serine), the former receives priority. [The structure is in (b) at right.] Sulfur receives priority over oxygen because it has the higher atomic number, and the priority is decided *at the first point of difference*. (See Sec. 4.2B of the text.) Notice that both (R)-cysteine and (S)-serine have the L configuration in the D,L-system.

(a) HS ... CO₂⁻ *R*
L-cysteine

(b) HO ... CO₂⁻ *S*
L-serine

27.7 (a)

$$pK_{a1} = 1.8 \qquad pK_{a3} = 9.2$$

H₃N⁺—CH—CO₂H ⇌(H₂O / H₃O⁺) H₃N⁺—CH—CO₂⁻ ⇌(H₂O / H₃O⁺) H₃N⁺—CH—CO₂⁻ ⇌(H₂O / H₃O⁺) H₂N—CH—CO₂⁻

(with CH₂ and imidazole side chains)

$pK_{a2} = 6.0$

A   *B*   *C*   *D*

(b) A plot of the fraction of each species shown in part (a) as a function of pH is shown in **Fig. SG27.1** on the next page. The pH at which any two species are present in equal concentrations is noted.

Figure SG27.1 A sketch of the fraction of each form of histidine present as a function of pH for the solution to Focused Problem 27.7(b).

27.8 (a) The amino group of tyrosine can be protonated, and both the carboxy group and the phenolic O—H group can be ionized.

(b) At pH 6, the net charge on tyrosine is zero. A pH value of 6 is below the pK_a of the conjugate acid of the amino group, and the amino group is therefore protonated; a pH value of 6 is above the pK_a of the carboxy group, which is therefore ionized; and a pH value of 6 is below the pK_a of the phenolic O—H group, which is therefore un-ionized.

(c) The structure of tyrosine in aqueous solution at pH 6 is shown at right.

27.9 (a) The isoelectric point of A-K-V-I-M (Ala-Lys-Val-Ile-Met) is approximately the same as lysine, since the lysine amino group is the only charged side chain in the peptide. Therefore, its isoelectric point is about 9.8. The isoelectric point of G-D-G-L-F (Gly-Asp-Gly-Leu-Phe) is about 2.8, because it contains an aspartic acid, which has an acidic side chain,.

(b) In the major neutral form of A-K-V-I-M, which is the form in which the peptide exists at its isoelectric point, the amino terminal amino group is unprotonated and the side chain on the Lys is in its conjugate-acid (protonated) form because has the higher pK_a of the two amino groups (Table 27.1, text p. 1478). This form would exist at a pH between the pK_a values of the two amino groups. That is, the pH at which this neutral form predominates is relatively basic. (See Sec. 27.3B.)

A–K–V–I–M

In the major neutral form of G-D-G-L-F, the side-chain carboxy group of Asp is un-ionized and the carboxy-terminal carboxy group is ionized because it has the lower pK_a of the two carboxy groups (Table 27.1, text p. 1478). This form would exist at a pH between the pK_a values of the two carboxy groups. That is, the pH at which this neutral form predominates is relatively acidic. (See Sec. 27.3B.)

G–D–G–L–F

27.10 The general rule of thumb is that if the peptide contains more acidic than basic groups it is an acidic peptide; if it contains more basic than acidic groups it is a basic peptide; and if the number of acidic and basic groups are equal, the peptide is neutral. (An acidic group is a group that is in its conjugate-base form at neutral pH, such as a carboxy group; a basic group is a group that is in its conjugate-acid form at neutral pH, such as an amino group.)

(a) The peptide is neutral because it contains one basic group (the terminal amino group, which is protonated at pH 6) and one acidic group (the terminal carboxy group, which is ionized at pH 6). Its net charge at pH 6 is 0.

(b) This peptide contains three basic groups: the α-amino group of the amino-terminal residue Leu and the side-chain amino groups of the Lys residues. The peptide contains one acidic group: the carboxy group of the carboxy-terminal Lys residue. Because the peptide contains more basic residues than acidic residues; it is a basic peptide. Its net charge at pH 6 is +2.

(c) Acetylation eliminates the basicity of the terminal amino group. The peptide contains two basic groups (the Arg residues) and two acidic groups (the Asp residue and the terminal carboxy group). Thus, the peptide is neutral, and its net charge is 0.

(d) This peptide contains three acidic groups: the side-chain carboxy groups of the Asp and Glu residues and the carboxy group of the carboxy-terminal Ile residue. The peptide contains two basic groups: the α-amino group of the amino-terminal residue Glu and the side-chain amino group of the Lys residue. Because the peptide contains more acidic residues than basic residues, it is an acidic peptide. Its net charge at pH 6 is −1.

27.11 (a) A resin containing a cationic group will serve as an anion exchanger. A quaternary ammonium salt is a suitable cationic group:

(b) The ion-exchange column is positively charged, as shown by the solution to part (a). At pH 6, the peptide R-K-N-G has a net charge of +2, is repelled by the column and will emerge first; the peptide A-V-G has a net charge of zero, is not retained by the column, and will emerge next, possibly very close to the first peptide; and the peptide D-N-E-G has a charge of −2, is most strongly retained by the column, and will emerge last.

27.12 (a) Alkylation of ammonia by α-bromophenylacetic acid will work, particularly since the alkyl halide is benzylic.

(continued next page)

$$PhCHCO_2H \xrightarrow{\text{NH}_3 \text{ (large excess)}} PhCHCO_2^-$$

with Br below the first structure and $^+NH_3$ below the product

α-bromophenyl-acetic acid   **α-phenylglycine**

The acetamidomalonate method would not work because it would require alkylation of a malonate anion by bromobenzene. Bromobenzene does not undergo S_N2 reactions. The Strecker synthesis would work.

$$PhCH{=}O + {}^+NH_4\,Cl^- + Na^{+\,-}CN \longrightarrow PhCHNH_2 \xrightarrow[\text{2) neutralize}]{\text{1) conc. HCl, H}_2\text{O heat}} PhCH\overset{+}{N}H_3$$

benzaldehyde   with CN below first product, CO_2^- below final product

α-phenylglycine

(b)   All three methods would work in principle for leucine.

Alkylation of ammonia:

$$(CH_3)_2CHCH_2CHCO_2H \xrightarrow{\text{NH}_3 \text{ (excess)}} (CH_3)_2CHCH_2CHCO_2^-$$

with Br below reactant, $^+NH_3$ below product

2-bromo-4-methylpentanoic acid   **leucine**

Acetamidomalonate method:

$$CH_3\overset{O}{\overset{\|}{C}}NHCH(CO_2Et)_2 \xrightarrow[\text{2) (CH}_3)_2\text{CHCH}_2\text{Br}]{\text{1) NaOEt, EtOH}} CH_3\overset{O}{\overset{\|}{C}}NHC(CO_2Et)_2 \xrightarrow[\text{2) NaOH (dilute)}]{\text{1) H}_3\text{O}^+, \text{H}_2\text{O, heat}} H_3\overset{+}{N}CHCO_2^-$$

with $CH_2CH(CH_3)_2$ below middle product, $CH_2CH(CH_3)_2$ below final product

leucine

Strecker synthesis:

$$(CH_3)_2CHCH_2CH{=}O + {}^+NH_4\,Cl^- + Na^{+\,-}CN \longrightarrow (CH_3)_2CHCH_2CHNH_2 \xrightarrow[\text{2) neutralize}]{\text{1) conc. HCl, H}_2\text{O heat}} (CH_3)_2CHCH_2CH\overset{+}{N}H_3$$

3-methylbutanal   with CN below first product, CO_2^- below final product

leucine

27.13   (a)

$$(CH_3)_2CHCH_2CHNH-\overset{O}{\underset{O}{\overset{\|}{\underset{\|}{S}}}}-\underset{}{\text{(benzene ring)}}-CH_3$$

with CO_2H below

N-tosylleucine

(b)

$$CH_3CH\overset{+}{N}H_3$$

with CO_2CH_3 below

alanine methyl ester

In (a), tosyl chloride forms an amide like any other acid chloride would that reacts with an amine. It is possible that the carboxylate group might form a mixed carboxylic–sulfonic anhydride; but this would be hydrolyzed immediately to the carboxylic acid on aqueous workup. In (b) the carboxylic acid group is esterified; it doesn't undergo aminolysis by an amino group as long as acidic conditions are maintained.

27.14   Polyglycine results from aminolysis of the ester group on one molecule by the amine on another:

$$\underset{\substack{\text{O}\\ \|}}{\text{H}_2\text{NCH}_2\overset{\text{O}}{\overset{\|}{\text{C}}}\text{OCH}_3} \quad \underset{\substack{\text{O}\\ \|}}{\text{H}_2\text{NCH}_2\overset{\text{O}}{\overset{\|}{\text{C}}}\text{OCH}_3} \xrightarrow{-\text{CH}_3\text{OH}} \text{H}_2\text{NCH}_2\overset{\text{O}}{\overset{\|}{\text{C}}}-\text{NHCH}_2\overset{\text{O}}{\overset{\|}{\text{C}}}\text{OCH}_3 \xrightarrow[\text{H}_2\text{NCH}_2\text{CO}_2\text{CH}_3]{-\text{CH}_3\text{OH}}$$

$$\text{H}_2\text{NCH}_2\overset{\text{O}}{\overset{\|}{\text{C}}}-\text{NHCH}_2\overset{\text{O}}{\overset{\|}{\text{C}}}-\text{NHCH}_2\overset{\text{O}}{\overset{\|}{\text{C}}}\text{OCH}_3 \longrightarrow \text{etc.}$$

In this reaction the amino group of glycine acts as a nucleophile. In acidic solution the amino group of glycine is protonated and cannot act as a nucleophile, so the reaction does not occur.

27.15 If the average yield of each step is Y, then the yield of the first step (assuming it is average) is Y; the yield of the second is $Y \cdot Y$; and that of the nth step is Y^n. Therefore,

$$Y^{369} = 0.17$$

$$369 \log Y = \log 0.17 = -0.7696$$

$$\log Y = -0.0021, \text{ or } Y = 0.995$$

Therefore, the average yield of each step is 99.5%!

 The high yield of each step demonstrates two points about peptide synthesis. First, it has been developed into a remarkably efficient process. The second point, however, is that even with yields in excess of 99%, the overall yields of large proteins prepared by this method will be extremely small. Furthermore, these will be contaminated by large numbers of impurities that will be difficult to separate from the desired material. (See the margin box on text p. 1501.)

27.16 The sidebar, "Solid-Phase Peptide Synthesis, on text p. 1495 shows that solid-phase peptide synthesis requires about three synthetic steps per residue. Therefore, about 300 steps are required for the synthesis of a protein containing 100 amino acid residues. If the average yield of each step is Y, then

$$Y^{300} = 0.50$$

$$300 \log Y = \log 0.50 = -0.3010$$

$$\log Y = -0.001003, \text{ or } Y = 0.998$$

Thus, an average yield of 99.8% per step would be required.

 If you used 100 steps for your calculation, the average yield is 99.3%. Either way, the yield must be nearly perfect at each step to achieve even a 50% overall yield. Notice that, because there are so many steps, a drop of a small fraction of a percent in average yield per step has a large effect on the overall yield.

27.17 (a) Because lysine contains two amino groups, both must be protected in order to prevent nucleophilic side reactions, such as the reaction of the amino group of one lysine molecule with the Py-BOP-activated carboxy group of another.

(b) The 20% piperidine step removes both Fmoc protecting groups. Therefore, both amino groups react in subsequent acylation reactions. (The Lys residue is drawn in more structural detail for clarity.)

$$
\begin{array}{ccc}
\overset{\text{O}}{\overset{\|}{}} & \overset{\text{O}}{\overset{\|}{}} & \overset{\text{O}}{\overset{\|}{}}\\
\text{Gly}-\text{NHCHC}-\text{Ala} & \text{H}_2\text{NCHC}-\text{Ala} & \text{Gly}-\text{NHCHC}-\text{Ala}\\
| & | & |\\
(\text{CH}_2)_4 & (\text{CH}_2)_4 & (\text{CH}_2)_4\\
| & | & |\\
\text{NH}_2 & \text{NH}-\text{Gly} & \text{NH}-\text{Gly}
\end{array}
$$

27.18 The S_N1 reaction occurs because the *tert*-butyl cation, a relatively stable tertiary carbocation, is formed as an intermediate. The acidic conditions convert the carbamate group into a good leaving group, which then undergoes acid–catalyzed decarboxylation to give the amine, which then protonates under the acidic conditions. (Pep = the remaining part of the peptide; A—H = trifluoroacetic acid.)

29.19 (a) Compound *A* results from the S_N2 coupling reaction to the resin (analogous to Eq. 27.21 on text p. 1393), and compound *B* results from deprotection of *A*:

Compound *C* is the product that results from coupling of the lysine derivative reacting at its carboxylic acid group with the free amino group of *B*, and compound *D* results from removal of the Fmoc protecting group, *but not the Boc group*, from compound *C*.

Compound *E* results from coupling of the carboxy terminus of Boc-Val with the free amino group of *D*. Peptide *P* results from both removal of *E* from the resin and removal of the Boc groups.

(from reaction of the *tert*-butyl cation of the Boc group with trifluoroacetic acid)

Note that the tripeptide *P* is in the di-cationic, amino-protonated form as a result of the acidic deprotection conditions. The neutral (zwitterionic) form would be obtained by neutralization.

(b)   The Boc group prevents the side-chain amino group of lysine from reacting with the carboxylic acid group of another molecule of itself under the coupling conditions. Also, it remains intact when the amino-terminal Fmoc group is removed with piperidine. The lysine side-chain thus remains protected in the subsequent coupling step as well.

(c)   Boc-Val introduces a protecting group that can be removed at the same time as the peptide is released from the resin. Use of Fmoc-Val would require an additional deprotection step with 20% piperidine.

27.20 (a)   As shown by Table 27.2 on text p. 1503, more than one mRNA code exists in some cases for each individual amino acid. So, for any given peptide, there are likely to be multiple mRNA sequences. The code for this peptide would start with either of the two codes for Phe, followed by any of the six codes for Arg, followed by any of the four codes for Gly, followed by either of the two codes for His, and terminated by the single code for Trp. Therefore, there are 96 possible codes ($2 \times 6 \times 4 \times 2 \times 1$) for this peptide.

mRNA codes					DNA codes				
Phe	Arg	Gly	His	Trp	Phe	Arg	Gly	His	Trp
UUU	CGU	GGU	CAU	UGG	AAA	GCA	CCA	GTA	ACC
UUC	CGC	GGC	CAC		AAG	GCG	CCG	GTG	
	CGA	GGA				GCT	CCT		
	CGG	GGG				GCC	CCC		
	AGA					TCT			
	AGG					TCC			

(b)   The DNA codes are complimentary to every RNA code, so there are 96 possible DNA codes. The possible codes are shown in the diagram for part (a).

(c)   The mRNA code for Trp, from 5' ⟶ 3', is UGG. The tRNA code is then ACC.

27.21 (a)   (b)

27.22 (a)   The amide group of Asn hydrolyzes to give ammonium ion and Asp. Consequently, Asn appears in amino acid analysis as Asp + $\overset{+}{N}H_4$..

(b)   Glu and Gln are not differentiated by amino acid analysis because the side-chain carboxamide group of Gln, like that of Asn in part (a), is hydrolyzed; that is, Gln is converted into Glu + $\overset{+}{N}H_4$.

27.23 The indole side-chain of tryptophan is very hydrocarbonlike, more so than the side chain of any other amino acid. [Recall (Sec. 26.2 of the text) that pyrrole and indole are neither basic nor acidic.] Therefore, we would expect to see AQC-Trp emerge on the far right of the amino acid analysis chromatogram, beyond AQC-Tyr. (This is in fact what is observed.)

27.24   The side chains of these amino acids as well as the α-amino groups are nucleophilic. Therefore, they are acylated as well by AQC-NHS. The structures of these derivatives are as follows, with the carboxylic acid group in its ionized form:

AQC-lysine   **AQC-cysteine**

27.25   Trypsin catalyzes the hydrolysis of peptides or proteins at the carbonyl group of arginine (R) or lysine (K) residues. Therefore, the peptide E-R-G-A-N-I-K-K-H-E-M would produce the following peptides upon trypsin-catalyzed hydrolysis:

   (1)   E-R   (2)   G-A-N-I-K   (3)   K   (4)   K-H-E-M   (5)   H-E-M

Peptides (2) and (4) originate by hydrolysis at the peptide bond between the two K residues. Peptides (3) that is, K, and (5) originate from hydrolysis at the peptide bond between the K and H residues to give G-A-N-I-K-K followed by hydrolysis at the C-terminal peptide bond of this peptide. However, peptide (5) does not originate from (4), because, when a lysine residue is at the amino terminus of a peptide, the hydrolysis of its peptide bond is not catalyzed by trypsin. (Trypsin is an endopeptidase.)

27.26   Trypsin catalyzes the hydrolysis of peptides or proteins at the carbonyl group of arginine (Arg, R) or lysine (Lys, K) residues, provided that these residues are (a) not at the amino end of the protein, and (b) not followed by a proline residue. Because *T1* contains an R residue, the carbonyl terminal of *T1* must be an R and *T1* must be (A,F,S)-R; because *T2* contains a K residue, the carbon terminal of *T2* must be a K and *T2* must be (G,I)-K; and thus *T3* must be at the carbonyl terminal of *Q* and the amino terminal cannot be a P.

$$Q \xrightarrow{\text{chymotrypsin}} \text{(A,N,R,S)-Y} \; + \; \text{K-F} \; + \; \text{G-I} \; + \; \text{(G,P)}$$

                                                        C1   *C2*   *C3*   *C4*

Chymotrypsin cleaves peptides at phenylalanine (Phe, F), tryptophan (Trp, W), tyrosine (Tyr, Y), and occasionally, at leucine (Leu, L) and isoleucine (Ile, I). Because *C1* contains a Y residue, the carbonyl terminal of *C1* must be a Y and *C1* must be (A,N,R,S)-Y; Because the dipeptide *C2* contains an F residue, the carbonyl terminal of *C2* must be an F and *C2* must be K-F; and because *C3* contains an I residue, the carbonyl terminal of *C3* must be an I and *C3* must be G-I, and thus *T2* must be G-I-K.

$$Q \xrightarrow{\text{chymotrypsin}} \text{(A,N,R,S)-Y} \; + \; \text{K-F} \; + \; \text{G,I} \; + \; \text{(G,P)}$$

                                                        C1   *C2*   *C3*   *C4*

Because *C2* is K-F and *T1* contains an F residue while *T2* contains a K residue, that means that those two peptides are linked as *T2-T1* to give G-I-K-F-(A,S)-R. With that information, *C1* must be (A,S)-R-N-Y. So, *T3* must be N-Y-(G,P) and *Q* is G-I-K-F-(A,S)-R-N-Y-(G,P), where the order of (A,S) and (G,P) are uncertain. (Although it is not mentioned in the text, chymotrypsin, like trypsin, does not catalyze the hydrolysis of peptides in which proline is at the C-terminal side of the cleavage point. Therefore, it is likely that the sequence of peptide *C4* is G-P.)

27.27 (a) b-Type fragmentation of the peptide N-F-E-S-G-K would produce the following m/z values. In this notation, each amino acid letter represents a residue (that is, —NH—CHR—CO—). Therefore, the *H*— must be added to indicate the additional hydrogen at the amino terminus H_2N—, and one mass unit must be added to the corresponding residue mass.

$$H—N \Rightarrow \qquad\qquad 1+114.0 = 115.0$$

$$H—N\text{-}F \Rightarrow \qquad\qquad 1+114.0 + 147.1 = 262.1$$

$$H—N\text{-}F\text{-}E \Rightarrow \qquad 1+114.0 + 147.1 + 129.0 = 391.1$$

$$H—N\text{-}F\text{-}E\text{-}S \Rightarrow \qquad 1+114.0 + 147.1 + 129.0 + 87.0 = 478.1$$

$$H—N\text{-}F\text{-}E\text{-}S\text{-}G \Rightarrow \qquad 1+114.0 + 147.1 + 129.0 + 87.0 + 57.0 = 535.1$$

$$H—N\text{-}F\text{-}E\text{-}S\text{-}G\text{-}K—OH \Rightarrow 1+114.0 + 147.1 + 129.0 + 87.0 + 57.0 + 128.1 + 17.0 = 620.2$$

(b) The fragmentation occurs in the following way.

The "H_2" at the amino terminus of each fragment that follows represents two protons added to the residue itself to give the protonated terminal amino group indicated in the problem. For example, the residue mass for K includes a —NH— from the amide nitrogen of the peptide bond; a K residue with an amino group at the amino terminus is then *H*—K; and a protonated amino-terminal K residue is H_2K, with a positive charge on the nitrogen. The following fragments should be observed:

$$[H_2K—OH]^+ \Rightarrow 147.1 = y_1$$

$$[H_2G—K—OH]^+ \Rightarrow 204.1 = y_2$$

$$[H_2S—G—K—OH]^+ \Rightarrow 291.1 = y_3$$

$$[H_2E—S—G—K—OH]^+ \Rightarrow 420.1 = y_4$$

$$[H_2F—E—S—G—K—OH]^+ \Rightarrow 567.2 = y_5$$

$$[H_2N—F—E—S—G—K—OH]^+ \Rightarrow 681.2 = M + 1$$

27.28 (a) Note that this is an "a-type" cleavage. The residue mass is the mass of a "b-type" fragment minus the mass of a carbonyl group (28 mass units).

$$m/z = 551.94$$

(b) The M + 1 ion (shown here protonated on the OH group of the carboxylic acid) loses water to give an acylium ion, which is the ion of interest. Although carboxylic acids are normally protonated on the carbonyl group, the small amount of the form shown here, once formed, is very unstable.

$$m/z = 710.97$$

(c)   Two conceivable mechanisms for forming H_2 are the following:

Mechanism (1):

$$m/z = 727.09$$

The mechanism could be tested by labeling the CH_2 group adjacent to the sulfur with deuterium. Then, if this fragmentation occurs, a loss of DH ($m/z = 3$) would confirm the hypothesis.

Mechanism (2):

$$m/z = 727.09$$

This mechanism could be tested with a peptide in which the α-hydrogen is replaced by D. The indicated fragmentation would result in the loss of 3 mass units.

It is relevant that the loss of H_2 occurs in many peptides; a C-terminal methionine is not necessary. This observation makes mechanism (2) more likely but does not rule out mechanism (1) as a contributor for this particular peptide.

27.29   (a)   The amino group is a nucleophile in an addition to the isothiocyanate.

(b)   The sulfur is the nucleophile in an intramolecular nucleophilic acyl substitution reaction. The following mechanism begins with the carbonyl-protonated thiourea derivative.

(c)   Water is a nucleophile in opening the thiazolinone, and nitrogen of the resulting thiourea is the nucleophile in closing the ring to the PTH. The mechanism below begins with the carbonyl-protonated thiazolinone.

27.30   (a)   As Eq. 27.58a on text p. 1518 as well as its mechanism in the solution to Focused Problem 27.29(a) shows, the Edman degradation depends on the presence of a free terminal amino group in the peptide. Because this group is blocked as an amide in acetylated peptides, such peptides cannot undergo the Edman degradation.

(b)   *N*-Acylation of peptides and proteins has no adverse effect on sequencing by MS–MS because the acetylamino group is just another amide bond. Thus, typical b-type and y-type cleavage can still occur.

27.31   (a)   The structure of a phosphotyrosine residue:

a phosphotyrosine residue
in a protein

(b)   Tyrosine contains an acidic phenolic group unlike serine and threonine. The equilibrium constants for the formation of ordinary aryl esters are considerably less favorable than the equilibrium constants for the formation of alkyl esters, and the same is true of phosphate esters. Because the conjugate base of a phenol is less basic and is therefore a poorer nucleophile and a better leaving group, the equilibrium constant for the formation of a phosphotyrosine residue from ATP would be less than the K_{eq} for the phosphorylation of a serine residue.

27.32   The structure of an L-phosphocysteine residue in a peptide: (Note that this has the R configuration.)

an L-phosphoserine residue in a peptide

27.33   The structure of an N-acetylgalactosamine conjugate formed by inversion of configuration with the threonine residue of an acceptor protein follows. Because UDP-N-acetylgalactosamine has the α-configuration at the anomeric carbon of the galactosamine, then the glycoprotein conjugate has the β-configuration.

27.34   (a)   A nonbonding electron pair on the ring oxygen of the sugar overlaps with the empty 2p orbital of the carbocation:

(b)   The Mn^{2+} ion likely stabilizes the phosphate anhydride group in the UDP-leaving group (see Display 25.52 on p. 1411, Sec. 25.7D, of the text).

27.35   Refer to the mechanism in Eqs. 24.34a–b on text p. 1349, and apply it to the glucose-Hb imine in Eq. 27.68 on text p. 1527. The net result is that the alcohol is converted into a carbonyl group.

27.36   As described in Sec. 19.12A, primary amines react with aldehydes to form imines. An example is shown in Eq. 27.67 on text p. 1527, with the aldehyde of glucose reacting with the amino terminus of hemoglobin. A simplified version follows:

The question asks why the nitrogen couldn't be supplied from an amide, such as from the side chain of an asparagine amino acid. The reaction mechanism for imide formation shows that an intermediate forms after loss of water through an E1 mechanism. This intermediate contains an ammonium ion; in the case of an amide, an analogous intermediate is very unstable because the positively charged nitrogen is bonded to a carbonyl carbon, which also has a significant positive charge due to the carbon–oxygen bond dipole. It is unlikely that this unstable structure forms, and so imide formation between aldehydes and amides is much less favorable than imide formation from aldehydes and amines, in which this unfavorable interaction does not exist.

27.37   In the first step of the Amadori rearrangement (see solution to Focused Problem 27.35), the C2 carbon is deprotonated. The resulting carbanion can re-protonate from the opposite side, inverting the stereochemistry at C2. This process converts the glucose derivative into a mannose derivative.

glucose-Hb imine (Schiff base)

mannose-Hb imine (Schiff base)
+ :ÖH₂

The reversible hydrolysis of the mannose imine yields the aldehyde form of mannose, which equilibrates into the cyclic form of mannose.

mannose–Hb imine (Schiff base)

+ H₂N̈—Hb
hemoglobin

D-mannopyranose
(either anomer)

27.38 Because the four chains of hemoglobin are held together by the same noncovalent forces responsible for the tertiary structures of the individual chains, the subunits of hemoglobin would dissociate and would unfold into two random-coil α chains and two random-coil β chains.

27.39 One end of aeruginosin-B resembles an arginine side chain; one can hypothesize the following correspondences:

aeruginosin-B

arginine

Aeruginosin-B has a positively charged guanidino group, like arginine, that can interact with the ionized Asp-189 in the binding site; and this is attached to a hydrocarbon chain that can interact favorably with the "walls" of the hydrophobic pocket.

27.40 In the following mechanism, the groups labeled "Asp" are aspartic acid residues of the enzyme.

27.41 Although the sulfur has two nonbonding electron pairs, the nitrogen of the thiazole is like the nitrogen of pyridine or imidazole and is the most basic site on the ring; therefore, it is likely to be the hydrogen-bond acceptor. An amide backbone N—H is the only hydrogen bond donor site on a peptide backbone.

27.42   We use the criteria for hydrogen-bond donors and acceptors discussed in Sec. 8.5C of the text. (Note that amide nitrogens are not very basic, and are not very good hydrogen-bond acceptors.)

Darunavir

Solutions to Skills Objectives Problems

27.43   (a)   The single-letter abbreviation sequence of S-peptide is K-E-S-A-A-A-K-F-E-R-Q-H-M-D-S-S-T-S-S-A.
(b)   The structures of residues 1, 5, 8, 9, 11 and 13:

| K | A | F | E | Q | M |

27.44   (a)   The structure of (2S,3R)-isomethionine:

(b)   Because sulfur receives higher priority than a carbonyl carbon, the priorities of the hydrogen and the side-chain are reversed over a typical L-amino acid, so this is a D-amino acid.
(c)   We established in the previous two parts that the 2S configuration corresponds to the D configuration, so the 2S,3S amino acid also has the D configuration.
(d)   In the D,L-system, this stereoisomer would have to be given a different name: perhaps alloisomethionine?

27.45   (a)   The acid–base equilibria for ornithine: (The pK_a values are by analogy to those for lysine; see Eq. 27.20, text p. 1487.

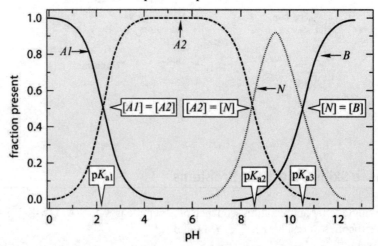

(b)   A plot of the fraction of each form from pH = 1 to pH = 12:

The curve for form *N* must cross the curves for *A2* and *B* at the points (8.5, 0.5) and (10.5, 0.5), respectively. The pK_a of form *N* differs by only 2 units from the pK_a of form *B*, and the peak for form *N* must occur at a pH that is halfway between the pK_a values of *N* and *B*. (Can you show that?) Therefore, the peak for *N* occurs only one pH unit away from its pK_a. With this constraint, form *N* can't quite reach a fraction percent = 1.0. (It peaks at fraction = 0.91. Can you show that?) However, if your curve looks essentially like this one but you have form *N* peaking at 1.0 you have the essential idea.

27.46 Because the amino acid has two acidic groups, one of them being a sulfonic acid, which is *much* more acidic than a carboxylic acid, cysteic acid has an isoelectric point lower than aspartic acid [option (a)].

27.47 Since lysozyme has many more basic residues (Lys and Arg) than acidic residues (Asp and Glu), lysozyme is expected to be (and is) a basic protein. Its isoelectric point should be >>6. (Its isoelectric point is about 12.)

27.48 The principle is that a peptide with a given charge migrates to the electrode of opposite charge; a peptide with zero charge does not migrate. Because Gly-Lys is a basic peptide, it has a net positive charge at pH 6, and it therefore migrates to the cathode, the negatively charged electrode. Because Gly-Asp is an acidic peptide, it has a net negative charge at pH 6, and it migrates to the anode, the positively charged electrode. Gly-Ala is a neutral peptide; it has net zero charge at pH 6, and therefore it does not migrate.

27.49 (a)   The peptide K-E-A-D-Y has two acidic residues; the peptide K-N-I-G-H-T has one basic residue; and the peptide P-A-I-N-T-E-R has one acidic and one basic residue. Therefore, the peptide K-N-I-G-H-T has the highest isoelectric point.

(b)   The charges at pH = 6 on the three peptides are *A*, –2; *B*, +1; and *C*, 0. An anion-exchange resin is positively charged. Therefore, the peptides will emerge in the order *B, C, A*.

27.50 (a)

Ph—CH=O + H₂N—CH—CO₂⁻ + Na⁺ ⁻C≡N ⟶ N≡C—CH—HN—CH—CO₂⁻ $\xrightarrow[\text{heat}]{\text{H}_2\text{O, H}_3\text{O}^+}$ $\xrightarrow{\text{neutralize}}$

(with Ph on one CH and isopropyl on the other chain)

⁻O₂C—CH—H₂N⁺—CH—CO₂⁻
(Ph; isopropyl)

(b)

$\xrightarrow[\text{ether}]{\text{Mg}}$ $\xrightarrow[\text{2) H}_2\text{O, H}_3\text{O}^+]{\text{1) CO}_2}$ ...CO₂H $\xrightarrow[\text{PBr}_3\text{ (cat.)}]{\text{Br}_2}$

...CO₂H (with Br) $\xrightarrow[\text{2) neutralize}]{\text{1) NH}_3\text{ (excess, heat)}}$ ...CO₂⁻ (with ⁺NH₃)

norleucine

(c)

CH₃CNH—CH(CO₂Et, with C=O) $\xrightarrow[\text{2) H}_2\text{C=CH—CN}]{\text{1) EtO}^-\text{/EtOH}}$ CH₃CNH—C—CH₂CH₂CN (with two CO₂Et groups, C=O) $\xrightarrow[\text{2) neutralize}]{\text{1) H}_2\text{O, H}_3\text{O}^+\text{ heat}}$ H₃N⁺—C—CH₂CH₂CO₂⁻ (with CO₂⁻)

glutamic acid

27.51 Every synthesis has the same final steps:

CH₃CNHCH(CO₂Et)₂ (with C=O) $\xrightarrow[\text{2) RBr}]{\text{1) NaOEt, EtOH}}$ CH₃CNHC(CO₂Et)₂ (with C=O, R) $\xrightarrow[\text{2) NaOH (dilute)}]{\text{1) H}_3\text{O}^+\text{, H}_2\text{O, heat}}$ H₃N⁺CHCO₂⁻ (with R)

The challenge in each case is to prepare the alkyl halide R—Br from the given starting material.

(a)

(CH₃)₂C=CH₂ $\xrightarrow{\text{DBr, peroxides}}$ (CH₃)₂CD—CH₂Br

methylpropene (isobutylene)

Alternatively, the same alkene could be hydroborated with BD₃, the resulting organoborane oxidized to the primary alcohol by alkaline H₂O₂, and the alcohol converted into the alkyl bromide with concentrated HBr and H₂SO₄ catalyst.

(b)

PhCH=O $\xrightarrow[\text{2) H}_3\text{O}^+]{\text{1) LiAlD}_4}$ PhCHD—OH $\xrightarrow{\text{HBr}}$ PhCHD—Br

benzaldehyde

(c)

CH₃O—(ring) $\xrightarrow[\text{2) H}_3\text{O}^+]{\text{1) ClCCH}_3\text{, AlCl}_3}$ CH₃O—(ring)—CCH₃ (with C=O) $\xrightarrow{\text{Br}_2\text{/HBr}}$ CH₃O—(ring)—CCH₂Br (with C=O)

anisole

Alkylate the acetamidomalonate ion with this halide. Then, in the final hydrolysis step, use concentrated HBr, which will also cleave the methyl ether. Recall that α-bromo carbonyl compounds are very reactive in S_N2 reactions (Sec. 22.3 of the text).

27.52   In part (a), the nitrogen of proline is benzoylated. In (b), the product of (a) is converted into the benzyl ester by alkylation of the potassium carboxylate salt. (Sec. 20.8B) In (c), the strongly acidic conditions convert the amide into methyl acetate and the amine salt an esterify both carboxylic acid groups.

(a) **N-benzoylproline**   (b) **N-benzoylproline benzyl ester**   (c) **aspartic acid dimethyl ester hydrogen sulfate**

In part (d), the amino group reacts with the anhydride and then, with heating, forms the imide (Eq. 21.68, Sec. 21.8B, text p. 1161). Addition of acid provides the un-ionized acid.

27.53   We assume that Fmoc-amino acids are available commercially. (If not, they are prepared by treating the amino acid with Fmoc-NHS ester as shown in Eq. 27.35, text p. 1496.) The Fmoc derivative of glutamic acid γ-*tert*-butyl ester is prepared in the same manner as other Fmoc-amino acids.

1.   The carboxylate salt of Fmoc-Ala is linked to the solid support, as shown in Eq. 27.38, p. 1497 of the text.

2.   The Fmoc group is removed with 20% piperidine in DMF, as shown in Eq. 27.39, p. 1498 of the text.

3.   The deprotected Ala-resin is allowed to react with PyBOP and Fmoc-glutamic acid γ-*tert*-butyl ester.

4.   The Fmoc group is removed with 20% piperidine in DMF to give Glu(γO*t*Bu)-Ala-resin. The amino group of the Glu residue is exposed, but the *tert*-butyl ester protecting group on the side-chain carboxylic acid group remains in place.

5.   The resin is treated with Fmoc-Val and PyBOP, and the resulting product is then deprotected to give Val-Glu(γO*t*Bu)-Ala-resin.

6.   This resin is treated with Fmoc-proline and PyBOP, and the resulting resin is deprotected with 20% piperidine to give Pro-Val-Glu(γO*t*Bu)-Ala-resin. (Alternatively, the resin is treated with Boc-Pro and PyBOP. Step 7 removes the protecting group at the same time that the peptide is detached from the resin.)

7.   The resin is now treated with 60% trifluoroacetic acid in dichloromethane. This removes the peptide from the resin and also removes the *tert*-butyl ester protecting group from the glutamic acid residue. The synthesis of Pro-Val-Glu-Ala is complete.

27.54   The problem is the unprotected leucine has a free amino group that, under the basic conditions, can also react as a nucleophile with the chloromethyl resin. Amino groups are more basic, and therefore more nucleophilic, than carboxylate groups, so this undesired reaction might well be the major one observed. This error can be avoided by attaching Fmoc-protected leucine to the resin and *then* deprotecting it.

27.55   (a)   The mRNA complementary sequence for the first residue of Sequence 2 is AAG, and Table 27.2 shows that this is the correct code for lysine. (The "Stop" sequence is not included.) Only Sequence 2 starts with the code for K, and we leave it to you to verify that the codes for the remaining amino acids are correct as well.

(b)   The mRNA complementary sequence for K is AAG, so the tRNA anticodon is UUC.

27.56   The peptide is hydrolyzed into the most acidic form of each of the constituent amino acids, except for Asn and Gln, which are hydrolyzed to Asp and Glu plus two equivalents of $\overset{+}{N}H_4$. Therefore, the products are His, Ala, Glu, Val, Leu, Arg, and Asp along with two equivalents each of Lys and Gly.

27.57   The peptides produced by the trypsin digestion of the S-peptide are Lys-Glu-Ser-Ala-Ala-Ala-Lys, Phe-Glu-Arg, and Gln-His-Met-Asp-Ser-Ser-Thr-Ser-Ser-Ala. Because trypsin is an endopeptidase, the peptide is not hydrolyzed at the amino-terminal lysine residue.

27.58   The amino-terminal lysine is removed from the peptide to give product *A*. Notice that the Edman reagent also reacts with the side-chain amino group of Lys. The peptide co-product is Glu-Ser-Ala-Ala-Ala-Lys*-Phe-Glu-Arg- Gln-His-Met-Asp-Ser-Ser-Thr-Ser-Ser-Ala, where the Lys* residue has a modified side-chain as shown in *B*.

phenylthiohydantoin (PTH) derivative of the
amino-terminal lysine

A

the Lys* residue on the product peptide

B

27.59   (a)   Trypsin should catalyze the cleavage of glucagon at Lys and Arg residues. The following five fragments are expected. (Remember that, by convention, peptides written in this way have their amino ends on the left.) We assume that all cleavages proceed to completion, although in practice intermediate cleavage fragments might be formed as well.

(1)   His-Ser-Gln-Gly-Thr-Phe-Thr-Ser-Asp-Tyr-Ser-Lys
(2)   Tyr-Leu-Asp-Ser-Arg

(3)   Arg

(4)   Ala-Gln-Asp-Phe-Val-Gln-Trp-Leu-Met-Asn-Thr

(5)   Arg-Ala-Gln-Asp-Phe-Val-Gln-Trp-Leu-Met-Asn-Thr

If cleavage occurs initially after the first Arg in the Arg-Arg sequence of glucagon, then the peptide (5) is formed, and the second Arg becomes the amino terminus of this peptide. Cleavage does not occur at the peptide bond of this Arg because trypsin is an endopeptidase. If cleavage of glucagon occurs after the second Arg, then Tyr-Leu-Asp-Ser-Arg-Arg is formed. Subsequent cleavage of this peptide at the Arg-Arg bond gives (2) and (3).

(b)   The products from one cycle of the Edman degradation are the PTH derivative of histidine (*A*) along with the peptide S-Q-G-T-F-T-S-D-Y-S-K*-Y-L-D-S-R-R-A-Q-D-F-V-Q-W-L-M-N-T, where K* is the side-chain-modified lysine residue (product *B* in solution to Problem 27.58).

phenylthiohydantoin (PTH) derivative of the
amino-terminal histidine (*A*)

27.60   MS–MS can separate components of a mixture by mass and then bring about fragmentation of each after the separation. CI–MS involves the same type of fragmentation mechanism, but does not separate the components of a mixture before fragmentation. This is important for sequencing because separation of the peptide mixture is required to give an unambiguous sequence for each peptide.

27.61   The sequence is read from high mass to low, taking the differences between successive masses. The first difference of 16 mass units is due to the mass difference between the fragments (M+1–H_2) and (M+1–H_2O). After that, the differences between successive masses correspond to the residue masses given in Table 27.3. Remember that the mass of the amino-terminal fragment is 1 unit more than its residue mass because this fragment has an H_2N— group, whereas the residue mass is calculated for an —NH—.

$$598.4 - 470.3 = 128.1 \Rightarrow \text{Gln}$$

$$470.3 - 371.2 = 99.1 \Rightarrow \text{Val}$$

$$371.2 - 300.2 = 71 \Rightarrow \text{Ala}$$

$$300.2 - 114.1 = 186.1 \Rightarrow \text{Trp}$$

$$114.1 \Rightarrow \text{Leu or Ile} + 1 \text{ mass unit}$$

The Edman degradation decides between Ile and Leu. The sequence (amino to carboxy) is therefore Leu-Trp-Ala-Val-Gln.

27.62   Number the residues of *Q* from the amino terminus as 1,2,3, … First, the Edman degradation of *Q* shows that leucine is the amino-terminal residue. The formation of only dipeptides, including the dipeptide Leu-Val, shows that DPAP catalyzes the hydrolysis of peptide *Q* at every even-numbered residue from the amino terminus. Thus, this enzyme also catalyzes the hydrolysis of peptide *R* at every other residue — the odd residues in the numbering of *Q*:

This analysis shows that Gly is residue 9, the carboxy-terminal residue of Q. To get the order of the dipeptides, work back and forth between the dipeptides derived from Q and those derived from R. Since Gly has to be at the carboxy terminus of one of the dipeptides from R, and since the only dipeptide that meets this criterion is Ala-Gly, then Ala is the next residue in from the carboxy terminus. If Ala is in position 8, then position 7, from the Q dipeptides, must be Gln; position 6, from the R dipeptides, must be Asp; and so on. Alternatively, work from the amino terminus of Q: Leu-Val of Q and Val-Arg of R establish the sequence Leu-Val-Arg; Arg-Gly of Q shows that the next residue is Gly; and so on. The final sequence of peptide R, then, is:

Leu-Val-Arg-Gly-Val-Asp-Gln-Ala-Gly

(Remember that the amino terminus by convention is on the left when peptides are written this way.) The ammonia arises from the hydrolysis of Gln to glutamic acid.

27.63 (a) Both types of glycosylation require either a Ser or a Thr residue in the active site.

(b) *N*-Glycosylation always proceeds with inversion of configuration, but different *O*-glycosylations can occur with different stereochemistry—in some cases with retention, and in others with inversion.

(c) The amide nitrogen of Asn, but *not* Gln, is the nucleophile only in *N*-glycosylation.

(d) The completed polysaccharide is delivered to the protein as a dolichol diphosphate derivative only in *N*-glycosylation.

(e) The polysaccharide is built up one residue at a time while attached to the protein in *O*-glycosylation.

(f) An *N*-acetylglucosamine is always the residue attached to the protein in *N*-glycosylation.

27.64 (a) The reaction by which histidine in proteins is phosphorylated:

The histidine is phosphorylated on N-3 of the imidazole ring—the nitrogen not bonded to hydrogen—because this is the "pyridine-like" nitrogen of the ring and is therefore the most basic nitrogen. The N-1 nitrogen is not basic because its nonbonding electron pair is part of the aromatic π-electron system.

(b) The dephosphorylation of histidine in a protein is catalyzed by a protein–histidine phosphorylase.

(c) Phosphorylation more commonly takes place on serine, threonine, tyrosine, and cysteine.

27.65 The two major types of secondary structure found in proteins are the α-helix and the β-sheet. The principal type of noncovalent interaction that maintains these structures is a hydrogen bond donated by a peptide backbone N—H to a carbonyl oxygen.

There are many subtle variations on the secondary structures discussed in the text, particularly the α-helix. These are beyond the scope of this chapter.

27.66   (a)   The carbonyl group of the Val residue accepts a hydrogen bond from the amide hydrogen four residues away in the carboxy-terminal direction—in this case, the Trp residue.

         (b)   The amide N—H of the valine residue donates a hydrogen bond to the carbonyl oxygen four residues away in the amino-terminal direction—that is, to the carbonyl of the Met residue.

         (c)   The two residues are eight residues apart. At 3.6 residues per helical turn (text Fig. 27.25), there are then 2.2 turns separating these residues. At 510 pm per turn, these residues are separated by about 1,100 pm.

27.67   Figure 27.11 on text p. 1529 shows that the amino acid side chains extend outward from the periphery of the α-helix and are actually rather close to each other in space. The pK_a of a protonated lysine side-chain amino group is about 10.5. At pH values below 10, the side-chain amino groups in poly(lysine) are protonated, and these side chains are positively charged. The peptide avoids the α-helical conformation at pH values below 10 to avoid the repulsive electrostatic interactions between the adjacent positively charged groups. At pH values above 11, the side-chain amino groups are unprotonated and hence uncharged. As a result, there are no charge–charge repulsions to destabilize the α-helix. (Conceivably the adjacent amino groups could form hydrogen bonds to each other, with one amino group acting as a donor and the other an acceptor. This could further stabilize the α-helix at high pH.)

     The same principles operate in poly(glutamic acid), but at the other end of the pH scale. In poly(glutamic acid), the side chains are negatively charged at pH values above the pK_a of the carboxy groups. Therefore, the repulsions between negative charges cause the helix to be destabilized, and it unfolds at high pH. At low pH, the carboxy groups are un-ionized and therefore uncharged. Consequently, the α-helix can form at low pH because there are no charge–charge repulsions to destabilize it.

27.68   (a)   The large enhancements in reaction rates brought about by enzyme catalysts are due to the large number of reacting groups brought into proximity within the enzyme–substrate complex. Usually, many of these reacting groups are involved in an enzyme catalyzed reaction. For these groups to "find" each other as separate molecules in solution simultaneously by random diffusion is highly improbable.

         (b)   The fact that even the smallest enzymes are large molecules suggests that large numbers of noncovalent interactions are required to position the catalytic groups in an enzyme active site precisely to fulfill their roles and to position substrates precisely to interact with these groups. If this is so, the synthesis of such large molecules is difficult to realize with conventional organic synthesis on a practical basis. This is not to say that there are no synthetic catalysts that bring about large rate enhancements. Transition-metal organometallic catalysts are examples.

27.69   Before trypsin can catalyze the hydrolysis of a peptide or synthetic substrate, it must bind the substrate noncovalently. This binding is where the specificity of trypsin catalysis is realized. Substrates bearing alkyl chains terminating with positively charged groups (such as ammonium and guanidinium groups) interact electrostatically with the ionized carboxylate group of an aspartic acid residue. Without such a positively charged group, substrates do not form strong enzyme-substrate complexes with trypsin.

27.70   Using the trypsin model as our guide, it would seem that a substrate with an ionized carboxylic acid group would require a *positively charged* group in the active site of V8 protease for strong binding specificity. (In fact, the amino-terminal amino group, which is protonated at physiological pH, is positioned within the active site of this enzyme in such a way that it interacts electrostatically with the carboxylate side chains of Glu and Asp residues in peptides.)

Solutions to Integrated Problems

27.71 (a) Aspartic acid is the most acidic amino acid. It has the lowest isoelectric point.

(b) Arginine is the most basic amino acid. It has the highest isoelectric point.

(c) Isoleucine and threonine can exist as diastereomers. Each has two asymmetric carbon stereocenters.

(d) Glycine has zero optical activity under all conditions because it is achiral.

(e) Asparagine and glutamine have carboxamide groups in their side chains. These amides are hydrolyzed to aspartic acid and glutamic acid in hot aqueous acid or base.

27.72 The amino-terminal residue is lysine. (See the solution to Problem 27.58.)

27.73 Dansyl chloride is a sulfonyl chloride that should react readily with nucleophiles, including the terminal amino groups of peptides. The amino-terminal residue of the peptide *P* is valine. Because dansyl-valine is obtained, the α-amino group of valine must have been free in the peptide; so, valine must have been the amino-terminal residue. The sequence of the other residues cannot be determined from the data given.

27.74 (a) A molecular mass of about 1000 is consistent with the actual composition (Ala$_4$,Arg$_2$,Gly$_2$,Ser$_2$); amino acid analysis gives only the *relative* amounts of each amino acid. The absence of a reaction with the Edman reagent suggests that there is no terminal amino group. A *cyclic* peptide would give these results. Cleavage of a cyclic peptide at two arginine residues would give two peptides with arginine at the C-terminus of each; but if the two peptides were identical, two equivalents of a single peptide would be obtained. The sequencing results are in accord with the following structure:

(b) Peptide *D* is an ordinary (noncyclic) peptide that results from two cleavages at the R residues. The MS–MS b-type fragmentation of the M + 1 ion of peptide *D* would be typical of any linear peptide. (The positive charge on the fragments represents the charge on the acylium ion, and the *H*— is the proton of the terminal H$_2$N—group that is not included in the residue masses in Table 27.3 on text p. 1518.)

$$H—A\text{-}I\text{-}G\text{-}A\text{-}R \bullet H^+—OH \Rightarrow m/z = 1.0 + 71.0 + 113.1 + 57.0 + 71.0 + 156.1 + 18.0 = 487.2 = M + 1$$

$$[H—A\text{-}I\text{-}G\text{-}A\text{-}R]^+ \Rightarrow m/z = 1.0 + 71.0 + 113.1 + 57.0 + 71.0 + 156.1 = 469.2$$

$$[H—A\text{-}I\text{-}G\text{-}A]^+ \Rightarrow m/z = 1.0 + 71.0 + 113.1 + 57.0 + 71.0 = 313.1$$

$$[H—A\text{-}I\text{-}G]^+ \Rightarrow m/z = 1.0 + 71.0 + 113.1 + 57.0 = 242.1$$

$$[H—A]^+ \Rightarrow m/z = 1.0 + 71.0 = 72.0$$

The cyclic peptide *C*, on the other hand, would undergo fragmentation at every possible peptide bond to give five different fragments.

$$\text{Peptide } C \xrightarrow[\text{of cyclic peptide}]{\text{fragmentation}} \begin{cases} H—A\text{-}I\text{-}G\text{-}A\text{-}R\text{-}A\text{-}I\text{-}G\text{-}A\text{-}R^+ \\ H—R\text{-}A\text{-}I\text{-}G\text{-}A\text{-}R\text{-}A\text{-}I\text{-}G\text{-}A^+ \\ H—A\text{-}R\text{-}A\text{-}I\text{-}G\text{-}A\text{-}R\text{-}A\text{-}I\text{-}G^+ \\ H—G\text{-}A\text{-}R\text{-}A\text{-}I\text{-}G\text{-}A\text{-}R\text{-}A\text{-}I^+ \\ H—I\text{-}G\text{-}A\text{-}R\text{-}A\text{-}I\text{-}G\text{-}A\text{-}R\text{-}A^+ \end{cases}$$

However, these all have identical masses because they differ only in the positions of the amino acids and not their identity. The m/z is M + 1, where M is the mass of the parent cyclic peptide; this peak occurs at $m/z = 973.4$

25.75 The presence of two amino-terminal residues suggests that either *(a)* insulin is an approximately equimolar mixture of two proteins, or *(b)* it is a single protein containing two peptide chains. In fact, the latter is correct; insulin is two polypeptide chains connected by disulfide bonds. Glycine is the amino-terminal residue of one chain, and phenylalanine is the amino-terminal residue of the other.

27.76 This amino acid is really a substituted malonic acid. When heated in acid, it, like most malonic acid derivatives, decarboxylates. In this case, the product is glutamic acid. (See Eq. 20.60, Sec. 20.11A, text p. 1114.)

27.77 The resin contains a large number of pendant phenyl groups. Phenylalanine also contains a phenyl group. The resin has dual characteristics. Not only is it an anion-exchange resin, but also it is like an HPLC resin in which the side-chains are phenyl instead of C-18 hydrocarbons. The "like-dissolves-like" principle suggests that Phe, the amino acid containing the phenyl side chain will be partitioned into the resin phase more extensively than Gly. (The phenyl rings of the amino acid molecules and the resin can interact noncovalently by edge-to-face interactions; see Sec. 15.8A of the text.) .The selective adsorption of phenylalanine is due to this effect.

27.78 (a) Reaction of chloromethylated poly(styrene) with trimethylamine would give the resin shown:

(b) This resin is an *anion-exchange resin*. (See the solution to Problem 27.11 in this chapter.) At pH 6, Arg is positively charged and is therefore repelled by the column; it emerges earliest from the column. Leu has zero charge at pH 6; it emerges next. Glu is negatively charged at pH 6 and is therefore attracted to the positively-charged groups on the column; it emerges last from the column.

27.79 (a) Alanine (like the other amino acids) is a different compound in HCl, NaOH, and neutral H_2O because its ionization state is different, and different compounds have different optical rotations. In acidic solution, the optical rotation is that of the acidic form; in base, the optical rotation is that of the basic form; and in water, the optical rotation is that of the neutral (zwitterion) form.

(b) Lysine has two amino groups, and acetylation of each (or both) is possible.

the two mono-*N*-acetylated lysine derivatives

(c)   The peptide is cleaved at the Arg-Ala bond at pH 8. However, in the presence of 8 *M* urea, trypsin, like most enzymes, is denatured. A denatured enzyme is devoid of catalytic activity because its tertiary structure (and therefore its active site) is disrupted. Thus, trypsin, when denatured, cannot catalyze cleavage of the peptide.

(d)   The 2-mercaptoethanol treatment ensures that disulfide bonds are reduced to thiols. Aziridine reacts much like an epoxide when the nitrogen is protonated. Aziridine is basic enough that a significant amount of it is protonated in aqueous solution. Reaction of protonated aziridine with the thiol group of a cysteine residue gives an amine:

The amino group of the modified residue is protonated at pH 8, the pH at which trypsin digestion is carried out. The side chain of the modified residue resembles the side chain of lysine in both length and charge, and trypsin hydrolyzes peptides at the modified residue much as it hydrolyzes peptides at lysine residues.

(e)   The two sulfoxides are diastereomers. They both have the same configuration at the α-carbon, but differ in configuration at the sulfur of the sulfoxide group. This sulfur is an asymmetric atom and stereocenter because it has four different "groups" attached: the CH_2, the CH_3, the O, and the nonbonding electron pair. Inversion at sulfur in a sulfoxide, unlike amine inversion, is very slow, so the individual diastereomers can be isolated. (See Sec. 6.9B, text p. 309.)

27.80   Amino acid *A* is 2,4-diaminobutanoic acid. The key piece of structural data is the Hofmann rearrangement, which converts the side-chain amide of glutamine into an amino group. Compound *A* is expected to be basic (and thus have a high isoelectric point) because it has two amino groups and only one carboxy group.

2,4-diaminobutanoic acid
(compund *A*)

27.81   (a)   At a pH of 8–9, the thiol group of cysteine is partially deprotonated (pK_a of the thiol group is 10.28; see Table 27.1, text p. 1478); the resulting thiolate is a good nucleophile and the iodine of iodoacetic acid is an excellent leaving group. Remember also that α-halo carbonyl compounds are very reactive in S_N2 reactions (Sec. 22.3D of the text). Masking the thiol group of any cysteine residues present in the protein is necessary to prevent the thiol groups from re-oxidizing to disulfides.

$$\boxed{\text{protein}}-CH_2-SH \underset{}{\overset{pH\ 8-9}{\rightleftharpoons}} \boxed{\text{protein}}-CH_2-S^- \xrightarrow{\overset{I-CH_2CO_2^-}{\curvearrowright}}$$

$$\boxed{\text{protein}}-CH_2-S-CH_2CO_2^- + I^-$$

cysteine residue

(b) Hydrogen peroxide oxidizes the disulfide bonds to sulfenic acids, then to sulfinic acids, and finally to sulfonic acids (see Fig. 11.2, text p. 607).

$$\boxed{\text{protein}}-CH_2-S-S-CH_2-\boxed{\text{protein}} \xrightarrow{H_2O_2} 2\ \boxed{\text{protein}}-CH_2-SO_3H \underset{}{\overset{pH\ 7}{\rightleftharpoons}}$$

cystine residue

$$2\ \boxed{\text{protein}}-CH_2-SO_3^-$$

27.82 The coenzyme involved in the conversion of aspartic acid residues to Asp-methyl ester residues is *S*-adenosylmethionine (SAM). (See Figure 12.1, text p. 650, for the structure of SAM.)

$$\underset{\underset{\underset{CO_2CH_3}{|}}{\overset{|}{CH_2}}}{-NH-CH-C\overset{O}{\overset{\|}{\diagdown}}}$$

Asp-methyl ester residue

27.83 Trypsin catalyzes the hydrolysis of peptides and proteins at the carbonyl group of arginine and lysine, amino acids with side chains containing strongly basic groups that are positively charged at physiological pH. Chymotrypsin, in contrast, catalyzes the hydrolysis of peptides and proteins at the carbonyl group of phenylalanine, tryptophan, tyrosine, leucine, and isoleucine. As explained in the text (see Sec 27.10) and illustrated in Fig. 27.19, text p. 1539, the positively charged group in the peptide or protein substrate forms a complex in the trypsin active site by electrostatic attractions and hydrogen bonding. The addition of a specific trypsin inhibitor such as benzamidinium chloride (Displays 27.70 and 27.71, text pp. 1538–1539) would block the trypsin active site without appreciably affecting chymotrypsin. (Considerably more potent trypsin inhibitors—some of which react *covalently* with residues in the trypsin active site—are also known and are used for this purpose.)

27.84 The interior of a membrane is a very hydrophobic environment. Therefore, amino acids containing side chains that are very nonpolar typically populate the surface of the protein that comes in contact with the membrane. (Refer to Table 27.1, text p. 1478–9, for the structures of the amino acid side chains.) Group *A* and *B* contain acidic and basic side chains, respectively, and would be more likely to be on the surface of a protein that comes into contact with water. Similarly, group *F* contains polar side chains, which are not hydrophobic. Group *D*, Pro, is rarely found in α-helix regions because of its cyclic, rigid, structure. Group *E* contains very hydrophobic side chains, and is thus the most likely group engage in favorable hydrocarbon-hydrocarbon interactions with the interior of the membrane. Group *C*, Gly, contains only hydrogen as a side chain and which is not as hydrophobic as the side chains in group *E*.

27.85 Part (a) is a reductive amination of formaldehyde by the amino group of the lysine residue. Because excess formaldehyde is present, the amino group of lysine is methylated twice. In part (b), The side-chain amino group of lysine serves as a nucleophile to open the anhydride. The resulting carboxy group is ionized at the pH of the reaction.

(a)

(b)

In part (c), the thiol of cysteine acts as a nucleophile in a conjugate-addition reaction to maleimide. The pH must be high enough to form a small amount of the conjugate-base thiolate anion of cysteine, which is the actual nucleophilic species. In part (d), the phenol ring of the tyrosine residue undergoes electrophilic substitution by the diazonium ion (see Sec. 23.10B, text p. 1304). Because the para position is occupied, substitution occurs at the position ortho to the hydroxy group.

(c)

(d)

27.86   The formation of the diketopiperazine is an intramolecular ester aminolysis. The mechanism below begins with the conjugate base of the peptide formed by loss of a proton from the terminal amino group.

You may have noticed that diet soft drinks that go on sale in the supermarket are not quite as sweet as really fresh ones. This is because some of the sweetener has degraded by the mechanism just depicted.

27.87   **(a)**   This is an aminolysis of the side-chain carboxylic group by the α-amino group. Normally, the reaction of an amine with a carboxylic acid is very slow because the amine is protonated and the carboxylic acid group is ionized, and the reaction requires heat. But there is always a little of the neutral form present, and the intramolecular aminolysis is accelerated by the proximity effect (Sec. 12.8). (We neglect stereochemistry in the second half of the mechanism that follows, but the pyroglutamic acid retains the L configuration of the starting glutamic acid.)

amino-terminal Glu residue

proton transfer

amino-terminal pyroGlu residue

+ H_2O

Essentially the same reaction occurs with amino-terminal glutamine, except that ammonia rather than water is lost.

(b) Because the terminal amino group of the peptide is blocked, formation of pyroGlu prevents Edman sequencing. However, a pyroGlu residue would not prevent sequencing by MS–MS.

(c) For the same reaction to occur with amino-terminal Asp or Asn, a four-membered ring would be formed. Because such a ring is very strained, this reaction doesn't occur.

27.88 Peptide *B* results from intramolecular nucleophilic reaction with the conjugate-base anion of the neighboring amide bond. This occurs mostly at Asn-Gly because glycine has no carbon side chain; in residues other than glycine, the carbon side chains can cause rate-retarding van der Waals repulsions in the ring-closure step.

+ H_2O

A

B

reacts with water to give ^-OH and NH_3

Derivative *B* is an imide, which is the nitrogen analog of an anhydride. Hydroxide can react with either carbonyl carbon of the imide ring to open the ring, thus generating either peptide *C* or peptide *D*. The mechanism for the formation of peptide *D* is as follows:

Peptide *C* results from the same type of reaction initiated by the reaction of ⁻OH at the other carbonyl group of the imide. (Dashed arrows.) We leave it to you to draw the curved-arrow notation.

27.89 (a)

(b)

(c)

The isotopically substituted reagents (⁺ND₄ Cl⁻, D₂O) are necessary in the first step because the pH of the reaction is sufficiently basic that the α-deuteriums of the starting trideuteroacetaldehyde would exchange for hydrogens if the analogous protic reagents ⁺NH₄ Cl⁻ and H₂O were used. (See Eq. 22.11 on text p. 1198.)

(d)

(e) First prepare 3-bromocyclopentene from cyclopentene:

3-bromocyclopentene

Then use it in the following synthesis. Because the α-hydrogen of the starting material [the product of part (d)] is about as acidic as an α-hydrogen of diethyl malonate, the techniques of the acetamidomalonate method can be used. Notice that 3-bromocyclopentene is an allylic halide and is therefore very reactive in S_N2 reactions. (See Sec. 17.4 of the text.)

(f)

27.90 Part (a) is like the reaction of an amine with an isocyanate; see Eq. 23.87 on text p. 1313. Part (b) is a modified Strecker synthesis in which methylamine is used in place of ammonia. Part (c) is a reductive amination in which the amine is the α-amino group of the amino acid.

(a)

(b)

(c)

(d) The methyl ester is saponified much more rapidly than the *tert*-butyl ester for two reasons. First, the methyl branches of the *tert*-butyl group impede the approach of hydroxide to the carbonyl group. Second, the *tert*-butyl "ester" is also an amide, and amides hydrolyze much more slowly than esters.

Part (e) is a diazotization to form the *N*-diazo compound, that is, the acyl azide. (See Eq. 23.86 on text p. 1313.) Part (f) is a Curtius rearrangement to the isocyanate. Part (g) is the urea formed by the addition of the amino group of the amino acid ester to the isocyanate.

H₂NCHCO₂CH₃

$$PhC(=O)-\overset{-}{N}-\overset{+}{N}\equiv N \xrightarrow[-N_2]{heat} PhN=C=O \xrightarrow[\text{valine methyl ester}]{} PhNHC(=O)NHCHCO_2CH_3$$

benzoyl azide
product of part (e)

phenyl isocyanate
product of part (f)

CH(CH₃)₂

**N-phenylaminocarbonylvaline
methyl ester**
product of part (g)

(h)   This is a formylamidomalonate reaction, a variation of the acetamidomalonate synthesis.

$$H_3\overset{+}{N}CHCO_2H + HCO_2H + EtOH$$

CH₂

$$H_2C=\overset{C}{\underset{CH_3}{}}$$

(i)   Hydrazine (H₂N—NH₂) displaces the ethoxy group of the ester to form a hydrazide *A*; diazotization forms the acyl azide *B*; heating this in ethanol gives a Curtius rearrangement in which the isocyanate is trapped by reaction with the solvent to give the ethyl carbamate *C*; and acidic hydrolysis generates the amino-protonated amino acid *D*.

$$N\equiv CCHC(=O)NHNH_2 \xrightarrow{NaNO_2/HCl} N\equiv CCHCN_3 \xrightarrow{EtOH, heat} N\equiv CCHNHC(=O)OEt \xrightarrow[heat]{HCl/H_2O} HO_2CCH\overset{+}{N}H_3\ Cl^-$$

CH(CH₃)₂   CH(CH₃)₂   CH(CH₃)₂   CH(CH₃)₂

A   *B*   *C*   *D*

27.91   (a)   The first step of the mechanism is formation of an imine. (This is discussed in Sec. 19.12A of the text.) Reaction of this imine with the conjugate base of the thiol gives an addition product which is eventually transformed into the product. The following mechanism begins with the imine.

(b)   In the first part of the mechanism, ⁻SH displaces chloride in an S_N2 reaction. Recall that α-halo carbonyl compounds are particularly reactive in S_N2 reactions (Sec. 22.3D of the text). Ammonia then reacts to forms an enamine. (Enamine formation is discussed in Sec. 19.12C of the text.) Although imine formation is favored, imines and enamines are in equilibrium just as aldehydes and enols are in equilibrium. Addition of both the —SH group and the —NH$_2$ group of this enamine to acetone (in a reaction much like acetal formation) and proton transfers give the first product *A*.

The isomerization leading to compound *A* is analogous to the conversion of an enol into a carbonyl compound. Compound *A* then reacts with cyanide ion in a variation of the Strecker synthesis. Hydrolysis in acid liberates acetone and the conjugate acid *B* of the α-cyano amine. The cyano group hydrolyzes in acid to give the conjugate acid of cysteine. (The mechanism of nitrile hydrolysis is given in Sec. 21.7C of the text.)

(c)   Formaldehyde and the amino group of tryptophan react to give the conjugate acid of an imine, which acts as an electrophile in an intramolecular electrophilic aromatic substitution reaction to give the product. The mechanism below begins with the conjugate acid of the imine.

(d) This reaction is a Curtius rearrangement (with concomitant loss of N_2) to give an isocyanate, which reacts intramolecularly with the side-chain hydroxy group. The mechanism of the Curtius rearrangement is given in Eq. 23.83 on text p. 1312. The mechanism below begins with the isocyanate.

27.92 (a) In the Z conformation, the two large groups Pep^N and the N-alkyl group Pep^C are trans, so van der Waals repulsions between these groups cannot occur. In the E conformation, these two groups are close enough that van der Waals repulsions can result. These raise the energy of the peptide; therefore, the Z conformation is energetically preferred.

(b) The proline nitrogen bears two N-alkyl groups, whereas the peptide-bond nitrogens of other residues have only one. Thus, in either the E or the Z conformation, an alkyl group is cis to the Pep^N group. Although the group that is cis to Pep^N in the E configuration is larger, there is a much smaller difference in energy between E and Z conformations for proline residues than there is for other residues; so E conformations in peptide bonds are sometimes found at proline residues in proteins and peptides.

Notice that the *E* conformation of proline in a peptide forms a turn or bend in the peptide chain. Proline residues with *E* conformations are often found within turns in the three-dimensional structures of proteins.

27.93   (a)   The two monomethyl esters result from ester formation at each of the two carboxy groups.

$$
\begin{array}{cc}
\underset{A}{
\begin{array}{c}
\text{CH}_3\text{CHNCHCO}^- \\
\big| \\
\text{CH}_2 \\
\big| \\
\text{C} \\
\text{O} \diagup\ \diagdown \text{OCH}_3
\end{array}
} &
\underset{B}{
\begin{array}{c}
\text{CH}_3\text{CHNCHCOCH}_3 \\
\big| \\
\text{CH}_2 \\
\big| \\
\text{C} \\
\text{O} \diagup\ \diagdown \text{O}^-
\end{array}
}
\end{array}
$$

(b)   According to Table 27.1, text p. 1479, the pK_a values of the two carboxy groups in aspartic acid are about 1.9 (for the α-carboxy group) and 3.6 (for the β-carboxy group). Assume that the pK_a values for the respective esters are similar. (Actually, they are both a little higher.) At pH 3.0, the carboxy group of compound *A* is largely ionized; at the same pH, the carboxy group of *B* is mostly un-ionized. Therefore, at pH 3.0, compound *B* is neutral, whereas compound *A* is negatively charged. Consequently, compound *A* will be retained by an anion-exchange column (which contains positively charged groups) and will be eluted more slowly than compound *B*. At pH 7, however, both compounds are negatively charged. Since they have the same charge, neither is retained by an anion-exchange column; consequently, these compounds are therefore not separated at pH 7.

27.94   The formula of compound *A* is the same as that of *N*-acetylaspartic acid minus the elements of water. Since treatment of a dicarboxylic acid with acetic anhydride is a good way to prepare a cyclic anhydride (see Eq. 20.50, text p. 1110), it is reasonable to suppose that compound *A* is the cyclic anhydride of aspartic acid. When this anhydride is treated with L-alanine, the amino group of alanine can react with either of the two carbonyl groups to give either of the two isomeric peptides *B* and *C*. Both would hydrolyze to give alanine and aspartic acid (as well as acetic acid).

27.95   The arylthiol group in peptide 1 is a leaving group. The *p*-nitro group reduces the basicity of the arylthiolate when the group is displaced, and the carboxylate substituent enhances water solubility of the leaving group and the peptide. The evidence in the problem *rules out* a direct reaction of the amino group of peptide 2 with the thioester of peptide 1; evidently the thiol is important.

   The amine and the conjugate-base thiolate of the side-chain thiol have similar basicities. However, in water, we expect the thiolate ion to be much more *nucleophilic* than the amine, because sulfur is a much weaker hydrogen-bond acceptor than the nitrogen. (Recall [Sec. 9.4E] that protic solvents reduce nucleophilicity by hydrogen bonding to the nucleophile, and this effect is most pronounced with nucleophilic atoms from the second period.) Therefore, the thiol reacts at the carbonyl group of peptide 1 to displace the arylthiolate leaving group and form another thioester T:

The amino group in thioester T is now positioned to be an *intramolecular* nucleophile. Recall (Sec. 12.8) that nucleophilic substitutions involving 5-membered rings are accelerated by many orders of magnitude. The amino group rapidly displaces the thiolate, which forms the thiol by protonation. Recall (Sec. 25.8C) that thioesters have evolved in biology because they are high-energy compounds; therefore, the equilibrium between the amide and the thioester *strongly* favors the amide. (This is why peptides don't self-cleave at cysteine residues.)

Another ingenious aspect of this ligation is that the nitro substituent, in addition to enhancing the leaving-group efficacy, makes the arylthiolate leaving group brightly colored; so, the ligation reaction has its own built-in "completion indicator." (The step in which the colored leaving group is expelled is rate-limiting.)

27.96 (a) With the mass M of peptide P equaling the mass of $(A + B) - 17$ mass units, the mass of C equaling the mass of $(T\text{-}K\text{-}N\text{-}L + N\text{-}S\text{-}L) - 17$ mass units, and the mass of D equaling the mass of $(V\text{-}T\text{-}K\text{-}N\text{-}L + N\text{-}S\text{-}L) - 17$ mass units, the evidence suggests that the N-S-L fragment is somehow connected in each of the fragments but minus 17 mass units. The residues common to A, C, and D are K, N, and L (lysine, asparagine, and leucine, respectively) and the 17 mass units could correspond to a loss of NH_3, indicating that perhaps lysine and/or asparagine may be the common factor. If asparagine undergoes a transamination with lysine, the resulting product would link the K-N with the loss of NH_3. Therefore, the structure of peptide P is:

L-T-V-T-K-N-L
|
N-S-L

peptide P

This peptide bond is unusual because it is a cross-link between the Lys and Asn side-chains of two peptide chains.

(b) A transamination ensues between the asparagine and lysine residues to produce the new amide linking the two peptides. We assume that an appropriate acid and base (shown here as AH and A⁻) are available to facilitate the proton transfers.

(c) The structure of the pilus evidently brings the nucleophilic amino group of the lysine and the carbonyl carbon of the asparagine into proximity. As discussed in Sec. 12.8, proximity can accelerate a substitution reaction by many orders of magnitude. This acceleration would turn an ordinarily sluggish reaction into a very rapid one.

Pericyclic Reactions

CHAPTER 28 SUMMARY

- Pericyclic reactions are concerted reactions that occur by cyclic electron shifts. Electrocyclic reactions, cycloadditions, and sigmatropic rearrangements are important pericyclic reactions.

- Electrocyclic reactions are stereochemically classified as conrotatory or disrotatory; cycloadditions and sigma-tropic rearrangements are classified as suprafacial or antarafacial.

- The stereochemical course of a pericyclic reaction is governed largely by the symmetry of the reactant HOMO (highest occupied molecular orbital), or, if there are two reacting components, by the relative

symmetries of the HOMO of one component and the LUMO (lowest unoccupied molecular orbital) of the other.

- Considerations of orbital symmetry lead to selection rules for pericyclic reactions. Whether a pericyclic reaction is allowed or forbidden depends on the number of electrons involved, the mode of activation (thermal or photochemical), and the stereochemical course of the reaction. The selection rules are summarized in Tables 28.1–28.3.

STUDY GUIDE LINKS

28.1   Frontier Orbitals

You've seen that the analysis of pericyclic reactions in the text has focused on frontier orbitals. The discussion of electrocyclic reactions focused on one of the frontier orbitals, the HOMO, and the analysis of cycloadditions focused on the HOMO of one component and the LUMO of the other. Yet each of the frontier orbitals houses only two of the many reacting π electrons. Because *all* π electrons in the reacting molecules are involved in pericyclic reactions, it can perhaps be appreciated that all of the π molecular orbitals are involved as well. Why focus, then, on frontier orbitals? This focus is justified because it has been shown that frontier-orbital interactions play the *major* role in determining the energies of transition states for concerted pericyclic reactions. However, these are not the only orbital interactions that occur; they are simply the *most important* interactions.

Some students try to associate frontier molecular orbital interactions with formation of *a particular bond* in the reactant or product. This temptation is understandable because of the association of a π molecular orbital with the double bond in a simple alkene, or the association of a σ molecular orbital with a particular single bond in saturated molecules. These early encounters with molecular orbitals fail to take account of the fact that all molecular orbitals within a molecule must have certain symmetry properties, namely, they must be symmetric or antisymmetric with respect to the symmetry elements of the molecule itself. (For example, if a molecule contains an internal mirror plane, molecular orbitals of the molecule must be either antisymmetric or symmetric with respect to that plane.) For this reason, a one-to-one correspondence in general does not exist between molecular orbitals and the localized bonds used to write Lewis structures. Molecular orbitals are orbitals of *molecules*, and changes in a particular molecular orbital may involve more than one bond, and vice versa.

28.2   Orbital Analysis of Sigmatropic Reactions

Sigmatropic reactions are concerted and do not involve free ions, but the orbital relationships are nevertheless understood by thinking of the two components of the transition state as ions of opposite charge. However, for the reaction in Eq. 28.36 on text p. 1574 we could just as well think of the reaction as the migration of a hydrogen atom across a 2,4-pentadien-1-yl radical, or as the migration of a hydrogen anion (hydride ion) across a 2,4-pentadien-1-yl cation; these different approaches give the same predictions. Let's verify this point by viewing the transition state of the [1,5] hydrogen migration as the migration of a hydrogen atom across a 2,4-pentadien-1-yl radical:

$$\left[\; \cdot CH_2 \quad\rightleftharpoons\quad CH_2 \cdot \; \right]^{\ddagger}$$

In this case, analyze the symmetries of the orbitals containing the unpaired electrons, that is, the SOMO of each radical. For a hydrogen atom, this orbital is just the 1s orbital. For the 2,4-pentadien-1-yl radical, this is π_3 (see Display 28.37 on text p. 1574). Because this molecular orbital is symmetric, the migration is predicted to occur suprafacially.

Now let's analyze the reaction as if it were the migration of a hydrogen anion (hydride ion, $H{:}^-$) across the termini of a 2,4-pentadien-1-yl cation and show that such an analysis makes the same prediction. In this case, the 1s orbital of the migrating hydride contains two electrons, so this orbital is classified as the HOMO of the hydride ion. It interacts with the LUMO of the 2,4-pentadien-1-yl cation, which, from Display 28.37 on text p. 1574, is π_3. Because this molecular orbital is symmetric, the migration hydride is predicted to occur suprafacially.

Notice that whether we deal with the migration of a proton across an anion, a radical across another radical, or an anion across a cation, the relevant orbitals of the species involved are the same: the 1s orbital of the hydrogen and π_3 of the pentadienyl system. Therefore, the predictions of orbital symmetry must also be the same.

REACTION REVIEW

I.   ELECTROCYCLIC REACTIONS

1. When an electrocyclic reaction takes place, the carbons at each end of the conjugated π-electron system undergo internal rotations in a concerted manner so that the 2p orbitals can overlap (and rehybridize) to form the σ bond that closes the ring.

2. These rotations can occur in two stereochemically distinct ways.

   a. In a conrotatory closure the two carbon atoms rotate in the same direction, either both clockwise or both counterclockwise.

conrotatory closure   or

b. In a disrotatory closure the two carbon atoms rotate in opposite directions so that either the upper lobes of the 2p orbitals overlap or the lower lobes of the 2p orbitals overlap.

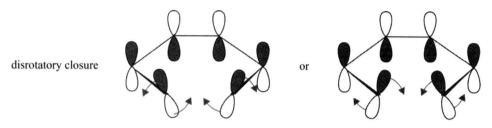

disrotatory closure   or

3. The HOMO of the conjugated alkene contains the π electrons of highest energy and governs the course of pericyclic reactions.

   a. When the ring closure takes place, the two 2p orbitals on the ends of the π system must overlap in phase.

   b. The wave peak on one carbon must overlap with the wave peak on the other, or a wave trough must overlap with a wave trough.

4. The relative orbital phase at the terminal carbon atoms of the HOMO (the orbital symmetry) determines whether the reaction is conrotatory or disrotatory.

   a. Conjugated alkenes with $4n$ π electrons (n = any integer) have antisymmetric HOMOs and undergo conrotatory ring closure. Conrotatory ring closure is allowed for systems with $4n$ π electrons; it is forbidden for systems with $4n + 2$ π electrons.

   b. Conjugated alkenes with $4n + 2$ π electrons have symmetric HOMOs and undergo disrotatory ring closure. Disrotatory ring closure is allowed for systems with $4n + 2$ π electrons; it is forbidden for systems with $4n$ π electrons.

5. For photochemical (excited-state) electrocyclic reactions, the selection rules are reversed; that is, conrotatory closure is allowed for a $4n + 2$ π-electron reaction and disrotatory closure is allowed for a $4n$ π-electron reaction.

6. Although the rules may be stated in terms of the number of π electrons, microscopic reversibility holds.

   a. The rules apply to both the forward and reverse of an electrocyclic reaction (that is, to both ring closure and ring opening).

   b. The rules have nothing to say about the position of equilibrium.

II. CYCLOADDITION REACTIONS

1. Cycloadditions are classified, first, by the number of electrons involved in the reaction with respect to each component.

   a. The number of electrons involved is determined by writing the reaction mechanism in the curved-arrow notation.

   b. The number of electrons contributed by each given reactant is equal to twice the number of curved arrows originating from that component (two electrons per arrow).

two electrons

two electrons

2. Cycloaddition reactions are also classified by their stereochemistry with respect to the plane of each reacting molecule and may in principle occur either across the same face, or across opposite faces, of the planes in each reacting component.

   a. If the reaction occurs across the same face of a π system, the reaction is said to be suprafacial with respect to that π system; that is, a syn-addition that occurs in a single mechanistic step.

   b. If the reaction bridges opposite faces of a π system, it is said to be antarafacial; that is, it is an anti-addition that occurs in one mechanistic step.

3. The reaction classification is written as a combination of numbers and letters. Thus, a [4s + 4a] cycloaddition involves 4 π electrons on each component and is suprafacial (s) on one component and antarafacial (a) on the other.

4. In order for a cycloaddition to occur, bonding overlap must take place between the 2p orbitals at the terminal carbons of each π electron system.

   a. The bonding overlap begins when the HOMO of one component interacts with the LUMO of the other.

   b. The LUMO of the other component is the empty orbital of lowest energy into which the electrons from the HOMO must flow.

   c. These two frontier MOs involved in the interaction must have matching phases if bonding overlap is to be achieved.

   d. It does not matter which component provides the HOMO and which provides the LUMO.

5. All-suprafacial cycloadditions are allowed thermally for systems in which the total number of reacting electrons is $4n + 2$, and they are allowed photochemically for systems in which the number is $4n$.

III. THERMAL SIGMATROPIC REACTIONS

A. STEREOCHEMISTRY OF SIGMATROPIC REACTIONS

1. Sigmatropic reactions can be classified by their stereochemistry according to whether the migrating bond moves over the same face, or between opposite faces, of a π-electron system.

   a. If the migrating bond moves across one face of a π system the reaction is said to be suprafacial (Eq. 28.31, text p. 1572).

   b. If the migrating bond moves from one face of a π system to the other the reaction is said to be antarafacial (Eq. 28.32, text p. 1572).

   c. When both ends of a σ bond migrate, the reaction can be suprafacial or antarafacial with respect to each π system (Eq. 28.33, text p. 1572).

2. The stereochemistry of a sigmatropic reaction is revealed experimentally only if the molecules involved have stereocenters at the appropriate carbons.

3. Molecular orbital theory provides the connection between the type of sigmatropic reaction and its stereochemistry.

B. [1,3] AND [1,5] SIGMATROPIC REARRANGEMENTS

1. The interaction of the LUMO of the migrating group with the HOMO of the π system, or vice-versa, controls the stereochemistry of the reaction.

   a. In the migration of a hydrogen, the orbital involved is a 1s orbital, which has no nodes.

   b. In the migration of a carbon, the orbital involved is a 2p orbital, which has one node.

   c. The shift of a carbon can occur in two stereochemically distinct ways:

i.   Migration with retention of configuration.

ii. Migration with inversion of configuration.

d. In the allyl anion, the HOMO is antisymmetric:

HOMO for the allyl anion

i.   The suprafacial [1,3] shift of a hydrogen is forbidden by orbital symmetry.

ii. The antarafacial [1,3] migration of hydrogen is allowed by orbital symmetry but is virtually nonexistent in organic chemistry because unreasonably long bonds would be required in the transition state.

iii. The suprafacial [1,3] shift of a carbon with inversion of configuration is allowed by orbital symmetry.

iv. The antarafacial [1,3] shift of a carbon with retention of configuration is allowed by orbital symmetry.

e. In the 2,4-pentadienyl anion, the HOMO is symmetric:

HOMO for the 2,4-pentadienyl anion

i.   The suprafacial [1,5] migration of a hydrogen is allowed by orbital symmetry.

ii. The antarafacial [1,5] migration of a hydrogen is forbidden by orbital symmetry.

iii. The suprafacial [1,5] migration of a carbon with retention of configuration is allowed by orbital symmetry.

iv. The antarafacial [1,5] migration of a carbon with inversion of configuration is allowed by orbital symmetry.

C. *SIGMATROPIC REARRANGEMENTS; COPE AND CLAISEN REARRANGEMENTS*

1. A [3,3] sigmatropic rearrangement is a reaction in which both ends of a σ bond change positions (migrate).

    a. The transition state of a [3,3] sigmatropic rearrangement can be visualized as the interaction of two allylic systems, one a cation and one an anion.

    b. The two MOs involved achieve bonding overlap when the [3,3] sigmatropic rearrangement occurs suprafacially on both components.

2. The Cope rearrangement is a [3,3] sigmatropic rearrangement in which a 1,5-diene isomerizes.

3. The oxyCope reaction is a variation of the Cope rearrangement that involves the initial formation of an enol.

    a. Isomerization of the enol into the corresponding carbonyl compound is a very favorable equilibrium that drives the reaction to completion.

b. The oxyCope reaction is strongly accelerated by conversion of the —OH group into its conjugate-base alkoxide (anionic oxyCope).

4. In the Claisen rearrangement, an ether that is both allylic and vinylic (or an allylic aryl ether) undergoes a [3,3] sigmatropic rearrangement. If both *ortho* positions of an aryl allylic ether are blocked by substituent groups, the *para*-substituted derivative is obtained by a sequence of two Claisen rearrangements followed by isomerization of the product to the phenol.

🔒🔑 SOLUTIONS TO PROBLEMS

Solutions to Focused Problems

28.1 (a) This is an electrocyclic reaction; six electrons are involved.

(b) This is a sigmatropic reaction; two electrons are involved.

(c) This is an intramolecular cycloaddition reaction; four electrons are involved.

(d)   This is a sigmatropic reaction; six electrons are involved.

(e)   This is an electrocyclic reaction; two electrons are involved.

28.2   The solution to Focused Problem 15.1(Fig. SG15.2, p. 437 in this manual) has a diagram of the 1,3,5-hexatriene molecular orbitals.

(a)   1,3,5-Hexatriene has six π molecular orbitals.

(b)   The symmetry of the 1,3,5-hexatriene MOs alternates; that is, π_1, π_3, and π_5^* are symmetric, and π_2, π_4^*, and π_6^* are antisymmetric.

(c)   The π_1, π_2, and π_3 MOs are bonding and the π_4^*, π_5^*, and π_6^* MOs are antibonding.

(d)   Because 1,3,5-hexatriene has six π electrons, the highest occupied molecular orbital (HOMO) is π_3; the lowest unoccupied molecular orbital (LUMO) is π_4^*.

(e)   As indicated in the solution to part (b), the HOMO (π_3) is symmetric; therefore, the phase of the HOMO at the terminal carbons is the same.

(f)   The phase of the LUMO (π_4^*) at the terminal carbons is different.

28.3   Because there are ten π electrons, there are five occupied MOs in this alkene, each containing two electrons. The MOs alternate in symmetry; the odd-numbered ones are symmetric, and the even-numbered ones are antisymmetric. Therefore, π_6 is antisymmetric (A); it is antibonding; and it is the LUMO and is thus one of the frontier orbitals.

28.4   (a)   The MO π_4 is nonbonding. [If there are an odd number n of molecular orbitals in an unbranched acyclic system, the nonbonding MO is always the one with the number $(n + 1)/2$.]

(b)   The symmetry of the MOs alternates; that is, π_1, π_3, π_5^*, and π_7^* are symmetric, and π_2, π_4, and π_6 are antisymmetric.

(c)   According to its resonance structures, this cation has sites of positive charge at alternating carbons:

In molecular orbital terms, the positive charge in the cation results from the absence of an electron in the LUMO; the LUMO *is* the nonbonding MO. Positive charge is shared at alternating carbons because the LUMO has nodes at the other carbons. That is, positive charge can only exist on carbons at which there is no node in the LUMO. If this is so, the LUMO must have three nodes: one at each carbon that does *not* share the positive charge. As shown in the solution to part (a), π_4 is the nonbonding MO and is therefore the LUMO, and it has three nodes, one at each carbon that does not share positive charge in the above structures.

28.5 (a) The resonance structures of the allyl radical show that the unpaired electron is shared between the terminal carbons:

(b) Figure 28.4, text p. 1559, shows that the unpaired electron of the allyl radical resides in the nonbonding MO, which has a node at the central carbon. The unpaired electron density therefore exists only at the carbons on either side of a node, that is, at the terminal carbons, as the resonance structures show.

28.6 *Reaction 1* is a disrotatory reaction involving $4n + 2$ electrons, and is allowed. Therefore it can occur readily by a concerted mechanism. The reaction is disrotatory because the hydrogens shown in the product are both on the outside of the starting material. (Draw these hydrogens in the starting material, if necessary.) In order for these hydrogens to end up on the same face of the product, the ends of the π-electron system must turn in opposite directions.

  Reaction 2 is a conrotatory reaction involving $4n + 2$ electrons, and is not allowed; it does not occur readily.

28.7 The two different conrotatory processes are as follows: (The curved arrows indicate atomic motion, not electron flow.)

The two processes are equally likely because they are enantiomeric; enantiomers have equal energies, and enantiomeric pathways have identical energies at all points and therefore identical rates (Sec. 7.7A).

28.8 The thermal ring opening of *trans*-3,4-dimethylcyclobutene must be a conrotatory process. The two possible conrotatory processes are as follows:

Product *A* has severe van der Waals repulsions between the methyl groups; product *B* does not. The repulsions raise the energy of product *A* and the transition state for its formation; consequently, product *B* is observed.

28.9 The opening of the cyclobutene is conrotatory and therefore gives (1*E*,3*Z*)-cyclodecadiene. The trans double bond does not introduce as much strain into the large ring as was present in the starting cyclobutene.

(1E,3Z)-cyclodecadiene

Here is one way to use models to visualize this transformation. To avoid having to build two models, construct a cyclodecane (the cyclic *alkane* with a ten-membered ring); don't worry about the missing double bond. Now connect two carbons to make a cis-fused four-membered ring. Identify the "up" hydrogens on these two carbons. Using a pencil, draw a single line across the bond that is a double bond in the starting material (even though it is a single bond in your model). Now draw two lines across each of the bonds that will become double bonds as a result of the reaction. Now break the cyclobutane ring and turn the carbons bearing the "up" hydrogens in a conrotatory manner. Examine the stereochemistry at the bonds marked with two lines; one should have an *E* configuration and the other a *Z* configuration. Voila!

28.10   The HOMO of the 2π-electron component is symmetric, and the LUMO of the 4π-electron component is symmetric. As the diagram below shows, the ends of these two MOs have matching phases in a $[4s + 2s]$ cycloaddition.

LUMO of 4π-electron component

in-phase overlap

HOMO of 2π-electron component

Interaction of the LUMO of the 4 π-electron component and the HOMO of the 2 π-electron component in a $[4s + 2s]$ cycloaddition gives in-phase overlap at both ends of the two π-electron systems.

28.11   Use the HOMO of the 4 π-electron component and the LUMO of the 2 π-electron component. The $[4a + 2s]$ process would involve the overlap shown in (a) below. The $[4s + 2a]$ process would involve the overlap shown in (b). In both cases an out-of-phase overlap between the π-electron systems is required at one end. Consequently, these processes are not allowed.

HOMO of 4π-electron component

HOMO of 4π-electron component

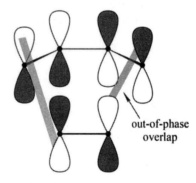

out-of-phase
overlap

out-of-phase
overlap

LUMO of 2π-electron component
$[4a + 2s]$

LUMO of 2π-electron component
$[4s + 2a]$

28.12   The $[8s + 2s]$ cycloaddition and its product:

28.13 The four products are the diastereomeric 1,2,3,4-tetramethylcyclobutanes. Each of the pure alkene stereoisomers can undergo an allowed photochemical [2s + 2s] cycloaddition in two distinguishable ways. *Cis*-2-butene reacts to give compounds *A* and *B*; *trans*-2-butene reacts to give compounds *B* and *C*; and the mixture of *cis*- and *trans*-2-butene can give these three compounds plus a fourth, compound *D*, that results from the [2s + 2s] cycloaddition of *cis*-2-butene to *trans*-2-butene.

28.14 Because exchanging the positions of two groups at a stereocenter changes the configuration of the stereocenter, exchange the positions of the T and the D in the starting material as shown in (a) below. This will lead to a product with the opposite, that is, the *R*, configuration at the indicated carbon. Two other starting materials that would give the same product as in part (a) are shown in (b) below.

(a)   (b)

28.15 Part (a) is a [1,4] sigmatropic rearrangement; part (b) is a [2,3] sigmatropic rearrangement; and part (c) is a [5,5] sigmatropic rearrangement. The following diagrams give the rationale for these assignments

(a)   migration of this end of the
bond does not occur, that is, it
"migrates" from carbon-1* to carbon-1*.

migration of this end of the
the bond occurs from
carbon-1 to carbon-4.

(b)   One end of this bond
moves from carbon-1
to carbon-3; the other end
moves from the sulfur (1*)
to the oxygen (2*).

(c)   This bond migrates
from carbons 1 and 1*
to carbons 5 and 5*,
respectively.

28.16   Reaction (1) occurs readily because it is a [1,5] sigmatropic rearrangement, which is allowed when the hydrogen migrates suprafacially over the π-electron system. This reaction pathway does not introduce significant strain or twist into the π-electron system. Reaction (2) does not occur because it is a [1,3] sigmatropic rearrangement, which is allowed only if the hydrogen migrates antarafacially. Such an antarafacial migration requires that the hydrogen simultaneously bridge the upper and lower faces of the π-electron system. This requirement cannot be met while at the same time maintaining normal bond lengths and angles.

28.17   If the migration were antarafacial, the opposite stereochemical result would have been observed; that is, the 3E stereoisomer of the product would have the S configuration at the asymmetric carbon stereocenter, and the 3Z stereoisomer of the product would have the R configuration at the asymmetric carbon stereocenter. This result was *not* observed.

28.18   (a)   Think of the five-carbon π-electron system as a 2,4-pentadienyl cation, whose MOs are shown in Fig. 28.5, text p. 1560, and think of the migrating group as a carbon anion in which the nonbonding electron pair resides in a 2p orbital. The orbitals involved in the rearrangement are the LUMO of the pentadienyl system and the HOMO of the carbon anion. The LUMO of the pentadienyl system, as shown by Fig. 28.5, is symmetric; that is, at each end of the system, the LUMO has the same phase on a given face. The HOMO of the carbon anion is simply the filled 2p orbital. The migration is allowed only if the carbon anion migrates suprafacially such that each end of the pentadienyl LUMO interacts with the same lobe of the carbon anion 2p orbital.

HOMO of migrating carbon anion

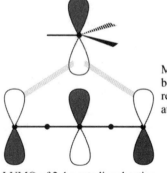

Migration occurs suprafacially on
both systems (and thus with
retention of configuration
at the migrating carbon)

LUMO of 2,4-pentadienyl cation

   Alternatively, you can think of this problem in terms of the 2,4-pentadienyl *anion* and a methyl *cation*. The orbitals involved are now the HOMO of the anion and the LUMO of the cation. But these are the same orbitals used in the foregoing solution (only their occupancies are different); therefore, the stereochemical result is the same.

(b) The first step (conversion of the starting material into compound *A*) is a [1,5] sigmatropic rearrangement. Because the stereochemistry of the methyl group on the migrating carbon does not change, the rearrangement occurs with retention of configuration. This is the result predicted for an allowed suprafacial process. The remaining two steps are [1,5] sigmatropic hydrogen migrations, which presumably occur in a suprafacial manner, although the stereochemical course of the reaction cannot be determined from the structures of the products. (The migrating hydrogen is shown with bold type as **H**.)

28.19 (a) This transformation involves two successive [1,5] sigmatropic hydrogen migrations—sort of a "hydrogen walk," as shown by the following curved arrows. These are allowed suprafacial processes. (See the solution to Problem 28.18b.)

Why does the equilibrium favor the two products over the starting material? See Sec. 4.5B, text p. 196.

(b) The "hydrogen walk" on 2,3-dimethyl-1,3-cyclopentadiene should proceed as shown in the following equation. Although a hydrogen that starts out on carbon-1 could migrate to four other carbons, only two of the possible products are unique. It can be argued that compounds *A* and *B* should be the predominant alkenes at equilibrium, and that compound *C* should be present in least amount. (Why? See Sec. 4.5B, text p. 196.) Remember that the selection rules have nothing to say about which compound is favored at equilibrium, only whether the equilibrium can be attained at a reasonable rate.

2,3-dimethyl- 1,3-cyclopentadiene	1,2-dimethyl- 1,3-cyclopentadiene	1,5-dimethyl- 1,3-cyclopentadiene
A	*B*	*C*

20.20 (a) Deduce the product by using the curved-arrow notation:

3,3-dimethyl-4-pentenal

(b)   Deduce the starting material by drawing the curved arrows for the reverse of a Claisen rearrangement:

28.21   Refer to the MOs of the allyl radical in Fig. 28.4, text p. 1559. We want to see the relationship between the SOMO of the two allyl radicals, which is π_2. This MO is antisymmetric. This antisymmetric MO of the two allyl radicals matches for a [3,3] sigmatropic reaction.

SOMO (π_2) of
the allyl radical

SOMO (π_2) of
the allyl radical

28.22   (a)   Compound A undergoes the oxyCope reaction because the allylic group attached to the carbon of the bicyclo[2.2.2]octene ring bearing the alcohol group has endo stereochemistry (see Sec. 15.3C, text p. 827) and is therefore positioned to interact with the double bond in the bicyclo[2.2.2]octene ring. The allylic group attached to the carbon bearing the alcohol group in compound B, on the other hand, has exo stereochemistry and cannot interact with the double bond in the bicyclo[2.2.2]octene ring.

an enolate ion

(b)

+ H₂

The solution to this part illustrates a useful two-step process for drawing the products of complicated rearrangements. It usually makes sense to draw the bond connections first and thus obtain a highly distorted-looking structure of the product. Then convert the distorted structure into a more conventional structure, using models if necessary. Problem 28.38(b) is another problem in which this technique will be useful.

28.23 An orbital symmetry analysis involves the interaction of π_2 of one allylic system with π_2 of the other as shown below. This antarafacial overlap of both allylic components gives bonding overlap but requires side-to-side (π) rather than head-to-head (σ) overlap. The formation of σ bonds would require severe distortions of both systems' so, for that reason, this type of reaction is not very common,

π_2

antarafacial overlap
on both components

π_2

28.24 (a) This rearrangement is s allowed because the migrating bond (emphasized below) is migrating suprafacially across both π-electron systems.

$\xrightarrow{[3,3]}$

(b) This is a carbocation rearrangement in which either methyl group migrates. (This is the first step of the "methyl walk" discussed in the solution to Problem 17.61(c) on p. 548 in this manual.)

This can also be thought of as an all-suprafacial [1,6] sigmatropic rearrangement, which is an allowed process.

28.25 As stated in the problem, previtamin D_2 is the same as previtamin D_3 except for the R-group. Comparing the structures of previtamin D_3 (or D_2) in Eq. 28.55 on text p. 1583, ergosterol in Eq. 28.58 on text p. 1584, and lumisterol in the problem reveals that the latter two compounds result from conrotatory electrocyclic ring closures, ergosterol being formed by a clockwise conrotatory motion (as viewed edge-on by the "eye" below) and the lumisterol from a counterclockwise conrotatory motion. Photochemical conrotatory electrocyclic reactions involving $4n + 2$ electrons are allowed. (The arrows indicate atomic motion.)

previtamin D₂

ergosterol   **lumisterol**

28.26 Heating promotes a disrotatory reaction; that is, the reaction should have a stereochemical course which is opposite to that of the photochemical reaction. (As in the previous problem, the arrows indicate atomic motion.)

previtamin D₂

A   *B*

28.27 First determine why irradiation of *A* or *B* does not give back previtamin D₂. One reason is mechanistic. Since *A* and *B* originate from *heating* previtamin D₂, then previtamin D₂ can only result from the exact reverse of the same reaction, which must occur as a disrotatory process. However, ring opening by irradiation, if it were to occur, would have to take place with opposite stereochemistry, that is, as a conrotatory process. Either of the two possible conrotatory processes would result in a product containing a trans double bond within a six-membered ring. Even irradiation does not provide enough energy for this to occur.

the two possible products of conrotatory ring opening

The only alternative to this process is a four-electron disrotatory process that gives a cyclobutene:

A   *C*

B $\xrightarrow{\text{light}}$ D

28.28 A [1,7] thermal sigmatropic hydrogen shift should be antarafacial. That is, a hydrogen from the methyl group migrates from a conformation in which it is "up" to the lower face of the ring, or a hydrogen from the methyl group migrates from a conformation in which it is "down" to the upper face of the ring, or both.

> Problem 28.58 (text p. 1591) describes an effort to elucidate the stereochemistry of the previtamin D rearrangement.

Solutions to Skills Objectives Problems

28.29 (a) This is a migration of hydrogen between carbons in a [1,5] sigmatropic rearrangement.

(b) This is the reverse of a [4s + 2s] cycloaddition.

(c) This is an electrocyclic reaction. (See the solution to Problem 28.61 for more on this reaction.)

28.30 (a) In a conjugated alkene of n carbons, the HOMO is number $n/2$, counting from the lowest-energy MO. Therefore, π_4 is the HOMO. The MO π_1 has no nodes, and each π MO of successively higher energy has one additional node. Therefore, π_4 has 3 nodes.

(b) Because π_1 is always symmetric regardless of n, and because the MOs alternate in symmetry. then π_4 must be antisymmetric.

(c) In this sketch of the three nodes in π_4, the carbons are the black dots and the nodes are the dashed lines. Because the phase of the MO alternates at each node, we can see that the parts of the MO at the two ends must have opposite phase; that is, peaks reflect into troughs at opposite ends. This verifies the conclusion of part (b), that this MO is antisymmetric.

(d) Because successive MOs alternate in symmetry, then π_5, the LUMO, must be symmetric.

28.31 (a) The molecule contains 6 carbons in its π-electron system, so it must have 6 π MOs.

(b) The three MOs of lowest energy are occupied, and each MO of successively higher energy has an additional node. Because the nodes are located symmetrically, the node for π_2 must go through the two central carbons. MO π_3 must have two nodes, also located symmetrically

π_1 $\qquad$ π_2 $\qquad$ π_3 (HOMO)

(Don't be concerned if the exact placement of the nodes in π_3 is not clear. As long as they were placed symmetrically, you have the right idea.).

(c)   The MO π_3 is symmetric.

28.32 (a)   The 2,4,6,8-nonatrien-1-yl system has nine MOs.

(b)   The orbital energy increases with each successive MO.

(c)   In an unbranched conjugated system, the SOMO is always the orbital number $(n + 1)/2$, where n = the number of carbons. The SOMO is orbital π_5 in this case. Alternatively, we can populate the orbitals with the nine π electrons (one from each carbon), as shown in the part (b), and the SOMO is the MO with the unpaired electron.

(d)   The nodes are most easily determined by the resonance structures for the ion. The nodes are all the carbons to which the unpaired electron is *not* delocalized. Therefore, the nodes are at carbons 2, 4, 6, and 8.

28.33 (a)   Because there are 7 atoms in its π-electron system, the benzyl cation has 7 π MOs.

(b)   The nonbonding MO is typically the one that is halfway up the MO "energy ladder." In this case, the NBMO is π_4. As in Problem 28.32(c), we can also populate the MOs with electrons. The NBMO in a cation is the first empty MO after the bonding MOs are filled. The 6 π electrons in the benzyl cation will fill bonding MOs π_1, π_2, and π_3. This means that the NBMO is π_4.

(c)   We deduce the nodes in this MO by noting the carbons to which the positive charge is *not* delocalized in the resonance structures.

resonance structures of the benzyl cation

The nodes in the NMBO of the benzyl cation are located at carbons 1, 3, and 5, respectively.

28.34   (a) and (b)   MO energy diagrams for the ground state and first excited state of 1,3,5-hexatriene:

(c)   The HOMOs of the two states have opposite symmetries. This means that any symmetry-based selection rules for their reactions will be reversed. For example, an electrocyclic reaction that is disrotatory in the ground state would be conrotatory in the excited state.

28.35   (a)   The stereochemistry of the first products *B1* and *B2* that would result from the two modes of cyclization are shown in the structures below.

(b)   The stereochemistry of the conrotatory and disrotatory products of the second reaction are shown for both. *B1* and *B2* as *C1, C2, C3,* and *C4*.

(c)   Compound *A* is a $4n$ π-electron system, so the cyclization must be conrotatory. The actual product is therefore *B1*. Because *B1* is a $4n + 2$ π-electron system, the reaction must be disrotatory. The actual product, then, is *C2*. Fortunately, this gives the more stable cis ring junction. (Both *B1* and *C2* are formed as racemates.)

(d)   The relevant frontier MO for an electrocyclic reaction is the HOMO, which, for a 6 π-electron system, is π_3. This MO is symmetric, as shown in the orbital diagram in the solution to Problem 28.34. Therefore, the electrocyclic reaction must be disrotatory. (See also Eq. 28.15, Study Problem 28.1, text p. 1564.)

26.36   (a)   This cycloaddition reaction involves 16 electrons; so, it must be suprafacial on one component and antarafacial on the other. Therefore, it must be a [14*s* + 2*a*] or a [14*a* + 2*s*] cycloaddition.

(b)   The heptafulvalene molecule is large enough that its π-electron system can twist without introducing too much strain or without losing too much π-electron overlap; therefore, the cycloaddition is a [14*a* + 2*s*] process. The product has the stereochemistry shown at right:

(c)   A 14 π-electron system, as a $4n + 2$ π-electron system, has a symmetric HOMO—in this case, π_7. Tetracyanoethylene accepts electrons into its LUMO, which, in a 2 π-electron system, is antisymmetric. Only the addition modes given in part (a) provide proper "MO matchup" for the reaction to proceed.

28.37 (a)   This seems to be a [1,3] sigmatropic reaction; one end of the bond to the migrating H stays on H, and the other end migrates to a carbon three carbons away. However, we should include all of the π electrons in the reaction, and, when we do, the reaction is actually a [1,7] sigmatropic reaction. Without considering orbital symmetry, the reaction could occur in a suprafacial or an antarafacial manner.

(b)   Toluene has added stability associated with aromaticity; compound *A*, although conjugated, is not aromatic. Because any equilibrium favors the more stable compound, the added stability of toluene causes it to be strongly favored in the equilibrium.

(c)   If the reaction were concerted, it would have to occur in an antarafacial manner. As shown in Eq. 28.39, Sec. 28.4A, text p. 1575, this is stereochemically impossible. (Eq. 28.59 shows a [1,3] sigmatropic reaction, but inclusion of the other four carbons doesn't make a difference in the stereochemistry of the reaction. Because six of the seven carbons involved are tied into a ring that is very nearly planar, the same stereochemical constraint exists as in a [1,3] sigmatropic reaction.)

(d)   The sigmatropic reaction can be envisioned as the migration of a hydrogen atom between the terminal carbons of the 1,3,5,7-heptatrien-1-yl radical. The relevant frontier orbitals are the SOMO of the hydrogen atom (that is, a 1s orbital), which is symmetric, and the SOMO of the radical (π_4) which is antisymmetric. Migration of a symmetric atom across an antisymmetric π-electron system must be antarafacial.

(e)   Compound *C* has two more carbons and therefore a *symmetric* SOMO (π_5). This [1,9] sigmatropic reaction should occur readily as a suprafacial process.

(f)   It is easiest to think of the reverse reaction to answer this question. In the trans-alkene, the terminal carbon is too far away to transmit a hydrogen atom to the ring carbon to form *C*. If this reverse reaction is impossible, by microscopic reversibility, the forward reaction is impossible also, That is, the terminal diene unit must be in an *s*-cis conformation for the migration to succeed; and when it does, the product alkene is locked into the cis stereochemistry.

28.38 (a)   This is a [3,3] sigmatropic reaction (a Cope rearrangement).

(b)   This a [3,3] sigmatropic rearrangement (a Claisen rearrangement).

(c)   This is a [4s + 2s] cycloaddition, that is, a Diels–Alder reaction. The exocyclic double bond is not involved in the reaction, and *endo* stereochemistry is assumed:

Two other reactions are allowed in principle that do involve the exocyclic double bond. One is a $[2s + 6a]$ cycloaddition to give product A, and the other is a less likely $[2a + 6s]$ cycloaddition to give product B. (The "squiggly bond" indicates that two stereochemistries are possible at this center; and these products are racemates.)

A

B

If you construct models of the starting materials, and if you bring them together so that the appropriate orbital interactions can occur, you will see that the transition states required to form these products (as well as the products themselves) are considerably strained. Therefore, the reactions that form these products are much slower than the Diels–Alder process.

(d)   This is an anionic oxyCope reaction of the alkoxide. The product of the reaction as shown is the enolate ion A, but addition of water will give the ketone B. Although the potassium ion is shown below in proximity to the anionic atoms, it is actually sequestered by the crown ether so that the oxy anion is "naked" and consequently bears a full negative charge. (See Fig. 8.11, Sec. 8.7C, text p. 437.) The conversion of this "naked" ion to a resonance-stabilized enolate ion undoubtedly provides a significant impetus for the reaction.

28.39   Statements (c) and (e) are the only true statements. As to statement (a), resonance structures by definition don't change with time. Statements (b) and (d) are both negated by the true statement (c).

Solutions to Integrated Problems

28.40   (a)   It is forbidden. Any electrocyclic reaction involving $4n$ electrons must be conrotatory.

(b)   Although a thermal [$8s + 4s$] cycloaddition is allowed, a photochemical one is not. Such a cycloaddition would have to be [$8s + 4a$] or [$8a + 4s$].

(c)   Methyl groups don't have an effect on the symmetry of the MOs. The HOMO is symmetric. (See the solution to Problem 28.34.)

28.41   The pericyclic selection rules say *absolutely nothing* about the position of equilibrium in each case. These rules refer to *rates* of reactions, not to equilibrium constants. Other considerations must be used to decide on the position of equilibrium. The point of the problem is for you to use what you have learned about the relative energies of molecules; the molecule of lower energy is favored in each equilibrium.

(a)   The right side of the equation is favored at equilibrium because the double bonds are conjugated with the phenyl rings; conjugation is a stabilizing effect.

(b)   The left side of the equation is favored at equilibrium because the double bonds have more alkyl branches. (Alkyl substitution stabilizes a double bond.)

(c)   The right side of the equation is favored at equilibrium because a C=O double bond is formed at the expense of a C=C double bond. (C=O bonds are more stable than isomeric hydroxy alkenes; see the heats of formation in Display 22.121 on text page 1255.)

(d)   The right side of the equation is favored at equilibrium because one product is aromatic and therefore particularly stable, and the other product is volatile (ethylene is a gas).

(e)   The left side of the equation is favored at equilibrium because the double bonds are conjugated not only with each other but also with the carbonyl group.

28.42   (a)   This reaction can be viewed as either a [1,9] or a [1,13] sigmatropic rearrangement. In either of these classifications the reaction is an allowed process that is suprafacial on the π-electron system and occurs with retention at the migrating carbon. The two final structures below differ only in the positions of double bonds in the benzene ring; that is, they are resonance structures and therefore represent the same compound.

(b)   Because a thermal suprafacial [1,9] or [1,13] methyl migration must occur with retention of configuration at the migrating carbon, the methyl group that migrates, when isotopically substituted, must have the S configuration in both starting material and product.

28.43   (a)   The formula indicates that an addition has taken place. This is an allowed photochemical [$2s + 2s$] cycloaddition.

(b) The anionic oxyCope reaction forms a product *A* that is both an enol and an enolate; on protonation, the diketone product *B* is formed.

28.44 (a) The structure of the ozonolysis product *C* shows that compound *B* is a cyclobutene, which must be formed in a disrotatory photochemical electrocyclic reaction. This defines the stereochemistry of *B*, which, in turn, defines the stereochemistry of *C*:

(b) Compound *D* is the trans-fused stereoisomer of the cyclobutene derivative, which undergoes conrotatory opening upon heating to give compound *A*.

The reason that compound *B* is inert under the same conditions is that conrotatory ring opening of *B* would give a stereoisomer of *A* in which one of the cyclohexene rings would contain a trans double bond within a six-membered ring. This introduces so much strain that the reaction does not occur, even though it is allowed by the selection rules.

28.45 As usual, it helps to analyze first the origins of the carbons. The carbonyl groups are a good place to start, as they can have originated only from the quinone. The carbons marked with an asterisk are derived from the quinone. The remaining carbons, marked with a dagger (†) are derived from 1,3-cyclopentadiene. The new bonds are shown with heavy shading.

The first reaction is a Diels–Alder reaction of a very reactive diene and dienophile. Notice that the Diels–Alder product must have endo stereochemistry for the next reaction to occur.

1,3-cyclopentadiene   ***p*-benzoquinone**   *X*

Compound *X* is all set up for a [2*s* + 2*s*] allowed photochemical cycloaddition to complete the formation of compound *Y*:

light
[2*s* + 2*s*]

X   *Y*

28.46 (a)   This is a [3,3] sigmatropic rearrangement, that is, a Claisen rearrangement.

(b)   This reaction superficially appears to involve internal rotation about both double bonds, a process that is virtually impossible. However, a more reasonable pericyclic mechanism is available: a sequence of two conrotatory electrocyclic reactions, the first one a ring closing, and the second one a ring opening. (The curved arrows refer to the clockwise rotations of groups and not to electron flow.)

(c)   This is a [3,3] sigmatropic rearrangement in which the product requires a little redrawing!

28.47 This is a [1,3] sigmatropic rearrangement with inversion in the migrating carbon, as required for the thermal [1,3] rearrangement.

28.48 Take a cue from the solution to Problem 28.35(c) of this manual, in which the last step involves a thermal reaction of 1,3,5-cyclooctatriene with two methyl groups. Replace the methyl groups with hydrogens and the result is the same. The resulting compound *B* undergoes a Diels–Alder reaction (that is, a [4s + 2s] cycloaddition) with the alkyne, and the product of that reaction, *C*, undergoes a reverse Diels–Alder to generate the final products.

1,3,5-cyclooctatriene
(compound *A*)   *B*

dimethyl phthalate   cyclobutene

C

28.49 (a) A Diels–Alder reaction is followed by the reverse of a different Diels–Alder reaction. The driving force for the reverse Diels–Alder is the formation of a product that is stabilized by aromaticity.

compound *A*

(b) The first transformation is a thermal conrotatory electrocyclic ring opening to give compound *Y*, and the second transformation is a thermal disrotatory electrocyclic ring closure.

(c)   Two successive photochemical disrotatory electrocyclic reactions account for the products.

The product, the very unstable antiaromatic cyclic diene 1,3-cyclobutadiene (see text Sec. 15.7E, text p. 851), was generated in 1972 by this sequence of reactions carried out at a temperature of 8 K in the laboratory of Professor Orville L. Chapman (1932–2004), then at Iowa State University, and later at the University of California, Los Angeles. This technique allowed the observation and characterization of 1,3-cyclobutadiene by infrared spectroscopy

28.50   Heating an allylic vinylic ether generally results in a [3,3] sigmatropic (Claisen) rearrangement.

Why compound *B* should have *E* stereochemistry is not obvious from the mechanism shown above, and an answer that gives either the *E* or the *Z* stereoisomer (or a mixture of both) is satisfactory. If you work Problem 28.55(a), you will find that the Claisen rearrangement proceeds through a chairlike conformation. If the ethyl group is placed in the more favorable equatorial position of the chairlike conformation, the *E* stereoisomer of substance *B* shown above is formed.

28.51   (a)   The intermediate *X* is formed from the starting material by an allowed thermal disrotatory ring closure. Because *X* is a conjugated diene, it reacts in a Diels–Alder reaction with maleic anhydride, a good dienophile, to give the product shown in the problem.

(b)   In this case maleic anhydride is used to trap the trienone intermediate *Y* formed in the first of what would, in the absence of maleic anhydride, be two successive [3,3] sigmatropic rearrangements. (See Eqs. 28.50, Sec. 28.4B, on text p. 1579.) Maleic anhydride reacts with the conjugated-diene unit of intermediate *Y*.

[reaction scheme: ortho-methyl allyl phenyl ether undergoing [3,3] to give Y]

[3,3]

reacts with maleic anhydride in a Diels–Alder reaction

Y

(c)   A [1,9] suprafacial sigmatropic reaction produces a nonaromatic intermediate *Z* which reacts with maleic anhydride to restore aromaticity in the product.

[reaction scheme]

[1,9]
250 °C

Z

maleic anhydride

28.52   Compound *A* is a secondary alcohol, and the aldehyde *B* is formed from an enol intermediate that results from a [3,3] sigmatropic (oxyCope) rearrangement.

[reaction scheme]

1) H_2C=CH—CH_2MgCl
2) H_3O^+

heat

2-methyl-2-propenal

A

an enol

$\equiv$ H_2C=CHCH$_2$CH$_2$CHCH
                         |
                         CH$_3$

O

B

 Evidently in this case the Grignard reagent undergoes mostly 1,2-addition to the α,β-unsaturated aldehyde; none of the 1,4-addition product is formed. (See text Sec. 22.11.) Conjugate addition of Grignard reagents can often be avoided by using highly purified magnesium that is free of copper and other transition metals.

28.53   The structure of the ozonolysis product *D* and the formula of elemicin show that elemicin has the structure *C*. Compound *B* and elemicin differ by one methyl group, and compound *B* ionizes in base. Consequently, a reasonable hypothesis is that one of the three oxygens of compound *B* is part of a phenol OH group, and that this is methylated by dimethyl sulfate treatment to give elemicin. Compound *A* is *not* a phenol because it is insoluble in base, but forms a phenol *B* when heated. This series of observations suggests that compound *A* is an allyl ether that undergoes a Claisen rearrangement when heated. Which oxygen is part of the allyl ether? If one of the two outer oxygens of *A* is an allyl ether, its rearrangement would result in transfer of the allyl group to the adjacent ortho position. Since the allyl group in *C* is not ortho to one of the methoxy groups, the rearrangement must have occurred from the central oxygen. Since the ortho positions are blocked, rearrangement occurs to the para position. The structures of *A* and *B* are thus defined.

28.54   Claisen rearrangements are key steps in both parts (a) and (b).

(a)

1-ethoxy-2-propylbenzene

(b)

28.55   (a)   Place the starting material in a chairlike conformation, carry out the [3,3] sigmatropic rearrangement, and compare the stereochemistry of the product with that obtained experimentally. (Models are useful to ensure that the asymmetric carbons have the proper configuration.)

One double bond is *E* and one is *Z*; this is identical to the product obtained experimentally.

This verifies that the product could be formed from a chairlike conformation. However, the boatlike conformation must also be examined:

Both double bonds are *Z*; this is different from the product obtained experimentally.

(A second boatlike conformation is possible, and it would lead to formation of the all-*E* product, which is also different from that observed experimentally.) Thus, of the two possible conformations, only a chairlike transition state fits the observations. You should use similar reasoning to verify that a chairlike conformation of the second reactant in the problem gives the all-*E* product observed and that a boatlike conformation does not.

 You may have noticed that the curved arrows in the two preceding equations go in opposite directions—counterclockwise in the first equation and clockwise in the second. The curved-arrow direction doesn't matter. You can readily verify that the same results are obtained by reversing the directions of the curved arrows.

(b) Given that germacrone assumes a chairlike conformation in its transition state when it reacts to give β-elemenone, deduce the structure of germacrone by mentally reversing the Cope rearrangement.)

β-elemenone   **germacrone**

28.56 A thermal electrocyclic reaction of the carbocation, which is formed by protonation of the ketone starting material, is followed by loss of a proton to give an enol, which spontaneously isomerizes to the corresponding ketone. The MOs of the pentadienyl cation are relevant here (Fig. 28.5, text p. 1560). Because four electrons are involved, the HOMO of this system is π_2, which is antisymmetric. This electrocyclic reaction is therefore predicted to be conrotatory, and the methyl groups in the product are therefore trans:

conjugate acid of the starting ketone

an enol

28.57 (a) A (presumably) conrotatory ring opening gives intermediate *A*. Although this compound is no longer aromatic, its formation is driven by the relief of strain in the four-membered ring. Compound *A* undergoes an *intramolecular* Diels–Alder reaction that restores aromaticity and gives the product with precisely the desired stereochemistry.

(b)   Oxidation of the secondary alcohol and cleavage of the ether gives estrone:

estrone

28.58   This is a [1,7] sigmatropic rearrangement, which, according to the selection rules, should be antarafacial. (See Problem 28.28, text p. 1584, and its solution in this chapter.) Migration of hydrogen gives *B* and migration of deuterium gives *C*. (Presumably, the predominance of *B* is due in part to a primary isotope effect, which should favor more rapid migration of hydrogen, all things being equal.) If the rearrangements are indeed antarafacial, the stereochemistry of the products should be as shown on the right. This was indeed the stereochemistry observed

This work was carried out to adduce evidence for the stereochemistry of the previtamin D rearrangement to vitamin D (Eq. 28.56, text p. 1583). Compound *A* was constructed to be a "model" for previtamin D. As noted in Problem 28.28, the structure of the previtamin itself does not reveal the stereochemistry of the rearrangement.

28.59   First deduce the structure of compound *C* from the structure of the Diels–Alder adduct. Then deduce the structures of compound *A* and *B* from the reactions. Compound *A* is the product of bromine addition; compound *B* is a conjugated diene that results from two successive β-eliminations of compound *A*. Compound *C* is formed from compound *B* by a disrotatory electrocyclic ring opening.

Compound *C* reacts as a diene with tetracyanoethylene in a Diels–Alder reaction to give the final product shown in the problem.

Compound *C* is interesting because, if it were planar, it would have antiaromatic character. Like 1,3,5,7-cyclooctatetraene (COT; text p. 772), it avoids antiaromaticity by avoiding planarity. However, this means that it reacts like an alkene.
    Compound *C* reacts as a diene with tetracyanoethylene rather than as a triene in a [6*a* + 2*s*] or [6*s* + 2*a*] cycloaddition. Why?

28.60   (a)   If compound *B* is aromatic, then it must be planar. If it is planar, the "inner" hydrogens (the ones shown explicitly in the problem) fall essentially on top of each other. (A model can be used to demonstrate this point.) The resulting van der Waals repulsions are so severe that compound *B* does not exist in spite of its aromaticity. (Compound *B* was also the subject of Problem 15.64 on text p. 863.)

(b)   The formation of compound *B* from compound *A* involves $4n + 2$ ($= 6$) electrons and is one of two possible allowed conrotatory photochemical electrocyclic ring-opening reactions. [The other is discussed in the solution to part (c).]

(c)   Compound *C* could be formed in two ways. First, compound *B* could be formed as an unstable intermediate, and then compound *C* could be formed from *B* by an allowed *thermal* disrotatory electrocyclic reaction. Such a reaction would be very rapid even at low temperature because of the instability of compound *B*. The second pathway is that the all-cis pentaene *D* at right could be formed from compound *A* in an allowed photochemical conrotatory electrocyclic reaction. Compound *C* could then be formed from compound *D* by a thermal disrotatory electrocyclic reaction.

compound *D*

In 1971, Stanford University chemists hypothesized that they might be able to identify the intermediate in the photolysis of *A* by carrying out the photolysis at very low temperature. When they subjected compound *A* to photolysis at −190 °C, they obtained *D*. Compound *D* is spontaneously transformed thermally into compound *C*. Therefore, compound *D* is a competent intermediate. The simplest interpretation of these observations is that compound *B* is not an intermediate, and the pathway for the observations recorded in the problem is the one involving compound *D*.

28.61   (a)   If the formation of benzene were concerted, it would have to be a thermal disrotatory electrocyclic ring-opening reaction involving four electrons—a disallowed concerted process.

(b)   The reaction that gives benzene is not allowed by the selection rules; therefore, it does not occur. In spite of the high degree of strain of Dewar benzene, and in spite of the great stability of benzene, Dewar benzene is effectively constrained to exist by the selection rules!

Dewar benzene was prepared by Stanford chemist Eugene van Tamelen and his group in 1963. They found that, despite its strain, it had a half-life of 2 days at room temperature, ultimately converting into benzene (probably by nonconcerted pathways).

Prismane, or Ladenburg benzene, is another benzene isomer that was discussed in Chapter 15 (Sec. 15.7A, p. 840). This highly strained compound was successfully prepared in 1973. It, too, is trapped into existence because its concerted transformation into benzene violates the selection rules for electrocyclic reactions. Because of its strain, it was referred to as a "caged tiger."

28.62   (a)   Bromine addition to the starting material is followed by four successive base-promoted β-elimination reactions to give compound *A*, which spontaneously forms compound *B* by an allowed disrotatory thermal electrocyclic reaction. Notice that compound *B* is a "tied-down" variation on hydrocarbon *B* in Problem 28.60. Because the offending inner hydrogens are absent in *B*, its continuous π-electron system can nearly achieve planarity.

(probably a mixture of stereoisomers)

(b)   As discussed in part (a), compound *B* contains a planar (or nearly planar) $4n + 2$ π-electron system. Consequently, it shows the typical ring-current effects expected of aromatic compounds in its NMR spectra. (See Fig. 16.2 on text p. 873 and Problem 16.5 on text p. 875.) The protons on the double bonds, like those in benzene, show typical aromatic chemical shifts near δ 7.1. The two methylene protons, however, occupy the region that is strongly shielded by the ring current. For this reason, these protons have a negative chemical shift.

δ (–0.5)

compound *B*

28.63 (a) This is a [2,3] sigmatropic reaction, because the wedged bond moves from O to S at one end (2 atoms) and from carbon 1 to carbon 3 on the other (3 atoms).

(b) Because three curved arrows can be used to show the reaction, it involves six electrons

(c) Because this is a 4*n* + 2-electron sigmatropic reaction, it should be suprafacial, as shown in the structures in part (a).

28.64 (a) The curved-arrow notation:

(b) As a two-electron electrocyclic reaction, it should be disrotatory—and it is. The two methyl groups, whose carbon–carbon bonds emerge up from the page in the starting material, are in the plane of the page in the product. Therefore, the methyl groups move in opposite directions—that is, in a disrotatory manner. (We can think of this reaction as the migration of a carbon anion [2 electrons] across two carbocations [0 electrons], or a carbocation [0 electrons] migrating across a π bond [2 electrons].)

(c) This reaction lies to the right for two reasons. First, the reaction relieves significant ring strain. Second, a resonance-stabilized allylic cation is formed.

28.65 Both of these reactions are diagrammed with three curved arrows. Therefore, they involve six electrons. Like the Diels–Alder reaction, they should be [4s + 2s] cycloadditions and occur suprafacially.

(a) This reaction is the first step in ozonolysis. (See Eq. 5.46, Sec. 5.8A, text p. 261.) The following reaction shows ozone approaching from the front face of the double bond. This must occur with retention of stereochemistry, so the trans methyl groups in the starting alkene are trans-anti in the product.

(b) In the following reaction the azide group serves as the four-electron component and the carbon–carbon double bond as the two-electron component. Exo stereochemistry is shown, but endo stereochemistry could also occur. (The last lactonization step will likely occur if the addition is exo because of the proximity of the OH group to the ester carbonyl group and the stringent conditions.)